BIOACTIVE PEPTIDES

BIOACTIVE PEPTIDES

EDITED BY
JOHN HOWL
SARAH JONES

CRC Press
Taylor & Francis Group
Boca Raton London New York

CRC Press is an imprint of the
Taylor & Francis Group, an **informa** business

CRC Press
Taylor & Francis Group
6000 Broken Sound Parkway NW, Suite 300
Boca Raton, FL 33487-2742

First issued in paperback 2019

ISBN-13: 978-1-4200-6114-7 (hbk)
ISBN-13: 978-0-367-38573-6 (pbk)

Library of Congress Cataloging-in-Publication Data

Bioactive peptides / editors, John Howl and Sarah Jones.
p. ; cm.
Includes bibliographical references and index.
ISBN 978-1-4200-6114-7 (hardcover : alk. paper)
1. Peptides--Physiological effect. I. Howl, John. II. Jones, Sarah, Dr. III. Title.
[DNLM: 1. Peptides--physiology. 2. Biological Factors--physiology. QU 68 B6133 2009]

QP552.P4B535 2009
612'.015756--dc22 2008043678

Visit the Taylor & Francis Web site at
http://www.taylorandfrancis.com

and the CRC Press Web site at
http://www.crcpress.com

Contents

Preface

It is widely accepted that the very first bioactive peptide to be discovered and named to reflect a physiological function was secretin (Bayliss and Starling, 1902). Some six decades later, structural analyses and the sequencing of tryptic peptides confirmed the heptacosapeptide sequence of porcine secretin (Mutt et al., 1965). Intriguingly, in the very same issue of the *Biochemistry* journal (volume 4, 1965), there was also a paper by Marshall and Merrifield (1965) entitled "Synthesis of Angiotensins by the Solid-Phase Method." Together, these seminal papers reflect the rapid developments in the understanding of bioactive peptides and their structural analogs that resulted from Merrifield's solid-phase synthetic strategy in the early 1960s. Moreover, the related disciplines of molecular genomics and proteomics have more recently accelerated the rate of discovery of bioactive peptides that, collectively, modulate all physiological systems and much human pathology. Many details of important biological mediators are included herein from sources that include the mammalian endocrine system, alternative secretory systems and venoms. There can be no doubt that, within the field of bioactive peptides, many more important discoveries await.

The study and diverse applications of bioactive peptides traverse many sub-disciplines within chemistry, biology, physics, and medicine. A cursory consideration of the affiliations of the contributing authors confirms that the study of bioactive peptides is globally compatible with many branches of scientific endeavor and supports many productive research avenues. When deciding upon the content of this volume, we were aware that we could not possibly include details of all, or even most, bioactive peptides in just one volume. Moreover, we deliberately selected topics that would provide the more casual reader with a broad flavor of the nature and applications of diverse peptides. Thus, we anticipate that both expert and learner alike will find this volume a particularly useful resource.

Part I provides details of bioactive peptides that predominantly function as ligands for G protein-coupled receptors (GPCRs). This is a major topic of current research since GPCRs include many common drug targets. Moreover, there is a resurgent interest in peptide therapeutics that currently includes several examples of such peptides. This growing family of bioactive peptides includes secretin and angiotensin as introduced above. Collectively, these chapters provide many details of some of the most competitive areas of current research and also illustrate the diverse nature and properties of bioactive peptides that bind common receptors selectively expressed on the surface of eukaryotic cells. It should become obvious when reading these chapters that the traditional classification of such peptides either as a hormone [a term first used by Starling (1905) to describe secretin], or as a neuropeptide is no longer fully justified. Thus, many of these important bioactive peptides serve multiple roles within both the nervous system and peripheral tissues. There can be no doubt that bioinformatics, molecular genomics and proteomics will identify many more examples of peptides that selectively bind GPCRs.

Not all peptides derive from the translation of genes. Prokaryotes and some lower order eukaryotes also produce bioactive peptides via non-ribosomal mechanisms. This fascinating area of science is the theme of Part II.

Although it is certain that mammalian physiological systems utilize bioactive peptides that have yet to be discovered or understood in detail, other vertebrates and non-vertebrates provide a rich and valuable source of bioactive peptides. Indeed, we have barely begun the task of cataloging these important biomolecules but, as indicated in Parts III and IV, many of these peptides have unique bioactivities that are ripe for further exploitation. Collectively, these six chapters provide details of host-defense peptides isolated from amphibians and antimicrobial peptides that clearly offer the potential for therapeutic development. Other viable sources of extremely potent bioactive peptides are invertebrate venoms and toxins as reported in Part IV. One objective in selecting these topics was to present a more balanced view of the current field of bioactive peptides. These contributions also cogently illustrate the fact that countless thousands of potentially important biological mediators await discovery.

In summary, we believe that the study of bioactive peptides, including the identification of new species, the synthesis of structural analogs, and diverse applications in biology and medicine, will remain in the forefront of scientific endeavor for many decades. Hopefully, this volume will serve to attract and excite tomorrow's researchers who will further develop some of the ideas and topics presented herein.

Professor John Howl
Dr. Sarah Jones
Wolverhampton, United Kingdom

REFERENCES

Bayliss, W.M. and Starling, E.H., The mechanism of pancreatic secretion, *J. Physiol.,* 28, 325–353, 1902.

Marshall, G.R. and Merrifield, R.B., Synthesis of angiotensins by the solid-phase method, *Biochemistry*, 4, 2394–2401, 1965.

Mutt, V., Magnusson, S., Jorpes, J.E. et al., Structure of porcine secretin. I. Degradation with trypsin and thrombin. Sequence of the tryptic peptides and the C-terminal residue, *Biochemistry,* 4, 2358–2362, 1965.

Starling, E.H., Croonian Lecture: On the chemical correlation of the functions of the body, I. *Lancet,* 2, 339–341, 1905.

Editors

John Howl is a professor of molecular pharmacology within the Research Institute in Healthcare Science, School of Applied Sciences, University of Wolverhampton, United Kingdom. Professor Howl received his PhD in the field of molecular pathology from the School of Biological Sciences, University of Birmingham, United Kingdom during 1988. From 1988 to 1997, he pursued research related to the pharmacology of neurohypophysial peptide hormones (vasopressin and oxytocin) and their cognate receptors and was supported by the Wellcome Trust and the British Heart Foundation within the School of Biochemistry, University of Birmingham, United Kingdom. During this time, Professor Howl developed a keen interest in the synthesis of chimeric peptide ligands that could bind multiple pharmacologically distinct receptors for both neurohypophysial hormones and kinins.

In 1997, Professor Howl joined the School of Health Sciences, University of Wolverhampton, United Kingdom, where he established the molecular pharmacology group. The study of peptides, including receptor ligands and, more recently, cell penetrating peptides, is an established focus of this group. He is a regular contributor to international conferences and co-author of numerous scientific articles describing the biological activities of novel peptides. Professor Howl is a member of the European Peptide Society and founder of Pantechnia, a company that provides both *de novo* materials and consultancy in the synthesis and applications of bioactive peptides.

Sarah Jones is a research fellow also within the Research Institute in Healthcare Sciences and has worked alongside Professor Howl for many years. As a former psychiatric nurse, akin to Jo Brand, Dr. Jones came late to science where she found solace in test tubes (rather than people) by embarking on a degree in biomedical science. Dr. Jones received her PhD in the field of molecular pharmacology of receptor mimetic peptides at the University of Wolverhampton and has continued her devotion to peptide biology ever since. Besides her role as product specialist to Pantechnia, her current research endeavors include the application of cell-penetrating peptides to modulate biological phenomena, with a particular emphasis on novel inducers of apoptosis. Sarah has produced numerous scientific publications within the field of bioactive peptides and is a member of the European Peptide Society.

Contributors

Jane V. Aldrich
Department of Medicinal Chemistry
School of Pharmacy
University of Kansas
Lawrence, Kansas

Margit A. Apponyi
Department of Chemistry
The University of Adelaide
South Australia, Australia

Grant W. Booker
School of Molecular and Biomedical Sciences
The University of Adelaide
South Australia, Australia

John H. Bowie
Department of Chemistry
The University of Adelaide
South Australia, Australia

Robert E. Carraway
Department of Physiology
University of Massachusetts Medical School
Worcester, Massachusetts

John A. Carver
Department of Chemistry
The University of Adelaide
South Australia, Australia

Cláudia Cavadas
Center for Neurosciences and Cell Biology, and Faculty of Pharmacy
University of Coimbra
Coimbra, Portugal

Mark C. Chappell
Hypertension and Vascular Disease Center
Department of Physiology and Pharmacology
Wake Forest University School of Medicine
Winston-Salem, North Carolina

Paul R. Dobner
Department of Molecular Genetics and Microbiology
University of Massachusetts Medical School
Worcester, Massachusetts

Jason R. Doyle
Australian Institute of Marine Science
Townsville MC
Queensland, Australia

Eric Grouzmann
Division of Clinical Pharmacology and Toxicology
Centre Hospitalier Universitaire Vaudois
Lausanne, Switzerland

Per Grybäck
Department of Nuclear Medicine and Diagnostic Radiology
Karolinska Institutet
Karolinska University Hospital
Stockholm, Sweden

Debbie L. Hay
School of Biological Sciences
University of Auckland
Auckland, New Zealand

Per M. Hellström
Department of Medicine
Karolinska Institutet
Karolinska University Hospital
Stockholm, Sweden

Richard A. Hughes
Department of Pharmacology
University of Melbourne
Victoria, Australia

Rebecca J. Jackway
Department of Chemistry
The University of Adelaide
South Australia, Australia

Lucia Kuhn-Nentwig
Zoological Institute
University of Bern
Bern, Switzerland

Ülo Langel
Department of Neurochemistry
Stockholm University
Stockholm, Sweden

Lyndon E. Llewellyn
Australian Institute of Marine Science
Townsville MC
Queensland, Australia

Eve M. Lutz
Strathclyde Institute of Pharmacy and Biomedical Sciences
University of Strathclyde
Glasgow, United Kingdom

Christopher MacKenzie
Strathclyde Institute of Pharmacy and Biomedical Sciences
University of Strathclyde
Glasgow, United Kingdom

Alison M. McDermott
College of Optometry
University of Houston
Houston, Texas

Thomas K. Monaghan
Strathclyde Institute of Pharmacy and Biomedical Sciences
University of Strathclyde
Glasgow, United Kingdom

Erik Näslund
Department of Clinical Sciences
Karolinska Institutet
Danderyd Hospital
Stockholm, Sweden

Brett A. Neilan
School of Biotechnology and Biomolecular Sciences
The University of New South Wales
Sydney, New South Wales, Australia

Wolfgang Nentwig
Zoological Institute
University of Bern
Bern, Switzerland

Suzanne E. Newton
School of Life Sciences
Kingston University London
Kingston-upon-Thames
Surrey, United Kingdom

Kjell Öberg
Department of Endocrine Oncology
Uppsala University Hospital
Uppsala, Sweden

Nigel M. Page
School of Life Sciences
Kingston University London
Kingston-upon-Thames
Surrey, United Kingdom

Leanne A. Pearson
School of Biotechnology and Biomolecular Sciences
The University of New South Wales
Sydney, New South Wales, Australia

Chantevy Pou
Strathclyde Institute of Pharmacy and Biomedical Sciences
University of Strathclyde
Glasgow, United Kingdom

David R. Poyner
School of Life and Health Sciences
Aston University
Birmingham, United Kingdom

Joseph A. Price
Department of Pathology
OSU Center for Health Sciences
Tulsa, Oklahoma

Tara L. Pukala
Department of Chemistry
The University of Adelaide
South Australia, Australia

Jens F. Rehfeld
Department of Clinical Biochemistry
Rigshospitalet
University of Copenhagen
Copenhagen, Denmark

Alexandra A. Roberts
School of Biotechnology and Biomolecular Sciences
The University of New South Wales
Sydney, New South Wales, Australia

John K. Robinson
Department of Psychology
Stony Brook University
Stony Brook, New York

Johan Runesson
Department of Neurochemistry
Stockholm University
Stockholm, Sweden

Joana Rosmaninho Salgado
Center for Neurosciences and Cell Biology
University of Coimbra
Coimbra, Portugal

Frances Separovic
School of Chemistry
Bio21 Institute
University of Melbourne
Victoria, Australia

Fazel Shabanpoor
Howard Florey Institute
University of Melbourne
Victoria, Australia

Ulla Eriksson Sollenberg
Department of Neurochemistry
Stockholm University
Stockholm, Sweden

Parvathy Subramaniam
School of Life Sciences
Kingston University London
Kingston-upon-Thames
Surrey, United Kingdom

Christian Trachsel
Zoological Institute
University of Bern
Bern, Switzerland

Michael J. Tyler
School of Earth and Environmental Sciences
The University of Adelaide
South Australia, Australia

John D. Wade
Howard Florey Institute
University of Melbourne
Victoria, Australia

Part I

Receptor Ligands

1 Angiotensins: From Endocrine to Intracrine Functions

Mark C. Chappell

CONTENTS

The renin–angiotensin–aldosterone system (RAAS) constitutes one of the most important hormonal systems in the physiological regulation of blood pressure. Indeed, dysregulation of the RAAS is considered a major factor in the development of cardiovascular disease, and blockade of this system offers an effective therapeutic regimen. Originally defined as a circulating or endocrine system, multiple tissues express a complete local RAAS and compelling evidence now favors intracellular sites of action for angiotensin II, the primary effector peptide of the system. The diversity of actions of the circulating and tissue RAAS further reflects the generation of additional biologically active angiotensin peptides that recognize distinct receptor subtypes.

1.1 INTRODUCTION

From the initial description of renin in the canine kidney by Tigerstedt and Bergman in the late 19th century (Tigerstedt and Bergman, 1898), the ongoing characterization of the bioactive peptides that comprise the renin–angiotensin–aldosterone system (RAAS) continues to reveal novel findings that have redefined the functional nature of this hormonal system. Historically, the biological actions of the circulating RAAS were viewed primarily with regard to alterations in blood pressure and inducing vasoconstriction, and to some extent, these remain the dominant aspects of the interaction of angiotensin II (Ang II) with the Ang II type 1 (AT1) receptor isotype. Abundant evidence now also defines the RAAS as a tissue system for which the components are expressed in essentially every organ and whose actions are implicated in numerous physiological events that influence renal, neuronal, cardiac, pancreatic, vascular, adrenal, pituitary, cognitive, aging, and reproductive functions

(Paul et al., 2006). Emerging data also reveals the RAAS to exhibit intracellular actions within multiple cell types, supporting the localization of various components of the RAAS including the AT1 receptor subtype on intracellular organelles. In addition to the array of actions associated with the Ang II–AT1 receptor pathway (Hunyady et al., 2000), there is also compelling evidence for other peptide products and additional receptor subtypes that may buffer or evoke antagonistic actions to this pathway (Carey, 2005; Ferrario et al., 2005).

1.2 SOURCES OF ANGIOTENSIN PEPTIDES

The ultimate precursor for angiotensin peptides in the circulation is angiotensinsogen, a 453-amino-acid protein that is constitutively secreted from various tissues. The principal source of circulating angiotensinsogen are hepatocytes, although other tissues including vascular smooth muscle cells, adipocytes, and the adrenal gland, which also express angiotensinsogen, may contribute to plasma levels. Circulating angiotensinsogen is hydrolyzed between residues 10 and 11 (Leu^{10}–Leu^{11} for rat and mouse; Leu^{10}–Val^{11} for dog and sheep; Leu^{10}–Ile^{11} for human) by the aspartyl protease renin to form angiotensin-(1–10) (Ang I) and (des-Ang I)-angiotensinsogen (Figure 1.1, Table 1.1). The N-terminal region of the protein-containing residues His^{6}–Pro^{7}–Phe^{8} is also highly conserved among species and plays a pivotal role in the appropriate recognition by renin to form Ang I (Nakagawa et al., 2007). Angiotensinsogen has long been considered the only known substrate for renin; however, other enzymes including tonin, kallikrein, and cathepsin D are capable of metabolizing the precursor to Ang I or Ang-(1–8) (Ang II) (Paul et al., 2006). The extent that other nonrenin enzymes participate in the processing of angiotensinogen to Ang I or Ang II is important given the recent development of nonpeptide inhibitors of renin for the clinical treatment of hypertension (Kushiro et al., 2006; Drummond et al., 2007; Gradman et al., 2008). Moreover, a prorenin receptor was recently cloned by Nguyen and colleagues with significant concentrations of the protein in the glomerulus and vascular smooth muscle cells (Nguyen et al., 2002). Receptor-bound prorenin exhibits increased proteolyitc activity, leading to Ang I formation (Batenburg et al., 2007), but prorenin may also induce distinct signaling pathways following binding, and renin inhibitors such as aliskiren do not attenuate prorenin binding (Ichihara et al., 2007). Sakoda and colleagues (2007) found that prorenin activated the MAP kinase pathway in human smooth muscle cells—actions similar to that for Ang II–AT1 receptor activation (Figure 1.1). In isolated mesangial cells, exogenous renin increased TGF-β expression and other matrix proteins including plasminogen activator inhibitor (PAI-1) and fibronectin, which was independent of formation of Ang II (Huang et al., 2006; Kaneshiro et al., 2007). Increased expression of PAI-1 and other matrix proteins may lead to progressive renal damage, which would not be inhibited by traditional agents that block the RAAS such as ACE inhibitors or AT1 receptor antagonists. Others report the predominant intracellular localization of the prorenin receptor in the perinuclear area (Schefe et al., 2006) that may interact with internalized renin (Peters et al., 2002) or the expression of nonsecreted forms of the enzyme (Clausmeyer et al., 1999).

The biologically inactive Ang I is further processed to directly yield the active peptides Ang II by angiotensin converting enzyme (ACE) or Ang-(1–7) through various neutral endopeptidases (NEP) including neprilysin, prolyl endopeptidase, and thimet oligopeptidase (Allred et al., 2000). ACE is a metallopeptidase primarily characterized as a dipeptidyl carboxypeptidase that cleaves the Phe^{8}–His^{9} bond of Ang I to form Ang II and the dipeptide His–Leu (Figure 1.1). ACE likely represents the predominant Ang II-forming enzyme in the circulation and various tissues under physiologic conditions; however, other peptidases including mast cell and smooth muscle chymases likely contribute to tissue levels of Ang II in pathological states (Guo et al., 2001; Sadjadi et al., 2005). In addition to the formation of Ang II, ACE participates in the metabolism of several other peptides that exhibit cardiovascular actions, including bradykinin, substance P, the hematopoietic fragment acetyl–Ser–Asp–Lys–Pro (Ac-SDKP), and Ang-(1–7) (Figure 1.1). Thus, the cardiovascular effects of ACE inhibitors likely reflect the inhibitory influence on Ang II formation, as well as attenuating the metabolism of other bioactive peptides. Indeed, we have shown that ACE inhibitors markedly

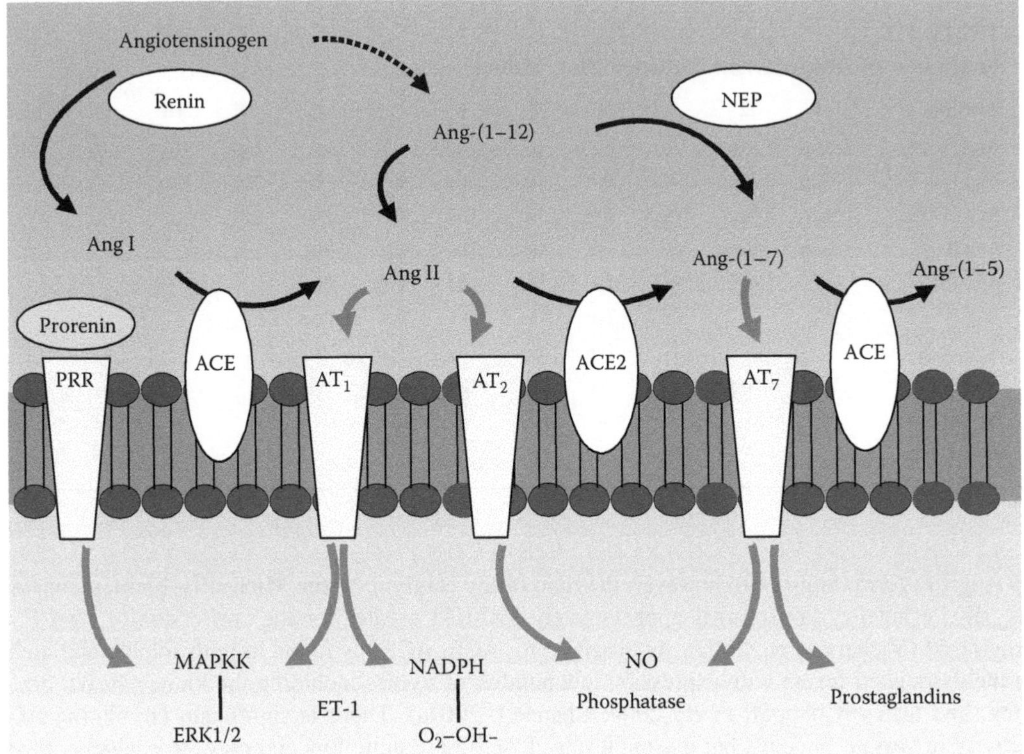

FIGURE 1.1 Cascade of the processing of angiotensin peptides and their interaction with AT_1, AT_2, and AT_7 receptor systems. Circulating renin cleaves the precursor angiotensinogen to Ang I. Angiotensin converting enzyme (ACE) hydrolyzes Ang I, releasing the dipeptide His–Leu to form Ang II, and ACE2 subsequently hydrolyzes Ang II to Ang-(1–7). ACE also metabolizes Ang-(1–7) to Ang-(1–5) and the dipeptide His–Pro. Ang I can also be directly cleaved by neprilysin (NEP) to form Ang-(1–7). Ang-(1–12) may be cleaved from Angiotensinogen and potentially processed (ACE, NEP) directly to Ang II or Ang-(1–7). Prorenin binds to the prorenin receptor (PRR) and stimulates the MAPK kinase (MAPKK)/ERK pathway. Ang II–AT_1 also stimulates MAPK kinase, endothelin-1 (ET-1) and an increase in reactive oxygen species superoxide (O_2-) and the hydroxyl radical (OH^-). Ang-(1–7) attenuates the actions of the Ang II–AT_1 receptor pathway through inhibition of the MAPK kinase pathway and the stimulation of cellular phosphatases, nitric oxide (NO), and prostaglandins. Activation of the AT_2 shares similar pathways to Ang-(1–7), including increased expression of NO and phosphatases.

increase the half-life of circulating Ang-(1–7) and that antagonism of Ang-(1–7) following ACE inhibition in hypertensive rats partially reverses the reduction in the blood pressure (Iyer et al., 1998, 2000; Yamada et al., 1998). ACE inhibitors can elicit cell-specific signaling by inducing conformational changes in membrane-bound ACE that result in the phosphorylation and nuclear trafficking of cJun kinase, a key transcriptional regulator (Benzing et al., 1999; Kohlstedt et al., 2002). ACE inhibitors or the angiotensin peptides Ang-(1–9) and Ang-(1–7) induce the association of ACE and the bradykinin B2 receptor to prevent the rapid downregulation of the ligand–receptor complex, thus potentiating the actions of bradykinin (Chen et al., 2005, 2006). Although ACE is predominantly localized at the plasma membrane with an extracellular orientation of both catalytic domains, there is evidence for cytosolic and nuclear forms of ACE within the kidney (Camargo de Andrade et al., 2006; Redublo Quinto et al., 2008).

A new homolog termed ACE2 was identified approximately 50 years after the discovery of ACE. ACE2 is also a metallopeptidase, but exhibits carboxypeptidase activity cleaving the Pro^7–Phe^8 bond of Ang II to generate Ang-(1–7) (Figure 1.1). ACE2 also hydrolyzes the His^9–Leu^{10} bond

TABLE 1.1
Sequence of Angiotensin Peptides (Rat, Mouse)

Position	1	2	3	4	5	6	7	8	9	10	11	12
Ang-(1–12)	Asp	Arg	Val	Tyr	Ile	His	Pro	Phe	His	Leu	Leu	Tyr
Ang I	Asp	Arg	Val	Tyr	Ile	His	Pro	Phe	His	Leu		
Ang-(1–9)	Asp	Arg	Val	Tyr	Ile	His	Pro	Phe	His			
Ang II	Asp	Arg	Val	Tyr	Ile	His	Pro	Phe				
Ang III		Arg	Val	Tyr	Ile	His	Pro	Phe				
Ang IV			Val	Tyr	Ile	His	Pro	Phe				
Ang-(4–8)				Tyr	Ile	His	Pro	Phe				
Ang-(1–7)	Asp	Arg	Val	Tyr	Ile	His	Pro					
Human	–	–	–	–	–	–	–	–	–	–	Ile	His
Sheep, dog	–	–	–	–	–	–	–	–	–	–	Val	Tyr

of Ang I to form Ang-(1–9); however, the ratio of the catalytic to the Michealis–Menten constant (k_{cat}/K_m, efficiency constant) is approximately 500-fold greater for the conversion of Ang II to Ang-(1–7) (Vickers et al., 2002). As observed for ACE, ACE2 is found in both soluble and membrane-associated forms with expression in a number of tissues including the kidney, heart, brain, lung, and testes (Chappell et al., 2004; Chappell, 2007a). There is significant circulating ACE activity in various species, but plasma levels of ACE2 are quite low and may vary widely among species with specific pathologies (Elased et al., 2006; Rice et al., 2006; Shaltout et al., 2006; Tikellis et al., 2008). In this regard, Lew and colleagues (2008) report that human blood contains an endogenous inhibitor of ACE2; however, following dissociation of the inhibitory substance, ACE2 activity remains very low. Thus, ACE2 is likely to be more important in the tissue generation of Ang-(1–7) from Ang II than in the circulation. Indeed, infusion of Ang I but not Ang II results in increased circulating levels of Ang-(1–7) that are sensitive to neprilysin inhibition (Yamamoto et al., 1992; Campbell et al., 1998). Moreover, an emerging number of studies strongly support the physiological importance of ACE2 as knockout models of the enzyme exhibit increased blood pressure (Gurley et al., 2006), exacerbation of myocardial damage (Yamamoto et al., 2006), alterations in cardiac function associated with increased oxidative stress (Crackower et al., 2002; Oudit et al., 2007), a greater degree of pulmonary injury (Kuba et al., 2005), and the development of renal damage (Oudit et al., 2006; Wong et al., 2007). Utilizing ACE2 knockout mice, we demonstrate a reduction in both the formation of Ang-(1–7) and the metabolism of Ang II in cardiac tissues (Garabelli et al., 2008). The acute inhibition of ACE2 also increases the half-life of Ang II in the isolated proximal tubules, which contain a very high density of the enzyme (Shaltout et al., 2006). Finally, Diz and colleagues (2008) report that central administration of an ACE2 inhibitor reduces baroreflex sensitivity, suggesting that ACE2 within the brain may play an important role in the regulation of cardiovascular peptides.

Although angiotensinogen is considered the precursor for the generation of angiotensin peptides with Ang I formation as the obligatory first step in the processing cascade, recent studies suggest the intriguing possibility of another peptide intermediate (Nagata et al., 2006; Jessup et al., 2008). Nagata and colleagues isolated a peptide that contains the first 12 amino acids of the N-terminal sequence of angiotensinogen referred to as Ang-(1–12) using antibodies specific to the C-terminal sequence of the peptide (Nagata et al., 2006). These investigators found that Ang-(1–12) was expressed in essentially all tissues that contain either Ang I or Ang II, with the highest concentrations in the intestine, heart, and kidney (Nagata et al., 2006). Differential expression of Ang-(1–12) was also demonstrated in the heart and kidney of the spontaneously hypertensive rat (SHR) and the normotensive control Wistar Kyoto strain (WKY) (Jessup et al., 2008). Moreover, the site of hydrolysis for the formation of Ang-(1–12) from rat angiotensinogen at residues Tyr^{12}–Tyr^{13} is distinct from the Leu^{10}–Leu^{11} sequence cleaved by renin to form Ang I. We also find that Ang-(1–12) is a substrate for ACE that results in the processing to Ang I and Ang II in serum and in the isolated heart, as well for neprilysin in the conversion to Ang-(1–7) and Ang-(1–4) in the kidney (Chappell et al., 2007b; Trask et al., 2008) (Figure 1.1). Conversion of Ang-(1–12) to Ang II by ACE is consistent with the increase in blood pressure following an acute infusion of Ang-(1–12) (Nagata et al., 2006). Therefore, Ang-(1–12) is likely not biologically active, but may serve as an intermediate precursor to Ang II or Ang-(1–7) and, importantly, the formation of the Ang-(1–12) may be independent of renin (Varagic et al., 2008).

1.3 MOLECULAR ACTIONS

Ang II primarily interacts with two G-protein coupled receptors identified as the AT1 and AT2 subtypes. In several rodent species including the rat and mouse, two isotypes of the AT1 termed AT1a and AT1b are present that arise from separate gene products (de Gasparo et al., 2000). Intense research has focused on the signaling mechanisms of the Ang II–AT1 receptor axis, with more recent attention to the generation of reactive oxygen species (ROS) (Griendling et al., 1994; Fukai et al., 1999; Higuchi et al., 2007). Indeed, the importance of this pathway underlies the relationship of oxidative stress and various pathologies that are not exclusively related to cardiovascular disease. Moreover, the generation of ROS is not simply a cytotoxic mechanism but likely constitutes an important intracellular signaling pathway for the RAAS in a variety of tissues including the brain, heart, kidney, and vasculature. Ang II activation can induce the complex assembly of NADPH units, resulting in the production of superoxide (SO) and downstream metabolites. ROS may arise from the Ang II-dependent stimulation of nitric oxide synthase (NOS) either from uncoupled NOS to yield SO or the complexing of nitric oxide (NO) and SO to form peroxynitrite (Touyz and Schiffrin, 1999; Kase et al., 2005). The Ang II–AT1 pathway also influences the level of ROS by inhibiting various scavenging components such as superoxide dismutase and catalase (Wilcox, 2005). Although the clinical benefits of reducing ROS remain equivocal, inhibitors of ROS or the enhanced expression of scavengers are effective in reversing Ang II-dependent alterations in blood pressure, vascular dysfunction, and tissue damage (Schnackenberg and Wilcox, 1999).

The stimulation of ROS by Ang II is dependent, in part, on the stimulation or transactiviaton of the epidermal growth factor (EGRF) receptor (Matsubara et al., 2000). In this regard, Ang II induces the activity of ADAM17, a membrane protease that acts locally to cleave the inactive or proform of EGRF (Mifune et al., 2005). This process essentially couples the activation of the G-protein receptor (AT1) to the autoactivation of tyrosine kinase receptor (EGRF) that increases the signaling potential of these systems. In addition, a second peptidergic system that both contributes to the actions of and is stimulated by Ang II is the potent vasoconstrictor endothelin. In this regard, Ang II–AT1 signaling increases the local release of endothelin through transcriptional regulation of the precursor preproendothelin and endothelin-converting enzyme (Schiffrin, 1999; Hsu et al., 2004). In contrast to the AT1 receptor, Ang II-dependent activation of the AT2 receptor subtype is closely linked to the stimulation of NO and the activation of cellular phosphatases (Takahasi et al., 1994; Gendron et al., 2002; Siragy et al., 2007). AT2-dependent signaling in the regulation of proliferation may also involve the inhibition of protein-kinase C (PKC) activity (Beaudry et al., 2006). The activation of these pathways supports the overall tenet that the AT2 subtype may counterbalance or buffer the actions of the AT1 receptor (Carey, 2005). Indeed, knockdown of the AT2 receptor enhances the AT1-dependent actions of Ang II (Siragy et al., 1999). The stimulation of NOS by AT2 activation may be indirect, involving the release of bradykinin (Tsutsumi et al., 1999), or direct, culminating in increased levels of cGMP and cGMP kinase (Siragy and Carey, 1997; Siragy et al., 2007). The generation of Ang-(1–7) and its associated signaling pathways are now recognized to antagonize the actions of Ang II–AT1 activation as well (Ferrario, 2006; Chappell, 2007a). Similar to the AT2 receptor, Ang-(1–7) stimulates NO production via enhanced phosphorylation of Akt and increased levels of cGMP that may or may not depend on the release of bradykinin (Weiss et al., 2001; Sampaio et al., 2007a). Ang-(1–7) also activates cellular phosphatases in various cells, culminating in the attenuation of MAP kinase activity (Su et al., 2006; Sampaio et al., 2007b). Furthermore, Ang-(1–7) directly increases the production of various prostaglandins that may contribute to the vasodilatory and antiproliferative actions of the peptide (Iyer et al., 2000). In contrast to either AT1 or AT2 receptor subtypes, which bind Ang II with high affinity, Ang-(1–7) activates at least one other unique receptor protein (Santos et al., 2003b). Originally cloned as an orphan receptor, the G-protein coupled Mas receptor exhibits little affinity for Ang II, but mediates a number of actions associated with Ang-(1–7) (Chappell, 2007a). In addition, several peptide antagonists selective for the Ang-(1–7) receptor have been developed that do not antagonize the actions of Ang II at either AT1 or AT2 sites (Santos et al., 1994, 2003a).

1.4 INTRACRINE ACTIONS

The RAAS is traditionally viewed as an endocrine system in which circulating renin and angiotensinogen initiate an enzymatic cascade to form either Ang II, Ang-(1–7), or related metabolites (Ang III, Ang IV) to target tissues. It is now apparent that multiple tissues contain the necessary RAAS components for the local generation of peptides that have actions on adjacent (paracrine) or the same cells (autocrine). These tissue pathways, whether ultimately expressing Ang II or Ang-(1–7), must recognize and activate their G-protein coupled receptors on the cell surface to initiate signaling. Emerging evidence in various tissues including the brain, heart, and kidney now suggest an intracellular or intracrine system in which receptors localized within the cell on distinct organelles can mediate the actions of Ang II (Lu et al., 1998; Re, 2003; Kumar et al., 2007). Intracellular administration of Ang II to cardiomyocytes, vascular smooth muscle, or proximal tubule epithelial cells results in the immediate increase in intracellular calcium (Haller et al., 1996; De Mello and Danser, 2000; Zhuo et al., 2006). Furthermore, the intracellular actions of Ang II have been blocked by the coadministration of the AT1 receptor antagonist losartan. We and other groups have reported a surprisingly high density of AT1 receptor sites in the nuclear fraction from both the cortical and medullary areas of the rat kidney (Licea et al., 2002; Pendergrass et al., 2006; Li and Zhuo, 2008). The density of Ang II receptors in the renal cortex is at least 20-fold higher than the density of sites reported in the rat liver (Booz et al., 1992; Tang et al., 1992). Nuclear immunoblots have revealed a 52-kDa band, suggesting a mature form of the AT1 receptor, and our studies demonstrate that the majority of ^{125}I-(Sarcosine^{1},threonine^{8})–Ang II (Sarthran) bound to the nucleus was inhibited by the AT1 receptor antagonists losartan and candesartan, but not the AT2 antagonist PD123319 or the Ang-(1–7) blocker D-Ala^{7}-Ang-(1–7) (Pendergrass et al., 2006). Utilizing flow cytometry coupled with an AT1 receptor antibody, our preliminary studies reveal that essentially all the nuclei in the renal cortex express the AT1 receptor subtype (Pendergrass et al., 2004). Moreover, our preliminary findings reveal that incubation of Ang II with isolated renal nuclei results in the immediate increase in dichlorofluroscein (DCF) fluorescence, a sensitive indicator of ROS (Pendergrass et al., 2008). These findings of a functional nuclear receptor for Ang II support the recent work by Zhou and colleagues demonstrating that Ang II stimulates the inflammatory genes MCP-1 and TGF-β_1 on isolated nuclei (Li and Zhuo, 2008), as well as an earlier study suggesting that Ang II increases gene expression for renin and angiotensinogen (Eggena et al., 1993).

The assessment of the intracellular functions for Ang II is far from complete at this time, as well as our understanding of the compartmentalization of the intracellular RAAS. With regard to the intracellular trafficking of the AT1 receptor, binding to Ang II induces rapid internalization of the ligand–receptor complex (Figure 1.2). Similar to other peptidergic systems, the internalized complex is sorted into either endosomal or lysosomal compartments (Hein et al., 1997; Imig et al., 1999; Hunyady et al., 2000). In the kidney, endosomal sorting of Ang II may protect the peptide from immediate metabolism and provide a mechanism for subsequent release into the intracellular compartment (Figure 1.2)

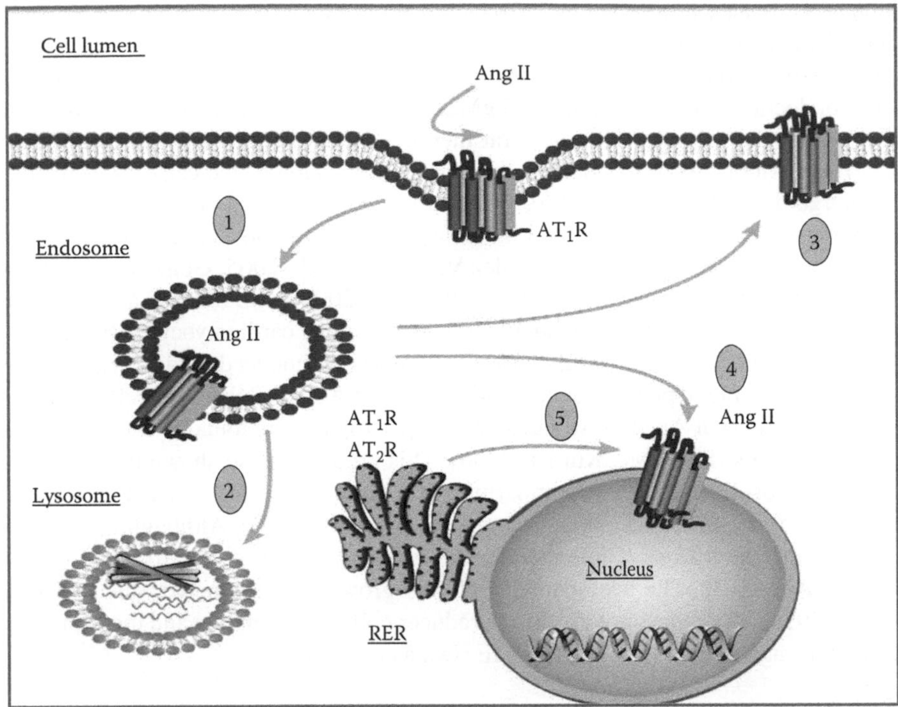

FIGURE 1.2 (See color insert following page 176.) Scheme for the internalization of the Ang II–AT1 receptor complex in the kidney. Ang II binds to the AT1 receptor (AT1R) on the cell surface and the complex is internalized. Sequestration into endosomes may provide a stable intracellular pool of Ang II (1) while transport to lysosomes will facilitate peptide metabolism (2). The AT1R recycles to the plasma membrane (3) or the Ang II–AT1R may traffic to the nucleus (4). The direct trafficking of the AT1R or the AT2R from the endoplasmic reticulum (RER) or Golgi apparatus to the nucleus may also occur (5).

(Imig et al., 1999). Following internalization, the AT1 receptor can traffic to the nuclear compartment or recycle to the plasma membrane (Thomas et al., 1996). Because the AT1 receptor contains a canonical nuclear localization sequence, we speculate that these receptors traffic directly to the nucleus following synthesis in the endoplasmic reticulum (Figure 1.2). Preliminary findings in the ACE knockout mice in which 80% of renal Ang II is reduced reveal no diminution in the AT1 receptor immunoreactive band in nuclear homogenates of the transgenic kidney (Modrall et al., 2003; Pendergrass et al., 2004). These data suggest the possibility that at least a portion of the AT1 receptors residing on the nucleus may not result from the agonist-dependent internalization of the plasma membrane receptor. In contrast, the AT2 receptor is generally not thought to undergo Ang II-induced internalization. Indeed, several studies suggest that cytosolic pools of the AT2 receptor may traffic to the plasma membrane following AT1 or dopamine receptor stimulation (de Godoy and Rattan, 2006; Salomone et al., 2007). The translocation of AT2 receptors to the cell membrane may provide a unique nongenomic mechanism to buffer the actions of Ang II–AT1 receptor activation. In contrast to the rat or mouse kidney, preliminary data in our laboratory reveals the localization of AT2 receptors in the nuclear fraction from renal cortex of both the fetal and adult sheep kidney, as well in the plasma membrane (Gwathmey et al., 2007). Receptor binding studies demonstrated the majority of ^{125}I-Sarthran bound to the nuclear fraction was competed for by the AT2 antagonist PD123319 (>60%), and the remaining sites by the AT1 antagonist losartan. However, essentially all Sarthran binding in the renal medulla of adult sheep was blocked by the AT1 antagonist losartan, but not the AT2 antagonist (Gwathmey et al., 2007). Apart from the issue of species differences in the expression of different Ang II receptor subtypes (Goldfarb et al., 1994), the functional significance of the differential distribution of these nuclear receptors in sheep kidney or the extent the nuclear AT2 sites contribute to the functional antagonism of the AT1 pathway awaits further study.

With regard to Ang-(1–7), we are unaware of any published studies on the intracellular actions of the peptide or localization of the Ang-(1–7)/Mas receptor on cellular organelles other than the plasma membrane. In primary cultures of rat mesangial cells, immunocytochemical staining for both Ang II and Ang-(1–7) was evident in the perinuclear region of the cells, as well as the cytoplasm (Camargo de Andrade et al., 2006). Whether receptor-mediated uptake of Ang-(1–7) occurs is also not known, although the transport protein megalin can promote uptake of this peptide as well as Ang II in the rat proximal tubule (Gonzalez-Villalobos et al., 2005; Gonzalez-Villalobos et al., 2006). Moreover, in the rat heart, Baker and colleagues (2004) have utilized a novel approach to express Ang II intracellularly using an adenoviral vector with a cardiomyocytes-specific promoter. These investigators find that intracellular Ang II promotes pronounced cardiac hypertrophy in the absence of changes in blood pressure or circulating levels of Ang II (Baker et al., 2004). Surprisingly, the intracellular actions of Ang II to stimulate cell hypertrophy were found to be independent of the AT1 or AT2 receptors (Baker and Kumar, 2006). This group has also shown the accumulation of Ang II in both cytosolic and nuclear compartments of cardiomyocytes exposed to high glucose, a known activator of the RAS and oxidative stress (Singh et al., 2007). Although the identity of the receptor mediating the actions of intracellular Ang II is unknown, it is unlikely to be due to the conversion to Ang-(1–7). The heptapeptide inhibits growth in cardiomyocytes and other cells (Tallant et al., 1999, 2005; Menon et al., 2007), reduces collagen formation and the expression of the proinflammatory agents TGF-β, ET-1, and LIF (Iwata et al., 2005).

1.5 STRUCTURAL ANALOGS

The original interest in peptide analogs of Ang II were as receptor antagonists to reduce blood pressure and included the sarcosine compounds with various substitutions at the Phe^8 position (Khosla, 1985). The development of potent and orally active ACE inhibitors was a major achievement in the treatment of cardiovascular disease and likely closed the chapter on Ang II peptide antagonists as therapeutic agents (Ondetti, 1982). However, *nonpeptide* AT1 receptor antagonists such as losartan, telmisartan, and olmesartan are now as widely utilized as ACE inhibitors to treat hypertension and reduce organ damage. Unlike the sarcosine compounds, this class of nonpeptide antagonists distinguished at least two Ang II receptor subtypes. AT1 receptor antagonism effectively blocks the actions of Ang II, but also results in significantly higher circulating levels of Ang II and Ang-(1–7) due to the disinhibition of renin. Thus, the cardiovascular actions of either ACE inhibition or AT1 receptor blockade involve the activation of either the AT2 or Ang-(1–7) receptors to some degree. In lieu of these actions, there is clear interest in the development of agonists to both the AT2 and Ang-(1–7) receptors to derive additional therapeutic benefit by activation of non-AT1 pathways. AVE0991 is the first nonpeptide agonist that mimics many actions of Ang-(1–7) including the stimulation of NO, reduction in tissue fibrosis, and reversal of vascular dysfunction, but does not recognize AT1 or AT2 receptors (Wiemer et al., 2002; Pinheiro et al., 2004; Benter et al., 2006; Santos and Ferreira, 2006; Ferreira et al., 2007). In addition to the oral availability of this compound, AVE0991 is not metabolized by ACE or other peptidases that contribute to the very short half-life of circulating Ang-(1–7) (Yamada et al., 1998). Altermans and colleagues have synthesized both peptide and nonpeptide agonists against the AT2 receptor (Wan et al., 2004a, 2004b; Wu et al., 2006). These compounds exhibit reasonable affinity towards the AT2 and stimulate neurite outgrowth in NG108 cells (Wu et al., 2006); however, the cardiovascular actions of these agents remain to be explored.

1.6 COMMON METHODOLOGIES AND ASSAY SYSTEMS

As is evident in this review, the emerging view of the RAAS is that of an exceedingly complex peptidergic system that exhibits a myriad array of biological actions. Indeed, one difficulty in the study of this system concerns the apparent effect to block one component and induce significant alterations in other components that may contribute to or attenuate the overall action. Although transgenic mice provide an undeniable opportunity to dissect out these complexities, developmental knockout models may induce their own unique phenotype. The methods described herein describe assays routinely employed in the laboratory to characterize the RAAS in the circulation and various tissue systems.

1.6.1 Measurement of Angiotensin Peptides

Radioimmunoassays (RIAs) continue to provide a sensitive and specific approach to measure angiotensin peptides in blood, urine, and tissue (Chappell et al., 1989). We currently utilize assays for Ang I, Ang II, Ang-(1–7), and Ang-(1–12). Each peptide shares the identical N-terminal sequence of angiotensinogen; therefore, antibodies are raised against the carboxy terminus by coupling to keyhole limpet hemocyanin (KLH) (Table 1.1). KLH is a strong immunogen, does not exhibit extensive homology with mammalian proteins, and the KLH-stimulated antibodies are also appropriate for immunocytochemical studies. The use of peptidase inhibitors in the collection of blood and the extraction of tissue prior to assay is critical to prevent the generation or metabolism *in vitro*. We employ the following inhibitors: 25 mM ethylenediaminetetraacetic acid (EDTA), 0.44 mM *O*-phenanthroline (PHEN), 1 mM 4-chloromercuribenzoic acid (PCMB), 0.12 mM pepstatin A, and 3 μM acetyl–His–Pro–Phe–Val–Statine–Leu–Phe, a specific rat renin inhibitor. The N-terminal metabolites of angiotensin (des–Asp^1, des–Asp^1–Arg^2, etc.) inevitably cross-react to some extent with the antibodies against the parent peptide; however, the endogenous content of these metabolites is generally quite low. Nevertheless, the immunoreactive data requires verification either by HPLC separation followed by RIA of each fraction or the use of mass spectroscopy (MS). As shown in Figure 1.3 (upper panel), the chromatograph reveals predominant immunoreactive peaks to Ang-(1–7), Ang-(4–8), and Ang II in renal extracts from wild-type mice. In the tissue ACE knockout mice (tACE$^{-/-}$), the peaks of Ang-(4–8) and Ang II are diminished (Figure 1.3, lower panel). These data are consistent with the predominant contribution of ACE activity to the generation of Ang II in the kidney (Modrall et al., 2003). We use HPLC, given the number of plasma and tissue samples that require multiple RIAs and that routine analysis of samples by MS is impractical at this time. Our current system employs a Waters NovaPak C18 narrow bore column and a heptafluorbutyric acid–acetonitrile (HFBA-ACN) solvent system performed on a Shimadzu CM2012LC (Shimadzu Scientific Instruments, Columbia, MD). The HFBA solvent (Sequanal Grade, Pierce, Rockford, IL) achieves baseline separation among 12 angiotensin peptides, exhibits high sample recovery, and its volatility reduces the potential interference with subsequent immunoassays. A 2-injector/2-column system to avoid the immunoreactive contribution of peptide standards (for establishment of elution times) to the endogenous content of samples is strongly recommended.

1.6.2 Measurement of Peptidase Activities

Of paramount importance to our understanding of the functional relationships among different angiotensin peptides, it is useful to establish the metabolism pathways involved in their formation and degradation. In the case of Ang II and Ang-(1–7), ACE, ACE2, and neprilysin play key roles in balancing the expression of these peptides (Figure 1.1) (Chappell, 2007a). Although a number of synthetic substrates are commercially available including those for ACE2, we utilize the endogenous peptide when possible to determine peptidase activity. We iodinate the peptide substrate by the chloramine T method to increase the sensitivity of detection and to determine the extent of uptake in intact cells (Chappell et al., 1990, 1994). Following incubation with serum, tissue, or cell preparations, the reaction products are separated by HPLC and the radioactive peaks quantified. This system is completely automated, utilizing a high-capacity automatic injector, and a flow-through gamma counter coupled to a data acquisition system (Shaltout et al., 2006). Using this approach in

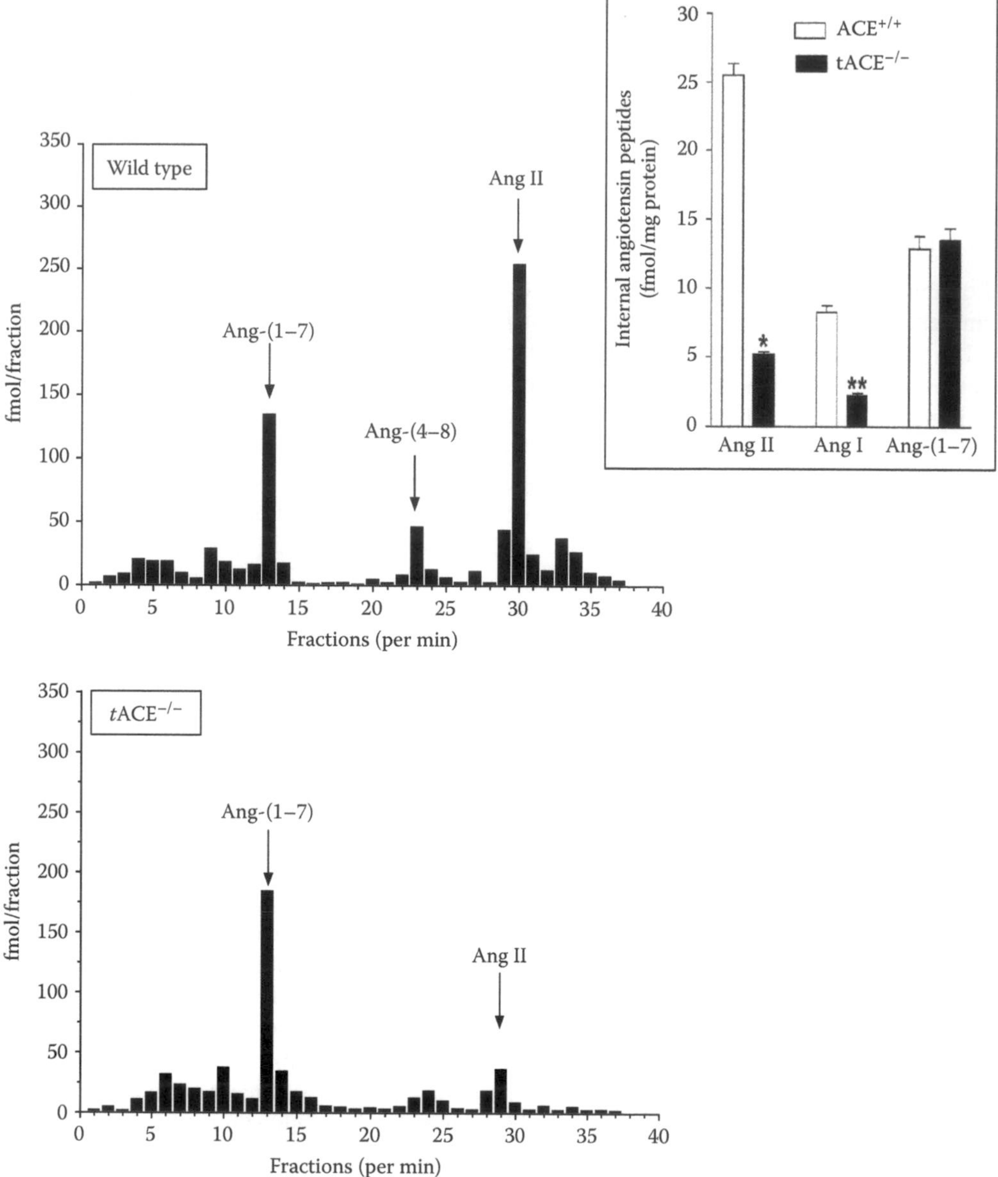

FIGURE 1.3 HPLC/RIA analysis of pooled kidney samples from wild-type and tissue ACE knockout ($tACE^{-/-}$) mice. The HPLC fractions were measured with Ang-(1–7) (fractions 1–20) and Ang II (fractions 21–40) RIAs. The arrows indicate the elution peak for Ang-(1–7), Ang-(4–8), and Ang II. Inset: intrarenal concentration of angiotensin peptides (fmol/mg protein) in wild-type ($n = 8$) and $tACE^{-/-}$ mice ($n = 8$); $*P < 0.001$; $**P < 0.01$. (Adapted from Modrall, J.G. et al., *Hypertension*, 43, 4849–4853, 2003.)

combination with selective peptidase inhibitors, we show abundant ACE2 activity in isolated proximal tubules of the sheep kidney and murine cardiac tissue that converts Ang II to Ang-(1–7), but does not hydrolyze Ang I (Shaltout et al., 2006; Garabelli et al., 2008). Moreover, we are able to characterize several competing peptidases for Ang II metabolism in proximal tubules and demonstrate that ACE2 is the major Ang-(1–7)-forming enzyme from Ang II (Figure 1.4).

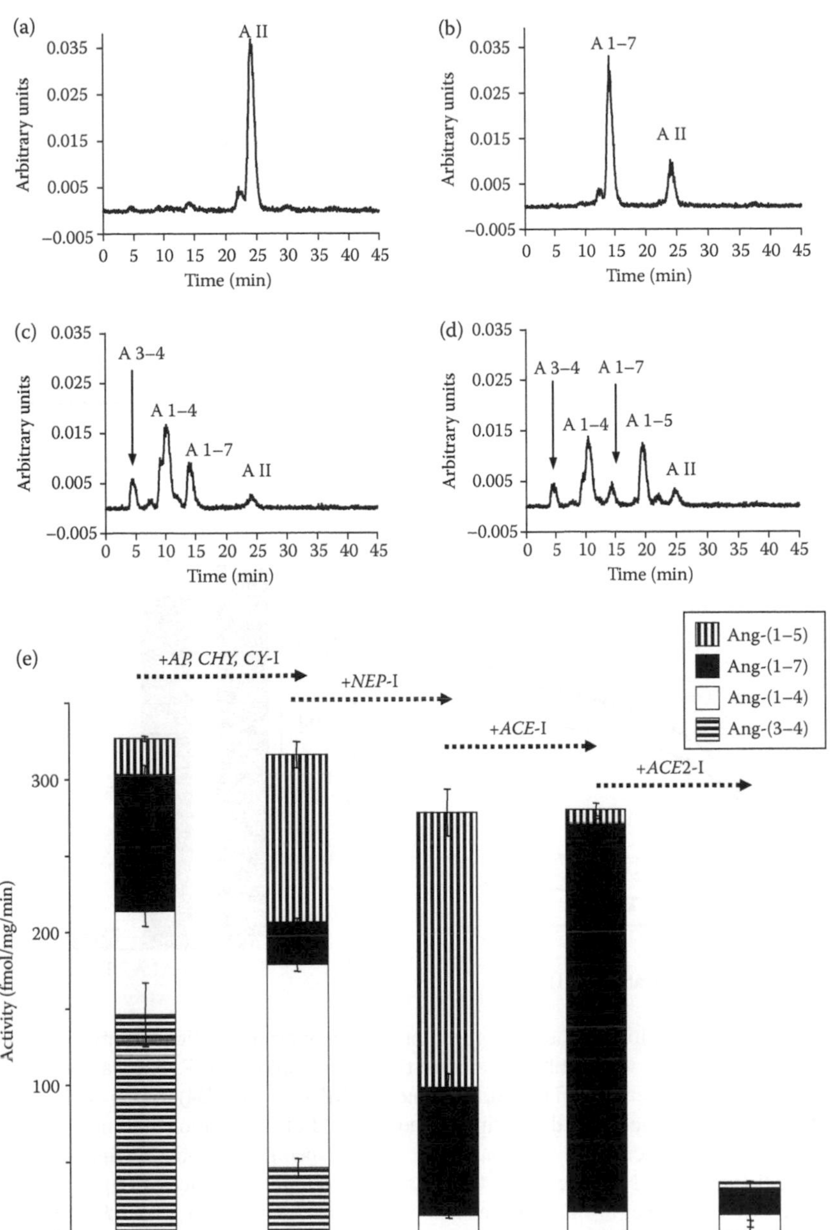

FIGURE 1.4 Metabolism of ^{125}I-Ang II by sheep proximal tubule membranes. ^{125}I-Ang II was incubated with 50 μg of proximal tubules membranes for 30 min at 37°C and the metabolites separated by HPLC. (a) A: ^{125}I-Ang II was not metabolized in the complete inhibitor cocktail. (b) Removal of the ACE2 inhibitor (ACE2-I) MLN4760 (MLN) revealed a prominent peak of Ang-(1–7) in the tubule membranes. (c) Further removal of the neprilysin inhibitor (NEP-I) SCH39370 (SCH) revealed peaks for Ang-(1–4) and Ang-(3–4), but a reduced peak for Ang-(1–7). (d) Removal of ACE inhibitor (ACE-I) lisinopril (LIS) revealed a prominent peak of Ang-(1–5) and reduced Ang-(1–7). (e) Quantification of the peptidase activities for ^{125}I-Ang II metabolism from the sheep proximal tubule membranes expressed as the rate of metabolism products formed in fmol/mg/min. Conditions: (a) no inhibitors; (b) inhibitor cocktail with aminopeptidase, chymase, and cysteine protease inhibitors (AP, CHY, CP-I); (c) addition of NEP-I; (d) addition of ACE-I; (e) addition of ACE2-I. Results are mean ± SEM; $n = 4$. (Modified from Shaltout, H.A. et al., *Am. J. Physiol. Renal Physiol.*, 292, F82–F91, 2006.)

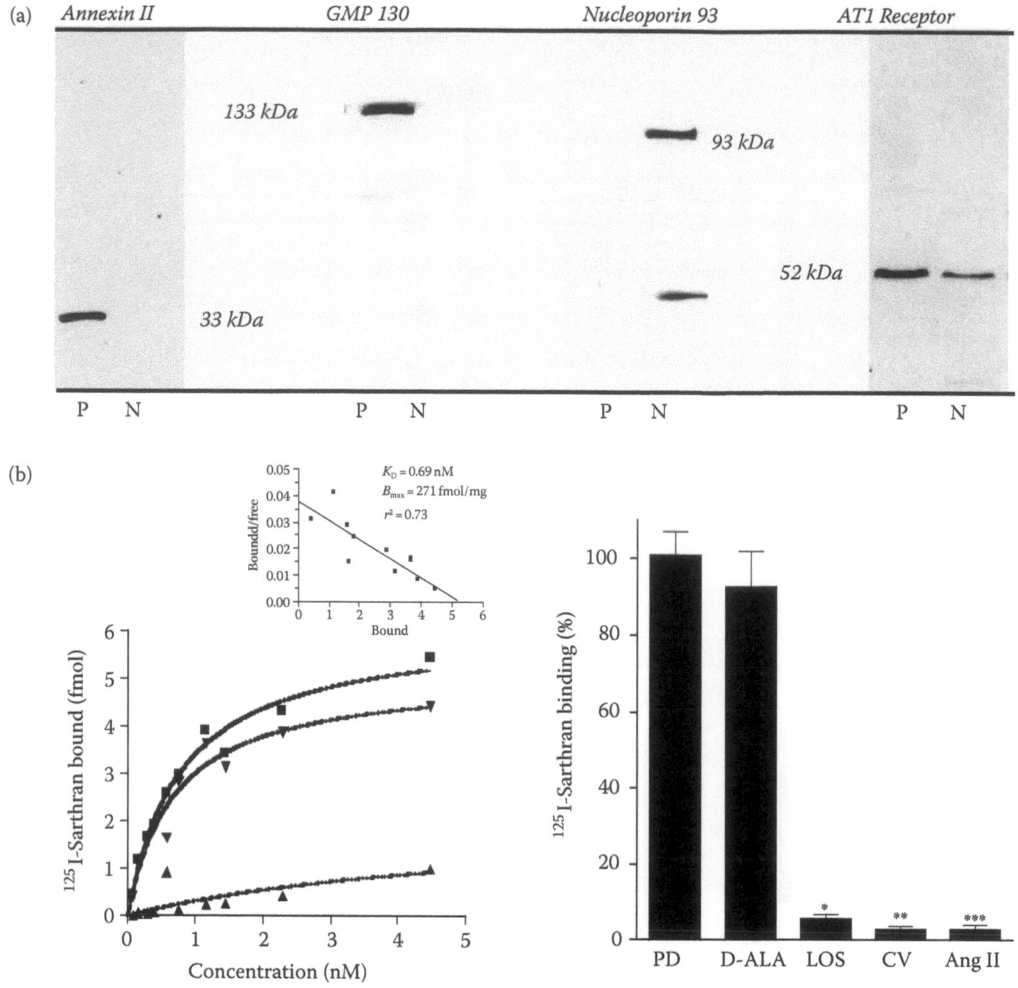

FIGURE 1.5 Characterization of the AT1 receptor in the nuclear fractions from rat renal cortex. (a) Immunoblot of full-length gel with antibodies against the AT1 receptor (AT1R, 52 kDa), nucleoporin (Nuc93, 93 kDa), annexin II (33 kDa), and Golgi membrane protein (GMP130, 133 kDa) for the nuclear (N) and plasma membrane (P) fractions obtained from the Optiprep gradient. (b) Left: Saturation binding and Scatchard analysis were performed with increasing concentrations of the specific receptor antagonist ^{125}I-(Sarcosine^{1}, threonine^{8})-Ang II (Sarthran). Nonspecific binding was obtained in the presence of 10 μM unlabeled Sarthran. The receptor density or number of binding sites and affinity are defined as B_{max} and K_D, respectively. Right: Competition studies with antagonists or Ang II utilized 0.5 nM ^{125}I-Sarthran. The antagonists included losartan (LOS), candesartan (CV), PD123319 (PD), (D-Ala^{7})-Ang-(1–7) (D-ALA), all used at a final concentration of 10 μM. Results are mean ± SEM; *$P < 0.05$ versus specific binding with respect to Sarthran, **$P < 0.05$ versus PD, ***$P < 0.05$ versus D-ALA ($n = 3$). (Modified from Pendergrass, K.D. et al., *Am. J. Physiol. Renal Physiol.*, 290, F1497–F1506, 2006.)

1.6.3 Isolation of Intracellular Components

For the isolation of intracellular components in renal tissue, we use an iodixanol (OptiPrep) density gradient. Tissues are initially homogenized in 20 mM Tricine KOH, 25 mM sucrose, 25 mM KCl, 5 mM $MgCl_2$, pH 7.8, with a Polytron Ultra Turrax T25 on ice and subsequently with a Teflon dounce homogenizer (Barnant Mixer Series 10, Barrington, IL; setting 3, 3 strokes). The homogenate is passed through a 100-μm mesh filter and centrifuged at 1000*g* for 10 min at 4°C. The resultant supernatant is centrifuged at 30,000*g* for 30 min to obtain the plasma membrane fraction. The crude nuclear pellet is diluted with 20% Optiprep diluted in the above buffer and layered on a 10, 20, 25, 30, and 35% gradient. The preparation is centrifuged at 10,000*g* for 20 min at 4°C and the isolated nuclei are obtained at the 30%/35% visible interface. This fraction is washed to remove the Optiprep medium (3000*g* for 10 min at 4°C) and compatible for immunoblot analysis as well as Ang II receptor binding. As shown in the immunoblot (Figure 1.5a), the nuclear fraction was devoid of markers for the plasma membrane (annexin) and the Golgi bodies (GMP 130) but positive for the nuclear pore protein nucleoporin 93. The AT1 receptor band was evident in both the nuclear (N) and plasma membrane (P) fractions. Saturation binding with ^{125}I-Sarthran revealed a high density of nuclear binding sites ($B_{max} > 200$ fmol/mg protein) that are sensitive to the AT1 receptor antagonists losartan (LOS) and candesartan (CV), but resistant to the AT2 and Ang-(1–7) antagonists PD123319 (PD) and (DAla^7)–Ang-(1–7) (D-ALA), respectively. These studies confirm the predominant expression of the AT1 receptor in the nuclear fraction of the rat kidney.

1.7 CLINICAL APPLICATIONS AND FUTURE DEVELOPMENTS

Therapeutic agents that block ACE or antagonize the AT1 receptor are very effective agents in the successful treatment of cardiovascular disease. Strategies to target AT2 or Ang-(1–7) receptors alone are likely to be limited without also attenuating to some degree the powerful and deleterious actions of the Ang II–AT1 receptor axis. The clinical relevance of the AT2 or Ang-(1–7) pathways may be of importance in therapies for noncardiovascular pathologies as well. Recent studies from Tallant and colleagues demonstrate that Ang-(1–7) reduces adenocarcinoma growth *in vivo* through the reduction in the cyclooxygenase-2/proinflammatory pathway (Menon et al., 2007). Indeed, the demonstration of the antiproliferative actions of Ang-(1–7) may explain the clinical observation that ACE inhibitor therapy is associated with a reduced incidence of certain types of cancer (Lever et al., 1998). In addition, Chow and colleagues (2008) have found that activation of AT2 receptors alone reduces the growth of the prostate cancer *ex vivo*. Finally, the emerging evidence for a functional intracellular RAS may require that future therapies also exhibit the ability to antagonize the AT1-dependent pathways, as well as activate AT2 or Ang-(1–7) systems within the cell.

ACKNOWLEDGMENTS

These studies were supported in part by grants from the National Institute of Health (HL-56973, HL-51952, HD047584, and HD017644) and the American Heart Association (AHA-151521 and AHA-355741). An unrestricted grant from the Unifi Corporation (Greensboro, NC) and the Farley-Hudson Foundation (Jacksonville, NC) is also acknowledged.

REFERENCES

Allred, A.J. et al., Pathways for angiotensin-(1–7) metabolism in pulmonary and renal tissues, *Am. J. Physiol.*, 279, F841–F850, 2000.

Baker, K.M. et al., Evidence of a novel intracrine mechanism in angiotensin II-induced cardiac hypertrophy, *Regul. Pept.*, 120, 5–13, 2004.

Baker, K.M. and Kumar, R., Intracellular angiotensin II induces cell proliferation independent of AT1 receptor, *Am. J. Physiol. Cell Physiol.*, 291, C995–C1001, 2006.

Batenburg, W.W. et al., Prorenin is the endogenous agonist of the (pro)renin receptor. Binding kinetics of renin and prorenin in rat vascular smooth muscle cells overexpressing the human (pro)renin receptor, *J. Hypertens.*, 25, 2441–2453, 2007.

Beaudry, H. et al., Involvement of protein kinase C alpha (PKC alpha) in the early action of angiotensin II type 2 (AT2) effects on neurite outgrowth in NG108–15 cells: AT2-receptor inhibits PKC alpha and p21ras activity, *Endocrinology*, 147, 4263–4272, 2006.

Benter, I.F. et al., Angiotensin-(1–7) prevents development of severe hypertension and end-organ damage in spontaneously hypertensive rats treated with L-NAME, *Am. J. Physiol. Heart Circ. Physiol.*, 290, H684–H691, 2006.

Benzing, T. et al., Angiotensin-converting enzyme inhibitor ramiprilat interferes with the sequestration of the B2 kinin receptor within the plasma membrane of native endothelial cells, *Circulation*, 99, 2034–2040, 1999.

Booz, G.W. et al., Angiotensin-II-binding sites on hepatocyte nuclei, *Endocrinology*, 130, 3641–3649, 1992.

Camargo de Andrade, M.C. et al., Expression and localization of N-domain ANG I-converting enzymes in mesangial cells in culture from spontaneously hypertensive rats, *Am. J. Physiol. Renal Physiol.*, 290, F364–F375, 2006.

Campbell, D.J. et al., Effects of neutral endopeptidase inhibition and combined angiotensin converting enzyme and neutral endopeptidase inhibition on angiotensin and bradykinin peptides in rats, *J. Pharmacol. Exp. Therap.*, 287, 567–577, 1998.

Carey, R.M., Cardiovascular and renal regulation by the angiotensin type 2 receptor. The AT_2 receptor comes of age, *Hypertension*, 45, 840–844, 2005.

Chappell, M.C., Emerging evidence for a functional angiotesin-converting enzyme 2-angiotensin-(1–7) mas receptor axis; more than regulation of blood pressure? *Hypertension*, 50, 596–599, 2007a.

Chappell, M.C. et al., Distinct processing pathways for the novel peptide angiotensin-(1–12) in the serum and kidney of the hypertensive mRen2.Lewis Rat, *Hypertension*, 50, E139 [P228], 2007b [Abstract].

Chappell, M.C. et al., Novel aspects of the renal renin-angiotensin system: angiotensin-(1-7), ACE2, and blood pressure regulation. *Contrib. Nephrol.*, 143, 77–89, 2004.

Chappell, M.C. et al., Conversion of angiotensin I to angiotensin-(1–7) by thimet oligopeptidase (EC 3.4.24.15) in vascular smooth muscle cells, *J. Vasc. Med. Biol.*, 5, 129–137, 1994.

Chappell, M.C. et al., Processing of angiotensin peptides by NG108–15 neuroblastoma X glioma hybrid cell line, *Peptides*, 11, 375–380, 1990.

Chappell, M.C. et al., Identification of angiotensin-(1–7) in rat brain: Evidence for differential processing of angiotensin peptides, *J. Biol. Chem.*, 264, 16518–16523, 1989.

Chen, Z. et al., Human ACE and bradykinin B_2 receptors form a complex at the plasma membrane, *FASEB*, 20, 2261–2270, 2006.

Chen, Z. et al., Hydrolysis of angiotensin peptides by human angiotensin I-converting enzyme and the resensitization of B_2 kinin receptors, *Hypertension*, 46, 1368–1373, 2005.

Chow, L. et al., Functional angiotensin II type 2 receptors inhibit growth factor signaling in LNCaP and PC3 prostate cancer cell lines, *Prostate*, 68, 651–660, 2008.

Clausmeyer, S., Sturzebecher, R., and Peters, J., An alternative transcript of the rat renin gene can result in a truncated prorenin that is transported into adrenal mitochondria, *Circ. Res.*, 84, 337–344, 1999.

Crackower, M.A. et al., Angiotensin-converting enzyme 2 is an essential regulator of heart function, *Nature*, 417, 822–828, 2002.

de Gasparo, M. et al., The angiotensin II receptors, *Pharmacol. Rev.*, 52, 415–472, 2000.

de Godoy, M.A. and Rattan, S., Translocation of AT1- and AT2-receptors by higher concentrations of angiotensin II in the smooth muscle cells of rat internal anal sphincter, *J. Pharmacol. Exp. Ther.*, 319, 1088–1095, 2006.

De Mello, W.C. and Danser, A.H., Angiotensin II and the heart: On the intracrine renin–angiotensin system, *Hypertension*, 35, 1183–1188, 2000.

Diz, D.I. et al., ACE2 inhibition in the solitary tract nucleus reduces baroreceptor reflex sensitivity for heart rate control, *Exp. Physiology*, 93, 694–700, 2008.

Drummond, W. et al., Antihypertensive efficacy of the oral direct renin inhibitor aliskiren as add-on therapy in patients not responding to amlodipine monotherapy, *J. Clin. Hypertens. (Greenwich)*, 9, 742–750, 2007.

Eggena, P. et al., Nuclear angiotensin receptors induce transcription of renin and angiotensinogen mRNA, *Hypertension*, 22, 496–501, 1993.

Elased, K.M. et al., New mass spectrometric assay for angiotensin-converting enzyme 2 activity, *Hypertension*, 47, 1010–1017, 2006.

Ferrario, C.M., Angiotensin-converting enzyme 2 and angiotensin-(1–7): An evolving story in cardiovascular regulation, *Hypertension*, 47, 515–521, 2006.

Ferrario, C.M., Trask, A.J., and Jessup, J.A., Advances in biochemical and functional roles of angiotensin converting enzyme 2 and angiotensin-(1–7) in the regulation of cardiovascular function, *Am. J. Physiol. Heart Circ. Physiol.*, 289, H2281–H2290, 2005.

Ferreira, A.J. et al., The nonpeptide angiotensin-(1–7) receptor Mas agonist AVE-0991 attenuates heart failure induced by myocardial infarction, *Am. J. Physiol. Heart Circ. Physiol.*, 292, H1113–H1119, 2007.

Fukai, T. et al., Modulation of extracellular superoxide dismutase expression by angiotensin II and hypertension, *Circ. Res.*, 85, 23–28, 1999.

Garabelli, P.J. et al., Distinct roles for angiotensin converting enzyme 2 and carboxypeptidase A in the processing of angiotensins in the murine heart, *Exp. Physiol.*, 93, 613–621, 2008.

Gendron, L. et al., Nitric oxide and cyclic GMP are involved in angiotensin II AT(2) receptor effects on neurite outgrowth in NG108–15 cells, *Neuroendocrinology*, 75, 70–81, 2002.

Goldfarb, D.A. et al., Angiotensin II receptor subtypes in the human renal cortex and renal cell carcinoma, *J. Urol.*, 151, 208–213, 1994.

Gonzalez-Villalobos, R. et al., Megalin binds and internalizes angiotensin-(1–7), *Am. J. Physiol. Renal Physiol.*, 290, F1270–F1275, 2006.

Gonzalez-Villalobos, R. et al., Megalin binds and internalizes angiotensin II, *Am. J. Physiol. Renal Physiol.*, 288, F420–F427, 2005.

Gradman, A.H., Pinto, R., and Kad, R., Current concepts: Renin inhibition in the treatment of hypertension, *Curr. Opin. Pharmacol.*, 8, 120–126, 2008.

Griendling, K.K. et al., Angiotensin II stimulates NADH and NADPH oxidase activity in cultured vascular smooth muscle cells, *Circ. Res.*, 74, 1141–1148, 1994.

Guo, C. et al., A novel vascular smooth muscle chymase is upregulated in hypertensive rats, *J. Clin. Invest.*, 107, 703–715, 2001.

Gurley, S.B. et al., Altered blood pressure responses and normal cardiac phenotype in ACE2-null mice, *J. Clin. Invest.*, 116, 2218–2225, 2006.

Gwathmey, T.M. et al., Nuclear angiotensin II type 2 (AT2) receptors are differentially expressed in the sheep kidney, *Hypertension*, 50, E97 [P6], 2007 [Abstract].

Haller, H. et al., Effects of intracellular angiotensin II in vascular smooth muscle cells, *Circ. Res.*, 79, 765–772, 1996.

Hein, L. et al., Intracellular trafficking of angiotensin II and its AT1 and AT2 receptors: Evidence for selective sorting of receptor and ligand, *Mol. Endocrinol.*, 11, 1266–1277, 1997.

Higuchi, S. et al., Angiotensin II signal transduction through the AT1 receptor: Novel insights into mechanisms and pathophysiology, *Clin. Sci. (London)*, 112, 417–428, 2007.

Hsu, Y.H. et al., Role of reactive oxygen species-sensitive extracellular signal-regulated kinase pathway in angiotensin II-induced endothelin-1 gene expression in vascular endothelial cells, *J. Vasc. Res.*, 41, 64–74, 2004.

Huang, Y. et al., Renin increases mesangial cell transforming growth factor-*B*1 matrix proteins through receptor-mediated, angiotensin II-independent mechanisms, *Kidney Int.*, 69, 105–113, 2006.

Hunyady, L. et al., Mechanisms and functions of AT(1) angiotensin receptor internalization, *Regul. Pept.*, 91, 29–44, 2000.

Ichihara, A. et al., The (pro)renin receptor and the kidney, *Semin. Nephrol.*, 27, 524–528, 2007.

Imig, J.D. et al., Renal endosomes contain angiotensin peptides, converting enzyme, and AT(1A) receptors, *Am. J. Physiol.*, 277, F303–F311, 1999.

Iwata, M. et al., Angiotensin-(1–7) bind to specific receptors on cardiac fibroblasts to initiate antifibrotic and antitrophic effects, *Am. J. Physiol. Heart Circ. Physiol.*, 289, H2356–H2363, 2005.

Iyer, S.N., Ferrario, C.M., Chappell, M.C., Angiotensin-(1–7) contributes to the antihypertensive effects of blockade of the renin–angiotensin system, *Hypertension*, 31, 356–361, 1998.

Iyer, S.N. et al., Evidence that prostaglandins mediate the antihypertensive actions of angiotensin-(1–7) during chronic blockade of the renin–angiotensin system, *J. Cardiovasc. Pharmacol.*, 36, 109–117, 2000.

Jessup, J.A. et al., Localization of the novel angiotensin peptide Angiotensin-(1-12) in the heart and kidney of hypertensive and normotensive rats, *Am. J. Physiol. Heart Circ. Physiol.*, 294, H2614–H2618, 2008.

Kaneshiro, Y. et al., Slowly progressive, angiotensin II-independent glomerulosclerosis in human (pro)renin receptor-transgenic rats, *J. Am. Soc. Nephrol.*, 18, 1789–1795, 2007.

Kase, H. et al., Supplementation with tetrahydrobiopterin prevents the cardiovascular effects of angiotensin II-induced oxidative and nitrosative stress, *J. Hypertens.*, 23, 1375–1382, 2005.

Khosla, M.C., Biological effects of angiotensin analogs and antagonists, in *The Renin–Angiotensin System*, Vol. II, Menard, J., Ed., Gower Medical Publishing Limited, London, p. 226–229, 1985.

Kohlstedt, K. et al., CK2 phosphorylates the angiotensin-converting enzyme and regulates its retention in the endothelial cell plasma membrane, *Circ. Res.*, 91, 749–756, 2002.

Kuba, K. et al., A crucial role of angiotensin converting enzyme 2 (ACE2) in SARS coronavirus-induced lung injury, *Nat. Med.*, 11, 875–879, 2005.

Kumar, R., Singh, V.P., and Baker, K.M., The intracellular renin–angiotensin system: A new paradigm, *Trends Endocrinol. Metab.*, 18, 208–214, 2007.

Kushiro, T. et al., Aliskiren, a novel oral renin inhibitor, provides dose-dependent efficacy and placebo-like tolerability in Japanese patients with hypertension, *Hypertension Res.*, 29, 997–1005, 2006.

Lever, A.F. et al., Do inhibitors of angiotensin-I-converting enzyme protect against risk of cancer? *Lancet*, 352, 179–184, 1998.

Lew, R.A. et al., ACE2 catalytic activity in plasma is masked by an endogenous inhibitor, *Exp. Physiol.*, 93, 685–693, 2008.

Li, X.C. and Zhuo, J.L., Intracellular angiotensin II induces in vitro transcription of TGF-β1, MCP-1, and NHE3 mRNAs in rat renal cortical nuclei via activation of nuclear AT1 receptors, *Am. J. Physiol. Cell Physiol.*, 294 (4), C1034–C1045, 2008.

Licea, H., Walters, M.R., and Navar, L.G., Renal nuclear angiotensin II receptors in normal and hypertensive rats, *Acta Physiol. Hung.*, 89, 427–438, 2002.

Lu, D. et al., Angiotensin II-induced nuclear targeting of the angiotensin type 1 (AT1) receptor in brain neurons, *Endocrinology*, 139, 365–375, 1998.

Matsubara, H. et al., Transactivation of EGF receptor induced by angiotensin II regulates fibronectin and TGF-beta gene expression via transcriptional and post-transcriptional mechanisms, *Mol. Cell Biochem.*, 212, 187–201, 2000.

Menon, J. et al., Angiotensin-(1–7) inhibits growth of a human lung adenocarcinoma xenografts in nude mice through a reduction in cyclooxygenase-2, *Cancer Res.*, 67, 2809–2815, 2007.

Mifune, M. et al., G protein coupling and second messenger generation are indispensable for metalloprotease-dependent, heparin-binding epidermal growth factor shedding through angiotensin II type-1 receptor, *J. Biol. Chem.*, 280, 26592–26599, 2005.

Modrall, J.G. et al., Depletion of tissue ace differentially influences the intrarenal and urinary expression of angiotensins, *Hypertension*, 43, 4849–4853, 2003.

Nagata, S. et al., Isolation and identification of proangiotensin-12, a possible component of the renin–angiotensin system, *Biochem. Biophys. Res. Commun.*, 350, 1026–1031, 2006.

Nakagawa, T. et al., The His–Pro–Phe motif of angiotensinogen is a crucial determinant of the substrate specificity of renin, *Biol. Chem.*, 388, 237–246, 2007.

Nguyen, G. et al., Pivotal role of the renin/prorenin receptor in angiotensin II production and cellular responses to renin, *J. Clin. Invest.*, 109, 1417–1427, 2002.

Ondetti, M.A., From peptides to peptidases: A chronicle of drug discovery, *Am. J. Cardiol.*, 49, 1390–1394, 1982.

Oudit, G.Y. et al., Loss of angiotensin-converting enzyme-2 leads to the late development of angiotensin II-dependent glomerulosclerosis, *Am. J. Pathol.*, 168, 1808–1820, 2006.

Oudit, G.Y. et al., Angiotensin II-mediated oxidative stress and inflammation mediate the age-dependent cardiomyopathy in ACE2 null mice, *Cardiovasc. Res.*, 75, 29–39, 2007.

Paul, M., Mehr, A.P., and Kreutz, R., Physiology of local renin–angiotensin systems, *Physiol. Rev.*, 86, 747–803, 2006.

Pendergrass, K.D. et al., Chronic depletion of renal angiotensin II does not influence the intracellular distribution of the renal AT1 receptor, *Hypertension*, 44 (4), 558[P220], 2004 [Abstract].

Pendergrass, K.D. et al., Differential expression of nuclear AT1 receptors and angiotensin II within the kidney of the male congenic mRen2.Lewis rat, *Am. J. Physiol. Renal Physiol.*, 290, F1497–F1506, 2006.

Pendergrass, K.D. et al., Nuclear AT1 receptors stimulate the formation of reactive oxygen species, *Hypertension*, 52, 4 [e48] (O67), 2008 [Abstract].

Peters, J. et al., Functional significance of prorenin internalization in the rat heart, *Circ. Res.*, 90, 1135–1141, 2002.

Pinheiro, S.V. et al., Nonpeptide AVE 0991 is an angiotensin-(1–7) receptor Mas agonist in the mouse kidney, *Hypertension*, 44, 490–496, 2004.

Re, R.N., Implications of intracrine hormone action for physiology and medicine, *Am. J. Physiol. Heart Circ. Physiol.* 284, H751–H757, 2003.

Redublo Quinto, B.M. et al., Expression of angiotensin I-converting enzymes and bradykinin B2 receptors in mouse inner medullary-collecting duct cells, *Int. Immunopharmacol.*, 8, 254–260, 2008.

Rice, G.I. et al., Circulating activities of angiotensin-converting enzyme, its homolog, angiotensin-converting enzyme 2, and neprilysin in a family study, *Hypertension*, 48, 914–920, 2006.

Sadjadi, J. et al., Angiotensin converting eyzyme-independent angiotensin II production by chymase is upregulated in the ischemic kidney in renovascular hypertension, *J. Surg. Res.*, 127, 65–69, 2005.

Sakoda, M. et al., (Pro)renin receptor-mediated activation of mitogen-activated protein kinases in human vascular smooth muscle cells, *Hypertens. Res.*, 30, 1139–1146, 2007.

Salomone, L.J. et al., Intrarenal dopamine D1-like receptor stimulation induces natriuresis via an angiotensin type-2 receptor mechanism, *Hypertension*, 49, 155–161, 2007.

Sampaio, W.O. et al., Angiotensin-(1–7) through receptor mas mediates endothelial nitric oxide synthase activation via Akt-dependent pathways, *Hypertension*, 49, 185–192, 2007a.

Sampaio, W.O. et al., Angiotensin-(1–7) counterregulates angiotensin II signaling in human endothelial cells, *Hypertension*, 50, 1093–1098, 2007b.

Santos, R.A. and Ferreira, A.J., Pharmacological effects of AVE 0991, a nonpeptide angiotensin-(1–7) receptor agonist, *Cardiovasc. Drug Rev.*, 24, 239–246, 2006.

Santos, R.A. et al., Characterization of a new selective antagonist for angiotensin-(1–7), D-pro7-angiotensin-(1–7), *Hypertension*, 41, 737–743, 2003a.

Santos, R.A. et al., Angiotensin-(1–7) is an endogenous ligand for the G protein-coupled receptor Mas, *Proc. Natl Acad. Sci. USA*, 100, 8258–8263, 2003b.

Santos, R.A. et al., Characterization of a new angiotensin antagonist selective for angiotensin-(1–7): Evidence that the actions of angiotensin-(1–7) are mediated by specific angiotensin receptors, *Brain Res. Bull.*, 35, 293–298, 1994.

Schefe, J.H. et al., A novel signal transduction cascade involving direct physical interaction of the renin/prorenin receptor with the transcription factor promyelocytic zinc finger protein, *Circ. Res.*, 99, 1355–1366, 2006.

Schiffrin, E.L., Role of endothelin-1 in hypertension, *Hypertension*, 34, 876–881, 1999.

Schnackenberg, C.G. and Wilcox, C.S., Two-week administration of tempol attenuates both hypertension and renal excretion of 8-Iso prostaglandin f2alpha, *Hypertension*, 33, 424–428, 1999.

Shaltout, H.A. et al., Angiotensin metabolism in renal proximal tubules, urine and serum of sheep: Evidence for ACE2-dependent processing of Angiotensin II, *Am. J. Physiol. Renal Physiol.*, 292, F82–F91, 2006.

Singh, V.P. et al., High-glucose-induced regulation of intracellular ANG II synthesis and nuclear redistribution in cardiac myocytes, *Am. J. Physiol. Heart Circ. Physiol.*, 293, H939–H948, 2007.

Siragy, H.M. and Carey, R.M., The subtype 2 (AT_2) angiotensin receptor mediates renal production of nitric oxide in conscious rats, *J. Clin. Invest.*, 100, 264–269, 1997.

Siragy, H.M., Inagami, T., and Carey, R.M., NO and cGMP mediate angiotensin AT2 receptor-induced renal renin inhibition in young rats, *Am. J. Physiol. Regul. Integr. Comp. Physiol.*, 293, R1461–R1467, 2007.

Siragy, H.M. et al., Sustained hypersensitivity to angiotensin II and its mechanism in mice lacking the subtype-2 (AT2) angiotensin receptor, *Proc. Natl Acad. Sci. USA*, 96, 6506–6510, 1999.

Su, Z., Zimpelmann, J., and Burns, K.D., Angiotensin-(1–7) inhibitis angiotensin II-stimulated phosphorylation of MAP kinases in proximal tubular cells, *Kidney Int.*, 69, 2212–2218, 2006.

Takahasi, K. et al., Protein tyrosine phosphatase inhibition by angiotensin II in rat pheochromocytoma cells through type 2 receptor, AT2, *Biochem. Biophys. Res. Commun.*, 198, 60–66, 1994.

Tallant, E.A., Diz, D.I., and Ferrario, C.M., State-of-the-art lecture. Antiproliferative actions of angiotensin-(1–7) in vascular smooth muscle, *Hypertension*, 34, 950–957, 1999.

Tallant, E.A., Ferrario, C.M., and Gallagher, P.E., Angiotensin-(1–7) inhibits growth of cardiac myocytes through activation of the mas receptor, *Am. J. Physiol. Heart Circ. Physiol.* 289, H1560–H1566, 2005.

Tang, S.S. et al., Characterization of nuclear angiotensin-II-binding sites in rat liver and comparison with plasma membrane receptors, *Endocrinology*, 131, 374–380, 1992.

Thomas, W.G., Thekkumkara, T.J., and Baker, K.M., Molecular mechanisms of angiotensin II (AT1A) receptor endocytosis. [Review], *Clin. Exp. Pharmacol. Physiol.*, Suppl. 3, S74–S80, 1996.

Tigerstedt, R. and Bergman, P.G., Niere und kreislant, *Skand Arch Physiol.*, 8, 223–271, 1898.

Tikellis, C. et al., ACE2 deficiency modifies renoprotection afforded by ACE inhibition in experimental diabetes, *Diabetes*, 57, 1018–1025, 2008.

Touyz, R.M. and Schiffrin, E.L., Ang II-stimulated superoxide production is mediated via phospholipase D in human vascular smooth muscle cells, *Hypertension*, 34, 976–982, 1999.

Trask, A.J. et al., Angiotensin-(1–12) is an alternate substrate for angiotensin peptide production in the heart. *Am. J. Physiol. Heart Circ. Physiol.*, 294, H2242–H2247, 2008.

Tsutsumi, Y. et al., Angiotensin II type 2 receptor overexpression activates the vascular kinin system and causes vasodilation, *J. Clin. Invest.*, 104, 925–935, 1999.

Varagic, J. et al., New angiotensins, *J. Renin Angiotensin Aldosterone Syst.*, 86, 663–671, 2008.

Vickers, C. et al., Hydrolysis of biological peptides by human angiotensin-converting enzyme-related carboxypeptidase, *J. Biol. Chem.*, 277, 14838–14843, 2002.

Wan, Y. et al., First reported nonpeptide AT1 receptor agonist (L-162,313) acts as an AT2 receptor agonist *in vivo*, *J. Med. Chem.*, 47, 1536–1546, 2004a.

Wan, Y. et al., Design, synthesis, and biological evaluation of the first selective nonpeptide AT2 receptor agonist, *J. Med. Chem.*, 47, 5995–6008, 2004b.

Weiss, D., Kools, J.J., and Taylor, W.R., Angiotensin II-induced hypertension accelerates the development of atherosclerosis in ApoE-deficient mice, *Circulation*, 103, 448–454, 2001.

Wiemer, G. et al., AVE 0991, a nonpeptide mimic of the effects of angiotensin-(1–7) on the endothelium, *Hypertension*, 40, 847–852, 2002.

Wilcox, C.S., Oxidative stress and nitric oxide deficiency in the kidney: A critical link to hypertension? *Am. J. Physiol. Regul. Integr. Comp. Physiol.*, 289, R913–R935, 2005.

Wong, D.W. et al., Loss of angiotensin-converting enzyme-2 (Ace2) accelerates diabetic kidney injury, *Am. J. Pathol.*, 171, 438–451, 2007.

Wu, X. et al., Selective angiotensin II AT2 receptor agonists: Arylbenzylimidazole structure–activity relationships, *J. Med. Chem.*, 49, 7160–7168, 2006.

Yamada, K. et al., Converting enzyme determines the plasma clearance of angiotensin-(1–7), *Hypertension*, 98, 496–502, 1998.

Yamamoto, K. et al., Deletion of angiotensin-converting enzyme 2 accelerates pressure overload-induced cardiac dysfunction by increasing local angiotensin II, *Hypertension*, 47, 718–726, 2006.

Yamamoto, K. et al., *In vivo* metabolism of angiotensin I by neutral endopeptidase (EC 3.4.24.11) in spontaneously hypertensive rats, *Hypertension*, 19, 692–696, 1992.

Zhuo, J.L. et al., Intracellular ANG II induces cytosolic Ca^{2+} mobilization by stimulating intracellular AT1 receptors in proximal tubule cells, *Am. J. Physiol. Renal Physiol.*, 290, F1382–F1390, 2006.

2 CT, CGRP, Amylin, and Adrenomedullins

Debbie L. Hay and David R. Poyner

CONTENTS

The calcitonin family of peptides includes calcitonin, calcitonin gene-related peptide (CGRP), adrenomedullin (AM), AM2 (also known as intermedin), and amylin, all sharing a common structure. Biological activity depends on the presence of a disulfide-bonded ring at the N-terminus of each peptide. This is followed by an amphipathic α-helix and there is a C-terminal amide. These peptides act at seven receptors produced by the interaction of two G-protein coupled receptors (the calcitonin receptor and the calcitonin receptor-like receptor) with three accessory proteins (receptor activity modifying proteins 1, 2, and 3). Thus, each peptide has a complicated pharmacology, with the ability to act at multiple receptors. There are few selective antagonists, although progress is now being made in this regard, at least for CGRP receptors. The use of genetic models, combined with traditional pharmacological techniques, has revealed roles for the peptides in a wide range of processes including bone formation, nutrient intake, vasodilation, cardioprotection, pain perception, neurogenic inflammation, angiogenesis, and lymphangiogenesis. Thus the family is of considerable clinical potential and the receptors for these peptides are important drug targets.

2.1 INTRODUCTION

2.1.1 The Calcitonin Family

Calcitonin (CT), amylin, CT gene-related peptide (CGRP), and adrenomedullin (AM) comprise the CT family of peptides. Recently, AM2 (intermedin) was added to this list (Roh et al., 2004; Takei et al., 2004). A series of other related peptides, known as CT receptor-stimulating peptides (CRSP) have also been identified in several species (Katafuchi et al., 2003a, 2003b, 2005; Katafuchi and Minamino, 2004). Although the members of this family share only low amino acid sequence homology, they have common structural features. Each peptide has two cysteine residues close to its N-terminus, which form a disulfide bridge; this region is critical for biological activity in all cases. An α-helical region is present towards the N-terminus of each of the peptides and an amidated residue at the carboxy terminus is also a family trait. An alignment of CT family peptides is shown in Table 2.1.

TABLE 2.1
Alignment of the CT Family of Peptides

hAM2	GC-VLGTCQVQNLSHRLWQLMGPAGRQDSAPVDPSSPHSY
rAM2	GC-VLGTCQVQNLSHRLWQLVRPSGRRDSAPVDPSSPHSY
hAM	GC-RFGTCTVQKLAHQIYQFTD-KDKDNVAPRSKISPQGY
rAM	GC-RFGTCTMQKLAHQIYQFTD-KDKDGMAPRNKISPQGY
rAmylin	KC-NTATCATQRLANFLVRSSN-NLGPVLPPTN-VGSNTY
hAmylin	KC-NTATCATQRLANFLVHSSN-NFGAILSSTN-VGSNTY
hβCGRP	AC-NTATCVTHRLAGLLSRSGG-MVKSNFVPTN-VGSKAF
hαCGRP	AC-DTATCVTHRLAGLLSRSGG-VVKNNFVPTN-VGSKAF
rβCGRP	SC-NTATCVTHRLAGLLSRSGG-VVKDNFVPTN-VGSKAF
rαCGRP	SC-NTATCVTHRLAGLLSRSGG-VVKDNFVPTN-VGSEAF
pCRSP3	SC-NTAICVTHKMAGWLSRSGS-VVKNNFMPIN-MGSKVL
pCRSP2	SC-NTASCVTHKMTGWLSRSGS-VAKNNFMPTN-VDSKIL
pCRSP1	SC-NTATCMTHRLVGLLSRSGS-MVRSNLLPTK-MGFKVF
rCT	-CGNLSTCMLGTYTQDLNKFHT-------FPQTSIGVGAP
hCT	-CGNLSTCMLGTYTQDFNKFHT-------FPQTAIGVGAP

Note: r, rat; h, human; p, porcine. For rat and human AM and AM2, the N-terminal extensions of the mature peptides are not shown.

2.1.2 Calcitonin

The CTs are 32 amino acids in length and the shortest members of this family. They have been classified into three groups; artiodactyls, primate/rodent, and teleost. The disulfide bridge of all members is located between residues 1 and 7 (Table 2.1) and a proline amide is present at the carboxy terminus (Sexton et al., 1999). Interestingly, salmon CT is considerably more potent than human CT at human CT receptors. A number of modified CT peptides have been generated (see discussion of structural analogs, Section 2.5.2); however, a thorough discussion is beyond the scope of this chapter. For a comprehensive analysis, the review of Sexton et al. (1999) should be consulted. The CT receptor mediates the actions of CT.

2.1.3 Amylin

Amylin or islet amyloid polypeptide (IAPP) is a 37-amino-acid peptide first purified from amyloid deposits in the pancreatic islets of type 2 diabetic patients (Cooper et al., 1987). Amylin probably has a hormonal role in the regulation of nutrient intake (Young et al., 1998). The sequences of human and rat amylin peptides are shown in Table 2.1. There are multiple subtypes of amylin receptors (AMY_1–AMY_3).

2.1.4 CGRP

αCGRP was first described in 1982 (Amara et al., 1982). It is a 37-amino-acid peptide, formed by alternate splicing of the CT gene with a disulfide bridge between residues 2 and 7. In some species (human, mouse, rat) a second version is found, βCGRP (Amara et al., 1985). In humans, the two differ by three amino acids; in the rat, there is only a single amino acid difference (Steenbergh et al., 1986). Generally, they have very similar biological activities. They are extremely potent vasodilators, although they have broader physiological or pharmacological effects. CGRP produces its main actions via the CGRP receptor (McLatchie et al., 1998).

Multiple forms of CGRP are found in teleost fish (Ogoshi et al., 2006). Although one of these arises from the same CT/CGRP gene as found in mammals, the duplication event in teleosts was unrelated to the gene duplication in mammals that has given rise to α- and βCGRP (and also the CRSPs, see below). For this reason, some authorities prefer to refer to the teleost CGRPs as CGRP-1 and -2 (Ogoshi et al., 2006).

2.1.5 CRSP

A number of mammalian species (pig, horse, cow, and dog) do not express βCGRP but instead express up to three forms of CRSP (Katafuchi and Minamino, 2004). The CRSPs are all produced as propeptides. The predicted mature CRSP-1 has 38 amino acids, and CRSP-2 and -3 have 37 amino acids. However, for CRSP-2 and -3, the presumed N-terminal cleavage site lacks the characteristic dibasic amino acid motif. A C-terminally extended form of CRSP-1 has been isolated with an extra glycine; significantly, this form of the peptide lacks the C-terminal amide found in all other members of the CT peptide family (Katafuchi et al., 2005). Furthermore, dog and bovine CRSP-1 cannot have this amide, as their precursors lack the necessary glycine immediately after the C-terminal cleavage site that is required for its formation (Katafuchi and Minamino, 2004). The peptides all show substantial homology to CGRP. In the pig (the only species so far known to express all three forms), CRSP-1, -2, and -3 have, respectively, 50%, 81%, and 86% identity compared to CGRP. It appears likely that the CRSP gene evolved at a relatively recent stage in mammalian evolution by duplication of the CGRP/CT gene (Ogoshi et al., 2006). However, in spite of this, the peptides have no actions on the CGRP receptor but instead stimulate the CT receptor (Katafuchi and Minamino, 2004).

2.1.6 AM

AM was discovered in 1993 (Kitamura et al., 1993). Human AM is a 52-amino-acid peptide. The first 14 amino acids are not homologous to anything found in CT, CGRP, or amylin. However, the remaining 38 residues show 24% identity to CGRP, with a disulfide bond between cysteines 16 and 20 and a C-terminal amide. Rat AM is homologous to the human peptide, although it has only 50 amino acids, having only 13 residues before the start of the disulfide. AM is a potent vasodilator and is essential for proper angiogenesis and lymphangiogenesis. It produces its actions via AM_1 and AM_2 receptors (McLatchie et al., 1998).

It is apparent from studies on other species, especially teleost fish, that there has been considerable expansion of the AM family; five distinct peptides have been identified in some fish, known as AM1 to 5 (Ogoshi et al., 2006). Of these, AM2 has been identified in humans. There is unfortunately some confusion as to its names; it was discovered independently by two groups and it is also known as intermedin (Roh et al., 2004; Takei et al., 2004). It appears to exist in two forms, one with 47 amino acids and a second with an extra 6 amino acids at the N-terminus. As with AM, the N-terminal sequence has no homology with CT, CGRP, or amylin, but thereafter there is a five-member disulfide bonded ring and a C-terminal amide. There is 28% identity with AM (Roh et al., 2004; Takei et al., 2004). A sequence corresponding to AM5 has been found in the EST databases from dog, cow, and pig. In primates the equivalent of this sequence is also present but it lacks two nucleotides, and it may not be transcribed (Ogoshi et al., 2006). It has been pointed out that, based on the current (albeit limited) data, the species that apparently express AM5 also express CRSP (Ogoshi et al., 2006). Nothing corresponding to AMs 3 or 4 has been identified in mammals.

AM2 will activate all the CT family receptors, albeit with modest potency (Roh et al., 2004; Hay et al., 2005). Its physiological role and cognate receptor remain to be established.

2.2 SOURCE OF PEPTIDES

2.2.1 CT AND αCGRP

CT is a circulating hormone, produced by the parafollicular (C-) cells of the thyroid. In contrast, CGRP is a neurotransmitter, found throughout the sensory nervous system and also in the central nervous system (CNS) and some motor nerves. The CT/αCGRP gene has six exons. Exons 1 to 4 are transcribed to give preproCT, a 141-amino-acid preprohormone in humans. Cleavage of the signal peptide gives proCT, which is 116 amino acids long. Within this, it is the equivalent of exon 4 that codes for CT and also a 21-amino-acid extension at its C-terminus sometimes called katacalcin or PDN-21. For CGRP, exons 1 to 3 and 5 to 6 are transcribed to give preproCGRP (Rosenfeld et al., 1983). This is 128 residues long; CGRP is located between residues 83 and 119 (Gkonos et al., 1986). βCGRP is produced as a 127-amino-acid preprohormone but from a distinct gene located adjacent to the CT/αCGRP gene on chromosome 11 of humans.

The concentration of αCGRP at the site of release by neuronal cells is likely to be much higher than in the blood. In male Sprague–Dawley rats, the fasted peripheral plasma concentration of endogenous CGRP-like immunoreactivity (CGRP-LI) was reportedly ~25 pM. In the portal vein the reported concentrations were ~41 pM and in skeletal muscle the amount of CGRP detected was 313 ± 40 pmol/kg (Rossetti et al., 1993). In other rat studies, endogenous concentrations were reported as ~5 pM (female Sprague–Dawley, Gangula et al., 2000a), ~21 pM (male Sprague–Dawley, Luo et al., 2004) or ~10–40 pM (male Wistar rats, Zaidi et al., 1985). In healthy human volunteers, reported circulating endogenous CGRP-LI concentrations were ~25 pM (Ando et al., 1992). In other studies, the reported concentrations were ~3 to 25 pM (Beglinger et al., 1988; Anand et al., 1991). Although generally in good agreement with one another, variations in values obtained between studies may relate to the assay used for measurement and the circumstances of sampling. Circulating CGRP-LI concentrations may be higher in certain disease states, for example, congestive heart failure (Anand et al., 1991; Shekhar et al., 1991). Resting plasma CT concentrations are ~5 to 10 pM (Schifter, 1993).

2.2.2 AMYLIN

Amylin is a 37-amino-acid hormone produced by β-cells within the islets of Langerhans in the pancreas as an 89-amino-acid precursor; the mature peptide is located between residues 34 and 70. The gene has three exons. Plasma amylin concentrations are typically around 5 pM (Mitsukawa et al., 1990; reviewed by Cooper, 1994).

2.2.3 ADRENOMEDULLIN

AM is a locally released paracrine factor. The most important physiological sites of production are the vascular endothelial cells, although it can also be produced from phaeochromocytoma tumors (Kitamura et al., 1993). It is initially produced as a 185-amino-acid preprohormone in humans from three consecutive exons in its gene; the peptide has a 21-residue signal peptide and the mature peptide begins at residue 96. Three other peptides are produced from the preprohormone: adrenotensin, proadrenomedullin N-terminal 20 peptide (PAMP), and proadrenomedullin45–92. Of these, PAMP has the best-characterized actions, inhibiting noradrenaline release from sympathetic nerves (Hinson et al., 2000). Adrenotensin has been reported to increase blood pressure (Hinson et al., 2000). It has been suggested that proadrenomedullin45–92 has similar actions to PAMP (Cuifen et al., 2008).

Circulating AM concentrations are ~2 pM, but this is likely to bear little resemblance to the concentration around endothelial cells (Letizia et al., 2002). Furthermore, AM bound to its binding protein, complement factor H, may underestimate the amount of circulating AM (Pio et al., 2002).

AM2 also appears to be a locally produced paracrine factor. It is produced from a 148-amino-acid precursor (Roh et al., 2004).

2.2.4 Degradation

The CT family of peptides are subject to proteolysis by various peptidases. For CGRP, the most significant appear to be neural endopeptidase, aminopeptidase N, carboxypeptidase, and the mast cell tryptase. Angiotensin converting enzyme appears to be of less significance (Kramer et al., 2006). For amylin, insulin-degrading enzyme is able to cause its breakdown. This is predominantly an intracellular enzyme; it could be responsible for the deposition of amyloid deposits derived from amylin in pancreatic β-cells (Bennett et al., 2000). Very little information is available about degradation of AM. Inhibition of neural endopeptidase can potentiate its activities, although this may not be a direct action (Rademaker et al., 2002).

2.3 MOLECULAR MECHANISM OF ACTION

2.3.1 Introduction

2.3.1.1 Ligand Binding

CT receptors belong to the B family of G protein-coupled receptors (GPCRs). This family contains receptors such as the secretin receptor and parathyroid hormone receptor, in addition to the CT receptor-like receptor (CLR, Section 2.3.1.2). The peptide-binding subfamily of family B receptors share conserved cysteines in their extracellular N-termini of around 100 to 160 amino acids (Hoare, 2005). Compared with family A GPCRs, which includes rhodopsin and adrenergic receptors, understanding of ligand binding mechanisms is relatively sparse but predicted to be rather different to the family A model (Hoare, 2005).

Given that these receptors have long N-termini and that they bind to large peptides (up to around 50 amino acids in length) it is expected that there are multiple sites of contact between the peptide and receptor. The limited dataset predicts a general mechanism for peptide binding to these receptors in which the C-terminal peptide region is thought to bind to the extracellular N-terminus of the receptor. This promotes the interaction of the N-terminal region of the peptide with the receptor juxtamembrane domain. Receptor activation and intracellular signaling ensue. Important support for the model was recently provided by the NMR structure of the mouse corticotrophin-releasing factor $2_{(b)}$ receptor and the crystal structure of the extracellular domain of the GIP receptor, bound to its cognate ligand (Grace et al., 2007; Malde et al., 2007). A significant body of evidence for this mechanism also comes from the study of receptor chimeras with CT receptors, for example, PTH/CT receptor chimeras (Bergwitz et al., 1996).

2.3.1.2 RAMPs

Although the receptor for CT is a conventional family B GPCR, the receptors for CGRP, AM, and amylin require additional proteins, from one of the receptor activity modifying proteins (RAMPs) (McLatchie et al., 1998; Hay et al., 2006 for review). There are three RAMPs in mammals, although considerably more are found in some fish species, matching the expansion of the CT peptide family. The RAMPs interact with the CT receptor to convert it to receptors for amylin (see Section 2.3.3 and Figure 2.1). In the case of CGRP and AM, a distinct but related family B GPCR, the CLR can interact with RAMP1 to give a CGRP receptor and RAMP2 or 3 to give AM receptors (see Sections 2.3.4, 2.3.5 and Figure 2.1); CLR by itself will bind no known endogenous ligand. Each RAMP has a short intracellular C-terminal tail and a single transmembrane spanning domain. This is then followed by an N-terminus of ~100 amino acids. It is predominantly the interaction of the N-terminus of the RAMP with the N-terminus of the GPCR that determines the pharmacology of the complex. It remains to be established whether the RAMPs directly interact with the bound ligands or whether they act indirectly by altering the conformation of the GPCR.

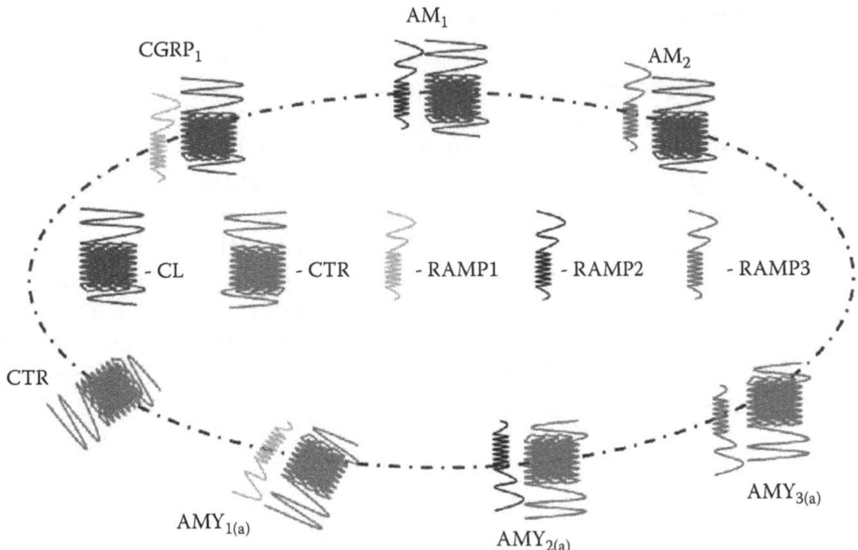

FIGURE 2.1 CT-family peptide receptors. The receptor subunits combine intracellularly to form complexes as indicated at the cell surface.

2.3.1.3 Receptor Activation

There is very little evidence for how ligand binding is converted to receptor activation for family B GPCRs. For family A GPCRs, a general scheme has been proposed in which there is bending of the transmembrane helices, particularly around a conserved proline in thc 6th transmembrane helix, resulting in the movement of the intracellular loops to open a binding pocket for G-proteins. In both the CGRP receptor and the CT receptor, mutation of a conserved proline in the 6th transmembrane helix also disrupts receptor activation, suggesting that at least elements of this mechanism may be conserved (Conner et al., 2005; Bailey and Hay, 2007).

2.3.2 CT Receptors

2.3.2.1 Mammalian CT Receptors

The first CT receptor was cloned in 1991 (Lin et al., 1991). This was a porcine receptor. The human homolog of this receptor, now known as $CT_{(b)}$, was reported a year later (Gorn et al., 1992; Poyner et al., 2002). A CT receptor with identical sequence (apart from 16 amino acids fewer in the first intracellular loop as a result of alternative splicing) is known as $CT_{(a)}$. These receptors have formerly been known as CTR1 and CTR2 or insert positive or negative, respectively. Other splice variants, such as those that lack the first 47 amino acids in the N-terminus, are also found in humans. These have been reviewed (Sexton et al., 1999; Purdue et al., 2002). $CT_{(a)}$ is the most widely distributed receptor and the most studied. In rodents, receptors analogous to $CT_{(a)}$ and $CT_{(b)}$ are found; however, in these receptors the insert is 37 amino acids and is found in the first extracellular loop. The receptor variants differ in their binding properties, their ability to couple to signal transduction pathways, and some have dominant negative activity. The physiological significance of the variants is unclear. Another variant of the human CT receptor has a T to C base mutation, encoding a leucine^{447} to proline change. There is apparently no effect on ligand binding or receptor function *in vitro*, but the polymorphism is associated with decreased fracture risk in post-menopausal women. The prevalence of the polymorphism differs between ethnic groups (Nakamura et al., 1997; Masi et al., 1998; Taboulet et al., 1998; Wolfe et al., 2003).

2.3.2.2 Fish CT Receptors and CT Receptors from Other Species

CT receptors from several mammalian species have been cloned and studied quite extensively, but fish CT receptors have only recently been cloned. The molecular identity of these receptors has long been sought, given the well-documented, high-affinity binding of salmon and eel CT to mammalian CT receptors. Mefugu CT receptor has recently been cloned and its structure reveals unique hormone binding domains (HBD) (Nag et al., 2007). This receptor bound salmon and eel CT but human CT did not displace radiolabeled salmon CT from this receptor. The extracellular N-terminus of this protein is ~470 amino acids in length and comprises four putative HBDs, each with ~50% amino acid identity to the human CT receptor. Interestingly, when truncated forms of the protein were tested, only one of these domains (HBD-D) was required for the receptor to bind ligand and activate cAMP. Mefugu, Torafugu, *Tetraodon*, and Stickleback have four HBDs, whereas Medaka have three. In Tetrapod animals, this has been reduced to one and forms the N-terminus, which we recognize in mammalian CT receptors.

2.3.3 Amylin Receptors

Amylin receptors have always been closely associated with CT receptors. This is because amylin receptors are multimeric complexes, formed from the interaction of the CT receptor with RAMPs. The CT receptor (or its splice variants) interact with three RAMPs to generate several subtypes of amylin receptor (e.g., $AMY_{1(a)-3(a)}$; Figure 2.1). These amylin receptor subtypes are pharmacologically distinct. $AMY_{1(a)}$ and $AMY_{3(a)}$ receptors are the best characterized. Both bind amylin with high affinity but $AMY_{1(a)}$ receptors also interact with CGRP with high affinity; $AMY_{3(a)}$ receptors show less preference for this peptide (Hay et al., 2005). The major splice variant of the CT receptor ($CT_{(b)}$, see above) also interacts with RAMPs to generate subtly different phenotypes (Tilakaratne et al., 2000). The precise consequences of RAMP interaction with the other CT receptor splice variants are unknown but there is certainly significant complexity in this system. The fish CT receptor also interacts with RAMPs to form amylin receptors (Nag et al., 2007).

2.3.4 CGRP Receptors

The CGRP receptor is a complex between CLR and RAMP1 (McLatchie et al., 1998). This has a high affinity for CGRP (~0.1 to 1 nM) and around a 10-fold lower affinity for AM and the antagonist $CGRP_{8-37}$ (Poyner et al., 2002). There are a number of nonpeptide antagonists (Salvatore et al., 2006, 2008), the best characterized of which is BIBN4096BS (Doods et al., 2000). This has a subnanomolar affinity for primate CGRP receptors, but binds substantially less well to other CGRP receptors. This species selectivity is due to a tryptophan at position 74 in primate RAMP1; in the rat it is a lysine residue (Mallee et al., 2002).

Until recently, the CLR/RAMP1 complex has been known as the $CGRP_1$ receptor. This was based on the observation that there are tissues where the action of CGRP could not be blocked with high affinity by antagonists such as $CGRP_{8-37}$, but linear agonists such as [$Cys^{2,7}$ acetamidomethyl]-CGRP showed some selectivity for these sites (Dennis et al., 1989; Poyner et al., 2002). It now seems likely that this $CGRP_2$ receptor phenotype is due to expression of receptors for other peptides that show appreciable affinity for CGRP, most notably the AM_2 and $AMY_{1(a)}$ receptors (Kuwasako et al., 2004; Hay et al., 2005; Hay et al., 2008).

2.3.5 AM Receptors

There are two AM receptors; the AM_1 receptor, formed by CLR and RAMP2, and the AM_2 receptor, formed by CLR and RAMP3. Both show high affinity for AM and moderate affinity for the antagonist AM_{22-52}. The AM_1 receptor has ~100-fold selectivity for AM over CGRP but the AM_2

receptor appears to have less discrimination (~50-fold) (McLatchie et al., 1998; Poyner et al., 2002). There are no reported nonpeptide antagonists at either AM receptor. In addition to differences in pharmacology, RAMP3 has a PDZ domain in its C-terminus; this means that in appropriate circumstances, the AM_2 receptor recycles after internalization, whereas the AM_1 receptor does not (Bomberger et al., 2005). It is generally unknown which AM receptor is responsible for the varied physiological or pathophysiological activities of AM.

2.4 BIOLOGICAL AND/OR PATHOLOGICAL MODE OF ACTION

2.4.1 Physiological and Pharmacological Roles of CT

The most well known action of CT is in the regulation of bone metabolism. The primary mechanism is that of inhibition of bone resorption by osteoclasts (Chambers and Magnus, 1982), occurring most significantly under conditions where bone turnover is high. CT protects against bone loss. On the other hand, the phenotypes displayed by relevant knockout mouse models suggest that bone regulation by CT might be more complex than originally envisaged. CT/CGRP$^{-/-}$ mice displayed increased bone formation (Hoff et al., 2002); however, αCGRP was also lacking in these mice, complicating the interpretation of this phenotype (see Section 2.4.3.4). Further back-crossing of these mice changes the phenotype, so caution should be applied to interpreting the earlier data from a mixed genetic background (Gagel et al., 2007). Interestingly, however, *calcr*$^{+/-}$ mice have increased bone formation and a corresponding increase in bone mass (Dacquin et al., 2004). A newly generated global CT receptor-deletion model, using the Cre-*lox*P system in which >94% of CT receptor was deleted sought to clarify the role of the CT system in the regulation of bone (Davey et al., 2008). These mice displayed only mildly increased bone formation under normal conditions. When challenged, however, in calcitriol ($1,25(OH)_2D_3$)-induced hypercalcaemia, serum total calcium was greatly increased in the global CT receptor knockout mice. This is consistent with a regulatory role for CT on bone primarily under conditions of calcium stress. Together, the data support a role for CT and its receptor in the formation and resorption aspects of bone metabolism under physiological conditions.

2.4.2 Physiological and Pharmacological Roles of Amylin

Amylin is coreleased with insulin from the β-cells of the pancreatic islets following nutrient stimulation (Ogawa et al., 1990). Amylin acts centrally (most likely in the *area postrema*) to regulate blood glucose levels by slowing the rate at which nutrients are delivered into the circulation by the gastrointestinal tract (Jodka et al., 1996). The amylin system is also involved in the regulation of food intake (Lutz et al., 1995; Morley and Flood, 1991; Roth et al., 2006). When islet β-cell mass is compromised, such as in diabetes, there is correspondingly less insulin and amylin that may be released into the circulation. Logically, therefore, dual treatment with both hormones offers promise in the treatment of diabetes. The recent clinical interest in amylin (see Section 2.6.2) has sprouted several useful reviews (e.g., Martin, 2006; Lutz, 2006; Wookey et al., 2006).

Amylin also plays another role in diabetes, in which oligomeric forms of amylin are actually toxic to the β-cell (Lorenzo et al., 1994) and therefore result in a decline in β-cell mass. In fact, amylin was isolated from the amyloid deposits commonly found in pancreata from type-2 diabetic patients (Cooper et al., 1987). Similar amyloid plaques are found in cats and monkeys (reviewed in Cooper, 1994). The role of these deposits in disease etiology is still unclear, in particular the nature of the toxic form, whether it is the soluble or insoluble oligomers that are causative (Jaikaran and Clark, 2001).

Male amylin-deficient mice are characterized by enhanced insulin secretion in response to glucose challenge, in parallel with reduced plasma glucose concentrations compared to wild-type mice. Interestingly, only changes in plasma glucose and not insulin were apparent in female mice

(Gebre-Medhin et al., 1998). This mouse model also displays a low bone mass due to increased bone resorption (Dacquin et al., 2004). As described above, *calcr* deficient mice show greater bone formation, suggesting that the CT receptor does not mediate the actions of amylin on bone.

2.4.3 Physiological and Pharmacological Roles of CGRP

2.4.3.1 Introduction

As one of the most abundant sensory neurotransmitters known, it is unsurprising that CGRP has been reported to have a wide range of pharmacological effects, including vasodilation, metabolic effects, and actions in the CNS. It is harder to establish its physiological role. Studies with knockout animals have not yielded dramatic phenotypes and there are few suitably selective pharmacological tools to study the consequences of blocking its receptors. Studies with knockout mice are complicated by the fact that there are two forms of the peptide; to date no study has examined the effect of deletion of both α- and βCGRP. However, it appears that at least in some knockout models of αCGRP loss, βCGRP levels either stay the same or are decreased (Gangula et al., 2000b; Schutz et al., 2004). In some models, both αCGRP and CT have been deleted. Nonetheless, by combining genetic and pharmacological studies, some patterns appear.

2.4.3.2 CGRP and the Cardiovascular System

CGRP is the main neurotransmitter released by capsacin-sensitive sensory nerve fibres; thus it is found throughout the cardiovascular system (CVS). It has positive chronotropic and inotropic effects on the heart and is a very potent vasodilator. It is released in myocardial ischemia, causing local vasodilation, and possibly reducing the effects of any infarction (Franco-Cereceda and Liska, 2000). It has been implicated in numerous animal models of hypertension, although there is no consistent evidence for changes in plasma CGRP concentrations in humans with the condition. Given that hypertension is likely to have multiple causes, this is perhaps not surprising. In some conditions such as primary hyperaldosteronism, there is an increase in CGRP, suggesting that this might be a compensatory mechanism. There is evidence for CGRP interacting with the renin–angiotensin–aldosterone system, endothelin production, and the sympathetic nervous system as well as directly on vascular cells to reduce blood pressure (Deng and Li, 2005).

The first knockout study, involving just the deletion of αCGRP, showed that there were no obvious changes in heart rate or blood pressure (Lu et al., 1999). In contrast, a study in which αCGRP and CT were both deleted indicated that there was an increase in blood pressure (Gangula et al., 2000b). Subsequent work has indicated an increase in plasma renin activity (Li et al., 2004). A pure αCGRP knockout demonstrated an increase in blood pressure, mainly due to increased sympathetic drive leading to a higher peripheral resistance. The ultimate site of action for CGRP in this model was the regions in the CNS that control blood pressure (Kurihara et al., 2003). Deletion of RAMP1 also caused an increase in blood pressure with no associated change in heart rate; whilst RAMP1 has other functions, in addition to forming CGRP receptors, inhibition of CGRP receptor function is by far the simplest explanation for this result (Tsujikawa et al., 2007). Thus there is support for the involvement of CGRP in blood pressure regulation from knockout animal models.

2.4.3.3 Pain and Inflammation

CGRP is distributed in sensory C fibers implicated in pain transmission. CGRP is released both at the peripheral termini of these fibers, where it is important in neurogenic inflammation, and also in the dorsal horn of the spinal cord, where it can modulate pain pathways, both directly and in synergy with other peptides such as substance P. There is good evidence to link the release of CGRP with the development of morphine tolerance (Trang et al., 2006). CGRP and its receptors are also distributed in higher centers in the brain involved in pain perception (Adwinikar et al., 2007). In a particularly elegant study, it has been shown that strain differences in mice to nociception correlates with their expression of αCGRP and that pharmacological manipulation of CGRP levels alters nociception thresholds (Mogil et al., 2005). In the CGRP/CT knockout animals, although the mice did not show heightened pain sensibility, if they were made arthritic, they showed reduced hyperalgesia as measured by hot plate and paw withdrawal tests (Zhang et al., 2001).

Release of CGRP from C-fibers acts at several levels in neurogenic inflammation. As part of its vasodilatory actions, it increases edema and hence recruitment of immune cells. There are also CGRP receptors on many immune cells such as macrophages, and T- and B-lymphocytes (McGillis and Figueiredo, 1996). RAMP1-deficient mice show reduced inflammatory cytokine production from bone marrow dendritic cells (Tsujikawa et al., 2007).

Given the role of CGRP in pain and neurogenic inflammation, it has been suggested that one of its key roles is in the neural defence mechanism, a system in which neuronal activation protects tissues against noxious stimuli. Indeed, a number of CGRP knockout models have shown greater sensitivity to injury (Huang et al., 2008; Ohno et al., 2008; Szabo et al., 2008), although in some models CGRP seems to enhance tissue damage (Aoki-Nagase et al., 2007).

2.4.3.4 CGRP and Bone Metabolism

Although CT has always attracted most interest in regulating bone metabolism, there is pharmacological evidence that CGRP might also have a role. *In vitro* it increases bone formation and osteoblast proliferation and decreases bone resorption and osteoclast formation. Overexpression of CGRP increases bone formation (Lerner, 2006).

Unexpectedly, the CGRP/CT knockout mice showed an increase in bone mass at ages up to 3 months, mainly due to increased bone formation (Hoff et al., 2002). Although unexpected, this is probably a consequence of the deletion of CT, as selective deletion of αCGRP caused a decrease in bone mass (Schinke et al., 2004). However, in a model of accelerated bone osteolysis (polyethylene particles), CGRP knockout animals showed reduced bone resorption (Wedemeyer et al., 2007). Thus there is good evidence for a role for CGRP in enhancing bone formation, but in some circumstances it might also promote osteolysis.

2.4.4 Physiological and Pharmacological Roles of Adrenomedullin

2.4.4.1 Cardiovascular Actions

Most attention has focused on the actions of AM in the CVS. It is well established as an extremely potent vasodilator. In addition, it increases cardiac output, causes naturesis and diuresis, promotes angiogenesis, inhibits apoptosis and protects against oxidants (Ishimitsu et al., 2006). Indeed, there is abundant evidence for increased release of AM or upregulation of its receptors in many cardiovascular disorders such as hypertension, heart failure, myocardial infarction, renal failure, and pulmonary hypertension (Ishimitsu et al., 2006).

Genetic models involving AM have confirmed the importance of the peptide for CVS function. Heterozygote AM^+/AM^- mice have elevated blood pressure (Shindo et al., 2001). AM gene transfer/overexpression improves blood pressure regulation in rats and protects against ischemia in mice (Eto, 2001; Imai et al., 2001). Knockout mice fed on a high salt diet show reduced sympathetic activity compared to wild type when challenged with intracerebroventricular hyperosmotic saline (Fujita et al., 2005).

2.4.4.2 AM and Cardioprotection

AM displays cardioprotective activity in knockout models (Shimosawa et al., 2002; Niu et al., 2003). These can at least in part be attributed to its antioxidant properties, arising from stimulation of nitric oxide production and inhibition of the production of reactive oxygen species (Nishimatsu et al., 2002; Matsui et al., 2004). The antioxidant effect of AM also explains a protective effect against insulin resistance and associated deleterious changes in the CVS (Xing et al., 2004).

2.4.4.3 Developmental Effects

There is now a substantial body of work demonstrating the importance of AM in angiogenesis and lymphangiogenesis. Homozygote AM^-/AM^- mice die *in utero* with abnormalities to their CVS (Caron and Smithies, 2001; Shindo et al., 2001). Further characterization has shown that the phenotype of these and AM receptor-deficient mice may be explained by abnormalities in blood and lymphatic vasculature (Fritz-Six et al., 2008; Ichikawa-Shindo et al., 2008). These genetic models point to the AM system as a key angiogenic pathway (Kahn, 2008). Mice deficient in AM show reduced vascularization (Iimuro et al., 2004). Reduced maternal expression of AM leads to defects in placental development and hence reduced fetal viability (Li et al., 2006). The underlying mechanism appears to be increases in AM expression in both fetal trophoblast cells and the uterine wall at implantation, and this is important for trophoblast invasion and hence placenta formation. Interestingly, the changes seen in placentas from AM knockout mice show changes similar to those brought about by preeclampsia in humans (Li et al., 2006).

AM effects on cell growth and differentiation are not just restricted to the CVS and lymph systems. There is evidence that it can prevent inappropriate airway smooth muscle proliferation in asthma (Yamamoto et al., 2007).

2.5 STRUCTURAL ANALOGS

2.5.1 Introduction

As is the case for most ligands for family B GPCRs, the N-termini of each endogenous peptide is responsible for activation of the receptor, but the bulk of the binding affinity arises from the C-terminus. Thus, N-terminally truncated peptides are antagonists. There are no special synthetic problems associated with the synthesis of these peptides or their analogs; they are all made by standard solid-phase peptide synthesis techniques. Only for CGRP receptors are selective ligands now becoming available due to the generation of small-molecule antagonists.

2.5.2 CT

2.5.2.1 Biophysical Studies

Structural analysis of human, salmon, and eel CT as well as analogs of eel and human CT have been performed in different solvents using a variety of methods (Meyer et al., 1991; Ogawa et al., 1994, 1998, 2006; Katahira et al., 1995). There is general agreement for the existence of an α-helical region within the sequences of these peptides and that differences in the hydrophobic surface of this may be correlated with the relative potencies of these peptides. A complete hydrophobic surface as found in eel and salmon CT may be responsible for their enhanced potency when compared to human CT (Ogawa et al., 2006).

2.5.2.2 Agonists and Antagonists

The most commonly used antagonist analog of CT is salmon CT_{8-32} (sCT_{8-32}; Feyen et al., 1992). This is a moderately high-affinity antagonist ($pK_B \approx 8$; Hay et al., 2005). As with all currently available antagonists of CT-family receptors, sCT_{8-32} has only limited selectivity and is also an effective antagonist of amylin receptors (Hay et al., 2005).

2.5.3 Amylin

Human amylin has a propensity to aggregate and therefore is unreliable as a pharmacological tool. Consequently, rat amylin is routinely used in pharmacological studies. Agonist analogs of human amylin have been generated, including pramlintide, in which Ala^{25}, Ser^{28}, and Ser^{29} have been substituted with proline residues. This has similar pharmacological properties to the unmodified peptide and rat amylin (Young et al., 1996). This peptide is used clinically (see Section 2.6.2.1).

sCT_{8-32} is also an antagonist of amylin receptors. $rAMY_{8-37}$ has been reported to antagonise various responses to amylin *in vivo* and in isolated tissue preparations (Cornish et al., 1999; Villa et al., 2000; Piao et al., 2004; Wang et al., 1999; Uezono et al., 2001). Full concentration analyses have rarely been performed with this antagonist and the reported concentration range required to block the effects of amylin in these studies is large, from 10 pM to 30 μM. In cell culture studies with transfected $AMY_{1(a)}$ and $AMY_{3(a)}$ receptors, $rAmy_{8-37}$ was very weak with pK_B values of around 5.8 (Hay et al., 2005). AC187, reportedly a selective amylin antagonist (Young et al., 1994) displays only modest (~10-fold) discrimination between CT and $AMY_{1(a)}$ receptors in transfected cells (Hay et al., 2005). Similar observations were made with AC413, like AC187, a modified form of sCT_{8-32} (Hay et al., 2005).

2.5.4 CGRP

2.5.4.1 Biophysical Studies

Full-length CGRP has been studied by NMR and circular dichroism. It shows evidence of the predicted α-helix between residues 8 and 18, but it is not otherwise particularly well structured in aqueous solution. However, its secondary structure is stabilized in the presence of agents such as trifluoroethanol. Under optimum conditions, it seems that the residues immediately following the α-helix may form either a β or a γ-turn. Little structure is then discernable until the C-terminus. Here there is evidence for a βI-, βII-, or a γ-turn centered around Val^{32}. Some workers have also reported a hydrogen bond between Val^{28} and Thr^{30} (see Poyner et al., 2000, for review).

2.5.4.2 Agonist Peptides

The most commonly used agonists are the endogenous CGRP peptides: rat and human α and β. These show only minor differences in their sequences and for the most part, their biological activities are similar. Although numerous other sequence variants are known, they are not readily available. Tyr^0-human αCGRP is commercially available and is used by some workers as a basis for radioligand synthesis. It shows some selectivity differences compared to normal CGRP at CGRP and amylin receptors, but they are generally quite small (Hay et al., 2005; Bailey and Hay, 2006). The main structural variants on CGRP are two forms where the cysteines at positions 2 and 7 have been derivatized, to stop disulfide bond formation. These were suggested as selective agonists at the $CGRP_2$ receptor, which is now understood to be an amalgam of (mainly) amylin receptors (see Sections 2.3.3 and 2.3.4). [$Cys^{2,7}$ acetamidomethyl]-CGRP is a weak partial agonist but [$Cys^{2,7}$ ethoxy]-CGRP is more potent. However, their apparent selectivity probably owes more to tissue-dependent factors than differences in receptor recognition (Bailey and Hay, 2006; Hay, 2007).

2.5.4.3 Antagonist Peptides

The chief peptide antagonist for CGRP is $CGRP_{8-37}$, although other fragments such as $CGRP_{12-37}$ have also been used (Dennis et al., 1989, 1990). This has a pKd at CGRP receptors of around 8. [Asp^{31},Pro^{34},Phe^{35}]$CGRP_{27-37}$ has nanomolar affinity and is the smallest peptide fragment with these properties, although the 29–37 equivalent is only fivefold less potent (Rist et al., 1999; Carpenter et al., 2001). It has been shown that the structure–activity relationship for the 29–37 and 27–37 analogs is different, suggesting that they might bind to the receptor in different ways (Boeglin et al., 2007); however, for both derivatives a βI-turn centered around $Gly^{33}Pro^{34}$ is essential for high-affinity binding.

2.5.5 Adrenomedullin

Human AM_{13-52}, AM_{15-52}, and rat AM_{11-50} show very similar agonist potencies to the full-length peptide and are routinely used for most studies. Human AM_{22-52} is the most common antagonist; it has a pKd of around 7 (see Poyner et al., 2002, for review). There is no significant published structural data on AM. A reported AM2/intermedin antagonist is the 17–47 fragment of the peptide (Roh et al., 2004; Chang et al., 2005).

2.6 CLINICAL APPLICATIONS

2.6.1 CT

There have been multiple clinical studies of CT (various forms) for the treatment of numerous bone disorders (reviews: Sexton et al., 1999; Zaidi et al., 2002). It has been most effective in stabilizing bone mineral density in postmenopausal osteoporosis and in treating Paget's disease. CT also appears to be quite effective for treating pain. The route of administration of CT is generally subcutaneous or intranasal. The main barrier to its use, like the other peptides in this family, is poor bioavailability. There is an ongoing effort to improve delivery of CT by improving formulations and trying different routes of delivery.

2.6.2 Amylin

2.6.2.1 Amylin Replacement Therapy in Diabetes

For many physiological processes, there is a growing appreciation that multiple factors act in concert to create balance in biological systems. Blood glucose regulation is no exception and it is now clear that insulin (glucose disposal) and glucagon (glucose appearance), the two players traditionally considered to be primary regulators, are but part of a bigger picture. Incretin peptides, such as glucagon-like peptide 1, are integral to this process, as is amylin. There are several, recent, informative studies on amylin from human trials with the synthetic human amylin analog Pramlintide ([$Pro^{25,28,29}$]-human amylin) in diabetic patients. These are reviewed in Kruger and Gloster (2004) and Schmitz et al. (2004), to give but two examples. This form of amylin does not aggregate (see above) and is therefore suitable as an amylin replacement therapy in humans. Pramlintide treatment is associated with suppression of postmeal glucagon secretion, slowed gastric emptying, reduced weight gain, and overall improvement in postprandial glucose control. These favorable clinical results resulted in the approval of Symlin™ by the U.S. FDA in 2005 and this drug is recommended for insulin-requiring diabetic patients as an adjunct therapy. The satiety effect of Pramlintide is encouraging, as diabetic patients find it particularly difficult to maintain their weight, compounding their health problems. Rodent studies have shown significant beneficial effects on obesity, and it is gratifying that these results are apparently mirrored in humans.

2.6.3 CGRP

As a peptide involved in neurogenic inflammation, CGRP has considerable clinical potential. The development of nonpeptide antagonists (Doods et al., 2000; Salvatore et al., 2006, 2008) has removed significant barriers to targeting the CGRP system in the clinic. It is now important to determine conditions where the antagonists can improve disease management.

Current attention is focused on migraine. There is good evidence that the release of CGRP from the trigeminal nerve is associated with migraine and so the use of CGRP antagonists ought to be an effective therapy. Phase II clinical trials have been conducted with both BIBN4096BS (given intravenously) and MK-0974 (orally active). These agents both show encouraging results compared to existing treatments in terms of their effectiveness at reducing symptoms and their side effects (Doods et al., 2007). There may be other uses for CGRP antagonists in treating conditions such as chronic pelvic pain, where there is inappropriate release of the peptide.

As CGRP often acts to promote tissue or cell protection, it has potential in treating various disorders. As there is no nonpeptide agonist, at present CGRP itself would need to be used for any therapy directed at the CGRP receptor. CGRP release plays a major part in the cardioprotective effect of ischemic preconditioning (Franco-Cereceda and Liska, 2000). In congestive heart failure, CGRP infusion increases cardiac output, mainly due to vasodilation. In chronic stable angina, it increases exercise tolerance. Beneficial effects of CGRP have been reported in Raynaud's disease, where it produces a long-lasting increase in hand skin blood flow and promotes healing of ulcers. Early clinical trials involving CGRP to treat the vasospasm that follows subarachnoid hemorrhage gave promising results, but they were not followed up as the peptide frequently caused hypotension (Franco-Cereceda and Liska, 2000). However, recent gene transfer experiments in animal models have produced positive results, suggesting that the peptide will be useful if it can be targeted appropriately. Gene transfer of CGRP in animal models has also been used successfully to treat pulmonary hypertension and prevent vascular restenosis after injury (Champion et al., 2000; Toyoda et al., 2000).

2.6.4 Adrenomedullin

There have been no serious attempts to use AM in the clinic. However, the peptide has significant potential for use in cardiovascular disease and other conditions. AM administration in man has been shown to be beneficial in patients with essential hypertension, pulmonary hypertension, renal impairment, and congestive heart failure (Troughton et al., 2000; McGregor et al., 2001; Nagaya and Kangawa, 2004). The insights into the role of AM in vascular (blood and lymph) development gained from *AM*, *calcrl*, and *ramp2* knockout mice (Caron and Smithies, 2001; Dackor et al., 2006; Ichikawa-Shindo et al., 2008) suggests that the AM system may be a viable target for conditions involving impairment in angiogenesis.

2.7 COMMON METHODOLOGIES AND ASSAY SYSTEMS

There are no specialized methods required to investigate members of the CT family. CT, CGRP, AM, and amylin are all available in radiolabeled form for use in binding assays. For investigation of CT binding, salmon CT is sometimes preferred to human as it demonstrates almost irreversible binding. For CGRP binding, either native CGRP iodinated at His^8 or the Tyr^0 derivative may be used; iodinated $CGRP_{8-37}$ may also be purchased, although the relatively low affinity of this peptide limits its usefulness (Conner et al., 2005). For amylin binding, the Bolton–Hunter derivative of the peptide is used. It should be noted that for all the peptides, the iodinated derivatives may not necessarily show the same pharmacological behavior as the parent ligand and that the binding is also frequently accompanied by high levels of nonspecific binding (Owji et al., 1995).

The most robust second messenger system coupled to receptor activation for all peptides is stimulation of adenylate cyclase. This may be measured in either cells or isolated tissues. However, if the aim of the study is to investigate the pharmacology of any given receptor subtype, then this is much easier to do using recombinant receptors expressed in cell lines. In native tissues there is often expression of multiple receptors for the different CT family peptides. Some of the antagonists such as BIBN4096BS, $CGRP_{8-37}$, or sCT_{8-32} have either good or reasonable selectivity and, depending on the other receptors present, the peptide agonists *may* show selectivity (Poyner et al., 2002; Hay et al., 2005; Bailey and Hay, 2006; Hay, 2007). However, there is no guarantee that these conditions can be met in intact tissues and the resulting data may be very difficult to interpret (Poyner et al., 1999; Takhshid et al., 2006). Accordingly, transfected cells are by the far the easiest system for examining the properties of peptides at defined receptors. Cos 7 cells offer particular advantages as they usually show very low endogenous RAMP expression (Hay et al., 2005; Bailey and Hay, 2006).

The detection of these peptides in the circulation is usually by radioimmunoassay (RIA) or enzyme-linked immunosorbant assay (ELISA). There are invariably differences in values obtained between studies, likely due the inherent variability of the methods used. In the case of measurements of αCGRP, the variability may reflect variable antibody affinity and/or cross-reactivity with βCGRP or possibly other CT-family peptides. Furthermore, the species specificity of the kit used may have implications for the final value obtained. For instance, in sheep and dogs, endogenous CGRP could not be measured because it was below the detection limit of the assay, probably due to the inability of the antibody to cross-react with canine or ovine CGRP (Braslis et al., 1988; Reasbeck et al., 1988). Likewise, these antibody-related factors may also influence the apparent tissue distribution of the peptide ascertained, where CGRP-LI, for example, is likely to comprise at least exogenous/endogenous αCGRP and βCGRP, as well as amylin in some tissues, for example, endocrine pancreas.

2.8 FUTURE DEVELOPMENTS

Considerable progress has been made in understanding the pharmacology of the CT family of peptides, and the availability of knockout models has given new insights into the pharmacology of CT, CGRP, and AM. It will be important to continue and extend this work to the other peptides in this family. A major challenge will be to understand at the molecular level how the peptides bind to their receptors and cause activation. In the case of AM and amylin, where multiple receptor subtypes exist, determining which is responsible for the physiology and pathophysiology of the peptides presents a considerable challenge. Allied to this is the search for new, more selective antagonists and agonists. These molecules will be fundamental tools for understanding the role of the peptides in pathophysiology and will have the potential to make an impact in the clinic.

REFERENCES

Adwanikar, H. et al., Spinal CGRP1 receptors contribute to supraspinally organized pain behavior and pain-related sensitization of amygdala neurons, *Pain*, 132, 53–66, 2007.

Amara, S.G. et al., Alternative RNA processing in calcitonin gene expression generates mRNAs encoding different polypeptide products, *Nature*, 298, 240–244, 1982.

Amara, S.G. et al., Expression in brain of a messenger RNA encoding a novel neuropeptide homologous to calcitonin gene-related peptide, *Science*, 229, 1094–1097, 1985.

Anand, I.S. et al., Cardiovascular and hormonal effects of calcitonin gene-related peptide in congestive heart failure, *J. Am. Coll. Cardiol.*, 17, 208–217, 1991.

Ando, K. et al., Vasodilating actions of calcitonin gene-related peptide in normal man: Comparison with atrial natriuretic peptide, *Am. Heart. J.*, 123, 111–116, 1992.

Aoki-Nagase, T. et al., Calcitonin gene-related peptide mediates acid-induced lung injury in mice, *Respirology*, 12, 807–813, 2007.

Bailey, R.J. and Hay, D.L. Pharmacology of the human CGRP1 receptor in Cos 7 cells, *Peptides*, 27, 1367–1375, 2006.

Bailey, R.J. and Hay, D.L. Agonist-dependent consequences of proline to alanine substitution in the transmembrane helices of the calcitonin receptor, *Br. J. Pharmacol.*, 151, 678–687, 2007.

Beglinger, C. et al., Calcitonin gene-related peptides I and II and calcitonin: Distinct effects on gastric acid secretion in humans, *Gastroenterology*, 95, 958–965, 1988.

Bennett, R.G., Duckworth, W.C., and Hamel, F.G., Degradation of amylin by insulin-degrading enzyme, *J. Biol. Chem.*, 275, 36621–36625, 2000.

Bergwitz, C. et al., Full activation of chimeric receptors by hybrids between parathyroid hormone and calcitonin. Evidence for a common pattern of ligand–receptor interaction, *J. Biol. Chem.*, 127, 26469–26472, 1996.

Boeglin, D. et al., Calcitonin gene-related peptide analogues with aza and indolizidinone amino acid residues reveal conformational requirements for antagonist activity at the human calcitonin gene-related peptide 1 receptor, *J. Med. Chem.*, 22, 1401–1408, 2007.

Bomberger, J.M. et al., Novel function for receptor activity-modifying proteins (RAMPs) in post-endocytic receptor trafficking, *J. Biol. Chem.*, 280, 9297–9307, 2005.

Braslis, K.G. et al., Pharmacokinetics and organ-specific metabolism of calcitonin gene-related peptide in sheep, *J. Endocrinol.*, 118, 25–31, 1988.

Caron, K.M. and Smithies, O., Extreme hydrops fetalis and cardiovascular abnormalities in mice lacking a functional Adrenomedullin gene, *Proc. Natl Acad. Sci. USA*, 98, 615–619, 2001.

Carpenter, K.A. et al., Turn structures in CGRP C-terminal analogues promote stable arrangements of key residue side chains, *Biochemistry*, 40, 8317–8325, 2001.

Chambers, T.J. and Magnus, C.J., Calcitonin alters behaviour of isolated osteoclasts, *J. Pathol.*, 136, 27–39, 1982.

Champion, H.C. et al., *In vivo* gene transfer of prepro-calcitonin gene-related peptide to the lung attenuates chronic hypoxia-induced pulmonary hypertension in the mouse, *Circulation*, 101, 923–930, 2000.

Chang, L.C. et al., Intermedin functions as a pituitary paracrine factor regulating prolactin release, *Mol. Endocrinol.*, 19, 2824–2838, 2005.

Conner, A.C. et al., A key role for transmembrane prolines in calcitonin receptor-like receptor agonist binding and signalling: Implications for family B G-protein-coupled receptors, *Mol. Pharmacol.*, 67, 20–31, 2005.

Cooper, G.J., Amylin compared with calcitonin gene-related peptide: Structure, biology, and relevance to metabolic disease, *Endocrinol. Rev.*, 15, 163–201, 1994.

Cooper, G.J. et al., Purification and characterization of a peptide from amyloid-rich pancreases of type 2 diabetic patients, *Proc. Natl Acad. Sci. USA*, 84, 8628–8632, 1987.

Cornish, J. et al., Comparison of the effects of calcitonin gene-related peptide and amylin on osteoblasts, *J. Bone Miner. Res.*, 14, 1302–1309, 1999.

Cuifen, Z. et al., Changes and distributions of peptides derived from proadrenomedullin in left-to-right shunt pulmonary hypertension of rats, *Circ. J.*, 72, 476–481, 2008.

Dackor, R.T. et al., Hydrops fetalis, cardiovascular defects, and embryonic lethality in mice lacking the calcitonin receptor-like receptor gene, *Mol. Cell. Biol.*, 26, 2511–2518, 2006.

Dacquin, R. et al., Amylin inhibits bone resorption while the calcitonin receptor controls bone formation *in vivo*, *J. Cell. Biol.*, 164, 509–514, 2004.

Davey, R.A. et al., The calcitonin receptor plays a physiological role to protect against hypercalcemia in mice, *J. Bone Miner. Res.*, 23, 1182–1193, 2008.

Deng, P.Y. and Li, Y.J., Calcitonin gene-related peptide and hypertension, *Peptides*, 26, 1676–1685, 2005.

Dennis, T. et al., Structure–activity profile of calcitonin gene-related peptide in peripheral and brain tissues. Evidence for receptor multiplicity, *J. Pharmacol. Exp. Ther.*, 251, 718–725, 1989.

Dennis, T. et al., hCGRP8-37, a calcitonin gene-related peptide antagonist revealing CT gene-related peptide receptor heterogeneity in brain and periphery, *J. Pharmacol. Exp. Ther.*, 254, 123–128, 1990.

Doods, H. et al., Pharmacological profile of BIBN4096BS, the first selective small molecule CGRP antagonist, *Br. J. Pharmacol.*, 129, 420–423, 2000.

Doods, H. et al., CGRP antagonists: Unravelling the role of CGRP in migraine, *Trends Pharmacol. Sci.*, 28, 580–587, 2007.

Eto, T., A review of the biological properties and clinical implications of adrenomedullin and proadrenomedullin N-terminal 20 peptide (PAMP), hypotensive and vasodilating peptides, *Peptides*, 22, 1693–1711, 2001.

Feyen, J.H. et al., N-terminal truncation of salmon calcitonin leads to calcitonin antagonists. Structure activity relationship of N-terminally truncated salmon calcitonin fragments *in vitro* and *in vivo*, *Biochem. Biophys. Res. Commun.*, 187, 8–13, 1992.

Franco-Cereceda, A. and Liska, L., Potential of CT gene-related peptide in coronary heart disease, *Pharmacology*, 60, 1–8, 2000.

Fritz-Six, K.L. et al., Adrenomedullin signaling is necessary for murine lymphatic vascular development, *J. Clin. Invest.*, 118, 40–50, 2008.

Fujita, M. et al., Sympatho-inhibitory action of endogenous adrenomedullin through inhibition of oxidative stress in the brain, *Hypertension*, 45, 1165–1172, 2005.

Gagel, R.F. et al., Deletion of calcitonin/CGRP gene causes a profound cortical resorption phenotype in mice, *J. Bone Miner. Res.*, 22, Suppl. 1, S35, 2007.

Gangula, P.R., Wimalawansa, S.J., and Yallampalli, C. Pregnancy and sex steroid hormones enhance circulating calcitonin gene-related peptide concentrations in rats, *Hum. Reprod.*, 15, 949–953, 2000a.

Gangula, P.R. et al., Increased blood pressure in alpha-calcitonin gene-related peptide/calcitonin gene knockout mice, *Hypertension*, 35, 470–475, 2000b.

Gebre-Medhin, S. et al., Increased insulin secretion and glucose tolerance in mice lacking islet amyloid polypeptide (amylin), *Biochem. Biophys. Res. Commun.*, 250, 271–277, 1998.

Gkonos, P.J. et al., Biosynthesis of calcitonin gene-related peptide and calcitonin by a human medullary thyroid carcinoma cell line, *J. Biol. Chem.*, 261, 14386–14391, 1986.

Gorn, A.H. et al., Cloning, characterization, and expression of a human calcitonin receptor from an ovarian carcinoma cell line, *J. Clin. Invest.*, 90, 1726–1735, 1992.

Grace, C.R. et al., Structure of the N-terminal domain of a type B1 G protein-coupled receptor in complex with a peptide ligand, *Proc. Natl Acad. Sci. USA*, 104, 4858–4863, 2007.

Hay, D.L. et al., Pharmacological discrimination of calcitonin receptor: Receptor activity-modifying protein complexes, *Mol. Pharmacol.*, 67, 1655–1665, 2005.

Hay, D.L., Poyner, D.R., and Quirion, R., International Union of Pharmacology. LXIX. The status of the calcitonim gene-related peptide subtype 2 receptor, *Pharmacological Rev.*, 60, 143–145, 2008.

Hay, D.L., Poyner, D.R., and Sexton, P.M., GPCR modulation by RAMPs, *Pharmacol. Ther.*, 109, 173–197, 2006.

Hinson, J.P., Kapas, S., and Smith, D.M., Adrenomedullin, a multifunctional regulatory peptide, *Endocrinol. Rev.*, 21, 138–167, 2000.

Hoare, S.R., Mechanisms of peptide and nonpeptide ligand binding to Class B G-protein-coupled receptors, *Drug Discov. Today*, 10, 417–427, 2005.

Hoff, A.O. et al., Increased bone mass is an unexpected phenotype associated with deletion of the calcitonin gene, *J. Clin. Invest.*, 110, 1849–1857, 2002.

Huang, R. et al., Deletion of the mouse α-calcitonin gene-related peptide gene increases the vulnerability of the heart to ischemia/reperfusion injury, *Am. J. Physiol. Heart Circ. Physiol.*, 294, H1291–H1297, 2008.

Ichikawa-Shindo, Y. et al., The GPCR modulator protein RAMP2 is essential for angiogenesis and vascular integrity, *J. Clin. Invest.*, 118, 29–39, 2008.

Iimuro, S. et al., Angiogenic effects of adrenomedullin in ischemia and tumor growth, *Circ. Res.*, 95, 415–423, 2004.

Imai, Y. et al., Evidence for the physiological and pathological roles of adrenomedullin from genetic engineering in mice, *Ann. NY Acad. Sci.*, 947, 26–33, 2001.

Ishimitsu, T. et al., Pathophysiologic and therapeutic implications of adrenomedullin in cardiovascular disorders, *Pharmacol. Ther.*, 111, 909–927, 2006.

Jaikaran, E.T. and Clark, A., Islet amyloid and type 2 diabetes: From molecular misfolding to islet pathophysiology, *Biochim. Biophys. Acta*, 1537, 179–203, 2001.

Jodka, C. et al., Amylin modulation of gastric emptying in rats depends upon an intact vagus nerve (abstract), *Diabetes*, 45, Suppl. 2, 235A, 1996.

Kahn, M.L., Blood is thicker than lymph, *J. Clin. Invest.*, 118, 23–26, 2008.

Katafuchi, T. and Minamino, N., Structure and biological properties of three calcitonin receptor-stimulating peptides, novel members of the calcitonin gene-related peptide family, *Peptides*, 25, 2039–2045, 2004.

Katafuchi, T. et al., Calcitonin receptor-stimulating peptide, a new member of the calcitonin gene-related peptide family. Its isolation from porcine brain, structure, tissue distribution, and biological activity, *J. Biol. Chem.*, 278, 12046–12054, 2003a.

Katafuchi, T., Hamano, K., and Kikumoto, K., Identification of second and third calcitonin receptor-stimulating peptides in porcine brain, *Biochem. Biophys. Res. Commun.*, 308, 445–451, 2003b.

Katafuchi, T. et al., Isolation and characterization of a glycine-extended form of calcitonin receptor-stimulating peptide-1: Another biologically active form of calcitonin receptor-stimulating peptide-1, *Peptides*, 26, 2616–2623, 2005.

Katahira, R. et al., Solution structure of a human calcitonin analog elucidated by NMR and distance geometry calculations, *Int. J. Pept. Protein Res.*, 45, 305–311, 1995.

Kitamura, K. et al., Adrenomedullin: A novel hypotensive peptide isolated from human pheochromocytoma, *Biochem. Biophys. Res. Commun.*, 192, 553–560, 1993.

Krämer, H.H. et al., Angiotensin converting enzyme has an inhibitory role in CGRP metabolism in human skin, *Peptides*, 27, 917–920, 2006.

Kruger, D.F. and Gloster, M.A., Pramlintide for the treatment of insulin-requiring diabetes mellitus: Rationale and review of clinical data, *Drugs*, 64, 1419–1432, 2004.

Kurihara, H. et al., Targeted disruption of adrenomedullin and alphaCGRP genes reveals their distinct biological roles, *Hypertens. Res.*, 26, Suppl., S105–S108, 2003.

Kuwasako, K. et al., Characterization of the human calcitonin gene-related peptide receptor subtypes associated with receptor activity-modifying proteins, *Mol. Pharmacol.* 65, 207–213, 2004.

Lerner, U.H., Deletions of genes encoding CT/alpha-CGRP, amylin, and CT receptor have given new and unexpected insights into the function of CT receptors and CT receptor-like receptors in bone, *J. Musculoskelet. Neuronal Interact.*, 6, 87–95, 2006.

Letizia, C. et al., High plasma adrenomedullin concentrations in patients with high-renin essential hypertension, *J. Renin Angiotensin Aldosterone Syst.*, 3, 126–129, 2002.

Li, J. et al., Activation of the renin–angiotensin system in alpha-CT gene-related peptide/CT gene knockout mice, *J. Hypertens.*, 22, 1345–1349, 2004.

Li, M. et al., Reduced maternal expression of adrenomedullin disrupts fertility, placentation, and fetal growth in mice, *J. Clin. Invest.*, 116, 2653–2662, 2006.

Lin, H.Y., Harris, T.L., and Flannery, M.S., Expression cloning of an adenylate cyclase-coupled calcitonin receptor, *Science*, 254, 1022–1024, 1991.

Lorenzo, A. and Yankner, B.A., Beta-amyloid neurotoxicity requires fibril formation and is inhibited by congo red, *Proc. Natl Acad. Sci. USA*, 91, 12243–12247, 1994.

Lu, J.T. et al., Mice lacking alpha-calcitonin gene-related peptide exhibit normal cardiovascular regulation and neuromuscular development, *Mol. Cell. Neurosci.*, 14, 99–120, 1999.

Luo, D. et al., Delayed preconditioning by cardiac ischemia involves endogenous calcitonin gene-related peptide via the nitric oxide pathway, *Eur. J. Pharmacol.*, 502, 135–141, 2004.

Lutz, T.A. et al., Amylin decreases meal size in rats, *Physiol. Behav.*, 58, 1197–1202, 1995.

Lutz, T.A., Amylinergic control of food intake, *Physiol. Behav.*, 89, 465–471, 2006.

Malde, A.K., Srivastava, S.S., and Coutinho, E.C., Understanding interactions of gastric inhibitory polypeptide (GIP) with its G-protein coupled receptor through NMR and molecular modeling, *J. Pept. Sci.*, 13, 287–300, 2007.

Mallee, J.J. et al., Receptor activity-modifying protein 1 determines the species selectivity of non-peptide CGRP receptor antagonists, *J. Biol. Chem.*, 277, 14294–14298, 2002.

Martin, C., The physiology of amylin and insulin: Maintaining the balance between glucose secretion and glucose uptake, *Diabetes Educ.*, 32, Suppl. 3, 101S–104S, 2006.

Masi, L. et al., Allelic variants of human calcitonin receptor: Distribution and association with bone mass in postmenopausal Italian women, *Biochem. Biophys. Res. Commun.*, 245, 622–626, 1998.

Matsui, H. et al., Adrenomedullin can protect against pulmonary vascular remodeling induced by hypoxia, *Circulation*, 109, 2246–2251, 2004.

McDonald, K.R. et al., Ablation of calcitonin/calcitonin gene-related peptide-alpha impairs fetal magnesium but not calcium homeostasis, *Am. J. Physiol. Endocrinol. Metab.*, 287, E218–E226, 2004.

McGillis, J.P. and Figueiredo, H.F., Sensory neuropeptides: The role of CT gene-related peptide (CGRP) in modulating inflammation and immunity, in *The Physiology of Immunity*, Marsh, J.A. and Kendall, M.D., Eds., CRC, Boca Raton, FL, p. 127–143, 1996.

McGregor, D.O. et al., Hypotensive and natriuretic actions of adrenomedullin in subjects with chronic renal impairment, *Hypertension*, 37, 1279–1284, 2001.

McLatchie, L.M. et al., RAMPs regulate the transport and ligand specificity of the calcitonin-receptor-like receptor, *Nature*, 393, 333–339, 1998.

Meyer, J.P. et al., Solution structure of salmon calcitonin, *Biopolymers*, 31, 233–241, 1991.

Mitsukawa, T. et al., Islet amyloid polypeptide response to glucose, insulin, and somatostatin analogue administration, *Diabetes*, 39, 639–642, 1990.

Mogil, J.S. et al., Variable sensitivity to noxious heat is mediated by differential expression of the CGRP gene, *Proc. Natl Acad. Sci. USA*, 102, 12938–12943, 2005.

Morley, J.E. and Flood, J.F., Amylin decreases food intake in mice, *Peptides*, 12, 865–869, 1991.

Nag, K. et al., Fish calcitonin receptor has novel features, *Gen. Comp. Endocrinol.*, 154, 48–58, 2007.

Nagaya, N. and Kangawa, K., Adrenomedullin in the treatment of pulmonary hypertension, *Peptides*, 25, 2013–2018, 2004.

Nakamura, M. et al., Allelic variants of human calcitonin receptor in the Japanese population, *Hum. Genet.*, 99, 38–41, 1997.

Nishimatsu, H. et al., Role of endogenous adrenomedullin in the regulation of vascular tone and ischemic renal injury: Studies on transgenic/knockout mice of adrenomedullin gene, *Circ. Res.*, 90, 657–663, 2002.

Niu, P. et al., Accelerated cardiac hypertrophy and renal damage induced by angiotensin II in adrenomedullin knockout mice, *Hypertens. Res.*, 26, 731–736, 2003.

Ogawa, A., Harris, V., and McCorkle, S.K., Amylin secretion from the rat pancreas and its selective loss after streptozotocin treatment, *J. Clin. Invest.*, 85, 973–976, 1990.

Ogawa, K., Nishimura, S., and Doi, M., Conformational analysis of elcatonin in solution, *Eur. J. Biochem.*, 222, 659–666, 1994.

Ogawa, K. et al., Conformation analysis of eel calcitonin—comparison with the conformation of elcatonin, *Eur. J. Biochem.*, 257, 331–336, 1998.

Ogawa, K. et al., Conformational analysis of human calcitonin in solution, *J. Pept. Sci.*, 12, 51–57, 2006.

Ogoshi, M. et al., Evolutionary history of the calcitonin gene-related peptide family in vertebrates revealed by comparative genomic analyses, *Peptides*, 27, 3154–3164, 2006.

Ohno, T. et al., Roles of calcitonin gene-related peptide in maintenance of gastric mucosal integrity and in enhancement of ulcer healing and angiogenesis, *Gastroenterology*, 134, 215–225, 2008.

Owji, A.A. et al., An abundant and specific binding site for the novel vasodilator adrenomedullin in the rat, *Endocrinology*, 136, 2127–2134, 1995.

Piao, F.L. et al., Amylin-induced suppression of ANP secretion through receptors for CGRP1 and salmon calcitonin, *Regul. Pept.*, 117, 159–166, 2004.

Pío, R. et al., Identification, characterization, and physiological actions of factor H as an adrenomedullin binding protein present in human plasma, *Microsc. Res. Tech.*, 57, 23–27, 2002.

Poyner, D.R. et al., Characterization of receptors for CT gene-related peptide and adrenomedullin on the guinea-pig vas deferens, *Br. J. Pharmacol.*, 126, 1276–1282, 1999.

Poyner, D.R. et al., Structure–activity relationships for CGRP in the CGRP family: Calcitonin gene-related peptide, amylin and adrenomedullin, *Brain*, 25–32, 2000.

Poyner, D.R. et al., International Union of Pharmacology. XXXII. The mammalian calcitonin gene-related peptides, adrenomedullin, amylin, and calcitonin receptors, *Pharmacol. Rev.*, 54, 233–246, 2002.

Purdue, B.W., Tilakaratne, N., and Sexton, P.M., Molecular pharmacology of the calcitonin receptor, *Receptors Channels*, 8, 243–255, 2002.

Rademaker, M.T. et al., Combined endopeptidase inhibition and adrenomedullin in sheep with experimental heart failure, *Hypertension*, 39, 93–98, 2002.

Reasbeck, P.G., Burns, S.M., and Shulkes, A., Calcitonin gene-related peptide: Enteric and cardiovascular effects in the dog, *Gastroenterology*, 95, 966–971, 1988.

Rist, B. et al., CGRP27-37 analogues with high affinity to the CGRP1 receptor show antagonistic properties in a rat blood flow assay, *Regul. Pept.*, 79, 153–158, 1999.

Roh, J. et al., Intermedin is a calcitonin/calcitonin gene-related peptide family peptide acting through the calcitonin receptor-like receptor/receptor activity-modifying protein receptor complexes, *J. Biol. Chem.*, 279, 7264–7274, 2004.

Rosenfeld, M.G. et al., Production of a novel neuropeptide encoded by the calcitonin gene via tissue-specific RNA processing, *Nature*, 304, 129–135, 1983.

Rossetti, L. et al., Multiple metabolic effects of CGRP in conscious rats: Role of glycogen synthase and phosphorylase, *Am. J. Physiol.*, 264, E1–E10, 1993.

Roth, J.D. et al., Antiobesity effects of the beta-cell hormone amylin in diet-induced obese rats: Effects on food intake, body weight, composition, energy expenditure, and gene expression, *Endocrinology*, 147, 5855–5864, 2006.

Salvatore, C.A. et al., Identification and pharmacological characterization of domains involved in binding of CGRP receptor antagonists to the calcitonin-like receptor, *Biochemistry*, 45, 1881–1887, 2006.

Salvatore, C.A. et al., Pharmacological characterization of MK-0974 [N-[(3R,6S)-6-(2,3-difluorophenyl)-2-oxo-1-(2,2,2-trifluoroethyl)azepan-3-yl]-4-(2-oxo-2,3-dihydro-1H-imidazo[4,5-b]pyridin-1-yl)-piperidine-1-carboxamide], a potent and orally active calcitonin gene-related peptide receptor antagonist for the treatment of migraine, *J. Pharmacol. Exp. Ther.*, 324, 416–421, 2008.

Schifter, S., A new highly sensitive radioimmunoassay for human calcitonin useful for physiological studies, *Clin. Chim. Acta*, 215, 99–109, 1993.

Schinke, T. et al., Decreased bone formation and osteopenia in mice lacking α-CT gene-related peptide, *J. Bone Miner. Res.*, 19, 2049–2056, 2004.

Schmitz, O., Brock, B., and Rungby, J., Amylin agonists: A novel approach in the treatment of diabetes, *Diabetes*, 53, Suppl. 3, S233–S238, 2004.

Schütz, B., Mauer, D., and Salmon, A.M., Analysis of the cellular expression pattern of beta-CGRP in alpha-CGRP-deficient mice, *J. Comp. Neurol.*, 476, 32–43, 2004.

Sexton, P.M., Findlay, D.M., and Martin, T.J., Calcitonin, *Curr. Med. Chem.*, 6, 1067–1093, 1999.

Shekhar, Y.C. et al., Effects of prolonged infusion of human alpha calcitonin gene-related peptide on hemodynamics, renal blood flow, and hormone levels in congestive heart failure, *Am. J. Cardiol.*, 67, 732–736, 1991.

Shimosawa, T. et al., Adrenomedullin, an endogenous peptide, counteracts cardiovascular damage, *Circulation*, 105, 106–111, 2002.

Shindo, T. et al., Vascular abnormalities and elevated blood pressure in mice lacking adrenomedullin gene, *Circulation*, 104, 1964–1971, 2001.

Steenbergh, P.H. et al., Structure and expression of the human calcitonin/CGRP genes, *FEBS Lett.*, 209, 97–103, 1986.

Szabó, A. et al., Investigation of sensory neurogenic components in a bleomycin-induced scleroderma model using transient receptor potential vanilloid 1 receptor- and calcitonin gene-related peptide-knockout mice, *Arthritis Rheum.*, 58, 292–301, 2008.

Taboulet, J. et al., Calcitonin receptor polymorphism is associated with a decreased fracture risk in post-menopausal women, *Hum. Mol. Genet.*, 7, 2129–2133, 1998.

Takei, Y. et al., Identification of novel adrenomedullin in mammals: A potent cardiovascular and renal regulator, *FEBS Lett.*, 556, 53–58, 2004.

Takhshid, M.A. et al., Characterization and effects on cAMP accumulation of adrenomedullin and CT gene-related peptide (CGRP) receptors in dissociated rat spinal cord cell culture, *Br. J. Pharmacol.*, 148, 459–468, 2006.

Tilakaratne, N. et al., Amylin receptor phenotypes derived from human calcitonin receptor/RAMP coexpression exhibit pharmacological differences dependent on receptor isoform and host cell environment, *J. Pharmacol. Exp. Ther.*, 294, 61–72, 2000.

Toyoda, K. et al., Gene transfer of calcitonin gene-related peptide prevents vasoconstriction after subarachnoid hemorrhage, *Circ. Res.*, 87, 818–824, 2000.

Trang, T. et al., Spinal modulation of calcitonin gene-related peptide by endocannabinoids in the development of opioid physical dependence, *Pain*, 126, 256–271, 2006.

Troughton, R.W. et al., Hemodynamic, hormone, and urinary effects of adrenomedullin infusion in essential hypertension, *Hypertension*, 36, 588–593, 2000.

Tsujikawa, K. et al., Hypertension and dysregulated proinflammatory cytokine production in receptor activity-modifying protein 1-deficient mice, *Proc. Natl Acad. Sci. USA*, 104, 16702–16707, 2007.

Uezono, Y., Nakamura, E., and Ueda, Y., Production of cAMP by adrenomedullin in human oligodendroglial cell line KG1C: Comparison with calcitonin gene-related peptide and amylin, *Brain Res. Mol. Brain Res.*, 97, 59–69, 2001.

Villa, I. et al., Effects of calcitonin gene-related peptide and amylin on human osteoblast-like cells proliferation, *Eur. J. Pharmacol.*, 409, 273–278, 2000.

Wang, Y., Xu, A., and Cooper, G.J., Amylin evokes phosphorylation of P20 in rat skeletal muscle, *FEBS Lett.*, 457, 149–152, 1999.

Wedemeyer, C. et al., Polyethylene particle-induced bone resorption in alpha-CT gene-related peptide-deficient mice, *J. Bone Miner. Res.*, 22, 1011–1019, 2007.

Wolfe, L.A. et al., *In vitro* characterization of a human calcitonin receptor gene polymorphism, *Mutat. Res.*, 522, 93–105, 2003.

Wookey, P.J., Lutz, T.A., and Andrikopoulos, S., Amylin in the periphery II: An updated mini-review, *The Scientific World J.*, 6, 1642–1655, 2006.

Xing, G. et al., Angiotensin II-induced insulin resistance is enhanced in adrenomedullin-deficient mice, *Endocrinology*, 145, 3647–3651, 2004.

Yamamoto, H. et al., Adrenomedullin insufficiency increases allergen-induced airway hyperresponsiveness in mice, *J. Appl. Physiol.*, 102, 2361–2368, 2007.

Young, A.A. et al., Selective amylin antagonist suppresses rise in plasma lactate after intravenous glucose in the rat. Evidence for a metabolic role of endogenous amylin, *FEBS Lett.*, 343, 237–241, 1994.

Young, A.A. et al., Preclinical pharmacology of pramlintide in the rat: Comparisons with human and rat amylin, *Drug Dev. Res.*, 37, 231–248, 1996.

Young, A. and Denaro, M., Roles of amylin in diabetes and in regulation of nutrient load, *Nutrition*, 14, 524–526, 1998.

Zaidi, M. et al., Circulating CGRP comes from the perivascular nerves, *Eur. J. Pharmacol.*, 117, 283–284, 1985.

Zaidi, M., Inzerillo, A.M., and Moonga, B.S., Forty years of calcitonin—where are we now? A tribute to the work of Iain Macintyre, FRS, *Bone*, 30, 655–663, 2002.

Zhang, L. et al., Arthritic calcitonin/alpha calcitonin gene-related peptide knockout mice have reduced nociceptive hypersensitivity, *Pain*, 89, 265–273, 2001.

3 Neuropeptide Y Family Peptides

Cláudia Cavadas, Joana Rosmaninho Salgado, and Eric Grouzmann

CONTENTS

3.1 INTRODUCTION

The neuropeptide Y family includes three different 36-amino-acid-long peptide members: the neuropeptide Y (NPY), which is expressed in the central and peripheral nervous systems, and the peptide YY (PYY) and pancreatic peptide (PP), which are gut endocrine peptides (Larhammar, 1996). NPY and PYY have 70% sequence similarity, whereas PP exhibits only 50% homology with NPY (Minth et al., 1984). Although very similar, these peptides exhibit distinct functions. All three peptides act on at least six types of G-protein-coupled receptors, termed Y_1, Y_2, Y_3, Y_4, Y_5, and y_6 (Michel et al., 1998).

A large number of publications and reviews have reported on the characterization of the different types of NPY receptors. The objective of this chapter is to reveal biological events that are associated with the activation of specific receptors in relationships with the metabolism of NPY by processing enzymatic machinery. We will first describe the established posttranslational modifications of NPY and its family produced by peptidases. We will then report on the bioactivity of these fragments on NPY receptors and, when known, define where these fragments are produced in the body and describe the posttranslational regulation of NPY processing by these peptidases. Finally, we will summarize the consequences of the metabolism of NPY family peptides by peptidases in the assessment of their clinical properties.

3.2 NPY SYNTHESIS

NPY is derived from a precursor that is 97 amino acids in length, the preproNPY (Minth et al., 1984). After removal of the 28-amino-acid signal sequence, the 69-amino-acid proNPY undergoes cleavage preferentially by proconverting enzymes PC1/3 then PC2 at a single paired basic site (^{38}Lys–^{39}Arg), resulting in the formation of $NPY_{1\text{-}39}$, which is further processed by a carboxypeptidase-like enzyme (CPE) to yield $NPY_{1\text{-}37}$, which in turn is used as a substrate by an amidating enzyme (peptidyl-glycine-alpha-amidating monooxygenase; PAM) to yield the active amidated $NPY_{1\text{-}36}$ (Brakch et al., 1997). ProNPY and some of the glycine extended forms of NPY have been found in plasma, cell media, and follicular fluid (Jorgensen et al., 1990) but in the absence of a specific assay for proNPY, a systematic quantification of this peptide in tissues and of its glycine extended form has not yet been carried out. The degree of processing of propeptides is highly variable. For example, proinsulin is efficiently processed, because in humans mature insulin concentrations are approximately fivefold higher than those of proinsulin (Kao et al., 1994). In contrast, prorenin is poorly cleaved into active renin (Lacombe et al., 2005). In this context, proopiomelanocortin is a good example of a prohormone that yields several important peptides with differing functions; its processing may be incomplete and is regulated in a tissue-specific way with respect to physiological requirements, having considerable implications in the pathophysiology of obesity and being related to changes in PC1/3 activity (Pan et al., 2006). Therefore, although studies on proNPY processing are lacking, the available data suggests that regulation of processing could be as important as transcriptional regulation as a means of controlling the synthesis of bioactive peptides.

3.3 NPY METABOLISM

Transcriptional control of NPY undoubtedly plays a role in regulating NPY family signaling, but posttranslational modifications of these peptides is also likely to be important in terminating or switching bioactivities toward specific receptors. There is little information on how NPY degradation by enzymes occurs in tissues, in part because reliable techniques are lacking to distinguish different NPY immunoreactive (NPYir) forms. NPY and PYY are substrates for dipeptidylpeptidase IV (DPPIV, EC 3.4.14.5), a serine protease aminopeptidase, also known as CD26, that releases the N-terminal dipeptide, Tyr–Pro, and thereby converts NPY and PYY into their $NPY_{3\text{-}36}$ and $PYY_{3\text{-}36}$ fragments with a $K_m = 8\,\mu M$ (Mentlein et al., 1993). NPY can also be degraded by aminopeptidase P (AmP, EC 3.4.11.9), a metalloprotease that removes the N-terminal tyrosine from NPY and PYY to generate $NPY_{2\text{-}36}$ and $PYY_{2\text{-}36}$ (Medeiros and Turner, 1994; Mentlein and Roos, 1996). NPY may be degraded by enzymes expressed in neurons, astrocytes, and microglia, as well as by purified urokinase-type plasminogen activator, plasmin, thrombin, and trypsin, through cleavage after Arg^{19}, Arg^{25}, Arg^{33}, and Arg^{35}; these patterns of degradation were confirmed with purified urokinase-type plasminogen activator, plasmin, thrombin, and trypsin. For further characterization of the neuropeptide-degrading serine proteinase activities from cell cultures, urokinase-type plasminogen activator was identified on microglia by immunostaining, whereas tissue-type plasminogen activator mRNA occurred in neurons and astrocytes, but not in microglia (Ludwig et al., 1996). However, it is still unclear whether another enzyme may also be involved in removing the tyrosynamide 36 of NPY, because an enzyme in human platelets that deamidates substance P has been

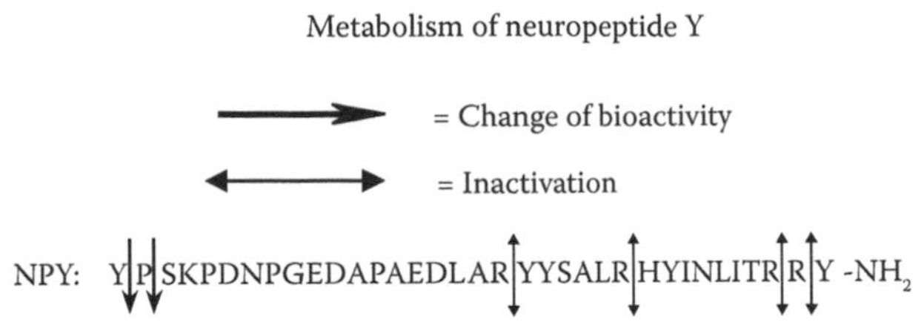

Metabolism of neuropeptide Y

⟶ = Change of bioactivity

⟷ = Inactivation

NPY: Y|P|SKPDNPGEDAPAEDLAR|YYSALR|HYINLITR|R|Y -NH$_2$

FIGURE 3.1 NPY post-translational modifications. NPY may be processed by aminopeptidases (aminopeptidase P and dipeptidylpeptidase IV) to produce peptides that no longer act on the Y_1 receptor but retain full activity with regard to the Y_2 and Y_5 receptors. NPY may be also degraded by endoproteases and carboxypeptidases to produce inactive peptides.

found (Jackman et al., 1990). This deamidase is inhibited by diisopropylfluorophosphate, inhibitors of chymotrypsin-type enzymes, and mercury compounds, while other inhibitors of catheptic enzymes, trypsin-like enzymes, and metalloproteases are ineffective. NPY degradation has been studied in the rat hippocampus using a combination of chromatographic techniques and nanospray mass spectrometry. The major component in the brain tissue corresponded to the authentic amidated form of $NPY_{1\text{-}36}$. The fate of NPY in the central nervous system was studied by subjecting pure peptide to the protease(s) present in hippocampal synaptosomes to reveal potential cleavage site(s). NPY was efficiently metabolized with a single cleavage between Leu^{30}–Ile^{31}. The enzyme revealed the properties of aspartic protease, being blocked by pepstatin, and having an optimum pH between 4 and 5. Recently, Frerker and colleagues reported (using MALDI-TOF mass spectrometry) that $NPY_{1\text{-}36}$ is exclusively degraded by DPPIV into $NPY_{3\text{-}36}$ in EDTA–plasma, but they did not provide the kinetics of NPY cleavage efficiency of DPPIV (Frerker et al., 2007). Beck-Sickinger and colleagues have studied (also using MALDI-TOF mass spectrometry) the metabolic stability of fluorescent N-terminal labeled NPY analogs incubated in human plasma and found that the 36th, 35th, and 33rd residues of NPY analogs may also be removed by unknown peptidases (Khan et al., 2007). NPY and PYY are also substrates for endopeptidase-24.11 ($K_m = 15.4\,\mu M$), which can cleave the Tyr^{20}–Tyr^{21} and Leu^{30}–Ile^{31} bonds, consistent with the known specificity of the enzyme. In striatal synaptic and renal brush border membranes, endopeptidase-24.11 is shown to comprise the major NPY hydrolyzing activity, but plays a lesser role in intestinal brush border membranes, while PYY produced in the intestinal L cells may be degraded in this tissue (Medeiros and Turner, 1994, 1996). A summary of cleavage sites of NPY is presented in Figure 3.1, indicating that the degradation pathway in the brain is different from that found in the periphery, and may have important consequences *in vivo* (Stenfors et al., 1997).

3.4 NPY RECEPTORS

The NPY receptor family includes the Y_1 receptor, first characterized as a postsynaptic receptor, the Y_2 receptor, the Y_3 receptor, which is a NPY-preferring receptor, the Y_4 receptor, first characterized as a PP receptor, the Y_5 receptor, and the y_6 that has been cloned, although this receptor is not functional in humans (Michel et al., 1998). All these receptors, except for the Y_3 receptor, have been cloned (Larhammar et al., 1992; Lundell et al., 1995; Rose et al., 1995; Weinberg et al., 1996).

3.4.1 Y_1 Receptor

The Y_1 receptor was the first PP-fold peptide receptor to be cloned and was identified as a rodent orphan organ receptor (Eva et al., 1990). Later, based on mRNA distribution, it was identified as a Y_1 receptor (Herzog et al., 1992). [Leu^{31},Pro^{34}]NPY was the first Y_1-receptor selective agonist to

TABLE 3.1
NPY Receptors

	Ligand-Binding Profile	Selective Agonists	Antagonists
Y_1 receptor	NPY ≈ PYY ≈ [Leu^{31},Pro^{34}]NPY > NPY_{2-36} > NPY_{3-36}PP > NPY_{13-36}	[Phe^7,Pro^{34}]NPY; [Leu^{31},Pro^{34}]NPY; [Pro^{34}]NPY; [Leu^{31},Pro^{34}]PYY; [Pro^{34}]PYY	BIBP3226; GW1229 (or GR231118 or 1229U91); SR120819A; BIBO3304; LY357897; J-115814; H394/84c; J-104870; GI264879A; H409/22;tetrahydrocarbazole derivatives
Y_2 receptor	NPY ≈ PYY ≥ NPY_{2-36} ≈ NPY $_{3-36}$ ≈ NPY $_{13-36}$ >> [Leu^{31},Pro^{34}]NPY	NPY_{3-36}, NPY_{13-36}; Ac-[Lys^{28},Glu^{32}] NPY_{25-36}; TASP-V;	BIIE0246; T_4-[NPY_{33-36}]$_4$; JNJ-5207787
Y_3 receptor	NPY ≈ NPY_{3-36} ≈ PP	—	—
Y_4 receptor	PP > PYY ≥ NPY > NPY_{2-36}	PP; GW1229 (or GR231118 or 1229U91); dimer of NPY_{32-36}	UR-AK49
Y_5 receptor	NPY ≈ PYY ≈ NPY_{2-36} > hPP > [D-Trp^{32}]NPY > NPY_{13-36} > rPP	FMS586; [Ala^{31},Aib^{32}]NPY; [hPP_{1-17},Ala^{31},Aib^{32}]NPY; [cPP_{1-7}, NPY_{19-23},Ala^{31},Aib^{32},Gln^{34}]hPP	CGP71683A; JCF109; NPY5RA-972; 3-[2-[6-(2-tertbutoxyethoxy) pyridin-3-yl]-1H-imidazol-4-yl]benzonitrile hydrochloride salt; GW438014A; L-152,804; pyrrolo[3,2-d] pyrimidine derivatives, 2-substituted 4-amino-quinazolin derivatives; α-substituted N-(sulfonamino)alkyl-β aminotetralins; cyclohexylurea 21c
y_6 receptor	NPY ≈ PYY ≈ [Leu^{31},Pro^{34}]NPY >> PP or PP > [Leu^{31},Pro^{34}] NPY > NPY ≈ PYY	—	—

Source: Berglund, M.M., Hipskind, P.A., and Gehlert, D.R., *Exp. Biol. Med.*, 228, 217–244, 2003; Silva, A.P., Xapelli, S., Grouzmann, E., and Cavadas, C., *Curr. Drug Targets*, 4, 331–347, 2005; Balasubramaniam, A. et al., *J. Med. Chem.*, 49, 2661–2665, 2006; Kakui, N. et al., *J. Pharmacol. Exp. Ther.*, 317, 562–570, 2006; Ziemek. R. et al., *J. Recept. Signal Transduct. Res.*, 27, 217–233, 2007; Li, G. et al., *Bioorg. Med. Chem. Lett.*, 18, 1146–1150, 2008.

be discovered (Fuhlendorff et al., 1990) (Table 3.1). The Y_1 receptor is fully activated by peptides substituted with a proline in position 34 (i.e., [Pro^{34}]NPY and [Leu^{31},Pro^{34}]NPY) and can be internalized after agonist stimulation (Fabry et al., 2000; Parker et al., 2002b; Pheng et al., 2003). The agonists and antagonists for the Y_1 receptor are listed in Table 3.1.

3.4.2 Y_2 Receptor

The Y_2 receptor is pharmacologically characterized by a high affinity for NPY and PYY, but, unlike the Y_1 receptor, it retains a high binding affinity for the C-terminal fragments of NPY (i.e., NPY_{2-36},

NPY_{3-36}, NPY_{13-36}) (Michel et al., 1998). The Y_2 receptor does not appear to be internalized after prolonged agonist stimulation, or does so very slowly (Parker et al., 2001). The agonists and antagonists for the Y_2 receptor subtype are listed in Table 3.1.

3.4.3 Y_3 Receptor

In contrast to the Y_1 and Y_2 receptors, the Y_3 receptor is characterized by its at least 10-fold lower affinity for PYY than for NPY (Michel et al., 1998). However, no NPY receptor that exhibits the pharmacology of the Y_3 receptor has been cloned. Pharmacological and binding studies suggest the existence of this NPY-preferring receptor in human adrenal chromaffin cells (Lee and Miller 1998; Cavadas et al., 2001), in the rat nucleus tractus solitarius (Grundemar et al., 1991), in rat cardiac membranes (Balasubramaniam et al., 1990), rat aortic endothelial cells (Nan et al., 2004), and in bovine chromaffin cells (Norenberg et al., 1995). In these reports, the inactivity of PYY was used as the criteria to infer Y_3 receptor activity. Recently, some evidence suggests that the uncloned Y_3 receptor could be a heterodimer of NPY receptors. Movafagh and colleagues (2006) showed NPY-induced migration of human endothelial cells through the activation of all Y_1, Y_2, and Y_5 receptors, although this response was significantly inhibited by any of the Y_1, Y_2, or Y_5 single antagonists, the activation of only one of those receptors was not enough to produce a migratory response. Moreover, PYY_{1-36} ($Y_1/Y_2/Y_5$, but not the Y_3 receptor agonist) did not produce the two migratory peaks. The authors propose that these receptors form heteromeric complexes, the $Y_1/Y_2/Y_5$ receptor oligomer, which may be the uncloned Y_3 receptor (Movafagh et al., 2006). The formation of NPY receptors heterodimerization was first proposed by Silva and colleagues (2003). They suggested that the activation of heterodimers between Y_1 and Y_2 or Y_2 and Y_5 receptors produced an inhibitory effect on stimulated glutamate release from rat synaptosomes, and the activation of such a complex would predominantly activate the Y_2 receptors over Y_1 and Y_5 receptors (Silva et al., 2003a). Because the authors did not evaluate the effect of PYY, the pharmacological profile of those oligomers (Y_1/Y_2 or Y_2/Y_5) cannot be compared to the putative Y_3 receptor. Gehlert and colleagues (2007) demonstrated that Y_1/Y_5 coexpression results in the formation of heterodimer Y_1/Y_5 which profoundly altering the pharmacology. However, PYY is still able to inhibit adenylyl cyclase activity and induced internalization of heterodimer Y_1/Y_5 (Gehlert et al., 2007), which means that, at least in the *in vitro* conditions tested, this oligomer is not related to the uncloned Y_3 receptor.

3.4.4 Y_4 Receptor

The Y_4 receptor was originally named the PP_1 receptor, because of its preference for PP as a ligand (Lundell et al., 1995). Unlike the other NPY receptors, the Y_4 receptors display considerable differences in sequence and ligand-binding affinity across mammalian species, and this feature could be related to the rate of receptor internalization (Parker et al., 2002a). In mammals, the affinity of PP for this receptor is much greater than the affinities of NPY and PYY (Bard et al., 1995; Lundell et al., 1995; Gregor et al., 1996b; Berglund et al., 2003). The Y_1 receptor antagonist, GW1229 (also known as 1229U91 or GR231118), is a potent Y_4 receptor agonist (Parker et al., 1998). Dimers of NPY_{32-36} are Y_4-receptor selective agonists with picomolar affinity (Balasubramaniam et al., 2006). Recently, the first nonpeptidic antagonist for human Y_4 receptor was discovered: the 3-cyclohexyl-*N*-[(3-1H-imidazol-4-ylpropylamino)(imino)methyl]propanamide (UR-AK49) (Ziemek et al., 2007).

3.4.5 Y_5 Receptor

The Y_5 receptor subtype has high affinity for NPY, PYY, [Leu[31],Pro[34]]-NPY or PYY substituted analogs, hPP, and long C-terminal fragments such as NPY_{3-36} and PYY_{3-36} (Gregor et al., 1996b; Hu et al., 1996; Borowsky et al., 1998; Michel et al., 1998). Different selective Y_5 receptor agonists and antagonists have been reported (see Table 3.1).

3.4.6 y_6 Receptor

The y_6 receptor was first cloned in rabbit (Matsumoto et al., 1996) and mouse (Gregor et al., 1996a). All y_6 sequences from primate species studied to date (i.e., chimpanzee, gorilla, marmoset, and human) encode for a truncated gene (Matsumoto et al., 1996; Weinberg et al., 1996). Although the gene for y_6 is present in most mammals, including mouse, neither mRNA nor the gene encoding y_6 have been detected in rats (Burkhoff et al., 1998).

3.5 IMPLICATIONS OF NPY AND PYY METABOLISM IN RECEPTOR SELECTIVITY

Compared with the native peptide, the N-terminally truncated fragments $NPY_{3\text{-}36}$ and $NPY_{2\text{-}36}$ had considerably reduced activities at the Y_1 receptor, but not reduced or scarcely reduced activity at the Y_2 and Y_5 receptors. These considerations have only recently been taken into account with the potential side effects that may be associated with the marketing of DPPIV enzyme inhibitors that prevent glucagon-like peptide 1 degradation for the treatment of type 2 diabetes (Grouzmann et al., 2007). However, there are few data showing a clinical relevance in preventing NPY metabolism with these inhibitors.

Because the Y_1 receptor is the major vasoconstrictive receptor, DPPIV probably largely inactivates the action of NPY. Therefore, DPPIV has been termed an NPY-converting enzyme, and several physiological functions have been proposed, including the regulation of vascular smooth muscle contractility and angiogenesis. In the presence of a DPPIV inhibitor, an enhanced sympathetic tone might be expected from direct postsynaptic effects of native NPY. This hypothesis has been assessed in a recent study reporting that male rats exhibit basal endogenous Y_1 receptor modulation in their hindlimb vascular bed. On the other hand, the systemic inhibition of DPPIV and AmP elicited a Y_1-receptor-dependent decrease in hindlimb vascular conductance in females. Thus, the lack of basal endogenous Y_1-receptor activation in female hindlimb vasculature is (at least partially) due to proteolytic processing of NPY (Jackson et al., 2005). Jackson and colleagues found that $PYY_{1\text{-}36}$ potentiates renovascular responses to angiotensin II in kidneys from spontaneously hypertensive rats, whereas $PYY_{3\text{-}36}$ has little effect. In addition, a DPPIV inhibitor augments the ability of peptide $PYY_{1\text{-}36}$ to enhance renovascular responses to angiotensin II (Jackson et al., 2005). The fact that DPPIV is expressed in preglomerular microvessels and glomeruli and kidneys metabolize locally arterial $PYY_{1\text{-}36}$ to $PYY_{3\text{-}36}$ via a mechanism blocked by an inhibitor indicates that this enzyme may be of crucial importance for regulating kidney microcirculation. Another study has shown that DPPIV inhibition elevates arterial blood pressure in spontaneously hypertensive rats (SHR) compared with normotensive Wistar–Kyoto rats when elevated blood pressure is reduced with antihypertensive drugs, provided that the sympathetic nervous system is functional. This effect is reversed by the use of a Y_1 antagonist. These results suggest vigilance, because DPPIV inhibitors are often used in diabetic patients with associated comorbidity factors such as hypertension (Jackson et al., 2008).

The administration of NPY has been found to be more potent in mutant Fischer 344 rats lacking endogenous DPPIV enzymatic activity in exerting anxiolytic-like effects (increased social interaction time in the social interaction test) and sedative-like effects (decreased motor activity in the elevated plus maze) that are reversed by a Y_1 receptor antagonist. These results provide direct evidence that NPY-mediated effects in the central nervous system are modulated by DPPIV-like enzymatic activity (Karl et al., 2003).

3.6 NPY AND FEEDING BEHAVIOR

It is now well established that feeding is centrally regulated by the hypothalamus. The NPY system for feeding regulation is indeed mostly located in the hypothalamus, including the arcuate nucleus (ARC), where NPY is synthesized, and the paraventricular (PVN), dorsomedial (DMN), and ventromedial (VMN) nuclei and perifornical area, where it is active (Kalra et al., 1999). So, the NPY neuron in the ARC of the hypothalamus and NPY receptors in the ARC and adjacent regions of the hypothalamus are believed to be critical components of a pathway that functions to stimulate food intake and decrease energy expenditure (Kalra et al., 1999; Schwartz et al., 2000).

Several lines of evidence suggest that NPY plays a key role in the control of appetite, body weight gain, and obesity (Inui, 1999; Kalra et al., 1999). Intracerebroventricular injection (i.c.v.) of NPY in mammals strongly stimulates food intake while inhibiting thermogenesis and lipolysis (Clark et al., 1984; Stanley et al., 1986; Larsen et al., 1999; Corp et al., 2001; Narnaware and Peter, 2001; Raposinho et al., 2001; Pelz and Dark, 2007). Food intake is increased several fold and the effect lasts 6 to 8 hours. Moreover, hypothalamic NPY and its mRNA are increased in genetic models of obesity, namely in the obese Zucker rat, in *db/db* or obese *ob/ob* mouse, and also during poor metabolic conditions such as when fasting (Morley et al., 1987; Sahu et al., 1988; McCarthy et al., 1991; McKibbin et al., 1991; Dryden et al., 1995; Erickson et al., 1996; Schwartz et al., 1996b). Interestingly, chronic i.c.v. administration of NPY leads to obesity in rodents fed *ad libitum* (Stanley et al., 1986; Vettor et al., 1994).

NPY synthesis and release by the NPY neuron are controlled by hormones that transmit the status of peripheral energy stores, including insulin, leptin, ghrelin, and peptide YY (Schwartz et al., 1992; Stephens et al., 1995; Dickson and Luckman, 1997; Baskin et al., 1999; Willesen et al., 1999; Kumarnsit et al., 2003). Leptin-deficient *ob/ob* mice as well as rats with insulin-deficient diabetes are hyperphagic and show increased NPY synthesis (White et al., 1990; Schwartz et al., 1996b). In addition, NPY knockout in leptin-deficient *ob/ob* mice resulted in reduced body weight (Erickson et al., 1996). Moreover, leptin infusion decreases the expression and secretion of NPY from the hypothalamus (Schwartz et al., 1996a). Similarly, insulin replacement in streptozotocin-treated rats normalizes the increase in hypothalamic NPY expression and reduces the hyperphagia that develops secondary to insulinopenia (McKibbin et al., 1992).

Over the last 10 years, many studies have aimed to identify the NPY receptor subtype involved in the orexigenic effects of this peptide. All cloned and functional receptors of NPY play a crucial role in energy homeostasis. Several Y_1 receptor antagonists were developed and tested for their ability to inhibit food intake in rodents. Injection (i.c.v.) of the Y_1 antagonist (1229U911 or GR231118) inhibits both NPY-induced food intake and physiological feeding behavior after an overnight fast (Kanatani et al., 1996, 1998). The effect of i.c.v. injection of 1229U911 was higher in obese *fa/fa* (Zucker) rats than in lean rats. In contrast, 1229U911 does not inhibit food intake in Wistar rats with diet-induced obesity (DIO). Although spontaneous feeding in wild-type mice by intraperitoneal injection of the Y_1 receptor antagonist J-115814 was reduced, neither 1229U911 nor BIBP3226 had an effect on NPY-induced feeding in mice, suggesting that in some strains NPY-induced feeding is mediated either by a combination of more than one NPY receptor subtype (dimerization of NPY receptors) or by a unique NPY receptor subtype different from the Y_1 receptor, which could be the y_6 receptor (Iyengar et al., 1999). Additionally, 1229U911 has agonistic properties for the Y_4 receptor, even though it does not seem to antagonize other receptors (Matthews et al., 1997). The Y_1 antagonists BIBO3304 and SR120562A also inhibit food intake induced by the i.c.v. infusion of NPY or by fasting (Wieland et al., 1998; Polidori et al., 2000). The i.c.v. injection of the Y_1 antagonist J-104870 significantly attenuates spontaneous food intake in Zucker fatty rats and to decrease by 74% of the NPY-induced feeding (Kanatani et al., 1999). Similarly, i.p. injection of the Y_1 receptor antagonist J-115814 reduced feeding induced by i.c.v. injections of NPY in satiated Sprague–Dawley rats and had no effect on NPY-induced food intake in NPY Y_1 knockout mice (Kanatani et al., 2001). All these results strongly suggest a role of the Y_1 receptor in mediating feeding behavior.

Because PP is the preferential Y_4 receptor agonist, this receptor might mediate some of the effects produced by PP, namely the involvement of the Y_4 receptor on food intake. Peripheral administration of PP decreases food intake in rodents and also in humans (Clark et al., 1984; Berntson et al., 1993; Batterham et al., 2003b; Stanley et al., 2004), but the central administration of PP increases food intake in rodents and dogs (Clark et al., 1984; Inui et al., 1991; Campbell et al., 2003). Systemic injection of PP has been shown to reduce food intake through the activation of vagal afferents (Asakawa et al., 2003). Moreover, the Y_4 agonist sub[–Tyr–Arg–Leu–Arg–Tyr–NH_2]$_2$ inhibited food intake in fasted mice in a dose-dependent manner (Balasubramaniam et al., 2006). Y_4 knockout mice showed increase plasma levels of PP, decreased body weight, and lower white adipose mass, especially in males (Sainsbury et al., 2002a).

The stimulatory and inhibitory effects of the Y_5 receptor agonists and the Y_5 receptor antagonists, respectively, on food intake also support the involvement of the Y_5 receptor subtype in eating behavior. The Y_5 agonist [D-Trp^{32}]NPY produced a similar increase in food intake as NPY administration in satiated rats (Hwa et al., 1999; Parker et al., 2000) and in obese Zucker rats (Wyss et al., 1998; Beck et al., 2007). However, other studies with mice could not demonstrate an increase of food intake by [D-Trp^{32}]NPY (Iyengar et al., 1999). Another Y_5 receptor-selective agonist, [Ala^{31},Aib^{32}]NPY, also stimulates feeding in rats (Cabrele et al., 2000). There are some reports regarding the antiobesity effects of long-term treatment with Y_5 antagonists. The Y_5 antagonist CGP71683A produces inhibition of food intake in lean satiated rats without undesirable side effects such as increased anxiety or decreased locomotor activity (Kask et al., 2001). Also, the Y_5 antagonist of FMS586 inhibits NPY-induced food intake, and interestingly, this compound also showed dose-dependent but transient suppression in natural feeding models of both overnight fasting-induced hyperphagia and spontaneous daily intake (Kakui et al., 2006). Oral or i.c.v. administration of the Y_5 receptor antagonist L-152,804 significantly inhibited food intake evoked by i.c.v. injected bovine PP, a moderately selective Y_4/Y_5 agonist in satiated rats. However L-152,804 did not significantly inhibit i.c.v. NPY-induced food intake (Kanatani et al., 2000). Two Y_5 antagonists, CGP71683A and GW438014A, suppressed body weight gain in DIO and genetically obese models (Criscione et al., 1998; Daniels et al., 2002), whereas NPY5RA-972 had no effect in DIO (Turnbull et al., 2002). However, another Y_5 antagonist (3,3-dimethyl-9-(4,4-dimethyl-2,6-dioxocyclohexyl)-1-oxo-1,2,3,4-tetrahydroxanthene) produces antiobesity effects in mice that were developing DIO by regulating both energy intake and expenditure (Ishihara et al., 2006; Mashiko et al., 2007). The Y_5 antagonist MK-0557 antagonized the effects of the Y_5 receptor agonist, D-Trp^{34}NPY on body-weight gain and hyperphagia in C57BL/6J mice, and significantly suppressed body-weight gain in DIO mice (Erondu et al., 2006). Interestingly, these results suggest that the antiobesity effects of the Y_5 antagonists are specific to the DIO model, because the Y_5 antagonist was not effective in genetically obese models, such as *db/db* mice and Zucker fatty rats (Mashiko et al., 2003; Ishihara et al., 2006). As for the Y_5 antagonists, repeated central administration of Y_5 antisense oligodeoxynucleotides significantly decreased spontaneous as well as NPY-induced food intake (Schaffhauser et al., 1997; Larsen et al., 1999). This Y_5 antagonist was tested (single-dose study, 1 mg) over a 52-week, multicenter, randomized, double-blind, placebo-controlled phase I trial involving 1661 overweight and obese patients. Although it had a favorable clinical safety profile, the degree of weight loss observed was not clinically meaningful (Erondu et al., 2006). These observations provide the first clinical insight into the human NPY-energy homeostatic pathway and suggest that solely targeting the Y_5 receptor in future drug development programs is unlikely to produce therapeutic efficacy.

The involvement of each player of NPY system on feeding behaviour has already been analysed in different transgenic NPY (overexpressing or knocking down NPY) and NPY receptors KO rodents, and those studies revealed that other systems may compensate the absence of one player in the system.

The overexpression of NPY leads to conflicting results related to food intake and feeding behavior. In fact, NPY-overexpressing mice have normal feeding or food intake (Inui et al., 1998; Ste Marie et al., 2005). However, with a sucrose-loaded diet, an increased body weight and transitory increased food intake was observed (Kaga et al., 2001). Some unexpected results were obtained with NPY knockout mice; in one study they were found to grow and eat normally, and their response to i.c.v. NPY administration was similar to that of wild-type animals (Erickson et al., 1996). These results could be explained by a compensatory mechanism involving other neuropeptides during development. In some conditions deletion of NPY led to a lower feeding response, such as after fasting, after exposure to a highly palatable diet, or in response to moderate insulin-induced hypoglycemia (Bannon et al., 2000; Sindelar et al., 2004, 2005). A deficiency of NPY also attenuated DIO in mice prone to obesity (Patel et al., 2006).

To try to clarify which NPY receptor subtype is involved in food intake, different knockout mice were developed. The Y_1 knockout mice have no difference or only a slight reduction in food intake, accumulate more fat, and after fasting eat less and have increased body fat, but they have a blunted feeding response after i.c.v. administration of NPY (Kushi et al., 1998; Pedrazzini et al., 1998).

Y_2 knockout mice show increased body weight, food intake, and fat deposition (Naveilhan et al., 1999), blunted response to leptin, normal response to NPY-induced food intake, and intact regulation of refeeding and body weight after fasting (Naveilhan et al., 1999). These mice also showed a blunted response to leptin but a normal response to NPY-induced food intake and intact regulation of refeeding and body weight after fasting (Naveilhan et al., 1999). Also in Y_2 knockout mice there is no reduction of food intake or weight gain after peripheral administration of the appetite stimulator $PYY_{3\text{-}36}$ (Batterham et al., 2002). In contrast, Y_2 knockout mice have a transient reduced body weight, despite an increase in food intake, probably due to an increase in mRNA levels for NPY, POMC, AgRP, and CART in the ARC (Sainsbury et al., 2002b).

Although a role for Y_5 in feeding behavior has been established through acute administration of selective agonists, Y_5 knockout mice feed normally, do not have a lean phenotype, and also develop late-onset mild obesity (Criscione et al., 1998; Marsh et al., 1998). Moreover, feeding induced by i.c.v. injection of NPY was not reduced in knockout Y_5 mice, as was observed in wild-type animals (Criscione et al., 1998), suggesting an activation of a compensatory orexigenic pathway during development.

The role of Y_1, Y_2, and Y_4 receptors in the NPY-induced obesity syndrome was elegantly investigated using recombinant adeno-associated viral vector (rAAV) overexpressing NPY in mice deficient of single or multiple receptors (Lin et al., 2006). Overexpressing NPY induced a smaller body weight increase in Y_1 knockout mice compared to control mice. The degree of body weight gain in rAAV-NPY-treated mice was smaller in $Y_1Y_2Y_4$ knockout mice compared to Y_1 knockout mice (Lin et al., 2006).

In conclusion, in all genetic disruption of NPY signaling (peptide and receptors), the consequences for feeding and body weight did not constitute a robust effect. It is not known whether there is compensation, synergy, or biological redundancy among the NPY receptors, or regulation of energy balance by other neuropeptides and/or regulators. Some studies have suggested that in the absence of NPY signaling compensatory mechanisms normalize the regulation of feeding and energy expenditure to maintain energy homeostasis (Luquet et al., 2005; Ishii et al., 2007). Moreover, the existence of other unidentified NPY receptors or the occurrence of oligomerization of NPY receptors should not be discounted. Therefore, drugs targeting different Y receptors should be considered in the development of new treatments for obesity.

3.7 $PYY_{3\text{-}36}$—A NEW THERAPEUTIC TARGET FOR TREATING EATING DISORDERS AND OBESITY

After meals, PYY is released from L-cells in the lower gastrointestinal tract, in proportion to the calories consumed and fat content in particular (Adrian et al., 1985), and is probably mediated via a neural reflex in response to meal ingestion (Fu-Cheng et al., 1997). PYY is locally converted into PYY_{3-36} by DPPIV in the capillaries supplying the L-cells of the intestine (Hansen et al., 1999). PYY_{3-36} is the main circulating form in the fasted and fed state, indicating that short-term regulation of DPPIV activity in the gastrointestinal tract is not a rate-limiting production step (Batterham et al., 2006; Korner et al., 2006). Batterham and colleagues found that peripheral administration of PYY_{3-36} reduced food intake in rodents and in normal-weight and obese humans (Batterham et al., 2002; 2003a). These findings stimulated research into PYY_{3-36} as a potential therapy for obesity, although these data were tempered by other groups that were not able to confirm the anorectic actions of PYY_{3-36}, although these effects were reproduced numerous times in rodents, nonhuman primates, and humans (see the review by Chandarana and Batterham, 2008). The mechanism of action by which PYY_{3-36} contributes to the reduced appetite and weight reduction observed in patients receiving the peptide chronically has not been fully elucidated. However, the observations that peripheral PYY_{3-36} administration to rodents increases the expression of *c*-fos within the ARC suggests that its effects should be mediated by a hypothalamic Y_2 receptor, a putative inhibitory presynaptic receptor, highly expressed on NPY neurons in the ARC to reduce food intake (Chandarana and Batterham, 2008). A phase II clinical trial aiming at reducing body weight in obese subjects has been initiated by Nastech, a pharmaceutical company, using a nasal formulation of PYY_{3-36} (http://www.clinicaltrials.gov.NCT00537420).

3.8 NPY AND NPY RECEPTORS IN ADIPOSE TISSUE: A PERIPHERAL ROLE OF NPY IN OBESITY

The increase in adipose mass that occurs in obesity is due to proliferation and differentiation of preadipocytes into mature adipocytes (Prins and O'Rahilly, 1997). The proliferation and differentiation of adipocytes is regulated by NPY receptors. Y_2 receptor activation increases mouse adipocyte proliferation (Kuo et al., 2007). However, the proliferation of the primary rat preadipocytes cell line of mouse preadipocytes (3T3-L1) is stimulated by Y_1 receptor activation (Yang et al., 2008). The importance of NPY for the regulation of adipocyte number has also been demonstrated using a Y_5 receptor antisense that induces adipocyte apoptosis (Gong et al., 2003). Lin and colleagues (2006), on the other hand, have shown that rAAV-NPY-treated mice show significant increases in both the absolute and relative weights of the various white adipose tissue depots studied, including inguinal, epididymal, mesenteric, and retroperitoneal (Lin et al., 2006). NPY may also inhibits lipolysis in these cells (Valet et al., 1990; Castan et al., 1992; Labelle et al., 1997; Turtzo et al., 2001). The inhibition of lipolysis by NPY was achieved in 3T3-L1 through the Y_5 receptor (Turtzo et al., 2001) or in rat adipose tissues primarily through Y_1 receptor activation (Labelle et al., 1997; Margareto et al., 2000).

In conclusion, adipocytes not only produce and secrete NPY but also express different NPY receptor. The presence of NPY receptors contributes to the modulation of adipocyte physiology: adipocyte proliferation, differentiation with an increase of intracellular lipid accumulation and lipolysis inhibition. NPY and its receptors in adypocytes are new putative targets to develop new antiobesity pharmacological tools.

3.9 THE NPY SYSTEM AS A THERAPEUTIC CANDIDATE IN EPILEPSY

The neuroprotective role of NPY in the central nervous system is well documented (see the review in Silva et al., 2005). NPY protects hippocampal and cortical neurons exposed to the prolonged stimulation of glutamate receptors (excitotoxic conditions; Silva et al., 2003b, 2005; Wu and Li, 2005; Domin et al., 2006; Xapelli et al., 2007) and has a protective role against methamphetamine-induced neuronal apoptosis in the mouse striatum (Thiriet et al., 2005). Moreover, in rat retina NPY prevents glutamate- and ecstasy-induced retinal neural cell death (Álvaro et al., 2007). Interest in the NPY system has gained particular relevance in pathological conditions characterized by neuronal hyperexcitability such as epilepsy (reviewed in Silva et al., 2002). The antiepileptic effects of NPY have been studied in several models of epilepsy *in vitro* and *in vivo*, and the results show an antiexcitatory role for this peptide (Erickson et al., 1996; Baraban et al., 1997; Woldbye et al., 1997). NPY application *in vivo* reduces seizure susceptibility in mice (Baraban et al., 1997; Erickson et al., 1996) and NPY knockout are more susceptible to seizures (Palmiter et al., 1998). On the other hand, NPY overexpression results in a decreased susceptibility to seizures (Vezzani et al., 2002).

Considerable effort has gone into identifying the receptor or receptors involved in the antiepileptic actions of NPY, and despite some controversy, these studies suggest that the Y_1, Y_2, and Y_5 receptors are putative candidates. Agonists for Y_1, Y_2, and Y_5 receptors have been found to reduce seizure-like activity in hippocampal neurons in culture (Reibel et al., 2001). *In vivo* study with the Y_1 antagonist BIBP 3226 suggests that the activation of the Y_1 receptor is proconvulsant, and the blockade of Y_1 receptors is anticonvulsant (Gariboldi et al., 1998). Another study using a model of genetic generalized epilepsy suggests that Y_2 receptors are more important than Y_1 and Y_5 receptors in mediating the effect of NPY to suppress seizures (Morris et al., 2007). The importance of the Y_2 receptor was also suggested by El Bahh et al. (2005), who showed that Y_2 receptors alone mediate all the antiexcitatory actions of NPY seen in the hippocampus, whereas the Y_5 receptors have no role *in vitro* or *in vivo* (El Bahh et al., 2005). These studies point out the Y_2 receptor agonists as putative useful anticonvulsants.

Nevertheless other studies suggest that NPY reduces seizures of hippocampal origin through the activation of the Y_5 receptors. For example, NPY Y_5 knockout mice do not exhibit spontaneous epileptic activity; however, they revealed an enhanced sensitivity to kainic acid-induced seizures (Marsh et al., 1999). Furthermore, slices prepared from these animals were insensitive to the actions of NPY (Marsh et al., 1999; Baraban, 2002). The Y_5 receptor appeared to mediate the antiepileptic effect of NPY in a kainate seizures model. Thus the mixed Y_1/Y_5 agonist [Leu^{31},Pro^{34}]NPY, mixed Y_2/Y_5-preferring agonists (NPY_{3-36}, PYY_{3-36}), and the Y_5-preferring agonist hPP mimicked the antiepileptic effects of NPY, whereas the Y_2-preferring agonist NPY_{13-36} had no effect (Woldbye et al., 1997). In generalized seizure models, by systemic injection of pentylenetetrazole (PTZ), the Y_2-preferring agonist NPY_{13-36} and the Y_1 antagonist BIBP3226 also significantly reduced seizure latency and severity in PTZ seizures induced in rats pretreated 30 days in advance with kainate, suggesting that Y_2 receptors can mediate antiepileptic effect against PTZ seizures, whereas Y_1 receptors appear to promote this type of seizure (Vezzani et al., 2000). In another study, i.c.v. administration of NPY_{13-36} showed no effect on PTZ seizures (Reibel et al., 2001). The type of receptors involved in the anticonvulsant role of NPY is model- and experimental-condition-dependent.

Gene therapy approaches with recombinant adeno-associated viral (rAAV) vector-induced NPY in the hippocampus have clearly demonstrated that NPY exerts powerful antiepileptic and antiepileptogenic effects in rats (Richichi et al., 2004; Foti et al., 2007). These effects were also obtained by bilateral piriform cortex infusions of AAV vectors that express and constitutively secrete full-length NPY (AAV-NPY) or the Y_2 agonist NPY_{13-36} (AAV-NPY_{13-36}) (Foti et al., 2007). These gene therapeutical approaches exert a relatively limited effect on synaptic plasticity and learning in the hippocampus, and therefore this approach could be considered as a viable alternative for epilepsy treatment (Sorensen et al., 2008). Additional investigations are required to demonstrate a therapeutic role of gene therapy in chronic models of seizures and to address in more detail safety concerns and possible side effects (Noe et al., 2007).

Clearly, NPY can have potent actions in models of epilepsy; the type of therapeutics candidate could be a nonpeptidic Y_2 and/or Y_5 agonist, but the relevance of this approach in human patients is still unknown.

3.10 NPY STIMULATES NEUROPROLIFERATION AND NEUROGENESIS

The mature mammalian nervous system arises from precise coordination of proliferation, differentiation, and migration of precursor cells during embryonic and early postnatal development. There are three major neurogenic niches that retain the ability to generate substantial numbers of new neurons in adult life: the subventricular zone (SVZ) lining the lateral ventricles, the subgranular zone (SGZ) of the dentate gyrus in the hippocampal formation, and the olfactory bulb (OB) (Lennington et al., 2003).

NPY has been implicated in the proliferation of neuronal precursor cells in olfactory regions and the SVZ. NPY knockout mice have a significant reduction in the number of dividing olfactory neuronal precursor cells and fewer olfactory neurons by adulthood (Hansel, 2001; Howell et al., 2005; Anitha et al., 2006). In other neuronal models, such as and rat enteric neurons, rat retinal neuronal progenitor cells, and stem cells of SVZ, NPY also increases proliferation (Anitha et al., 2006; Agasse et al., 2008; Alvaro et al., 2008).

The proliferative role of NPY for neural cells is likely mediated through the Y_1 receptor (Hansel, 2001; Howell et al., 2005; Agasse et al., 2008; Alvaro et al., 2008) and also by the Y_2 receptors (Stanic et al., 2008). In rat retinal cells, the proliferative effect induced by NPY arises through the activation of several NPY receptors, such as Y_1, Y_2, and Y_5 receptors (Alvaro et al., 2008). The studies that have investigated the intracellular mechanisms involved in the stimulation of neuronal cells proliferation by NPY suggest the involvement of extracellular signal-regulated kinase (Erk1/2) c-Jun-NH_2-terminal kinase/stress-activated protein kinase (JNK/SAPK), nitric oxide (NO), and protein kinase C (Hansel, 2001; Howell et al., 2005; Agasse et al., 2008).

Besides its involvement in regulating the proliferation of precursors in the SVZ and RMS, NPY also promotes cell differentiation into distinct neuronal cells (Howell et al., 2005; Agasse et al., 2008; Stanic et al., 2008). The use of NPY as a neurogenesis activator opens new avenues for the development of stem-cell-based therapies for neurodegenerative diseases.

3.11 NPY IN THE REGULATION OF CATECHOLAMINE SECRETION BY ADRENAL CHROMAFFIN CELLS

The adrenal medulla is a target organ during stress, and mediates catecholamine release. Catecholamines, by increasing heart rate, blood pressure, and glucose consumption, help our bodies to respond to stress. As well as its presence in nerve endings and the adrenal cortex, NPY is produced by adrenonomedullary chromaffin cells of different species, including human and mice (Allen et al., 1983; Cavadas et al., 2001, 2006; Renshaw and Hinson, 2001; Spinazzi et al., 2005). The effect of NPY on adrenal cortex has been well discussed elsewhere (Renshaw et al., 2000; Renshaw and Hinson, 2001; Spinazzi et al., 2005).

The role of NPY in the regulation of the secretion of catecholamine by adrenal chromaffin cells is subject to debate, depending on the species and experimental conditions applied. NPY stimulated catecholamine release from intact rat adrenal capsular tissue (Renshaw et al., 2000), although an inhibitory effect of NPY on catecholamine secretion in rat adrenomedullary primary cell cultures was also observed (Shimoda et al., 1993). Moreover, there is a weak inhibitory effect of NPY on catecholamine release from bovine chromaffin cells, evoked by the addition of a cholinergic agonist (Higuchi et al., 1988; Hexum and Russett, 1989). However, in perfused bovine adrenal gland, NPY stimulated the secretion of catecholamine in the presence of cholinergic agonists (Higuchi et al., 1988; Hexum and Russett, 1989). The exact subtype(s) of the NPY receptor(s) involved in the modulation of bovine catecholamine secretion remain(s) undefined, although the presence of NPY Y_1 and the Y_3 receptors has been reported (Norenberg et al., 1995; Zheng et al., 1997; Zhang et al., 2000). In human chromaffin cells, we demonstrated that human chomaffin cells in culture expressed mRNA for the Y_1, Y_2, Y_4, and Y_5 receptors (Cavadas et al., 2001). These receptors are functional, as various receptor-specific agonists elicit an increase in intracellular calcium (Cavadas et al., 2001).

The stimulatory effect of NPY on catecholamine release from human chromaffin cells was also obtained with hPP, $NPY_{13\text{-}36}$, and $NPY_{3\text{-}36}$, but PYY did not produce any effect, suggesting that the Y_3 receptor modulates the NPY effect (Cavadas et al., 2001). Another important observation in human and mice chromaffin cells in culture is that NPY is constitutively released from these cells (Cavadas et al., 2001, 2006). Moreover, no receptor-specific antagonists (not Y_1, nor Y_2, nor Y_5 antagonists) were able to reduce constitutive catecholamine release; however, an NPY-immunoneutralizing antibody markedly reduced constitutive catecholamine release. Because NPY is coreleased with norepinephrine (NE) from nerve endings and acts as an important modulator of sympathetic function by potentiating the catecholamine vasoconstrictor activity through the Y_1 receptor, we speculate that NPY could behave as an amplifier of catecholamine release by acting on the Y_3 receptor.

In mouse chromaffin cells in culture that also express mRNA for Y_1, Y_2, Y_4, and Y_5 receptors, NPY evokes catecholamine release (Cavadas et al., 2006; Rosmaninho-Salgado et al., 2007b). Moreover, NPY evokes catecholamine release by the activation of the Y_1 receptor, in a Ca^{2+}-dependent manner, either by activating MAPK or leading to production of NO, which is linked to the activation of protein kinase C (PKC) and guanylyl cyclase (GC) (Rosmaninho-Salgado et al., 2007b).

The importance of Y_1 receptors in regulating the release and synthesis of catecholamine was investigated using Y_1 knockout mice (Cavadas et al., 2006). In fact, Y_1 knockout mice have a higher adrenal content and constitutive release of catecholamine and catecholamine plasma concentrations (Figure 3.2). This unexpected high turnover of adrenal catecholamine in these animals

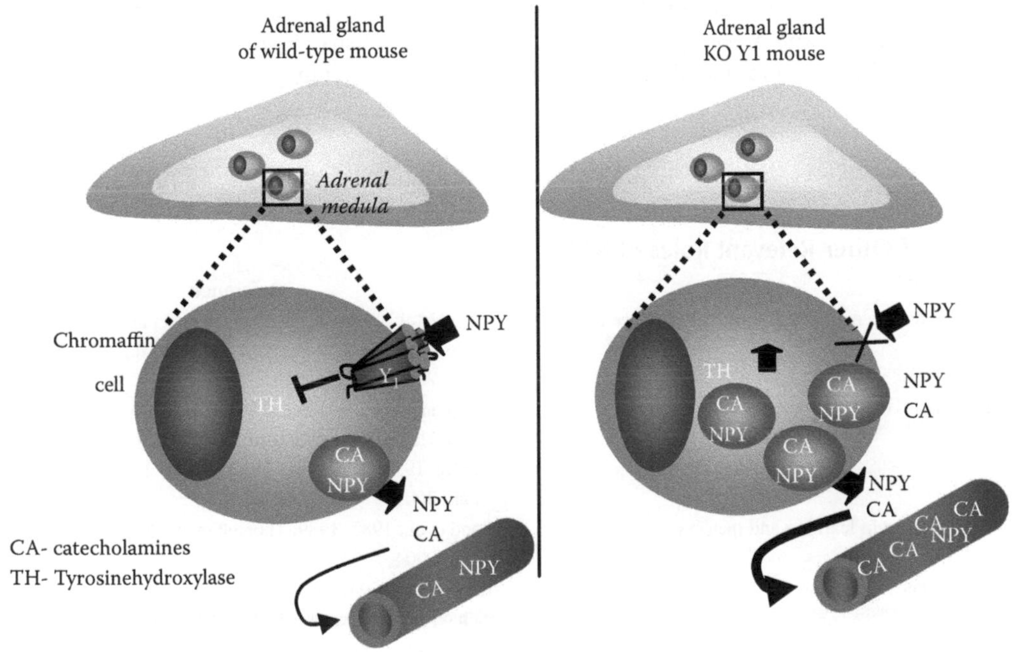

FIGURE 3.2 NPY plays an important role in the fight-or-flight response as it regulates the synthesis and release of catecholamines from the adrenal medulla. Upon activation of the sympathetic nervous system, chromaffin cells in the adrenal gland release NPY together with the catecholamines (CA). In wild-type mice, NPY evokes catecholamine release through the activation of the Y_1 receptor. In Y_1 knockout mice (KO) the baseline levels of catecholamines are elevated in both the blood plasma and adrenal glands in yhese animals. Stimulated or not, chromaffin cells from these animals released higher amounts of the catecholamines than cells from wild-type mice. Exogenous NPY did not further increase catecholamine release from the receptor-deficient cells. Although adrenal glands in the mutant mice were morphologically normal, their tyrosine hydroxylase (TH) activity was greatly increased. TH catalyzes the rate-limiting step in the production of catecholamines, and NPY inhibits the enzyme in a pathway mediated by the Y_1 receptor. This indicates that NPY has dual roles as hormone and neurotransmitter in the regulation of the sympathetic nervous system.

was explained by the enhancement of adrenal tyrosine hydroxylase (TH) activity (Cavadas et al., 2006). These results emphasize the important role of NPY in controlling catecholamine synthesis and secretion in the adrenal gland of mice, by directly acting on TH, and fine-tuning of adreno-sympathetic tone.

Interestingly, the proinflamatory cytokine interleukin 1β (IL-1β) stimulates NPY and catecholamine release from mouse chromaffin cells and the NPY released stimulates further release of catecholamines through the activation of NPY Y_1 receptor (Rosmaninho-Salgado et al., 2007a). These data suggest that in pathophysiological conditions where there is an increase of plasma IL-1β, for example, in sepsis, rheumatoid arthritis, stress, or hypertension, it will induce an increase in catecholamine release from the adrenals that will induce vasoconstriction, a rise in blood pressure, and in peripheral vascular resistance.

3.12 NPY AND CANCER

NPY receptors are present in different tumors and may open new strategies for controlling tumor growth. In breast cancer it is observed that the NPY Y_1 receptor is present in almost all cases, whereas Y_2 is only expressed in 24% of cases. In normal breast NPY Y_2 expression is predominant, suggesting that NPY receptor expression switches from the Y_2 receptor, present in normal tissue, to the Y_1 receptor in breast tumor cells (Reubi et al., 2001). Some results suggest that estrogen is likely to be one of the mediators responsible for the induction of Y_1 expression on breast cancer cells (Kuang et al., 1998; Amlal et al., 2006). Moreover, 80% of nephroblastomas expressed Y_1 and Y_2 receptors in moderate to high density. Y_1 receptors were also highly expressed in intratumoral blood vessels. In the non-neoplasic kidney, Y_1 receptors are identified

TABLE 3.2
Résumé of Other Relevant Roles of NPY

Roles of NPY	Reference
Pain: NPY has been shown to increase or reduce pain	Tracey et al., 1995; White 1997; Wang et al., 2000; Shi et al., 2006; Brumovsky et al., 2007; Smith et al., 2007; Taylor et al., 2007; Thomsen et al., 2007
NPY is anxiolytic	Heilig and Murison 1987; Broqua et al., 1995; Thorsell et al., 1999; Heilig 2004; Inui et al., 1998; Wahlestedt et al., 1993; Heilig, 1995; Kask et al., 2001; Heilig, 2004; Karlsson et al., 2008
NPY is involved in learning and memory	Flood et al., 1987, 1989; Thorsell et al., 2000; Redrobe et al., 2004
NPY modulates the circadian clock	Biello et al., 1994; Gribkoff et al., 1998; Harrington and Schak, 2000; Fukuhara et al., 2001; Yannielli and Harrington, 2001; Maywood et al., 2002; Soscia and Harrington, 2004
NPY mediates angiogenesis	Zukowska-Grojec et al., 1998; Ekstrand et al., 2003; Lee et al., 2003a, 2003b; Zukowska et al., 2003; Movafagh et al., 2006; Abe et al., 2007
NPY is a vasoconstrictor	Wahlestedt et al., 1985; Fallgren et al., 1993; Malmstrom and Lundberg, 1995; Schuerch et al., 1998; Potter and Tripovic, 2006
The NPY receptor system is implicated in bone formation	Sainsbury et al., 2003; Allison and Herzog, 2006; Baldock et al., 2007; Lundberg et al., 2007; Mikic et al., 2008

in small arteries as well as in tubulli corresponding to proximal and distal nephron segments (Korner et al., 2005).

NPY is expressed in tumors derived from the autonomic nervous system, such as neuroblastoma, pheochromocytoma, and also in Ewing's sarcoma family of tumors (ESFT) (O'Hare and Schwartz, 1989; van Valen et al., 1992). The adrenal cortical tumors express the Y_1 receptor with high frequency, pheochromocytomas and paragangliomas express Y_1 and Y_2 receptors with moderate frequency, and neuroblastic tumors express Y_2 with high frequency (Reubi et al., 2001). Also, in about 56% of renal cell carcinomas, the Y_1 receptor has been detected in moderate density.

NPY has also been identified in pituitary tumors, but there is no relationship between its presence and the headache described in pituitary tumor patients (Levy et al., 2006).

Thyroid tumor cells contain and secrete NPY, and high plasma concentrations of NPY have been found in patients with medullary thyroid carcinoma. Moreover, the concentration of NPY in tumors is 50-fold higher than in adjacent normal thyroid tissue (Connell et al., 1987).

The role of NPY in tumor increase is subject to debate. In fact, NPY induces cell proliferation on PCa and PC3 prostate cell lines, but inhibits cell proliferation in other prostate cell line, such as LNCaP and DU145 cell lines (Ruscica et al., 2007). In SK-N-BE cells, NPY induces cell proliferation through both Y_2 and Y_5 receptors. In contrast, in SK-N-MC cells, NPY decreases cell proliferation and Y_1/Y_5 agonist mimicked NPY inhibitory effect (Kitlinska et al., 2005). Moreover, it has also been suggested that NPY released from neuroblastomas enhances their growth by stimulation of tumor cell proliferation and/or angiogenesis (Kitlinska et al., 2002, 2005; Amlal et al., 2006).

3.13 CONCLUSION

The NPY peptide family exerts pleiotropic effects that are at the edge of the endocrinology, immune, and cardiovascular systems (see also Table 3.2). However, 26 years after its discovery, NPY had not yet led to therapeutic drugs, probably bccause of its pivotal role in homeostasis.

ACKNOWLEDGMENTS

J. Rosmaninho Salgado is supported by Portuguese Science Foundation (SFRH/BPD/31547/2006).

REFERENCES

Abe, K., Tilan, J.U., and Zukowska, Z., NPY and NPY receptors in vascular remodeling, *Curr. Top. Med. Chem.*, 7, 1704–1709, 2007.

Adrian, T.E. et al., Human distribution and release of a putative new gut hormone, peptide YY, *Gastroenterology*, 89, 1070–1077, 1985.

Agasse, F. et al., Neuropeptide Y promotes neurogenesis in murine subventricular zone, *Stem Cells*, 6, 1636–1645, (Dayton, OH), 2008.

Allen, J.M., Adrian, T.E., Polak, J.M., and Bloom, S.R., Neuropeptide Y (NPY) in the adrenal gland, *J. Auton. Nerv. Syst.*, 9, 559–563, 1983.

Allison, S.J. and Herzog, H., NPY and bone, *EXS.*, 95, 171–182, 2006.

Alvaro, A.R. et al., Neuropeptide Y stimulates retinal neural cell proliferation—involvement of nitric oxide, *J. Neurochem.*, doi:10.1111/j.1471–4159.2008.05334.x, 2008.

Álvaro, A.R. et al., Neuropeptide Y protects retinal neural cells against cell death induced by ecstasy, *Neuroscience*, doi:10.1016/j.neuroscience.2007.1012.1027, 2007.

Amlal, H., Faroqui, S., Balasubramaniam, A., and Sheriff S., Estrogen up-regulates neuropeptide Y Y1 receptor expression in a human breast cancer cell line, *Cancer Res.*, 66, 3706–3714, 2006.

Anitha, M. et al., Glial-derived neurotrophic factor modulates enteric neuronal survival and proliferation through neuropeptide Y, *Gastroenterology*, 131, 1164–1178, 2006.

Asakawa, A. et al., Characterization of the effects of pancreatic polypeptide in the regulation of energy balance, *Gastroenterology*, 124, 1325–1336, 2003.

Balasubramaniam, A., Sheriff, S., Rigel, D.F., and Fischer, J.E., Characterization of neuropeptide Y binding sites in rat cardiac ventricular membranes, *Peptides*, 11, 545–550, 1990.

Balasubramaniam, A. et al., Neuropeptide Y (NPY) Y4 receptor selective agonists based on NPY(32–36): development of an anorectic Y4 receptor selective agonist with picomolar affinity, *J. Med. Chem.*, 49, 2661–2665, 2006.

Baldock, P.A. et al., Novel role of Y1 receptors in the coordinated regulation of bone and energy homeostasis, *J. Biol. Chem.*, 282, 19092–19102, 2007.

Bannon, A.W. et al., Behavioral characterization of neuropeptide Y knockout mice, *Brain Res.*, 868, 79–87, 2000.

Baraban, S.C., Antiepileptic actions of neuropeptide Y in the mouse hippocampus require Y5 receptors, *Epilepsia*, 43, Suppl. 5, 9–13, 2002.

Baraban, S.C. et al., Knock-out mice reveal a critical antiepileptic role for neuropeptide Y, *J. Neurosci.*, 17, 8927–8936, 1997.

Bard, J.A., Walker, M.W., Branchek, T.A., and Weinshank, R.L., Cloning and functional expression of a human Y4 subtype receptor for pancreatic polypeptide, neuropeptide Y, and peptide YY, *J. Biol. Chem.*, 270, 26762–26765, 1995.

Baskin, D.G., Breininger, J.F., and Schwartz, M.W., Leptin receptor mRNA identifies a subpopulation of neuropeptide Y neurons activated by fasting in rat hypothalamus, *Diabetes*, 48, 828–833, 1999.

Batterham, R.L. et al., Inhibition of food intake in obese subjects by peptide YY3–36, *N. Engl. J. Med.*, 349, 941–948, 2003a.

Batterham, R.L. et al., Pancreatic polypeptide reduces appetite and food intake in humans, *J. Clin. Endocrinol. Metab.*, 88, 3989–3992, 2003b.

Batterham, R.L. et al., Critical role for peptide YY in protein-mediated satiation and body-weight regulation, *Cell Metab.*, 4, 223–233, 2006.

Batterham, R.L. et al., Gut hormone PYY(3–36) physiologically inhibits food intake, *Nature*, 418, 650–654, 2002.

Beck, B., Richy, S., and Stricker-Krongrad, A., Responsiveness of obese Zucker rats to [D-Trp34]-NPY supports the targeting of Y5 receptor for obesity treatment, *Nutr. Neurosci.*, 10, 211–214, 2007.

Berglund, M.M., Hipskind, P.A., and Gehlert, D.R., Recent developments in our understanding of the physiological role of PP-fold peptide receptor subtypes, *Exp. Biol. Med.*, 228, 217–244, 2003.

Berntson, G.G. et al., Pancreatic polypeptide infusions reduce food intake in Prader–Willi syndrome, *Peptides*, 14, 497–503, 1993.

Biello, S.M., Janik, D., and Mrosovsky, N., Neuropeptide Y and behaviorally induced phase shifts, *Neuroscience*, 62, 273–279, 1994.

Borowsky, B. et al., Molecular biology and pharmacology of multiple NPY Y5 receptor species homologs, *Regul. Pept.*, 75–76, 45–53, 1998.

Brakch, N. et al., Role of prohormone convertases in pro-neuropeptide Y processing: Coexpression and *in vitro* kinetic investigations, *Biochemistry*, 36, 16309–16320, 1997.

Broqua, P. et al., Behavioral effects of neuropeptide Y receptor agonists in the elevated plus-maze and fear-potentiated startle procedures, *Behav. Pharmacol.*, 6, 215–222, 1995.

Brumovsky, P. et al., Neuropeptide tyrosine and pain, *Trends Pharmacol. Sci.*, 28, 93–102, 2007.

Burkhoff, A., Linemeyer, D.L., and Salon, J.A., Distribution of a novel hypothalamic neuropeptide Y receptor gene and its absence in rat, *Brain Res. Mol. Brain Res.*, 53, 311–316, 1998.

Cabrele, C. et al., The first selective agonist for the neuropeptide YY5 receptor increases food intake in rats, *J. Biol. Chem.*, 275, 36043–36048, 2000.

Campbell, R.E. et al., Orexin neurons express a functional pancreatic polypeptide Y4 receptor, *J. Neurosci.*, 23, 1487–1497, 2003.

Castan, I. et al., Identification and functional studies of a specific peptide YY-preferring receptor in dog adipocytes, *Endocrinology*, 131, 1970–1976, 1992.

Cavadas, C. et al., NPY regulates catecholamine secretion from human adrenal chromaffin cells, *J. Clin. Endocrinol. Metab.*, 86, 5956–5963, 2001.

Cavadas, C. et al., Deletion of the neuropeptide Y (NPY) Y1 receptor gene reveals a regulatory role of NPY on catecholamine synthesis and secretion, *Proc. Natl Acad. Sci. USA*, 103, 10497–10502, 2006.

Chandarana, K. and Batterham, R., Peptide YY, *Curr. Opin. Endocrinol. Diabetes Obes.*, 15, 65–72, 2008.

Clark, J.T., Kalra, P.S., Crowley, W.R., and Kalra, S.P., Neuropeptide Y and human pancreatic polypeptide stimulate feeding behavior in rats, *Endocrinology*, 115, 427–429, 1984.

Connell, J.M. et al., Neuropeptide Y in multiple endocrine neoplasia: Release during surgery for phaeochromocytoma, *Clin. Endocrinol. (Oxf)*, 26, 75–84, 1987.

Corp, E.S., McQuade, J., Krasnicki, S., and Conze, D.B., Feeding after fourth ventricular administration of neuropeptide Y receptor agonists in rats, *Peptides*, 22, 493–499, 2001.

Criscione, L. et al., Food intake in free-feeding and energy-deprived lean rats is mediated by the neuropeptide Y5 receptor, *J. Clin. Invest.*, 102, 2136–2145, 1998.

Daniels, A.J. et al., Food intake inhibition and reduction in body weight gain in lean and obese rodents treated with GW438014A, a potent and selective NPY-Y5 receptor antagonist, *Regul. Pept.*, 106, 47–54, 2002.

Dickson, S.L. and Luckman, S.M., Induction of c-fos messenger ribonucleic acid in neuropeptide Y and growth hormone (GH)-releasing factor neurons in the rat arcuate nucleus following systemic injection of the GH secretagogue, GH-releasing peptide-6, *Endocrinology*, 138, 771–777, 1997.

Domin, H., Kajta, M., and Smialowska, M., Neuroprotective effects of MTEP, a selective mGluR5 antagonists and neuropeptide Y on the kainate-induced toxicity in primary neuronal cultures, *Pharmacol. Rep.*, 58, 846–858, 2006.

Dryden, S., Pickavance, L., Frankish, H.M., and Williams, G., Increased neuropeptide Y secretion in the hypothalamic paraventricular nucleus of obese (fa/fa) Zucker rats, *Brain Res.*, 690, 185–188, 1995.

Ekstrand, A.J. et al., Deletion of neuropeptide Y (NPY) 2 receptor in mice results in blockage of NPY-induced angiogenesis and delayed wound healing, *Proc. Natl Acad. Sci. USA*, 100, 6033–6038, 2003.

El Bahh, B. et al., The anti-epileptic actions of neuropeptide Y in the hippocampus are mediated by Y2 and not Y5 receptors, *Eur. J. Neurosci.*, 22, 1417–1430, 2005.

Erickson, J.C., Clegg, K.E., and Palmiter, R.D., Sensitivity to leptin and susceptibility to seizures of mice lacking neuropeptide Y, *Nature*, 381, 415–421, 1996.

Erondu, N. et al., Neuropeptide Y5 receptor antagonism does not induce clinically meaningful weight loss in overweight and obese adults, *Cell Metab.*, 4, 275–282, 2006.

Eva, C. et al., Molecular cloning of a novel G protein-coupled receptor that may belong to the neuropeptide receptor family, *FEBS Lett.*, 271, 81–84, 1990.

Fabry, M. et al., Monitoring of the internalization of neuropeptide Y on neuroblastoma cell line SK-N-MC, *Eur. J. Biochem.* 267, 5631–5637, 2000.

Fallgren, B., Arlock, P., and Edvinsson, L., Neuropeptide Y potentiates noradrenaline-evoked vasoconstriction by an intracellular calcium-dependent mechanism, *J. Auton. Nerv. Syst.*, 44, 151–159, 1993.

Flood, J.F., Hernandez, E.N., and Morley, J.E., Modulation of memory processing by neuropeptide Y, *Brain Res.*, 421, 280–290, 1987.

Flood, J.F., Baker, M.L., Hernandez, E.N., and Morley, J.E., Modulation of memory processing by neuropeptide Y varies with brain injection site, *Brain Res.*, 503, 73–82, 1989.

Foti, S., Haberman, R.P., Samulski, R.J., and McCown, T.J., Adeno-associated virus-mediated expression and constitutive secretion of NPY or NPY13–36 suppresses seizure activity *in vivo*, *Gene Ther.*, 14, 1534–1536, 2007.

Frerker, N. et al., Neuropeptide Y (NPY) cleaving enzymes: Structural and functional homologues of dipeptidyl peptidase 4, *Peptides*, 28, 257–268, 2007.

Fu-Cheng, X. et al., Mechanisms of peptide YY release induced by an intraduodenal meal in rats: Neural regulation by proximal gut, *Pflugers Arch.*, 433, 571–579, 1997.

Fuhlendorff, J. et al., [Leu31, Pro34]neuropeptide Y: A specific Y1 receptor agonist, *Proc. Natl Acad. Sci. USA*, 87, 182–186, 1990.

Fukuhara, C. et al., Neuropeptide Y rapidly reduces Period 1 and Period 2 mRNA levels in the hamster suprachiasmatic nucleus, *Neurosci. Lett.*, 314, 119–122, 2001.

Gariboldi, M. et al., Anticonvulsant properties of BIBP3226, a non-peptide selective antagonist at neuropeptide Y Y1 receptors, *Eur. J. Neurosci.*, 10, 757–759, 1998.

Gehlert, D.R., Schober, D.A., Morin, M., and Berglund, M.M., Co-expression of neuropeptide Y Y1 and Y5 receptors results in heterodimerization and altered functional properties, *Biochem. Pharmacol.*, 74, 1652–1664, 2007.

Gong, H.X. et al., Lipolysis and apoptosis of adipocytes induced by neuropeptide Y-Y5 receptor antisense oligodeoxynucleotides in obese rats, *Acta Pharmacologica Sinica*, 24, 569–575, 2003.

Gregor, P. et al., Molecular characterization of a second mouse pancreatic polypeptide receptor and its inactivated human homologue, *J. Biol. Chem.*, 271, 27776–27781, 1996a.

Gregor P. et al., Cloning and characterization of a novel receptor to pancreatic polypeptide, a member of the neuropeptide Y receptor family, *FEBS Lett.*, 381, 58–62, 1996b.

Gribkoff, V.K. et al., Phase shifting of circadian rhythms and depression of neuronal activity in the rat suprachiasmatic nucleus by neuropeptide Y: Mediation by different receptor subtypes, *J. Neurosci.*, 18, 3014–3022, 1998.

Grouzmann, E., Monod, M., Landis, B.N., and Lacroix, J.S., Adverse effects of incretin therapy for type 2 diabetes, *JAMA*, 298, 1759–1760; author reply 1760, 2007.

Grundemar, L., Wahlestedt, C., and Reis, D.J., Neuropeptide Y acts at an atypical receptor to evoke cardiovascular depression and to inhibit glutamate responsiveness in the brainstem, *J. Pharmacol. Exp. Ther.*, 258, 633–638, 1991.

Hansel, D., Eipper, B.A., and Ronnet, G.V., Neuropeptide Y functions as a neuroproliferative factor, *Nature*, 410, 940–944, 2001.

Hansen, L., Deacon, C.F., Orskov, C., and Holst, J.J., Glucagon-like peptide-1-(7-36)amide is transformed to glucagon-like peptide-1-(9-36)amide by dipeptidyl peptidase IV in the capillaries supplying the L cells of the porcine intestine, *Endocrinology*, 140, 5356–5363, 1999.

Harrington, M.E. and Schak, K.M., Neuropeptide Y phase advances the *in vitro* hamster circadian clock during the subjective day with no effect on phase during the subjective night, *Can. J. Physiol. Pharmacol.*, 78, 87–92, 2000.

Heilig, M., Antisense inhibition of neuropeptide Y (NPY)-Y1 receptor expression blocks the anxiolytic-like action of NPY in amygdala and paradoxically increases feeding, *Regul. Pept.* 59, 201–205, 1995.

Heilig, M., The NPY system in stress, anxiety and depression, *Neuropeptides*, 38, 213–224, 2004.

Heilig, M. and Murison, R., Intracerebroventricular neuropeptide Y suppresses open field and home cage activity in the rat, *Regul. Pept.*, 19, 221–231, 1987.

Herzog, H. et al., Cloned human neuropeptide Y receptor couples to two different second messenger systems, *Proc. Natl Acad. Sci. USA*, 89, 5794–5798, 1992.

Hexum, T.D. and Russett, L.R., Stimulation of cholinergic receptor mediated secretion from the bovine adrenal medulla by neuropeptide Y, *Neuropeptides*, 13, 35–41, 1989.

Higuchi, H., Costa, E., and Yang, H.Y., Neuropeptide Y inhibits the nicotine-mediated release of catecholamines from bovine adrenal chromaffin cells, *J. Pharmacol. Exp. Ther.*, 244, 468–474, 1988.

Howell, O.W. et al., Neuropeptide Y stimulates neuronal precursor proliferation in the post-natal and adult dentate gyrus, *J. Neurochem.*, 93, 560–570, 2005.

Hu, Y. et al., Identification of a novel hypothalamic neuropeptide Y receptor associated with feeding behavior, *J. Biol. Chem.*, 271, 26315–26319, 1996.

Hwa, J.J. et al., Activation of the NPY Y5 receptor regulates both feeding and energy expenditure, *Am. J. Physiol.*, 277, R1428–R1434, 1999.

Inui, A., Neuropeptide Y feeding receptors: Are multiple subtypes involved? *Trends Pharmacol. Sci.*, 20, 43–46, 1999.

Inui, A. et al., Neuropeptide regulation of feeding in dogs, *Am. J. Physiol.*, 261, R588–R594, 1999.

Inui, A. et al., Anxiety-like behavior in transgenic mice with brain expression of neuropeptide Y, *Proc. Assoc. Am. Physicians*, 110, 171–182, 1998.

Ishihara, A. et al., A neuropeptide Y Y5 antagonist selectively ameliorates body weight gain and associated parameters in diet-induced obese mice, *Proc. Natl Acad. Sci. USA*, 103, 7154–7158, 2006.

Ishii, T., Muranaka, R., Tashiro, O., and Nishimura, M., Chronic intracerebroventricular administration of anti-neuropeptide Y antibody stimulates starvation-induced feeding via compensatory responses in the hypothalamus, *Brain Res.*, 1144, 91–100, 2007.

Iyengar, S., Li, D.L., and Simmons, R.M., Characterization of neuropeptide Y-induced feeding in mice: Do Y1-Y6 receptor subtypes mediate feeding? *J. Pharmacol. Exp. Ther.*, 289, 1031–1040, 1999.

Jackman, H.L. et al., A peptidase in human platelets that deamidates tachykinins. Probable identity with the lysosomal "protective protein", *J. Biol. Chem.*, 265, 11265–11272, 1990.

Jackson, D.N., Milne, K.J., Noble, E.G., and Shoemaker, J.K., Neuropeptide Y bioavailability is suppressed in the hindlimb of female Sprague–Dawley rats, *J. Physiol.*, 568, 573–581, 2005.

Jackson, E.K., Dubinion, J.H., and Mi, Z., Effects of dipeptidyl peptidase iv inhibition on arterial blood pressure, *Clin. Exp. Pharmacol. Physiol.* 35, 29–34, 2008.

Jorgensen, J.C., O'Hare, M.M., and Andersen, C.Y., Demonstration of neuropeptide Y and its precursor in plasma and follicular fluid, *Endocrinology*, 127, 1682–1688, 1990.

Kaga, T. et al., Modest overexpression of neuropeptide Y in the brain leads to obesity after high-sucrose feeding, *Diabetes*, 50, 1206–1210, 2001.

Kakui, N. et al., Pharmacological characterization and feeding-suppressive property of FMS586 [3-(5,6,7,8-tetrahydro-9-isopropyl-carbazol-3-yl)-1-methyl-1-(2-pyridin-4-yl-ethyl)-urea hydrochloride], a novel, selective, and orally active antagonist for neuropeptide Y Y5 receptor, *J. Pharmacol. Exp. Ther.*, 317, 562–570, 2006.

Kalra, S.P. et al., Interacting appetite-regulating pathways in the hypothalamic regulation of body weight. *Endocr. Rev.*, 20, 68–100, 1999.

Kanatani, A. et al., Potent neuropeptide Y Y1 receptor antagonist, 1229U91: Blockade of neuropeptide Y-induced and physiological food intake, *Endocrinology*, 137, 3177–3182, 1996.

Kanatani, A. et al., NPY-induced feeding involves the action of a Y1-like receptor in rodents, *Regul. Pept.*, 75–76, 409–415, 1998.

Kanatani, A. et al., The novel neuropeptide Y Y(1) receptor antagonist J-104870: A potent feeding suppressant with oral bioavailability, *Biochem. Biophys. Res. Commun.*, 266, 88–91, 1999.

Kanatani, A. et al., A typical Y1 receptor regulates feeding behaviors: Effects of a potent and selective Y1 antagonist, J-115814, *Mol. Pharmacol.*, 59, 501–505, 2001.

Kanatani, A. et al., L-152,804: Orally active and selective neuropeptide Y Y5 receptor antagonist, *Biochem. Biophys. Res. Commun.*, 272, 169–173, 2000.

Kao, P.C., Taylor, R.L., and Service, F.J., Proinsulin by immunochemiluminometric assay for the diagnosis of insulinoma, *J. Clin. Endocrinol. Metab.*, 78, 1048–1051, 1994.

Karl, T., Hoffmann, T., Pabst, R., and von Horsten, S., Behavioral effects of neuropeptide Y in F344 rat substrains with a reduced dipeptidyl-peptidase IV activity, *Pharmacol. Biochem. Behav.*, 75, 869–879, 2003.

Karlsson, R.M. et al., The neuropeptide Y Y1 receptor subtype is necessary for the anxiolytic-like effects of neuropeptide Y, but not the antidepressant-like effects of fluoxetine, in mice, *Psychopharmacology*, 195, 547–557, 2008.

Kask, A., Nguyen, H.P., Pabst, R., and Von Horsten, S., Neuropeptide Y Y1 receptor-mediated anxiolysis in the dorsocaudal lateral septum: Functional antagonism of corticotropin-releasing hormone-induced anxiety, *Neuroscience*, 104, 799–806, 2001.

Khan, I.U., Reppich, R., and Beck-Sickinger, A.G., Identification of neuropeptide Y cleavage products in human blood to improve metabolic stability, *Biopolymers*, 88, 182–189, 2007.

Kitlinska, J. et al., Neuropeptide Y-induced angiogenesis in aging, *Peptides*, 23, 71–77, 2002.

Kitlinska, J. et al., Differential effects of neuropeptide Y on the growth and vascularization of neural crest-derived tumors, *Cancer Res.*, 65, 1719–1728, 2005.

Korner, J. et al., Differential effects of gastric bypass and banding on circulating gut hormone and leptin levels, *Obesity (Silver Spring)*, 14, 1553–1561, 2006.

Korner, M., Waser, B., and Reubi, J.C., Neuropeptide Y receptors in renal cell carcinomas and nephroblastomas, *Int. J. Cancer*, 115, 734–741, 2005.

Kuang, W.W., Thompson, D.A., Hoch, R.V., and Weigel, R.J., Differential screening and suppression subtractive hybridization identified genes differentially expressed in an estrogen receptor-positive breast carcinoma cell line, *Nucleic Acids Res.*, 26, 1116–1123, 1998.

Kumarnsit, E., Johnstone, L.E., and Leng, G., Actions of neuropeptide Y and growth hormone secretagogues in the arcuate nucleus and ventromedial hypothalamic nucleus, *Eur. J. Neurosci.*, 17, 937–944, 2003.

Kuo, L.E. et al., Neuropeptide Y acts directly in the periphery on fat tissue and mediates stress-induced obesity and metabolic syndrome, *Nature Med.*, 13, 803–811, 2007.

Kushi, A. et al., Obesity and mild hyperinsulinemia found in neuropeptide Y-Y1 receptor-deficient mice, *Proc. Natl Acad. Sci. USA*, 95, 15659–15664, 1998.

Labelle, M. et al., Tissue-specific regulation of fat cell lipolysis by NPY in 6-OHDA-treated rats, *Peptides*, 18, 801–808, 1997.

Lacombe, M.J., Mercure, C., Dikeakos, J.D., and Reudelhuber, T.L., Modulation of secretory granule-targeting efficiency by *cis* and *trans* compounding of sorting signals, *J. Biol. Chem.*, 280, 4803–4807, 2005.

Larhammar, D., Evolution of neuropeptide Y, peptide YY and pancreatic polypeptide. *Regul. Pept.*, 62, 1–11, 1996.

Larhammar, D. et al., Cloning and functional expression of a human neuropeptide Y/peptide YY receptor of the Y1 type, *J. Biol. Chem.*, 267, 10935–10938, 1992.

Larsen, P.J. et al., Activation of central neuropeptide Y Y1 receptors potently stimulates food intake in male rhesus monkeys, *J. Clin. Endocrinol. Metab.*, 84, 3781–3791, 1999.

Lee, C.C. and Miller, R.J., Is there really an NPY Y3 receptor? *Regul. Pept.*, 75–76, 71–78, 1998.

Lee, E.W., Grant, D.S., Movafagh, S., and Zukowska, Z., Impaired angiogenesis in neuropeptide Y (NPY)-Y2 receptor knockout mice, *Peptides*, 24, 99–106, 2003a.

Lee, E.W. et al., Neuropeptide Y induces ischemic angiogenesis and restores function of ischemic skeletal muscles, *J. Clin. Invest.*, 111, 1853–1862, 2003b.

Lennington, J.B., Yang, Z., and Conover, J.C., Neural stem cells and the regulation of adult neurogenesis, *Reprod. Biol. Endocrinol.*, 1, 99, 2003.

Levy, M.J. et al., The relationship between neuropeptide Y expression and headache in pituitary tumours, *Eur. J. Neurol.*, 13, 125–129, 2006.

Li, G. et al., Discovery of novel orally active ureido NPY Y5 receptor antagonists, *Bioorg. Med. Chem. Lett.*, 18, 1146–1150, 2008.

Lin, E.J. et al., Combined deletion of Y1, Y2, and Y4 receptors prevents hypothalamic neuropeptide Y overexpression-induced hyperinsulinemia despite persistence of hyperphagia and obesity, *Endocrinology*, 147, 5094–5101, 2006.

Ludwig, R. et al., Metabolism of neuropeptide Y and calcitonin gene-related peptide by cultivated neurons and glial cells, *Brain Res. Mol. Brain Res.*, 37, 181–191, 1996.

Lundberg, P. et al., Greater bone formation of Y2 knockout mice is associated with increased osteoprogenitor numbers and altered Y1 receptor expression, *J. Biol. Chem.*, 282, 19082–19091, 2007.

Lundell, I. et al., Cloning of a human receptor of the NPY receptor family with high affinity for pancreatic polypeptide and peptide YY, *J. Biol. Chem.*, 270, 29123–29128, 1995.

Luquet, S., Perez, F.A., Hnasko, T.S., and Palmiter, R.D., NPY/AgRP neurons are essential for feeding in adult mice but can be ablated in neonates, *Science*, 310, 683–685, 2005.

Malmstrom, R.E. and Lundberg, J.M., Endogenous NPY acting on the Y1 receptor accounts for the long-lasting part of the sympathetic contraction in guinea-pig vena cava: Evidence using SR 120107A, *Acta Physiol. Scand.*, 155, 329–330, 1995.

Margareto, J. et al., A new NPY-antagonist strongly stimulates apoptosis and lipolysis on white adipocytes in an obesity model, *Life Sci.*, 68, 99–107, 2000.

Marsh, D.J., Hollopeter, G., Kafer, K.E., and Palmiter, R.D., Role of the Y5 neuropeptide Y receptor in feeding and obesity, *Nature Med.*, 4, 718–721, 1998.

Marsh, D.J., Baraban, S.C., Hollopeter, G., and Palmiter, R.D., Role of the Y5 neuropeptide Y receptor in limbic seizures, *Proc. Natl Acad. Sci. USA*, 96, 13518–13523, 1999.

Mashiko, S. et al., A pair-feeding study reveals that a Y5 antagonist causes weight loss in diet-induced obese mice by modulating food intake and energy expenditure, *Mol. Pharmacol.*, 71, 602–608, 2007.

Mashiko, S. et al., Characterization of neuropeptide Y (NPY) Y5 receptor-mediated obesity in mice: Chronic intracerebroventricular infusion of D-Trp(34)NPY, *Endocrinology*, 144, 1793–1801, 2003.

Matsumoto, M. et al., Inactivation of a novel neuropeptide Y/peptide YY receptor gene in primate species, *J. Biol. Chem.*, 271, 27217–27220, 1996.

Matthews, J.E. et al., Pharmacological characterization and selectivity of the NPY antagonist GR231118 (1229U91) for different NPY receptors, *Regul. Pept.*, 72, 113–119, 1997.

Maywood, E.S., Okamura, H., and Hastings, M.H., Opposing actions of neuropeptide Y and light on the expression of circadian clock genes in the mouse suprachiasmatic nuclei, *Eur. J. Neurosci.*, 15, 216–220, 2002.

McCarthy, H.D. et al., Hypothalamic neuropeptide Y receptor characteristics and NPY-induced feeding responses in lean and obese Zucker rats, *Life Sci.*, 49, 1491–1497, 1991.

McKibbin, P.E., McCarthy, H.D., Shaw, P., and Williams, G., Insulin deficiency is a specific stimulus to hypothalamic neuropeptide Y: A comparison of the effects of insulin replacement and food restriction in streptozocin-diabetic rats, *Peptides*, 13, 721–727, 1992.

McKibbin, P.E. et al., Altered neuropeptide Y concentrations in specific hypothalamic regions of obese (fa/fa) Zucker rats. Possible relationship to obesity and neuroendocrine disturbances, *Diabetes*, 40, 1423–1429, 1991.

Medeiros, M.D. and Turner, A.J., Processing and metabolism of peptide-YY: Pivotal roles of dipeptidylpeptidase-IV, aminopeptidase-P, and endopeptidase-24.11, *Endocrinology*, 134, 2088–2094, 1994.

Medeiros Mdos, S. and Turner, A.J., Metabolism and functions of neuropeptide Y, *Neurochem. Res.*, 21, 1125–1132, 1996.

Mentlein, R. and Roos, T., Proteases involved in the metabolism of angiotensin II, bradykinin, calcitonin gene-related peptide (CGRP), and neuropeptide Y by vascular smooth muscle cells, *Peptides*, 17, 709–720, 1996.

Mentlein, R., Dahms, P., Grandt, D., and Kruger, R., Proteolytic processing of neuropeptide Y and peptide YY by dipeptidyl peptidase IV, *Regul. Pept.*, 49, 133–144, 1993.

Michel, M.C. et al., XVI. International Union of Pharmacology recommendations for the nomenclature of neuropeptide Y, peptide YY, and pancreatic polypeptide receptors, *Pharmacol. Rev.*, 50, 143–150, 1998.

Mikic, B., Zhang, M., Webster, E., and Rossmeier, K., Effect of Y_2 receptor deletion on whole bone structural behavior in mice, *Anat. Rec. (Hoboken)*, 291, 14–18, 2008.

Minth, C.D., Bloom, S.R., Polak, J.M., and Dixon, J.E., Cloning, characterization, and DNA sequence of a human cDNA encoding neuropeptide tyrosine, *Proc. Natl Acad. Sci. USA*, 81, 4577–4581, 1984.

Morley, J.E., Hernandez, E.N., and Flood, J.F., Neuropeptide Y increases food intake in mice, *Am. J. Physiol.*, 253, R516–R522, 1987.

Morris, M.J. et al., Neuropeptide Y suppresses absence seizures in a genetic rat model primarily through effects on Y receptors, *Eur. J. Neurosci.*, 25, 1136–1143, 2007.

Movafagh, S. et al., Neuropeptide Y induces migration, proliferation, and tube formation of endothelial cells bimodally via Y1, Y2, and Y5 receptors, *FASEB J.*, 20, 1924–1926, 2006.

Nan, Y.S. et al., Neuropeptide Y enhances permeability across a rat aortic endothelial cell monolayer, *Am. J. Physiol. Heart Circ. Physiol.*, 286, H1027–1033, 2004.

Narnaware, Y.K. and Peter, R.E., Neuropeptide Y stimulates food consumption through multiple receptors in goldfish, *Physiol. Behav.*, 74, 185–190, 2001.

Naveilhan, P. et al., Normal feeding behavior, body weight and leptin response require the neuropeptide Y Y2 receptor, *Nature Med.*, 5, 1188–1193, 1999.

Noe, F. et al., Gene therapy in epilepsy: The focus on NPY, *Peptides*, 28, 377–383, 2007.

Norenberg, W. et al., Inhibition of nicotinic acetylcholine receptor channels in bovine adrenal chromaffin cells by Y3-type neuropeptide Y receptors via the adenylate cyclase/protein kinase A system, *Naunyn Schmiedebergs Arch. Pharmacol.*, 351, 337–347, 1995.

O'Hare, M.M. and Schwartz, T.W., Expression and precursor processing of neuropeptide Y in human pheochromocytoma and neuroblastoma tumors, *Cancer Res.*, 49, 7010–7014, 1989.

Palmiter, R.D. et al., Life without neuropeptide Y, *Recent Prog. Horm. Res.*, 53, 163–199, 1998.

Pan, H. et al., The role of prohormone convertase-2 in hypothalamic neuropeptide processing: A quantitative neuropeptidomic study, *J. Neurochem.*, 98, 1763–1777, 2006.

Parker, E.M. et al., [D-Trp(34)] neuropeptide Y is a potent and selective neuropeptide Y Y(5) receptor agonist with dramatic effects on food intake, *Peptides*, 21, 393–399, 2000.

Parker, E.M. et al., GR231118 (1229U91) and other analogues of the C-terminus of neuropeptide Y are potent neuropeptide Y Y1 receptor antagonists and neuropeptide Y Y4 receptor agonists, *Eur. J. Pharmacol.*, 349, 97–105, 1998.

Parker, M.S., Lundell, I., and Parker, S.L., Internalization of pancreatic polypeptide Y4 receptors: Correlation of receptor intake and affinity, *Eur. J. Pharmacol.*, 452, 279–287, 2000a.

Parker, S.L. et al., Cloned neuropeptide Y (NPY) Y1 and pancreatic polypeptide Y4 receptors expressed in Chinese hamster ovary cells show considerable agonist-driven internalization, in contrast to the NPY Y2 receptor, *Eur. J. Biochem.*, 268, 877–886, 2001.

Parker, S.L. et al., Agonist internalization by cloned Y1 neuropeptide Y (NPY) receptor in Chinese hamster ovary cells shows strong preference for NPY, endosome-linked entry and fast receptor recycling, *Regul. Pept.* 107, 49–62, 2002b.

Patel, H.R. et al., Neuropeptide Y deficiency attenuates responses to fasting and high-fat diet in obesity-prone mice, *Diabetes*, 55, 3091–3098, 2006.

Pedrazzini, T. et al., Cardiovascular response, feeding behavior and locomotor activity in mice lacking the NPY Y1 receptor, *Nature Med.*, 4, 722–726, 1998.

Pelz, K.M. and Dark, J., I.C.V. NPY Y1 receptor agonist but not Y5 agonist induces torpor-like hypothermia in cold-acclimated Siberian hamsters, *Am. J. Physiol. Regul. Integr. Comp. Physiol.*, 292, R2299–R2311, 2007.

Pheng, L.H. et al., Agonist- and antagonist-induced sequestration/internalization of neuropeptide Y Y1 receptors in HEK293 cells, *Br. J. Pharmacol.*, 139, 695–704, 2003.

Polidori, C., Ciccocioppo, R., Regoli, D., and Massi, M., Neuropeptide Y receptor(s) mediating feeding in the rat: Characterization with antagonists, *Peptides*, 21, 29–35, 2000.

Potter, E.K. and Tripovic, D., Modulation of sympathetic neurotransmission by neuropeptide Y Y2 receptors in rats and guinea pigs, *Exp. Brain Res.*, 173, 346–352, 2006.

Prins, J.B. and O'Rahilly, S., Regulation of adipose cell number in man, *Clin. Sci. (London)*, 92, 3–11, 1997.

Raposinho, P.D. et al., Chronic administration of neuropeptide Y into the lateral ventricle of C57BL/6J male mice produces an obesity syndrome including hyperphagia, hyperleptinemia, insulin resistance, and hypogonadism, *Mol. Cell Endocrinol.*, 185, 195–204, 2001.

Redrobe, J.P., Dumont, Y., Herzog, H., and Quirion, R., Characterization of neuropeptide Y, Y(2) receptor knockout mice in two animal models of learning and memory processing, *J. Mol. Neurosci.*, 22, 159–166, 2004.

Reibel, S. et al., Neuropeptide Y and epilepsy: Varying effects according to seizure type and receptor activation, *Peptides*, 22, 529–539, 2001.

Renshaw, D. and Hinson, J.P., Neuropeptide Y and the adrenal gland: A review, *Peptides*, 22, 429–438, 2001.

Renshaw, D. et al., Actions of neuropeptide Y on the rat adrenal cortex, *Endocrinology*, 141, 169–173, 2000.

Reubi, J.C., Gugger, M., Waser, B., and Schaer, J.C., Y(1)-mediated effect of neuropeptide Y in cancer: Breast carcinomas as targets, *Cancer Res.*, 61, 4636–4641, 2001.

Richichi, C. et al., Anticonvulsant and antiepileptogenic effects mediated by adeno-associated virus vector neuropeptide Y expression in the rat hippocampus, *J. Neurosci.*, 24, 3051–3059, 2004.

Rose, P.M. et al., Cloning and functional expression of a cDNA encoding a human type 2 neuropeptide Y receptor, *J. Biol. Chem.*, 270, 29038, 1995.

Rosmaninho-Salgado, J. et al., Neuropeptide Y regulates catecholamine release evoked by interleukin-1beta in mouse chromaffin cells, *Peptides*, 28, 310–314, 2007a.

Rosmaninho-Salgado, J. et al., Intracellular signalling mechanisms mediating catecholamine release upon activation of NPY Y1 receptors in mouse chromaffin cells, *J. Neurochem.*, 103, 896–903, 2007b.

Ruscica, M., Dozio, E., Motta, M., and Magni, P., Modulatory actions of neuropeptide Y on prostate cancer growth: Role of MAP kinase/ERK 1/2 activation, *Adv. Exp. Med. Biol.*, 604, 96–100, 2007.

Sahu, A., Kalra, P.S., and Kalra, S.P., Food deprivation and ingestion induce reciprocal changes in neuropeptide Y concentrations in the paraventricular nucleus, *Peptides*, 9, 83–86, 1988.

Sainsbury, A. et al., Y4 receptor knockout rescues fertility in ob/ob mice, *Genes Dev.*, 16, 1077–1088, 2002a.

Sainsbury, A. et al., Synergistic effects of Y2 and Y4 receptors on adiposity and bone mass revealed in double knockout mice, *Mol. Cell Biol.*, 23, 5225–5233, 2003.

Sainsbury, A. et al., Important role of hypothalamic Y2 receptors in body weight regulation revealed in conditional knockout mice, *Proc. Natl Acad. Sci. USA*, 99, 8938–8943, 2002b.

Schaffhauser, A.O. et al., Inhibition of food intake by neuropeptide Y Y5 receptor antisense oligodeoxynucleotides, *Diabetes*, 46, 1792–1798, 1997.

Schuerch, L.V., Linder, L.M., Grouzmann, E., and Haefeli, W.E., Human neuropeptide Y potentiates α1-adrenergic blood pressure responses *in vivo*, *Am. J. Physiol.*, 275, H760–H766, 1998.

Schwartz, M.W. et al., Identification of targets of leptin action in rat hypothalamus, *J. Clin. Invest.*, 98, 1101–1106, 1996a.

Schwartz, M.W. et al., Central nervous system control of food intake, *Nature*, 404, 661–671, 2000.

Schwartz, M.W. et al., Inhibition of hypothalamic neuropeptide Y gene expression by insulin, *Endocrinology*, 130, 3608–3616, 1992.

Schwartz, M.W. et al., Specificity of leptin action on elevated blood glucose levels and hypothalamic neuropeptide Y gene expression in ob/ob mice, *Diabetes*, 45, 531–535, 1996b.

Shi, T.J. et al., Deletion of the neuropeptide Y Y1 receptor affects pain sensitivity, neuropeptide transport and expression, and dorsal root ganglion neuron numbers, *Neuroscience*, 140, 293–304, 2006.

Shimoda, K. et al., Antiserum against neuropeptide Y enhances the nicotine-mediated release of catecholamines from cultured rat adrenal chromaffin cells, *Neurochem. Int.*, 23, 71–77, 1993.

Silva, A.P., Cavadas, C., and Grouzmann, E., Neuropeptide Y and its receptors as potential therapeutic drug targets, *Clin. Chimica Acta Int. J. Chem.*, 326, 3–25, 2002.

Silva, A.P., Carvalho, A.P., Carvalho, C.M., and Malva, J.O., Functional interaction between neuropeptide Y receptors and modulation of calcium channels in the rat hippocampus, *Neuropharmacol.*, 44, 282–292, 2003a.

Silva, A.P., Xapelli, S., Grouzmann, E., and Cavadas, C., The putative neuroprotective role of neuropeptide Y in the central nervous system. *Curr. Drug Targets*, 4, 331–347, 2005.

Silva, A.P. et al., Activation of neuropeptide Y receptors is neuroprotective against excitotoxicity in organotypic hippocampal slice cultures, *FASEB J.*, 17, 1118–1120, 2003b.

Sindelar, D.K., Palmiter, R.D., Woods, S.C., and Schwartz, M.W., Attenuated feeding responses to circadian and palatability cues in mice lacking neuropeptide Y, *Peptides*, 26, 2597–2602, 2005.

Sindelar, D.K. et al., Neuropeptide Y is required for hyperphagic feeding in response to neuroglucopenia, *Endocrinology*, 145, 3363–3368, 2004.

Smith, P.A. et al., Spinal mechanisms of NPY analgesia, *Peptides*, 28, 464–474, 2007.

Sorensen, A.T. et al., NPY gene transfer in the hippocampus attenuates synaptic plasticity and learning. *Hippocampus*, 18 (6), 564–574, 2008.

Soscia, S.J. and Harrington, M.E., Neuropeptide Y attenuates NMDA-induced phase shifts in the SCN of NPY Y1 receptor knockout mice in vitro, *Brain Res.*, 1023, 148–153, 2004.

Spinazzi, R., Andreis, P.G., and Nussdorfer, G.G., Neuropeptide-Y and Y-receptors in the autocrine-paracrine regulation of adrenal gland under physiological and pathophysiological conditions (review), *Int. J. Mol. Med.*, 15, 3–13, 2005.

Stanic, D. et al., Peptidergic influences on proliferation, migration, and placement of neural progenitors in the adult mouse forebrain, *Proc. Natl. Acad. Sci. USA*, 105 (9), 3610–3615, 2008.

Stanley, B.G., Kyrkouli, S.E., Lampert, S., and Leibowitz, S.F., Neuropeptide Y chronically injected into the hypothalamus: a powerful neurochemical inducer of hyperphagia and obesity, *Peptides*, 7, 1189–1192, 1986.

Stanley, S., Wynne, K., and Bloom, S., Gastrointestinal satiety signals III. Glucagon-like peptide 1, oxyntomodulin, peptide YY, and pancreatic polypeptide, *Am. J. Physiol. Gastrointest. Liver Physiol.*, 286, G693–G697, 2004.

Ste Marie, L., Luquet, S., Cole, T.B., and Palmiter, R.D., Modulation of neuropeptide Y expression in adult mice does not affect feeding, *Proc. Natl Acad. Sci. USA*, 102, 18632–18637, 2005.

Stenfors, C., Hellman, U., and Silberring, J., Characterization of endogenous neuropeptide Y in rat hippocampus and its metabolism by nanospray mass spectrometry, *J. Biol. Chem.*, 272, 5747–5751, 1997.

Stephens, T.W. et al., The role of neuropeptide Y in the antiobesity action of the obese gene product, *Nature*, 377, 530–532, 1995.

Taylor, B.K. et al., Neuropeptide Y acts at Y1 receptors in the rostral ventral medulla to inhibit neuropathic pain, *Pain*, 131, 83–95, 2007.

Thiriet, N. et al., Neuropeptide Y protects against methamphetamine-induced neuronal apoptosis in the mouse striatum, *J. Neurosci.*, 25, 5273–5279, 2005.

Thomsen, M. et al., Involvement of Y(5) receptors in neuropeptide Y agonist-induced analgesic-like effect in the rat hot plate test, *Brain Res.*, 1155, 49–55, 2007.

Thorsell, A., Carlsson, K., Ekman, R., and Heilig, M., Behavioral and endocrine adaptation, and up-regulation of NPY expression in rat amygdala following repeated restraint stress, *Neuroreport*, 10, 3003–3007, 1999.

Thorsell, A. et al., Behavioral insensitivity to restraint stress, absent fear suppression of behavior and impaired spatial learning in transgenic rats with hippocampal neuropeptide Y overexpression, *Proc. Natl Acad. Sci. USA*, 97, 12852–12857, 2000.

Tracey, D.J., Romm, M.A., and Yao, N.N., Peripheral hyperalgesia in experimental neuropathy: Exacerbation by neuropeptide Y, *Brain Res.*, 669, 245–254, 1995.

Turnbull, A.V. et al., Selective antagonism of the NPY Y5 receptor does not have a major effect on feeding in rats, *Diabetes*, 51, 2441–2449, 2002.

Turtzo, L.C., Marx, R., and Lane, M.D., Cross-talk between sympathetic neurons and adipocytes in coculture, *Proc. Natl Acad. Sci. USA*, 98, 12385–12390, 2001.

Valet, P. et al., Neuropeptide Y and peptide YY inhibit lipolysis in human and dog fat cells through a pertussis toxin-sensitive G protein, *J. Clin. Invest.*, 85, 291–295, 1990.

van Valen, F., Winkelmann, W., and Jurgens, H., Expression of functional Y1 receptors for neuropeptide Y in human Ewing's sarcoma cell lines, *J. Cancer Res. Clin. Oncol.*, 118, 529–536, 1992.

Vettor, R. et al., Induction and reversibility of an obesity syndrome by intracerebroventricular neuropeptide Y administration to normal rats, *Diabetologia*, 37, 1202–1208, 1994.

Vezzani, A. et al., Plastic changes in neuropeptide Y receptor subtypes in experimental models of limbic seizures, *Epilepsia*, 41, Suppl. 6, S115–S121, 2000.

Vezzani, A. et al., Seizure susceptibility and epileptogenesis are decreased in transgenic rats overexpressing neuropeptide Y, *Neuroscience*, 110, 237–243, 2002.

Wahlestedt, C., Edvinsson, L., Ekblad, E., and Hakanson, R., Neuropeptide Y potentiates noradrenaline-evoked vasoconstriction: Mode of action, *J. Pharmacol. Exp. Ther.*, 234, 735–741, 1985.

Wahlestedt, C. et al., Modulation of anxiety and neuropeptide Y-Y1 receptors by antisense oligodeoxynucleotides, *Science (New York)*, 259, 528–531, 1993.

Wang, J.Z., Lundeberg, T., and Yu, L., Antinociceptive effects induced by intra-periaqueductal grey administration of neuropeptide Y in rats, *Brain Res.*, 859, 361–363, 2000.

Weinberg, D.H. et al., Cloning and expression of a novel neuropeptide Y receptor, *J. Biol. Chem.*, 271, 16435–16438, 1996.

White, D.M., Intrathecal neuropeptide Y exacerbates nerve injury-induced mechanical hyperalgesia, *Brain Res.*, 750, 141–146, 1997.

White, J.D., Olchovsky, D., Kershaw, M., and Berelowitz, M., Increased hypothalamic content of preproneuropeptide-Y messenger ribonucleic acid in streptozotocin-diabetic rats, *Endocrinology*, 126, 765–772, 1990.

Wieland, H.A. et al., Subtype selectivity of the novel nonpeptide neuropeptide Y Y1 receptor antagonist BIBO 3304 and its effect on feeding in rodents, *Br. J. Pharmacol.*, 125, 549–555, 1998.

Willesen, M.G., Kristensen, P., and Romer, J., Co-localization of growth hormone secretagogue receptor and NPY mRNA in the arcuate nucleus of the rat, *Neuroendocrinology*, 70, 306–316, 1999.

Woldbye, D.P. et al., Powerful inhibition of kainic acid seizures by neuropeptide Y via Y5-like receptors, *Nature Med.*, 3, 761–764, 1997.

Wu, Y.F. and Li, S.B., Neuropeptide Y expression in mouse hippocampus and its role in neuronal excitotoxicity, *Acta Pharmacologica Sinica*, 26, 63–68, 2005.

Wyss, P., Levens, N., and Stricker-Krongrad, A., Stimulation of feeding in lean but not in obese Zucker rats by a selective neuropeptide Y Y5 receptor agonist, *Neuroreport*, 9, 2675–2677, 1998.

Xapelli, S., Silva, A.P., Ferreira, R., and Malva, J.O., Neuropeptide Y can rescue neurons from cell death following the application of an excitotoxic insult with kainate in rat organotypic hippocampal slice cultures, *Peptides*, 28, 288–294, 2007.

Yang, K. et al., Neuropeptide Y is produced in visceral adipose tissue and promotes proliferation of adipocyte precursor cells via the Y1 receptor, *FASEB J.*, 7, 2452–2464, 2008.

Yannielli, P.C. and Harrington, M.E., Neuropeptide Y in the mammalian circadian system: Effects on light-induced circadian responses, *Peptides*, 22, 547–556, 2001.

Zhang, P., Zheng, J., Vorce, R.L., and Hexum, T.D., Identification of an NPY-Y1 receptor subtype in bovine chromaffin cells, *Regul. Pept.*, 87, 9–13, 2000.

Zheng, J., Zhang, P., and Hexum, T.D., Neuropeptide Y inhibits chromaffin cell nicotinic receptor-stimulated tyrosine hydroxylase activity through a receptor-linked G protein-mediated process, *Mol. Pharmacol.*, 52, 1027–1033, 1997.

Ziemek. R. et al., Determination of affinity and activity of ligands at the human neuropeptide Y Y4 receptor by flow cytometry and aequorin luminescence, *J. Recept. Signal Transduct. Res.*, 27, 217–233, 2007.

Zukowska-Grojec, Z., Karwatowska-Prokopczuk, E., Fisher, T.A., and Ji, H., Mechanisms of vascular growth-promoting effects of neuropeptide Y: Role of its inducible receptors, *Regul. Pept.*, 75–76, 231–238, 1998.

Zukowska, Z., Grant, D.S., and Lee, E.W., Neuropeptide Y: A novel mechanism for ischemic angiogenesis, *Trends Cardiovasc. Med.*, 13, 86–92, 2003.

4 Pituitary Adenylate Cyclase-Activating Polypeptide

Eve M. Lutz, Chantevy Pou, Thomas K. Monaghan, and Christopher MacKenzie

CONTENTS

4.1 INTRODUCTION

The pituitary adenylate cyclase-activating polypeptide (PACAP) is a 38-amino-acid neuropeptide that was first identified from ovine hypothalamic extracts because of its potent ability to stimulate cAMP production in rat anterior pituitary cells (Miyata et al., 1989). Its amino (N)-terminal region shares a high degree of amino acid sequence similarity with the 28-amino-acid vasoactive intestinal peptide (VIP), with which it has several overlapping as well as distinct functions. PACAP and VIP have distinct distributions in the brain (Girard et al., 2006), but in the periphery both peptides are coexpressed in the same neuronal cell bodies and fibers innervating a number of peripheral organs (Fahrenkrug and Hannibal, 2004). PACAP is a pleiotropic neuropeptide, having a range of activities in the nervous system, including acting as a hypothalamic hormone, a neurotransmitter, a neuromodulator, and a neurotrophic factor. It is also a vasodilator and modulator of endocrine, cardiac, urogenital, and immune functions. PACAP is widely distributed in the brain and is expressed in certain peripheral organs, notably in the endocrine pancreas, adrenal medulla, gonads, and cardiac, respiratory, and urogenital tracts. Its actions are mediated through three related guanine nucleotide-binding protein (G protein)-coupled receptors (GPCRs), the $VPAC_1$, $VPAC_2$, and PAC_1 receptors, two of which PACAP shares with VIP. The distribution and actions of PACAP, VIP, and their receptors are described in great detail in several reviews (Rawlings and Hezareh, 1996; Arimura, 1998; Sherwood et al., 2000; Vaudry et al., 2000; Zhou et al., 2002; Ganea et al., 2003; Conconi et al., 2006; Laburthe et al., 2007; Ghzili et al., 2008) and a book edited by Hubert Vaudry and Akira Arimura (Vaudry and Arimura, 2003), to which the reader is directed for further information.

4.2 SOURCE OF PEPTIDES

PACAP is a member of a family of structurally similar peptides that includes VIP, secretin, peptide histidine-methionine (PHM, in human) or peptide histidine-isoleucine (PHI, other mammals and birds), glucagon, glucagon-like peptide (GLP)-1, GLP-2, growth hormone-releasing hormone (GRF), and glucose-dependent insulinotropic polypeptide (GIP) (Figure 4.1). In addition, nonmammalian

PACAP/secretin/glucagon family of peptides

	1 5 10 15 20 25 30 35 40 45
Human peptides	
PACAP-38	H S D G I F T D S Y S R Y R K Q M A V K K Y L A A V L G K R Y K Q R V K N K -NH_2
PACAP-27	H S D G I F T D S Y S R Y R K Q M A V K K Y L A A V L -NH_2
VIP	H S D A V F T D N Y T R L R K Q M A V K K Y L N S I L N -NH_2
Secretin	H S D G T F T S E L S R L R R G A R L Q R L L Q G L V G -NH_2
PHM	H A D G V F T S D F S K L L G Q L S A K K Y L E S L M -NH_2
Glucagon	H S Q G T F T S D Y S K Y L D S R R A Q D F V Q W L M N T
GLP-1	H A E G T F T S D V S S Y L E G Q A A K E F I A W L V K G R -NH_2
GLP-2	H A D G S F S D E M N T I L D N L A A R D F I N W L I Q K I Y D
GIP	Y A E G T F I S D Y S I A M D K I H Q Q D F V N W L L A Q K G K K N D W K H N I T Q
GRF	Y A D A I F T N S Y R K V L G Q L S A R K L L Q D I M S R Q Q G E S N Q E R G A R A R L G -NH_2
PRP	D V A H G I L N E A Y R K V L D Q L S A G K H L Q S L V A
Lizard peptides	
Helodermin	H S D A I F T E E Y S K L L A K L A L Q K Y L A S I L G S R T S P P P P -NH_2
Exendin-1	H S D A T F T A E Y S K L L A K L A L Q K Y L E S I L G S S T S P R P P S S
Exendin-2	H S D A T F T A E Y S K L L A K L A L Q K Y L E S I L G S S T S P R P P S
Exendin-3	H S D G T F T S D L S K Q M E E E A V R L F I E W L K N G G P S S G A P P P S -NH_2
Exendin-4	H G E G T F T S D L S K Q M E E E A V R L F I E W L K N G G P S S G A P P P S -NH_2

FIGURE 4.1 The PACAP/secretin/glucagon family of peptides. Alignment of amino acid sequences of the different members of the PACAP/secretin/glucagon family of peptides in human and the related lizard peptides found in the venom of the Gila monster *Heloderma suspectum* and *Heloderma horridum* (Raufman, 1996). Exendin-1 is also known as helospectin-1 and exendin-2 as helospectin-2. Residues that are homologous with the PACAP-38 sequence are highlighted in gray. (Adapted from Kieffer, T.J. and Habener, J.F., *Endocrinol. Rev.*, 20, 876–913, 1999.)

bioactive peptides with structural similarities to this family of peptides have been isolated from the venom of the Gila monster, *Heloderma suspectum* and *Heloderma horridum* (Raufman, 1996). These include helodermin, which shares 53% sequence identity with mammalian PACAP, and the exendins (including helospectin), of which exendin-4 shares 53% sequence identity with human GLP-1 (Pohl and Wank, 1998). The 28-amino-acid VIP is the closest related family member to PACAP, with 68% homology to PACAP-27 (Miyata et al., 1990). The PACAP-related peptide (PRP), which is encoded by the same gene as PACAP, is more structurally similar to GRF than to PACAP, probably reflecting a distant ancestor where GRF and PACAP were encoded by the same gene (Sherwood et al., 2000; Tam et al., 2007).

The genes encoding members of the PACAP family of peptides [apart from helodermin and the exendins (Pohl and Wank, 1998)] have similar structures (reviewed by Sherwood et al., 2000) and are believed to have evolved from one ancestral gene (Campbell and Scanes, 1992). PACAP is encoded in a single exon (exon 5) of the gene. The human gene, *ADCYAP1*, is located on chromosome 18 at position 18p11.32 (Chang et al., 1993). Like other members of this peptide family, PACAP is initially expressed as a large precursor peptide, preproPACAP (Figure 4.2), which is sequentially cleaved to release the biologically active 38-residue peptide termed PACAP-38 (Okazaki et al., 1992; Vaudry et al., 2000). The preproPACAP contains potential prohormone convertase (PC) cleavage sites (Hook et al., 2008), two of which flank the PACAP-38 sequence (Figure 4.2). In addition,

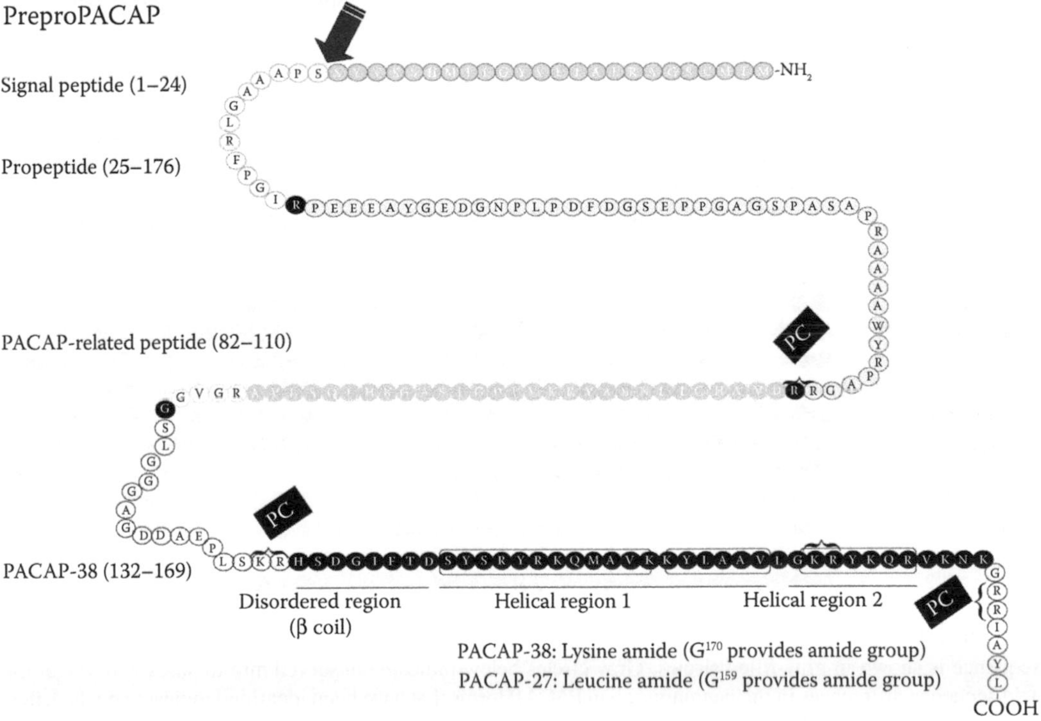

FIGURE 4.2 Amino-acid sequence of human preproPACAP. The human PACAP gene *ADCYAP1*, is organised into 5 exons and encodes a 176-amino acid precursor peptide preproPACAP. The amino acid residues are encoded by exons 2-5, residues at exon boundaries are indicated by black-filled circles. The gray-filled circles at the N-terminus indicate the signal peptide sequence and the predicted site of cleavage is indicated by the large arrow. The 29-amino acid PRP is indicated by the light gray-filled circles and the 38-amino acid PACAP sequence by dark gray-filled circles. Dibasic motifs are indicated by the symbol { and potential prohormone convertase cleavage sites by the black-boxes labelled PC.

PACAP-38 contains an internal site, Gly[28]Lys[29]Arg[30], which when cleaved produces the 27-amino-acid carboxyl (C)-terminal truncated form, PACAP-27 (Miyata et al., 1990). Coexpression of cDNAs encoding preproPACAP with PC1 or PC2 in rat GH4C1 cells have shown that both enzymes can process the precursor to produce PACAP-38 and PACAP-27 (Li et al., 1999). The preproPACAP is also efficiently processed by PC4, which is the principal PC form expressed in the testis and ovary (Li et al., 2000; Basak et al., 2004). Both PACAP-38 and PACAP-27 are C-terminally amidated; in the precursor peptide each is flanked on the C-terminus with a glycine that can be utilized by peptidyl glycine α-amidating monooxygenase (Eipper et al., 1992; Hansel et al., 2001). It has been determined by radioimmunoassay and HPLC analysis that PACAP-38 is the predominantly expressed form of PACAP in rats, comprising more than 90% of total PACAP (Arimura et al., 1991; Arimura, 1998). The mammalian preproPACAP precursor also contains the 29-amino-acid PRP (encoded by exon 4 of the gene) and a signal sequence to direct PACAP and PRP to vesicles for secretion. PRP does not appear to have a physiological function in mammals (Tam et al., 2007).

The peptide sequence of PACAP is highly conserved across several vertebrate species, with all the characterized mammalian forms having identical sequences (Sherwood et al., 2000) (Figure 4.3). The forms identified in chicken (McRory et al., 1997), frog (Hu et al., 2000), and lizard (Valiante et al., 2007) each vary only by one amino acid from the mammalian peptide. The fish PACAPs vary more considerably, principally in the C-terminus amino-acid residues 28 to 38 (Figure 4.3). Two genes encoding PACAP-27-like peptides have been isolated from a protochordate (sea squirt), and one of these encodes a peptide that only has one amino-acid residue different from mammalian PACAP-27 (McRory and Sherwood, 1997). Other members of the PACAP/secretin/glucagon family have been found to vary more considerably in peptide sequence and in length (Sherwood et al., 2000).

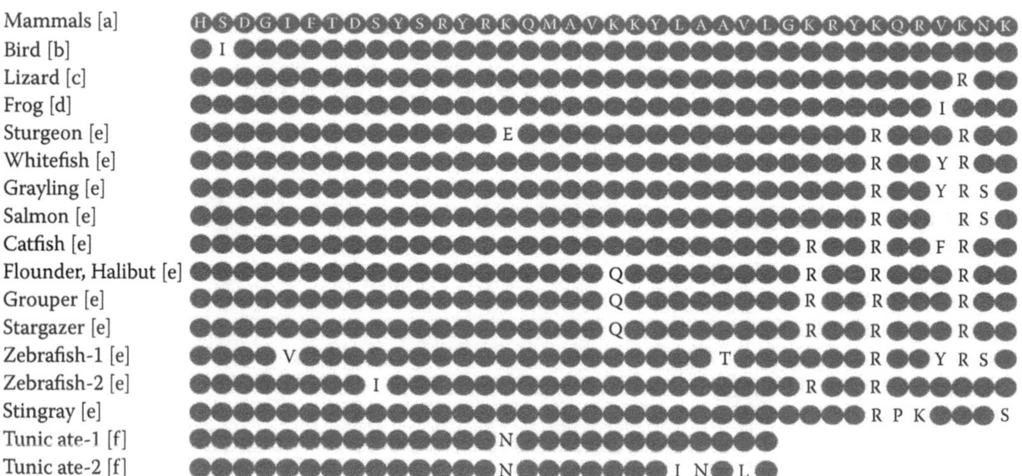

FIGURE 4.3 Comparison of PACAP amino-acid sequences from different species. Mammalian amino-acid sequence is shown in gray-filled circles. Gray circles below indicate conserved amino acids with the amino acid sequence differences in the nonmammalian PACAP forms that have been identified indicated by the letters. Published sequences include (a) the mammalian form reported in sheep (Kimura et al., 1990), human (Kimura et al., 1990; Ohkubo et al., 1992), rat (Ogi et al., 1990), and mouse (Yamamoto et al., 1998); the nonmammalian forms identified in (b) chicken (McRory et al., 1997) and turkey (Yoo et al., 2000); (c) lizard (Pohl and Wank, 1998; Valiante et al., 2007); (d) frog (Alexandre et al., 2000; Hu et al., 2000); (e) several species of fish including sturgeon, whitefish, grayling, flounder, and halibut (Adams et al., 2002), salmon (Parker et al., 1997), catfish (McRory et al., 1995), grouper (Jiang et al., 2003), stargazer (Matsuda et al., 1997), zebrafish (Fradinger and Sherwood, 2000; Wang et al., 2003), and stingray (Matsuda et al., 1998); and (f) protochordate forms (McRory and Sherwood, 1997).

The observation of the ability of PACAP to trigger depolarization and to modulate K^+ currents at the Drosophila neuromuscular junction along with PACAP-like immunoreactivity in the Drosophila CNS and neuromuscular junction has led to speculation as to the existence of an insect PACAP-like peptide (Zhong and Peña, 1995). The subsequent isolation and characterization of the *amnesiac* gene has shown that it encodes an adenylyl-cyclase-activating polypeptide (Feany and Quinn, 1995). Comparison of the putative peptides encoded by the *amnesiac* gene with PACAP reveals some sequence similarity (Hashimoto et al., 2002), although it is more consistent with a convergent evolution of peptide structure/activities.

4.3 MOLECULAR MECHANISM OF ACTION

4.3.1 Solution Structure of PACAP

The solution structure of PACAP has been determined by two-dimensional nuclear magnetic resonance (NMR) spectroscopy and circular dichroism (CD) spectroscopy (Inooka et al., 1992; Wray et al., 1993). PACAP-38 and PACAP-27 are very similar to the other secretin/glucagon peptides, adopting an unstructured random coil in aqueous solution that is shifted to an α-helical structure in aqueous organic solvent mixtures and in phospholipid micelles that are used to mimic the hydrophobic receptor/membrane environment (Wray et al., 1993; Blankenfeldt et al., 1996; Tsueshita et al., 2002; Onoue et al., 2008). PACAP-38 displays three distinct domains in the hydrophobic environment, an initial disordered N-terminus sequence of eight amino-acid residues followed by two α-helical regions (Ser^9–Val^{26} and Gly^{28}–Arg^{34}) the first of which is interrupted by a discontinuity occurring at $Lys^{20}Lys^{21}$ (Wray et al., 1993). PACAP-27 is similar in structure in that the truncation does not disturb the N-terminal random coil structure and first α-helical region (Inooka et al., 1992; Wray et al., 1993). By comparison, the solution structure of VIP has been reported to show that the α-helical structure begins two residues closer to the N-terminus of the VIP peptide (Theriault et al., 1991; Tan et al., 2006).

Receptor binding and activation studies using N- and C-terminal truncated derivatives of PACAP and of VIP, along with VIP/PACAP chimeric peptides, have determined that the N-terminal random coil structure confers receptor selectivity and activation and that the C-terminal α-helical conformation is necessary for receptor binding (Gourlet et al., 1995, 1996b; Blankenfeldt et al., 1996; Onoue et al., 2004b, 2008). Recent NMR studies have confirmed that a C-terminal α-helical conformation is adopted by the receptor-bound peptide (Inooka et al., 2001; Tan et al., 2006; Sun et al., 2007).

4.3.2 PACAP Receptors

Two classes of high-affinity PACAP binding sites were characterized initially in tissues and termed Type I binding sites, which preferentially bind PACAP ($K_d \approx 0.5\,nM$) compared to VIP ($K_d > 500\,nM$), and Type II binding sites, which bind PACAP and VIP with similar affinity ($K_d \approx 1\,nM$) (Shivers et al., 1991; Vaudry et al., 2000). Type I binding sites were found predominantly in the CNS and endocrine tissues, whereas Type II were found predominantly in peripheral organs such as lung and liver (Shivers et al., 1991; Vaudry et al., 2000). Subsequently, three genes were cloned that encode PACAP receptors, the $VPAC_1$, $VPAC_2$, and PAC_1 receptors (Ishihara et al., 1992; Hosoya et al., 1993; Lutz et al., 1993). These are seven transmembrane (TM) GPCRs that share ~50% sequence homology with each other and belong to the Group II family of GPCRs that also includes receptors for secretin, glucagon, GLP-1, GIP, GRF, and the calcitonin, corticotrophin-releasing hormone, and parathyroid hormone receptors (Harmar, 2001). The $VPAC_1$ and $VPAC_2$ receptors both bind PACAP and VIP with equal affinity and their expression overlaps that of Type II PACAP binding sites, whereas the PAC_1 receptor binds PACAP with 1000-fold greater affinity than VIP and is expressed similarly to that of Type I PACAP binding sites (Vaudry et al., 2000). The PAC_1 receptor is generally described as the PACAP-preferring receptor. The current receptor nomenclature along with that previously used in the literature is listed in Table 4.1.

TABLE 4.1
PACAP Receptors Belonging to the Group II GPCR Family and Splice Variants

Current Receptor Nomenclature[a]	Previous Nomenclature	Gene	Splice Variants[b]	Additional Information (Reference)
PAC_1	PACAP-R $PACAP_1$-R	*ADCYAP1R1*	$PAC_{1\text{-null}}$	$PACAP_1$–Rs [c]; PACAPR–null [d]; PAC1normal [f]; full NT + null [g]
	PACAP type I receptor PACAP/VIP receptor I (PVR1)		$PAC_{1\text{-hop1}}$ $PAC_{1\text{-hop2}}$	$PACAP_1$–Rhop1/hop2 [c]; PACAPR–SV–2 [d]; full NT + hop [g]
			$PAC_{1\text{-hip}}$	$PACAP_1$–Rhip [c]; PACAPR–SV–1 [d]; full NT + hip [g]
			$PAC_{1\text{-hiphop1}}$ $PAC_{1\text{-hiphop2}}$	$PACAP_1$–Rhiphop1/hiphop2 [c]; PACAPR–SV–3 [d]; full NT + hiphop [g]
			$PAC_{1\text{-}\delta}$5,6null	$PACAP_1$–Rvs [e]; PAC1short [f]; NTdeV^{89}–S^{109} + null [g]
			$PAC_{1\text{-}\delta}$5,6hop	NTdeV^{89}–S^{109} + hop [g]
			$PAC_{1\text{-}\delta}$5,6hip	NTdeV^{89}–S^{109} + hip [g]
			$PAC_{1\text{-}\delta}$5,6hiphop	NTdeV^{89}–S^{109} + hiphop [g]
			$PAC_{1\text{-}\delta}$5null	NTdeV^{89}–I^{95} + null [g]
			$PAC_{1\text{-}\delta}$5hop	NTdeV^{89}–I^{95} + hop [g]
			$PAC_{1\text{-}\delta}$5hip	NTdeV^{89}–I^{95} + hip [g]
			$PAC_{1\text{-}\delta}$5hiphop	NTdeV^{89}–I^{95} + hiphop [g]
			$PAC_{1\text{-}\delta}$4,5,6null	PAC1veryshort [f]; NTdeG^{53}–S^{109} + null [g]
			$PAC_{1\text{-}\delta}$4,5,6hop	NTdeG^{53}–S^{109} + hop [g]
			$PAC_{1\text{-}\delta}$4,5,6hip	NTdeG^{53}-S^{109} + hip [g]
			$PAC_{1\text{-}\delta}$4,5,6hiphop	NTdeG^{53}–S^{109} + hiphop [g]
			$PAC_{1\text{-}\delta}$5,6,16,17	NTdeV^{89}–S^{109}, R^{350}–T^{468} [g]; 5TM receptor missing TM6, ic3, TM7 and CT due to frameshift

$VPAC_1$	VIP receptor/ classic VIP receptor VIP_1 receptorVIP-PACAP Type II receptor PACAP/VIP receptor 2 (PVR2) VIP1/PACAP receptor	*VIPR1*	$hVPAC_1$ δ10,11,12	VPAC1 TM5 [k]; hVPAC1de307–394
$VPAC_2$	Helodermin-preferring VIP receptor VIP_2 receptor PACAPR-3	*VIPR2*	$mVPAC_2$ δ12 $hVPAC_2$ δ11	mVPAC2de367–380 [h,i]; SD VPAC2 [j]; missing part of TM7 hVPAC2de325–438 [i]; 5TM receptor missing TM6, ic3, TM7 and CT due to frameshift
	PACAP/VIP receptor 3 (PVR3) VIP2/PACAP receptor		$hVPAC_2$ δ10,11	VPAC2 TM5 [k]; hVPAC2de294–367

Source: Rawlings, S.R. and Hezareh, M., *Endocrinol. Rev.*, 17, 4–29, 1996.

Note: NT, amino terminus; s, short; vs, very short; SV, splice variant; δ, deletion exons; h, human; m, mouse; TM, transmembrane domain; de, deletion of amino acids; SD, short-deletion; TM5, transmembrane domain 1–5 variant; CT, carboxyl-terminal tail.

[a] The VPAC and PAC_1 nomenclature are as in Harmar et al. (1998).

[b] Nomenclature for the PAC_1 receptor splice variants as in Lutz et al. (2006). The hip and hop designations for the additional 28 amino-acid cassettes that can be inserted into ic3 were originally used by Spengler et al. for the rat PAC_1 receptor ic3 variants (Spengler et al., 1993) and subsequently by Lutz et al. for the human PAC_1 receptor ic3 variants (Lutz et al., 2006). The hop1 and hop2 ic3 variants of the PAC_1 receptor arises through alternate usage of adjacent 3′ splice acceptor sites in exon 15 that encodes the hop cassette.

[c] Spengler et al., 1993.

[d] Pisegna and Wank, 1996.

[e] Pantaloni et al., 1996.

[f] Dautzenberg et al., 1999.

[g] Lutz et al., 2006.

[h] Grinninger et al., 2004.

[i] Miller et al., 2006.

[j] Huang et al., 2006.

[k] Bokaei et al., 2006.

All three PACAP receptors are coupled to adenylyl cyclase (AC) activation via G_s (McCulloch et al., 2002). In addition, these receptors have been shown to differentially couple to phospholipase C (PLC) activation via the pertussis toxin-insensitive G proteins, $G_{q/11}$ (PAC_1 receptor) and G_{16} ($VPAC_2$ receptor), or the pertussis toxin-sensitive G proteins, Gi/o ($VPAC_1$ and $VPAC_2$ receptors) (Hezareh et al., 1996; MacKenzie et al., 1996, 2001; Langer et al., 2001; McCulloch et al., 2002). However, coupling to PLC activation appears to be cell-type specific. For instance PAC_1 receptors expressed in rat sympathetic neurons couple to PLC activation in addition to AC activation (Braas and May, 1999; Beaudet et al., 2000), whereas PAC_1 receptors expressed in pancreatic β-islet cells (Borboni et al., 1999; Jamen et al., 2002) and in human neuroblastoma SH-SY5Y cells (which can be differentiated into sympathetic neuron-like cells; Monaghan et al., 2008) couple to AC activation but not to PLC. However, although PLC activation is not involved, PAC_1 receptor activation mediates increases in intracellular Ca^{2+} levels in both these cell types, likely via a cAMP-mediated pathway (Borboni et al., 1999; Jamen et al., 2002; Dickson et al., 2006). The hop1 isoform of the PAC_1 receptor (see Figure 4.5) and the $VPAC_1$ and $VPAC_2$ receptors contain an additional site for docking the small G-protein ADP-ribosylation factor (ARF), which is believed to facilitate phospholipase D (PLD) activation (McCulloch et al., 2001). Figure 4.4 shows a generalized schematic diagram of the PACAP-activated intracellular signaling pathways mediated by the VPAC and PAC_1 receptors.

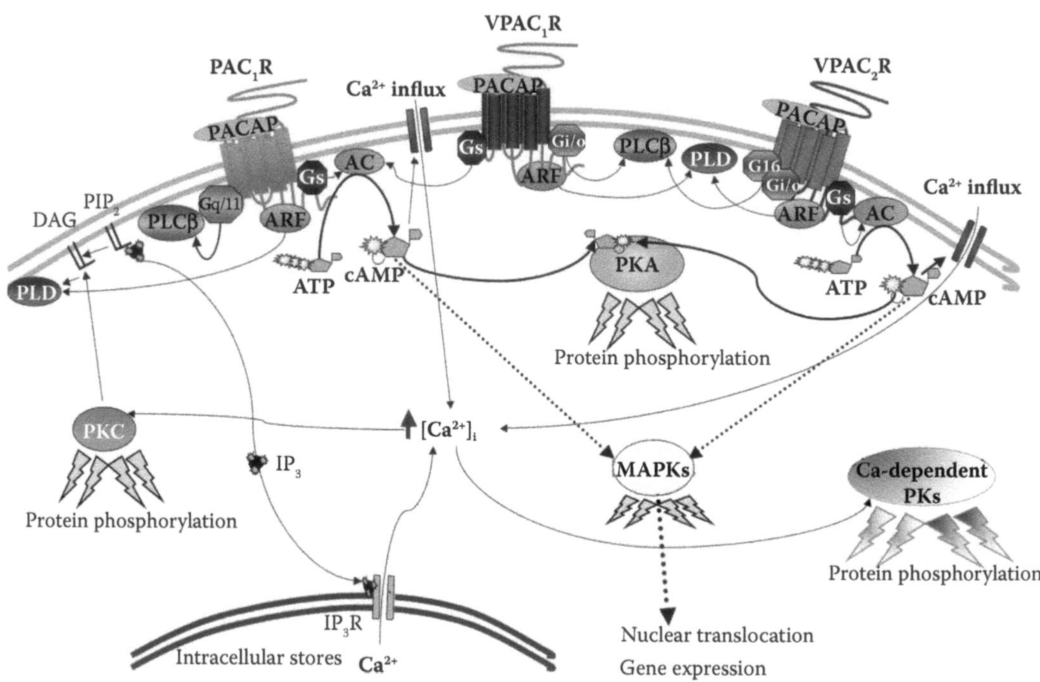

FIGURE 4.4 Generalized schematic diagram of intracellular signaling pathways activated by PACAP. The activation of cAMP production through VPAC and PAC_1 receptor coupling to Gs is the primary mediator of the cellular effects of PACAP; however, these receptors are capable of coupling to other heterotrimeric and small G protein-mediated pathways. PAC_1 receptors are also able to couple to Gq/11, and thus can mediate activation of PLC and the production of the phospholipid second messengers DAG and IP_3. The VPAC receptors may activate PLC through the βγ subunits of Gi/o. In addition, PLD activation occurs through PKC activation and also through the small G protein ARF. The particular pathways that mediate the cellular actions of PACAP are utterly dependent on the cellular context. Abbreviations: PAC_1R, PACAP type I receptor; VPAC, VIP/PACAP receptor; AC, adenylyl cyclase; PLC, phospholipase C; PLD, phospholipase D; ARF, the small G protein ADP ribosylation factor; PKA, cAMP-dependent protein kinase; PKC, protein kinase C; Ca-dependent PKs, calcium-dependent protein kinases; MAPKs, mitogen activated protein-kinases; PIP_2, phosphatidylinositol 4,5-bisphosphate; DAG, diacylglycerol; IP_3, inositol 1,4,5-trisphosphate.

The genes encoding the Group II receptors are organized into several exons, and there is a high degree of conservation with respect to the position of intron/exon boundaries. The VPAC receptor encoding genes, *VIPR1* and *VIPR2*, are similarly organized into 13 exons, with the methionine start codon located in exon 1 and the carboxyl tail and stop codon encoded in exon 13 (Sreedharan et al., 1995; Lutz et al., 1999b). The PAC_1 receptor gene, *ACDYAP1R1*, is organized into 18 exons, with 5 exons having no equivalent in the genes encoding the VPAC receptors (Lutz et al., 2006). These include exon 1, which is noncoding, exons 5 and 6, which encode an additional 21 amino acids in the N-terminal extracellular (NT) domain, and exon 14 (hip), which encodes an additional 28-amino-acid cassette and exon 15 (hop) that encodes an additional 28- or 27-amino acid cassette, depending on the use of adjacent 5′ splice acceptor sites. The hip and/or hop cassettes can be inserted into the third intracellular loop (ic3) domain near to the TM6 boundary (Figure 4.5). Multiple isoforms of the PAC_1 receptor with different NT and ic3 domain combinations can be generated as a result of alternative splicing of the gene in two separate regions (Lutz et al., 2006). The first region spans exons 4, 5, and 6 and gives rise to receptors with the full NT domain (encoded by exons 2 to 8), the NT domain missing 7 amino acids (δ5), the NT domain missing 21 amino acids (δ5,6), or the NT domain missing 57 amino acids (δ4,5,6). The second spans exons 14 to 17 and can give rise to several different ic3 isoforms containing the hip, the hop, or the hiphop inserts or null (for no insert)

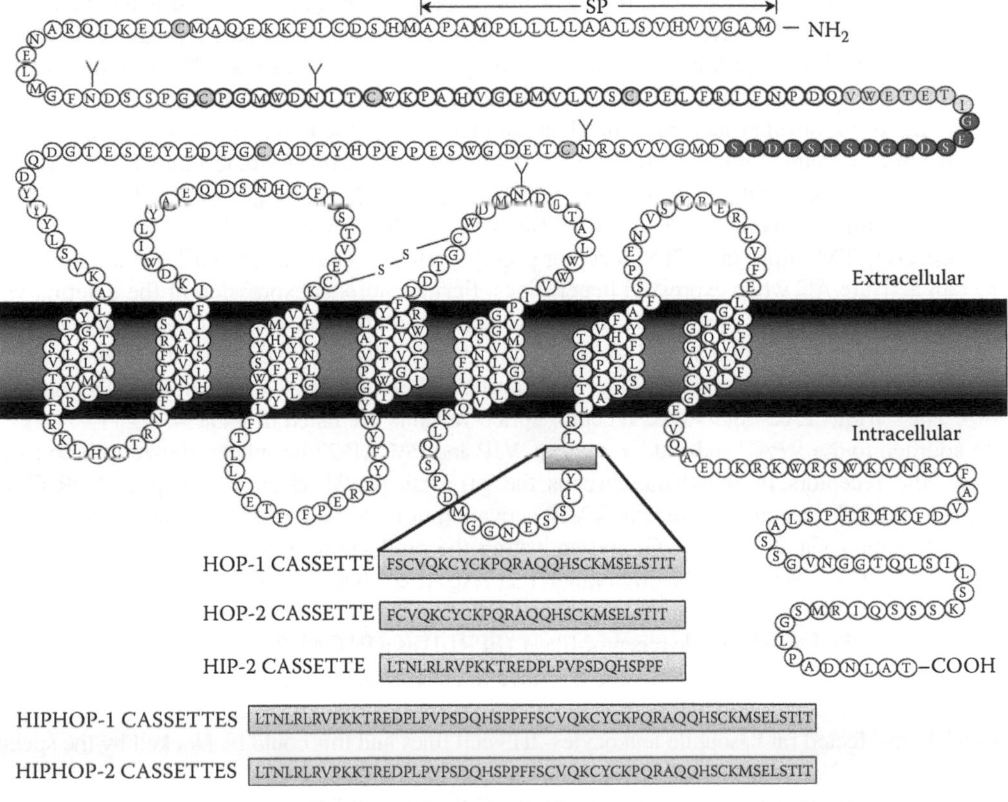

FIGURE 4.5 (See color insert following page 176.) Schematic diagram of the human PAC_1 receptor. The predicted 468-amino-acid sequence of the null PAC_1 receptor isoform, $hPAC_1$-null, is shown along with the amino-acid cassettes that can be inserted into ic3 shown beneath. The signal peptide sequence (SP) is indicated by the arrows, and cleavage is predicted to occur between A^{20} and M^{21}. The potential N-glycosylation sites in the NT domain are indicated by the symbol Y. Amino acid sequences encoded by exon 4 are outlined in red, the additional 21-amino-acid sequences encoded by exons 5 and 6 are indicated in yellow and orange, respectively. The transmembrane spanning domains were predicted by hydrophobicity plot.

(Spengler et al., 1993; Pantaloni et al., 1996; Pisegna and Wank, 1996; Dautzenberg et al., 1999) as well as a variant missing exons 14 to 17 that encodes a 5TM rather than a 7TM receptor, which does not activate AC when expressed in COS 7 cells and stimulated with various concentrations of PACAP-38 (Lutz et al., 2006). The predominant PAC_1 receptor isoforms in the adult brain appear to be the NT domain + ic3 combinations of full NT + null or full NT + hop PAC_1 receptors ($PAC_{1\text{-null}}$ and $PAC_{1\text{-hop}}$), whereas in fetal brain the NT domain isoforms missing the 21 amino acids encoded by exons 5 and 6 along with the null or hop ic3 domain PAC_1 receptors ($PAC_{1\text{-}\delta5,6\text{null}}$ and $PAC_{1\text{-}\delta5,6\text{hop}}$) are also highly expressed (Lutz et al., 2006). Interestingly, VIP is more potent at the δ5,6 NT PAC_1 variants compared to the full NT variants, and there is some suggestion that these may mediate both VIP and PACAP actions during development (Lutz et al., 2006). In addition to the PAC_1 receptor splice variants, a TM4 PAC_1 receptor variant that differs from the typical PAC_1 receptor by discrete sequence differences in the TM2 and TM4 domains was described in rat β-islet cells. Interestingly, the TM4 PAC_1 variant does not couple to AC or PLC activation, but does stimulate Ca^{2+} influx through an L-type Ca^{2+} channel when activated by PACAP (Chatterjee et al., 1996). However, the presence of this variant could not be confirmed in a separate study of mouse and rat pancreatic β-islets, which were found to express $VPAC_1$, $VPAC_2$ and $PAC_{1\text{-null}}$, $PAC_{1\text{-hop}}$ and $PAC_{1\text{-}\delta5,6\text{null}}$ variants (Jamen et al., 2002).

Variants of the $VPAC_1$ and $VPAC_2$ receptors have also been identified, but appear to be expressed in a limited number of cell types, in particular in lymphocytes and in certain cancer cell lines. A variant of the $VPAC_2$ receptor that is highly selective for VIP was cloned from guinea pig tenia coli. This differs by two amino acids (L40F, L41F) in the NT domain to that of the typical $VPAC_2$ receptor that was cloned from gastric smooth muscle (Teng et al., 2001; Zhou et al., 2006). It has not been determined if the $VPAC_2$ and NT variant are encoded by separate genes, or whether some alternative splicing event or other molecular mechanism gives rise to the two receptor forms. In addition, different splice variants for the $VPAC_1$ and $VPAC_2$ receptors missing TM6, ic3, and TM7; TM6, ic3, TM7, and the carboxyl tail; or part of TM7, which arise through skipping of exons 10, 11, and/or 12 have been identified (Grinninger et al., 2004; Bokaei et al., 2006; Miller et al., 2006). Three of these would encode 5TM rather than 7TM receptor proteins (Bokaei et al., 2006; Miller et al., 2006), but none can activate AC when expressed in cell lines. Furthermore, coexpression of the deletion variant $VPAC_2$ receptor, mouse (m)VPAC2de367-380, along with the full-length $mVPAC_2$ receptor was shown to diminish VIP-stimulated IL-4 production in Th2 cells, suggesting that the deletion variant may act as a dominant negative inhibitor modulating VIP actions on T-cell function (Huang et al., 2006). The various PAC_1 and VPAC receptor splice variants are listed in Table 4.1.

In addition to the VPAC and PAC_1 receptors, VIP and PACAP-27 have been shown to bind to and activate other receptors. In smooth muscle cells, the natriuretic peptide clearance receptor (NPR-C) was identified as a VIP receptor that mediates VIP activation of nitric oxide (NO) production and smooth muscle relaxation via coupling to Gi and activating the production of cGMP (Murthy et al., 1998; Murthy et al., 2000). Moreover, the observation that PACAP-27, but not PACAP-38 or VIP, specifically stimulated intracellular calcium mobilization and ERK phosphorylation in human neutrophils led to the identification of the formyl peptide receptor-like 1 (FPRL1) as a PACAP-27 receptor (Kim et al., 2006). It has been suggested that PACAP-27 could induce the chemotactic migration of neutrophils through activation of this receptor. PACAP-27 was found to selectively stimulate intracellular calcium increase in FPRL1-transfected rat basophile leukocytes-2H3 cell lines and this could be blocked by the specific FPRL1 antagonist WRWWWW (WRW4) (Kim et al., 2006). The selective FPRL-1 agonist, WKYTIVm appears to share some structural similarity with the C-terminus of PACAP-27.

4.4 BIOLOGICAL AND/OR PATHOLOGICAL MODE OF ACTION

PACAP is expressed throughout the central and peripheral nervous systems and has been shown to have a wide array of actions (Vaudry et al., 2000). A large number of immunohistochemical and *in situ* hybridization studies have shown that PACAP is mainly localized in neurons (Arimura, 1998; Vaudry et al., 2000). PACAP is also expressed in chromaffin cells in the adrenal medulla (Mazzocchi et al., 2002a, 2002b). In addition, some lymphoid cells have been shown to produce and release PACAP (Abad et al., 2002–2003) and it is particularly highly expressed in the testis (Arimura et al., 1991). Like VIP, PACAP has potent vasodilatory and vasorelaxant actions on smooth muscle (Warren et al., 1992; Naruse et al., 1993; Bhogal et al., 1994). However, PACAP induces contraction of gallbladder smooth muscle whereas VIP remains relaxant only. This may reflect the coexpression of the VIP/PACAP $VPAC_1$ and $VPAC_2$ receptors that are coupled to Gs- but not Gq-mediated signaling pathways and PACAP-selective PAC_1 receptor variants that are coupled to both (Wei et al., 2007). PACAP has numerous other functions, such as modulating the secretion of gastric acid in the gastrointestinal tract (Piqueras et al., 2004), regulating protein expression during spermatogenesis (Kononen et al., 1994), and increasing thyroxin secretion from the thyroid (Chen et al., 1993). PACAP is a circadian regulator (Fukuhara et al., 1998; Zhou et al., 2002). In the pineal gland it stimulates melatonin release (Simonneaux et al., 1993, 1998), and in the retinal ganglion PACAP is colocalized with glutamate in neurones mediating light information to the circadian clock in the suprachiasmatic nucleus (SCN) (Hannibal and Fahrenkrug, 2004). In the brain, PACAP is most highly expressed in the hypothalamus (Arimura et al., 1991), where it has been shown to modulate the depolarization of magnocellular neurons (Uchimura et al., 1996) in addition to glutamatergic signaling in the SCN (Chen et al., 1999). Moreover, PACAP is released from the hypothalamus into the portal blood (Dow et al., 1994) and has been shown to stimulate the release of a number of pituitary hormones from clonal pituitary-derived cell lines and primary cell cultures, including adrenocorticotropic hormone (ACTH), growth hormone (GH), prolactin (Hart et al., 1992; Propato-Mussafiri et al., 1992), oxytocin, and vasopressin (Lutz-Bucher et al., 1996). In addition, PACAP stimulates catecholamine release from the adrenal medulla (Pellegri et al., 1998; Mazzocchi et al., 2002a, 2002b) and potentiates glucose-dependent insulin release from pancreatic β-islet cells (Yamada et al., 2004). There is also evidence for PACAP acting as a neurotransmitter in the antrum of the stomach (Tornoe et al., 2001).

4.4.1 Role of PACAP in the Immune System

PACAP, like VIP, has particularly interesting modulatory effects on both innate and adaptive immunity (Ganea et al., 2003; Pozo, 2003; Abad et al., 2006). PACAP has been shown to regulate the release of histamine from mast cells (Odum et al., 1998) and the activation of monocytes (Zein et al., 2006), and modulate the expression of inflammatory cytokines (Martinez et al., 1998). PACAP in general has been shown to be anti-inflammatory, reducing expression of the proinflammatory mediators: tumor necrosis factor-α (TNFα) (Delgado et al., 1999a, 1999c), interleukin-6 (IL-6) (Delgado et al., 1999a), and IL-12 (Delgado et al., 1999b) from stimulated macrophages, and increasing expression of the anti-inflammatory IL-10 (Bozza et al., 1998). Likewise PACAP is an important anti-inflammatory factor in the CNS (Abad et al., 2006), where it has been suggested to act as a microglial deactivating factor (Delgado et al., 2003). It has been shown to inhibit LPS-induced microglial production of TNFα, IL-1β, IL-6, NO, and reactive oxygen species (Kim et al., 2000; Delgado et al., 2003; Yang et al., 2006). In addition, PACAP-mediated inhibition of chemokine production has been shown to lead to a significant reduction in the chemotactic recruitment of peripheral leukocytes (neutrophils, macrophages, and lymphocytes) by activated microglia (Delgado et al., 2002). The $VPAC_1$ receptor, and to a lesser extent the PAC_1 receptor, appears to mediate the actions of PACAP on immune cells; however, expression of the $VPAC_2$ receptor is upregulated following lymphocyte and macrophage activation (Lara-Marquez et al., 2001; Ganea et al., 2003).

4.4.2 Role of PACAP during Development

Much research has focused on the role of PACAP as a neurotrophic and neuroprotective factor mediating neuronal progenitor cell proliferation, differentiation, and survival (Waschek, 2002; Zhou et al., 2002; Dejda et al., 2005; Brenneman, 2007; Falluel-Morel et al., 2007). During development, expression of PACAP and receptors have been observed in areas of the brain undergoing neurogenesis (Sheward et al., 1998; Zhou et al., 1999; Basille et al., 2000, 2006; Watanabe et al., 2007) and also in germinative regions in the adult brain (Köves et al., 1991; Nielsen et al., 1998; Jaworski and Proctor, 2000; Mercer et al., 2004; Basille et al., 2006). PACAP has been shown to have a role in neural tube patterning (Waschek et al., 1998) and to stimulate neurogenesis (DiCicco-Bloom et al., 2000; Mercer et al., 2004; Ohta et al., 2006). In the cerebellar cortex PACAP acts as an antimitogenic factor (Suh et al., 2001): primary neuronal cultured rat cerebellar neuroblasts (Gonzalez et al., 1997; Falluel-Morel et al., 2007), basal forebrain neurons (Yuhara et al., 2001), and superior cervical ganglion neurones (DiCicco-Bloom et al., 2000) respond to PACAP by forming neurites and halting apoptosis. Particular PAC_1 receptor splice variants have been shown to mediate the pro- or antimitogenic effects of PACAP on sympathetic neuroblasts and on cortical precursor neurons, respectively (Nicot and DiCicco-Bloom, 2001). PACAP has also been shown to regulate neurotransmitter phenotype, increasing the number of cholinergic cells in cultured rat basal forebrain (Takei et al., 2000) while increasing the number of catecholaminergic cells in cultured rat superior cervical ganglion (May and Braas, 1995).

4.4.3 Role of PACAP in Neuronal Injury and Regeneration

PACAP expression is upregulated at sites of neuronal injury (Zhang et al., 1995, 1996; Moller et al., 1997; Skoglösa et al., 1999; Armstrong et al., 2004; Boeshore et al., 2004; Pettersson et al., 2004) and this may be mediated by chronic depolarization, inflammatory mediators, and/or the loss of target-derived factors such as neurotrophin-3 or bone morphogenetic protein (Brandenburg et al., 1997; Wallin et al., 2001; Armstrong et al., 2003, 2004; Pavelock et al., 2007). This upregulated expression is likely to be important, as PACAP may be acting directly on injured neurons as a neuroprotective and neurotrophic factor, thus aiding survival and regeneration (Gonzalez et al., 1997; Yuhara et al., 2001; Waschek, 2002). For instance, focal ischemia has been shown to cause an NMDA receptor-dependent upregulation in the expression of PACAP in cortical pyramidal cells and PACAP has been shown to protect cortical cells in culture from mild hypoxic/ischemic damage (Stumm et al., 2007). PACAP may also function as an autocrine factor, causing injured neurons and astrocytes to express other neurotrophic factors and inflammatory mediators that aid neuronal survival and regeneration, illustrated by the observation that glutamate-challenged mouse cerebral neurons secrete brain-derived neurotrophic factor (BDNF) in response to PACAP (Pellegri et al., 1998). Blocking BDNF actions on traumatized rat cerebral neurons attenuates PACAP-induced neuroprotection (Frechilla et al., 2001), suggesting that part of PACAPs neuroprotective effects are mediated by BDNF. PACAP can also induce expression of glial cell-line-derived neurotrophic factor following facial nerve axotomy (Kimura et al., 2004) and PACAP-stimulated IL-6 secretion is required for maximum PACAP-induced protection of hippocampal neurons from apoptosis after ischemia (Shioda et al., 1998). More recently, PACAP has been shown to induce expression of the activity-dependent neuroprotective protein in glial cells (Zusev and Gozes, 2004; Nakamachi et al., 2008), a factor that was first discovered as a novel VIP-response gene that confers high potency neuroprotection (Gozes and Brenneman, 2000).

PACAP and VIP have been shown to be involved in modulating pain states in models of chronic nerve injury (Dickinson and Fleetwood-Walker, 1999; Mabuchi et al., 2004), and it appears that this is mediated via PAC_1 and $VPAC_2$ receptors (Dickinson et al., 1999; Jongsma et al., 2001; Garry et al., 2005). Along with changes in peptide expression, changes in receptor expression were found in that $VPAC_2$ was upregulated and $VPAC_1$ was downregulated in spinal cord, but the levels of the PAC_1 receptor did not change (Dickinson et al., 1999).

4.4.4 Gene Knockout Studies

A number of groups have generated mouse gene knockout lines through targeted disruption of the PACAP or the PAC_1 or $VPAC_2$ receptor genes (Girard et al., 2006; Hashimoto et al., 2006; Sherwood et al., 2007). $VPAC_1$ receptor knockout lines have also been made, but the results on these have not yet been published. There have been some reported differences in phenotypes that have been ascribed to the different methods used for gene knockout along with differences in genetic backgrounds; however, these studies have highlighted important functions for PACAP in feeding, lipid and carbohydrate metabolism, thermoregulation, stress responses, fertility, neuronal development, circadian and cognitive functions such as memory, emotion, and psychomotor abilities, and the development of inflammatory and neuropathic pain (Girard et al., 2006; Hashimoto et al., 2006; Bechtold et al., 2008). Recently it was shown that the reduced fertility observed in female PACAP-null mice is due to implantation failure rather than serious deficits occurring during puberty, in the estrous cycle, ovarian histology, ovulation, or fertilization of eggs (Isaac and Sherwood, 2008). PACAP-null mice are born at a normal Mendelian ratio, but have increased early postnatal mortality rate compared to their wild-type littermates, (Gray et al., 2001; Hashimoto et al., 2001). This appears to be due to increased cold sensitivity and lack of thermoregulation, implicating PACAP as a central regulator of energy homeostasis (Adams et al., 2008). Although there do not appear to be gross anatomical changes in the nervous system (Hamelink et al., 2002; Ogawa et al., 2005), a recent report has shown that PACAP-null mice display altered cerebellar development due to delayed maturation of granule cells and increased apoptosis (Allais et al., 2007). The functional overlap between PACAP and VIP could allow VIP to compensate for the loss of PACAP, possibly masking the effects of loss of PACAP during development. However, recent evidence from Girard and colleagues (2006) indicates there are no compensatory changes in brain VIP mRNA levels or PACAP receptors' mRNA levels in PACAP-null mice compared to wild-type counterparts, and that in fact there appear to be developmental delays in their expression. The Girard et al. study also highlighted the behavioral differences between PACAP-null, VIP-null, and wild-type counterparts. PACAP-null mice have also been reported to suffer greater neuronal cell death following middle cerebral artery occlusion (MCAO) (Chen et al., 2006; Ohtaki et al., 2006). PACAP-38 administration into the lateral ventricle of heterozygous and PACAP-null mice immediately after MCAO was shown to reduce infarct volume, cell loss, and neurological deficit (Ohtaki et al., 2006), underlining a neuroprotective role for PACAP.

PAC_1 receptor-null mice also show increased postnatal mortality that is due to pulmonary hypertension and right heart failure after birth (Otto et al., 2004). The role of the PAC_1 receptor in mediating the anti-inflammatory actions of PACAP in an experimental model of endotoxic shock was confirmed using PAC_1 receptor-null mice (Martinez et al., 2002). Further elucidation of the mechanism underlying the protective role of PACAP, VIP, and the PAC_1 receptor in endotoxic shock demonstrated that neutrophil recruitment along with regulation of adhesion molecule expression and coagulation-related molecule fibrinogen synthesis were involved (Martinez et al., 2005).

4.5 STRUCTURAL ANALOGS

A number of PACAP analogs have been created through sequential deletion of the N- or C-termini of both PACAP-38 and PACAP-27, amino acid substitutions, cyclization, N-terminal modifications, modification at both N- and C-termini, or creating peptide chimeras (Gourlet et al., 1991, 1995, 1996b, 1998b; Robberecht et al., 1992a, 1992b; Vandermeers et al., 1992; Bitar et al., 1994; Fishbein et al., 1994; Wei et al., 2007; Bourgault et al., 2008; Onoue et al., 2008). Comparison of the potency of PACAP to its N-terminally and C-terminally shortened analogs has indicated that the N-terminal residues are important for receptor-mediated activation of cAMP production, whereas the C-terminal region confers receptor binding (Robberecht et al., 1992a, 1992b; Gourlet et al., 1996b). Furthermore, the residues 28 to 38 appear to facilitate binding to the PAC_1 receptor (Vaudry et al., 2000). In support of this is the observation that although VIP is generally a poor agonist at the PAC_1 receptor, by extending VIP1-27 with the PACAP28–38 sequence binding affinity and the ability to activate AC for the chimeric peptide VIP(1–27)/PACAP(28–38) at the PAC_1 receptor are improved (Gourlet et al., 1996a). Because PACAP-27, which is also missing the second α-helical region found in the

38-amino-acid form (Figure 4.2), does not display reduced affinity or potency at the PAC_1 receptor compared to PACAP-38, this has highlighted that there are other structural refinements that VIP is lacking (Onoue et al., 2008).

PACAP analogs have been designed specifically to improve metabolic stability (Bourgault et al., 2008) in view of observations that PACAP is vulnerable to rapid proteolytic degradation *in vivo*. In a recent study PACAP-38 was shown to have a very short halflife in human plasma, <5 min, where it was cleaved to generate PACAP1–21 and PACAP22–38 by an unidentified protease, along with different C-terminal fragments that are generated by plasma carboxypeptidases (Bourgault et al., 2008). Furthermore, the cell surface protease dipeptidyl peptidase IV (DPPIV, also known as T-cell surface antigen CD26) cleaves PACAP at the N-terminal serine, generating PACAP3–38 or PACAP3–27 and further N-terminal deleted forms (Lambeir et al., 2001; Zhu et al., 2003; Green et al., 2006) that have antagonistic activity (Robberecht et al., 1992a). PACAP-27 is degraded by neutral endopeptidase EC 3.4.24.11, an enzyme that is highly expressed in lung, kidney, and the CNS, to generate PACAP1–22 and PACAP1–25, of which the latter was shown to retain some activity at the $VPAC_1$ receptor (Gourlet et al., 1997b). In that study PACAP-38 was shown to be resistant to cleavage by the neutral endopeptidase, and it was observed that even partial C-terminal extension of the PACAP1–27 sequence to PACAP1–29 or to PACAP1–32 was protective against this enzyme activity.

Like PACAP, many of its analogs that have been found to have agonist or antagonist actions are not receptor-selective (Gourlet et al., 1995; Moody et al., 2002). For instance, a truncated version of PACAP, PACAP6–38 (Robberecht et al., 1992a, 1992b) that is commonly used as a PAC_1 receptor antagonist also has potent antagonist action at the $VPAC_2$ receptor (Gourlet et al., 1995; Dickinson et al., 1997; Moro et al., 1999; Tatsuno et al., 2001). A range of selective agonists and antagonists that have been reported by different groups are listed in Table 4.2.

Although there are no PACAP analogs that show selectivity for the PAC_1 receptor, the naturally occurring 6.8kDa vasodilator peptide, maxadilan, originally discovered in the saliva of sand flies (*Lutzomia lingipalpis*) has been shown to be a selective PAC_1 receptor agonist (Lerner and Shoemaker, 1992; Moro and Lerner, 1997). The maxadilan gene was cloned and sequenced and found to encode a prepropeptide that undergoes cleavage and amidation to a 61-amino-acid mature peptide. Surprisingly, the peptide sequence of maxadilan shows very little sequence similarity to PACAP (Moro and Lerner, 1997), containing four cysteine residues that form two disulfide bonds between positions 1 and 5 and positions 14 and 51, respectively (Figure 4.6). Although a solution structure for maxadilan has not been published yet, computational analysis of the peptide sequence by the Chou and Fasman method has predicted two α-helical regions; this conformation might confer ability to bind to the PAC_1 receptor (Moro et al., 1999). In addition, threonine residues within the predicted β-strand portion of the peptide (amino acids 23 to 46), in particular T^{33}, have been shown to be important for PAC_1 receptor activation (Reddy et al., 2006). A truncated version of maxadilan in which residues 24 to 42 were deleted (maxadilan de24–42, max.d.4 or M65) was found to be a selective antagonist of the PAC_1 receptor (Moro et al., 1999; Tatsuno et al., 2001). It will be interesting to compare a solution structure of the maxadilan bound-PAC_1 receptor to that of the PACAP-bound PAC_1 receptor.

A number of the VPAC receptor agonists and antagonists that have been developed are VIP analogs, and these tend to have poor affinity and activity at the PAC_1 receptor. In addition, the observation that secretin is a partial agonist at $VPAC_1$ receptors but is inactive at $VPAC_2$ receptors led to the development of the secretin analog, [R^{16}]chicken secretin, which acts as a $VPAC_1$ agonist at the rat receptor, although it is a much poorer agonist at the human $VPAC_1$ receptor and still maintains activity at the secretin receptor (Gourlet et al., 1997c). A number of GRF analogs and VIP/GRF chimeric peptides have been shown to have agonist or antagonist actions at VPAC receptors (Waelbroeck et al., 1985; Gourlet et al., 1997a, 1997c; Rekasi et al., 2000) and VIP/GRF chimeric peptides such as the agonist [$K^{15}R^{16}L^{27}$]VIP((1–7)/GRF(8–27) and antagonist PG 97–269 (Gourlet et al., 1997c) show high selectivity and affinity for the $VPAC_1$ receptor. Helodermin binds with high

TABLE 4.2
VPAC and PAC_1 Receptor Agonists and Antagonists

		$VPAC_1$		$VPAC_2$		PAC_1		
Agonists		**Affinity (IC_{50}) nM**	**cAMP (EC_{50}) nM**	**Affinity (IC_{50}) nM**	**cAMP (EC_{50}) nM**	**Affinity (IC_{50}) nM**	**cAMP (EC_{50}) nM**	**Reference**
$[K^{15},R^{16},L^{27}]$VIP(1–7) GRF(8–27)	tCHO h/rRs	1/2	1/1	≥30,000	No effect	≥30,000	No effect	Gourlet et al., 1997c
$[R^{16}]$chicken secretin[a]	tCHO h/rRs	60/1	Nr	5000/10,000	Nr	20,000	Nr	
$[Ala^{11,22,28}]$VIP	tCHO hRs	7.4	0.4	2352	1222	No effect	No effect	Nicole et al., 2000
$[Tyr^9Dip^{18}]$VIP(1–28)	tCHO, tHEK–293 rRs	0.11	0.23	53	74	3000	>1000	Tams et al., 2000
$[Ala^{2,8,9,11,19,24,25,27,28}]$VIP	T47D/tPANC1 (hVPAC1), SupT1/tPANC1 (hVPAC2)	1.9/1.5	2.9/5.6	468/105	1029/1334	Nd	Nd	Igarashi et al., 2002a, 2005)
$[Ala^{2,8,9,11,19,22,24,15,27,28}]$VIP		11.5/13.5	0.68/1.9	>30,000	23,710/16,560	Nd	Nd	Igarashi et al., 2005
Ro 25–1553	tCHO hVPACRs, rPAC1	100	100–500	1	1	3000	300	Gourlet et al., 1997d
Ro 25–1392	tCHO hVPACRs	>1,000	>100	9.6	3	Nd	Nd	Xia et al., 1997
BAY 55–9837	tCHO hRs	8700	100	50	0.4	No effect	No effect	Tsutsumi et al., 2002
R3–P55	tCHO hRs	7000	100	40	0.2	No effect	No effect	Yung et al., 2003
BAYQ9Q28	tCHO hRs	>10,000	233	32	0.3	>10,000	>1000	Pan et al., 2007
BAYQ9Q28C32	tCHO hRs	>10,000	>1,000	20	0.38	>10,000	>1000	
Hexanoyl$[A^{19},K^{27,28}]$VIP	tCHO hVPACRs, rPAC1R	800	30,000	1	0.2	Nd	1000	Langer et al., 2004
rMROM	tCHO hRs	>10,000	300	50	0.6	No effect	No effect	Yu et al., 2004
rMBAY	tCHO hRs	>10,000	200	53	0.8	No effect	No effect	Yu et al., 2007

continued

TABLE 4.2 (continued)

		VPAC$_1$		VPAC$_2$		PAC$_1$		
Agonists		**Affinity (IC_{50}) nM**	**cAMP (EC_{50}) nM**	**Affinity (IC_{50}) nM**	**cAMP (EC_{50}) nM**	**Affinity (IC_{50}) nM**	**cAMP (EC_{50}) nM**	**Reference**
PG99–465	tCHO hRs	Nd	8 (full ag.)	Nd	4.9 (part. ag.)	Nd	71.3 (full ag.)	Dickson et al., 2006
Maxadilan	tCHO r/mRs	No effect	No effect	No effect	No effect	0.055	0.008 (null)/ 0.015 (hop)	Tatsuno et al., 2001
	PC12 (rPAC1), tCOS hRs	No effect	Nd	No effect	No effect	4.2	0.3	Moro et al., 1999
	tCHO hRs	Nd	No effect	Nd	No effect	Nd	0.054	Dickson et al., 2006
Antagonists		**Binding IC_{50} nM**	**cAMP IC_{50} nM**	**Binding IC_{50} nM**	**cAMP IC_{50} nM**	**Binding IC_{50} nM**	**cAMP IC_{50} nM**	
PACAP6–38	NB-OK-1 (hPAC1)	—	—	—	—	2	1.5	Robberecht et al., 1992a
	tCHO hVPACRs, rPAC1	600	Nd	40	Nd	30	Nd	Gourlet et al., 1995
	tCHO rRs	Nd	>3000	Nd	170	Nd	14	Dickinson et al., 1997
PG97–269	tCHO, native cell lines h/r	2/10	2/15	3000/2000	No effect	~30,000	No effect	Gourlet et al., 1997a
[Tyr^9,Dip^{18}]VIP(6–28)	tCHO, tHEK-293 rRs	18	16	96	94	6000	>500	Tams et al., 2000
PG99–465	tCHO hVPACRs	200	Part. ag.	2	2	Nd	Nd	Moreno et al., 2000
	tCHO hRs	Nd	Aw VIP	Nd	22.9	Nd	Aw VIP	Dickson et al., 2006
M65	tCHO r/mRs	No effect	No effect	No effect	No effect	0.574	300	Tatsuno et al., 2001
	PC12 (rPAC1), tCOS hRs	No effect	No effect	Nd	No effect	2	300	Moro et al., 1999
	tCHO hRs	Nd	No effect	Nd	No effect	Nd	150	Dickson et al., 2006

Note: Nr, not reported; Nd, not determined; h, human; r, rat; m, mouse; R, receptor; ag, agonist; part, partial; Aw, additive with.

[a] Has significant activity at the secretin receptor.

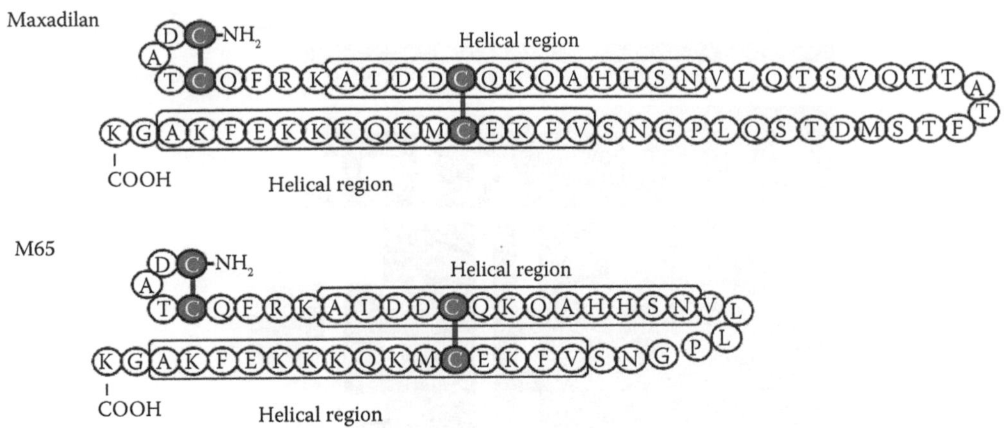

FIGURE 4.6 Amino-acid sequence and predicted structure of maxadilan and the deletion mutant M65. (Adapted from Moro, O. et al., *J. Biol. Chem.*, 274, 23103–23110, 1999.)

affinity and potently activates cAMP production at both human and rat $VPAC_2$ receptors (Svoboda et al., 1994; Gourlet et al., 1998b; Lutz et al., 1999a). By comparison, it shows >300-fold reduced affinity for binding to the human $VPAC_1$ receptor (Gourlet et al., 1998b) and is a poor agonist at the rat PAC_1 receptor (Lutz et al., 1999a). However, helodermin has been shown to bind with high affinity to the rat $VPAC_1$ receptor (Gourlet et al., 1998b).

Amino acid substitutions in the VIP molecule have been found to generate more selective analogs for the VPAC receptors. For instance, replacement of Tyr^{22} in the VIP sequence is selective against the $VPAC_2$ receptor (Gourlet et al., 1998c). Alanine-scan (O'Donnell et al., 1991; Nicole et al., 2000; Igarashi et al., 2002a, 2005) and D-amino acid scan (Igarashi et al., 2002a) have produced $VPAC_1$-selective analogs such as the $VPAC_1$-selective agonist $[Ala^{11,22,28}]$ VIP (Nicole et al., 2000). Although highlighting important residues for binding to $VPAC_2$ receptors, no $VPAC_2$-selective analogs were identified in these studies (Nicole et al., 2000; Igarashi et al., 2002b).

VIP is particularly vulnerable to a number of peptidases (Onoue et al., 2008), with VIP degradation products tending to have poor bioactivity and low affinity at VIP/PACAP receptors (Bolin et al., 1995; Gourlet et al., 1998b). Leukocytes produce and release both truncated and N-terminally extended VIP peptides (Goetzl et al., 1988), and T-lymphocytes have a cell-surface protease activity that generates several VIP fragments, of which the VIP4–28 form predominates (Goetzl et al., 1989). This was subsequently reported to have potent $VPAC_1$ agonist and $VPAC_2$ antagonist activity (Summers et al., 2003). Interestingly, the $VPAC_2$ receptor is more sensitive to C-terminal deletions of VIP and PACAP than the $VPAC_1$ receptor, supporting the notion that this receptor requires the α-helical C-terminus for peptide binding (Gourlet et al., 1998b). Substitution of the valine at position 18 with alanine was found to increase VIP halflife (O'Donnell et al., 1991). Arginine substitutions for lysines at positions 15, 20, and 21 and leucine for methionine at position 17, along with extension of the VIP1–28 sequence with GRR produced a high-affinity metabolically stable (but nonselective) VPAC agonist, IK312532 (Onoue et al., 2004a). Peptide modifications such as cyclization, acylation, glycosylation, and C-terminal PEGylation have been used to increase metabolic stability and in some cases have also resulted in increased peptide affinity and receptor selectivity (O'Donnell et al., 1994; Bolin et al., 1995; Gozes et al., 1995; Gourlet et al., 1997d; Xia et al., 1997; Juarranz et al., 1999; Pan et al., 2007, 2008; Bourgault et al., 2008; Dangoor et al., 2008). The acetylated cyclic peptides, Ro 25-1392 and Ro 25-1553, which were originally developed by Roche as a treatment for asthma (Bolin et al., 1995; Schmidt et al., 2001) were found to be potent $VPAC_2$ receptor-selective agonists (Gourlet et al., 1997d; Xia et al., 1997). These have been used as the basis

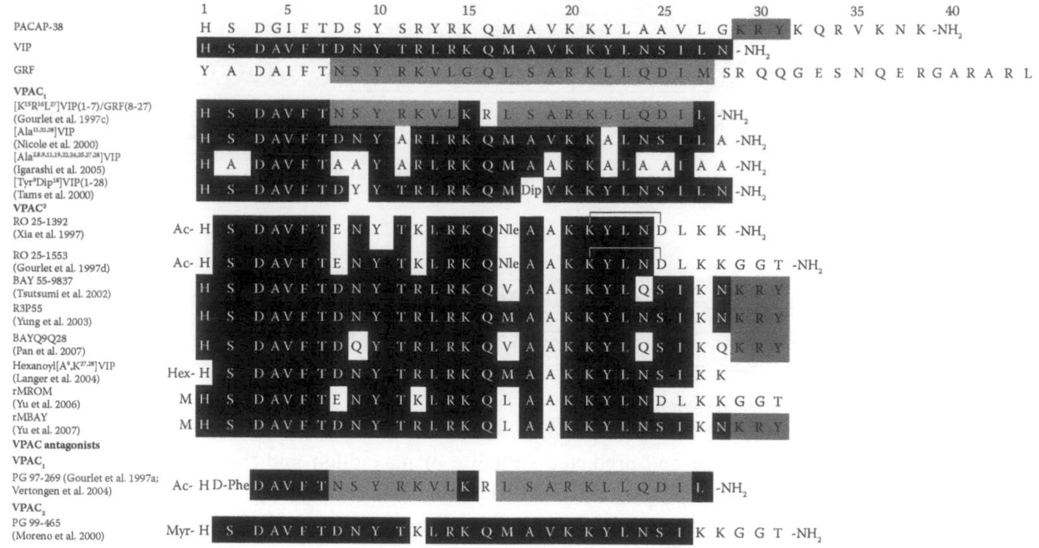

FIGURE 4.7 Amino-acid sequences of VPAC-selective ligands. The VIP sequence backbone is highlighted in black, the GRF-specific residues in light gray and PACAP-38-specific residues in dark gray. The β-lactam bridge between K[20] and D[24] in the Roche compounds is indicated by the bar. The tyrosine at position 10 in Ro 25-1392 is O-methylated. Abbreviations: Ac, Acetyl; Hex, hexanoyl; Dip, diphenylalanine; Nle, norleucine; Myr, myristoyl.

for developing further selective $VPAC_2$ receptor agonists and antagonists (Moreno et al., 2000; Yung et al., 2003; Yu et al., 2006). However, the purported $VPAC_2$ receptor-selective antagonist PG99-465 that was derived from Ro 25-1553 (Moreno et al., 2000) was subsequently shown to have full agonist activity at the $VPAC_1$ and PAC_1 receptors (Dickson et al., 2006). Thus there is at present no selective $VPAC_2$ receptor antagonist. The observation that fatty acyl modifications of the N-terminal histidine residue increase affinity and selectivity for $VPAC_2$ over $VPAC_1$ (Gourlet et al., 1998a; Moreno et al., 2000) led to the development of the high-affinity $VPAC_2$ receptor-selective agonist hexanoyl[$A^{19}K^{27,28}$]VIP (Langer et al., 2004). Figure 4.7 shows the structures of analogs that have been reported to be selective at either the $VPAC_1$ or $VPAC_2$ receptors, and their pharmacological properties at the three receptors are listed in Table 4.2.

4.6 COMMON METHODOLOGIES AND ASSAY SYSTEMS

Many tissues and cell lines contain a mixture of VPAC and PAC_1 receptors and a number of studies have utilized these to investigate the biological functions of PACAP (Vaudry et al., 2000). *In vivo* assays based on intravenous injection or infusion of peptide to measure effects on mean arterial blood pressure along with measurement of adrenal catecholamine secretion (Lamouche and Yamaguchi, 2001) or pancreatic insulin secretion (Tsutsumi et al., 2002) are also used. A number of clonal cell lines that predominantly express one or other of the VPAC and PAC_1 receptors can be used to investigate receptor-specific actions in a "native" environment. These are listed in Table 4.3. For instance, an assay based on PC12 cell proliferation/differentiation has been used to characterize PACAP analogs (Bourgault et al., 2008). We typically use transiently or stably transfected cell lines expressing one of the VPAC or PAC_1 receptor isoforms for evaluation of ligand binding and the activation of second messenger pathways. Chinese hamster ovary (CHO) cells are the most commonly used cell line for expressing these receptors (Robberecht and Waelbroeck, 1998), but may not be ideal in certain circumstances as this cellular environment has been found to alter receptor behavior with respect to ligand binding and activation of second messenger pathways (Robberecht

TABLE 4.3
Cell Lines Expressing VPAC and PAC_1 Receptor Isoforms

Receptor	Cell Line	Species and Cell Type	Reference
$VPAC_1$	HT-29	Human adenocarcinoma (intestinal epithelium)	Sreedharan et al., 1993; Igarashi et al., 2002b
	LoVo	Human adenocarcinoma (colon)	Gourlet et al., 1997a, 1997d
	HCT 15	Human adenocarcinoma (colon)	Gourlet et al., 1997d
	T47D	Human breast cancer cells	Waschek et al., 1995; Igarashi et al., 2002b
$VPAC_2$	Sup-T1	Human lymphoblastoma	Svoboda et al., 1994; Igarashi et al., 2002b
	GH3	Rat anterior pituitary somatomammotroph cells	MacKenzie et al., 2001
	GH4C1	Rat anterior pituitary somatomammotroph cells	Rawlings et al., 1995
	L2	Rat alveolar epithelial cells	Onoue et al., 2004a
PAC_1	SH-SY5Y	Human neuroblastoma cells	Lutz et al., 2006
	PC12	Rat phaeochromocytoma cells	Ravni et al., 2006a
	AR4-2J	Rat pancreatic tumour acinar cells	Pisegna and Wank, 1993

and Waelbroeck, 1998; Igarashi et al., 2002a). Other cell types that are used as heterologous expression systems are monkey kidney fibroblast COS-7 cells (Morrow et al., 1993; Lutz et al., 1999a), NIH Swiss mouse embryo fibroblast NIH-3T3 cells (Pisegna and Wank, 1996), human embryonic kidney HEK-293 cells (Jabranc-Ferrat et al., 2000), and human pancreatic tumor PANC1 cells (Igarashi et al., 2002a, 2005). In addition, the *Xenopus laevis* melanophore assay based on pigment granule translocation induced by second messengers has been utilized to investigate PAC_1 receptor-selective agonists and antagonists (Pereira et al., 2002; Reddy et al., 2006). Melanophore cell lines stably expressing the rat PAC_1, human $VPAC_1$ or human $VPAC_2$ receptors were created and used in these studies; however, it should be noted that these cells endogenously express a VIP/PACAP receptor (Marotti et al., 1999; Pereira et al., 2002) that may be similar to that cloned from *Rana ridibunda* (Alexandre et al., 1999). Needless to say, interpretation of results from heterologous cell expression systems should be done with the greatest of care (Robberecht and Waelbroeck, 1998).

4.6.1 Ligand Binding Assays

Ligand binding assays and competition displacement binding assays performed with cell membranes or whole cells, using various iodinated peptide ligands ^{125}I-PACAP-27, ^{125}I-PACAP-38, ^{125}I-VIP, ^{125}I-[AcHis1]PACAP-27, ^{125}I-maxadilan, ^{125}I-Helodermin, and ^{125}I-Ro 25-1553 have been described in the literature (Robberecht et al., 1992a; Van Rampelbergh et al., 1996; Moro and Lerner, 1997; Gourlet et al., 1998c; Moreno et al., 2000; MacKenzie et al., 2001). In addition, a nonradioactive, fluorescence-based assay using a fluorescent derivative of PACAP-27 (fluor-PACAP) has also been described (Germano et al., 2001). This has also been used recently for fluorescence activated cell sorting (FACS) to isolate pure populations of gastric ECL cells (Oh et al., 2005).

4.6.1.1 Membrane Binding Assay Procedure

A confluent flask of cells (80 cm^2) is placed on ice, the medium discarded and the cell layer gently washed with 10 mL of ice-cold HBSS. Cells are scraped into 1 mL of Buffer A [50 mM Tris HCl pH 7.4, 1 mM EGTA (ethyleneglycol-*bis*(b-aminoethyl ether)-*N,N,N′,N′*-tetraacetic acid)) containing 1 mM AEBSF (4-(2-aminoethyl)benzenesulfonyl fluoride), 2 μg/mL aprotinin, 4 μg/mL leupeptin, 2 μg/mL pepstatin, 1 mM Na_3VO_4, 1 mM NaF, 50 μg/mL soybean trypsin inhibitor and transferred into a clean tube on ice. Another 1 mL volume of Buffer A is added to the flask and the process repeated. The cells are homogenized using an Ystral cell homogenizer for 30 sec, then centrifuged at 1000*g* for 5 min at 4°C to pellet intact cells, nuclei and so on. The supernatant is transferred to a new tube and centrifuged at 12,000*g* for 30 min at 4°C to pellet the cell membranes. The membranes are resuspended in 1 to 2 mL of assay buffer (50 mM Tris HCl, pH 7.4, bacitracin 0.5 mg/mL, AEBSF 2 μg/mL, BSA 1% (w/v)].

The assay is carried out in screw cap tubes in a total volume of 500 μL per tube at 37°C in a shaking water bath. Each tube contains 300 μL of assay buffer, 50 μL of 10× peptide and 50 μL of radioiodinated peptide. A high dose of displacing peptide (e.g., 10 μM) is used to determine nonspecific binding. The reaction is started by the rapid addition of 100 μL of membrane suspension and the samples incubated for 10 to 15 min before the reactions are stopped by centrifugation at 12,000*g* for 30 min at 4°C. The supernatant is removed by aspiration, 500 μL of assay buffer is added to each tube, and the samples recentrifuged as above but for 15 min. The supernatant from the samples is then aspirated off again and the bottom of the eppendorf tube (which contains the pellet) is clipped into a pony vial and the amount of radioactivity determined by gamma counting.

4.6.1.2 Whole Cell Binding Assay Procedure

For determining cell-surface binding, cells seeded into 12-well plates and grown for 24 to 96 h are washed twice and incubated for 60 min on ice in Medium 199 containing 0.2% BSA (w/v) and radioiodinated peptide in the absence or presence of increasing concentrations of unlabeled peptide. A high dose of displacing peptide (e.g., 10 μM) is used to determine nonspecific binding. Unbound radioactivity is removed by washing cells three times with EBSS containing 0.1% BSA (w/v). Bound radioactivity is removed by an acid-wash procedure, 5 min incubation with 0.2 M acetic acid/0.5 M NaCl, and the amount of radioactivity in the solution determined by gamma counting.

For ligand-receptor internalization assays, cells seeded into 12-well plates are washed twice and incubated at 37°C in a waterbath for 10 min in Medium 199 containing 0.2% BSA (w/v), 30 μg/mL bacitracin and 1 μg/mL aprotinin with radioiodinated peptide in the absence or presence of increasing concentrations of unlabeled peptide. A high dose of displacing peptide (e.g., 10 μM) is used to determine nonspecific binding. Unbound radioactivity is removed by washing cells three times with EBSS containing 0.1% BSA (w/v). Bound radioactivity is removed by acid wash (0.2 M glacial acetic acid/0.5 M NaCl) for 5 min on ice. The amount of internalized radioactivity is determined by solubilizing the cell layer by treatment with 0.5 M NaOH overnight. The amount of radioactivity in the samples is determined by gamma counting.

4.6.2 Second Messenger Assays

4.6.2.1 AC and cAMP Assays

Several assay methods have been developed that are used to measure AC activation in membranes (Moreno et al., 2000) and to quantitate cAMP production in whole cells, including radioimmunoassay using a cAMP-specific antibody and iodinated cAMP tracer (Morrow et al., 1993). In addition, nonradioactive enzyme immunoassay-based kits are commercially available, although these tend to be expensive. We routinely use the Wong method described in the next section as this is relatively inexpensive to perform, but have also used the enzyme complementation assay described later as we have found this more rapid and less complex to perform, allowing the preparation and analysis of a larger number of samples per experiment.

4.6.2.1.1 Measurement of [^{3}H]cAMP Production in Whole Cells

Cells seeded into 24-well plates and grown for 24 to 96 h are prelabeled overnight with 1 μCi per well of [2-^{3}H]-adenine (25 Ci/mmol) in 0.5 mL labeling medium [modified Eagles medium (MEM) containing 1% fetal bovine serum (v/v)]. The following morning, cells are washed in MEM-BSA (0.25% w/v) and preincubated with 500 μM of the nonspecific phosphodiesterase inhibitor

3-isobutyl-1-methylxanthine (IBMX) for 15 min before agonist stimulation. If other inhibitors are to be used, they are also added at this time. For a typical dose response, cells are stimulated for 15 min with various concentrations of peptide agonist and the assay stopped by aspirating the medium and replacing with 5% ice-cold trichloroacetic acid. The plates are stored on ice for a minimum of 1 h, but can be stored at –20°C until the column chromatography is to be performed. The [^{3}H]ATP/ADP and [^{3}H]cyclic AMP are isolated from each sample by sequential chromatography on anion exchange and alumina columns described by Wong (1994). The amount of radioactivity in the flow-through (representing [^{3}H]ATP + [^{3}H]ADP) and eluate (representing [^{3}H]cyclic AMP) are measured by scintillation counting. The levels of cAMP production are calculated as the ratio of [[^{3}H]cAMP]/([[^{3}H]cAMP] + [[^{3}H]ATP + [^{3}H]ADP]).

4.6.2.1.2 Nonradioactive Enzyme Complementation Assay that can be Scaled for High-Throughput Screening

The production of cAMP is measured by the β-galactosidase enzyme fragment complementation based HitHunter™ cAMP XS+ chemiluminescence assay (DiscoveRx, Birmingham, UK) according to manufacturer's instructions. Briefly, cells are grown to ~80% confluency in 75 cm^2 flasks, then quiesced for 2 h in OptiMEM (Invitrogen, Paisley, UK) before being scraped into phosphate buffered saline (PBS) and the cell numbers determined using a hemocytometer. Cells then are centrifuged and resuspended at a density of 1.0 × 10^5 cells per 20 µL in assay medium (MEM containing 0.25% BSA (w/v) and 500 µM IBMX) for 15 min before starting the assay. Cells are assayed in suspension in a 96-well plate by adding 20 µL of the cell suspension to each well into which has already been added 5 µL of a ×5 concentration solution of the compounds to be tested. In addition, a separate set of wells containing 25 µL of different concentrations of a standard cAMP solution in assay medium are set up to generate the standard curve. The plate is incubated at 37°C in 5% CO_2, and the assay stopped after 15 min by consecutive addition of 5 µL of antibody solution then 30 µL of the solution containing 1 part Galacton-Star: 5 parts Emerald-II; 19 parts Lysis buffer. To this solution is added 1 part enzyme donor fragment (ED) solution. Following this, the plate is incubated for 1 h at room temperature before adding to each well 30 µL of a solution containing the enzyme acceptor fragment (EA) solution and gently mixing. The plate is incubated a further hour at room temperature and luminescence is then detected using a luminescence plate reader.

4.6.2.2 Inositol Phosphates Assay

For measuring [^{3}H]inositol phosphate ([^{3}H]InsP) production, cells seeded into 12-well plates and grown for 24 to 96 h are prelabeled overnight with 1 µCi per well of [2-^{3}H]-myoinositol (20 Ci/mmol) in 1 mL labeling medium [Earle's Balanced salt solution (EBSS) with 10 mM glucose, 10 mM HEPES, pH 7.4]. The following morning, cells are washed twice in EBSS/glucose/HEPES-BSA [0.2% (w/v)] and preincubated with 10 mM lithium chloride (LiCl) for 10 min before agonist stimulation. If inhibitors are to be used, they are added at this time also. For a typical dose response, cells are stimulated for 60 min with various concentrations of peptide agonist and the assay stopped by aspirating the medium and adding 1 mL ice-cold 10 mM formic acid per well. The plates are stored on ice for a minimum of 1 h before the [^{3}H]InsPs are isolated by anion exchange chromatography as described previously (MacKenzie et al., 2001; Mitchell et al., 2003) adapted from a method originally described by Berridge and colleagues (1983). Briefly, each sample is applied to a 1 mL column of Dowex anion exchange resin (1 × 8 resin, formate form, 200–400 mesh, Sigma-Aldrich, Poole, UK) and allowed to drip through. Following this, the columns are washed with 15 mL of ultrapure (UHP) H_2O. A stepwise gradient of ammonium formate is used to elute the [^{3}H]InsP by first washing the column with 5 mL 50 mM ammonium formate followed by 10 mL of 1 M ammonium formate/0.1 M formic acid. This fraction contains the [^{3}H]InsP and is collected in scintillation vials. After this step, the columns are regenerated by washing with 5 mL of 2 M ammonium formate/0.1 M formic acid followed by 15 mL of UHP H_2O. The amount of radioactivity for each sample is determined by scintillation counting.

4.6.2.3 Intracellular Ca^{2+} Assay

Fluorescence-based high-throughput assay methods have been developed to measure changes in intracellular Ca^{2+} levels (Dickson et al., 2006; Bourgault et al., 2008). Different assay kits are commercially available. The following is an assay method adapted from Lembo et al. (1999) based on the Molecular Devices FLIPR technology.

4.6.2.3.1 High-Throughput Calcium Flux Assay Using a Fluorometric Imaging Plate Reader (FLIPR)

Cells are seeded into 96-well black-walled culture plates at a density of ~5 × 10^4 cells per well in complete cell culture medium one day prior to the experiment. On the day of the experiment cells are washed twice with assay buffer [Hanks' balanced salt solution, 20 mM HEPES (pH7.6), 2.5 mM probenecid, and 0.1% BSA (w/v)] and loaded for 1h at 37°C in 100 µL assay buffer with 1 mM Fluo-3-AM (Molecular Probes) (dissolved in 10% pluronic acid first before adding to assay buffer). Cells are then washed twice with the assay buffer before analysis. Various concentrations of peptide agonist (50 µL of 3× final peptide concentration) are then added to each well and calcium flux measured in all 96 wells simultaneously with a fluorometric imaging plate reader (FLIPR) (Molecular Devices, Sunnyvale, CA), excitation 488 nm, emission 526 nm (Ca^{2+}-bound), according to the manufacturer's instructions. To determine inhibition, Fluo-3-AM loaded cells are preincubated for 10 min with antagonist (by adding 50 µL of 3× antagonist/well) before adding the agonist (50 µL of 4× final concentration/well). Data are expressed as arbitrary florescence units *vs.* time.

4.6.3 Peptide Synthesis Methodologies

PACAP, VIP, and their peptide analogs are typically produced by solid-phase chemical synthesis methods. However, the Ro 25-1392 compound is no longer obtainable and it is apparently too difficult to synthesize (Yung et al., 2003). The limitations of standard peptide synthetic methodologies can also be restrictive for the number of analogs that can be generated for testing, although a spot synthesis protocol was described for the complete substitutional analysis of VIP–dye conjugates (Bhargava et al., 2002). Peptides can be produced as recombinant proteins, such as GST-fusion proteins (Moro and Lerner, 1997; Yung et al., 2003), and these are amenable to high-throughput mutagenesis. However, these tend to suffer from low yields due to poor solubility and/or difficulty in cleaving and recovering the peptide. Intein-based recombinant methods (Chong et al., 1998) have also been used to generate peptides and peptide mutants (Yu et al., 2004, 2006, 2007; Reddy et al., 2006).

4.7 PERSPECTIVES AND FUTURE DEVELOPMENTS

It is 20 years since PACAP was first discovered, and it has certainly proved to be a very interesting and versatile peptide. PACAP is expressed in neuronal, endocrine, reproductive, and immune tissues and has been shown to have a wide array of biological actions in the CNS and periphery, including regulating proliferation, migration, differentiation, and cell survival. The recent discovery implicating alterations in the PACAP/PAC_1 receptor signaling pathway in schizophrenia has revealed a new function for this peptide in humans (Hashimoto et al., 2007) and it is likely that there are still more to be discovered. It is clear that the use of microarray and proteomic analysis combined with gene knockouts/knockins will lead to greater insights into the physiological roles of PACAP, helping to elucidate the molecular mechanisms that underlie these processes. The few studies that have been carried out have revealed some of the signaling and genetic pathways involved in the neuroprotective and neurotrophic actions of PACAP on rat PC12 cells (Vaudry et al., 2002; Grumolato et al., 2003; Ravni et al., 2006b), mouse adrenal gland, and bovine chromafffin cells (Samal et al., 2007) and primary sympathetic cells in culture (Braas et al., 2007), the neuroprotective actions of PACAP following stroke (Chen et al., 2006) and the role of PACAP in photic entrainment (Fahrenkrug et al., 2005). Many more studies, particularly using human tissues and cells, will be needed in order to understand the important functions of this peptide and in order to exploit its potential in the treatment of human disease.

4.7.1 Neurodifferentiation Factor for Embryonic Stem Cells

Mouse embryonic stem (ES) cells have been shown to express both PAC_1 and $VPAC_2$ receptors (Cazillis et al., 2004; Hirose et al., 2005) and the developmental role of PACAP is being examined with these (Hirose et al., 2006; Falluel-Morel et al., 2007). The observation that PACAP induces neurodifferentiation of ES cells has raised the exciting possibility of its potential use as a neurodifferentiation factor to produce neuronal cell types from ES cells for cell replacement therapies. It will be of particular interest to discover whether human embryonic stem cells also express PACAP receptors and to determine their responses to PACAP.

4.7.2 Neuroprotection in Acute and Chronic CNS Disorders

PACAP has promise for the treatment of acute (brain trauma, stroke) and chronic CNS disorders. First, PACAP-38 crosses the blood–brain barrier through a saturable transport mechanism involving peptide transport system-6 (PTS-6) (Banks et al., 1993). Second, PACAP has been shown *in vivo* to prevent the death of CA1 hippocampal cells when administered intracerebroventricularly or intravenously, even when administered 24 h postinjury (Banks et al., 1996; Uchida et al., 1996). It has been determined that part of PACAPs neuroprotective actions are mediated by the attenuation of microglial activation (Suk et al., 2004).

4.7.3 Anti-Inflammatory Agent

Observations relating to thc anti-inflammatory effects of PACAP may prove useful in developing new anti-inflammatory drugs (Pozo, 2003; Abad et al., 2006). Certainly, PACAP has been shown to have beneficial effects in a number of inflammatory disorders, including multiple sclerosis (Kato et al., 2004), rheumatoid arthritis (Abad et al., 2001), Crohn's disease and bowel cancer (Nemetz et al., 2008), multiple myeloma and other kidney diseases (Li et al., 2008), asthma and chronic obstructive pulmonary disease (Groneberg et al., 2006), and pulmonary hypertension (Otto et al., 2004). The neurotrophic and neuroprotective actions of PACAP in combination with its anti-inflammatory actions may have particular relevance for chronic neurodegenerative diseases, such as Parkinson's disease (Reglődi et al., 2004; Chung et al., 2005; Reglődi et al., 2006), for which both direct neurotrophic actions and indirect through the release of other neurotrophic factors along with attenuation of inflammatory signaling by microglial cells may be useful (Yang et al., 2006).

4.7.4 Receptor Isoform-Specific Agonists and Antagonists

The use of receptor isoform-selective agonists and antagonists will not only help determine the specific physiological roles each has in mediating PACAPs actions, but will also enable better targeted therapies to be developed. For instance, PACAP and VIP have a number of shared functions in the periphery that are likely to be mediated through their common VPAC receptors. The therapeutic potential of VIP has been exploited for some years as many tumors overexpress VIP/PACAP receptors, and in particular the $VPAC_1$ receptor (Reubi, 2000; Reubi et al., 2000). VIP receptor scintigraphy (Virgolini, 1997) and VIP–dye conjugates (living tissue is relatively transparent to near-infrared light; Bhargava et al., 2002) have been developed for better tumor imaging *in vivo* and for tumor-targeting radiotherapy treatment. The prominent role of PACAP in glucose and energy homeostasis (Nakata and Yada, 2007) and in particular the $VPAC_2$ receptor in insulin but not glucose release, has highlighted the potential for targeting this receptor for treating type 2 diabetes (Tsutsumi et al., 2002). Certainly, the role of the PAC_1 receptor in mediating the anti-inflammatory actions of PACAP in septic shock has identified this as a potential therapeutic target (Martinez et al., 2006). Furthermore, the expression of particular receptor isoforms, such as the δ5,6 NT PAC_1 receptor forms that are expressed in neuroblastoma and retinoblastoma, may prove useful as targets for the treatment of these diseases (Lutz et al., 2006).

4.7.5 Metabolic Stability Issues

For optimal therapeutic benefit, the problem of metabolic stability of PACAP and its analogs still needs to be addressed. One solution has been to exploit methods of delivery that protect against peptidase degradation, such as incorporation into lipid micelles (Krishnadas et al., 2003). Lentiviral delivery of the VIP gene has been used successfully to treat rheumatoid arthritis in a mouse model of the disease (Delgado et al., 2008). Other therapeutic approaches, such as the use of neutralizing monoclonal antibodies to target PACAP and $VPAC_1$ receptor have been shown to aid platelet recovery in animal models of myelosuppressive therapy (Freson et al., 2008). Interestingly, PACAP-38 (but not PACAP-27) binds to ceruloplasmin in human plasma (Tams et al., 1999). Although the physiological role of this has not been ascertained, it may have significance for transporting PACAP to distal targets and for protecting the peptide from degradation. However, it is the development of metabolically stable peptide mimetics and antagonists to target specific receptor subtypes that may be key to overcoming these issues (Abad et al., 2006), and this advance in the PACAP story will certainly be benefited by the current and future efforts to determine receptor structures (Sun et al., 2007; Ceraudo et al., 2008).

ACKNOWLEDGMENTS

This is dedicated to the memory of two very important people: Professor Akira Arimura, who made the initial discovery of PACAP and whose extensive and tireless work has helped elicit many of its biological functions, and Melanie Johnson, whose insight, advice, and good humor kept us on track these many years. She is very much missed.

REFERENCES

Abad, C., Martinez, C., Leceta, J., Gomariz, R.P., and Delgado, M., Pituitary adenylate cyclase-activating polypeptide inhibits collagen-induced arthritis: an experimental immunomodulatory therapy, *J. Immunol.*, 167, 3182–3189, 2001.

Abad, C. et al., Pituitary adenylate-cyclase-activating polypeptide expression in the immune system, *Neuroimmunomodulation*, 10, 177–186, 2002–2003.

Abad, C., Gomariz, R.P., and Waschek, J.A., Neuropeptide mimetics and antagonists in the treatment of inflammatory disease: focus on VIP and PACAP, *Curr. Top. Med. Chem.*, 6, 151–163, 2006.

Adams, B.A. et al., Pituitary adenylate cyclase-activating polypeptide and growth hormone-releasing hormone-like peptide in sturgeon, whitefish, grayling, flounder, and halibut: cDNA sequence, exon skipping, and evolution, *Regul. Pept*, 109, 27–37, 2002.

Adams, B.A. et al., Feeding and metabolism in mice lacking pituitary adenylate cyclase-activating polypeptide (PACAP), *Endocrinology*, 149, 1571–1580, 2008.

Alexandre, D., Anouar, Y., Jégou, S., Fournier, A., and Vaudry, H., A cloned frog vasoactive intestinal polypeptide/pituitary adenylate cyclase-activating polypeptide receptor exhibits pharmacological and tissue distribution characteristics of both $VPAC_1$ and $VPAC_2$ receptors in mammals, *Endocrinology*, 140, 1285–1293, 1999.

Alexandre, D., Vaudry, H., Jégou, S., and Anouar, Y., Structure and distribution of the mRNAs encoding pituitary adenylate cyclase-activating polypeptide and growth hormone-releasing hormone-like peptide in the frog, *Rana ridibunda*, *J. Comp. Neurol.*, 421, 234–246, 2000.

Allais, A. et al., Altered cerebellar development in mice lacking pituitary adenylate cyclase-activating polypeptide, *Eur. J. Neurosci.*, 25, 2604–2618, 2007.

Arimura, A. et al., Tissue distribution of PACAP as determined by RIA: highly abundant in the rat brain and testes, *Endocrinology*, 129, 2787–2789, 1991.

Arimura, A., Perspectives on pituitary adenylate cyclase activating polypeptide (PACAP) in the neuroendocrine, endocrine, and nervous systems, *Jpn J. Physiol.*, 48, 301–331, 1998.

Armstrong, B.D. et al., Lymphocyte regulation of neuropeptide gene expression after neuronal injury, *J. Neurosci. Res.*, 74, 240–247, 2003.

Armstrong, B.D. et al., Induction of neuropeptide gene expression and blockade of retrograde transport in facial motor neurons following local peripheral nerve inflammation in severe combined immunodeficiency and BALB/c mice, *Neuroscience*, 129, 93–99, 2004.

Banks, W.A., Kastin, A.J., Komaki, G., and Arimura, A., Passage of pituitary adenylate cyclase activating polypeptide1-27 and pituitary adenylate cyclase activating polypeptide1-38 across the blood–brain barrier, *J. Pharm. Exp. Therap.*, 267, 690–696, 1993.

Banks, W.A. et al., Transport of pituitary adenylate cyclase-activating polypeptide across the blood–brain barrier and the prevention of ischemia-induced death of hippocampal neurons, *Ann. NY Acad. Sci.*, 805, 270–277, 1996.

Basak, S., Chretien, M., Mbikay, M., and Basak, A., *In vitro* elucidation of substrate specificity and bioassay of proprotein convertase 4 using intramolecularly quenched fluorogenic peptides, *Biochem. J.*, 380, 505–514, 2004.

Basille, M. et al., Comparative distribution of pituitary adenylate cyclase-activating polypeptide (PACAP) binding sites and PACAP receptor mRNAs in the rat brain during development, *J. Comp. Neurol.*, 425, 494–509, 2000.

Basille, M. et al., Ontogeny of PACAP receptors in the human cerebellum: perspectives of therapeutic applications, *Regul. Pept.*, 137, 27–33, 2006.

Beaudet, M.M., Parsons, R.L., Braas, K.M., and May, V., Mechanisms mediating pituitary adenylate cyclase-activating polypeptide depolarization of rat sympathetic neurons, *J. Neurosci.*, 20, 7353–7361, 2000.

Bechtold, D.A., Brown, T.M., Luckman, S.M., and Piggins, H.D., Metabolic rhythm abnormalities in mice lacking VIP-$VPAC_2$ signaling, *Am. J. Physiol. Regul. Integr. Comp. Physiol.*, 294, R344–R351, 2008.

Berridge, M.J. et al., Changes in the levels of inositol phosphates after agonist-dependent hydrolysis of membrane phosphoinositides, *Biochem. J.*, 212, 473–482, 1983.

Bhargava, S. et al., A complete substitutional analysis of VIP for better tumor imaging properties, *J. Mol. Recog.*, 15, 145–153, 2002.

Bhogal, R. et al., The effects of PACAP and VIP on guinea pig tracheal smooth muscle *in vitro*, *Peptides*, 15, 1237–1241, 1994.

Bitar, K.G., Somogyvári-Vigh, A., and Coy, D.H., Cyclic lactam analogues of ovine pituitary adenylate cyclase activating polypeptide (PACAP): discovery of potent type II receptor antagonists, *Peptides*, 15, 461–466, 1994.

Blankenfeldt, W. et al., NMR spectroscopic evidence that helodermin, unlike other members of the secretin/VIP family of peptides, is substantially structured in water, *Biochemistry*, 35, 5955–5962, 1996.

Boeshore, K.L. et al., Novel changes in gene expression following axotomy of a sympathetic ganglion: a microarray analysis, *J. Neurobiol.*, 59, 216–235, 2004.

Bokaei, P.B. et al., Identification and characterization of five-transmembrane isoforms of human vasoactive intestinal peptide and pituitary adenylate cyclase-activating polypeptide receptors, *Genomics*, 88, 791–800, 2006.

Bolin, D.R., Michalewsky, J., Wasserman, M.A., and O'Donnell, M., Design and development of a vasoactive intestinal peptide analog as a novel therapeutic for bronchial asthma, *Biopolymers*, 37, 57–66, 1995.

Borboni, P. et al., Molecular and functional characterization of pituitary adenylate cyclase-activating polypeptide (PACAP-38)/vasoactive intestinal polypeptide receptors in pancreatic beta-cells and effects of PACAP-38 on components of the insulin secretory system, *Endocrinology*, 140, 5530–5537, 1999.

Bourgault, S. et al., Novel stable PACAP analogs with potent activity towards the PAC1 receptor, *Peptides*, doi:10.1016/j.peptides.2008.01.022., 2008.

Bozza, M. et al., The PACAP-type I receptor agonist maxadilan from sand fly saliva protects mice against lethal endotoxemia by a mechanism partially dependent on IL-10, *Eur. J. Immunol.*, 28, 3120–3127, 1998.

Braas, K.M. and May, V., Pituitary adenylate cyclase-activating polypeptides directly stimulate sympathetic neuron Neuropeptide Y release through PAC_1 receptor isoform activation of specific intracellular signaling pathways, *J. Biol. Chem.*, 274, 27702–27710, 1999.

Braas, K.M. et al., Microarray analyses of pituitary adenylate cyclase activating polypeptide (PACAP)-regulated gene targets in sympathetic neurons, *Peptides*, 28, 1856–1870, 2007.

Brandenburg, C.A., May, V., and Braas, K.M., Identification of endogenous sympathetic neuron pituitary adenylate cyclase-activating polypeptide (PACAP): depolarization regulates production and secretion through induction of multiple propeptide transcripts, *J. Neurosci.*, 17, 4045–4055, 1997.

Brenneman, D.E., Neuroprotection: A comparative view of vasoactive intestinal peptide and pituitary adenylate cyclase-activating polypeptide, *Peptides*, 28, 1720–1726, 2007.

Campbell, R.M. and Scanes, C.G., Evolution of the growth hormone-releasing factor (GRF) family of peptides, *Growth Regul.*, 2, 175–191, 1992.

Cazillis, M. et al., VIP and PACAP induce selective neuronal differentiation of mouse embryonic stem cells, *Eur. J. Neurosci.*, 19, 798–808, 2004.

Ceraudo, E. et al., The vasoactive intestinal peptide (VIP) α-helix up to C terminus interacts with the N-terminal ectodomain of the human VIP/pituitary adenylate cyclase-activating peptide 1 receptor: photoaffinity, molecular modeling, and dynamics, *Mol. Endocrinol.*, 22, 147–155, 2008.

Chang, E. et al., Generation of a human chromosome 18-specific YAC clone collection and mapping of 55 unique YACs by FISH and fingerprinting, *Genomics*, 17, 393–402, 1993.

Chatterjee, T.K., Sharma, R.V., and Fisher, R.A., Molecular cloning of a novel variant of the pituitary adenylate cyclase-activating polypeptide (PACAP) receptor that stimulates calcium influx by activation of L-type calcium channels, *J. Biol. Chem.*, 271, 32226–32232, 1996.

Chen, D. et al., Pituitary adenylate cyclase-activating peptide: A pivotal modulator of glutamatergic regulation of the suprachiasmatic circadian clock, *Proc. Natl Acad. Sci. USA*, 96, 13468–13473, 1999.

Chen, W. et al., Stimulatory action of pituitary adenylate cyclase-activating polypeptide (PACAP) on thyroid gland, *Biochem. Biophys. Res. Commun.*, 194, 923–929, 1993.

Chen, Y. et al., Neuroprotection by endogenous exogenous PACAP following stroke, *Regul. Pept.*, 137, 4–19, 2006.

Chong, S. et al., Utilizing the C-terminal cleavage activity of a protein splicing element to purify recombinant proteins in a single chromatographic step, *Nucleic Acids Res.*, 26, 5109–5115, 1998.

Chung, C.Y. et al., Cell type-specific gene expression of midbrain dopaminergic neurons reveals molecules involved in their vulnerability and protection, *Hum. Mol. Genet.*, 14, 1709–1725, 2005.

Conconi, M.R., Spinazzi, R., and Nussdorfer, G.G., Endogenous ligands of PACAP/VIP receptors in the autocrine–paracrine regulation of the adrenal gland, *Int. Rev. Cytol.*, 249, 1–51, 2006.

Dangoor, D. et al., Novel glycosylated VIP analogs: synthesis, biological activity, and metabolic stability, *J. Pept. Sci.*, 14, 321–328, 2008.

Dautzenberg, F.M., Mevenkamp, G., Wille, S., and Hauger, R.L., N-terminal splice variants of the type I PACAP receptor: isolation, characterization, and ligand binding/selectivity determinants, *J. Neuroendocrinol.*, 11, 941–949, 1999.

Dejda, A., Sokolowska, P., and Nowak, J.Z., Neuroprotective potential of three neuropeptides PACAP, VIP, and PHI, *Pharmacol. Rep.*, 57, 307–320, 2005.

Delgado, M. et al., Vasoactive intestinal peptide (VIP) and pituitary adenylate cyclase-activation polypeptide (PACAP) protect mice from lethal endotoxemia through the inhibition of TNF-alpha and IL-6, *J. Immunol.*, 162, 1200–1205, 1999a.

Delgado, M., Munoz-Elias, E.J., Gomariz, R.P., and Ganea, D., VIP and PACAP inhibit IL-12 production in LPS-stimulated macrophages. Subsequent effect on IFNγ synthesis by T cells, *J. Neuroimmunol.*, 96, 167–181, 1999b.

Delgado, M. et al., Vasoactive intestinal peptide and pituitary adenylate cyclase-activating polypeptide inhibit endotoxin-induced TNF-α production by macrophages: *in vitro* and *in vivo* studies, *J. Immunol.*, 162, 2358–2367, 1999c.

Delgado, M., Jonakait, G.M., and Ganea, D., Vasoactive intestinal peptide and pituitary adenylate cyclase-activating polypeptide inhibit chemokine production in activated microlia, *Glia*, 39, 148–161, 2002.

Delgado, M., Leceta, J., and Ganea, D., Vasoactive intestinal peptide and pituitary adenylated cyclase-activating polypeptide inhibit the production of inflammatory mediators by activated microglia, *J. Leukoc. Biol.*, 73, 155–164, 2003.

Delgado, M. et al., *In vivo* delivery of lentiviral vectors expressing vasoactive intestinal peptide complementary DNA as gene therapy for collagen-induced arthritis, *Arthritis Rheum.*, 58, 1026–1037, 2008.

DiCicco-Bloom, E. et al., Autocrine expression and ontogenetic function of the PACAP ligand/receptor system during sympathetic development, *Dev. Biol.*, 219, 197–213, 2000.

Dickinson, T., Fleetwood-Walker, S.M., Mitchell, R., and Lutz, E.M., Evidence for roles of vasoactive intestinal polypeptide (VIP) and pituitary adenylate cyclase activating polypeptide (PACAP) receptors in modulating the responses of rat dorsal horn neurons to sensory inputs, *Neuropeptides*, 31, 175–181, 1997.

Dickinson, T. and Fleetwood-Walker, S.M., VIP and PACAP: Very important in pain? *Trends Pharmacol. Sci.*, 20, 324–329, 1999.

Dickinson, T., Mitchell, R., Robberecht, P., and Fleetwood-Walker, S.M., The role of VIP/PACAP receptor subtypes in spinal somatosensory processing in rats with an experimental peripheral mononeuropathy, *Neuropharmacology*, 38, 167–180, 1999.

Dickson, L. et al., A systematic comparison of intracellular cyclic AMP and calcium singalling highlights complexities in human VPAC/PAC receptor pharmacology, *Neuropharmacology*, 51, 1086–1098, 2006.

Dow, R.C., Bennie, J., and Fink, G., Pituitary adenylate cyclase-activating peptide-38 (PACAP)-38 is released into hypophysial portal blood in the normal male and female rat, *J. Endocrinol.*, 142, R1–R4, 1994.

Eipper, B.A., Stoffers, D.A., and Mains, R.E., The biosynthesis of neuropeptides: Peptide alpha-amidation, *Annu. Rev. Neurosci.*, 15, 57–85, 1992.

Fahrenkrug, J. and Hannibal, J., Neurotransmitters co-existing with VIP or PACAP, *Peptides*, 25, 393–401, 2004.

Fahrenkrug, J., Hannibal, J., Honnoré, B., and Vorum, H., Altered calmodulin response to light in the suprachiasmatic nucleus of PAC1 receptor knockout mice revealed by proteomic analysis, *J. Mol. Neurosci.*, 25, 251–258, 2005.

Falluel-Morel, M. et al., The neuropeptide pituitary adenylate cyclase-activating polypeptide exerts anti-apoptotic and differentiating effects during neurogenesis: Focus on cerebellar granule neurones and embryonic stem cells, *J. Neuroendocrinol.*, 19, 321–327, 2007.

Feany, M.B. and Quinn, W.G., A neuropeptide gene defined by the *Drosophila* memory mutant *amnesiac*, *Science*, 268, 869–873, 1995.

Fishbein, V.A. et al., Chimeric VIP-PACAP analogue but not VIP pseudopeptides function as VIP receptor antagonists, *Peptides*, 15, 95–100, 1994.

Fradinger, E.A. and Sherwood, N.M., Characterization of the gene encoding both growth hormone-releasing hormone (GRF) and pituitary adenylate cyclase-activating polypeptide (PACAP) in the zebrafish, *Mol. Cell. Endocrinol.*, 165, 211–219, 2000.

Frechilla, D. et al., BDNF mediates the neuroprotective effecto of PACAP-38 on rat cortical neurons, *NeuroReport*, 12, 919–923, 2001.

Freson, K. et al., PACAP and its receptor VPAC1 regulate megakaryocyte maturation: Therapeutic implications, *Blood*, 111, 1885–1893, 2008.

Fukuhara, C. et al., Pituitary adenylate cyclase-activating polypeptide rhythm in the rat pineal gland, *Neurosci. Lett.*, 241, 115–118, 1998.

Ganea, D., Rodriguez, R., and Delgado, M., Vasoactive intestinal peptide and pituitary adenylate cyclase-activating polypeptide: Players in innate and adaptive immunity, *Cell Mol. Biol.* (*Noisy-le-grand*), 49, 127–142, 2003.

Garry, E.M. et al., Activation of p38 and p42/44 MAP kinase in neuropathic pain: Involvement of $VPAC_2$ and NK_2 receptors and mediation by spinal glia, *Mol. Cell. Neurosci.*, 30, 523–537, 2005.

Germano, P.M. et al., Characterization of the pharmacology, signal transduction, and internalization of the fluorescent PACAP ligand, fluor-PACAP, on NIH/3T3 cells expressing PAC1, *Peptides*, 22, 861–866, 2001.

Ghzili, H. et al., Role of PACAP in the physiology and pathology of the sympathoadrenal system, *Front. Neuroendocrinol.*, 29, 128–141, 2008.

Girard, B.A. et al., Noncompensation in peptide/receptor gene expression and distinct behavioral phenotypes in VIP- and PACAP-deficient mice, *J. Neurochem.*, 99, 499–513, 2006.

Goetzl, E.J., Sreedharan, S.P., and Turck, C.W., Structurally distinctive vasoactive intestinal peptides from rat basophilic leukemia cells, *J. Biol. Chem.*, 263, 9083–9086, 1988.

Goetzl, E.J. et al., Unique patterrn of cleavage of vasoactive intestinal peptide by human lymphocytes, *Immunology*, 66, 554–548, 1989.

Gonzalez, B.J. et al., Pituitary adenylate cyclase-activating polypeptide promotes cell survival and neurite outgrowth in rat cerebellar neuroblasts, *Neuroscience*, 78, 419–430, 1997.

Gourlet, P. et al., The activation of adenylate cyclase by pituitary adenlyate cyclase activating polypeptide (PACAP) via helodermin-preferring VIP receptors in human SUP-T1 lymphoblastic membranes, *Biochim. Biophys. Acta*, 1066, 245–251, 1991.

Gourlet, P. et al., Fragments of pituitary adenylate cyclase activating polypeptide discriminate between type I and II recombinant receptors, *Eur. J. Pharmacol.*, 287, 7–11, 1995.

Gourlet, P. et al., Addition of the (28–38) peptide sequence of PACAP to the VIP sequence modifies peptide selectivity and efficacy, *Int. J. Pept. Protein Res.*, 48, 391–396, 1996a.

Gourlet, P. et al., C-terminally shortened pituitary adenylate cyclase-activating peptides (PACAP) discriminate PACAP I, PACAP II–VIP1, and PACAP II–VIP2 recombinant receptors, *Regul. Pept.*, 62, 125–130, 1996b.

Gourlet, P. et al., *In vitro* properties of a high affinity selective antagonist of the VIP1 receptor, *Peptides*, 18, 1555–1560, 1997a.

Gourlet, P., Vandermeers, A., Robberecht, P., and Deschodt-Lanckman, M., Vasoactive intestinal peptide (VIP) and pituitary adenylate cyclase-activating peptide (PACAP-27, but not PACAP-38) degradation by the neutral endopeptidase EC 3.4.24.11, *Biochem. Pharmacol.*, 54, 509–515, 1997b.

Gourlet, P. et al., Development of high affinity selective VIP1 receptor agonists, *Peptides*, 18, 1539–1545, 1997c.

Gourlet, P. et al., The long-acting vasoactive intestinal polypeptide agonist RO25-1553 is highly selective of the VIP_2 receptor subclass, *Peptides*, 18, 403–408, 1997d.

Gourlet, P. et al., Interaction of lipophilic VIP derivatives with recombinant VIP1/PACAP and VIP2/PACAP receptors, *Eur. J. Pharmacol.*, 354, 105–111, 1998a.

Gourlet, P. et al., Analogues of VIP, helodermin, and PACAP discriminate between rat and human VIP_1 and VIP_2 receptors, *Ann. NY Acad. Sci.*, 865, 247–252, 1998b.

Gourlet, P. et al., Vasoactive intestinal peptide modification at position 22 allows discrimination between receptor subtypes, *Eur. J. Pharmacol.*, 348, 95–99, 1998c.

Gozes, I. et al., Superactive lipophilic peptides discriminate multiple vasoactive intestinal peptide receptors, *J. Pharm. Exp. Therap.*, 273, 161–167, 1995.

Gozes, I. and Brenneman, D.E., A new concept in the pharmacology of neuroprotection, *J. Mol. Neurosci.*, 14, 61–68, 2000.

Gray, S.L., Cummings, K.J., Jirik, F.R., and Sherwood, N.M., Targeted disruption of the pituitary adenylate cyclase-activating polypeptide gene results in early postnatal death associated with dysfunction of lipid and carbohydrate metabolism, *Mol. Endocrinol.*, 15, 1739–1747, 2001.

Green, B.D., Irwin, N., and Flatt, P.R., Pituitary adenylate cyclase-activating peptide (PACAP): assessment of dipeptidyl peptidase IV degradation, insulin-releasing activity, and antidiabetic potential, *Peptides*, 27, 1349–1358, 2006.

Grinninger, C. et al., A natural variant Type II G protein-coupled receptor for vasoactive intestinal peptide with altered function, *J. Biol. Chem.*, 279, 40259–40262, 2004.

Groneberg, D.A., Rabe, K.F., and Fischer, A., Novel concepts of neuropeptide-based drug therapy: Vasoactive intestinal peptide and its receptors, *Eur. J. Pharmacol.*, 533, 182–194, 2006.

Grumolato, L. et al., Microarray and suppression subtractive hybridizatin analyses of gene expression in pheochromocytoma cells reveal pleiotropic effects of pituitary adenylate cyclase-activating polypeptide on cell proliferation, survival, and adhesion, *Endocrinology*, 144, 2368–2379, 2003.

Hamelink, C. et al., Pituitary adenylate cyclase-activating polypeptide is a sympathoadrenal neurotransmitter involved in catecholamine regulation and glucohomeostasis, *Proc. Natl Acad. Sci. USA*, 99, 461–466, 2002.

Hannibal, J. and Fahrenkrug, J., Target areas innervated by PACAP-immunoreactive retinal ganglion cells, *Cell Tissue Res.*, 316, 99–113, 2004.

Hansel, D.E., May, V., Eipper, B.A., and Ronnett, G.V., Pituitary adenylyl cyclase-activating peptides and α-amidation in olfactory neurogenesis and neuronal survival *in vitro*, *J. Neurosci.*, 21, 4625–4636, 2001.

Harmar, A.J. et al., International Union of Pharmacology. XVIII. Nomenclature of receptors for vasoactive intestinal peptide and pituitary adenylate cyclase-activating polypeptide, *Pharmacol. Rev.*, 50, 265–270, 1998.

Harmar, A.J., Family-B G-protein-coupled receptors, *Genome Biol.*, 2, 3013.3011–3013.3010, 2001.

Hart, G.R., Gowing, H., and Burrin, J.M., Effects of a novel hypothalamic peptide, pituitary adenlyate cyclase-activating polypeptide, on pituitary hormone release in rats, *J. Endocrinol.*, 134, 33–41, 1992.

Hashimoto, H. et al., Altered psychomotor behaviors in mice lacking pituitary adenylate cyclase-activating polypeptide (PACAP), *Proc. Natl Acad. Sci. USA*, 98, 13355–13360, 2001.

Hashimoto, H., Shintani, N., and Baba, A., Higher brain functions of PACAP and a homologous Drosophila memory gene amnesiac: insights from knockouts and mutants, *Biochem. Biophys. Res. Commun.*, 297, 427–432, 2002.

Hashimoto, H., Shintani, N., and Baba, A., New insights into the central PACAPergic system from phenotypes in PACAP- and PACAP receptor-knockout mice, *Ann. NY Acad. Sci.*, 1070, 75–89, 2006.

Hashimoto, R. et al., Pituitary adenylate cyclase-activating polypeptide is associated with schizophrenia, *Mol. Psychiatry*, 12, 1026–1032, 2007.

Hezareh, M., Schlegel, W., and Rawlings, S.R., PACAP and VIP stimulate Ca^{2+} oscillations in rat gonadotrophs through the PACAP/VIP type 1 receptor (PVR1) linked to a pertussis toxin-insensitive G-protein and the activation of phospholipase C-beta, *J. Neuroendocrinol.*, 8, 367–374, 1996.

Hirose, M. et al., Differential expression of mRNAs for PACAP and its receptors during neural differentiation of embryonic stem cells, *Regul. Pept.*, 126, 109–113, 2005.

Hirose, M. et al., Inhibition of self-renewal and induction of neural differentiation by PACAP in neural progenitor cells, *Ann. NY Acad. Sci.*, 1070, 342–347, 2006.

Hook, V. et al., Proteases for processing proneuropeptides into peptide neurotransmitters and hormones, *Annu. Rev. Pharmacol. Toxicol.*, 48, 393–423, 2008.

Hosoya, M. et al., Molecular cloning and functional expression of rat cDNAs encoding the receptor for pituitary adenylate cyclase activating polypeptide (PACAP), *Biochem. Biophys. Res. Commun.*, 194, 133–143, 1993.

Hu, Z. et al., Molecular cloning of growth hormone-releasing hormone/pituitary adenylyl cyclase-activating polypeptide in the frog *Xenopus laevis*: brain distribution and regulation after castration, *Endocrinology*, 141, 3366–3376, 2000.

Huang, M.C. et al., Differential signaling of T cell generation of IL-4 by wild-type and short-deletion variant of type 2 G protein-coupled receptor for vasoactive intestinal peptide (VPAC2), *J. Immunol.*, 176, 6640–6646, 2006.

Igarashi, H. et al., Elucidation of vasoactive intestinal peptide pharmacophore for VPAC(1) receptors in human, rat, and guinea pig, *J. Pharm. Exp. Therap.*, 301, 37–50, 2002a.

Igarashi, H. et al., Elucidation of the vasoactive intestinal peptide pharmacophore for $VPAC_2$ receptors in human and rat and comparison to the pharmacophore for $VPAC_1$ receptors, *J. Pharm. Exp. Therap.*, 303, 445–460, 2002b.

Igarashi, H. et al., Development of simplified vasoactive intestinal peptide analogs with receptor selectivity and stability for human vasoactive intestinal peptide/pituitary adenylate cyclase-activating polypeptide receptors, *J. Pharm. Exp. Therap.*, 315, 370–381, 2005.

Inooka, H. et al., Pituitary adenylate cyclase activating polypeptide (PACAP) with 27 residues. Conformation determined by 1H NMR and CD spectroscopies and distance geometry in 25% methanol solution, *Int. J. Pept. Protein Res.*, 40, 456–464, 1992.

Inooka, H. et al., Conformation of a peptide ligand bound to its G-protein coupled receptor, *Nat. Struct. Biol.*, 8, 161–165, 2001.

Isaac, E.R. and Sherwood, N.M., Pituitary adenylate cyclase-activating polypeptide (PACAP) is important for embryo implantation in mice, *Mol. Cell. Endocrinol.*, 280, 13–19, 2008.

Ishihara, T. et al., Functional expression and tissue distribution of a novel receptor for vasoactive intestinal polypeptide, *Neuron*, 8, 811–819, 1992.

Jabrane-Ferrat, N., Pollock, A.S., and Goetzl, E.J., Inhibition of expression of the type I G protein-coupled receptor for vasoactive intestinal peptide (VPAC1) by hammerhead ribozymes, *Biochemistry*, 39, 9771–9777, 2000.

Jamen, F. et al., Pituitary adenylate cyclase-activating polypeptide receptors mediating insulin secretion in rodent pancreatic islets are coupled to adenylate cyclase but not to PLC, *Endocrinology*, 143, 1253–1259, 2002.

Jaworski, D.M. and Proctor, M.D., Developmental regulation of pituitary adenylate cyclase-activating polypeptide and PAC_1 receptor mRNA expression in the rat central nervous system, *Dev. Brain Res.*, 120, 27–39, 2000.

Jiang, Y., Li, W.S., Xie, J., and Lin, H.R., Sequence and expression of a cDNA encoding both pituitary adenylate cyclase activating polypeptide and growth hormone-releasing hormone in a grouper (*Epinephelus coioides*), *Sheng Wu Hua Xue Yu Sheng Wu Wu Li Xue Bao (Shanghai)*, 35, 864–872, 2003.

Jongsma, H. et al., Markedly reduced chronic nociceptive response in mice lacking the PAC_1 receptor, *NeuroReport*, 12, 2215–2219, 2001.

Juarranz, M.G. et al., Different vasoactive intestinal polypeptide receptor domains are involved in the selective recognition of two $VPAC_2$-selective ligands, *Mol. Pharmacol.*, 56, 1280–1287, 1999.

Kato, H. et al., Pituitary adenylate cyclase-activating polypeptide (PACAP) ameliorates experimental autoimmune encephalomyelitis by suppressing the functions of antigen presenting cells, *Mult. Scler.*, 10, 651–659, 2004.

Kieffer, T.J. and Habener, J.F., The glucagon-like peptides, *Endocrinol. Rev.*, 20, 876–913, 1999.

Kim, W.-K. et al., Vasoactive intestinal peptide and pituitary adeylyl cyclase-activaing polypeptide inhibit tumor necrosis factor-α production in injured spinal cord and in activated microglia via a cAMP-dependent pathway, *J. Neurosci.*, 20, 3622–3630, 2000.

Kim, Y. et al., Pituitary adenylate cyclase-activating polypeptide 27 is a functional ligand for formyl peptide receptor-like 1, *J. Immunol.*, 176, 2969–2975, 2006.

Kimura, C. et al., A novel peptide which stimulates adenylate cyclase: molecular cloning and characterization of the ovine and human cDNAs, *Biochem. Biophys. Res. Commun*, 166, 81–89, 1990.

Kimura, H., Kawatani, M., Ito, E., and Ishikawa, K., PACAP facilitate the nerve regeneration factors in the facial nerve injury, *Regul. Pept.*, 123, 135–138, 2004.

Kononen, J. et al., Stage-specific expression of pituitary adenylate cyclase-activating polyeptide (PACAP) mRNA in the rat seminiferous tubules, *Endocrinology*, 135, 2291–2294, 1994.

Krishnadas, A., Önyüksel, H., and Rubinstein, I., Interactions of VIP, secretin, and PACAP(1-38) with phospholipids: a biological paradox revisited, *Curr. Pharm. Des.*, 9, 1005–1012, 2003.

Köves, K., Arimura, A., Görcs, T.G., and Somogyvári-Vigh, A., Comparative distribution of immunoreactive pituitary adenylate cyclase activating polypeptide and vasoactive intestinal polypeptide in rat forebrain, *Neuroendocrinol.*, 54, 159–169, 1991.

Laburthe, M., Couvineau, A., and Tan, V., Class II G protein-coupled receptors for VIP and PACAP: structure, models of activation, and pharmacology, *Peptides*, 28, 1631–1639, 2007.

Lambeir, A.M. et al., Kinetic study of the processing by dipeptidyl-peptidase IV/CD26 of neuropeptides involved in pancreatic insulin secretion, *FEBS Lett.*, 507, 327–330, 2001.

Lamouche, S. and Yamaguchi, N., Role of PAC(1) receptor in adrenal catecholamine secretion induced by PACAP and VIP *in vivo*, *Am. J. Physiol. Regul. Integr. Comp. Physiol.*, 280, R510–R518, 2001.

Langer, I., Perret, J., Vertongen, P., Waelbroeck, M., and Robberecht, P., Vasoactive intestinal peptide (VIP) stimulates $[Ca^{2+}]_i$ and cyclic AMP in CHO cells expressing Galpha16, *Cell Calcium*, 30, 229–234, 2001.

Langer, I., Gregoire, F., De Neef, P., Vertongen, P., and Robberecht, P., Hexanoylation of a $VPAC_2$ receptor-preferring ligand markedly its selectivity and potency, *Peptides*, 25, 275–278, 2004.

Lara-Marquez, M., O'Dorisio, M., O'Dorisio, T., Shah, M., and Karacay, B., Selective gene expression and activation-dependent regulation of vasoactive intestinal peptide receptor type 1 and type 2 in human T cells, *J. Immunol.*, 166, 2522–2530, 2001.

Lembo, P.M. et al., The receptor for the orexigenic peptide melanin-concentrating hormone is a G-protein-coupled receptor, *Nat. Cell Biol.*, 1, 267–271, 1999.

Lerner, E.A. and Shoemaker, C.B., Maxadilan. Cloning and functional expression of the gene encoding this potent vasodilator peptide, *J. Biol. Chem.*, 267, 1062–1066, 1992.

Li, M., Shuto, Y., Somogyvári-Vigh, A., and Arimura, A., Prohormone convertases 1 and 2 process ProPACAP and generate matured, bioactive PACAP38 and PACAP27 in transfected rat pituitary GH4C1 cells, *Neuroendocrinology*, 69, 217–226, 1999.

Li, M., Mbikay, M., and Arimura, A., Pituitary adenylate cyclase-activating polypeptide precursor is processed solely by prohormone convertase 4 in the gonads, *Endocrinology*, 141, 3723–3730, 2000.

Li, M., Maderdrut, J.L., Lertora, J.J., Arimura, A., and Batuman, V., Renoprotection by pituitary adenylate cyclase-activating polypeptide in multiple myeloma and other kidney diseases, *Regul. Pept.*, 145, 24–32, 2008.

Lutz, E.M. et al., The VIP_2 receptor: Molecular characterisation of a cDNA encoding a novel receptor for vasoactive intestinal peptide, *FEBS Lett.*, 334, 3–8, 1993.

Lutz, E.M. et al., Domains determining agonist selectivity in chimaeric VIP_2 ($VPAC_2$)/PACAP (PAC_1) receptors, *Br. J. Pharmacol.*, 128, 934–940, 1999a.

Lutz, E.M. et al., Structure of the human VIPR2 gene for vasoactive intestinal peptide receptor type 2, *FEBS Lett.*, 458, 197–203, 1999b.

Lutz, E.M. et al., Characterization of novel splice variants of the PAC_1 receptor in human neuroblastoma cells: Consequences for signaling by VIP and PACAP, *Mol. Cell. Neurosci.*, 31, 193–201, 2006.

Lutz-Bucher, B., Monnier, D., and Koch, B., Evidence for the presence of receptors for pituitary adenylate cyclase-activating polypeptide in the neurohypophsis that are positively coupled to cyclic AMP formation and nuerohypophyseal hormone secretion, *Neuroendocrinol.*, 64, 153–161, 1996.

Mabuchi, T. et al., Pituitary adenylate cyclase-activating polypeptide is required for the development of spinal sensitization and induction of neuropathic pain, *J. Neurosci.*, 24, 7283–7291, 2004.

MacKenzie, C.J., Lutz, E.M., McCulloch, D.A., Mitchell, R., and Harmar, A.J., Phospholipase C activation by VIP1 and VIP2 receptors expressed in COS 7 cells involves a pertussis toxin-sensitive mechanism, *Ann. NY Acad. Sci.*, 805, 579–584, 1996.

MacKenzie, C.J. et al., Mechanisms of phospholipase C activation by the vasoactive intestinal polypeptide/pituitary adenylate cyclase-activating polypeptide type 2 receptor, *Endocrinology*, 142, 1209–1217, 2001.

Marotti, L.A.J., Jayawickreme, C.K., and Lerner, M.R., Functional characterization of a receptor for vasoactive-intestinal-peptide-related peptides in cultured dermal melanophores from *Xenopus laevis*, *Pigment Cell Res.*, 12, 89–97, 1999.

Martinez, C. et al., Vasoactive intestinal peptide and pituitary adenylate cyclase-activating polypeptide modulate endotoxin-induced IL-6 production by murine peritoneal macrophages, *J. Leukoc. Biol.*, 63, 591–601, 1998.

Martinez, C. et al., Anti-inflammatory role in septic shock of pituitary adenylate cyclase-activating polypeptide receptor, *Proc. Natl Acad. Sci. USA*, 99, 1053–1058, 2002.

Martinez, C. et al., Analysis of the role of the PAC1 receptor in neutrophil recruitment, acute-phase response, and nitric oxide production in septic shock, *J. Leukoc. Biol.*, 77, 729–738, 2005.

Martinez, C. et al., PAC1 receptor: Emerging target for septic shock therapy, *Ann. NY Acad. Sci.*, 1070, 405–410, 2006.

Matsuda, K. et al., Isolation and structural characterization of pituitary adenylate cyclase activating polypeptide (PACAP)-like peptide from the brain of a Teleost, Stargazer, *Uranoscopus japonicus*, *Peptides*, 18, 723–727, 1997.

Matsuda, K. et al., Purification and primary structure of pituitary adenylate cyclase activating polypeptide (PACAP) from the brain of an elasmobranch, stingray, *Dasyatis ajahei*, *Peptides*, 19, 1489–1495, 1998.

May, V. and Braas, K.M., Pituitary adenylate cyclase-activating polypeptide (PACAP) regulation of sympathetic neuron neuropeptide Y and catecholamine expression, *J. Neurochem.*, 65, 978–987, 1995.

Mazzocchi G. et al., Pituitary adenlyate cyclase-activating polypeptide and PACAP receptor expression and function in the rat adrenal gland, *Int. J. Mol. Med.*, 9, 233–243, 2002a.

Mazzocchi, G. et al., Expression and function of vasoactive intestinal peptide, pituitary adenylate cyclase-activating polypeptide, and their receptors in the human adrenal gland, *J. Clin. Endocrinol. Metab.*, 87, 2575–2580, 2002b.

McCulloch, D. et al., ADP-ribosylation factor-dependent phospholipase D activation by VPAC receptors and a PAC(1) receptor splice variant, *Mol. Pharmacol.*, 59, 1523–1532, 2001.

McCulloch, D.A. et al., Additional signals from VPAC/PAC family receptors, *Biochem. Soc. Trans.*, 30, 441–446, 2002.

McRory, J. and Sherwood, N.M., Two protochordate genes encode pituitary adenylate cyclase-activating polypeptide and related family members, *Endocrinology*, 138, 2380–2390, 1997.

McRory, J.E., Parker, D.B., Ngamvongchon, S., and Sherwood, N.M., Sequence and expression of cDNA for pituitary adenylate cyclase activating polypeptide (PACAP) and growth hormone-releasing hormone (GHRH)-like peptide in catfish, *Mol. Cell Endocrinol.*, 108, 169–177, 1995.

McRory, J.E., Parker, R.L., and Sherwood, N.M., Expression and alternative processing of a chicken gene encoding both growth hormone-releasing hormone and pituitary adenylate cyclase-activating polypeptide, *DNA Cell Biol.*, 16, 95–102, 1997.

Mercer, A. et al., PACAP promotes neural stem cell proliferation in adult mouse brain, *J. Neurosci. Res.*, 76, 205–215, 2004.

Miller, A.L. et al., Functional splice variants of the type II G protein-coupled receptor (VPAC2) for vasoactive intestinal peptide in mouse and human lymphocytes, *Ann. NY Acad. Sci.*, 1070, 422–426, 2006.

Mitchell, R. et al., ADP-ribosylation factor-dependent phospholipase D activation by the M_3 muscarinic receptor, *J. Biol. Chem.*, 278, 33818–33830, 2003.

Miyata, A. et al., Isolation of a novel 38 residue-hypothalmic polypeptide which stimulates adenylate cyclase in pituitary cells, *Biochem. Biophys. Res. Commun.*, 164, 567–574, 1989.

Miyata, A. et al., Isolation of a neuropeptide corresponding to the N-terminal 27 residues of the pituitary adenlyate cyclase activating polypeptide with 38 residues (PACAP38), *Biochem. Biophys. Res. Commun.*, 170, 643–648, 1990.

Moller, K. et al., The effects of axotomy and preganglionic denervation on the expression of pituitary adenylate cyclase activating peptide (PACAP), galanin and PACAP type 1 receptors in the rat superior cervical ganglion, *Brain Res.*, 775, 166–182, 1997.

Monaghan, T.K. et al., Neurotrophic actions of PACAP-38 and LIF on human neuroblastoma SH-SY5Y cells, *J. Mol. Neurosci.*, 36, 45–56, 2008.

Moody, T.W., Jensen, R.T., Fridkin, M., and Gozes, I., (N-stearyl, norleucine17)VIP hybrid is a broad spectrum vasoactive intestinal peptide receptor antagonist, *J. Mol. Neurosci.*, 18, 29–35, 2002.

Moreno, D. et al., Development of selective agonists and antagonists for the human vasoactive intestinal polypeptide $VPAC_2$ receptor, *Peptides*, 21, 1543–1549, 2000.

Moro, O. and Lerner, E.A., Maxadilan, the vasodilator from sand flies, is a specific pituitary adenylate cyclase activating peptide type I receptor agonist, *J. Biol. Chem.*, 272, 966–970, 1997.

Moro, O. et al., Functional characterization of structural alterations in the sequence of the vasodilatory peptide maxadilan yields a pituitary adenylate cyclase-activating peptide type 1 receptor-specific antagonist, *J. Biol. Chem.*, 274, 23103–23110, 1999.

Morrow, J.A. et al., Molecular cloning and expression of a cDNA encoding a receptor for pituitary adenylate cyclase activating polypeptide (PACAP), *FEBS Lett.*, 329, 99–105, 1993.

Murthy, K.S., Teng, B.-Q., Jin, J.-G., and Makhlouf, G.M., G protein-dependent activation of smooth muscle eNOS via natriuretic peptide clearance receptor, *Am. J. Physiol. Cell Physiol.*, 275, C1409–C1416, 1998.

Murthy, K.S. et al., G_{i-1}/G_{i-2}-dependent signaling by single-transmembrane natriuretic peptide clearance receptor, *Am. J. Physiol. Gastrointest. Liver Physiol.*, 278, G974–G980, 2000.

Nakamachi, T. et al., Pituitary adenylate cyclase-activating polypeptide (PACAP) type 1 receptor (PAC1R) co-localizes with activity-dependent neuroprotective protein (ADNP) in the mouse brains, *Regul. Pept.*, 145, 88–95, 2008.

Nakata, M. and Yada, T., PACAP in the glucose and energy homeostasis: physiological role and therapeutic potential, *Curr. Pharm. Des.*, 13, 1105–1112, 2007.

Naruse, S., Suzuki, T., Ozaki, T., and Nokihara, K., Vasodilator effect of pituitary adenylate cyclase activating polypeptide (PACAP) on femoral blood flow in dogs, *Peptides*, 14, 505–510, 1993.

Nemetz, N. et al., Induction of colitis and rapid development of colorectal tumors in mice deficient in the neuropeptide PACAP, *Int. J. Cancer*, 122, 1803–1809, 2008.

Nicole, P. et al., Identification of key residues for interaction of vasoactive intestinal peptide with human $VPAC_1$ and $VPAC_2$ receptors and development of a highly selective $VPAC_1$ receptor agonist. Alanine scanning and molecular modeling of the peptide, *J. Biol. Chem.*, 275, 24003–24012, 2000.

Nicot, A. and DiCicco-Bloom, E., Regulation of neuroblast mitosis is determined by PACAP receptor isoform expression, *Proc. Natl Acad. Sci. USA*, 98, 4758–4763, 2001.

Nielsen, H.S., Hannibal, J., and Fahrenkrug, J., Expression of pituitary adenylate cyclase activating polypeptide (PACAP) in the postnatal and adult rat cerebellar cortex, *NeuroReport*, 9, 2639–2642, 1998.

O'Donnell, M. et al., Structure–activity studies of vasoactive intestinal polypeptide, *J. Biol. Chem.*, 266, 6389–6392, 1991.

O'Donnell, M. et al., Ro 25-1553: A novel, long-acting vasoactive intestinal peptide agonist. Part 1: *In vitro* and *in vivo* bronchodilator studies, *J. Pharmacol. Exp. Therap.*, 270, 1282–1288, 1994.

Odum, L., Petersen, L.J., Skov, P.S., and Ebskov, L.B., Pituitary adenylate cyclase activating polypeptide (PACAP) is localized in human dermal neurons and causes histamine release from skin mast cells, *Inflamm. Res.*, 47, 488–492, 1998.

Ogawa, T. et al., Monoaminergic neuronal development is not affected in PACAP-gene-deficient mice, *Regul. Pept.*, 126, 103–108, 2005.

Ogi, K. et al., Molecular cloning and characterization of cDNA for the precursor of rat pituitary adenylate cyclase activating polypeptide (PACAP), *Biochem. Biophys. Res. Commun.*, 173, 1271–1279, 1990.

Oh, D.S. et al., PACAP regulation of secretion and proliferation of pure populations of gastric ECL cells, *J. Mol. Neurosci.*, 26, 85–97, 2005.

Ohkubo, S. et al., Primary structure and characterization of the precursor to human pituitary adenylate cyclase activating polypeptide, *DNA Cell Biol.*, 11, 21–30, 1992.

Ohta, S., Gregg, C., and Weiss, S., Pituitary adenylate cyclase-activating polypeptide regulates forebrain neural stem cells and neurogenesis *in vitro* and *in vivo*, *J. Neurosci. Res.*, 84, 1177–1186, 2006.

Ohtaki, H. et al., Pituitary adenylate cyclase-activating polypeptide (PACAP) decreases ischemic neuronal cell death in association with IL-6, *Proc. Natl Acad. Sci. USA*, 103, 7488–7493, 2006.

Okazaki, K. et al., Expression of human pituitary adenylate cyclase activating polypeptide (PACAP) cDNA in CHO cells and characterization of the products, *FEBS Lett.*, 298, 49–56, 1992.

Onoue, S. et al., Long-acting analogue of vasoactive intestinal peptide, $[R^{15,20,21},L^{17}]$-VIP-GRR (IK312532), protects rat alveolar L2 cells from the cytotoxicity of cigarette smoke, *Regul. Pept.*, 123, 193–199, 2004a.

Onoue, S. et al., α-helical structure in the C-terminus of vasoactive intestinal peptide: Functional and structural consequences, *Eur. J. Pharmacol.*, 485, 307–316, 2004b.

Onoue, S., Misaka, S., and Yamada, S., Structure–activity relationship of vasoactive intestinal peptide (VIP): Potent agonists and potential clinical applications, *Naunyn-Schmeideberg's Arch. Pharmacol.*, 10.1002/jps.21329, 2008.

Otto, C. et al., Pulmonary hypertension and right heart failure in pituitary adenylate cyclase-activating polypeptide type 1 receptor-deficient mice, *Circulation*, 110, 3245–3251, 2004.

Pan, C.Q. et al., Engineering novel VPAC2-selective agonists with improved stability and glucose-lowering activity, *J. Pharm. Exp. Therap.*, 320, 900–906, 2007.

Pan, C.Q., Hamren, S., Roczniak, S., Tom, I., and DeRome, M., Generation of PEGylated VPAC1-selective antagonists that inhibit proliferation of a lung cancer cell line, *Peptides*, 29, 479–486, 2008.

Pantaloni, C. et al., Alternative splicing in the N-terminal extracellular domain of the pituitary adenylate cyclase-activating polypeptide (PACAP) receptor modulates receptor selectivity and relative potencies of PACAP-27 and PACAP-38 in phospholipase C activation, *J. Biol. Chem.*, 271, 22146–22151, 1996.

Parker, D.B. et al., Exon skipping in the gene encoding pituitary adenylate cyclase-activating polypeptide in salmon alters the expression of two hormones that stimulate growth hormone release, *Endocrinology*, 138, 1414–1423, 1997.

Pavelock, K.A. et al., Bone morphogenetic protein down-regulation of neuronal pituitary adenylate cyclase-activating polypeptide and reciprocal effects on vasoactive intestinal peptide expression, *J. Neurochem.*, 100, 603–616, 2007.

Pellegri, G., Magistretti, P.J., and Martin, J.L., VIP and PACAP potentiate the action of glutamate on BDNF expression in mouse cortical neurones, *Eur. J. Neurosci.*, 10, 272–280, 1998.

Pereira, P. et al., Maxadilan activates PAC1 receptors expressed in Xenopus laevis xelanophores, *Pigment Cell Res.*, 15, 461–466, 2002.

Pettersson, L.M., Dahlin, L.B., and Danielsen, N., Changes in expression of PACAP in rat sensory neurons in response to sciatic nerve compression, *Eur. J. Neurosci.*, 20, 1838–1848, 2004.

Piqueras, L., Taché, Y., and Martinez, V., Peripheral PACAP inhibits gastric acid secretion through somatostatin release in mice, *Br. J. Pharmacol.*, 142, 67–78, 2004.

Pisegna, J.R. and Wank, S.A., Molecular cloning and functional expression of the pituitary adenylate cyclase-activating polypeptide type I receptor, *Proc. Natl Acad. Sci. USA*, 90, 6345–6349, 1993.

Pisegna, J.R. and Wank, S.A., Cloning and characterization of the signal transduction of four splice variants of the human pituitary adenylate cyclase activating polypeptide receptor, *J. Biol. Chem.*, 271, 17267–17274, 1996.

Pohl, M. and Wank, S.A., Molecular cloning of the helodermin and exendin-4 cDNAs in the lizard, *J. Biol. Chem.*, 273, 9778–9784, 1998.

Pozo, D., VIP- and PACAP-mediated immunomodulation as prospective therapeutic tools, *Trends Mol. Med.*, 9, 211–217, 2003.

Propato-Mussafiri, R., Kanse, S.M., Ghatei, M.A., and Bloom, S.R., Pituitary adenylate cyclase-activating polypeptide releases 7B2, adrenocorticotrophin, growth hormone, and prolactin from the mouse and rat clonal pituitary cell lines AtT-20 and GH3, *J. Endocrinol.*, 132, 107–113, 1992.

Raufman, J.-P., Bioactive peptides from lizard venoms, *Regul. Pept.*, 61, 1–18, 1996.

Ravni, A. et al., The neurotrophic effects of PACAP in PC12 cells: Control by multiple transduction pathways, *J. Neurochem.*, 98, 321–329, 2006a.

Ravni, A. et al., Cycloheximide treatment to identify components of the transitional transcriptome in PACAP-induced PC12 differentiation, *J. Neurochem.*, 98, 1229–1241, 2006b.

Rawlings, S.R. et al., Differential expression of pituitary adenylate cyclase-activating polypeptide/vasoactive intestinal polypeptide receptor subtypes in clonal pituitary somatotrophs and gonadotrophs, *Endocrinology*, 136, 2088–2098, 1995.

Rawlings, S.R. and Hezareh, M., Pituitary adenylate cyclase-activating polypeptide (PACAP) and PACAP/vasoactive intestinal polypeptide receptors: Actions on the anterior pituitary gland, *Endocrinol. Rev.*, 17, 4–29, 1996.

Reddy, V.B., Iuga, A.O., Kounga, K., and Lerner, E.A., Functional analysis of recombinant mutants of maxadilan with a PAC1 receptor expressing melanophore cell line, *J. Biol. Chem.*, 281, 16197–16201, 2006.

Reglődi, D. et al., Pituitary adenlyate cyclase activating polypeptide protects dopaminergic neurons and improves behavioral deficits in a rat model of Parkinson's disease, *Behav. Brain Res.*, 151, 303–312, 2004.

Reglődi, D. et al., Comparative study of the effects of PACAP in young, aging, and castrated males in a rat model of Parkinson's disease, *Ann. NY Acad. Sci.*, 1070, 518–524, 2006.

Rekasi, Z. et al., Antagonistic actions of analogs related to growth hormone-releasing hormone (GHRH) on receptors for GHRH and vasoactive intestinal peptide on rat pituitary and pineal cells *in vitro*, *Proc. Natl Acad. Sci. USA*, 97, 1218–1223, 2000.

Reubi, J.C., *In vitro* evaluation of VIP/PACAP receptors in healthy and diseased human tissues. Clinical implications, *Ann. NY Acad. Sci.*, 921, 1–25, 2000.

Reubi, J.C. et al., Vasoactive intestinal peptide/pituitary adenylate cyclase-activating peptide receptor subtypes in human tumours and their tissues of origin, *Cancer Res.*, 60, 3105–3112, 2000.

Robberecht, P. et al., Structural requirements for the occupancy of pituitary adenylate-cyclase-activating-peptide (PACAP) receptors and adenylate cyclase activation in human neuroblastoma NB-OK-1 cell membranes. Discovery of PACAP(6-38) as a potent antagonist, *Eur. J. Biochem.*, 207, 239–246, 1992a.

Robberecht, P. et al., Receptor occupancy and adenylate cyclase activation in AR4-2J rat pancreatic acinar cell membranes by analogs of pituitary adenylate cyclase activating peptides amino-terminally shortened or modified at position 1, 2, 3, 20, or 21, *Mol. Pharmacol.*, 42, 347–355, 1992b.

Robberecht, P. and Waelbroeck, M., A critical view of the methods for characterization of the VIP/PACAP receptor subclasses, *Ann. NY Acad. Sci.*, 865, 157–163, 1998.

Samal, B. et al., Meta-analysis of microarray-derived data from PACAP-deficient adrenal gland *in vivo* and PACAP-treated chromaffin cells identifies distinct classes of PACAP-regulated genes, *Peptides*, 28, 1871–1882, 2007.

Schmidt, D.T. et al., The effect of the vasoactive intestinal polypeptide agonist Ro 25-1553 on induced tone in isolated human airways and pulmonary artery, *Naunyn-Schmeideberg's Arch. Pharmacol.*, 364, 314–320, 2001.

Sherwood, N.M., Krueckl, S.L., and McRory, J.E., The origin and function of the pituitary adenylate cyclase-activating polypeptide (PACAP)/glucagon superfamily, *Endocrin. Rev.*, 21, 619–670, 2000.

Sherwood, N.M. et al., Knocked down and out: PACAP in development, reproduction, and feeding, *Peptides*, 28, 1680–1687, 2007.

Sheward, W.J., Lutz, E.M., Copp, A.J., and Harmar, A.J., Expression of PACAP, and PACAP type 1 (PAC_1) receptor mRNA during development of the mouse embryo, *Dev. Brain Res.*, 109, 245–253, 1998.

Shioda, S. et al., PACAP protects hippocampal neurons against apoptosis: involvement of JNK/SAPK signaling pathway, *Ann. NY Acad. Sci.*, 865, 111–117, 1998.

Shivers, B.D., Görcs, T.J., Gottschall, P.E., and Arimura, A., Two high affinity binding sites for pituitary adenylate cyclase-activating polypeptide have different tissue distributions, *Endocrinology*, 128, 3055–3065, 1991.

Simonneaux, V., Ouichou, A., and Pévet, P., Pituitary adenylate cyclase-activating polypeptide (PACAP) stimulates melatonin synthesis from rat pineal gland, *Brain Res.*, 603, 148–152, 1993.

Simonneaux, V. et al., Pharmacological, molecular, and functional characterization of vasoactive intestinal polypeptide/pituitary adenylate cyclase-activating polypeptide receptors in the rat pineal gland, *Neuroscience*, 85, 887–896, 1998.

Skoglösa, Y. et al., Regulation of pituitary adenylate cyclase activating polypeptide and its receptor type 1 after traumatic brain injury: comparison with brain-derived neurotrophic factor and the induction of neuronal cell death, *Neuroscience*, 90, 235–247, 1999.

Spengler, D. et al., Differential signal transduction by 5 splice variants of the PACAP receptor, *Nature*, 365, 170–175, 1993.

Sreedharan, S.P., Patel, D.R., Huang, J.X., and Goetzl, E.J., Cloning and functional expression of a human neuroendocrine vasoactive intestinal peptide receptor, *Biochem. Biophys. Res. Commun.*, 193, 546–553, 1993.

Sreedharan, S.P., Huang, J.-X., Cheung, M.-C., and Goetzl, E.J., Structure, expression, and chromosomal localization of the type I human vasoactive intestinal peptide receptor gene, *Proc. Natl Acad. Sci. USA*, 92, 2939–2943, 1995.

Stumm, R. et al., Pituitary adenylate cyclase-activating polypeptide is up-regulated in cortical pyramidal cells after focal ischemia and protects neurons from mild hypoxic/ischemic damage, *J. Neurochem.*, 103, 1666–1681, 2007.

Suh, J. et al., PACAP is an anti-mitogenic signal in developing cerebral cortex, *Nature Neurosci.*, 4, 123–124, 2001.

Suk, K., Park, J.-H., and Lee, W.-H., Neuropeptide PACAP inhibits hypoxic activation of brain microglia: A protective mechanism against microglial neurotoxicity in ischemia, *Brain Res.*, 1026, 151–156, 2004.

Summers, M.A. et al., A lymphocyte-generated fragment of vasoactive intestinal peptide with VPAC1 agonist activity and VPAC2 antagonist effects, *J. Pharm. Exp. Therap.*, 306, 638–645, 2003.

Sun, C. et al., Solution structure and mutational analysis of pituitary adenylate cyclase-activating polypeptide binding to the extracellular domain of PAC1-R_s, *Proc. Natl Acad. Sci. USA*, 104, 7875–7880, 2007.

Svoboda, M. et al., Molecular cloning and functional characterization of a human VIP receptor from SUP-T1 lymphoblasts, *Biochem. Biophys. Res. Commun.*, 205, 1617–1624, 1994.

Takei, N. et al., Pituitary adenylate cyclase-activating polypeptide promotes the survival of basal forebrain cholinergic neurons *in vitro* and *in vivo*: comparison with effects of nerve growth factor, *Eur. J. Neurosci.*, 12, 2273–2280, 2000.

Tam, J.K.V., Lee L.T.O., and Chow, B.K.C., PACAP-related peptide (PRP)-molecular evolution and potential functions, *Peptides*, 28, 1920–1929, 2007.

Tams, J.W., Johnsen, A.H., and Fahrenkrug, J., Identification of pituitary adenylate cyclase-activating polypeptide 1–38-binding factor in human plasma, as ceruloplasmin, *Biochem. J.*, 341, 271–276, 1999.

Tams, J.W., Jorgensen, R.M., Holm, A., and Fahrenkrug, J., Creation of a selective antagonist and agonist of the rat $VPAC_1$ receptor using a combinatorial approach with vasoactive intestinal peptide 6–23 as template, *Mol. Pharmacol.*, 58, 1035–1041, 2000.

Tan, Y.V. et al., Peptide agonist docking in the N-terminal ectodomain of a class II G protein-coupled receptor, the VPAC1 receptor. Photoaffinity, NMR, and molecular modeling, *J. Biol. Chem.*, 281, 12792–12798, 2006.

Tatsuno, I. et al., Maxadilan specifically interacts with PAC1 receptor, which is a dominant form of PACAP/VIP family receptors in cultured rat cortical neurons, *Brain Res.*, 889, 138–148, 2001.

Teng, B.Q., Grider, J.R., and Murthy, K.S., Identification of a VIP-specific receptor in guinea pig tenia coli, *Am. J. Physiol. Gastrointest. Liver Physiol.*, 281, G718–G725, 2001.

Theriault, Y., Boulanger, Y., and St-Pierre, S., Structural determination of the vasoactive intestinal peptide by two-dimensional H-NMR spectroscopy, *Biopolymers*, 31, 459–464, 1991.

Tornoe, K. et al., PACAP 1–38 as neurotransmitter in the porcine antrum, *Regul. Pept.*, 101, 109–121, 2001.

Tsueshita, T., Gandhi, S., Önyüksel, H., and Rubinstein, I., Phospholipids modulate the biophysical properties and vasoactivity of PACAP-(1–38), *J. Appl. Physiol.*, 93, 1377–1383, 2002.

Tsutsumi, M. et al., A potent and highly selective $VPAC_2$ agonist enhances glucose-induced insulin release and glucose disposal. A potential therapy for Type 2 diabetes, *Diabetes*, 51, 1453–1460, 2002.

Uchida, D. et al., Prevention of ischemia-induced death of hippocampal neurons by pituitary adenylate cyclase activating polypeptide, *Brain Res.*, 736, 280–286, 1996.

Uchimura, D., Katafuchi, T., Hori, T., and Yanaihara, N., Facilitatory effects of pituitary adenylate cyclase activating polypeptide (PACAP) on neurons in the magnocellular portion of the rat hypothalamic paraventricular nucleus (PVN) *in vitro*, *J. Neuroendocrinol.*, 8, 137–143, 1996.

Valiante, S. et al., Molecular characterization and gene expression of the pituitary adenylate cyclase-activating polypeptide (PACAP) in the lizard brain, *Brain Res.*, 1127, 66–75, 2007.

Van Rampelbergh, J. et al., Properties of the PACAP I, PACAP II VIPI and chimeric N-terminal PACAP/VIP1 receptors: Evidence for multiple receptor states, *Mol. Pharmacol.*, 50, 1596–1605, 1996.

Vandermeers, A. et al., Antagonistic properties are shifted back to agonistic properties by further N-terminal shortening of pituitary adenylate-cyclase-activating peptides in human neuroblastoma NB-OK-1 cell membranes, *Eur. J. Biochem.*, 208, 815–819, 1992.

Vaudry, D. et al., Pituitary adenylate cyclase-activating polypeptide and its receptors: From structure to functions, *Pharmacol. Rev.*, 52, 269–324, 2000.

Vaudry, D. et al., Analysis of the PC12 cell transcriptome after differentiation with pituitary adenylate cyclase-activating polypeptide (PACAP), *J. Neurochem.*, 83, 1272–1284, 2002.

Vaudry, H. and Arimura, A., Eds., *Pituitary Adenylate Cyclase-Activating Polypeptide*, Vol. 20, Kluwer Academic Publishers, 2003.

Vertongen, P. et al., AcHis^1[D-Phe^2,K^{15},R^{16},L^{27}]VIP(3-7)/GRF(8-27)—a $VPAC_1$ receptor antagonist—is an inverse agonist on two constutively active truncated $VPAC_1$ receptors, *Peptides*, 25, 1943–1949, 2004.

Virgolini, I., Mack Forster Award Lecture. Receptor nuclear medicine: vasointestinal peptide and somatostatin receptor scintigraphy for diagnosis and treatment of tumour patients, *Eur. J. Clin. Invest.*, 27, 793–800, 1997.

Waelbroeck, M. et al., Interaction of growth hormone-releasing factor (GRF) and 14 GRF analogs with vasoactive intestinal peptide (VIP) receptors of rat pancreas. Discovery of (N-Ac-Tyr^1,D-Phe^2)-GRF(1-29)-NH_2 as a VIP antagonist, *Endocrinology*, 116, 2643–2649, 1985.

Wallin, H.J. et al., Exogenous NT-3 and NGF differentially modulate PACAP expression in adult sensory neurons, suggesting distinct roles in injury and inflammation, *Eur. J. Neurosci.*, 14, 267–282, 2001.

Wang, Y., Anderson, O.L., and Ge, W., Cloning, regulation of messenger ribonucleic acid expression, and function of a new isoform of pituitary adenylate cyclase-activating polypeptide in the zebrafish ovary, *Endocrinology*, 144, 4799–4810, 2003.

Warren, J.B. et al., Pituitary adenylate cyclase activating polypeptide is a potent vasodilator in humans, *J. Cardiovasc. Pharmacol.*, 20, 83–87, 1992.

Waschek, J.A., Richards, M.L., and Bravo, D.T., Differential expression of VIP/PACAP receptor genes in breast, intestinal, and pancreatic cell lines, *Cancer Lett.*, 92, 143–149, 1995.

Waschek, J.A. et al., Neural tube expression of pituitary adenylate cyclase-activating peptide (PACAP) and receptor: Potential role in patterning and neurogenesis, *Proc. Natl Acad. Sci. USA*, 95, 9602–9607, 1998.

Waschek, J.A., Multiple actions of pituitary adenylyl cyclase activating peptide in nervous system development and regeneration, *Dev. Neurosci.*, 24, 14–23, 2002.

Watanabe, J. et al., Localization, characterization, and function of pituitary adenylate cyclase-activating polypeptide during brain development, *Peptides*, 28, 1713–1719, 2007.

Wei, M. et al., Identification of key residues that cause differential gallbladder response to PACAP and VIP in the guinea pig, *Am. J. Physiol. Gastrointest. Liver Physiol.*, 292, G76–G83, 2007.

Wong, Y.H., G_i assays in transfected cells, *Methods Enzymol.*, 238, 81–94, 1994.

Wray, V., Kakoschke, C., Nokihara, K., and Naruse, S., Solution structure of pituitary adenylate cyclase activating polypeptide by nuclear magnetic resonance spectroscopy, *Biochemistry*, 32, 5832–5841, 1993.

Xia, M. et al., Novel cyclic peptide agonist of high potency and selectivity for the type II vasoactive intestinal peptide receptor, *J. Pharm. Exp. Therap.*, 281, 629–633, 1997.

Yamada, H., Watanabe, M., and Yada, T., Cytosolic Ca^{2+} responses to sub-picomolar and nanomolar PACAP in pancreatic beta-cells are mediated by $VPAC_2$ and PAC_1 receptors, *Regul. Pept.*, 123, 147–153, 2004.

Yamamoto, K. et al., Cloning and characterization of the mouse pituitary adenylate cyclase-activating polypeptide (PACAP) gene, *Gene*, 211, 63–69, 1998.

Yang, S. et al., Pituitary adenylate cyclase-activating polypeptide (PACAP) 38 and PACAP4–6 are neuroprotective through inhibition of NADPH oxidase: Potent regulators of microglia-mediated oxidative stress, *J. Pharm. Exp. Therap.*, 319, 595–603, 2006.

Yoo, S.J. et al., Molecular cloning and characterization of alternatively spliced transcripts of the turkey pituitary adenylate cyclase-activating polypeptide, *Gen. Comp. Endocrinol.*, 120, 326–335, 2000.

Yu, R.-J., Hong, A., Dai, Y., and Gao, Y., Intein-mediated rapid purification of recombinant human pituitary adenylate cyclase activating polypeptide, *Acta Biochim. Biophys. Sin.* (*Shanghai*), 36, 759–766, 2004.

Yu, R.-J. et al., Intein-mediated rapid purification and characterization of a novel recombinant agonist for $VPAC_2$, *Peptides*, 27, 1359–1366, 2006.

Yu, R.-J. et al., A novel recombinant, $VPAC_2$-selective agonist enhancing insulin release and glucose disposal, *Acta Pharmacol. Sin.*, 28, 526–533, 2007.

Yuhara, A. et al., PACAP has a neurotrophic effect on cultured basal forebrain cholinergic neurons from adult rats, *Brain Res. Dev. Brain Res.*, 131, 41–45, 2001.

Yung, S.L. et al., Generation of highly selective VPAC2 receptor agonists by high throughput mutagenesis of vasoactive intestinal peptide and pituitary adenylate cyclase-activating peptide, *J. Biol. Chem.*, 278, 10273–10281, 2003.

Zein, N.E., Corazza, F., and Sariban, E., The neuropeptide pituitary adenylate cyclase activating protein is a physiological activator of human monocytes, *Cell Signal*, 18, 162–173, 2006.

Zhang, Q. et al., Expression of pituitary adenylate cyclase-activating polypeptide in dorsal root ganglia following axotomy: Time course and coexistence, *Brain Res.*, 705, 149–158, 1995.

Zhang, Y.-Z. et al., Pituitary adenylate cyclase activating peptide expression in the rat dorsal root ganglia: Up-regulation after peripheral nerve injury, *Neuroscience*, 74, 1099–1110, 1996.

Zhong, Y. and Peña, L.A., A novel synaptic transmission mediated by a PACAP-like neuropeptide in Drosophila, *Neuron*, 14, 527–536, 1995.

Zhou, C.-J. et al., PACAP and its receptors exert pleiotropic effects in the nervous system by activating multiple signaling pathways, *Curr. Protein Pept. Sci.*, 3, 423–439, 2002.

Zhou, C.J. et al., Pituitary adenylate cyclase-activating polypeptide receptors during development: Expression in the rat embryo at primitive streak stage, *Neuroscience*, 93, 375–391, 1999.

Zhou, H., Huang, J., and Murthy, K.S., Molecular cloning and functional expression of a VIP-specific receptor, *Am. J. Physiol. Gastrointest. Liver Physiol.*, 291, G728–G734, 2006.

Zhu, L. et al., The role of dipeptidyl peptidase IV in the cleavage of glucagon family peptides: *In vivo* metabolism of pituitary adenylate cyclase activating polypeptide-(1–38), *J. Biol. Chem.*, 278, 22418–22423, 2003.

Zusev, M. and Gozes, I., Differential regulation of activity-dependent neuroprotective protein in rat astrocytes by VIP and PACAP, *Regul. Pept.*, 123, 33–41, 2004.

5 Opioid Peptides

Jane V. Aldrich

CONTENTS

Since the discovery of the opioid peptides in the mid-1970s, extensive research has been performed to characterize the activities of the endogenous opioid peptides and the more recently discovered related peptide nociceptin/orphanin FQ. Intensive investigation of the structure–activity relationships (SAR) of the different classes of peptides has resulted in a variety of peptides, with both agonist and antagonist activity, that are selective for different receptor types and are useful pharmacological tools to study the functions of the opioid and nociceptin/orphanin FQ receptors. Recently, peptide analogs that exhibit improved pharmacokinetic profiles have shown promising activity *in vivo* with a few of these peptides making it into clinical trials. This chapter provides an overview of the endogenous peptides, their mechanism of action and biological activities, their SAR, and progress toward their potential clinic applications.

5.1 INTRODUCTION

Two discoveries during the mid-1970s revolutionized our understanding of narcotic analgesics and how they produce their effects. The discovery of the first opioid peptides, the enkephalins (Hughes et al., 1975), opened up a whole new area of research on opioid ligands. The characterization and classification of multiple types of opioid receptors, the mu (μ) and kappa (κ) receptors (Gilbert and Martin, 1976; Martin et al., 1976), and following the discovery of the enkephalins the delta (δ) opioid receptors (Lord et al., 1977), formed the foundation of our current understanding of opioid pharmacology.

Since the discovery of the endogenous opioid peptides, thousands of analogs have been synthesized and evaluated for their opioid activity. A number of these peptides have become standard pharmacological tools used to study opioid receptors and have led to a more detailed understanding of the functions of these receptors. Recently, significant advances have been made in the development of opioid peptide analogs with improved pharmacokinetic profiles for potential clinical application.

The identification of the endogenous opioid peptides is a relatively recent development in the area of narcotic analgesics and opioid receptor ligands. The effects of opium, from which morphine is isolated, have been known for thousands of years, and the first analog of morphine (heroin) was synthesized in the late 1800s. The SARs of morphine and other nonpeptide opiates were studied extensively in the 1950s and 1960s (see Aldrich and Vigil-Cruz, 2003 for a review) in an effort to identify better analgesics. Since the identification of the opioid peptides, the exploration of the SARs of peptide and nonpeptide opioid receptor ligands have proceeded in parallel, with the results of the two areas providing complementary information on the functions of opioid receptors and their ligands.

5.2 SOURCE OF PEPTIDES

Opioid peptides have been isolated from both mammalian and amphibian sources. These peptides share a common structural motif, with an N-terminal Tyr separated from a Phe residue by one or two amino acids (Figure 5.1). The C-terminal sequences of these peptides vary substantially both in sequence and length (Table 5.1).

5.2.1 Mammalian Opioid Peptides

The search for the endogenous ligands for opioid receptors led first to the discovery of the enkephalins (Hughes et al., 1975), followed shortly thereafter by dynorphin A (Cox et al., 1975; Teschemacher et al., 1975) and β-endorphin (Li and Chung, 1976). These mammalian opioid peptides share a common N-terminal tetrapeptide sequence, Tyr–Gly–Gly–Phe, but differ in their C-terminal sequences (Table 5.1) and also in the opioid receptor type with which they preferentially interact. While the enkephalins exhibit some preference for interacting with δ opioid receptors, the dynorphins preferentially interact with κ opioid receptors (β-endorphin possesses high affinity for both μ

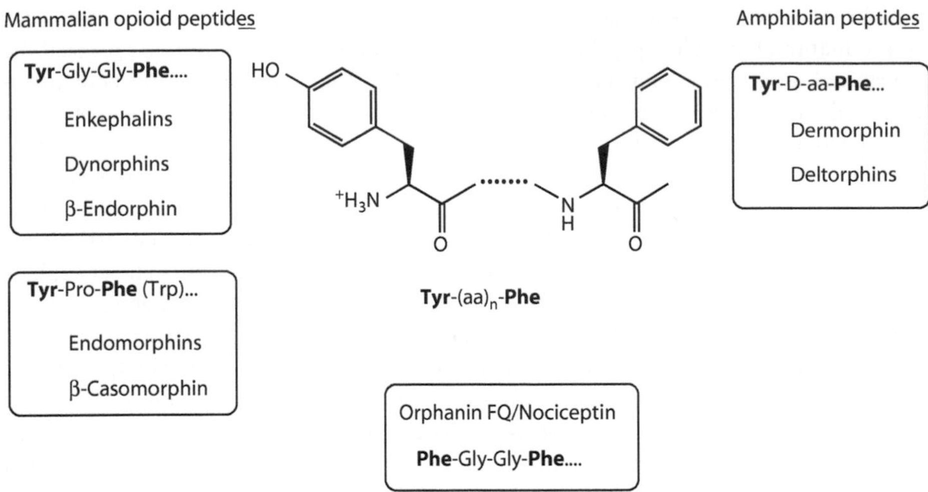

FIGURE 5.1 Common structural motif of opioid peptides and N-terminal sequences of different classes of opioid peptides.

TABLE 5.1
Endogenous Opioid Peptides

Peptide	Sequence
Mammalian Peptides	
Proenkephalin peptides	
Leu-enkephalin	Tyr–Gly–Gly–Phe–Leu
Met-enkephalin	Tyr–Gly–Gly–Phe–Met
Prodynorphin peptides	
Dynorphin A	Tyr–Gly–Gly–Phe–Leu–Arg–Arg–Ile–Arg–Pro–Lys–Leu–Lys–Trp–Asp–Asn–Gln
Dynorphin B	Tyr–Gly–Gly–Phe–Leu–Arg–Arg–Gln–Phe–Lys–Val–Val–Thr
α-Neoendorphin	Tyr–Gly–Gly–Phe–Leu–Arg–Lys–Tyr–Arg–Pro–Lys
POMC opioid peptides	
β_h-endorphin	Tyr–Gly–Gly–Phe–Met–Thr–Ser–Glu–Lys–Ser–Gln–Thr–Pro–Leu–Val–Thr–Leu–Phe–Lys–Asn–Ala–Ile–Ile–Lys–Asn–Ala–Tyr–Lys–Lys–Gly–Glu
Endomorphins	
Endomorphin-1	Tyr–Pro–Trp–PheNH_2
Endomorphin-2	Tyr–Pro–Phe–PheNH_2
Nociceptin/orphanin FQ	Phe–Gly–Gly–Phe–Thr–Gly–Ala–Arg–Lys–Ser–Ala–Arg–Lys–Leu–Ala–Asn–Gln
Amphibian Peptides	
Dermorphin	Tyr–D-Ala–Phe–Gly–Tyr–Pro–SerNH_2
Deltorphins	
Deltorphin (dermenkephalin, deltorphin A)	Tyr–D-Met–Phe–His–Leu–Met–AspNH_2
Deltorphin I (deltorphin C)	Tyr–D-Ala–Phe–Asp–Val–Val–GlyNH_2
Deltorphin II (deltorphin B)	Tyr–D-Ala–Phe–Glu–Val–Val–GlyNH_2

and δ opioid receptors). This led Goldstein to apply the "message–address" concept (Schwyzer, 1977) to the mammalian opioid peptides (Chavkin and Goldstein, 1981); in this proposal the common N-terminal tetrapeptide sequence constitutes the "message" responsible for activating opioid receptors, while the unique C-terminal sequences function as the "address" component to direct the peptides to particular opioid receptor types.

The classical mammalian opioid peptides are derived from three distinct precursor proteins, with multiple bioactive peptides derived from each protein (see Höllt, 1986; Herz et al., 1993a for reviews). β-Endorphin is derived from proopiomelanocortin (POMC), along with adrenocorticotropic hormone (ACTH), melanocyte-stimulating hormone (α-MSH), and β-lipotropin. The enkephalins are derived from proenkephalin A, along with several extended methionine-enkephalin derivatives. The dynorphins are derived from prodynorphin (also called proenkephalin B).

A new class of opioid peptides, the endomorphins, was reported in 1997 (Zadina et al., 1997). These peptides do not share the classical "message" sequence with other mammalian opioid peptides (Figure 5.1; Table 5.1). Also, in contrast to the classical mammalian peptides, the endomorphins show high selectivity for their preferential receptor, the μ opioid receptor. The precursor protein for the endomorphins has yet to be identified.

A peptide related to opioid peptides, the 17-residue peptide nociceptin/orphanin FQ (N/OFQ), was also isolated during the mid-1990s by two research groups following the identification of the opioid receptor-like (ORL-1) receptor (Mollereau et al., 1994), which is closely related to the opioid receptors (see Section 5.3.1). This peptide, which is closely related to dynorphin, was named nociceptin by one group because in the initial studies this peptide was reported to produce hyperalgesia (Meunier et al., 1995), and orphanin FQ by the other research group because it was the ligand for the orphan receptor (FQ are the N- and C-terminal residues, respectively, of the peptide; Reinscheid et al., 1995). The N-terminal sequence of N/OFQ is similar to classical opioid peptides (Figure 5.1; Table 5.1), but the presence of Phe rather than Tyr in position 1 is an important factor in the peptide's high selectivity for ORL-1 receptors over opioid receptors. The precursor protein for this peptide, pronociceptin (Meunier et al., 1995), also contains the additional related biologically active peptides orphanin/nociceptin II (Florin et al., 1997) and nocistatin (Okuda-Ashitaka et al., 1998).

In addition to these five classes of peptides, other peptides of mammalian origin with opioid activity have been identified. In particular, β-casomorphin (Tyr–Pro–Phe–X–Y–Pro–Ile, where X–Y = Val–Glu for the human peptide and Pro–Gly for the bovine peptide), obtained by the enzymatic digestion of the milk protein casein (Brantl et al., 1979; Henschen et al., 1979), is a weak opioid agonist that has been used as a lead peptide for further structural modification.

5.2.2 Amphibian Opioid Peptides

Potent and selective opioid peptides were isolated from the skin of South American Phyllomedusinae hylid frogs in the 1980s (see Erspamer, 1992 for a review). In contrast to most mammalian opioid peptides, the amphibian opioid peptides exhibit high selectivity for opioid receptors, with dermorphin (Table 5.1) preferentially binding to μ opioid receptors, while the deltorphins exhibit high selectivity for δ opioid receptors (Erspamer et al., 1989). The unique feature of these amphibian skin opioid peptides is the sequence between the important aromatic residues. In contrast to the enkephalins and other mammalian opioid peptides that contain the Gly–Gly dipeptide sequence between Tyr and Phe, the amphibian opioid peptides contain a single D-amino acid, which apparently arises from a post-translational conversion of the L-amino acid to its D-isomer (Richter et al., 1987). (See Negri et al., 2000 for a review of the pharmacology of these amphibian opioid peptides.)

5.2.3 Other Sources

In addition to natural sources of opioid peptides, novel peptides and peptidomimetics with high affinity for opioid receptors and selectivity for different opioid receptor types have been identified from combinatorial libraries (see Section 5.5.6; Dooley and Houghten, 1999; Aldrich and Vigil-Cruz, 2003 for reviews).

5.3 MOLECULAR MECHANISM OF ACTION

5.3.1 Opioid and ORL-1 (NOP) Receptors

Opioid agonists produce their biological effects by activating opioid receptors. A major breakthrough came when first the δ opioid receptor (Evans et al., 1992; Kieffer et al., 1992), followed shortly thereafter by the μ (Chen et al., 1993a) and κ (Yasuda et al., 1993) opioid receptors, were successfully cloned in the early 1990s. Since then, the cloning of all three opioid receptor types from several different species, including human, have been reported (see Knapp et al., 1995; Satoh and Minami, 1995; Dhawan et al., 1996; Kieffer, 1997; Gaveriaux-Ruff and Kieffer, 1999 for reviews). Knockout mice have been generated lacking individual opioid receptors, as well as combinatorial opioid receptor knockout mice lacking two or all three receptors (see Kieffer, 1999; Gaveriaux-Ruff and Kieffer, 2002; Kieffer and Gaveriaux-Ruff, 2002 for reviews), which have been used to study the roles of the opioid receptors in various physiological processes and in the pharmacological activities of opioid receptor ligands.

During attempts to clone the opioid receptors, the related ORL-1 receptor, which has high sequence homology to the opioid receptors, was identified by several research groups (see Calò et al., 2000c; Harrison and Grandy, 2000; Meunier et al., 2000; Mogil and Pasternak, 2001 for reviews). This receptor, now referred to as the N/OFQ peptide (NOP) receptor by the International Union of Pharmacology (Foord et al., 2005), does not display affinity for classical opioid ligands including naloxone. While distinctly different from opioid receptors, the NOP receptor interacts with the opioid receptor system in the regulation of analgesia and other physiological effects (see Calò et al., 2000c, 2002a; Zaveri, 2003 and references cited therein).

The opioid receptors belong to the family of G-protein coupled receptors (GPCRs). Based on the model for this family of receptors, the receptors contain extracellular regions including the N-terminus and extracellular loops, seven putative transmembrane (TM) regions, and intracellular regions including the C-terminus and intracellular loops (see Chaturvedi et al., 2000; Zollner and Stein, 2007 for reviews). Comparison of the sequences (Chen et al., 1993b) indicates the highest sequence homology between the μ, δ, and κ opioid receptors in TM2, TM3, and TM7. TM2 and TM3 each contain a conserved Asp residue; the conserved Asp in TM3 is thought to interact with protonated amine groups on opioid ligands. There are also high similarities in the intracellular loops; the third intracellular loop is thought to be involved in interactions with G-proteins. The second and third extracellular loop, TM1, and TM4–6 are less conserved. The largest structural diversity occurs in the extracellular N-terminus. Potential sites for possible post-translational modification have been identified on the receptors. Potential glycosylation sites are located in the N-terminal sequence. Possible sites for phosphorylation are located in the C-terminus and in intracellular loops; a palmitoylation site is also located in the C-terminus. Conserved Cys residues in the first and second extracellular loops are involved in a disulfide linkage. A variety of chimeric receptors between different opioid receptor types have been prepared in attempts to identify regions of the receptors that are involved in ligand recognition and receptor type selectivity (see Law et al., 1999; Chaturvedi et al., 2000 for reviews).

5.3.2 Effector Systems for Opioid Receptors

Opioid receptors are coupled to G-proteins and produce most of their effects through these proteins (see Childers, 1993; Cox, 1993; Zollner and Stein, 2007 for reviews). G-proteins mediate the interaction of opioid and other receptors with a variety of effector systems, including adenylyl cyclase and ion channels. Opioid inhibition of adenylyl cyclase is the best studied of the effector systems that have been implicated in the transduction mechanisms for opioid receptors; the inhibition of adenyl cyclase following agonist activation of the NOP receptor has also been demonstrated (Harrison and Grandy, 2000). There is also some evidence that μ and δ opioid receptors can stimulate adenylyl cyclase in certain tissues (see Childers, 1993; Fowler and Fraser, 1994 for reviews). Opioid receptors (North, 1993; Zollner and Stein, 2007) and the NOP receptor (Harrison and Grandy, 2000) can also be coupled to ion channels via G proteins. All four receptor types can decrease voltage-dependent Ca^{+2} current. The coupling of opioid receptors to calcium channels involves a G-protein, and activation of μ and δ opioid receptors can also increase K^+ conductance apparently through a G-protein. Agonist binding to opioid receptors also appears to activate the extracellular signal regulated kinase (ERK) cascade (Childers, 1993; Akil et al., 1998; Zollner and Stein, 2007).

5.3.3 Proposed Opioid Receptor Subtypes and Opioid Receptor Dimerization

Subtypes of each of the three opioid receptors were proposed based on evidence from pharmacological assays prior to cloning of the opioid receptors (see Porreca and Burks, 1993; Fowler and Fraser, 1994 for excellent reviews). There is considerable evidence from both functional assays as well as binding studies supporting the existence of two δ opioid receptor subtypes, δ_1 and δ_2 (Traynor and Elliott, 1993; Fowler and Fraser, 1994; Zaki et al., 1996). A key factor in the characterization of these δ opioid receptor subtypes was the availability of ligands selective for each of the proposed subtypes, including several peptides (see Section 5.5.1). Delta receptor subtypes were also proposed by Rothman and colleagues that were differentiated by whether they are or are not associated with a μ–δ opioid receptor complex (δ_{cx} and δ_{ncx}, respectively; see Rothman et al., 1993 for a review). In δ opioid receptor knockout mice the binding of tritiated ligands postulated to be selective for both receptor subtypes is absent (Zhu et al., 1999), indicating that the proposed subtypes are not due to different gene products. Multiple types of μ opioid receptors have also been proposed. Pasternak and colleagues initially proposed two subtypes of μ opioid receptors (Wolozin and Pasternak, 1981), the μ_1 site, which was suggested to be a common high affinity site for both nonpeptide opioids and opioid peptides, and the μ_2 site, which was proposed to correspond to the "traditional" μ-binding site (Pasternak and Wood, 1986; Fowler and Fraser, 1994). Following the cloning of the opioid receptors Pasternak and colleagues characterized multiple splice variants of the μ opioid receptor (see Pasternak, 2001a, 2001b; 2004 for reviews). Multiple subtypes of κ opioid receptors have also been postulated. Kappa$_1$ (κ_1) and κ_2 opioid receptors were differentiated based on their sensitivity and insensitivity, respectively, to the nonpeptide κ opioid receptor selective agonists U50,488 and U69,593 (Zukin et al., 1988).

Cloning has consistently identified only a single gene product for each opioid receptor type, so the molecular basis for opioid receptor subtypes has been unclear. Recently, a possible explanation for receptor subtypes, receptor dimerization, has appeared in the literature. Receptor homodimerization has been described for both cloned δ (Cvejic and Devi, 1997) and κ opioid receptors (Jordan and Devi, 1999), as well as heterodimerization between μ and δ (George et al., 2000; Gomes et al., 2000), between κ and δ (Jordan and Devi, 1999), and recently between μ and κ opioid receptors (Wang et al., 2005). The κ–δ opioid receptor heterodimers exhibit a ligand binding profile that is virtually identical to that previously reported for the proposed κ_2 receptor subtype (Jordan and Devi, 1999). In addition, evidence has been reported for opioid receptor oligomerization with a variety of other GPCRs, including receptors for somatostatin and substance P (see Gupta et al., 2006 for a review). Heterodimerization of opioid receptors has been proposed to lead to altered signaling pathways via different G-proteins or via non-G-protein mechanisms, at least for μ and δ opioid receptors (George et al., 2000; Levac et al., 2002; Fan et al., 2005; Hasbi et al., 2007; Rozenfeld et al., 2007; Rozenfeld and Devi, 2007). Opioid receptor dimerization is an active area of current research (Gupta et al., 2006), and such dimers represent possible targets for drug development (see Section 5.8).

5.4 BIOLOGICAL AND PATHOLOGICAL MODES OF ACTION

The exploration of the biological effects of opioids is an active area of research. A detailed discussion of all these effects is beyond the scope of this chapter, and therefore it will focus primarily on effects that are relevant for clinical or potential clinical applications of opioids. Readers are referred to a comprehensive two volume series on opioids (Herz et al., 1993a, 1993b), plus more recent reviews (Ossipov et al., 1999; Stein, 1999; Gutstein and Akil, 2001) for detailed discussion of opioid pharmacology and physiology. In addition, a series of comprehensive reviews is published annually in the journal *Peptides*, summarizing all aspects of opioid pharmacology and physiology (see Bodnar and Klein, 2006; Bodnar, 2007 for the most recent reviews).

5.4.1 Analgesia

The principal biological activity that has been exploited clinically is the analgesic activity of opioid agonists acting primarily at μ opioid receptors. Opioid analgesia is mediated by opioid receptors both in the brain and spinal cord; peripheral opioid receptors have also been implicated in analgesia, particularly in cases of inflammation (see Barber and Gottschlich, 1992; Stein, 1993; Stein et al., 1999 for reviews).

N/OFQ and NOP receptors are also involved in the response to painful stimuli, but their effects are complex (see Grisel and Mogil, 2000; Harrison and Grandy, 2000; Mogil and Pasternak, 2001; Chiou et al., 2007 for reviews). Effects reported for N/OFQ have ranged from hyperalgesia (hence the name nociceptin), analgesia, antianalgesic activity, or a combination of these effects (see Grisel and Mogil, 2000; Harrison and Grandy, 2000; Xu et al., 2000; Mogil and Pasternak, 2001). A number of factors appear to influence these different findings, particularly the route of administration (see Mogil and Pasternak, 2001 for a detailed discussion). The antianalgesic effects following supraspinal (intracerebroventricular, i.c.v.) administration are the most consistently observed effects, with N/OFQ acting as a functional antagonist of opioid receptors and blocking the analgesia produced by a wide variety of opioids (Harrison and Grandy, 2000; Mogil and Pasternak, 2001). Supraspinal administration of NOP receptor antagonists and antisense oligonucleotides consistently produce antinociceptive responses (see Chiou et al., 2007 for a detailed review). At the spinal level there are conflicting reports for the role of N/OFQ and NOP receptors in pain states; the predominant effect of N/OFQ appears to be inhibitory, resulting in analgesia and/or antihyperalgesia/antiallodynia (see Harrison and Grandy, 2000; Xu et al., 2000; Mogil and Pasternak, 2001). Particularly in rat models of inflammation and nerve injury (Harrison and Grandy, 2000; Xu et al., 2000), several studies have reported antihyperalgesic or antiallodynic activity for N/OFQ. The results for several NOP receptor antagonists suggest that endogenous spinal N/OFQ may play a protective role in inflammatory, but not acute, pain states (Chiou et al., 2007). The activity of N/OFQ in neuropathic pain following nerve injury could have important therapeutic implications because morphine appears to have reduced effectiveness in treating this type of pain. N/OFQ has also been implicated in the development of tolerance to morphine (Harrison and Grandy, 2000; Mogil and Pasternak, 2001), and blockade of NOP receptors reverses tolerance induced by low doses of morphine (Chiou et al., 2007).

5.4.2 Other Effects

In addition to analgesia, opioid agonists display a plethora of biological effects (Herz et al., 1993a, 1993b). For clinically used opioid analgesics, alterations of the nervous, respiratory, and gastrointestinal systems contribute to the side effects of these drugs, which include respiratory depression, sedation, and constipation. Some of these effects are mediated by opioid receptors located in the central nervous system (CNS, e.g., respiratory depression and sedation), while others result from activation of peripheral opioid receptors (e.g., inhibition of gut motility and constipation).

Ligands with activity at δ and κ opioid receptors, which are not associated with the side effect profile of μ opioid receptor agonists, are of interest because of their effects on μ opioid agonist activity, depression, and drug abuse. Delta opioid receptor ligands, both agonists and antagonists, can modulation the activity of μ opioid receptor agonists such as morphine (see Aldrich and Vigil-Cruz, 2003; Schiller, 2005). Thus δ opioid agonists can potentiate the analgesic activity of μ receptor agonists (Porreca et al., 1992; He and Lee, 1998), and δ opioid receptor antagonists can decrease the development of tolerance and dependence to morphine (Abdelhamid et al., 1991; Fundytus et al., 1995). Thus δ opioid receptor ligands, particularly antagonists, could have important therapeutic applications to minimize the deleterious effects of morphine; there is also considerable interest in developing compounds that exhibit both μ opioid agonist and δ receptor antagonist activity (see Sections 5.5.4.2 and 5.8).

Ligands for both δ and κ opioid receptors exhibit antidepressant activity. Delta opioid receptor agonists display antidepressant activity (see Jutkiewicz, 2006 for a review), and the antidepressant activity of inhibitors of endogenous opioid peptide metabolism has been attributed to the activation of these receptors (Noble and Roques, 2007). Antagonists at κ opioid receptors also exhibit antidepressant activity (Mague et al., 2003), and this is thought to be a factor in their ability to prevent reinstatement of cocaine abuse.

Ligands for both δ and κ opioid receptors have also been examined for their effects on drugs of abuse (see Aldrich and Vigil-Cruz, 2003 and references cited therein). Delta receptor antagonists can attenuate a number of the effects of cocaine and methamphetamine (Coop and Rice, 2000). Because κ opioid agonists decrease mesolimbic dopamine levels that can counteract the increases in extracellular dopamine caused by cocaine, they have been studied extensively for their ability to attenuate cocaine self-administration (Schenk et al., 1999; Mello and Negus, 2000). The dysphoric effects of many centrally acting κ opioid receptor agonists, however, may limit their therapeutic application. Interestingly, κ opioid receptor antagonists (Beardsley et al., 2005), including the peptide antagonist arodyn (Carey et al., 2007), have been shown to block stress-induced reinstatement of cocaine-seeking behavior, activity that is thought to be related to their antidepressant effects (Beardsley et al., 2005). Antagonism of κ opioid receptors has also shown beneficial effects in the treatment of opiate addiction (Rothman et al., 2000).

N/OFQ and the NOP receptor are also involved in a number of other physiological effects (Harrison and Grandy, 2000; Mogil and Pasternak, 2001; Chiou et al., 2007). Two of the most significant areas are the effect of N/OFQ and NOP receptor ligands on anxiety and depression (see Gavioli and Calo, 2006; Chiou et al., 2007 for recent reviews). N/OFQ administered i.c.v. produces anxiolytic-like effects in several different anxiety paradigms, and there is substantial evidence for a role of endogenous N/OFQ in stress and anxiety (Gavioli and Calò, 2006; Chiou et al., 2007). In contrast, NOP receptor antagonists show antidepressant effects (Gavioli and Calo, 2006; Chiou et al., 2007); these effects were not observed in NOP knockout mice (Gavioli et al., 2003), suggesting that endogenous N/OFQ and NOP receptors play a role in depression. N/OFQ also plays a modulatory role on the reward system in the brain (Harrison and Grandy, 2000; Mogil and Pasternak, 2001), and i.c.v. injection of N/OFQ has been shown to have an inhibitory effect on several drugs of abuse (Chiou et al., 2007) and alcohol (Cowen and Lawrence, 2006). In addition, N/OFQ and NOP receptors have a number of other effects, both centrally and peripherally mediated, including effects on food intake, learning and memory, motor activity, and the cardiovascular system (Chiou et al., 2007).

The peptide nocistatin, which is obtained from the same precursor protein as N/OFQ but is structurally unrelated, appears to be a functional antagonist of N/OFQ. It reverses a number of the central modulatory effects of N/OFQ on pain, etc., but does not alter the cardiovascular effects of N/OFQ (Chiou et al., 2007).

5.4.3 Metabolism of Opioid Peptides and Effects on Activity

The study of the biological activity of endogenous opioid peptides and N/OFQ is complicated by their metabolic lability. In plasma and brain tissue the half-life of most of these peptides is typically less than a couple of minutes (Roques et al., 1993a). A major metabolic pathway for endogenous mammalian opioid peptides is removal of the N-terminal Tyr residue by an aminopeptidase, especially aminopeptidase-N (Roques et al., 1993a). Removal of the N-terminal Tyr eliminates the ability of the peptides to bind to opioid receptors. However, cleavage of the peptides, particularly the longer opioid peptides such as the dynorphins and endorphins, by other proteolytic enzymes can result in peptide fragments that still have opioid activity, and these active fragments can have altered preferences for interaction with different opioid receptors compared to the parent peptide.

While proteolytic cleavage of the N-terminal Tyr residue of opioid peptides abolishes affinity for opioid receptors, in the case of dynorphin A the resulting des-Tyr peptide has nonopioid effects (see Caudle and Mannes, 2000; Lai et al., 2001; Hauser et al., 2005 for reviews). Intrathecal (i.t.) administration of dynorphin A and des-Tyr fragments causes motor impairment and hind limb paralysis. Nerve injury results in an increase in spinal dynorphin concentration, which has been correlated with the development of neuropathic pain. Glutamate receptors, in particular NMDA (N-methyl-D-aspartate) receptors, have been implicated in the paradoxical neurotoxic effects of dynorphin A (Caudle and Mannes, 2000; Lai et al., 2001; Hauser et al., 2005), although the detailed mechanisms of how these effects occur are not well understood.

5.5 STRUCTURAL ANALOGS

As noted in the introduction, since the discovery of the first opioid peptides in the mid-1970s thousands of analogs of these peptides have been synthesized and evaluated for opioid receptor affinity and opioid activity. These have included peptide ligands selective for all three opioid receptor types, both agonists and antagonists, and analogs of each of the different endogenous opioid peptides (Figure 5.1). Space precludes a detailed discussion of the SARs of opioid peptides here, so readers are referred to a detailed review (Aldrich and Vigil-Cruz, 2003, which contains extensive tabular data), a special issue of *Biopolymers (Peptide Science)* (Schiller, 1999), other recent reviews (Janecka et al., 2004; Janecka and Kruszynski, 2005), and the references cited therein for additional information. The overview here will concentrate on key peptide ligands used to study opioid receptors, and some examples of recent developments in opioid peptide ligand design.

While the naturally occurring opioid peptides are linear, a variety of cyclic analogs of opioid peptides have been synthesized (see Sections 5.5.1, 5.5.2, and 5.5.5 and Hruby, 2001; Aldrich and Vigil-Cruz, 2003; Janecka and Kruszynski, 2005 for reviews). Cyclization restricts the conformational flexibility of peptides, facilitating examination of possible bioactive conformations of the peptides. Conformationally constrained amino acid derivatives have also been incorporated into a number of these opioid peptides, particularly in the cyclic peptides (Hruby, 2001; Aldrich and Vigil-Cruz, 2003; Janecka and Kruszynski, 2005). These have focused on derivatives of phenylalanine and tyrosine, including β-methyl substituted derivatives and cyclic analogs, in order to explore the possible relative spatial orientations of the side chains of these key aromatic residues in the peptides. Readers are referred to the reviews above and references cited therein for details concerning the conformational analysis and possible bioactive conformations of these conformationally constrained opioid peptide analogs.

5.5.1 ENKEPHALIN ANALOGS

The enkephalins have been the most extensively modified of the opioid peptides, and thousands of analogs of these pentapeptides have been prepared (see Morley, 1980; Udenfried and Meienhofer, 1984; Hruby and Gehrig, 1989; Schiller, 1993; Hruby and Agnes, 1999; Aldrich and Vigil-Cruz, 2003 for reviews). The naturally occurring enkephalins exhibit some selectivity for δ opioid receptors, but these peptides are rapidly degraded by a variety of peptidases. Therefore one major goal of structural modification of these small peptides was to increase metabolic stability.

Early SAR studies (see Morley, 1980 for a review) identified important structural features of the enkephalins and which positions could be readily modified. The importance of Tyr^1 for opioid activity was apparent from the large decreases in potency when this residue was substantially modified. A D-amino acid in position 2 was one of the early modifications examined in order to decrease the cleavage of the Tyr–Gly bond by aminopeptidases (Hambrook et al., 1976). Incorporation of a D-amino acid at this position increases potency at both μ and δ opioid receptors and is found in the vast majority of enkephalin derivatives; an L-amino acid at position 2 significantly decreases potency at both receptors. In the enkephalins the 3 position is very intolerant to substitution, and therefore enkephalin derivatives generally retain Gly at this position. There are significant differences between μ and δ opioid receptors, however, in the structural requirements for residues in positions 4 and 5, and modifications in these positions have been used to impart selectivity for one of these opioid receptor types. For μ opioid receptor selective peptides the C-terminus can be amidated, reduced, or eliminated completely with retention of μ receptor affinity and often with a substantial increase in μ opioid receptor selectivity. One of the most commonly used μ opioid receptor selective ligands is DAMGO ([D-Ala^2,NMePhe^4,glyol]enkephalin, Table 5.2; Handa et al., 1981) which contains a reduced C-terminus. This peptide is available in tritiated form and is used extensively in radioligand binding assays for μ opioid receptors.

Differences in the SAR of enkephalin analogs for interaction with δ and μ opioid receptors have been exploited to develop δ opioid receptor selective derivatives. Early studies of δ opioid receptors used DADLE ([D-Ala^2,D-Leu^5]enkephalin, Table 5.2), which shows a slight preference for δ over

TABLE 5.2
Synthetic Enkephalin and Dynorphin Analogs Discussed in the Text

Peptide	Sequence	Receptor Preference[a]
Enkephalin Analogs		
DAMGO	Tyr–D-Ala–Gly–NMePhe–glyol	μ
DADLE	Tyr–D-Ala–Gly–Phe–D-Leu	(δ)[b]
DSLET/DTLET	Tyr–X–Gly–Phe–Leu–Thr, X = D-Ser or D-Thr	δ
DSTLBULET	Tyr–D-Ser(*t*Bu)–Gly–Phe–Leu–Thr	δ
BUBU/BUBUC	Tyr–X–Gly–Phe–Leu–Thr(*t*Bu), X = D-Ser(*t*Bu)/Cys(*t*Bu)–	δ
DPDPE/DPLPE	Tyr–*cyclo*[D-Pen–Gly–Phe–D/L-Pen]	δ
Biphalin	(Tyr–D-Ala–Gly–Phe–NH–)$_2$	μ and δ
ICI 174,864	(allyl)$_2$Tyr–Aib–Aib–Phe–Leu	δ antagonist
Dynorphin (Dyn) A Analogs		
E2078	NMeTyr–Gly–Gly–Phe–Leu–Arg–NMeArg–D-LeuNHEt	κ
N-alkyl[D-Pro^{10}]-Dyn A-(1–11)	RNTyr–Gly–Gly–Phe–Leu–Arg–Arg–Ile–Arg–D-Pro–Lys R = allyl or CPM	κ
[L/D-Ala^3]DynA-(1–11)NH$_2$	Tyr–Gly–L/D-Ala–Phe–Leu–Arg–Arg–Ile–Arg–Pro–LysNH$_2$	κ
[Pro^3]DynA-(1–11)NH$_2$	Tyr–Gly–Pro–Phe–Leu–Arg–Arg–Ile–Arg–Pro–LysNH$_2$	κ antagonist
Dynantin	(2*S*)Mdp–Gly–Gly–Phe–Leu–Arg–Arg–Ile–Arg–Pro–LysNH$_2$[c]	κ antagonist
Arodyn	AcPhe–Phe–Phe–Arg–Leu–Arg–Arg–D-Ala–Arg–Pro–LysNH$_2$	κ antagonist
Cyclodyn	*cyclo*[Tyr–Gly–Trp–Trp–Glu]–Arg–Arg–Ile–Arg–Pro–LysNH$_2$	κ antagonist
[N-BenzylTyr^1,*cyclo*(D-Asp^5,Dap^8)]-Dyn A-(1–11)NH$_2$	NBzTyr–Gly–Gly–Phe–*cyclo*[D-Asp–Arg–Arg–Dap]–Arg–Pro–LysNH$_2$[c]	κ antagonist

[a] Peptides are agonists unless otherwise indicated.
[b] The receptor type in parentheses indicates that the selectivity for that receptor is low.
[c] Mdp = 2-methyl-3-(2′,6′-dimethyl-4-hydroxyphenyl)propanoic acid (see Figure 5.2), Dap = 2,3-diaminopropionic acid.

μ opioid receptors. A more hydrophilic D-amino acid, such as D-Ser or D-Thr, incorporated into position 2 imparts δ opioid receptor selectivity, while μ opioid receptors prefer a more hydrophobic residue in this position (Fournié-Zaluski et al., 1981). At the C-terminus δ opioid receptors prefer a free carboxylic acid, and lengthening of the peptide with a residue such as Thr can result in increased δ opioid receptor selectivity. Thus the extended analogs DSLET and DTLET ([D-Ser^2,Leu^5,Thr^6]- and [D-Thr^2,Leu^5,Thr^6]enkephalin, Table 5.2; Zajac et al., 1983; Gacel et al., 1988) exhibit higher selectivity for this receptor. Retaining bulky *t*-butyl ether protecting groups on the residues in positions 2 and/or 6 resulted in the analogs DSTBULET and BUBU ([D-Ser(*t*Bu)^2,Leu^5,Thr^6]- and [D-Ser(*t*Bu)^2,Leu^5,Thr(*t*Bu)^6]enkephalin, respectively; Gacel et al., 1988) and the related derivative BUBUC ([D-Cys(*t*Bu)^2,Leu^5,Thr(*t*Bu)^6]enkephalin; Gacel et al., 1990) which exhibit decreased affinity for μ receptors and thus enhanced selectivity for δ opioid receptors (>1000-fold for the latter peptide). The N-terminal dialkylated derivative ICI 174,864 was an early antagonist identified for δ opioid receptors (Cotton et al., 1984); this peptide exhibits inverse agonist activity in GTPase and GTPγS binding assays (Costa and Herz, 1989; Mullaney et al., 1996). This peptide has been largely replaced by the more potent and selective tetrapeptide antagonist TIPP (see Section 5.5.4.2).

Delta selective enkephalin analogs have also been prepared by cyclization via disulfide bond formation. The initial cyclic enkephalin analogs containing D- or L-Cys in positions 2 and 5 exhibit only a slight preference for δ over μ opioid receptors, but introduction of methyl groups on the β-carbons of the cysteine residues by incorporation of penicillamine (Pen) in one or both of these positions markedly enhances δ opioid receptor selectivity. The most δ-selective compounds in the series were the *bis*-penicillamine derivatives DPDPE and DPLPE (*cyclo*[D-Pen^2,D-Pen^5]- and *cyclo*[D-Pen^2,L-Pen^5]enkephalin, respectively, Table 5.2; Mosberg et al., 1983). A large number of DPDPE analogs were subsequently synthesized, some of which exhibit enhanced δ opioid receptor affinity and/or selectivity (see Hruby and Agnes, 1999 for a review); modifications included halogenation of Phe^4 (Toth et al., 1990) which can also increase blood–brain barrier permeability (Weber et al., 1991; Gentry et al., 1999), and C-terminal extension with Phe (Bartosz-Bechowski et al., 1994). Recently analogs of *cyclo*[D-Cys^2,D/L-Cys^5]enkephalin in which the disulfide bond is replaced by an olefin prepared by ring closing metathesis have been reported (Berezowska et al., 2007; Mollica et al., 2007).

Several enkephalin derivatives have been used to differentiate postulated δ opioid receptor subtypes. The proposed δ_1 opioid receptor subtype was characterized by its preferential stimulation by the enkephalin analogs DPDPE and DADLE, while DSLET was postulated to preferentially stimulate the δ_2 receptor subtype (see Traynor and Elliott, 1993; Fowler and Fraser, 1994; Zaki et al., 1996).

Various approaches have been examined to enhance *in vivo* activity and facilitate the transport of enkephalin derivatives into the CNS. Glycopeptide derivatives have been prepared by incorporation of a glycosylated amino acid (e.g., Ser(β-D-glucose); Polt et al., 1994) at the C-terminus to enhance penetration of the blood–brain barrier (see Polt et al., 2005 for a review). A prodrug derivative of DPDPE in which PEG (poly(ethylene glycol)) is attached to the N-terminus has been reported to also exhibit enhanced analgesia compared to DPDPE following intravenous (i.v.) administration; this prodrug, which has very weak affinity for δ opioid receptors, is converted to DPDPE *in vivo* (Witt et al., 2001).

A number of dimeric enkephalin analogs have been prepared (see Aldrich and Vigil-Cruz, 2003; Janecka et al., 2004 for reviews). Of these, the best studied is biphalin (Lipkowski et al., 1982) (Table 5.2) in which the two tetrapeptides are connected by a hydrazine spacer. This peptide, which exhibits comparable affinity at both μ and δ opioid receptors, exhibits potent analgesic activity *in vivo* (Horan et al., 1993). Recently the first cyclic analogs of biphalin were reported (Mollica et al., 2006). PEG has also been conjugated to biphalin to increase potency following systemic (i.v. or subcutaneous, s.c.) administration (Huber et al., 2003).

5.5.2 Dynorphin Analogs

Dynorphin has not been as well studied as other smaller opioid peptides. Although several peptides with high affinity for κ opioid receptors are obtained from prodynorphin (see Table 5.1), SAR studies have focused almost exclusively on derivatives of dynorphin A (see Naqvi et al., 1998; Hruby and Agnes, 1999; Aldrich and Vigil-Cruz, 2003 for reviews). Dynorphin A is a heptadecapeptide, but dynorphin A-(1–13) accounts for essentially all of the activity of the larger peptide (Goldstein et al., 1981); typically dynorphin A-(1–13) or A-(1–11) have been used as the parent peptide for further modification.

Studies of dynorphin A have been complicated by its metabolic lability (Leslie and Goldstein, 1982). In addition to inactivation by peptidase cleavage in the N-terminus, cleavage in the C-terminus yields shorter active peptides that may have different receptor selectivity profiles from the parent peptide. Dynorphin A-(1–8), which is the predominant form of dynorphin A present in rat brain (Weber et al., 1982; Cone et al., 1983), is less selective for κ opioid receptors than the longer peptides (Goldstein, 1984). C-Terminal amidation enhances the metabolic stability of dynorphin A-(1–13) (Leslie and Goldstein, 1982), and therefore this modification is typically incorporated into dynorphin A analogs. An analog of dynorphin A-(1–8) E2078 ([NMeTyr^1,NMeArg^7,D-Leu^8]dynorphin A-(1–8)NHEt, Table 5.2; Yoshino et al., 1990a, 1990b) containing modifications to stabilize the peptide to metabolism has been studied extensively *in vivo* (see Section 5.6.1); this peptide exhibits a slight preference for κ opioid receptors in binding assays.

Although dynorphin A and its fragments preferentially interact with κ opioid receptors, they also exhibit nanomolar affinity for μ and δ opioid receptors. Therefore a major focus has been to identify dynorphin A analogs that exhibit high κ opioid receptor selectivity. Interestingly the most κ-receptor-selective derivatives have been prepared by modifications in the N-terminal "message" sequence rather than in the unique C-terminal "address" sequence (Table 5.2). N-Terminal monoalkylation results in agonists with a marked enhancement in κ opioid receptor selectivity by decreasing μ opioid receptor affinity (Choi et al., 1992); N,N-dialkylation results in analogs with antagonist activity (Gairin et al., 1988; Choi et al., 1997; Soderstrom et al., 1997). Substitution of either Ala or D-Ala in position 3 also results in agonist analogs with high κ opioid receptor selectivity (Lung et al., 1995); incorporation of Pro in this positions results in a highly selective, but relatively weak, κ opioid receptor antagonist (Schlechtingen et al., 2000).

A variety of cyclic analogs of dynorphin A have been prepared (see Hruby and Agnes, 1999; Aldrich and Vigil-Cruz, 2003 for reviews). Although cyclization in a number of different positions in dynorphin A are well tolerated by κ opioid receptors, the selectivity of these derivatives is generally only modest, with the exception of the cyclic antagonist [N-benzylTyr^1,*cyclo*(D-Asp,Dap^8)]-dynorphin A-(1–11)NH_2 (Patkar et al., 2005).

Early attempts to prepare κ opioid receptor selective antagonists by modification of dynorphin A met with limited success, but in recent years antagonist analogs with improved pharmacological profiles have been identified (Table 5.2). These include analogs without a basic amine terminus (Wan et al., 1999; Lu et al., 2001; Bennett et al., 2002), Pro^3 derivatives (Schlechtingen et al., 2000, 2003), and recently cyclic derivatives with antagonist activity (Vig et al., 2003; Patkar et al., 2005).

5.5.3 Nociceptin/Orphanin FQ Analogs

Shortly after the identification of the endogenous ligand N/OFQ, several research groups began exploring the SAR of this peptide. Several reviews (Calò et al., 2000a, 2000c; Guerrini et al., 2000; Zaveri, 2003; Chiou et al., 2007) have discussed the details of the SAR of N/OFQ, so the discussion here will focus only on some of the key findings.

Initial studies focused on identifying important residues for NOP receptor interaction. Truncation studies were performed to identify the minimum sequence of N/OFQ required for NOP receptor affinity and activation. These studies revealed that, like opioid peptides, the N-terminal aromatic residue was important for biological activity (Dooley and Houghten, 1996; Butour et al., 1997). In contrast to dynorphin A, where shorter fragments retain opioid activity (Chavkin and Goldstein,

1981), 13 of the 17 residues of N/OFQ appear to be required for NOP receptor affinity and activation (Dooley and Houghten, 1996; Butour et al., 1997). The amide derivative of OFQ-(1–13), which is a considerably more potent agonist than the acid derivative apparently due to decreased metabolism (Guerrini et al., 2000), is often used as the parent structure in further SAR studies. An alanine scan has identified Phe^1, Phe^4, and Arg^8 as critical residues in the sequence (Dooley and Houghten, 1996; Reinscheid et al., 1996). Phe^1 can be replaced by tyrosine (Reinscheid et al., 1996; Butour et al., 1997), resulting in an analog that retains affinity and potency at NOP receptors, but this peptide also exhibits increased affinity and activity at opioid receptors, particularly κ and μ opioid receptors (Calò et al., 2000a; Guerrini et al., 2000). Unlike the opioid peptides, an aromatic amino acid is not required in position 1, and Phe^1 can be replaced by the aliphatic residues cyclohexylalanine (Cha, Figure 5.2) and Leu (Guerrini et al., 1997). In contrast, an aromatic residue is required in position 4, and replacement of Phe^4 with an aliphatic residue results in loss of activity (Guerrini et al., 1997). An Arg in position 8 appears to be critical, and replacement with Lys results in large decreases (>100-fold) in affinity and potency (Guerrini et al., 2000).

N/OFQ exhibits the greatest similarity to dynorphin A, but N/OFQ exhibits very low affinity for κ opioid receptors and conversely Dyn A exhibits low affinity for NOP receptors. A series of chimeric peptides between N/OFQ and Dyn A were prepared in order to explore the structural reasons for the selectivity of N/OFQ for NOP receptors over κ opioid receptors (Lapalu et al., 1997); the chimera Dyn A-(1–5)/ N/OFQ-(6–17) was able to bind and activate both NOP and κ opioid receptors.

Recent reports have described N/OFQ analogs with enhanced potency and in some cases longer duration of action. Introduction of fluorine on the para position of the phenyl ring of Phe^4 in N/OFQ-(1–13)NH_2 resulted in a peptide (Table 5.3) with enhance potency and a longer duration of action (Bigoni et al., 2002; Rizzi et al., 2002b). Introduction of basic residues in positions 14 and 15 of N/OFQ resulted in the more potent agonist [Arg^{14},Lys^{15}]N/OFQ (Table 5.3; Okada et al., 2000) that exhibits longer duration of action *in vivo* (Rizzi et al., 2002c). Combining these two modifications resulted in an even more potent analog UFP-102 (Table 5.3; Carra et al., 2005); Aib (α-aminoisobutyric acid) was subsequently incorporated into positions 7 and 11 of this peptide (Peng et al., 2006). Derivatives cyclized in the C-terminal sequence that retain high NOP receptor affinity and potency have also been reported (Table 5.3; Ambo et al., 2001; Charoenchai et al., 2008).

Early studies of the pharmacology of the NOP receptor system were hindered by the lack of antagonists for this receptor. Therefore there was considerable excitement in the field when the first report of an antagonist appeared in the literature (Guerrini et al., 1998). The reduced amide derivative of N/OFQ [Phe^1(ΨCH_2NH)Gly^2]N/OFQ-(1–13)NH_2, which was synthesized to protect the peptide from metabolism by aminopeptidases, was initially reported to be an antagonist of N/OFQ-(1–13)NH_2 in smooth muscle preparations (Guerrini et al., 1998). Subsequent examination of this compound in a variety of assays, however, indicated that the activity observed depended on the

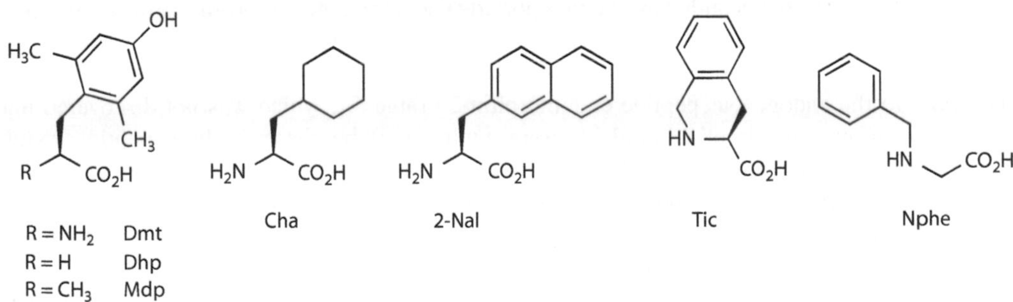

FIGURE 5.2 Structures of Tyr and Phe analogs discussed in the text (derivatives of L-amino acids are shown).

TABLE 5.3
Selected Nociceptin/Orphanin FQ Analogs

Peptide	Sequence
Agonists	
[Arg^{14},Lys^{15}]N/OFQ	Phe–Gly–Gly–Phe–Thr–Gly–Ala–Arg–Lys–Ser–Ala–Arg–Lys–Arg–Lys–Asn–Gln
[Phe(*p*-F)^4]N/OFQ-(1–13)NH_2	Phe–Gly–Gly–Phe(*p*-F)–Thr–Gly–Ala–Arg–Lys–Ser–Ala–Arg–LysNH_2
UFP-102	Phe–Gly–Gly–Phe(*p*-F)–Thr–Gly–Ala–Arg–Lys–Ser–Ala–Arg–Lys–Arg–Lys–Asn–GlnNH_2
cyclo[Cys^{10},Cys^{14}]N/OFQ-(1–14)NH_2	Phe–Gly–Gly–Phe–Thr–Gly–Ala–Arg–Lys–*cyclo*[Cys–Ala–Arg–Lys–Cys]-NH_2
Antagonists	
[Phe$^1\Psi$(CH_2NH)Gly^2]N/OFQ-(1–13)NH_2[a]	Phe(ΨCH_2NH)Gly–Gly–Phe–Thr–Gly–Ala–Arg–Lys–Ser–Ala–Arg–LysNH_2
[Nphe^1]N/OFQ-(1–13)NH_2	Nphe–Gly–Gly–Phe–Thr–Gly–Ala–Arg–Lys–Ser–Ala–Arg–LysNH_2
UFP-101	Nphe–Gly–Gly–Phe–Thr–Gly–Ala–Arg–Lys–Ser–Ala–Arg–Lys–Arg–Lys–Asn–GlnNH_2

[a] Partial agonist in several assays (see text).

assay, and that while this peptide was an antagonist in some assays, it was a partial or full agonist in a number of other assays (see Calò et al., 2000a; Zaveri, 2003; Chiou et al., 2007 for detailed reviews). Subsequently the N-substituted glycine analog [Nphe^1]N/OFQ-(1–13)NH_2 (see Figure 5.2 for structure of Nphe) was reported to be a NOP receptor antagonist (Calò et al., 2000b). Although the potency of this compound is weak ($pA_2 > 6$ in most assays), it is an antagonist in all of the assays examined to date (see Calò et al., 2000a, 2002a; Zaveri, 2003; Chiou et al., 2007 for reviews). Incorporating the Nphe^1 modification into [Arg^{14},Lys^{15}]N/OFQ resulted in the more potent antagonist UFP-101 (Calò et al., 2002b) which has been studied extensively *in vivo* (Calo et al., 2005). Combining several modifications (Phe(*p*-F)^4, Arg^{14}, Lys^{15}, and Aib^{7,11}) resulted in a potent long-lasting antagonist (Peng et al., 2006).

5.5.4 Opioid Peptides with the Tyr–Pro–Phe Sequence and Related Peptides

5.5.4.1 β-Casomorphin and Endomorphin Analogs

β-Casomorphin was identified almost 30 years ago (Brantl et al., 1979; Henschen et al., 1979), and a number of analogs of this peptide have been reported. The tetrapeptide morphiceptin (Chang et al., 1981) and its more potent analog PL017 (Table 5.4; Chang et al., 1983), both of which exhibit high selectivity for μ opioid receptors, have been used most extensively. These peptides are closely related structurally to the endogenous peptide endomorphin-2 (Table 5.1, which was not discovered until over a decade after morphiceptin and PL017 were synthesized), but these synthetic peptides exhibit lower μ opioid receptor affinity and selectivity than the endogenous tetrapeptide. *Cis/trans* isomerization occurs around the amide bond to the proline nitrogen in all of these peptides; although the major conformation is the *trans* amide bond (75%), the *cis* conformation represents a minor but significant conformer (25%) for these peptides (Goodman and Mierke, 1989; Podlogar et al., 1998). Schiller and colleagues reported endomorphin-2 and morphiceptin analogs containing a pseudoproline derivative which exists almost exclusively in the *cis* conformation in place of Pro^2 (Keller et al., 2001); the analogs of both peptides retain μ opioid receptor affinity and agonist activity, strongly

TABLE 5.4
Selected β-Casomorphin and TIPP Analogs

Peptide	Sequence	Receptor Preference
β-Casomorphin (bovine)	Tyr–Pro–Phe–Pro–Gly–Pro–Ile	μ
Morphiceptin	Tyr–Pro–Phe–ProNH_2	μ
PL017	Tyr–Pro–NMePhe–D-ProNH_2	μ
Antanal-1	Dmt–Pro–Trp–D-2-NalNH_2	μ antagonist
Antanal-2	Dmt–Pro–Phe–D-2-NalNH_2	μ antagonist
TIPP	Tyr–Tic–Phe–Phe	δ antagonist
TIPP[Ψ]	Tyr–TicΨ[CH_2NH]Phe–Phe[a]	δ antagonist
TICP	Tyr–Tic–Cha–Phe	δ antagonist
TICP[Ψ]	Tyr–TicΨ[CH_2NH]Cha–Phe	δ antagonist
DIPP	Dmt–Tic–Phe–Phe	δ antagonist/μ agonist
DIPPNH_2[Ψ]	Dmt–TicΨ[CH_2NH]Phe–PheNH_2	δ antagonist/μ agonist
Dmt–Tic	Dmt–Tic	δ antagonist

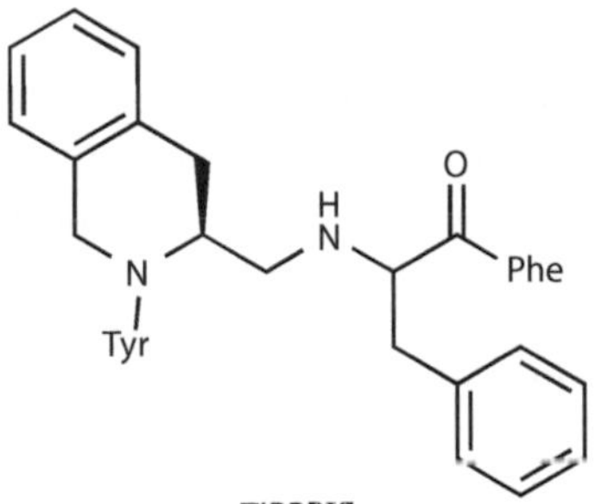

TIPP[Ψ]

suggesting that the *cis* conformation around the Tyr–Pro amide bond is the bioactive form (see Gentilucci and Tolomelli, 2004 for a review of conformational studies of the endomorphins).

A number of analogs of the endomorphins have been prepared and examined for μ opioid receptor affinity and opioid activity (see Janecka et al., 2007 for a recent review). It is clear from these studies that the structural modifications in the endomorphins that are tolerated by the μ opioid receptor are quite limited, with the vast majority of changes resulting in decreased μ opioid receptor affinity. Some replacements for the C-terminal amide (e.g., an alcohol or ester) are tolerated, and substitution of Dmt (2′,6′-dimethyltyrosine, Figure 5.2) in position 1 enhances affinity (Bryant et al., 2003). Introduction of a D-naphthylalanine (Nal, Figure 5.2) into position 4 of endomorphin-2 resulted in analogs with μ opioid receptor antagonist activity (Kruszynski et al., 2005); additional incorporation of Dmt into position 1 gave [Dmt[1],D-2-Nal[4]]endomorphin-1 and 2 (named antanal-1 and 2, respectively), which exhibit enhanced affinity and antagonist potency at μ opioid receptors, but also enhanced δ opioid receptor affinity (Fichna et al., 2007).

5.5.4.2 TIPP and Related Peptides

Exploration of a series of tetrapeptides consisting solely of aromatic residues led Schiller and colleagues to identify TIPP (Tyr–Tic–Phe–Phe, Tic = 1,2,3,4-tetrahydroisoquinoline-3-carboxylic acid, Figure 5.2; Schiller et al., 1992). TIPP is a potent δ opioid receptor antagonist in the mouse vas deferens (MVD) smooth muscle preparation (see Section 5.7.2) with very high selectivity for these receptors (in other assays, i.e., adenylyl cyclase, TIPP, and its analog TIPP[Ψ] (Table 5.4) exhibit partial agonist activity; Martin et al., 2001, 2002). The tripeptide derivatives TIP and $TIPNH_2$ are also δ opioid receptor antagonists in the MVD. In contrast, the tetrapeptide Tyr–D-Tic–Phe–$PheNH_2$ with D-Tic in position 2 is a potent agonist that is selective for μ opioid receptors. The amide derivative of TIPP, in addition to δ receptor antagonist activity in the MVD, is a full agonist in the guinea pig ileum (GPI) smooth muscle preparation (see Section 5.7.2), and was the first compound reported to be a mixed μ opioid receptor agonist/δ opioid receptor antagonist (Schiller et al., 1992).

During examination of TIPP by nuclear magnetic resonance (NMR) spontaneous degradation via diketopiperazine formation occurred (Marsden et al., 1993). This led Schiller and colleagues to prepare TIPP[Ψ] and the tripeptide TIP[Ψ] (Schiller et al., 1993) containing a reduced peptide bond between Tic^2 and Phe^3 (Table 5.4). These analogs exhibit increased δ opioid receptor antagonist potency in the MVD and higher δ receptor affinity compared to the parent peptides and exceptional δ opioid receptor selectivity. This modification also enhances the metabolic stability of the peptides.

A variety of structural modifications have been made to all positions of TIPP (see Schiller et al., 1999b, 1999c; Aldrich and Vigil-Cruz, 2003 for reviews), and these can have a profound effect on the activity profile of these peptides. Numerous substitutions have been made in position 3, and the results indicate that an aromatic residue is not required in this position for δ receptor antagonist activity. Thus the cyclohexylalanine (Cha) analogs of both TIPP and TIPP[Ψ] (TICP and TICP[Ψ], respectively, Table 5.4) are approximately 10-fold more potent as δ receptor antagonists than the parent peptides (Schiller et al., 1996).

The substitution of Dmt in position 1 has been examined in a number of opioid peptides and can substantially increase opioid receptor affinity and opioid activity (see Lazarus et al., 1998; Bryant et al., 2003 for reviews). Incorporation of Dmt into TIPP yielded DIPP (Dmt–Tic–Phe–Phe) which is an extremely potent δ opioid receptor antagonist and, like $TIPPNH_2$, is also a full agonist in the GPI (Schiller et al., 1994a, 1994c). The nonpeptide δ opioid receptor antagonist naltrindole can prevent the development of morphine tolerance and dependence in mice (Abdelhamid et al., 1991), and therefore compounds with mixed μ agonist/δ antagonist activity could have therapeutic potential. To enhance μ opioid agonist activity Schiller and colleagues also incorporated Dmt into position 1 of $TIPPNH_2$ and TIPP[Ψ]NH_2; the resulting $DIPPNH_2$[Ψ] (Table 5.4) was the first compound with balanced μ agonist/δ antagonist properties (Schiller et al., 1994b, 1999a). $DIPPNH_2$[Ψ] is a potent analgesic following i.c.v. administration (three times more potent than morphine), produces less acute tolerance than morphine, and results in no physical dependence following chronic administration (Schiller et al., 1999a).

Based on the hypothesis that the "message" domain in these δ receptor antagonist peptides consisted of only the Tyr–Tic dipeptide, Temussi and colleagues synthesized Tyr–L/D-Tic–NH_2 (Temussi et al., 1994), and Lazarus and colleagues subsequently explored the SARs of Tyr–Tic in considerable detail (see Bryant et al., 1998; Lazarus et al., 1998; Schiller et al., 1999b for reviews, including extensive tables of analogs). The discovery that replacement of Tyr by Dmt in Tyr–Tic and related tripeptides substantially increased opioid receptor affinity (Salvadori et al., 1995) resulted in Dmt–Tic becoming the template for extensive structural modification (see Bryant et al., 1998, 2003; Aldrich and Vigil-Cruz, 2003 for reviews). Derivatives have been reported with a wide range of activity, including δ opioid agonist (Balboni et al., 2002), δ inverse agonist (Labarre et al., 2000), and μ opioid antagonist activity (Van den Eynde et al., 2005) in addition to δ opioid antagonist and μ opioid agonist activity.

5.5.5 Opioid Peptides with the Tyr–D-AA–Phe Sequence

As noted above, the amphibian peptides dermorphin and the deltorphins contain a D-amino acid between the Tyr and Phe residues. The SAR of these unusual and highly selective opioid peptides have been studied in detail, with amino acid substitutions examined in every position, and hundreds of analogs prepared (see Erspamer, 1992; Heyl and Schullery, 1997; Lazarus et al., 1999; Aldrich and Vigil-Cruz, 2003 for reviews). The stereochemistry of the amino acid at position 2 is critical for the opioid activity of these peptides, and peptides containing an L-amino acid in this position have very low opioid receptor affinity and potency (Amiche et al., 1989; Kreil et al., 1989; Lazarus et al., 1989). The deltorphins in particular have been used in a number of pharmacological studies of δ opioid receptors, and tritiated deltorphin II is commercially available. Deltorphin II has been postulated to preferentially stimulate the δ_2 receptor subtype, and therefore has been used to study δ opioid receptor subtypes (Traynor and Elliott, 1993; Fowler and Fraser, 1994; Zaki et al., 1996).

A number of other synthetic μ opioid receptor selective peptides with the Tyr–D-aa–Phe sequence have been reported. Linear peptides include tetrapeptide amides such as TAPP (Tyr–D-Ala–Phe–Phe–NH_2, Table 5.5; Schiller et al., 1989); this peptide can be considered an analog of endomorphin-2, although it was synthesized several years before the discovery of endomorphins. Incorporation of D-Arg in position 2 of tetrapeptide analogs of dermorphin (Table 5.5) resulted in peptides that are potent opioids in antinociceptive assays. Interestingly, the potent opioid peptide TAPS (Tyr–D-Arg–Phe–Sar; Sar = sarcosine) causes respiratory stimulation rather than respiratory depression (Paakkari et al., 1993) and antagonizes the respiratory depression caused by dermorphin; based on these results, TAPS was postulated to be a μ_1 agonist and a μ_2 antagonist *in vivo* (Paakkari et al., 1993). Combination of D-Arg in position 2 with a second basic residue in position 4 resulted in DALDA (Table 5.5; Schiller et al., 1989), which exhibits exceptional μ opioid receptor selectivity. Replacement of Tyr[1] in DALDA by Dmt resulted in [Dmt[1]]DALDA (Schiller et al., 2000) with 10-fold higher affinities for both μ and δ opioid receptors. This derivative is an extraordinarily potent analgesic (3000 times more potent than morphine following i.t. administration; Shimoyama et al., 2001) that shows promising pharmacological and pharmacokinetic properties (see Section 5.6.1). The cyclic tetrapeptide Tyr-*cyclo*[D-Orn–Phe–Asp]NH_2 (Schiller et al., 1985) also exhibits high selectivity for μ opioid receptors. The ring size in these cyclic lactams affects receptor selectivity, with derivatives with a larger ring size being nonselective for μ vs. δ opioid receptors.

For the peptides cyclized through a disulfide or dithioether linkage, the receptor selectivity depends on the linkage. The cyclic tetrapeptide JOM-13 (Table 5.5) is a δ-receptor selective agonist (Mosberg et al., 1988), but incorporation of an ethyl group between the two sulfurs and amidation of the C-terminus results in the μ opioid receptor selective peptide JOM-6 (Table 5.5; Mosberg et al., 1988). Conformationally constrained analogs of Tyr and Phe were incorporated into positions 1 and 3, respectively, of JOM-13 to study possible bioactive conformations for the side chains of these important residues when bound to δ opioid receptors (see Mosberg, 1999; Aldrich and Vigil-Cruz, 2003). The conformation of JOM-13 that was compatible with the receptor binding pocket in a computational model of the δ opioid receptor was different from that predicted based on comparison of JOM-13 to other δ opioid receptor ligands (Mosberg, 1999). These results are an indication that receptor interactions may differ for different ligands, and caution against a simplistic view of a single common binding mode for different ligands (Mosberg, 1999). Molecular modeling also suggested that hydrogen bonding between the phenol of Tyr[1] of the peptide and the His from TM6 in the opioid receptor binding site was less important for binding to μ than δ opioid receptors

TABLE 5.5
Selected Synthetic Opioid Peptides with the Tyr–D-aa–Phe Sequence

Peptide	Sequence	Receptor Preference
TAPP	Tyr–D-Ala–Phe–PheNH_2	μ
TAPS	Tyr–D-Arg–Phe–Sar	μ
DALDA	Tyr–D-Arg–Phe–LysNH_2	μ
[Dmt[1]]DALDA	Dmt–D-Arg–Phe–LysNH_2	μ
JOM-6	S CH_2CH_2 S (linking Cys and Pen) Tyr–D-Cys–Phe–D-PenNH_2	μ
JOM-13	Tyr–*cyclo*[D-Cys–Phe–D-Pen]	δ

(Pogozheva et al., 1998). This led Mosberg and colleagues to prepare the Phe[1] analog of JOM-6, which retains high affinity for μ opioid receptors but greatly reduced affinity for δ receptors, consistent with the expected results from the modeling (Mosberg et al., 1998); this peptide is a potent full agonist in the GPI. Indeed, an aromatic residue is not required in position 1 of this peptide, and the Cha[1] analog of JOM-6 retains moderate μ receptor affinity and agonist potency (McFadyen et al., 2000).

5.5.6 Other Peptides with High Affinity for Opioid Receptors

Peptides with affinity for opioid receptors that have sequences completely different from those of the endogenous opioid peptides have also been identified. Analogs of somatostatin that are μ opioid receptor antagonists have been prepared (see Hruby and Agnes, 1999 for a review). Somatostatin exhibits low affinity for opioid receptors, and the potent somatostatin analog SMS-201,995 (D-Phe–*cyclo*[Cys–Phe–D-Trp–Lys–Thr–Cys]–Thr-ol) was found to be an antagonist at μ opioid receptors (Maurer et al., 1982). Further structural modification yielded the analogs CTP, CTOP, and CTAP (Table 5.6; Pelton et al., 1985; 1986), which exhibit greatly reduced affinity for somatostatin receptors and enhanced affinity and selectivity for μ opioid receptors. These peptides are frequently used in pharmacological studies of μ opioid receptors.

Mixture-based combinatorial peptide libraries have been extensively explored by Houghten and colleagues and have led to the identification of a variety of peptides with affinity for opioid receptors (see Dooley and Houghten, 1999; Aldrich and Vigil-Cruz, 2003 for reviews). A variety of novel acetylated peptides with sequences unrelated to known opioid peptides have been identified for all three opioid receptors from combinatorial libraries. Acetylated hexapeptides, termed acetalins (Table 5.6), have been identified that are potent μ opioid receptor antagonists in the GPI (Dooley et al., 1993). Screening of a tetrapeptide library containing both L- and D-amino acids, including a number of unnatural amino acids, led to the identification of peptides with nanomolar affinity for each receptor type (Dooley et al., 1998). The most unusual results were those obtained for the κ opioid receptor, where the resulting peptides contained all D-amino acids (e.g., D-Phe–D-Phe–D-Nle–D-ArgNH$_2$; Table 5.6) and exhibit full agonist activity. Subsequent modification of the C-terminal amide resulted in FE200665 and FE200666 (Table 5.6), which are peripherally selective κ opioid receptor agonists that have analagesic and anti-inflammatory properties (Binder et al., 2001).

TABLE 5.6
Other Peptides with Affinity for Opioid and NOP Receptors

Peptides	Sequence	Receptor Preference
CTP, CTOP, and CTAP	D-Phe–*cyclo*[Cys–Phe–D-Trp–X–Thr–Cys]ThrNH$_2$ X = Lys, Orn, or Arg	μ antagonists
Acctalins	AcArg–Phe–Met–Trp–Met–XNH$_2$ X = Arg, Lys, or Thr	μ antagonists
	D-Phe–D-Phe–D-Nle–D-ArgNH$_2$	κ agonist
FE200665	D-Phe–D-Phe–D-Nle–D-ArgNH-4-picolyl	κ agonist
FE200666	D-Phe–D-Phe–D-Leu–D-Orn–morpholine amide	κ agonist
	AcArg–Tyr–Tyr–Arg–X–LysNH$_2$, X = Trp or Ile	NOP partial agonists
ZP120	AcArg–Tyr–Tyr–Arg–Trp–(Lys)$_7$NH$_2$	NOP partial agonist
OS461	$H_2NC(NH)NH(CH_2)_6$–Tyr–Tyr–Arg–TrpNH$_2$	NOP agonist
OS462	$H_2NC(NH)NH(CH_2)_6$–Dmt–Tyr–NRArgNH$_2$ R = (*R*)-1-(2-naphthyl)ethyl	NOP agonist
OS500	$H_2NC(NH)NH(CH_2)_6$–Dmt–Dmt–NRArgNH$_2$ R = (*R*)-1-(2-naphthyl)ethyl	NOP agonist

Peptidomimetic ligands for opioid receptors have also been identified from combinatorial libraries (Dooley and Houghten, 1999; Aldrich and Vigil-Cruz, 2003).

Combinatorial libraries have also been used to identify peptidic and peptidomimetic ligands for the NOP receptor that are not structurally related to N/OFQ (Zaveri, 2003; Chiou et al., 2007). Houghten and colleagues identified acetylated hexapeptides with high affinity for the NOP receptor, with AcArg–Tyr–Tyr–Arg–X–LysNH_2 (X = Trp or Ile; Table 5.6) having the highest affinity (Dooley et al., 1997); these peptides were initially reported to be potent antagonists, but later were found in most assays to exhibit partial agonist activity (Calò et al., 2000a; Chiou et al., 2007). Subsequent modification by attachment of six lysine residues at the C-terminus resulted in ZP120 (Table 5.6; Rizzi et al., 2002a; Kapusta et al., 2005), which, like the hexapeptides, exhibits partial agonist activity; *in vivo* this peptide is more potent, exhibits a longer duration of action and is restricted to the periphery (Rizzi et al., 2002a; Kapusta et al., 2005). A combinatorial library of conformationally constrained peptides resulted in the identification of the peptidomimetic III-BTD, which exhibits moderate affinity but only modest selectivity for NOP receptors; this compound acts as an antagonist at NOP receptors, but exhibits partial agonist activity at opioid receptors (Becker et al., 1999). Recently the effects of the three novel agonists OS461, OS462, and OS500—N-terminally modified tri- and tetrapeptide derivatives that exhibit subnanomolar affinity for NOP receptors—on food intake were described (Economidou et al., 2006).

5.5.7 New Approaches to the Design of Opioid Peptide Analogs

New approaches to the design of opioid peptide analogs have recently been explored, including replacement of key functionalities in Tyr^1 that have been thought to be critical for opioid receptor interaction. Schiller and colleagues have developed an approach for converting agonist opioid peptides into antagonists by replacing the basic N-terminal amine with a hydrogen or a methyl group. They initially introduced the novel Dmt analogs Dhp and Mdp (Figure 5.2) in place of Tyr^1 in dynorphin A-(1–11) amide to yield the potent κ opioid receptor selective antagonist dynantin (Lu et al., 2001). Subsequently they demonstrated that this is a general approach that is applicable to the preparation of a wide variety of opioid peptide derivatives with antagonist activity (Schiller et al., 2003). Recent reports have also demonstrated that the phenol of Tyr^1 in various opioid peptides can be replaced with a carboxamide group (Dolle et al., 2004) or other functionalities (Weltrowska et al., 2005) with retention of opioid receptor affinity and agonist potency.

5.6 CLINICAL APPLICATIONS

Narcotic analgesics such as morphine are used primarily for the treatment of severe pain. In addition to analgesia, these clinically used nonpeptide opioid agonists, which interact primarily with μ opioid receptors, display a plethora of biological effects. The most serious side effects associated with the majority of clinically used narcotic analgesics are respiratory depression, addiction liability, and the development of tolerance. Other side effects, for example, sedation and cognitive impairment, constipation, and nausea and vomiting, can also limit the clinical use of these agents. Therefore there has been considerable interest in identifying analgesics without these side effects.

5.6.1 Opioid and NOP Peptides

In order to use opioid peptide analogs as potential analgesics the peptides must be stable to metabolic degradation and be able to reach the targeted receptors. The incorporation of D- and unnatural amino acids, cyclization, amidation, modifications to peptide bonds, etc. which have been studied extensively in the different classes of opioid peptides (see Section 5.5) stabilize the peptides to metabolism, and therefore a number of opioid peptide analogs which exhibit activity following *in vivo* administration have been identified. In selected cases (i.e., inflammation) antinociception can be produced by activation of peripheral opioid receptors, and peripherally restricted opioid agonists, including a few peptides, have been examined as novel analgesic agents (see DeHaven-Hudkins and Dolle, 2004 for a review). Recently one of these peripherally selective peptides, the κ opioid receptor selective agonist FE200665 (CR665), has entered into clinical trials (Vanderah et al., 2008).

Narcotic analgesic effects are generally centrally mediated, however, and peptides must be able to cross the blood–brain barrier (BBB) in order to reach opioid receptors in the central nervous system. Several different approaches have been utilized to enhance the penetration of the BBB by opioid peptides (see Witt and Davis, 2006 for a recent review). Increasing lipophilicity by halogenation has been shown to increase the BBB permeability of several opioid peptides. Glycosylation has also been explored to increase the BBB penetration of opioid peptides (see Polt et al., 2005 for a review), and glycosylated opioid peptides exhibit analgesia after systemic (i.v.) administration. Interestingly, the glycosylated opioid peptide [D-Thr^2,Leu^5,Ser(β-D-Glc)^6]enkephalin amide does not produce the characteristic increases in locomotor activity or the stereotypic Straub tail response that morphine does (Polt et al., 2005). Some cationic peptides, including the dynorphin derivative E-2078 and the μ opioid receptor selective peptide Tyr–D-Arg–Phe–βAla, have been reported to penetrate the BBB through absorptive-mediated endocytosis (see Witt and Davis, 2006).

In recent years several opioid peptide analogs have shown promising *in vivo* activity profiles, suggesting that opioid peptides could ultimately find clinical use. [Dmt^1]DALDA is active after systemic administration (see Schiller, 2005 for a review), suggesting that this peptide can cross the BBB, and exhibits longer duration of antinociceptive action than morphine. Unlike morphine, [Dmt^1]DALDA does not produce respiratory depression after i.t. administration, so that it could have an improved safety profile. [Dmt^1]DALDA can also cross a Caco-2 cell monolayer (Zhao et al., 2003), suggesting that it may have reasonable oral absorption. The dimeric enkephalin analog biphalin is also active following systemic administration with effects comparable to morphine, but it produces less physical dependence (Yamazaki et al., 2001). The stabilized dynorphin analog E-2078 produces analgesia following systemic (i.v. and s.c.) administration (Nakazawa et al., 1990; Yoshino et al., 1990a, 1990b), apparently via spinal κ opioid receptors (Nakazawa et al., 1991); in humans its analgesic effect is comparable to the clinically used nonpeptide opioid pentazocine (Fujimoto and Momose, 1995). The dynorphin analog SK-9709 (Tyr–D-Ala–Phe–Leu–ArgΨ[CH_2NH]ArgNH_2) also exhibits antinociceptive activity following systemic (s.c.) administration (Hiramatsu et al., 2001). Recently the results of a phase II clinical trial of a dynorphin A-(1-13) derivative CJC-1008, which contains a maleimidopropionyl group that promotes its covalent attachment to serum albumin, in patients with postherpatic neuralgia were reported (Wallace et al., 2006); the peptide exhibited analgesic activity in these patients that lasted for up to 24 hours.

A number of clinical indications have been suggested for ligands for NOP receptors (see Bignan et al., 2005; Chiou et al., 2007), and clinical trials are being conducted on several NOP receptor antagonists (Bignan et al., 2005). This includes the peptide antagonist ZP-120, which has been filed for phase II clinical trials for the treatment of acute decompensated heart failure (Bignan et al., 2005). N/OFQ has also been reported to be in phase II clinical trials for the treatment of urinary incontinence (Bignan et al., 2005).

5.6.2 Peptidase Inhibitors

All of the peptides discussed so far in this chapter interact directly with opioid receptors. An alternative approach to producing analgesia using opioid peptides is to inhibit the metabolism of the endogenous peptides, thus increasing their concentration and occupancy of opioid receptors (see Roques et al., 1993a; Aldrich and Vigil-Cruz, 2003; Noble and Roques, 2007 for reviews). The two major enzymes involved in metabolism of the opioid peptides are the aminopeptidases, especially the bestatin-sensitive aminopeptidase-N (APN, EC 3.4.11.2) and the neutral endopeptidase-24.11 (NEP or neprilysin, EC 3.4.24.11; Roques et al., 1993b). Both of these enzymes have a broad substrate

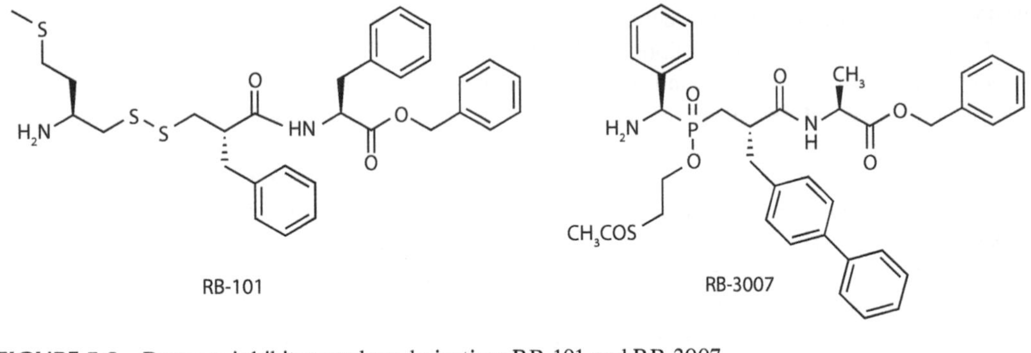

FIGURE 5.3 Protease inhibitor prodrug derivatives RB-101 and RB-3007.

specificity and cleave other peptides in addition to opioid peptides. NEP in particular exhibits much lower activity towards longer opioid peptides, and thus the opioid effects resulting from inhibition of this enzyme are probably mainly due to interfering with the metabolism of the enkephalins. Studies of peptidase inhibitors indicate the importance of inhibiting both enzymes in order to significantly increase the concentrations of the endogenous opioid peptides (Roques et al., 1993a; Roques, 2000). Both APN and NEP, as well as the other enzymes involved in enkephalin metabolism, are zinc metallopeptidases. Thus it has been possible to design mixed inhibitors capable of blocking multiple enzymes that more effectively protect the opioid peptides from metabolism (Roques et al., 1993a; Roques, 2000; Aldrich and Vigil-Cruz, 2003; Noble and Roques, 2007). One challenge was identifying inhibitors that can cross the BBB. An approach that successfully improved the bioavailability of the enzyme inhibitors was to link potent inhibitors of NEP and APN containing thiol groups by a disulfide bond (Fournié-Zaluski et al., 1992). The disulfide bond, while stable in plasma, is readily cleaved in the brain, and thus these compounds function as prodrugs. The dual inhibitor prodrug RB-101 (Figure 5.3) is very active in antinociceptive assays after either i.v. or s.c. administration of low doses (Noble et al., 1992; see Jutkiewicz, 2007; Noble and Roques, 2007 for recent reviews). In addition RB101 produces antidepressant and anxiolytic effects without the typical opioid analgesic side effects (respiratory depression, tolerance, dependence, abuse potential; see Jutkiewicz, 2007; Noble and Roques, 2007), suggesting a promising approach to the development of a new class of therapeutic agents for the treatment of pain and other disease states. The first true dual inhibitors of these two enzymes with nanomolar affinity for both enzymes have been developed (Chen et al., 1998, 2000); prodrug derivatives of these aminophosphinic compounds (e.g., RB-3007; Figure 5.3) produce an antinociceptive response following i.v. or intraperitoneal (i.p.) administration that lasted longer than RB101 (~2 h; Chen et al., 2001; Le Guen et al., 2003). Because these peptidase inhibitors may also inhibit the metabolism of other neuropeptides that are susceptible to metabolism by APN and NEP, it will be important to examine them for possible nonopioid effects.

5.7 COMMON METHODOLOGIES AND ASSAY SYSTEMS

A variety of assays are routinely used to evaluate the pharmacological activity of opioid peptides. These include radioligand binding assays to measure affinity, functional assays based on second messenger systems and isolated tissues, and *in vivo* assays for antinociceptive activity.

5.7.1 Radioligand Binding Assays

New compounds are routinely evaluated for their affinity for each of the opioid receptor types in radioligand binding assays. Tritiated ligands selective for each receptor type are available. Thus [^{3}H]DAMGO is most often used as the radioligand in binding assays for μ opioid receptors, while

[^{3}H]DPDPE is commonly used to determine the affinity of compounds for δ opioid receptors. The highly selective κ opioid receptor nonpeptide ligand [^{3}H]U69,593 is commonly used to determine κ opioid receptor affinity, and ^{125}I- and ^{3}H-labeled N/OFQ are available for use in radioligand binding assays for NOP receptors. Other tritiated opioid ligands that are available for use in binding assays include the δ opioid receptor selective ligands [^{3}H]deltorphin II and [^{3}H]naltrindole (a nonpeptide antagonist), [^{3}H]DADLE, and the nonselective nonpeptide antagonists [^{3}H]diprenorphine and [^{3}H]-aloxone. With the cloning of opioid receptors, cells such as Chinese hamster ovary (CHO) or human embryonic kidney (HEK) cells into which individual receptor types have been cloned are now typically used in binding assays. Earlier assays used rodent brain homogenates that contain all three receptor types, although in varying amounts depending upon the species (Aldrich and Vigil-Cruz, 2003). Binding assays in brain homogenate are complicated by the presence of multiple types of opioid receptors.

5.7.2 *In Vitro* Functional Assays

Cells expressing cloned opioid receptors are also used in functional assays to determine the efficacy and potency of opioid peptides. These assays measure the effects of these ligands on signal transduction systems. Because opioid receptors are G-protein coupled receptors, measuring the stimulation of binding of the radiolabeled nonhydrolyzable GTP analog [^{35}S]GTPγS (Traynor and Nahorski, 1995; Selley et al., 1997) following interaction of a compound with the receptor can be used to determine the compound's efficacy and potency. Inhibition of adenylyl cyclase is also frequently used as a functional assay to evaluate the efficacy and potency of opioid ligands at opioid receptors (Soderstrom et al., 1997; Dooley et al., 1998).

Smooth muscle preparations have been used extensively to characterize opioid ligands (see Leslie, 1987; Smith and Leslie, 1993 for reviews), especially prior to the cloning of the opioid receptors. The electrically stimulated GPI myenteric plexus-longitudinal muscle and the MVD preparations have been the tissues most extensively used. In these tissues opioids inhibit smooth muscle contraction by inhibiting the release of neurotransmitters. Both of these tissues contain more than one opioid receptor type, complicating the analysis of the results of these assays. The GPI contains both μ and κ opioid receptors with little if any functional δ opioid receptors, while the MVD contains all three opioid receptor types. Vas deferens from other species appear to contain predominantly a single receptor population and have been used to characterize opioids; the rabbit vas deferens appears to contain only κ opioid receptors (Oka et al., 1980), while the hamster vas deferens contains only δ opioid receptors (McKnight et al., 1985).

5.7.3 *In Vivo* Evaluation in Antinociceptive Assays

A variety of antinociceptive assays utilizing different noxious stimuli and different animal species have been used to examine the activity of potential analgesics. Animal models for pain include models of acute pain (e.g., hot plate, tail flick, paw pressure, and abdominal constriction assays), persistent pain (e.g., the formalin test), chronic pain (e.g., adjuvant-induced arthritis), and neuropathic pain (e.g., nerve ligation; see Table 2.1 of Cowen, 1999). The different types of noxious stimuli commonly used in these antinociceptive assays include heat (e.g., hot plate and tail flick assays), pressure (e.g., tail pinch and paw pressure assays), chemical (abdominal constriction or writhing assay and the formalin test), and electrical (tail shock vocalization) stimuli. Among the most commonly used assays are the hot plate, tail flick, and abdominal constriction assays. In the hot plate test, the latency to various behavioral responses (e.g., forepaw or hindpaw licking) is measured when the animal is placed on a hot plate, typically set at 55°C, while the tail flick measures the time for the animal to flick its tail away from radiant heat focused on the tail or withdraw its tail from water typically heated to 55°C. In the abdominal constriction (writhing) assay the animal is injected intraperitoneally (i.p.) with an irritant (typically acetic acid or phenylquinone) and the dose of a test compound required to abolish the abdominal constrictions is determined. Based on comparison of the potencies of standard opioids in several pain models, it was concluded that the mouse abdominal constriction assay (using 0.4% acetic acid) was the most sensitive to opioids, while the mouse hot plate test (at 55°C) was the least sensitive and the rat tail flick test was intermediate (Cowen, 1999). For a detailed discussion of these antinociceptive assays readers are referred to excellent reviews (Tjølsen and Hole, 1997; Cowen, 1999) that describe these procedures and discuss methodological issues with their execution.

Activation of all three types of opioid receptors can produce antinociception, but there are significant differences in the effects of activating different receptor types depending upon the noxious stimulus used and the animal species (see Millan, 1990; Porreca and Burks, 1993 for excellent reviews; see particularly Table 1 of Porreca and Burks, 1993 for a summary of supraspinal opioid receptor involvement in antinociceptive assays). While μ opioid agonists are active against all types of noxious stimuli, the activity of κ opioid agonists depends on the type of stimulus. The κ agonists are active against chemical and pressure stimuli, but they are inactive against electrical stimulation and their efficacy against thermal stimuli is dependent on the intensity (Millan, 1989). Delta agonists may be active against all four types of stimuli in mice, depending on the route of administration (Porreca and Burks, 1993), but there are species differences. It has been reported that while δ opioid agonists are effective in both the tail-flick and hot-plate assays in mice, in rats they are active in the hot-plate, but not tail-flick assays (Heyman et al., 1988).

The antinociceptive activity of opioid ligands can be evaluated following administration by a variety of routes, including by supraspinal (i.c.v.), spinal (i.t.), and peripheral (i.v., s.c., i.p., or oral) administration. Opioid peptide analogs are typically examined initially following i.c.v. or i.t. administration to assess the antinociceptive or antagonist activity of the peptides without complications related to peripheral metabolism or the ability of the peptides to cross the BBB. The demonstration of antinociceptive or antagonist activity following peripheral administration has been demonstrated for several opioid peptide analogs that are stabilized to metabolism (see Section 5.6.1), suggesting that they can cross the BBB.

5.8 FUTURE DEVELOPMENTS

The identification of systemically active opioid peptide derivatives is an exciting advance in the area of opioid peptide research and is a major step toward the development of opioid peptides for potential clinical use. Differences in the pharmacological profiles of opioid peptides and morphine, particularly differences in the side effect profiles (see Section 5.6), suggest that some opioid peptides could have significant advantages over morphine and other currently used nonpeptide narcotic analgesics in terms of safety. However, the delivery of such peptide therapeutic agents remains a challenge. The recent report that [Dmt^1]DALDA can cross the Caco-2 cell model of the intestinal barrier (Zhao et al., 2003) raises the intriguing possibility that some opioid peptide analogs could be active following oral administration. This is a very exciting development, but it is likely that oral bioavailability will be limited to only very few opioid peptide derivatives. Therefore, alternative methods for systemic administration of opioid peptides will be important for their application as therapeutic agents; the development of methods that do not involve injection (e.g., inhalation; Brugos et al., 2004) are likely to increase the acceptance of therapeutic agents that are not taken orally.

A very promising approach to improving the therapeutic profile for narcotic analgesics is the development of compounds that target both μ and δ opioid receptors (see Schiller, 2005; Ananthan, 2006 for reviews). The early study demonstrating that a selective δ opioid receptor antagonist reduced the development of morphine tolerance and dependence (Abdelhamid et al., 1991) prompted considerable interest in this approach to potential new analgesic agents. As noted in Section 5.5.4.2, modification of TIPPNH$_2$, the first compound with a mixed μ agonist/δ antagonist profile (Schiller et al., 1992), resulted in the analog DIPPNH$_2$[Ψ] (Schiller et al., 1999a), which exhibits a balanced profile between the two activities (see Schiller, 2005; Ananthan, 2006). The ability of this peptide to cross the BBB is limited, however, and therefore new peptide analogs with improved penetration into the CNS still need to be developed. One approach that is being explored is the preparation of a chimeric peptide between a TIPP derivative (TICP[Ψ]) and [Dmt[1]]DALDA to impart BBB permeability (Weltrowska et al., 2004). The δ opioid receptor antagonist Dmt–Tic has also been linked to both the μ opioid peptide agonist endomorphin 2 (Salvadori et al., 2007) and a nonpeptide opioid agonist (Neumeyer et al., 2006) to produce ligands targeting multiple opioid receptors. Developing peptides that can interact with other receptors [e.g., cholecystokinin (Hruby et al., 2006) or neurokinin receptors; Yamamoto et al., 2007] in addition to opioid receptors is another approach to improving the pharmacological profile that is being explored.

A recent development that could lead to new approaches in opioid ligand design and the identification of completely new types of opioid ligands, both peptide and nonpeptide, is the characterization of opioid receptor dimers. This is particularly true for receptor heterodimers, both between different opioid receptor types and between opioid receptors and other GPCRs (see Section 5.3.3). The different preferences of δ opioid receptor selective peptides for postulated δ opioid receptor subtypes (see Section 5.3.3) suggest that it should be possible to identify opioid ligands that selectively target such receptor heterodimers. Clearly this increases the potential complexity of opioid receptor pharmacology, but also opens up whole new approaches for opioid ligand design. This is clearly a very exciting time to be involved in the study of these remarkable peptides!

ACKNOWLEDGMENTS

The author thanks Dr. Mark Del Borgo, Dr. Nicolette Ross, Bhaswati Sinha, and Katherine Prevatt-Smith for their assistance with preparation of the manuscript. Research in the author's laboratory on opioid peptides is supported by the National Institute on Drug Abuse.

REFERENCES

Abdelhamid, E.E. et al., Selective blockage of delta-opioid receptors prevents the development of morphine-tolerance and dependence in mice, *J. Pharmacol. Exp. Ther.*, 258, 299–303, 1991.

Akil, H. et al., Endogenous opioids: Overview and current issues, *Drug Alcohol Depend.*, 51, 127–140, 1998.

Aldrich, J.V. and Vigil-Cruz, S.C., Narcotic analgesics, in *Burger's Medicinal Chemistry and Drug Discovery*, Abraham, D.J., Ed., John Wiley & Sons, Inc., New York, 6, 2003, p. 329–481.

Ambo, A. et al., Structure–activity studies on nociceptin analogues: ORL1 receptor binding and biological activity of cyclic disulfide-containing analogues of nociceptin peptides, *J. Med. Chem.*, 44, 4015–4018, 2001.

Amiche, M. et al., Dermenkephalin (Tyr–D-Met–Phe–His–Leu–Met–Asp–NH_2): A potent and fully specific agonist for the δ opioid receptor, *Mol. Pharmacol.*, 35, 774–779, 1989.

Ananthan, S., Opioid ligands with mixed mu/delta opioid receptor interactions: An emerging approach to novel analgesics, *AAPS J.*, 8, E118–E125, 2006.

Balboni, G. et al., Potent delta-opioid receptor agonists containing the Dmt–Tic pharmacophore, *J. Med. Chem.*, 45, 5556–5563, 2002.

Barber, A. and Gottschlich, R., Opioid agonists and antagonists: An evaluation of their peripheral actions in inflammation, *Med. Res. Rev.*, 12, 525–562, 1992.

Bartosz-Bechowski, H. et al., Cyclic enkephalin analogs with exceptional potency at peripheral delta opioid receptors, *J. Med. Chem.*, 37, 146–150, 1994.

Beardsley, P.M. et al., Differential effects of the novel kappa opioid receptor antagonist, JDTic, on reinstatement of cocaine-seeking induced by footshock stressors *vs.* cocaine primes and its antidepressant-like effects in rats, *Psychopharmacol. (Berl.)*, 183, 118–126, 2005.

Becker, J.A.J. et al., Ligands for κ-opioid and ORL1 receptors identified from a conformationally constrained peptide combinatorial library, *J. Biol. Chem.*, 274, 27513–27522, 1999.

Bennett, M.A., Murray, T.F., and Aldrich, J.V., Identification of arodyn, a novel acetylated dynorphin A-(1-11) analogue, as a κ opioid receptor antagonist, *J. Med. Chem.*, 45, 5617–5619, 2002.

Berezowska, I. et al., Dicarba analogues of the cyclic enkephalin peptides H-Tyr–*c*[D-Cys–Gly–Phe–D (or L)-Cys]NH_2 retain high opioid activity, *J. Med. Chem.*, 50, 1414–1417, 2007.

Bignan, G.C., Connolly, P.J., and Middleton, S.A., Recent advances towards the discovery of ORL-1 receptor agonists and antagonists, *Exp. Opin. Ther. Patents*, 15, 357–388, 2005.

Bigoni, R. et al., Pharmacological characterisation of [(pX)Phe^4]nociceptin(1–13)amide analogues. 1. *In vitro* studies, *Naunyn-Schmiedeberg's Arch. Pharmacol.*, 365, 442–449, 2002.

Binder, W. et al., Analgesic and anti-inflammatory effects of two novel κ-opioid peptides, *Anesthesiology*, 94, 1034–1044, 2001.

Bodnar, R.J., Endogenous opiates and behavior: 2006, *Peptides*, 28, 2435–2513, 2007.

Bodnar, R.J. and Klein, G.E., Endogenous opiates and behavior: 2005, *Peptides*, 27, 3391–3478, 2006.

Brantl, V. et al., Novel opioid peptides derived from casein (β-casomorphins). I. Isolation from bovine casein peptone, *Hoppe-Seylers Z. Physiol. Chem.*, 360, 1211–1216, 1979.

Brugos, B., Arya, V., and Hochhaus, G., Stabilized dynorphin derivatives for modulating antinociceptive activity in morphine tolerant rats: Effect of different routes of administration, *AAPS J.*, 6, e36, 2004.

Bryant, S.D. et al., New delta-opioid antagonists as pharmacological probes, *Trends Pharmacol. Sci.*, 19, 42–46, 1998.

Bryant, S.D. et al., Dmt and opioid peptides: A potent alliance, *Biopolymers*, 71, 86–102, 2003.

Butour, J.-L. et al., Recognition and activation of the opioid receptor-like ORL1 receptor by nociceptin, nociceptin analogs, and opioids, *Eur. J. Pharmacol.*, 321, 97–103, 1997.

Calò, G. et al., Nociceptin/orphanin FQ receptor ligands, *Peptides*, 21, 935–947, 2000a.

Calò, G. et al., Characterization of [Nphe^1]NC(1–13)NH_2, a new selective nocieptin receptor antagonist, *Br. J. Pharmacol.*, 129, 1183–1193, 2000b.

Calò, G. et al., Review: Pharmacology of nociceptin and its receptor: A novel therapeutic target, *Br. J. Pharmacol.*, 129, 1261–1283, 2000c.

Calò, G. et al., Pharmacological profile of nociceptin/orphanin FQ receptors, *Clin. Exp. Pharmacol. Physiol.*, 29, 223–228, 2002a.

Calò, G. et al., [Nphe^1,Arg^{14},Lys^{15}]nociceptin-NH_2, a novel potent and selective antagonist of the nociceptin/orphanin FQ receptor, *Br. J. Pharmacol.*, 136, 303–311, 2002b.

Calò, G. et al., UFP-101, a peptide antagonist selective for the nociceptin/orphanin FQ receptor, *CNS Drug Rev.*, 11, 97–112, 2005.

Carey, A.N. et al., Reinstatement of cocaine place-conditioning prevented by the peptide kappa-opioid receptor antagonist arodyn, *Eur. J. Pharmacol.*, 569, 84–89, 2007.

Carra, G. et al., [(pF)Phe^4,Arg^{14},Lys^{15}]N/OFQ-NH_2 (UFP-102), a highly potent and selective agonist of the nociceptin/orphanin FQ receptor, *J. Pharmacol. Exp. Ther.*, 312, 1114–1123, 2005.

Caudle, R.M. and Mannes, A.J., Dynorphin: Friend or foe? *Pain*, 87, 235–239, 2000.

Chang, K.-J. et al., Morphiceptin (NH_4–Tyr–Pro–Phe–Pro–$CONH_2$): A potent and specific agonist for morphine (μ) receptors, *Science*, 212, 75–77, 1981.

Chang, K.-J. et al., Potent morphiceptin analogs: Structure activity relationships and morphine-like activities, *J. Pharmacol. Exp. Ther.*, 227, 403–408, 1983.

Charoenchai, L. et al., High affinity conformationally constrained nociceptin/orphanin FQ(1–13) amide analogues, *J. Med. Chem.*, 51, 4385–4387, 2008.

Chaturvedi, K. et al., Structure and regulation of opioid receptors, *Biopolymers*, 55, 334–346, 2000.

Chavkin, C. and Goldstein, A., A specific receptor for the opioid peptide dynorphin: Structure–activity relationships, *Proc. Natl Acad. Sci. USA*, 78, 6543–6547, 1981.

Chen, H. et al., Aminophosphinic inhibitors as transition state analogues of enkephalin-degrading enzymes: A class of central analgesics, *Proc. Natl Acad. Sci. USA*, 95, 12028–12033, 1998.

Chen, H. et al., Phosphinic derivatives as new dual enkephalin-degrading enzyme inhibitors: Synthesis, biological properties, and antinociceptive activities, *J. Med. Chem.*, 43, 1398–1408, 2000.

Chen, H. et al., Long lasting antinociceptive properties of enkephalin degrading enzyme (NEP and APN) inhibitor prodrugs, *J. Med. Chem.*, 44, 3523–3530, 2001.

Chen, Y. et al., Molecular cloning and functional expression of a μ-opioid receptor from rat brain, *Mol. Pharmacol.*, 44, 8–12, 1993a.

Chen, Y. et al., Molecular cloning of a rat κ opioid receptor reveals sequence similarities to the μ and δ opioid receptors, *Biochem. J.*, 295, 625–628, 1993b.

Childers, S.R., Opioid receptor-coupled second messenger systems, in *Opioids I.*, Herz, A., Akil, H., and Simon, E.J., Ed., Springer-Verlag, Berlin, 104/I, 1993, p. 189–216.

Chiou, L.C. et al., Nociceptin/orphanin FQ peptide receptors: Pharmacology and clinical implications, *Curr. Drug Targets*, 8, 117–135, 2007.

Choi, H. et al., N-Terminal alkylated derivatives of [D-Pro^{10}]dynorphin A-(1–11) are highly selective for κ-opioid receptors, *J. Med. Chem.*, 35, 4638–4639, 1992.

Choi, H. et al., Synthesis and opioid activity of [D-Pro[10]]dynorphin A-(1–11) analogues with N-terminal alkyl substitution, *J. Med. Chem.*, 40, 2733–2739, 1997.

Cone, R.I. et al., Regional distribution of dynorphin and neo-endorphin peptides in rat brain, spinal cord, and pituitary, *J. Neurosci.*, 3, 2146–2152, 1983.

Coop, A. and Rice, K.C., Role of δ-opioid receptors in biological processes, *Drug News Perspect.*, 13, 481–487, 2000.

Costa, T. and Herz, A., Antagonists with negative intrinsic activity at delta opioid receptors coupled to GTP-binding proteins, *Proc. Natl Acad. Sci. USA*, 86, 7321–7325, 1989.

Cotton, R. et al., ICI 174864: A highly selective antagonist for the opioid δ-receptor, *Eur. J. Pharmacol.*, 97, 331–332, 1984.

Cowen, A., Animal models of pain, in *Novel Aspects of Pain Management: Opioids and Beyond*, Sawynok, J. and Cowen, A., Ed., Wiley-Liss, New York, 1999, p. 21–47.

Cowen, M.S. and Lawrence, A.J., Alcoholism and neuropeptides: An update, *CNS Neurol. Disord. Drug Targets*, 5, 233–239, 2006.

Cox, B.M., Opioid receptor-G protein interactions: Acute and chronic effects of opioids, in *Opioids I*, Herz, A., Akil, H., and Simon, E.J., Ed., Springer-Verlag, Berlin, 104/I, 1993, p. 145–188.

Cox, B.M. et al., A peptide-like substance from pituitary that acts like morphine. 2. Purification and properties, *Life Sci.*, 16, 1777–1782, 1975.

Cvejic, S. and Devi, L., Dimerization of the delta opioid receptor: Implications for a function in receptor internalization, *J. Biol. Chem.*, 272, 26959–26964, 1997.

DeHaven-Hudkins, D.L. and Dolle, R.E., Peripherally restricted opioid agonists as novel analgesic agents, *Curr. Pharm. Des.*, 10, 743–757, 2004.

Dhawan, B.N. et al., International Union of Pharmacology. XII. Classification of opioid receptors, *Pharmacol. Rev.*, 48, 567–592, 1996.

Dolle, R.E. et al., (4-Carboxamido)phenylalanine is a surrogate for tyrosine in opioid receptor peptide ligands, *Bioorg. Med. Chem. Lett.*, 14, 3545–3548, 2004.

Dooley, C.T. and Houghten, R.A., Orphanin FQ: Receptor binding and analog structure activity relationships in rat brain, *Life Sci.*, 59, 23–29, 1996.

Dooley, C.T. and Houghten, R.A., New opioid peptides, peptidomimetics, and heterocyclic compounds from combinatorial libraries, *Biopolymers*, 51, 379–390, 1999.

Dooley, C.T. et al., Acetalins: Opioid receptor antagonists determined through the use of synthetic peptide combinatorial libraries, *Proc. Natl Acad. Sci. USA*, 90, 10811–10815, 1993.

Dooley, C.T. et al., Binding and *in vitro* activities of peptides with high affinity for the nociceptin/orphanin FQ receptor, ORL1, *J. Pharmacol. Exp. Ther.*, 283, 735–741, 1997.

Dooley, C.T. et al., Selective ligands for the μ, δ, and κ opioid receptors identified from a single mixture based tetrapeptide positional scanning combinatorial library, *J. Biol. Chem.*, 273, 18848–18856, 1998.

Economidou, D. et al., Effect of novel NOP receptor ligands on food intake in rats, *Peptides*, 27, 775–783, 2006.

Erspamer, V., The opioid peptides of the amphibian skin, *Int. J. Devel. Neuroscience*, 10, 3–30, 1992.

Erspamer, V. et al., Deltorphins: A family of naturally occurring peptides with high affinity and selectivity for δ opioid binding sites, *Proc. Natl Acad. Sci. USA*, 86, 5188–5192, 1989.

Evans, C.J. et al., Cloning of a delta opioid receptor by functional expression, *Science*, 258, 1952–1955, 1992.

Fan, T. et al., A role for the distal carboxyl tails in generating the novel pharmacology and G protein activation profile of mu and delta opioid receptor hetero-oligomers, *J. Biol. Chem.*, 280, 38478–38488, 2005.

Fichna, J. et al., Synthesis and characterization of potent and selective mu-opioid receptor antagonists, [Dmt[1], D-2-Nal[4]]endomorphin-1 (Antanal-1) and [Dmt[1], D-2-Nal[4]]endomorphin-2 (Antanal-2), *J. Med. Chem.*, 50, 512–520, 2007.

Florin, S. et al., Orphan neuropeptide NocII, a putative pronociceptin maturation product, stimulates locomotion in mice, *Neuroreport*, 8, 705–707, 1997.

Foord, S.M. et al., International Union of Pharmacology. XLVI. G protein-coupled receptor list, *Pharmacol. Rev.*, 57, 279–288, 2005.

Fournié-Zaluski, M.-C. et al., Structural requirements for specific recognition of μ and δ opiate receptors, *Mol. Pharmacol.*, 20, 484–491, 1981.

Fournié-Zaluski, M.-C. et al., Mixed inhibitor-prodrug as a new approach towards systemically active inhibitors of enkephalin degrading enzymes, *J. Med. Chem.*, 35, 2473–2481, 1992.

Fowler, C.J. and Fraser, G.L., Invited review: Mu-, delta-, and kappa-opioid receptors and their subtypes. A critical review with emphasis on radioligand binding experiments, *Neurochem. Int.*, 24, 401–426, 1994.

Fujimoto, K. and Momose, T., Analgesic efficacy of E-2078 (dynorphin analog) in patients following abdominal surgery, *Jpn. J. Anesth.*, 44, 1233–1237, 1995.

Fundytus, M.E. et al., Attenuation of morphine tolerance and dependence with the highly selective delta-opioid receptor antagonist TIPP[psi], *Eur. J. Pharmacol.*, 286, 105–108, 1995.

Gacel, G. et al., Development of conformationally constrained linear peptides exhibiting a high affinity and pronounced selectivity for δ opioid receptors, *J. Med. Chem.*, 31, 1891–1897, 1988.

Gacel, G. et al., Synthesis, biochemical, and pharmacological properties of BUBUC, a highly selective and systemically active agonist for *in vivo* studies of δ opioid receptors, *Peptides*, 11, 983–988, 1990.

Gairin, J.E. et al., N,N-Diallyl-tyrosyl substitution confers antagonist properties on the κ-selective opioid peptide [D-Pro^{10}]dynorphin A-(1–11), *Br. J. Pharmacol.*, 95, 1023–1030, 1988.

Gaveriaux-Ruff, C. and Kieffer, B.L., Opioid receptors: Gene structure and function, in *Opioids in Pain Control: Basic and Clinical Aspects*, Stein, C., Ed., Cambridge University Press, Cambridge, 1999, p. 1–20.

Gaveriaux-Ruff, C. and Kieffer, B.L., Opioid receptor genes inactivated in mice: The highlights, *Neuropeptides*, 36, 62–71, 2002.

Gavioli, E.C. and Calò, G., Antidepressant- and anxiolytic-like effects of nociceptin/orphanin FQ receptor ligands, *Naunyn-Schmiedeberg's Arch. Pharmacol.*, 372, 319–330, 2006.

Gavioli, E.C. et al., Blockade of nociceptin/orphanin FQ-NOP receptor signalling produces antidepressant-like effects: Pharmacological and genetic evidences from the mouse forced swimming test, *Eur. J. Neurosci.*, 17, 1987–1990, 2003.

Gentilucci, L. and Tolomelli, A., Recent advances in the investigation of the bioactive conformation of peptides active at the μ-opioid receptor. Conformational analysis of endomorphins, *Curr. Top. Med. Chem.*, 4, 105–121, 2004.

Gentry, C.L. et al., The effect of halogenation on blood–brain barrier permeability of a novel peptide drug, *Peptides*, 20, 1229–1238, 1999.

George, S.R. et al., Oligomerization of μ- and δ-opioid receptors. Generation of novel functional properties, *J. Biol. Chem.*, 275, 26128–26135, 2000.

Gilbert, P.E. and Martin, W.R., The effects of morphine- and nalorphine-like drugs in the nondependent, morphine-dependent, and cyclazocine-dependent chronic spinal dog, *J. Pharmacol. Exper. Ther.*, 198, 66–82, 1976.

Goldstein, A., Biology and chemistry of the dynorphin peptides, in *The Peptides: Analysis, Synthesis, Biology. Volume 6. Opioid Peptides: Biology, Chemistry, and Genetics*, Udenfried, S. and Meienhofer, J., Ed., Academic Press, Orlando, 1984, p. 95–145.

Goldstein, A. et al., Porcine pituitary dynorphin: Complete amino acid sequence of the biologically active heptadecapeptide, *Proc. Natl Acad. Sci. USA*, 78, 7219–7223, 1981.

Gomes, I. et al., Heterodimerization of mu and delta opioid receptors: A role in opiate synergy, *J. Neurosci.*, 20, RC110, 2000.

Goodman, M. and Mierke, D.F., Configurations of morphiceptins by ^{1}H and ^{13}C NMR spectroscopy, *J. Am. Chem. Soc.*, 111, 3489–3496, 1989.

Grisel, J.E. and Mogil, J.S., Effects of supraspinal orphanin FQ/nociceptin, *Peptides*, 21, 1037–1045, 2000.

Guerrini, R. et al., Address and message sequences for the nociceptin receptor: A structure-activity study of nociceptin-(1–13)-peptide amide, *J. Med. Chem.*, 40, 1789–1793, 1997.

Guerrini, R. et al., A new selective antagonist of the nociceptin receptor, *Br. J. Pharmacol.*, 123, 163–165, 1998.

Guerrini, R. et al., Structure–activity relationships of nociceptin and related peptides: Comparison with dynorphin A, *Peptides*, 21, 923–933, 2000.

Gupta, A., Decaillot, F.M., and Devi, L.A., Targeting opioid receptor heterodimers: Strategies for screening and drug development, *AAPS J.*, 8, E153–E159, 2006.

Gutstein, H.B. and Akil, H., Opioid analgesics, in *Goodman and Gilman's The Pharmacological Basis of Therapeutics*, Hardman, J., Limbird, L., and Gilman, A., Ed., Mc-Graw-Hill Medical Publishing Division, New York, 2001, p. 569–619.

Hambrook, J.M. et al., Mode of deactivation of the enkephalins by rat and human plasma and rat brain homogenates, *Nature*, 262, 782–783, 1976.

Handa, B.K. et al., Analogues of β-LPH_{61-64} possessing selective agonist activity at μ-opiate receptors, *Eur. J. Pharmacol.*, 70, 531–540, 1981.

Harrison, L.M. and Grandy, D.K., Opiate modulating properties of nociceptin/orphanin FQ, *Peptides*, 21, 151–172, 2000.

Hasbi, A. et al., Trafficking of preassembled opioid mu-delta heterooligomer-Gz signaling complexes to the plasma membrane: Coregulation by agonists, *Biochemistry*, 46, 12997–13009, 2007.
Hauser, K.F. et al., Pathobiology of dynorphins in trauma and disease, *Front Biosci.*, 10, 216–235, 2005.
He, L. and Lee, N.M., Delta opioid receptor enhancement of mu opioid receptor-induced antinociception in spinal cord, *J. Pharmacol. Exp. Ther.*, 285, 1181–1186, 1998.
Henschen, A. et al., Novel opioid peptides derived from casein (β-casomorphins). II. Structure of active components from bovine casein peptone, *Hoppe-Seylers Z. Physiol. Chem.*, 360, 1217–1224, 1979.
Herz, A., Akil, H., and Simon, E.J., Eds., Opioids I, *Handbook of Experimental Pharmacology*, Springer-Verlag, Berlin, 104/I, 1993a.
Herz, A., Akil, H., and Simon, E.J., Eds., Opioids II, *Handbook of Experimental Pharmacology*, Springer-Verlag, Berlin, 104/II, 1993b.
Heyl, D.L. and Schullery, S.E., Developments in the structure–activity relationships for the δ-selective opioid peptides of amphibian skin, *Curr. Med. Chem.*, 4, 117–150, 1997.
Heyman, J.S. et al., Can supraspinal δ receptors mediate antinocieption? *Trends Pharmacol. Sci.*, 9, 134–138, 1988.
Hiramatsu, M. et al., Long-lasting antinociceptive effects of a novel dynorphin analogue, Tyr–D-Ala–Phe–Leu–Arg-Ψ(CH_2NH)Arg–NH_2, in mice, *Br. J. Pharmacol.*, 132, 1948–1956, 2001.
Höllt, V., Opioid peptide processing and receptor selectivity, *Annu. Rev. Pharmacol. Toxicol.*, 26, 59–77, 1986.
Horan, P.J. et al., Antinociceptive profile of biphalin, a dimeric enkephalin analog, *J. Pharmacol. Exp. Ther.*, 265, 1446–1454, 1993.
Hruby, V.J., Design in topographical space of peptide and peptidomimetic ligands that affect behavior. A chemist's glimpse at the mind—body problem, *Acc. Chem. Res.*, 34, 389–397, 2001.
Hruby, V.J. and Agnes, R.S., Conformation–activity relationships of opioid peptides with selective activities at opioid receptors, *Biopolymers*, 51, 391–410, 1999.
Hruby, V.J. and Gehrig, C.A., Recent developments in the design of receptor specific opioid peptides, *Med. Res. Rev.*, 9, 343–401, 1989.
Hruby, V.J. et al., New paradigms and tools in drug design for pain and addiction, *AAPS J.*, 8, E450–E460, 2006.
Huber, J.D. et al., Conjugation of low molecular weight poly(ethylene glycol) to biphalin enhances antinociceptive profile, *J. Pharm. Sci.*, 92, 1377–1385, 2003.
Hughes, J. et al., Identification of two related pentapeptides from the brain with potent opiate agonist activity, *Nature*, 258, 577–579, 1975.
Janecka, A., Fichna, J., and Janecki, T., Opioid receptors and their ligands, *Curr. Top. Med. Chem.*, 4, 1–17, 2004.
Janecka, A. and Kruszynski, R., Conformationally restricted peptides as tools in opioid receptor studies, *Curr. Med. Chem.*, 12, 471–481, 2005.
Janecka, A., Staniszewska, R., and Fichna, J., Endomorphin analogs, *Curr. Med. Chem.*, 14, 3201–3208, 2007.
Jordan, B.A. and Devi, L.A., G-Protein-coupled receptor heterodimerization modulated receptor function, *Nature*, 399, 697–700, 1999.
Jutkiewicz, E.M., The antidepressant-like effects of delta-opioid receptor agonists, *Mol. Interv.*, 6, 162–169, 2006.
Jutkiewicz, E.M., RB101-mediated protection of endogenous opioids: Potential therapeutic utility? *CNS Drug Rev.*, 13, 192–205, 2007.
Kapusta, D.R. et al., Functional selectivity of nociceptin/orphanin FQ peptide receptor partial agonists on cardiovascular and renal function, *J. Pharmacol. Exp. Ther.*, 314, 643–651, 2005.
Keller, M. et al., Pseudoproline-containing analogues of morphiceptin and endomorphin-2: Evidence for a *cis* Tyr–Pro amide bond in the bioactive conformation, *J. Med. Chem.*, 44, 3896–3903, 2001.
Kieffer, B.L., Molecular aspects of opioid receptors, in *The Pharmacology of Pain*, Dickenson, A. and Besson, J.-M., Eds., Springer, Berlin, 130, 1997, p. 281–303.
Kieffer, B.L., Opioids: First lessons from knockout mice, *Trends Pharmacol. Sci.*, 20, 19–26, 1999.
Kieffer, B.L. and Gaveriaux-Ruff, C., Exploring the opioid system by gene knockout, *Prog. Neurobiol.*, 66, 285–306, 2002.
Kieffer, B.L. et al., The δ-opioid receptor: Isolation of a cDNA by expression cloning and pharmacological characterization, *Proc. Natl Acad. Sci. USA*, 89, 12048–12052, 1992.
Knapp, R.J. et al., Molecular biology and pharmacology of cloned opioid receptors, *FASEB J.*, 9, 516–525, 1995.

Kreil, G. et al., Deltorphin, a novel amphibian skin peptide with high selectivity and affinity for δ opioid receptors, *Eur. J. Pharmacol.*, 162, 123–128, 1989.

Kruszynski, R. et al., Novel endomorphin-2 analogs with mu-opioid receptor antagonist activity, *J. Pept. Res.*, 66, 125–131, 2005.

Labarre, M. et al., Inverse agonism by Dmt–Tic analogues and HS 378, a naltrindole analogue, *Eur. J. Pharmacol.*, 406, R1–R3, 2000.

Lai, J. et al., Neuropathic pain: The paradox of dynrophin, *Mol. Interv.*, 1, 160–167, 2001.

Lapalu, S. et al., Comparison of the structure–activity relationships of nociceptin and dynorphin A using chimeric peptides, *FEBS Lett.*, 417, 333–336, 1997.

Law, P.Y., Wong, Y.H., and Loh, H.H., Mutational analysis of the structure and function of opioid receptors, *Biopolymers*, 51, 440–455, 1999.

Lazarus, L.H. et al., Dermorphin gene sequence peptide with high affinity and selectivity for δ-opioid receptors, *J. Biol. Chem.*, 264, 3047–3050, 1989.

Lazarus, L.H. et al., Design of δ-opioid peptide antagonists for emerging drug applications, *Drug Discov. Today*, 3, 284–294, 1998.

Lazarus, L.H. et al., What peptides these deltorphins be, *Prog. Neurobiol.*, 57, 377–420, 1999.

Le Guen, S. et al., Further evidence for the interaction of mu- and delta-opioid receptors in the antinociceptive effects of the dual inhibitor of enkephalin catabolism, RB101(S). A spinal c-Fos protein study in the rat under carrageenin inflammation, *Brain Res.*, 967, 106–112, 2003.

Leslie, F.M., Methods used for study of opioid receptors, *Pharmacol. Rev.*, 39, 197–249, 1987.

Leslie, F.M. and Goldstein, A., Degradation of dynorphin-(1–13) by membrane-bound rat brain enzymes, *Neuropeptides*, 2, 185–196, 1982.

Levac, B.A., O'Dowd, B.F., and George, S.R., Oligomerization of opioid receptors: Generation of novel signaling units, *Curr. Opin. Pharmacol.*, 2, 76–81, 2002.

Li, C.H. and Chung, D., Isolation and structure of an untriakontapeptides with opiate activity from camel pituitary glands, *Proc. Natl Acad. Sci. USA*, 73, 1145–1148, 1976.

Lipkowski, A.W., Konecka, A.M., and Sroczynska, I., Double enkephalins—synthesis, activity on guinea-pig ileum, and analgesic effect, *Peptides*, 3, 697–700, 1982.

Lord, J.A.H. et al., Endogenous opioid peptides: Multiple agonists and receptors, *Nature*, 267, 495–499, 1977.

Lu, Y. et al., [2′,6′-Dimethyltyrosine]dynorphin A(1-11)NH_2 analogues lacking an N-terminal amino group: Potent and selective κ opioid antagonists, *J. Med. Chem.*, 44, 3048–3053, 2001.

Lung, F.-D.T. et al., Highly κ receptor-selective dynorphin A analogues with modifications in position 3 of dynorphin A(1–11)-NH_2, *J. Med. Chem.*, 38, 585–586, 1995.

Mague, S.D. et al., Antidepressant-like effects of κ-opioid receptor antagonists in the forced swim test in rats, *J. Pharmacol. Exp. Ther.*, 305, 323–330, 2003.

Marsden, B.J., Nguyen, T.M.-D., and Schiller, P.W., Spontaneous degradation via diketopiperazine formation of peptides containing a tetrahydoisoquinoline-3-carboxylic acid residue in the 2-position of the peptide sequence, *Int. J. Pept. Protein Res.*, 41, 313–316, 1993.

Martin, N.A. et al., Agonist, antagonist, and inverse agonist characteristics of TIPP (H–Tyr–Tic–Phe–Phe–OH), a selective delta-opioid receptor ligand, *J. Pharmacol. Exp. Ther.*, 301, 661–671, 2002.

Martin, N.A., Terruso, M.T., and Prather, P.L., Agonist activity of the δ-antagonists TIPP and TIPP-Ψ in cellular models expressing endogenous or transfected δ-opioid receptors, *J. Pharmacol. Exp. Ther.*, 298, 240–248, 2001.

Martin, W.R. et al., The effects of morphine- and nalorphine-like drugs in the nondependent and morphine-dependent chronic spinal dog, *J. Pharmacol. Exp. Ther.*, 197, 517–532, 1976.

Maurer, R. et al., Opiate antagonistic properties of an octapeptide somatostatin analog, *Proc. Natl Acad. Sci. USA*, 79, 4815–4817, 1982.

McFadyen, I.J. et al., Tetrapeptide derivatives of [D-Pen^2,D-Pen^5]-enkephalin (DPDPE) lacking an N-terminal tyrosine residue are agonists at the μ-opioid receptor, *J. Pharmacol. Exp. Ther.*, 295, 960–966, 2000.

McKnight, A.T. et al., The opioid receptors in the hamster vas deferens are of the δ-type, *Neuropharmacol.*, 24, 1011–1017, 1985.

Mello, N. and Negus, S.S., Interactions between kappa opioid agonists and cocaine, *Ann. NY Acad. Sci.*, 909, 104–132, 2000.

Meunier, J., Mouledous, L., and Topham, C.M., The nociceptin (ORL1) receptor: Molecular cloning and functional architecture, *Peptides*, 21, 893–900, 2000.

Meunier, J.-C. et al., Isolation and structure of the endogenous agonist of opioid receptor-like ORL_1 receptor, *Nature*, 377, 532–535, 1995.

Millan, M.J., Kappa-opioid receptor-mediated antinociception in the rat .1. Comparative actions of *mu*-opioids and *kappa*-opioids against noxious thermal, pressure, and electrical stimuli, *J. Pharmacol. Exp. Ther.*, 251, 334–341, 1989.

Millan, M.J., Kappa-opioid receptors and analgesia, *Trends Pharmacol. Sci.*, 11, 70–76, 1990.

Mogil, J.S. and Pasternak, G.W., The molecular and behavioral pharmacology of the orphanin FQ/nociceptin peptide and receptor family, *Pharmacol. Rev.*, 53, 381–415, 2001.

Mollereau, C. et al., ORL1, a novel member of the opioid receptor family—cloning, functional expression, and localization, *FEBS Lett.*, 341, 33–38, 1994.

Mollica, A. et al., Synthesis and biological activity of the first cyclic biphalin analogues, *Bioorg. Med. Chem. Lett.*, 16, 367–372, 2006.

Mollica, A. et al., Synthesis of stable and potent delta/mu opioid peptides: Analogues of H–Tyr–*c*[D-Cys–Gly–Phe–D-Cys]–OH by ring-closing metathesis, *J. Med. Chem.*, 50, 3138–3142, 2007.

Morley, J.S., Structure–activity relationships of enkephalin-like peptides, *Annu. Rev. Pharmacol. Toxicol.*, 20, 81–110, 1980.

Mosberg, H.I., Complementarity of delta opioid ligand pharmacophore and receptor models, *Biopolymers*, 51, 426–439, 1999.

Mosberg, H.I., Ho, J.C., and Sobczyk-Kojiro, K., A high affinity, mu-opioid receptor-selective enkephalin analogue lacking an N-terminal tyrosine, *Bioorg. Med. Chem. Lett.*, 8, 2681–2684, 1998.

Mosberg, H.I. et al., Bis-penicillamine enkephalins possess highly improved selectivity toward δ opioid receptors, *Proc. Natl Acad. Sci. USA*, 80, 5871–5874, 1983.

Mosberg, H.I. et al., Cyclic disulfide- and dithioether-containing opioid tetrapeptides: Development of a ligand with high delta opioid receptor selectivity and affinity, *Life Sci.*, 43, 1013–1020, 1988.

Mullaney, I., Carr, I.C., and Milligan, G., Analysis of inverse agonism at the delta opioid receptor after expression in Rat 1 fibroblasts, *Biochem. J.*, 315, Pt. 1, 227–234, 1996.

Nakazawa, T. et al., Analgesia produced by E-2078, a systemically active dynorphin analog, in mice, *J. Pharmacol. Exp. Ther.*, 252, 1247–1254, 1990.

Nakazawa, T. et al., Spinal kappa-receptor-mediated analgesia of E-2078, a systemically active dynorphin analog, in mice, *J. Pharmacol. Exp. Ther.*, 256, 76–81, 1991.

Naqvi, T., Haq, W., and Mathur, K.B., Structure–activity relationship studies of dynorphin A and related peptides, *Peptides*, 19, 1277–1292, 1998.

Negri, L., Melchiorri, P., and Lattanzi, R., Pharmacology of amphibian opiate peptides, *Peptides*, 21, 1639–1647, 2000.

Neumeyer, J.L. et al., New opioid designed multiple ligand from Dmt-Tic and morphinan pharmacophores, *J. Med. Chem.*, 49, 5640–5643, 2006.

Noble, F. and Roques, B.P., Protection of endogenous enkephalin catabolism as natural approach to novel analgesic and antidepressant drugs, *Expert Opin. Ther. Targets*, 11, 145–159, 2007.

Noble, F. et al., Inhibition of the enkephalin-metabolizing enzymes by the first systematically active mixed inhibitor prodrug RB101 induces potent analgesic responses in mice and rats, *J. Pharmacol. Exp. Ther.*, 261, 181–190, 1992.

North, R.A., Opioid actions on membrane ion channels, in *Opioids I*, Herz, A., Akil, H., and Simon, E.J., Eds., Springer-Verlag, Berlin, 104/I, 1993, p. 773–797.

Oka, T. et al., Rabbit vas deferens: A specific bioassay for opioid κ-receptor agonists, *Eur. J. Pharmacol.*, 73, 235–236, 1980.

Okada, K. et al., Highly potent nociceptin analog containing the Arg–Lys triple repeat, *Biochem. Biophys. Res. Commun.*, 278, 493–498, 2000.

Okuda-Ashitaka, E. et al., Nocistatin, a peptide that blocks nociceptin action in pain transmission, *Nature*, 392, 286–289, 1998.

Ossipov, M. et al., Recent advances in the pharmacology of opioids, in *Novel Aspects of Pain Management: Opioids and Beyond*, Sawynok, J. and Cowan, A., Eds., Wiley-Liss, New York, 1999, p. 49–71.

Paakkari, P. et al., Dermorphin analog Tyr–D-Arg^2–Phe-sarcosine-induced opioid analgesia and respiratory stimulation: The role of mu_1-receptors? *J. Pharmacol. Exp. Ther.*, 266, 544–550, 1993.

Pasternak, G.W., Incomplete cross tolerance and multiple mu opioid peptide receptors, *Trends Pharmacol. Sci.*, 22, 67–70, 2001a.

Pasternak, G.W., Insights into mu opioid pharmacology. The role of mu opioid receptor subtypes, *Life Sci.*, 68, 2213–2219, 2001b.

Pasternak, G.W., Multiple opiate receptors: *Deja vu* all over again, *Neuropharmacol.*, 47, Suppl. 1, 312–323, 2004.

Pasternak, G.W. and Wood, P.J., Minireview: Multiple mu opiate receptors, *Life Sci.*, 38, 1889–1898, 1986.

Patkar, K.A. et al., [N^{α}-BenzylTyr[1],*cyclo*(D-Asp[5],Dap[8])]-dynorphin A-(1–11)NH_2 cyclized in the "address" domain is a novel kappa-opioid receptor antagonist, *J. Med. Chem.*, 48, 4500–4503, 2005.

Pelton, J.T. et al., Conformationally restricted analogs of somatostatin with high μ-opiate receptor specificity, *Proc. Natl Acad. Sci. USA*, 82, 236–239, 1985.

Pelton, J.T. et al., Design and synthesis of conformationally constrained somatostatin analogues with high potency and specificity for μ opioid receptors, *J. Med. Chem.*, 29, 2370–2375, 1986.

Peng, Y.L. et al., Novel potent agonist [(pF)Phe[4],Aib[7],Aib[11],Arg[14],Lys[15]]N/OFQ-NH_2 and antagonist [Nphe[1], (pF)Phe[4],Aib[7],Aib[11],Arg[14],Lys[15]]N/OFQ-NH_2 of nociceptin/orphanin FQ receptor, *Regul. Pept.*, 134, 75–81, 2006.

Podlogar, B.P. et al., Conformational analysis of the endogenous mu-opioid agonist endomorphin-1 using NMR spectroscopy and molecular modeling, *FEBS Lett.*, 439, 13–20, 1998.

Pogozheva, I.D., Lomize, A.L., and Mosberg, H.I., Opioid receptor three-dimensional structures from distance geometry calculations with hydrogen bonding constraints, *Biophys. J.*, 75, 612–634, 1998.

Polt, R., Dhanasekaran, M., and Keyari, C.M., Glycosylated neuropeptides: A new vista for neuropsychopharmacology? *Med. Res. Rev.*, 25, 557–585, 2005.

Polt, R. et al., Glycopeptide enkephalin analogues produce analgesia in mice: Evidence for penetration of the blood–brain barrier, *Proc. Natl Acad. Sci. USA*, 91, 7114–7118, 1994.

Porreca, F. and Burks, T.F., Supraspinal opioid receptors in antinociception, in *Opioids II*, Herz, A., Akil, H., and Simon, E.J., Eds., Springer-Verlag, Berlin, 104/II, 1993, p. 21–51.

Porreca, F. et al., Modulation of mu-mediated antinociception in the mouse involves opioid delta-2 receptors, *J. Pharmacol. Exp. Ther.*, 263, 147–152, 1992.

Reinscheid, R.K. et al., Orphanin FQ: A neuropeptide that activates an opioidlike G protein-coupled receptor, *Science*, 270, 792–794, 1995.

Reinscheid, R.K. et al., Structure–activity relationship studies of the novel neuropeptide orphanin FQ, *J. Biol. Chem.*, 271, 14163–14168, 1996.

Richter, K., Egger, R., and Kreil, G., D-Alanine in the frog skin peptide dermorphin is derived from L-alanine in the precursor, *Science*, 238, 200–202, 1987.

Rizzi, A. et al., Pharmacological characterization of the novel nociceptin/orphanin FQ receptor ligand, ZP120: *In vitro* and *in vivo* studies in mice, *Br. J. Pharmacol.*, 137, 369–374, 2002a.

Rizzi, A. et al., Pharmacological characterisation of [(pX)Phe[4]]nociceptin(1-13)NH_2 analogues. 2. *In vivo* studies, *Naunyn-Schmiedeberg's Arch. Pharmacol.*, 365, 450–456, 2002b.

Rizzi, D. et al., [Arg[14],Lys[15]]nociceptin, a highly potent agonist of the nociceptin/orphanin FQ receptor: *In vitro* and *in vivo* studies, *J. Pharmacol. Exp. Ther.*, 300, 57–63, 2002c.

Roques, B.P., Novel approaches to targeting neuropeptide systems, *Trends Pharmacol. Sci.*, 21, 475–483, 2000.

Roques, B.P. et al., Peptidase inactivation of enkephalins: Design of inhibitors and biochemical, pharmacological, and clinical applications, in *Opioids I*, Herz, A., Akil, H., and Simon, E.J., Eds., Springer-Verlag, Berlin, 104/I, 1993a, p. 547–584.

Roques, B.P. et al., Neutral endopeptidase 24.11 structure, inhibition, and experimental and clinical pharmacology, *Pharmacol. Rev.*, 45, 87–147, 1993b.

Rothman, R.B. et al., An open-label study of a functional opioid κ antagonist in the treatment of opioid dependence, *J. Substance Abuse Treat.*, 18, 277–281, 2000.

Rothman, R.B., Holaday, J.W., and Porreca, F., Allosteric coupling among opioid receptors: Evidence for an opioid receptor complex, in *Opioids I*, Herz, A., Akil, H., and Simon, E.J., Eds., Springer-Verlag, Berlin, 104/I, 1993, 217–237.

Rozenfeld, R. and Devi, L.A., Receptor heterodimerization leads to a switch in signaling: Beta-arrestin2-mediated ERK activation by mu-delta opioid receptor heterodimers, *FASEB J.*, 21, 2455–2465, 2007.

Rozenfeld, R. et al., An emerging role for the delta opioid receptor in the regulation of mu opioid receptor function, *ScientificWorldJournal*, 7, 64–73, 2007.

Salvadori, S. et al., δ opioidmimetic antagonists: Prototypes for designing a new generation of ultraselective opioid peptides, *Mol. Med.*, 1, 678–689, 1995.

Salvadori, S. et al., A new opioid designed multiple ligand derived from the μ opioid agonist endomorphin-2 and the delta opioid antagonist pharmacophore Dmt-Tic, *Bioorg. Med. Chem.*, 15, 6876–6881, 2007.

Satoh, M. and Minami, M., Molecular pharmacology of the opioid receptors, *Pharmacol. Ther.*, 68, 343–364, 1995.

Schenk, S., Partridge, B., and Shippenberg, T.S., U69,593, a kappa-opioid agonist, decreases cocaine self-administration and decreases cocaine-produced drug-seeking, *Psychopharmacol.*, 144, 339–346, 1999.

Schiller, P.W., Development of receptor-selective opioid peptide analogs as pharmacologic tools and as potential drugs, in *Opioids I*, Herz, A., Akil, H., and Simon, E.J., Eds., Springer-Verlag, Berlin, 104/I, 1993, p. 681–710.

Schiller, P.W., Ed., Peptide and peptidomimetic ligands of opioid receptors, *Biopolymers*, 51 (6), 377–455, 1999.

Schiller, P.W., Opioid peptide-derived analgesics, *AAPS J.*, 7, E560–E565, 2005.

Schiller, P.W. et al., A novel cyclic opioid peptide analog showing high preference for μ-receptors, *Biochem. Biophys. Res. Commun.*, 127, 558–564, 1985.

Schiller, P.W. et al., Dermorphin analogues carrying an increased positive net charge in their "message" domain display extremely high μ opioid receptor selectivity, *J. Med. Chem.*, 32, 698–703, 1989.

Schiller, P.W. et al., Differential stereochemical requirements of μ vs δ opioid receptors for ligand binding and signal transduction: Development of a class of potent and highly δ-selective peptide antagonists, *Proc. Natl Acad. Sci. USA*, 89, 11871–11875, 1992.

Schiller, P.W. et al., TIPP[Ψ]: A highly potent and stable pseudopeptide δ opioid receptor antagonist with extraordinary δ selectivity, *J. Med. Chem.*, 36, 3182–3187, 1993.

Schiller, P.W. et al., TIPP opioid peptides: Development of extraordinarily potent and selective δ antagonists and observation of astonishing structure–instrinsic activity relationships, in *Peptides: Chemistry, Structure, and Biology*, Hodges, R.S. and Smith, J.A., Eds., ESCOM, Leiden, 1994a, p. 514–516.

Schiller, P.W. et al., A highly potent TIPP-NH_2 analog with balanced mixed mu agonist/delta antagonist properties, *Regul. Pept.*, 54, 257–258, 1994b.

Schiller, P.W. et al., TIPP analogs—highly selective δ opioid antagonists with subnanomolar potency and first known compounds with mixed μ agonist/δ antagonist properties, *Regul. Pept.*, S63–S64, 1994c.

Schiller, P.W. et al., Structure–agonist/antagonist activity relationships of TIPP analogs, in *Peptides: Chemistry, Structure, Biology*, Kaumaya, T.P. and Hodges, R.S., Eds., Mayflower Scientific Ltd., West Midlands, England, 1996, p. 609–611.

Schiller, P.W. et al., The opioid μ agonist/δ antagonist DIPP-NH_2[Ψ] produces a potent analgesic effect, no physical dependence, and less tolerance than morphine in rats, *J. Med. Chem.*, 42, 3520–3526, 1999a.

Schiller, P.W. et al., The TIPP opioid peptide family: Development of δ antagonists, δ agonists, and mixed μ agonist/δ antagonists, *Biopolymers*, 51, 411–425, 1999b.

Schiller, P.W. et al., Subtleties of structure–agonist versus antagonist relationships of opioid peptides and peptidomimetics, *J. Recept. Signal Transduct. Res.*, 19, 573–588, 1999c.

Schiller, P.W. et al., Synthesis and *in vitro* opioid activity profiles of DALDA analogues, *Eur. J. Med. Chem.*, 35, 895–901, 2000.

Schiller, P.W. et al., Conversion of δ-, κ-, and μ-receptor selective opioid peptide agonists into δ-, κ-, and μ-selective antagonists, *Life Sci.*, 73, 691–698, 2003.

Schlechtingen, G. et al., [Pro^3]Dyn A(1–11)NH_2: A dynorphin analogue with high selectivity for the κ opioid receptor, *J. Med. Chem.*, 43, 2698–2702, 2000.

Schlechtingen, G. et al., Structure–activity relationships of dynorphin A analogues modified in the address sequence, *J. Med. Chem.*, 46, 2104–2109, 2003.

Schwyzer, R., ACTH: A short introductory review, *Ann. NY Acad. Sci.*, 297, 3–26, 1977.

Selley, D.E. et al., μ-opioid receptor-stimulated guanosine-5′-*O*-(γ-thio)-triphosphate binding in rat thalamus and cultured cell lines: Signal transduction mechanisms underlying agonist efficacy, *Mol. Pharmacol.*, 51, 87–96, 1997.

Shimoyama, M. et al., Antinociceptive and respiratory effects of intrathecal H-Tyr–D-Arg–Phe–Lys–NH_2 (DALDA) and [Dmt^1] DALDA, *J. Pharmacol. Exp. Ther.*, 297, 364–371, 2001.

Smith, J.A.M. and Leslie, F.M., Use of organ systems for opioid bioassay, in *Opioids I*, Herz, A., Akil, H., and Simon, E.J., Eds., Springer-Verlag, Berlin, 104/I, 1993, p. 53–78.

Soderstrom, K. et al., N-Alkylated derivatives of [D-Pro^{10}]dynorphin A-(1–11) are high affinity partial agonists at the cloned rat κ-opioid receptor, *Eur. J. Pharmacol.*, 338, 191–197, 1997.

Stein, C., Peripheral mechanisms of opioid analgesia, in *Opioids II*, Herz, A., Akil, H., and Simon, E.J., Eds., Springer-Verlag, Berlin, 104/II, 1993, p. 91–103.

Stein, C., Ed., *Opioids in Pain Control: Basic and Clinical Aspects*, Cambridge University Press, Cambridge, 1999.

Stein, C., Cabot, P.J., and Schafer, M., Peripheral analgesia: Mechanisms and clinical implications, in *Opioids in Pain Control: Basic and Clinical Aspects*, Stein, C., Ed., Cambridge University Press, Cambridge, 1999, p. 96–108.

Temussi, P.A. et al., Selective opioid dipeptides, *Biochem. Biophys. Res. Commun.*, 198, 933–939, 1994.

Teschemacher, H. et al., A peptide-like substance from pituitary that acts like morphine. 1. Isolation, *Life Sci.*, 16, 1771–1776, 1975.

Tjølsen, A. and Hole, K., Animal models for analgesia, in *The Pharmacology of Pain*, Dickenson, A. and Besson, J.-M., Eds., Springer, Berlin, 130, 1997, p. 1–20.

Toth, G. et al., [D-Pen[2],D-Pen[5]]enkephalin analogues with increased affinity and selectivity for δ opioid receptors, *J. Med. Chem.*, 33, 249–253, 1990.

Traynor, J.R. and Elliott, J., δ-Opioid receptor subtypes and cross-talk with μ-receptors, *Trends Pharmacol. Sci.*, 14, 84–86, 1993.

Traynor, J.R. and Nahorski, S.R., Modulation by μ-opioid agonists of guanosine-5′-*O*-(3-[^{35}S]thio)triphosphate binding to membranes from human neuroblastoma SH-SY5Y cells, *Mol. Pharmacol.*, 47, 848–854, 1995.

Udenfried, S. and Meienhofer, J., Eds., *Opioid Peptides: Biology, Chemistry, and Genetics*, *The Peptides: Analysis, Synthesis, Biology*, Academic Press, Orlando, 1984.

Van den Eynde, I. et al., A new structural motif for mu-opioid antagonists, *J. Med. Chem.*, 48, 3644–3648, 2005.

Vanderah, T.W. et al., Novel D-amino acid tetrapeptides produce potent antinociception by selectively acting at peripheral kappa-opioid receptors, *Eur. J. Pharmacol.*, 583, 62–72, 2008.

Vig, B.S., Murray, T.F., and Aldrich, J.V., A novel N-terminal cyclic dynorphin A analogue *cyclo*^{N,5}[Trp[3],Trp[4], Glu[5]]-dynorphin A-(1–11)NH_2 that lacks the basic N-terminus, *J. Med. Chem.*, 46, 1279–1282, 2003.

Wallace, M.S. et al., A Phase II, multicenter, randomized, double-blind, placebo-controlled crossover study of CJC-1008—a long-acting, parenteral opioid analgesic—in the treatment of postherpetic neuralgia, *J. Opioid Manag.*, 2, 167–173, 2006.

Wan, Q., Murray, T.F., and Aldrich, J.V., A novel acetylated analogue of dynorphin A-(1–11) amide as a κ opioid receptor antagonist, *J. Med. Chem.*, 42, 3011–3013, 1999.

Wang, D. et al., Opioid receptor homo- and heterodimerization in living cells by quantitative bioluminescence resonance energy transfer, *Mol. Pharmacol.*, 67, 2173–2184, 2005.

Weber, E., Evans, C.J., and Barchas, J.D., Predominance of the amino-terminal octapeptide fragment of dynorphin in rat brain regions, *Nature*, 299, 77–79, 1982.

Weber, S.J. et al., Distribution and analgesia of [^{3}H][D-Pen[2],D-Pen[5]]enkephalin and two halogenated analogs after intravenous administration, *J. Pharmacol. Exp. Ther.*, 259, 1109–1117, 1991.

Weltrowska, G. et al., A chimeric opioid peptide with mixed mu agonist/delta antagonist properties, *J. Pept. Res.*, 63, 63–68, 2004.

Weltrowska, G. et al., Cyclic enkephalin analogs containing various para-substituted phenylalanine derivatives in place of Tyr1 are potent opioid agonists, *J. Pept. Res.*, 65, 36–41, 2005.

Witt, K.A. and Davis, T.P., CNS drug delivery: Opioid peptides and the blood–brain barrier, *AAPS J.*, 8, E76–E88, 2006.

Witt, K.A. et al., Pharmacodynamic and pharmacokinetic characterization of poly(ethylene glycol) conjugation to met-enkephalin analog [D-Pen[2],D-Pen[5]]-enkephalin (DPDPE), *J. Pharmacol. Exp. Ther.*, 298, 848–856, 2001.

Wolozin, B.L. and Pasternak, G.W., Classification of multiple morphine and enkephalin binding sites in the central nervous system, *Proc. Natl Acad. Sci. USA*, 78, 6181–6185, 1981.

Xu, X. et al., Nociceptin/orphanin FQ in spinal nociceptive mechanisms under normal and pathological conditions, *Peptides*, 21, 1031–1036, 2000.

Yamamoto, T. et al., Design, synthesis, and biological evaluation of novel bifunctional C-terminal-modified peptides for delta/mu opioid receptor agonists and neurokinin-1 receptor antagonists, *J. Med. Chem.*, 50, 2779–2786, 2007.

Yamazaki, M. et al., The opioid peptide analogue biphalin induces less physical dependence than morphine, *Life Sci.*, 69, 1023–1028, 2001.

Yasuda, K. et al., Cloning and functional comparison of κ and δ opioid receptors from mouse brain, *Proc. Natl Acad. Sci. USA*, 90, 6736–6740, 1993.

Yoshino, H. et al., Synthesis and structure–activity relationships of [MeTyr[1],MeArg[7]]-dynorphin A(1-8)-OH analogues with substitution at position 8, *Chem. Pharm. Bull.*, 38, 404–406, 1990a.

Yoshino, H. et al., Synthesis and structure–activity relationships of dynorphin A-(1–8) amide analogues, *J. Med. Chem.*, 33, 206–212, 1990b.

Zadina, J.E. et al., A potent and selective endogenous agonist for the μ-opiate receptor, *Nature*, 386, 499–502, 1997.

Zajac, J.M. et al., Deltakephalin, Tyr–D-Thr–Gly–Phe–Leu–Thr: A new highly potent and fully specific agonist for opiate δ-receptors, *Biochem. Biophys. Res. Commun.*, 111, 390–397, 1983.

Zaki, P.A. et al., Opioid receptor types and subtypes: The delta receptor as a model, *Annu. Rev. Pharmacol. Toxicol.*, 36, 379–401, 1996.

Zaveri, N., Peptide and nonpeptide ligands for the nociceptin/orphanin FQ receptor ORL1: Research tools and potential therapeutic agents, *Life Sci.*, 73, 663–678, 2003.

Zhao, K. et al., Transcellular transport of a highly polar 3+ net charge opioid tetrapeptide, *J. Pharmacol. Exp. Ther.*, 304, 425–432, 2003.

Zhu, Y. et al., Retention of supraspinal delta-like analgesia and loss of morphine tolerance in δ opioid receptor knockout mice, *Neuron*, 24, 243–252, 1999.

Zollner, C. and Stein, C., Opioids, *Handb. Exp. Pharmacol.*, 177, 31–63, 2007.

Zukin, R.S. et al., Characterization and visualization of rat and guinea pig brain κ opioid receptors: Evidence for κ_1 and κ_2 receptors, *Proc. Natl Acad. Sci. USA*, 85, 4061–4065, 1988.

6 Cholecystokinins and Gastrins

Jens F. Rehfeld

CONTENTS

Cholecystokinins (CCK) and gastrins are homologous peptides, and are both ligands for the gastrin/CCK-B receptor, whereas the CCK-A receptor binds only sulfated CCK-peptides. CCK peptides are mainly produced in intestinal endocrine I-cells and in cerebral neurons. CCK peptides regulate pancreatic enzyme secretion and growth, gallbladder contraction, intestinal motility, satiety, and inhibit gastric acid secretion. Moreover, they are potent neurotransmitters in the brain and in the periphery. CCK peptides are derived from proCCK and have the active site, –Tyr(SO_4)–Met–Gly–Trp–Met–Asp–Phe–NH_2, at their C-terminus. The hormonal forms in plasma are CCK-58, CCK-33, CCK-22, and CCK-8, whereas CCK-8 and CCK-5 are the major neurotransmitter forms. The scarcity of specific CCK assays has limited the knowledge about CCK in disease. Gastrin peptides are synthesized in antroduodenal G-cells, from where they are released to blood to regulate gastric acid secretion and mucosal growth. Small amounts are expressed further down the intestinal tract, in the fetal pancreas, in a few cerebral and peripheral neurons, in the pituitary gland, and in spermatozoes. Gastrin peptides are derived from progastrin and have the C-terminal active site, –Tyr (SO_4)–Gly–Trp–Met–Asp–Phe–NH_2. The major gastrin forms in tissue and plasma are gastrin-34 and gastrin-17. Gastrin peptides are secreted in excessive amounts from gastrinomas and expressed at a lower level in common cancers.

6.1 INTRODUCTION

Cholecystokinin (CCK) and gastrin peptides are members of a peptide family with a common C-terminal bioactive sequence that has been exceedingly well preserved during evolution (Larsson and Rehfeld, 1977; Johnsen, 1998). It is possible that the common ancestor resembles a dityrosyl-sulfated peptide, cionin, which has been isolated from the central ganglion of protochordates (Johnsen and Rehfeld, 1990). The family also includes the frogskin peptides, caerulein and phyllocaerulein (Anastasi et al., 1967, 1969), but CCK and gastrin peptides are the only members of the family found in mammals. Protochordate and nonmammalian vertebrate members of the family are more CCK- than gastrin-like, and it is only at the level of elasmobranch that separate CCK and gastrin genes are expressed during evolution (Johnsen et al., 1997). Elasmobranchs are also the first organisms in evolution to produce hydrochloric acid in the stomach.

CCK was discovered in 1928 as a gallbladder-contracting hormone in extracts of the small intestine (Ivy and Oldberg, 1928). The last decades, however, have shown that CCK, in addition to having hormonal cholecystokinetic and pancreozymic actions, is also a growth factor for the pancreas and a potent neurotransmitter in both the central and peripheral nervous systems (for reviews, see Jorpes and Mutt, 1973; Rehfeld, 1989, 2004). Its long history has made the literature about CCK comprehensive but inconsistent, because early physiological studies used impure CCK preparations, with little attention being paid to species differences and to physiological dosing. Also, until recently, most assays used to measure CCK have lacked specificity and sensitivity (Rehfeld, 1984, 1998a, 1998b).

Gastrin was discovered more than 100 years ago as an acid-stimulating factor in the mucosa of the antral parts of the stomach (Edkins, 1905). Identification of gastrin and CCK peptides in the 1960s revealed that their C-terminal pentapeptide sequences are identical and constitute the active site of both hormones (Gregory and Tracy, 1964; Gregory et al., 1964; Mutt and Jorpes, 1968, 1971). Over the last few decades, the concept of gastrin as a simple peptide from the upper digestive tract has changed, like that of CCK. Now gastrin is known to occur in multiple molecular forms, and like that of CCK, the gastrin gene is expressed in a cell-specific manner in neurons, endocrine cells, and other epithelial cells outside the gastrointestinal tract. Moreover, gastrin has been shown to stimulate growth of the gastric mucosa cells (Johnson and Guthrie, 1974; Johnson, 1976).

All biological effects of gastrin and CCK peptides reside in the conserved common C-terminal tetrapeptide amide sequence (Figure 6.1). Modification of this sequence grossly reduces or abolishes the biological effects (Morley et al., 1965). The N-terminal extensions of the common C-terminal sequence increase the biological potency and the specificity for receptor binding. Of particular importance is the tyrosyl residue in position six of mammalian gastrins and position seven of CCK peptides (as counted from the C-terminal phenylalanyl amide; Figure 6.1). The tyrosyl residue is partly sulfated in gastrins (Gregory and Tracy, 1964; Gregory et al., 1964), and more completely sulfated in CCKs (Bonetto et al., 1999). The gastrin/CCK-B receptor binds sulfated and unsulfated ligands equally well, whereas the CCK-A receptor requires sulfation of the ligand. The gastrins are, consequently, defined as peptides that stimulate gastric acid secretion and have the C-terminal sequence Tyr–X–Trp–Met–Asp–Phe–NH_2, and CCK peptides are defined as gallbladder-contracting peptides with the C-terminal sequence Tyr–Met–X–Trp–Met–Asp–Phe–NH_2 (where X in most mammalian species is a glycyl residue).

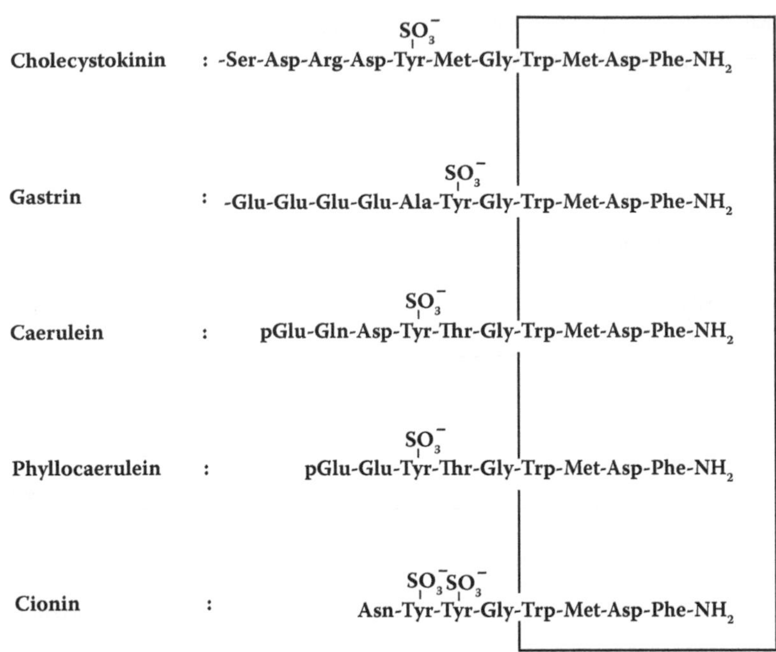

FIGURE 6.1 The homologous bioactive sequences of CCK, gastrin, caerulein, phyllocaerulein, and cionin. The evolutionary preserved bioactive tetrapeptide amide is boxed.

6.2 CHOLECYSTOKININ

6.2.1 Cellular Synthesis of CCK Peptides

The transcriptional unit of the CCK gene is 7 kilobases interrupted by two introns (Deschenes et al., 1985). The first of the three exons is small and noncoding. Except for the homology with the gastrin gene in the coding region and in the 5′-untranslated region (5′ UTR), the structure of the CCK gene displays no unique features. Conserved regulatory elements have been identified in the first 100 bp of the human promoter, including an E-box element, a combined cAMP response element (CRE)/12-O-tetradeconoylphorbol-13-acetate response element (TRE), and a guanine-cytosine (GC)-rich region (Nielsen et al., 1996; Hansen, 2001). Whereas the function of the E-box and the GC-rich region is not fully clarified (Rourke et al., 1999), the combined CRE/TRE sequence plays an important role in the regulation of transcription. The CRE/TRE binds the transcription factor CREB, which is activated by phosphorylation by several signaling pathways, including cAMP, fibroblast growth factor (FGF), pituitary adenylate cyclase-activating polypeptide (PACAP), calcium, hydrolysates, and peptones to ultimately induce CCK transcription (Hansen et al., 1999, 2004; Deavall et al., 2000; Bernard et al., 2001; Gevrey et al., 2002). Only one CCK mRNA molecule has been found, and the CCK peptides are thus fragments of the same primary mRNA product. The mRNA has 750 bases, 345 of which are protein coding (Deschenes et al., 1984; Gubler et al., 1984). The concentrations of CCK mRNA in cerebrocortical tissue are similar to those of the duodenal mucosa (Gubler et al., 1984), and at least in the brain there is a rapid synthesis of CCK peptides (Goltermann et al., 1980).

The primary translational product, preproCCK, has a sequence of 115 amino-acid residues (Figure 6.2). The first part is a 20-amino-acid signal peptide. The second part, with considerable species variation, is a spacer. The bioactive CCK peptides are derived from the subsequent 58-amino-acid residue sequence (Jorpes and Mutt, 1966; Mutt and Jorpes, 1968; Dockray et al., 1978; Rehfeld and Hansen, 1986; Reeve et al., 1986), and the species variation is small in this

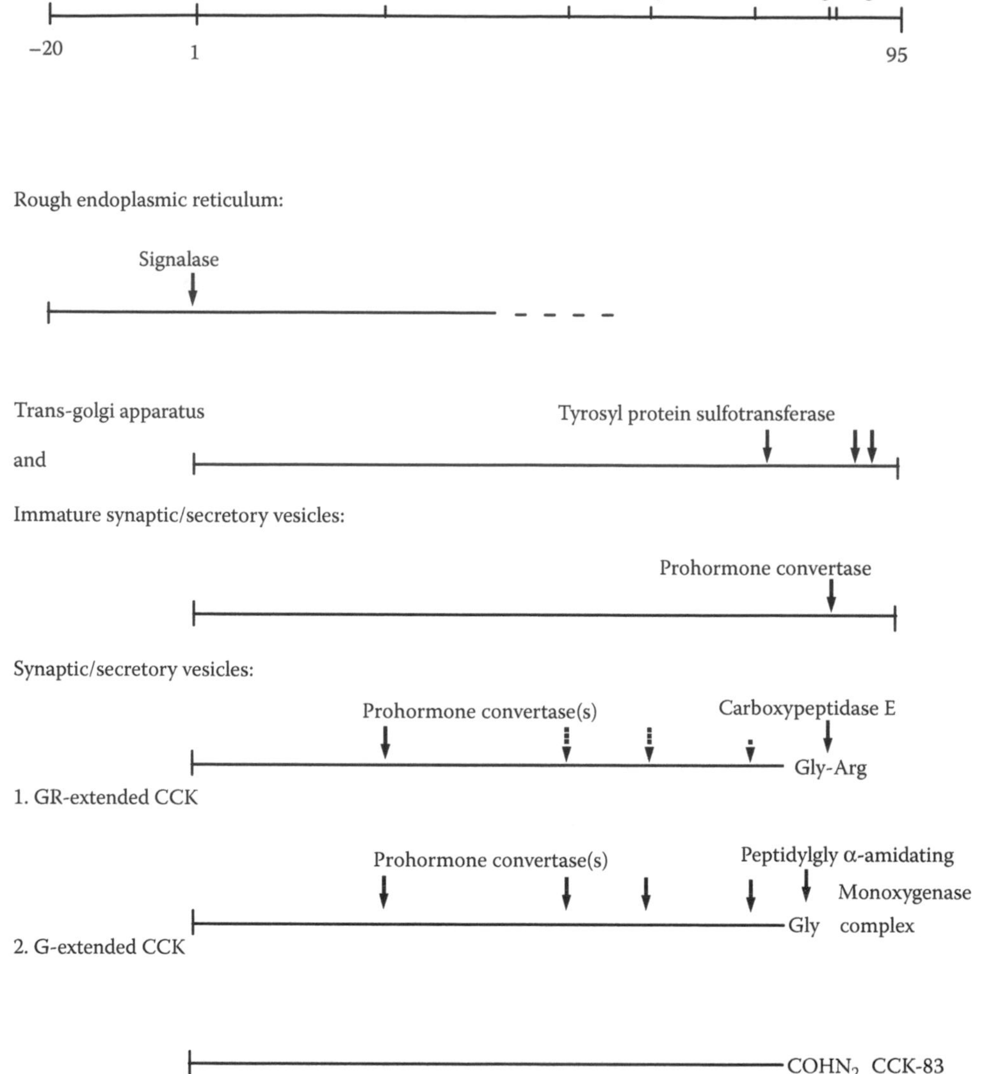

FIGURE 6.2 Co- and post-translational processing of procholecystokinin (proCCK). Mono- and dibasic cleavage sites are indicated (Arg, Lys, Arg-Arg).

sequence. The processing of proCCK is cell-specific due to different expression of prohormone convertases (Rehfeld et al., 2008). Thus, endocrine cells contain a mixture of medium-sized CCKs, whereas neurons mainly release the small CCK-8 and CCK-5 (Rehfeld, 1978a; Rehfeld and Hansen, 1986). The cell-specific synthesis is essential for the CCK functions in the different compartments of the body. The endoproteolysis of proCCK occur mainly at monobasic sites (Figure 6.2). Tyr-77 is O-sulfated, which determines the affinity for the receptors. The known bioactive CCKs have 58, 39, 33, 22, and 8 amino-acid residues (Figure 6.2).

In the small intestine CCK peptides are synthesized in endocrine I-cells (Buffa et al., 1976) whose apical membrane is in direct contact with the intestinal lumen and whose basal region contains secretory granules with CCK peptides. CCK is also synthesized in pituitary corticotrophs and melanotrophs and in a few adrenal medullary cells (Rehfeld, 1978a; Larsson and Rehfeld, 1979a). In the pituitary cells CCK constitutes a small fraction of the hormones. Tumors originating from pituitary corticotrophs, however, produce substantial amounts of CCK (Rehfeld, 1987; Rehfeld et al., 1987).

The brain produces the most CCK (Rehfeld, 1978a; Larsson and Rehfeld, 1979a). Moreover, cerebral CCK neurons are more abundant than neurons of any other neuropeptide (Larsson and Rehfeld, 1979a; Crawley, 1985; Rehfeld and Hansen, 1986). Although most peptidergic neurons occur in subcortical regions, CCK is expressed in the highest concentrations in neocortical regions (Larsson and Rehfeld, 1979a; Crawley, 1985; Fallon and Seroogy, 1985). The perikarya of the cortical CCK nerves are distributed in layers II–VI, with the highest frequency in layers II and III (Larsson and Rehfeld, 1979a; Hendry et al., 1983). CCK in mesencephalic dopamine neurons projecting to the limbic area of the forebrain (Hökfelt et al., 1980) has aroused particular clinical interest because these neurons are supposed to be involved in schizophrenia.

Outside the brain the colon contains numerous CCK nerves, whereas the jejunum and ileum are sparsely innervated. Colonic CCK fibers occur mainly in the circular muscle layer, which they penetrate to form a neural plexus in the submucosa (Larsson and Rehfeld, 1979a). In accordance with these locations, CCK peptides have been shown to excite colonic smooth muscles and to release acetylcholine from neurons in both plexus myentericus and submuca (Vizi et al., 1973). Ganglionic cell bodies in the pancreas are also surrounded by CCK nerves (Larsson and Rehfeld, 1979b). Moreover, in man, pig, and cat, CCK nerve terminals surround pancreatic islets (Rehfeld et al., 1980). The origin of intestinal and pancreatic CCK nerve fibers is uncertain. Finally, afferent vagal nerve fibers also contain CCK (Dockray et al., 1981; Rehfeld and Lundberg, 1983).

6.2.2 Cellular Release of CCK Peptides

CCK in the circulation originates mainly from the intestinal endocrine cells. Measurement of its release to blood could not be examined until specific assays were developed (Rehfeld, 1984, 1998a, 1998b; Liddle et al., 1985). These assays have confirmed most classic ideas about CCK. Thus, protein- and fat-rich food is the most important stimulus (Liddle et al., 1985; Rehfeld, 1998a). Of the constituents, protein and L-amino acids as well as digested fat cause significant CCK release (Himeno et al., 1983; Liddle et al., 1985). Carbohydrates only release small amounts of CCK (Liddle et al., 1985), but hydrochloric acid also stimulates release (Himeno et al., 1983).

The release from neurons has been examined directly in brain slices and synaptosomes (Dodd et al., 1980; Emson et al., 1980). Potassium-induced depolarization causes a calcium-dependent release of CCK-8. Similarly, depolarization releases CCK peptides from the hypothalamic dopamine neurons that innervate the intermediate lobe of the pituitary (Rehfeld et al., 1984). It is not known to what extent neuronal CCK overflows to plasma. By analogy with other neuropeptides, CCK of neuronal origin might constitute a small part of the circulating CCK.

6.2.3 CCK Peptides in Circulation

By comparison with identified forms in tissue, it has been possible to deduce the molecular nature of CCK in plasma. The picture has varied (Rehfeld, 1998a), partly due to species differences, and partly because the molecular pattern along the small intestine varies (Maton et al., 1984) so that venous blood from the duodenum contains more CCK-8 than blood from the distal gut (Rehfeld et al., 1982). Furthermore, the distribution may vary during stimulation. In man, CCK-22 and CCK-33 predominate in plasma, but CCK-8 and CCK-58 are also present (Rehfeld, 1998a; Rehfeld et al., 2001).

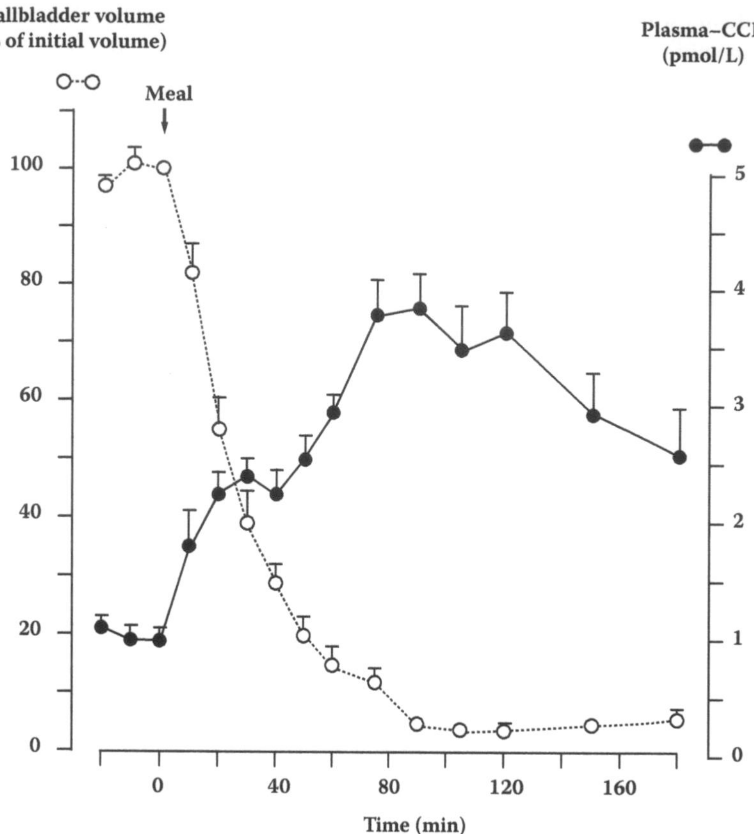

FIGURE 6.3 Plasma CCK concentrations and gallbladder volume during a meal in normal human subjects ($n = 8$).

In the basal state, the concentration of CCK in plasma is 1 pmol/L or less. The concentration increases within 20 min to 3 to 5 pmol/L during maximal stimulation, and then declines gradually to basal levels (Figure 6.3). In comparison with most other pancreatic and gastrointestinal hormones (Horness et al., 1980), the concentrations of CCK in plasma are low. When food-induced CCK in plasma is mimicked by the infusion of exogenous CCK, the same degree of gallbladder contraction and release of enzymes occurs as is seen during meals (Horness et al., 1980; Anagnostides et al., 1985; Kerstens et al., 1985). Therefore, circulating CCK is sufficient to account for gallbladder contraction and pancreatic enzyme secretion during meals (Figure 6.3).

Because the cholecystokinetic and pancreozymic potency of CCK-33 and CCK-8 on a molar base are identical (Solomon et al., 1984) it may seem less important what I-cells release during digestion. On the other hand CCK-58, -33, and -22 are cleared from blood at a significantly slower rate than CCK-8. It is therefore important to know the molecular pattern of CCK in plasma.

6.2.4 Action of CCK Peptides

6.2.4.1 The CCK Receptors

The cellular effects of CCK peptides are mediated via two receptors (Kopin et al., 1992; Wank et al., 1992). The "alimentary" CCK-A receptor (Wank et al., 1992) mediates gallbladder contraction, relaxation of the sphincter of Oddi, pancreatic growth and enzyme secretion, delay of gastric emptying, and inhibition of gastric acid secretion via fundic somatostatin (Chen et al., 2004). CCK-A receptors have also been found in the anterior pituitary, the myenteric plexus, and areas of the midbrain (Honda et al., 1993; You et al., 1996). The CCK-A receptor binds CCK peptides that are amidated and sulfated with high affinity, whereas the affinity for nonsulfated CCK peptides and gastrins is negligible.

The CCK-B receptor (the "brain" receptor) is the predominant CCK receptor in the brain (Kopin et al., 1992; Pisegna et al., 1992). It is less selective than the CCK-A receptor, as it also binds nonsulfated CCK, gastrins, and short C-terminal fragments with high affinity. Data on the gastrin receptor cloned from parietal cells (Kopin et al., 1992) show that the gastrin and CCK-B receptor are identical (Pisegna et al., 1992). The gastrin/CCK-B receptor is also abundantly expressed in the pancreas of man (Lee et al., 1993; Saillan-Barreau et al., 1999).

6.2.4.2 Gallbladder Emptying, Bile Release, Pancreatic Secretion, and Growth

CCK peptides stimulate hepatic secretion mainly as bicarbonate from hepatic ductular cells (Shaw and Jones, 1978) and act on gallbladder muscles with a potency correlated to the low plasma concentrations of sulfated CCK (Figure 6.3). From the liver and gallbladder, bile is released into the duodenum via CCK-mediated rhythmic contraction and relaxation of muscles in the common bile duct and the sphincter of Oddi. CCK regulates the secretion of pancreatic enzymes so potently that it seems sufficient to account for all enzyme secretion (Anagnostides et al., 1985; Kerstens et al., 1985; Solomon et al., 1984). CCK is also capable of releasing several small intestinal enzymes such as alkaline phosphatase (Dyck et al., 1974), disaccharidase (Dyck et al., 1974), and enterokinase (Götze et al., 1978). In addition, CCK stimulates the synthesis of digestive enzymes such as pancreatic amylase, chymotrypsinogen, and trypsinogen (Rothman and Wells, 1967; Bragado et al., 2000; Williams, 2001).

Although the effect of CCK on exocrine pancreatic secretion was for many years considered restricted to enzyme secretion, it is now well established that CCK can also stimulate fluid and bicarbonate secretion. The effect on bicarbonate secretion is in itself weak, but because CCK potentiates secretin-induced bicarbonate secretion from the pancreas in the same way as secretin potentiates CCK-induced enzyme release (Debas and Grossman, 1978), the effect of CCK peptides on bicarbonate and fluid secretion is sufficiently potent under physiological circumstances (Figure 6.4). There are species differences in the mechanism of the CCK effect on pancreatic exocrine secretion. Hence, it is now generally assumed that CCK in man stimulates pancreatic enzyme secretion through a cholinergic pathway that is considerably less significant in rodents (Soudah et al., 1992; Ji et al., 2001; Owyang and Logsdon, 2004).

In man and pig, CCK-33 and -8 are weak insulin and glucagon secretagogues (Rehfeld et al., 1980; Jensen et al., 1981), whereas CCK-5 may be more potent (Rehfeld et al., 1980). In contrast, CCK-8 in low concentrations releases insulin and glucagon in dog and rat (Otsuki et al., 1979; Hermansen, 1984). The species differences seem to be due to the innervation in man and pig of pancreatic islets with terminals that release small molecular forms of CCK (Rehfeld et al., 1980), whereas rat and dog islets have no such innervation (Larsson and Rehfeld, 1979b; Rehfeld et al., 1980). Moreover, islet cells in man and pig also express the gastrin/CCK-B receptor (Lee et al., 1993; Saillan-Barreau et al., 1999), whereas rat islet cells express mainly the CCK-A receptor (Monstein et al., 1996).

By 1967, Rothman and Wells had already noted that CCK increased pancreatic weight and enzyme synthesis. Also the maximal output of bicarbonate and protein from the hypertrophic pancreas was increased (Petersen et al., 1978). Although secretin in itself is without trophic effects, the combination of secretin and CCK showed additional trophic effects on ductular cells with a subsequent increase of secretin-induced bicarbonate output (Petersen et al., 1978).

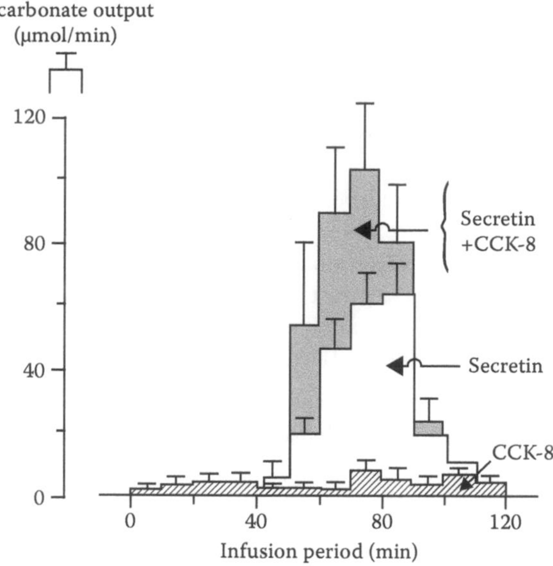

FIGURE 6.4 Pancreatic bicarbonate output during infusion of a physiological dose of cholecystokinin (CCK-8), a physiological dose of secretin, or physiological doses of secretin plus CCK in normal human subjects ($n = 6$).

6.2.4.3 Gastrointestinal Motility and Blood Flow

CCK peptides contribute to the control of motility of the digestive tract. The effect on gastric motility varies considerably between species, and it is still uncertain whether circulating CCK affects the motility of the stomach under physiological conditions. In contrast to the stomach, the distal part of the gut is abundantly innervated with CCK neurons (Larsson and Rehfeld, 1979a; Schultzberg et al., 1980). It is therefore likely that an increase of intestinal motor activity by exogenous CCK (Gutierrez et al., 1974) reflects control of intestinal muscles by CCK. Neuronal CCK probably acts both indirectly via acetylcholine release from postganglionic parasympathetic nerves and directly on muscle cells. The observation that CCK peptides stimulate intestinal blood flow is in harmony with the CCK nerve terminals around blood vessels in the basal lamina propria and the submucosa of all parts of the intestinal tract (Larsson and Rehfeld, 1979a).

6.2.4.4 Satiety

In 1973, Gibbs, Young, and Smith discovered that exogenous CCK inhibits food intake (Gibbs et al., 1973). The effect was dose dependent and specific in the sense that it mimicked the satiety induced by food and was not seen with other gut peptides known then. The effect could be demonstrated in several mammals. Vagotomy studies indicate that peripheral CCK induces satiety via gastric receptors, relaying the effect into afferent vagal fibers (Smith et al., 1981). The satiety signal then reaches the hypothalamus from the vagus via the nucleus tractus solitarius and area postrema.

6.2.4.5 Inhibition of Gastric Acid Secretion

The effect of CCK on gastric acid secretion has, until recently, been uncertain. On the one hand it has been suggested that CCK is an acid inhibitor released from the intestine during meals. On the other hand, the results from CCK infusions have been inconsistent. Recently, the question was solved in gastrin/CCK double knockout mice (Chen et al., 2004), in which it was shown that circulating CCK is a potent acid inhibitor that stimulates somatostatin release from fundic D-cells via CCK-A receptors. The local somatostatin in the fundic mucosa then inhibits acid secretion from parietal cells (Chen et al., 2004). Thus, CCK is a potent enterogastrone.

6.3 GASTRIN

6.3.1 Cellular Synthesis of Gastrin Peptides

The cloning of mammalian gastrin and CCK genes shows that the genes are structurally similar, both in the overall exon–intron organization and in certain peptide coding sequences (Boel et al., 1983; Wiborg et al., 1984; Deschenes et al., 1984, 1985). The gastrin gene spans 4.1kb chromosomal DNA and contains two introns of 3041 and 130bp, respectively. Antral G-cells generate a single mRNA of 0.7kb, which encodes the 101-amino-acid preprogastrin in man and mouse, whereas other mammalian preprogastrins have 104 amino acids due to a prolonged C-terminal flanking peptide (Figure 6.5). The first exon encodes the 5′-untranslated region (for reviews, see Rehfeld and van Solinge, 1994; Rehfeld, 1998c).

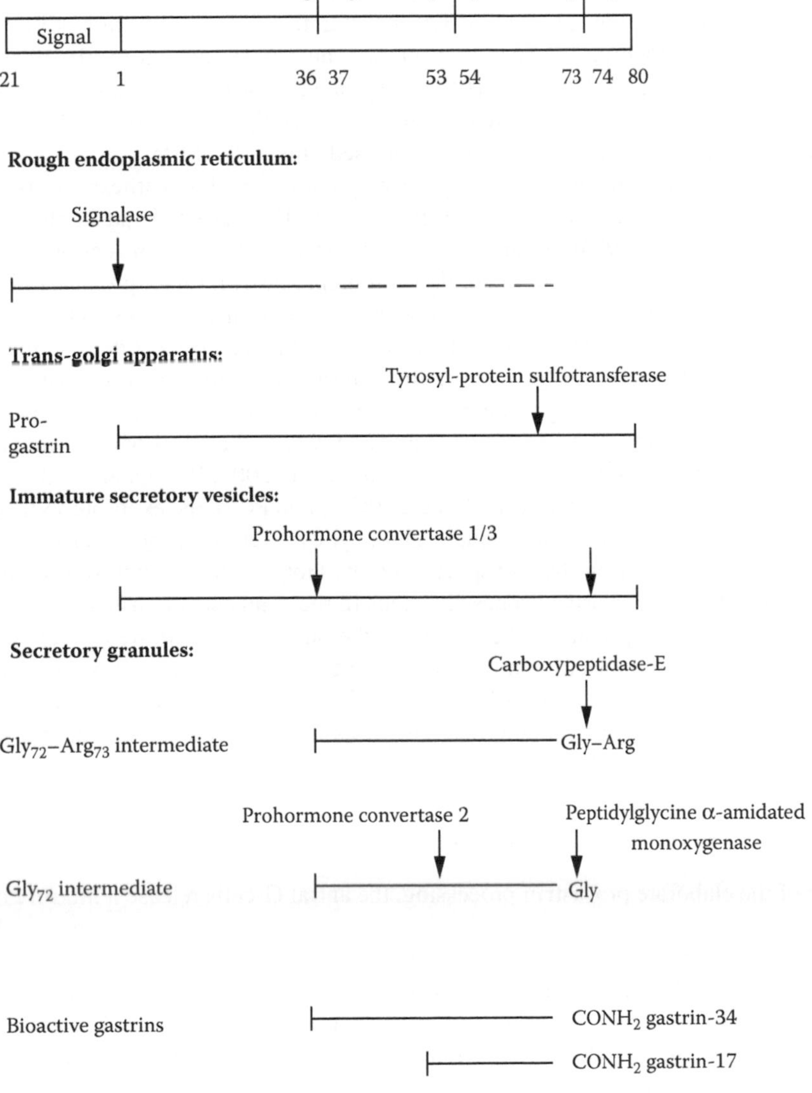

FIGURE 6.5 Co- and post-translational processing of progastrin to the predominant gastrins, gastrin-34 and gastrin-17. Dibasic sites are indicated (Arg-Arg and Lys-Lys).

Several studies have identified important regulatory domains in the gastrin gene promoter (Brand and Fuller, 1988; Wang and Brand, 1990; Merchant et al., 1991; Bachwich et al., 1992; Bundgaard et al., 1995; Ford et al., 1997; Hansen et al., 1999; for review, see Merchant et al., 2004). Hence, a cell-specific regulatory element has been located in the cap-exon I region of the human gastrin gene, and a pancreatic islet cell specific regulatory domain in the gastrin promoter containing adjacent positive and negative DNA elements has also been identified (Brand and Fuller, 1988; Wang and Brand, 1990). This regulatory domain may be a switch controlling the transient transcription of the gastrin gene in the pancreatic islets during fetal and neonatal development (Brand and Fuller, 1988; Wang and Brand, 1990). The gastrin gene transcription is stimulated by epidermal growth factor (EGF) and inhibited by somastostatin (Merchant et al., 1991; Bachwich et al., 1992; Bundgaard et al., 1995; Ford et al., 1997; Hansen et al., 1999). The EGF-responsive element is of particular relevance for the understanding of the growth-promoting and oncological significance of gastrin.

So far, no tissue-specific splicing or use of alternative promoters has been described. Therefore, the tissue-specific molecular pattern of gastrin peptides is due to differences in the post-translational processing rather than alternative RNA splicing.

In the developing rat colon, gastrin mRNA concentrations increase from birth to adult, apparently without a corresponding increase in peptide synthesis (Lüttichau et al., 1993). Therefore, expression of the gastrin gene also appears to be regulated at the translational level. The expression of the gastrin gene is ontogenetically regulated. Consequently, expression of the gastrin gene in tumors may involve factors that are normally expressed only in fetal life.

The antral G-cells are the main site of gastrin synthesis and biosynthesis studies have so far focused on antral tissue (Gregory et al., 1983; Brand et al., 1984; Hilsted and Rehfeld, 1987; Jensen et al., 1989; Rehfeld et al., 1995). After translation of gastrin mRNA in the endoplasmic reticulum and removal of the N-terminal signal peptide from preprogastrin, intact progastrin is transported to the Golgi apparatus. In the trans-Golgi network, O-sulfation of the tyrosyl-66 residue neighbouring the active site and the first endoproteolytic convertase cleavage at two monobasic processing sites occur. From the trans-Golgi network the N-terminal fragments of progastrin are sorted to the constitutive secretory pathway, whereas vesicles carry the C-terminally derived processing intermediates along the regulated secretory pathway towards the basal part of the G-cells where the gastrin peptides are stored in granules (Bundgaard et al., 2004; Bundgaard and Rehfeld, 2008). The cleavage by prohormone convertases 1 and 2 (PC1/3 and PC2), the exoproteolytic carboxypeptidase E trimming as well as the subsequent glutamyl cyclization at the N-termini of gastrin-34 and gastrin-17 continue during the transport from the Golgi to the immature secretory granules (Figure 6.5). The last and decisive processing steps in the synthesis of gastrin then occurs during storage in the secretory granules where the amidation enzyme removes glyoxylate from the glycine-extended intermediates to complete the synthesis of bioactive carboxyamidated peptides. Notably, amidation of gastrin is an all-or-none activation process, and is carefully controlled. Activation of the enzymatic amidation process requires copper, oxygen, and ascorbic acid as cofactors and a pH of around 5.

6.3.2 Cellular Release of Gastrin Peptides

As a result of the elaborate progastrin processing, the antral G-cells release a mixture of progastrin products from the secretory granules to blood. In man a few percent are nonamidated precursors, mainly glycine-extended gastrins, but more than 95% are α-amidated bioactive gastrins. Of these, 85% is gastrin-17, 5–10% gastrin-34, and the rest is a mixture of gastrin-71, gastrin-14, and the short gastrin-6. Approximately half the amidated gastrins are tyrosyl-sulfated. Owing to gross differences in metabolic clearance rates, the distribution pattern of gastrins in peripheral plasma changes, so that larger gastrins with their long half-lives predominate over gastrin-17 and shorter gastrins. Hence, gastrin-34 is the predominant form of gastrin in peripheral blood.

Increased gastrin synthesis changes the molecular pattern in plasma further, as seen in achlorhydria where the translational activity of gastrin mRNA in G-cells seems to be so high that the processing enzymes cannot keep up with the maturation (Jensen et al., 1989). Consequently, G-cells release more incompletely processed nonamidated progastrin products when the synthesis is increased. Also, the carboxyamidated gastrins are less sulfated and the N-terminus of progastrin cleaved to a lesser degree.

Gastrin peptides are also released from cell types other than the antroduodenal G-cells. Quantitatively, these other cells contribute little to circulating gastrin, partly because the secretion seems to serve local purposes, partly because the biosynthetic processing is cell-specific. So far, expression of progastrin has also been encountered in the ileum and the colon (Lüttichau et al., 1993; Friis-Hansen and Rehfeld, 1994), in endocrine cells in the fetal and neonatal pancreas (Larsson et al., 1976; Bardram et al., 1990), in pituitary corticotrophs and melanotrophs (Rehfeld, 1978b; Larsson and Rehfeld, 1981), in oxytocinergic hypothalamopituitary neurons (Rehfeld et al., 1984), in a few cerebellar and vagal neurons (Unväs-Wallensten et al., 1977; Rehfeld, 1991), in the bronchial mucosa (Rehfeld et al., 1989), in postmenopausal ovaria (van Solinge et al., 1993), and in human spermatogenic cells (Schalling et al., 1990). The concentrations and presumably also the synthesis in the extra-antal tissues are far below that of the antral "main factory." The function of gastrin synthesized outside the antroduodenal mucosa is not yet fully known, but a possibility is the paracrine or autocrine regulation of growth. It is also possible that the low expression is without significant function in the adult, but a relic of a more comprehensive fetal synthesis for local stimulation of growth. A third possibility is that the low cellular and, hence, low tissue concentration is due to constitutive rather than regulated secretion.

Although the extra-antral synthesis of gastrin may be without function in the adult organism, recognition of the phenomenon has biomedical interest. Hence, tumors originating from cells that express the gastrin gene at low levels in adult organisms may produce gastrin in significant amounts in tumors. Well-known examples are pancreatic, duodenal, and ovarian gastrinomas, which give rise to the Zollinger–Ellison syndrome (Bardram et al., 1990; van Solinge et al., 1993).

6.3.3 Gastrin Peptides in Circulation

The molecular nature of gastrin in plasma and the physiological variations during meals have been well studied during the last decades (for reviews, see Rehfeld, 1973; Walsh, 1994). The predominant circulating forms in most mammals are gastrin-34 and gastrin-17, which both occur in sulfated and nonsulfated forms. In addition, small amounts of gastrin-71 and gastrin-14 (Rehfeld, 1973) may be present. Only the cat differs, because gastrin in peripheral feline blood is all gastrin-14 (Rehfeld and Uvnäs-Wallensten, 1978). During hypergastrinemia in hypo- and achlorhydria, during treatment with antacids, and in gastrinoma patients, the pattern changes to be dominated by large molecular forms, gastrin-71 and gastrin-34 (Jensen et al., 1989). The shift is partly due to attenuated endoproteolytic processing of progastrin during G-cell hypersecretion (Jensen et al., 1989), and partly to slower clearance of the larger forms. Thus, gastrin-34, in man, has a halflife in circulation of 40 min, contrasting with 4 min for gastrin-17 peptides (Walsh et al., 1976).

In the basal state, the concentration of gastrin in plasma varies between 10 and 20 pmol/L. During stimulation with a protein-rich meal, the concentrations increase within few minutes to reach a peak of ~50 pmol/L after 15 to 20 min. The gastrin concentrations are at the same level as most other gastrointestinal and pancreatic hormones (Horness et al., 1980). Notably, however, they are 10 to 20-fold above those of CCK in plasma (Rehfeld, 1998a; Rehfeld et al., 2001). This difference is important for receptor selection. Hence, gastrin/CCK-B type receptors in the periphery are in physiological terms the receptors for gastrin only, because the 10 to 20-fold lower CCK concentrations in plasma cannot compete with gastrin. Thus, in normal mammals CCK effects in the periphery are elicited essentially only via CCK-A receptors that do not bind the gastrins. The described gastrin concentrations in plasma are sufficient to account for regulation of gastric acid secretion via gastrin receptor-binding on ECL-cells and subsequent release of histamine.

6.3.4 Action of Gastrin Peptides

6.3.4.1 The Gastrin Receptor

The gastrin/CCK-B receptor is, as described above, the predominant receptor for gastrin and CCK peptides in the central nervous system (Kopin et al., 1992; Lee et al., 1993; Wank, 1995). It is expressed with particularly high density in the cerebral cortex (Lee et al., 1993), where it binds both sulfated and nonsulfated gastrin and CCK peptides as well as short C-terminal fragments of CCK (CCK-5), all with similarly high affinity. The gastrin/CCK-B receptor is also abundantly expressed on ECL-cells in the stomach (Chen et al., 1998; Reubi et al., 1997) and at a lower level in the pancreas of man and pig (Reubi et al., 1997; Saillan-Barreau et al., 1999; Reubi, 2003). In physiological terms, the CCK-B receptor expressed outside of the nervous system only uses gastrins as ligands. Here, it is simply the gastrin receptor.

6.3.4.2 Gastric Acid Secretion

The main effect and purpose of gastrin is to stimulate gastric acid secretion. Gastrin was discovered and defined by its effect on gastric acid secretion (Edkins, 1905). Accordingly, separate gastrin peptides, distinct from CCK, did not occur during evolution until elasmobranches were developed. The elasmobranch stomach was apparently also the first to secrete hydrochloric acid (for review, see Johnsen, 1998). The fundamental link between gastrin and gastric acid secretion has recently been further emphasized in genetically modified animals. For instance, the stomach in gastrin knockout mice does not secrete acid, not even after stimulation with secretagogues such as histamine and carbachol (Koh et al., 1997; Friis-Hansen et al., 1998) (Figure 6.6). Only infusion of gastrin for 1 to 2 weeks revives the acid-producing machinery with responses to histamine and carbachol (Friis-Hansen et al., 1998).

Gastric acid is essential for initiation of food digestion. In addition, however, acid constitutes a defence barrier against microbial invasion of the gut, which again appears to help prevention of cancer in the stomach. Hence, after 1.5 years of age, gastrin knockout mice develop adenocarcinomas in the stomach (Zavros et al., 2002; Friis-Hansen et al., 2006). Consequently, through its stimulation of gastric acid, gastrin seems necessary for the maintenance of life in mammals.

Gastric acid secretion is regulated by gastrin via the gastrin/CCK-B receptor on ECL-cells, which again release histamine to stimulate parietal cells in the fundic mucosa. Parietal cells also express gastrin receptors, but at a lower level. Therefore, most of the effect of gastrin is indirect via ECL-cells and histamine (for reviews, see Chen et al., 1998; Håkanson et al., 2004; Prinz et al., 2004). Careful examination of gastrin release during meals and subsequent infusions in physiological doses of gastrin-17 and gastrin-34 have shown that normal gastrin release to circulation during meals is sufficient to account for the gastric phase of acid secretion (for review, see Walsh, 1994). Gastrin interacts with a number of other stimulatory and inhibitory substances in the fine-tuning of gastric acid secretion (for review, see Lloyd and Walsh, 1994). However, further discussion of the mechanisms is outside the scope of this review.

6.3.4.3 Mucosal Cell Growth

Four decades ago it was shown that repeated doses of pentagastrin to rats resulted in parietal cell hyperplasia (Crean et al., 1969). Numerous studies have since confirmed that gastrin regulates the growth of cells in the fundic mucosa (Johnson and Guthrie, 1974; for reviews, see Johnson, 1976, 1987). ECL-cells are particularly sensitive targets for the trophic effect of gastrin. Thus, prolonged hypergastrinemia of moderate degree in rats leads to carcinoid tumors of ECL-cell origin, that is, ECLomas. Up to 25% of female rats and a lower fraction of male rats developed ECLomas after lifelong moderate hypergastrinemia (Håkanson and Sundler, 1990; Havu et al., 1990). With

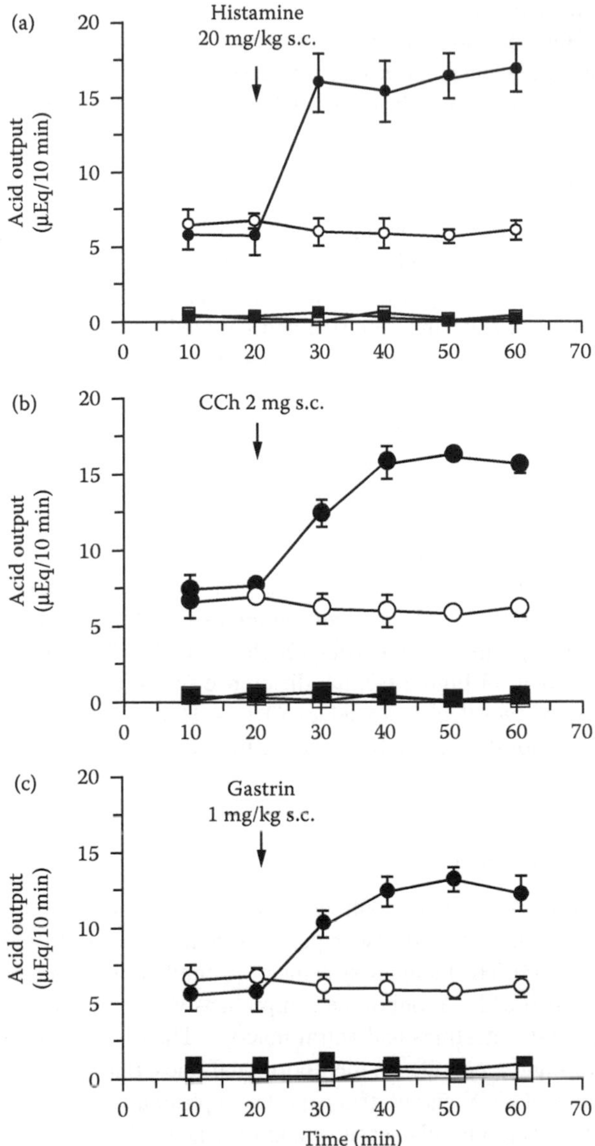

FIGURE 6.6 Gastric acid secretion in wild-type (circles) and gastrin knockout mice (squares). Basal secretion was measured for 60 min in one group of mice (open symbols). Other groups (filled symbols) were stimulated by subcutaneous injections of histamine (a), carbachol (b), or gastrin (c). Each group consisted of six mice. (Data from Friis-Hansen et al., *Am. J. Physiol.*, 274, G561–G568, 1998.)

a complete lack of gastrin (knockout mice), the stomach still contains ECL and parietal cells, but they are immature with a grossly abnormal morphology (Koh et al., 1997; Friis-Hansen et al., 1998). Presumably gastrin is therefore necessary also for maturation of the cells to exert normal secretion.

Gastrin has been suggested to stimulate the growth of several epithelial cells also outside the fundic mucosa (for reviews, see Johnson, 1976, 1987). These suggestions have, in combination with low-level expression of gastrins in the bronchial, colorectal, ovarian, and pancreatic cancers (Rehfeld et al., 1989; van Solinge et al., 1993; Rehfeld and van Solinge, 1994; Goetze et al., 2000), been used to discuss the possibility of gastrins playing important roles in the formation of major cancers

(Rehfeld and van Solinge, 1994; Rengifo-Cam and Singh, 2004; Rehfeld, 2006). It is obviously an important discussion. But a decisive carcinogenetic role of gastrin in man still remains to be proven (Rehfeld, 2006; Rehfeld and Goetze, 2006). So far, most evidence is based on studies of cell cultures and experimental rodent models, and the evidence has often been controversial.

6.3.5 Structural Analogs

As shown in Figure 6.1, naturally occurring homologs have been identified in extracts of frogskin (Anastasi et al., 1967, 1969) and protochordean ganglionic tissue (Johnsen and Rehfeld, 1990). The frogskin peptides, caerulein and phyllocaerulein, have been used as CCK analogs in examination of the function of pancreatic exocrine secretion and gallbladder emptying in man with suspected pancreatic and gallbladder diseases. The potency of caerulein and phyllocaerulein is similar to that of sulfated CCK-8.

A synthetic pentapeptide, BOC (butyloxycarbonyl)–Ala–Trp–Met–Asp–Phe–NH_2, also called pentagastrin®, has been widely used as a gastrin analog in examination of stimulated gastrin acid secretion. Because the C-cells in medullary thyroid carcinomas express the gastrin/CCK-B receptor abundantly (Reubi et al., 1997), pentagastrin has also been used to stimulate calcitonin secretion in early diagnosis of thyroid carcinomas.

It has been suggested that sulfakinins, that is, small carboxyamidated and tyrosyl-sulfated peptides isolated from the cockroach *Leucophaea maderae* (Nachman et al., 1986) or deduced from cDNA cloned from *Drosophila melanogaster* (Nichols et al., 1988), might be early homologs of CCK. However, as shown in Figure 6.7, the difference in the C-terminal active site sequences comprises substitutions of two decisive residues (Trp → His and Asp → Arg). Such differences render a phylogenetic relationship unlikely and make the sulfakinins useless as ligands for the gastrin/CCK receptors.

6.3.6 Common Assay Systems

Bioassay systems based on gallbladder contraction in cats and gastric acid secretion in dogs were established for purification of CCK and gastrin peptides in the 1950s and 1960s (Gregory and Tracy, 1964; Mutt and Jorpes, 1968). The bioassay systems were cumbersome and insensitive. But they had no alternatives and they sufficed to monitor the comprehensive purifications out of large volumes of extracts from porcine small intestines and antral mucosa. They were, however, far too unspecific and insensitive for defining the peptides in their hormonal roles, that is, for measurement of the low picomolar concentrations of CCK and gastrin peptides in plasma.

In the mid 1980s, however, a highly sensitive and specific CCK bioassay for measurements also in plasma was developed by Liddle et al. (1984, 1985). The assay was based on amylase release from dispersed rat pancreatic acinar cells. In spite of excellent reliability parameters, this assay found only little use due to its considerable labour intensiveness and consequently high costs per sample analysis.

Today, CCK and gastrin peptides are all measured by radioimmunoassay systems (RIA). Generally, development of gastrin RIAs has been straightforward. Production of monospecific high-titer and high-affinity gastrin antibodies in rabbits has not been difficult (Rehfeld et al., 1972), and the reproducible preparation of pure mono-iodinated gastrin tracers has made gastrin RIAs robust and simple (Stadil and Rehfeld, 1972). The only problem has been proper recognition of the RIA epitope specificity in relation to the molecular heterogeneity of gastrin in plasma. Hence, some gastrin RIAs fail to measure changed molecular patterns during hypersecretion, as seen in gastrin producing cancers. Consequently, the diagnostic specificity of such RIAs is inadequate (for review, see Rehfeld, 2008).

In contrast to gastrin RIAs, it has been exceedingly difficult to establish CCK RIAs of sufficient specificity and sensitivity (for reviews, see Rehfeld, 1984, 1998b). Finally, a decade ago, a high-titer

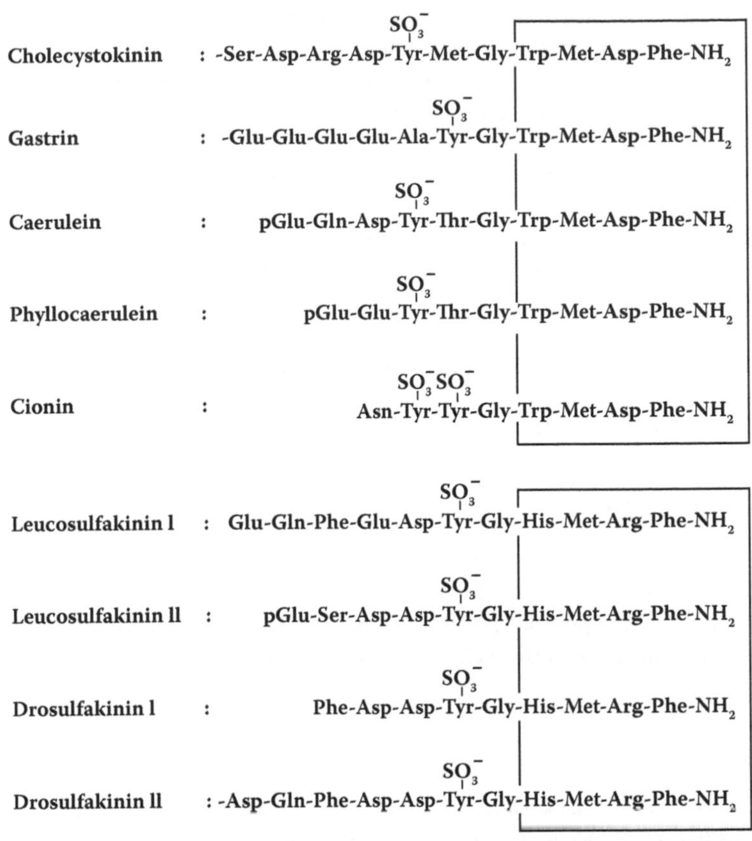

FIGURE 6.7 Comparison of the active site structures of true members of the CCK–gastrin family (upper part; see also Figure 6.1) with those of the sulfakinins (lower part of figure).

and high-affinity antiserum was raised, which, under suitable RIA conditions, could measure the femtomolar concentrations of CCK peptides in circulation without cross-reactivity with any gastrins (Rehfeld, 1998a; Rehfeld et al., 2001). CCK RIA kits based on this antiserum are now generally available and in use for physiological, pharmacological, and clinical studies.

6.3.7 Future Developments

CCK and gastrin peptides are prominent vertebrate members of a family of biologically active peptides that, in phylogenetic terms, date back more than 600 million years. They were also among the first mammalian peptide hormones to be discovered in the early 20th century. Originally they were conceived as local gut hormones, but recently their ubiquitous roles as general intercellular messengers have been recognized (for review, see Rehfeld, 1998c).

The biology of the CCK and gastrin peptides is in many respects still enigmatic, and important aspects of their roles remain to be explored. Thus, the following questions are still open and require answers:

1. In view of the widespread, high-level expression of the CCK gene in cerebral neurons, what is the role of CCK peptides in central neuronal pathways and brain functions?
2. In this context, what are the pathogenetic roles of the cerebral CCK system in neuropsychiatric diseases?

3. To what extent is the gastrin gene expressed in neurons?
4. What is the role of neuronal gastrin peptides?
5. In view of the promiscuous expression of the gastrin gene in common neoplasias (lung, gastric, pancreatic, colorectal, and ovarian cancers), what is the role of gastrin peptides in carcinogenesis?
6. If the gastrins do indeed play a growth-promoting role in cancers, is it possible then to target gastrin peptides in cancer treatment and prevention?
7. At a basal level, do progastrin and/or proCCK release other bioactive peptides (with separate receptors) than the known carboxyamidated peptides that use the CCK-A and CCK-B receptors?

Many more specific questions can be asked. Taken together, the questions reveal the substantial need for continuous and intensive research in CCK and gastrin systems in the future. The present understanding of their roles merely represent the tip of the iceberg.

ACKNOWLEDGMENTS

The skilful secretarial assistance of Diana Skovgaard is gratefully acknowledged. The studies from the laboratory of the author that has contributed to the picture of cholecystokinin and gastrin described here have been supported by grants from the Danish Medical Research Council, the Danish Cancer Union, the Danish Biotechnology Program for Peptide Research, and the Lundbeck Foundation.

REFERENCES

Anagnostides, A.A. et al., Human pancreatic and biliary responses to physiological concentrations of cholecystokinin octapeptide, *Clin. Sci. Mol. Med.*, 69, 259–263, 1985.

Anastasi, A., Erspamer, V., and Endean, R., Isolation and structure of caerulein, an active decapeptide from the skin of *Hyla caerula*, *Experientia*, 23, 699–700, 1967.

Anastasi, A. et al., Structure and pharmacological actions of phyllocaerulein, a caerulein-like nonapeptide. Its occurrence in extracts of the skin of *Phyllomedusa sauvagei* and related Phyllomedusa species, *Br. J. Pharmacol.*, 37, 198–206, 1969.

Bachwich, D., Merchant, J., and Brand, S.J., Identification of a cis-regulatory element mediating somatostatin inhibition of epidermal growth factor-stimulated gastrin gene transcription, *Mol. Endocrinol.*, 6, 1175–1184, 1992.

Bardram, L., Hilsted, L., and Rehfeld, J.F., Progastrin expression in mammalian pancreas, *Proc. Natl Acad. Sci. USA*, 87, 298–302, 1990.

Bernard, C. et al., Peptones stimulate intestinal cholecystokinin gene transcription via cyclic adenosine monophosphate response element-binding factors, *Endocrinology*, 142, 721–729, 2001.

Boel, E. et al., Molecular cloning of human gastrin cDNA. Evidence for evolution of gastrin by gene duplication, *Proc. Natl Acad. Sci. USA*, 80, 2866–2869, 1983.

Bonetto, V. et al., Isolation and characterization of sulfated and nonsulfated forms of cholecystokinin-58 and their action on gallbladder contraction, *Eur. J. Biochem.*, 264, 336–340, 1999.

Bragado, M.J., Tashiro, M., and Williams, J.A., Regulation of the initiation of pancreatic digestive enzyme protein synthesis by cholecystokinin in rat pancreas *in vivo*, *Gastroenterology*, 119, 1731–1739, 2000.

Brand, S.J. and Fuller, P.J., Differential gastrin gene expression in rat gastrointestinal tract and pancreas during neonatal development, *J. Biol. Chem.*, 263, 5341–5347, 1988.

Brand, S.J. et al., Biosynthesis of tyrosine O-sulfated gastrins in rat antral mucosa, *J. Biol. Chem.*, 59, 13246–13252, 1984.

Buffa, R., Solcia, E., and Go, V.L.W., Immunohistochemical identification of the cholecystokinin cell in the intestinal mucosa, *Gastroenterology*, 70, 528–530, 1976.

Bundgaard, J.R. et al., A distal Sp1-element is necessary for maximal activity of the human gastrin gene promoter, *FEBS Letts.*, 369, 225–228, 1995.

Bundgaard, J.R., Birkedal, H., and Rehfeld, J.F., Progastrin is directed to the regulated secretory pathway by synergistically acting basic and acidic motifs, *J. Biol. Chem.*, 279, 5488–5493, 2004.

Bundgaard, J.R. and Rehfeld, J.F., Distinct linkage between post-translational processing and differential secretion of progastrin derivatives in endocrine cells, *J. Biol. Chem.*, 283, 4014–4027, 2008.

Chen, D. et al., Altered control of gastric acid secretion in gastrin-cholecystokinin double mutant mice, *Gastroenterology*, 126, 476–487, 2004.

Chen, D. et al., Novel aspects of gastrin-induced activation of histidine decarboxylase in rat stomach ECL cells, *Regul. Peptides*, 77, 169–175, 1998.

Crawley, J.N., Comparative distribution of cholecystokinin and other neuropeptides: Why is this peptide different from all other peptides? *Ann. NY Acad. Sci.*, 448, 1–8, 1985.

Crean, G.P., Marshall, M.W., and Rumsey, R.D., Parietal cell hyperplasia induced by the administration of pentagastrin to rats, *Gastroenterology*, 57, 147–155, 1969.

Deavall, D.G. et al., Control of CCK gene transcription by PACAP in STC-1 cells, *Am. J. Physiol.*, 279, G605–G612, 2000.

Debas, H.T. and Grossman, M.I., Pure cholecystokinin: Pancreatic protein and bicarbonate response, *Digestion*, 9, 469–481, 1978.

Deschenes, R.J. et al., A gene encoding rat cholecystokinin: Isolation, nucleotide sequence, and promoter activity, *J. Biol. Chem.*, 260, 1280–1286, 1985.

Deschenes, R.J. et al., Cloning and sequence analysis of cDNA encoding rat preprocholecystokinin, *Proc. Natl Acad. Sci. USA*, 81, 726–730, 1984.

Dockray, G.J. et al., Isolation, structure, and biological activity of two cholecystokinin octapeptides from sheep brain, *Nature*, 274, 711–713, 1978.

Dockray, G.J. et al., Transport of cholecystokinin-octapeptide-like immunoreactivity towards the gut in afferent vagal fibers in cat and dog, *J. Physiol.*, 314, 501–511, 1981.

Dodd, P.R., Edwardson, J.A., and Dockray, G.J., The depolarisation-induced release of cholecystokinin octapeptide from rat synaptosomes and brain slices, *Regul. Peptides*, 1, 17–29, 1980.

Dyck, W.P. et al., Hormonal stimulation of intestinal disaccharidase release in the dog, *Gastroenterology*, 66, 533–538, 1974.

Dyck, W.P., Martin, G.A., and Ratliff, C.R., Influence of secretin and cholecystokinin on intestinal alkaline phosphatase secretion, *Gastroenterology*, 64, 599–602, 1973.

Edkins, J.S., On the chemical mechanism of gastric secretion, *Proc. Roy. Soc. (London)*, 76, 376, 1905.

Emson, P.C., Lee, C.M., and Rehfeld, J.F., Cholecystokinin octapeptide, vesicular localization, and calcium dependent release from rat brain *in vitro*, *Life Sci.*, 26, 2157–2163, 1980.

Fallon, J.H. and Seroogy, K.B., The distribution and some connections of cholecystokinin neurons in the rat brain, *Ann. NY Acad. Sci.*, 448, 121–132, 1985.

Ford, M.G. et al., EGF receptor activation stimulates endogenous gastrin gene expression in canine G cells and human gastric cell cultures, *J. Clin. Invest.*, 99, 2762–2771, 1997.

Friis-Hansen, L. and Rehfeld, J.F., Ileal expression of gastrin and cholecystokinin, *FEBS Letts.*, 343, 115–119, 1994.

Friis-Hansen, L. et al., Gastric inflammation, intestinal metaplasia, and tumor development in gastrin deficient mice, *Gastroenterology*, 131, 246–258, 2006.

Friis-Hansen, L. et al., Impaired gastric acid secretion in gastrin-deficient mice, *Am. J. Physiol.*, 274, G561–G568, 1998.

Gevrey, J.C. et al., Co-requirement of cyclic AMP- and calcium-dependent protein kinases for transcriptional activation of cholecystokinin gene by protein hydrolysates, *J. Biol. Chem.*, 277, 22407–22413, 2002.

Gibbs, J., Young, R.C., and Smith, G.P., Cholecystokinin elicits satiety in rats with open gastric fistulas, *Nature*, 245, 323–325, 1973.

Goetze, J.P. et al., Closing the gastrin loop in pancreatic carcinomas: Coexpression of gastrin and its receptor in solid human pancreatic adenocarcinoma, *Cancer*, 88, 2487–2494, 2000.

Goltermann, N., Rehfeld, J.F., and Petersen, H.R., *In vivo* biosynthesis of cholecystokinin in rat cerebral cortex, *J. Biol. Chem.*, 255, 6181–6185, 1980.

Götze, H., Götze, J., and Adelson, J.W., Studies on intestinal enzyme secretion: The action of cholecystokinin, pentagastrin, and bile, *Res. Exp. Med.*, 173, 17–25, 1978.

Gregory, R.A. et al., Isolation from porcine antral mucosa of a hexapeptide corresponding to the C-terminal sequence of gastrin, *Peptides*, 4, 319–323, 1983.

Gregory, H. et al., The antral hormone gastrin. I. Structure of gastrin, *Nature*, 204, 931–933, 1964.

Gregory, R.A. and Tracy, H.J., The constitution and properties of two gastrins extracted from hog antral mucosa, *Gut*, 5, 103–117, 1964.

Gubler, U. et al., Cloned cDNA to cholecystokinin mRNA predicts an identical preprocholecystokinin in pig brain and gut, *Proc. Natl Acad. Sci. USA*, 81, 4307–4310, 1984.

Gutierrez, J.G., Chey, W.Y., and Dinoso, V.P., Actions of cholecystokinin and secretin on the motor activity of the small intestine in man, *Gastroenterology*, 67, 35–41, 1974.

Håkanson, R. and Sundler, F., Proposal mechanism of induction of gastric carcinoids: The gastrin hypothesis, *Eur. J. Clin. Invest.*, 20, S65–S71, 1990.

Håkanson, R., Surve, V.V., and Chen, D., ECL cells in a physiological context, in *Gastrin in the New Millennium*, Merchant, J.L., Buchan, A.M.J., and Wang, T.C., Eds., Cure Foundation, Los Angeles, 2004, p. 161–181.

Hansen, T.v.O., Cholecystokinin gene transcription: Promoter elements, transcription factors, and signaling pathways, *Peptides*, 22, 1201–1211, 2001.

Hansen, T.v.O. et al., Composite action of three GC/GT boxes in the proximal promoter region is important for gastrin gene transcription, *Mol. Cell. Endocrinol.*, 155, 1–8, 1999.

Hansen, T.v.O., Rehfeld, J.F., and Nielsen, F.C., Function of the C-36 to T polymorphism in the human cholecystokinin gene promoter, *Mol. Psychiatry*, 5, 443–447, 2000.

Hansen, T.v.O., Rehfeld, J.F., and Nielsen, F.C., KCL and forskolin synergistically up-regulate cholecystokinin gene expression via coordinate activation of CREB and the co-activator CBP, *J. Neurochem.*, 89, 15–23, 2004.

Hansen, T.v.O., Rehfeld, J.F., and Nielsen, F.C., Mitogen-activated protein kinase and protein kinase A signaling pathways stimulate cholecystokinin gene transcription via activation of cyclic adenosine 3′, 5′-monophosphate response element-binding protein, *Mol. Endocrinol.*, 13, 466–475, 1999.

Havu, N. et al., Enterochromaffin-like cell carcinoids in the rat gastric mucosa following long-term administration of ranitidine, *Digestion*, 45, 189–195, 1990.

Hendry, S.H.C., Jones, E.G., and Beinfeld, M.C., Cholecystokinin-immunoreactive neurons in rat and monkey cerebral cortex make symmetric synapses and have intimate associations with blood vessels, *Proc. Natl Acad. Sci. USA*, 80, 2400–2404, 1983.

Hermansen, K., Effects of CCK-4, non-sulfated CCK-8 and sulfated CCK-8 on pancreatic somatostatin, insulin, and glucagons secretion in the dog, *Endocrinology*, 114, 1770–1778, 1984.

Hilsted, L. and Rehfeld, J.F., Alpha-carboxyamidation of antral progastrin: Relation to other post-translational modifications, *J. Biol. Chem.*, 262, 16953–16957, 1987.

Himeno, S. et al., Plasma cholecystokinin responses after ingestion of liquid meal and intraduodenal infusion of fat, amino acids, or hydrochloric acid in man: Analysis with region specific radioimmunoassays, *Am. J. Gastroenterol.*, 78, 703–707, 1983.

Hökfelt, T. et al., Evidence for co-existence of dopamine and CCK in mesolimbic neurons, *Nature*, 285, 476–478, 1980.

Honda, T. et al., Differential gene expression of CCK-A and CCK-B receptors in the rat brain, *Mol. Cell. Neurosci.*, 4, 143–154, 1993.

Horness, P.J. et al., Simultaneous recording of the gastroentero-pancreatic hormone response to food in man, *Metabolism*, 29, 777–779, 1980.

Ivy, A.C. and Oldberg, E., A hormone mechanism for gallbladder contraction and evacuation, *Am. J. Physiol.* 86, 559–613, 1928.

Jensen, S.L. et al., Secretory effects of cholecystokinins on the isolated, perfused porcine pancreas, *Acta Physiol. Scand.*, 111, 225–231, 1981.

Jensen, S. et al., Progastrin processing during antral G-cell-hypersecretion in humans, *Gastroenterology*, 96, 1063–1070, 1989.

Ji, B. et al., Human pancreatic acinar cells lack functional responses to cholecystokinin and gastrin, *Gastroenterology*, 121, 1380–1390, 2001.

Johnsen, A.H., Phylogeny of the cholecystokinin/gastrin family, *Front. Neuroendocrinol.*, 19, 73–99, 1998.

Johnsen, A.H. et al., Elasmobranches express separate cholecystokinin and gastrin genes, *Proc. Natl Acad. Sci. USA*, 94, 1022–1026, 1997.

Johnsen, A.H. and Rehfeld, J.F., Cionin: A disulfotyrosyl hybrid of cholecystokinin and gastrin from the neural ganglion of the protochordate ciona intestinalis, *J. Biol. Chem.*, 265, 3054–3058, 1990.

Johnson, L.R., Regulation of gastrointestinal growth, in *Physiology of the Gastrointestinal Tract*, Johnson, L.R. et al., Eds., Raven Press, New York, 1987, p. 301–333.

Johnson, L.R., The trophic action of gastrointestinal hormones, *Gastroenterology*, 70, 278–288, 1976.

Johnson, L.R. and Guthrie, P.D., Mucosal DNA synthesis: A short term index of the trophic action of gastrin, *Gastroenterology*, 67, 453–459, 1974.

Jorpes, J.E. and Mutt, V., Cholecystokinin and pancreozymin, one single hormone? *Acta Physiol. Scand.*, 66, 196–202, 1966.

Jorpes, J.E. and Mutt, V., Secretin, cholecystokinin, and pancreozymin, in *Handbook of Experimental Pharmacology*, Jorpes, J.E. and Mutt, V., Eds., Springer Verlag, New York, Vol. 34, 1973, p. 1–179.

Kerstens, P.J. et al., Physiological plasma concentrations of cholecystokinin stimulate pancreatic enzyme secretion and gallbladder contraction in man, *Life Sci.*, 36, 565–569, 1985.

Koh, T.J. et al., Gastrin deficient mice show altered gastric differentiation and decreased colonic proliferation, *Gastroenterology*, 113, 1015–1025, 1997.

Kopin, A.S. et al., Expression cloning and characterization of the canine parietal cell gastrin receptor, *Proc. Natl Acad. Sci. USA*, 89, 3605–3609, 1992.

Larsson, L.I. and Rehfeld, J.F., Evidence for a common evolutionary origin of gastrin and cholecystokinin, *Nature*, 269, 335–338, 1977.

Larsson, L.I. and Rehfeld, J.F., Localization and molecular heterogeneity of cholecystokinin in the central and peripheral nervous system, *Brain Res.*, 165, 201–218, 1979a.

Larsson, L.I. and Rehfeld, J.F., Peptidergic and adrenergic innervation of pancreatic ganglia, *Scand. J. Gastroenterol.*, 14, 433–437, 1979b.

Larsson, L.I. and Rehfeld, J.F., Pituitary gastrins occur in corticotrophs and melanotrophs, *Science*, 213, 768–770, 1981.

Larsson, L.I. et al., Pancreatic gastrin in foetal and neonatal rats, *Nature*, 262, 609–611, 1976.

Lee, Y.M. et al., The human brain cholecystokinin-B/gastrin receptor, *J. Biol. Chem.*, 268, 8164–8169, 1993.

Liddle, R.A., Goldfine, I.D., and Williams, J.A., Bioassay of cholecystokinin in rats: Effects of food, trypsin inhibitor, and alcohol, *Gastroenterology*, 87, 542–549, 1984.

Liddle, R.A. et al., Cholecystokinin bioactivity in human plasma: Molecular forms, responses to feeding, and relationship to gallbladder contraction, *J. Clin. Invest.*, 75, 1144–1152, 1985.

Lloyd, K.C.K. and Walsh, J.H., Gastric secretion, in *Gut Peptides*, Walsh, J.H. and Dockray, G.J., Eds., Raven Press, New York, 1994, p. 633–654.

Lüttichau, H.R. et al., Developmental expression of the gastrin and cholecystokinin genes in rat colon, *Gastroenterology*, 104, 1092–1098, 1993.

Maton, P.N., Selden, A.C., and Chadwick, V.S., Differential distribution of molecular forms of cholecystokinin in human and porcine small intestinal mucosa, *Regul. Peptides*, 8, 9–19, 1984.

Merchant, J.L., Demediuk, B., and Brand, S.J., A GC-rich element confers epidermal growth factor responsiveness to transciption from the gastrin promoter, *Mol. Cell. Biol.*, 11, 2686–2696, 1991.

Merchant, J.L., Tucker, T.P., and Zavros, Y., Inducible regulation of gastrin gene expression, in *Gastrin in the New Millennium*, Merchant, J.L., Buchan, A.M.J., and Wang, T.C., Eds., Cure Foundation, Los Angeles, 2004, p. 55–69.

Monstein, H.J. et al., Cholecystokinin-A and -B receptor mRNA expression in the gastrointestinal tract and pancreas of the rat and man, *Scand. J. Gastroent.*, 31, 383–390, 1996.

Morley, J.S., Tracy, H.J., and Gregory, R.A., Function relationships of the active C-terminal tetrapeptide sequence of gastrin, *Nature*, 207, 1356–1359, 1965.

Mutt, V. and Jorpes, J.E., Structure of porcine cholecystokinin–pancreozymin. 1. Cleavage with thrombin and with trypsin, *Eur. J. Biochem.*, 6, 156–162, 1968.

Mutt, V. and Jorpes, J.E., Hormonal polypeptides of the upper intestine, *Biochem. J.*, 125, 57P–58P, 1971.

Nachman, R.J. et al., Leucosulfakinin, a sulfated insect neuropeptide with homology to gastrin and cholecystokinin, *Science*, 234, 71–73, 1986.

Nichols, R., Schneuwly, S.A., and Dixon, J.E., Identification and characterization of a Drosophila homologue to the vertebrate neuropeptide cholecystokinin, *J. Biol. Chem.*, 263, 12167–12170, 1988.

Nielsen, F.C. et al., Transcriptional regulation of the human cholecystokinin gene: Composite action of upstream stimulatory factor, Sp1, and members of the CREB/ATF-AP-1 family of transcription factors, *DNA Cell Biol.*, 15, 53–63, 1996.

Otsuki, M. et al., Discrepancies between the doses of cholecystokinin or caerulein stimulating exocrine and endocrine responses in the perfused isolated rat pancreas, *J. Clin. Invest.*, 63, 478–484, 1979.

Owyang, C. and Logsdon, C.D., New insights into neurohormonal regulation of pancreatic secretion, *Gastroenterology*, 127, 957–969, 2004.

Petersen, H., Solomon, T., and Grossman, M.I., Effect of chronic pentagastrin, cholecystokinin, and secretin on pancreas of rats, *Am. J. Physiol.*, 234, E286–E293, 1978.

Pisegna, J.R. et al., Molecular cloning of the human brain and gastric cholecystokinin receptor: Structure, functional expression, and chromosomal localization, *Biochem. Biophys. Res. Commun.*, 189, 296–303, 1992.

Prinz, C., Gastric enterochromaffin-like cells: The major target of gastrin in the GI tract, in *Gastrin in the New Millennium*, Merchant, J.L., Buchan, A.M.J., and Wang, T.C., Eds, Cure Foundation, Los Angeles, 2004, p. 183–187.

Reeve, J.R. et al., New molecular forms of cholecystokinin. Microsequence analysis of forms previously characterized by chromatographic methods, *J. Biol. Chem.*, 261, 16392–16397, 1986.

Rehfeld, J.F., Gastrins in serum, *Scand. J. Gastroenterol.*, 8, 577–583, 1973.

Rehfeld, J.F., Immunochemical studies on cholecystokinin. II. Distribution and molecular heterogeneity in the central nervous system and small intestine of man and hog, *J. Biol. Chem.*, 253, 4022–4030, 1978a.

Rehfeld, J.F., Localisation of gastrins to neuro- and adenohypophysis, *Nature*, 272, 771–773, 1978b.

Rehfeld, J.F., How to measure cholecystokinin in plasma? *Gastroenterology*, 87, 434–438, 1984.

Rehfeld, J.F., Procholecystokinin processing in the normal human anterior pituitary, *Proc. Natl Acad. Sci. USA*, 84, 3019–3024, 1987.

Rehfeld, J.F., Cholecystokinin, in *Handbook of Physiology: The Gastrointestinal System*, Makhlouf, G.M., Ed., Amer. Physiol. Soc., Bethesda, MD, Vol. II, 1989, p. 337–358.

Rehfeld, J.F., Progastrin and its products in the cerebellum, *Neuropeptides*, 20, 239–245, 1991.

Rehfeld, J.F., Accurate measurement of cholecystokinin in plasma, *Clin. Chem.*, 44, 991–1001, 1998a.

Rehfeld, J.F., How to measure cholecystokinin in tissue, plasma, and cerebrospinal fluid, *Regul. Peptides*, 78, 31–39, 1998b.

Rehfeld, J.F., The new biology of gastrointestinal hormones, *Physiol. Rev.*, 78, 1087–1108, 1998c.

Rehfeld, J.F., Cholecystokinin, *Best Practice Res. Clin. Endocrinol. Metab.*, 18, 569–586, 2004.

Rehfeld, J.F., Gastrin and cancer, in *The Handbook of Biologically Active Peptides*, Kastin, A.J. and Moody, T.W., Eds., Elsevier, New York, 2006, p. 486–491.

Rehfeld, J.F., The art of measuring gastrin in plasma: A dwindling, diagnostic discipline? *Scand. J. Clin. Lab. Invest.*, 66, 1–9, 2008

Rehfeld, J.F., Stadil, F., and Rubin, B., Production and evaluation of antibodies for the radioimmunoassay of gastrin, *Scand. J. Clin. Lab. Invest.*, 30, 221–232, 1972.

Rehfeld, J.F. and Uvnäs-Wallensten, K., Gastrins in cat and dog: Evidence for a biosynthetic relationship between the large molecular forms of gastrin and heptadecapeptide gastrin, *J. Physiol.*, 283, 379–396, 1978.

Rehfeld, J.F. et al., Neural regulation of pancreatic hormone secretion by the C-terminal tetrapeptide of CCK, *Nature*, 284, 33–38, 1980.

Rehfeld, J.F., Holst, J.J., and Jensen, S.L., The molecular nature of vascularly released cholecystokinin from the isolated, perfused porcine duodenum, *Regul. Peptides*, 3, 15–18, 1982.

Rehfeld, J.F. and Lundberg, J., Cholecystokinin in feline vagal and sciatic nerves: Concentration, molecular forms, and transport velocity, *Brain Res.*, 275, 341–347, 1983.

Rehfeld, J.F. et al., Gastrin and cholecystokinin in pituitary neurons, *Proc. Natl Acad. Sci. USA*, 81, 1902–1905, 1984.

Rehfeld, J.F. and Hansen, H.F., Characterization of preprocholecystokinin products in the porcine cerebral cortex: Evidence of different neuronal processing pathways, *J. Biol. Chem.*, 261, 5832–5840, 1986.

Rehfeld, J.F. et al., Pituitary tumors containing cholecystokinin, *N. Engl. J. Med.*, 316, 1244–1247, 1987.

Rehfeld, J.F., Bardram, L., and Hilsted, L., Gastrin in bronchogenic carcinomas: Constant expression but variable processing of progastrin, *Cancer Res.*, 49, 2840–2843, 1989.

Rehfeld, J.F. and van Solinge, W.W., The tumor biology of gastrin and cholecystokinin, *Adv. Cancer Res.*, 63, 295–347, 1994.

Rehfeld, J.F., Hansen, C.P., and Johnsen, A.H., Post-polyGlu cleavage and degradation modified by O-sulfated tyrosine: A novel post-translational processing mechanism, *EMBO J.*, 14, 389–396, 1995.

Rehfeld, J.F. et al., The predominant cholecystokinin in human plasma and intestine is cholecystokinin-33, *J. Clin. Endocrinol. Metab.*, 86, 251–258, 2001.

Rehfeld, J.F. and Goetze, J.P., Gastrin vaccination against gastrointestinal and pancreatic cancer, *Scand. J. Gastroent.*, 41, 122–123, 2006.

Rehfeld, J.F. et al., Differential expression of PC1/3, PC2, and PC5/6 governs the cell-specific processing of procholecystokinin, *Endocrinology*, 149, 1600–1609, 2008.

Rengifo-Cam, W. and Singh, P., Role of progastrins and gastrins and their receptors in GI and pancreatic cancers: Target for treatment, *Curr. Pharmaceut. Design*, 10, 2345–2358, 2004.

Reubi, J.C., Peptide receptors as molecular targets for cancer diagnosis and therapy, *Endocrinol. Rev.*, 24, 389–427, 2003.

Reubi, J.C., Schaer, J.C., and Waser, B., Cholecystokinin (CCK)-A and CCK-B/gastrin receptors in human tumors, *Cancer Res.*, 57, 1377–1386, 1997.

Rothman, S.S. and Wells, H., Enhancement of pancreatic enzyme synthesis by pancreozymin, *Am. J. Physiol.*, 213, 215–218, 1967.

Rourke, I.J. et al., Negative cooperativity between juxtaposed E-box and cAMP/TRA responsive elements in the cholecystokinin gene promoter, *FEBS Lett.*, 448, 15–18, 1999.

Saillan-Barreau, C. et al., Evidence for a functional role of the cholecystokinin-B/gastrin receptor in human fetal and adult pancreas, *Diabetes*, 48, 2015–2021, 1999.

Schalling, M. et al., Expression and localization of gastrin mRNA and peptide in human spermatogenic cells, *J. Clin. Invest.*, 86, 660–669, 1990.

Schultzberg, M. et al., Distribution of peptide and catecholamine-containing neurons in the gastrointestinal tract of rat and guinea-pig, *Neuroscience*, 5, 689–744, 1980.

Shaw, R.A. and Jones, R.S., The cholerectic action of cholecystokinin in dogs, *Surgery*, 84, 622–625, 1978.

Smith, G.P. et al., Abdominal vagotomy blocks the satiety effect of cholecystokinin in the rat, *Science*, 213, 1036–1037, 1981.

Solomon, T.E. et al., Bioactivity of cholecystokinin analogues: CCK-8 is not more potent than CCK-33, *Am. J. Physiol.*, 247, G105–G111, 1984.

Soudah, H.C. et al., Cholecystokinin at physiological levels evokes pancreatic enzyme secretion via a cholinergic pathway, *Am. J. Physiol.*, 263, G102–G107, 1992.

Stadil, F. and Rehfeld, J.F., Preparation of ^{125}I-labelled synthetic human gastrin for radioimmunoanalysis, *Scand. J. Clin. Lab. Invest.*, 30, 361–369, 1972.

Uvnäs-Wallensten, K. et al., Heptadecapeptide gastrin in the vagal nerve, *Proc. Natl Acad. Sci. USA*, 74, 5707–5710, 1977.

van Solinge, W.W., Ødum, L., and Rehfeld, J.F., Ovarian cancer express and process progastrin, *Cancer Res.*, 53, 1823–1828, 1993.

Vizi, S.E. et al., Evidence that acetylcholine released by gastrin and related peptides contributes to their effect on gastrointestinal motility, *Gastroenterology*, 64, 268–277, 1973.

Walsh, J.H., Gastrin, in *Gut Peptides*, Walsh, J.H. and Dockray, G.J., Eds., Raven Press, New York, 1994, p. 75–121.

Walsh, J.H. et al., Clearance and acid-stimulating action of human big and little gastrins in duodenal ulcer subjects, *J. Clin. Invest.*, 57, 1125–1131, 1976.

Wang, T.C. and Brand, S.J., Islet cell-specific regulatory domain in the gastrin promoter contains adjacent positive and negative DNA elements, *J. Biol. Chem.*, 265, 8908–8914, 1990.

Wank, S.A., Cholecystokinin receptors, *Am. J. Physiol.*, 269, G628–G646, 1995.

Wank, S.A. et al., Purification, molecular cloning, and functional expression of the cholecystokinin receptor from rat pancreas, *Proc. Natl Acad. Sci. USA*, 89, 3125–3129, 1992.

Wiborg, O. et al., Structure of a human gastrin gene, *Proc. Natl Acad. Sci. USA*, 81, 1067–1069, 1984.

Williams, J.A., Intracellular signaling mechanisms activated by cholecystokinin-regulating synthesis and secretion of digestive enzymes in pancreatic acinar cells, *Ann. Rev. Physiol.*, 63, 77–97, 2001.

You, Z.B. et al., Modulation of neurotransmitter release by cholecystokinin in the neostriatum and substantia nigra of the rat. Regional and receptor specificity, *Neuroscience*, 74, 793–804, 1996.

Zavros, Y. et al., Genetic or chemical hypochlorhydria is associated with inflammation that modulates parietal and G-cell populations in mice, *Gastroenterology*, 122, 119–133, 2002.

7 The Chemistry and Biology of Insulin-Like Peptide 3, a Novel Member of the Insulin Superfamily

Fazel Shabanpoor, Richard A. Hughes, Frances Separovic, and John D. Wade

CONTENTS

7.1 INTRODUCTION

In the early 1990s, the peptide hormone, insulin-like peptide 3 (INSL3), was independently discovered by two different groups as a result of differential cloning of pig (Adham et al., 1993) and mice (Pusch et al., 1996) testis-specific genes. cDNA analysis showed the probable expression product to consist of a preprohormone containing a signal peptide and B, C, and A domains. This, together with the presence of six cysteine residues spaced in a manner similar to that of insulin, led to the conclusion that this gene product was a novel member of the insulin family of peptides. INSL3 was originally designated Ley-IL (Leydig cell insulin-like peptide), because its mRNA was initially found in Leydig cells of the testis (Burkhardt et al., 1994). It has also been referred to as RLF (relaxin-like-factor) due to its relaxin-like activity in mouse interpubic ligament bioassay (Büllesbach and Schwabe, 1995).

7.2 RELAXIN–INSULIN SUPERFAMILY OF PEPTIDE HORMONES

INSL3 is a member of the relaxin–insulin superfamily of peptide hormones. In the human, this family of peptide hormones comprises relaxins 1, 2, and 3 and INSLs 3, 4, 5, and 6, as well as insulin

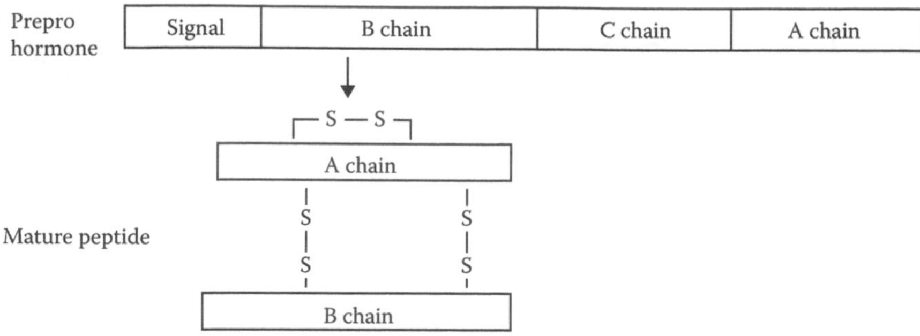

FIGURE 7.1 Schematic representation of expression of the relaxin–insulin family peptide expression as preprohormone, and their cleavage and folding into mature peptide hormones.

and insulin-like growth factors (IGF) I and II (Hudson et al., 1981; Conklin et al., 1999; Lok et al., 2000; Bathgate et al., 2002). The members of this family have high structural, but low sequence similarity (Zimmermann et al., 1997). They are each expressed as an immature preprohormone with four functional segments: (1) an N-terminal signal peptide for secretion, (2) a conserved B chain, (3) a nonconserved C peptide, and (4) a C-terminal A-chain (Hsu, 2003). The cotranslational loss of the N-terminal signal peptide results in the conversion of preprohormone to prohormone. Subsequent formation of two disulfide bonds between the A- and B-chains and one disulfide bond within the A-chain followed by proteolytic cleavage of the C peptide gives rise to a mature heterodimeric A–B peptide (Figure 7.1) (Adham et al., 1993; Hsu, 2003). The relaxins and INSLs were grouped with insulin and IGFs as structural homologs based on their primary and secondary characteristics (James et al., 1977; Hsu, 2003; Fu et al., 2004).

7.3 PHYSIOLOGICAL ROLES OF INSL3

INSL3 is a 6 kDa novel circulating hormone that is expressed exclusively in high levels by the Leydig cells of adult testes in a differentiation-dependent manner (Ivell and Bathgate, 2002a; Ivell and Einspanier, 2002b). It acts as a marker for fully differentiated adult-type Leydig cells (Ivell and Einspanier, 2002b). In the female, INSL3 is expressed in the ovarian follicle and in the corpus luteum, but at lower levels compared to that of the male (Tashima et al., 1995; Roche et al., 1996). The peptide circulates in the blood at a concentration of 1 ng/mL in men, but the circulating level remains below 100 pg/mL in women (Ivell et al., 2005).

Like relaxin-2, INSL3 has both reproductive and nonreproductive roles. Identification of the reproductive roles of INSL3 has been facilitated by the analysis of INSL3 knockout mice, in which testicular descent during development has been shown to be severely retarded due to loss of INSL3-mediated gubernaculum growth (i.e., they show bilateral cryptorchidism—Greek for hidden gonad) (Zimmermann et al., 1999; Spiess et al., 1999; Nef and Parada, 1999). In humans, cryptorchidism is the most common congenital disorder during sexual differentiation and affects approximately 3 to 5% of newborn male infants (Lim et al., 2001; Marin et al., 2001; Canto et al., 2003). Mutation of the INSL3 gene has also been shown to be the cause of cryptorchidism (Canto et al., 2003). The cryptorchid phenotype, where the testis is retained in the abdomen, results in disrupted spermatogenesis and infertility.

Recently, it has also been shown that INSL3 acts as a paracrine factor in mediating gonadotropin actions (Kawamura et al., 2004). Luteinizing hormone (LH), released by the anterior pituitary gland, stimulates Leydig insulin-like 3 (INSL3) transcripts in ovarian theca and testicular Leydig cells. INSL3, in turn, binds to RXFP2 receptors expressed in germ cells to activate the inhibitory G protein, thus leading to decreases in cAMP production. This leads to initiation of meiotic progression of arrested oocytes in preovulatory follicles *in vitro* and *in vivo* and suppresses male germ cell apoptosis *in vivo* (Kawamura et al., 2004).

A recent study of a group of men receiving a male contraceptive (a combination of testosterone and progesterone, which acts as an inhibitor of pituitary gonadotropins) showed that most men are azoospermic or severely oligospermic (≤1 million sperm/mL) (Amory et al., 2007). However, 10 to 20% of men have persistent sperm production despite profound gonadotropin suppression. This was found to be associated with a higher serum INSL3 concentration, which prevents complete suppression of spermatogenesis in those men on hormonal contraceptive regimens. This finding suggests that INSL3 antagonists could have potential for development as male contraceptives.

The expression of INSL3 has been found to be upregulated in hyperplastic and neoplastic human thyrocytes, suggesting that it may play a role in human thyroid carcinoma (Klonisch et al., 2005). The expression of human INSL3 has also been detected in neoplastic Leydig cells and mammary epithelial cells, which also suggests that it may be involved in tumor biology (Klonisch et al., 2005).

7.4 RECEPTORS FOR RELAXIN-FAMILY PEPTIDES

The receptors for most peptides of the relaxin family have only recently been identified. Owing to the high structural similarity of relaxin and insulin, it had been predicted that the receptors for the relaxins and INSLs would, like the insulin receptor, belong to the receptor tyrosine kinase family. However, it has been shown that, in fact, relaxin binds to and activates two previously orphan G-protein couple receptors through a G_s-cAMP-dependent pathway (Hsu et al., 2002). These two receptors, originally referred to as LGR7 (leucine-rich G-protein coupled receptor 7) and LGR8, were recently renamed based upon an International Union of Pharmacology recommendation as relaxin family peptide (RXFP) receptors 1 and 2, respectively (Bathgate et al., 2006). The LGRs (leucine-rich repeat-containing G protein-coupled receptors) are GPCRs with 7-transmembrane spanning domains, a long extracellular domain containing a series of leucine-rich repeats (LRRs), and an N-terminal ectodomain (Hsu et al., 2002; Sudo et al., 2003) (Figure 7.2). They belong to a family of glycoprotein receptors in class A GPCRs, which includes receptors for follicle stimulating and luteinizing hormones, and thyroid-stimulating hormone (Hsu et al., 2003).

The RXFP receptors have 60% sequence identity (Hsu et al., 2002) and a common domain structure of these receptors is the presence of an N-terminal low-density lipoprotein (LDL) region which is not found in other LGR family members (Sherwood, 2004). These are further distinguished from other glycoprotein hormone receptors by this unique low-density lipoprotein receptor-like cysteine-rich motif (low-density lipoprotein class A module, LDLa) (Hsu, 2003) which is located near the N-terminus of the extracellular region. The LDLa domain has been postulated to be involved in receptor activation but not in primary ligand binding (Adham and Agoulnik, 2004). Recent structural and functional studies on mutated full-length RXFP1 have shown that a hydrophobic surface within the N-terminal region of the LDLa module is essential for activation of the RXFP1 receptor signaling cascade in response to relaxin stimulation (Hopkins et al., 2007). Recent studies using truncated and chimeric receptors have revealed two binding sites on RXFPs 1 and 2. The leucine-rich repeats within the ectodomain have been shown to be the primary, high-affinity binding site and a secondary, low-affinity site within the transmembrane region (Halls et al., 2005) (Figure 7.2).

A direct interaction between INSL3 and RXFP2 receptors has also been demonstrated through different studies such as receptor-binding studies, changes in cAMP production, and ligand–receptor cross-linking studies (Claasz et al., 2002; Hsu et al., 2002; Kumagai et al., 2002). In order to define the primary INSL3 binding site in LRR of RXFP2, our group has used alanine scanning to show that RXFP2 Asp-227 is crucial for binding INSL3 Arg-B16, whereas RXFP2 Phe-131 and Gln-133 are involved in INSL3 Trp-B27 binding (Scott et al., 2007).

The RXFP1 transcript has been found in the uterus, ovary, adrenal gland, prostate, skin, testis, brain, and heart (Hsu et al., 2002). RXFP2 (also known as GREAT in rodent—G-protein coupled

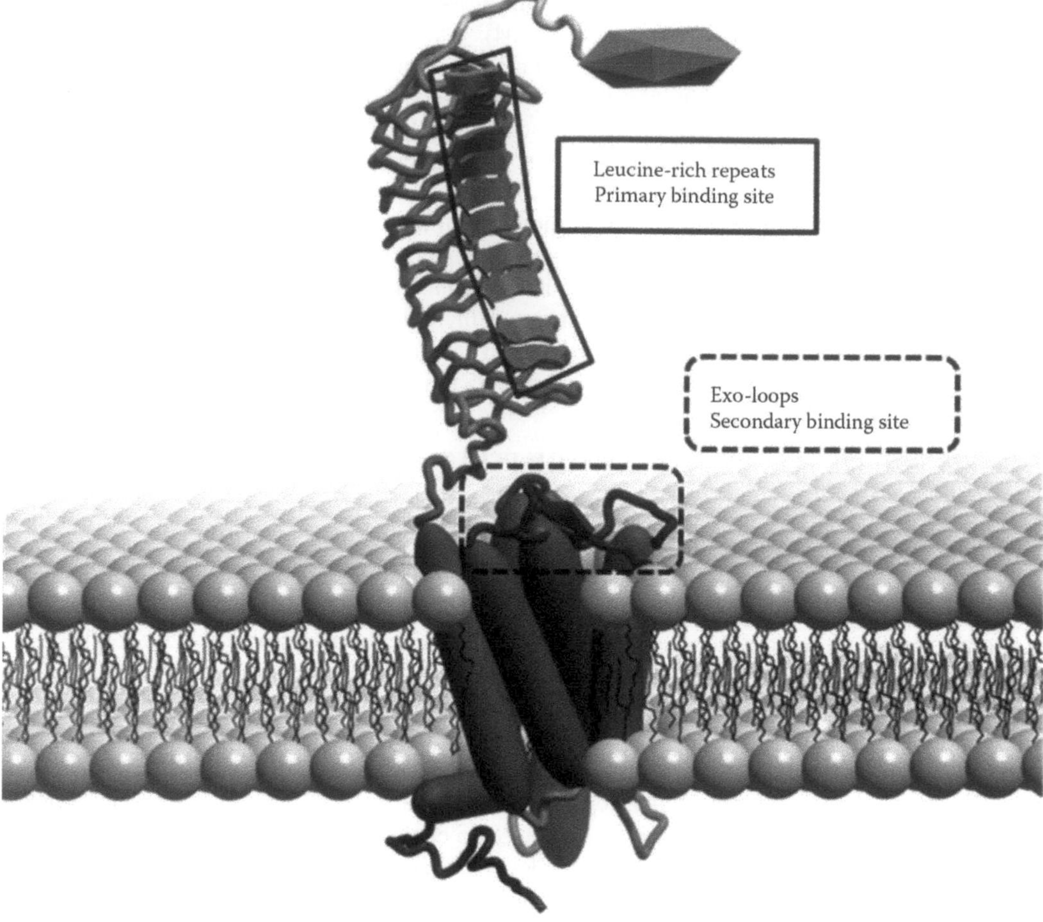

FIGURE 7.2 Proposed model of LGR7/8 with the primary, high-affinity binding site being located in the leucine-rich repeats (LRRs), and the secondary, low-affinity binding site being on the exoloops of the transmembrane domain. (Adapted from Scott, D.J., PhD thesis, Howard Florey Institute, Melbourne, Australia, 2007.)

receptor affecting testicular descent) expression has been observed in several organs and tissues, such as the gubernaculum, testis, brain, skeletal muscle, uterus, peripheral blood cell, thyroid, and bone marrow (Overbeek et al., 2001; Hsu et al., 2002; Gorlov et al., 2002). Relaxin-2 binds to and activates both RXFP1 and RXFP2; however, it has a lower affinity for RXFP2 (Hsu et al., 2002, 2003; Sudo et al., 2003; Hsu, 2003). In contrast, INSL3 has been shown to bind exclusively to and activate RXFP2 to induce an intracellular increase in cAMP, both in transfected cells overexpressing cloned RXFP2 receptor and in gubernacular and prostate cells with naturally expressed receptors (Overbeek et al., 2001).

Phenotypically, male mice deficient for RXFP2 (Bogatcheva et al., 2003; Feng et al., 2004; Foresta and Ferlin, 2004) have been shown to be similar to male INSL3-deficient mice in that both exhibit bilateral cryptorchidism (Nef and Parada, 1999; Zimmermann et al., 1999). On the other hand, female homozygous INSL3$^{-/-}$ mice display impaired fertility associated with an abnormal estrous cycle, whereas female RXFP2 receptor null mice remain fertile, suggesting that INSL3 is involved in female fertility but that there may be redundant signaling systems involved (Nef and Parada, 1999). In contrast, transgenic overexpression of INSL3 in female mice causes the ovaries to descend into the inguinal region due to an overdeveloped gubernaculum (Adham et al., 2002).

Although INSL3 plays a role in gonadal and other physiological processes, the intracellar signaling pathways involved are still unclear. Most of the signaling studies have thus far focused on cAMP pathways. The level of cAMP has been shown to either increase or decrease based on coupling of RXFP2 to different G proteins. INSL3 stimulation of HEK cells expressing recombinant RXFP2 and gubernaculum cells (which endogenously express RXFP2) causes an increase in cAMP as a result of G_s-mediated activation of adenylate cyclase in these cells (Kumagai et al., 2002). RXFP2 couples to G_i/G_o proteins in testicular germ cells and oocytes, which upon stimulation by INSL3 cause a decrease in cAMP that is prevented by pertussis toxin (Kawamura et al., 2004).

7.5 TERTIARY STRUCTURE OF HUMAN INSL3

The three-dimensional structure of relaxin-2 (Figure 7.3) has been determined by X-ray crystallography (Eigenbrot et al., 1991). The A-chain (24 amino acids) consists of two α-helices connected via a short loop. The B-chain (30 amino acids) consists of an α-helix and a strand. As shown from earlier studies, these two chains are connected via two interchain and one intra-A-chain disulfide bond.

Until recently, the tertiary structure of INSL3 was unknown. Due to the high sequence similarity between INSL3 and relaxin-2, it was assumed that INSL3 adopts a relaxin-like conformation, in particular in the region confined by the completely conserved cysteine bonds with the same pattern of inter- and intrachain disulfide bonds (Büllesbach and Schwabe, 2002). In the past, most structure–activity studies of INSL3 have been carried out based on a homology model created using the X-ray crystal structure of relaxin-2 as template (Del Borgo et al., 2005, 2006; Shabanpoor et al., 2007).

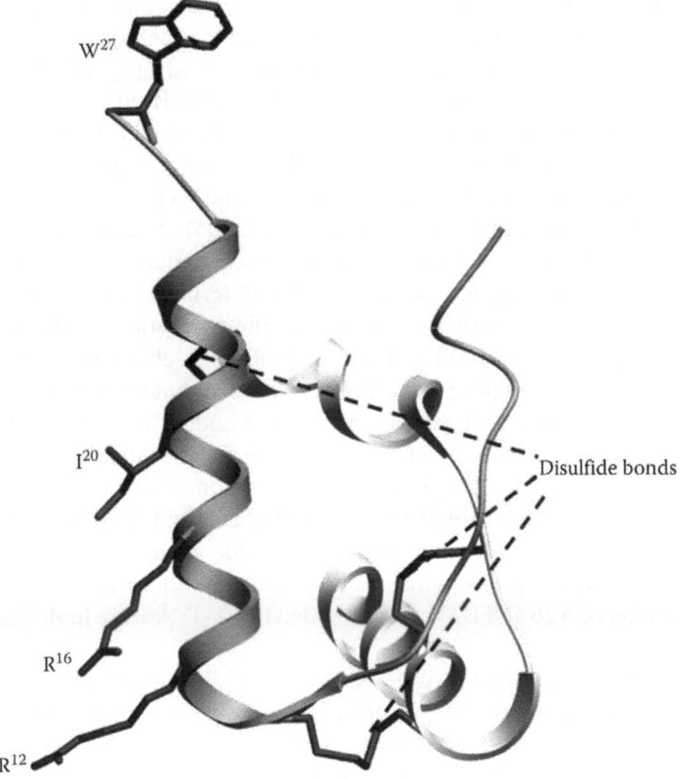

FIGURE 7.3 X-ray crystal structure of human relaxin-2 (A-chain in silver and B-chain in gray). Inter- and intradisulfide bonds are shown in black. The key receptor binding residues (Arg^{12}, Arg^{16}, Ile^{19}, and Trp^{27}) located on the B-chain α-helix are highlighted in black.

Recently, the tertiary structure of human INSL3 was solved by solution-phase NMR spectroscopy (Rosengren et al., 2006). As had been predicted, INSL3, and indeed all other members of the insulin family for which structures are known, adopts a very similar insulin core structure with differences being mostly around the termini. In relaxin-2, the A-chain has a well-defined structure whereas in INSL3 the first four residues at the N-terminus are disordered (Rosengren et al., 2006). The C-terminus of relaxin-2 ends with a cysteine, which is involved in the disulfide bond with the B-chain. This disulfide bond locks the A-chain into its helical conformation. Unlike relaxin-2, which has a clear helical structure at the C-terminus, the INSL3 A-chain C-terminus extends beyond cysteine by two extra residues, Pro^{A25} and Tyr^{A26}, and is structurally disordered (Rosengren et al., 2006). Another important structural distinction between relaxin-2 and INSL3 is the number of helical turns in their B-chain. The helical segment of relaxin B-chain consists of 15 residues, but in INSL3 the B-chain helix is shorter by three residues.

7.6 STRUCTURE–ACTIVITY RELATIONSHIP STUDIES OF INSL3

The interaction of relaxin-2 with its receptor RXFP1 has been shown to be dependent on the presence of a motif (Arg^{12}–X–X–X–Arg^{16}–X–X–Ile^{19}, where X is any amino acid) in the mid-region of the B-chain helix (Büllesbach and Schwabe, 1988, 2002). In addition to these three residues (Arg^{12}, Arg^{16}, and Ile^{19}) in the binding motif, Trp^{27} toward the C-terminus of B-chain has also been shown to be important for the binding of relaxin to RXFP1 (Büllesbach and Schwabe, 1999).

A number of structure–activity studies of INSL3 using site-directed mutagenesis techniques have been reported. Büllesbach and Schwabe (2004) have prepared a series of C- and N-terminally truncated INSL3 analogs. These authors showed that C-terminal truncation of the B-chain beyond Trp^{27} resulted in only 1% receptor binding (Büllesbach and Schwabe, 1999) whereas a B-chain construct ending in tryptophan retained 5.4% of the receptor binding. Amidation of the C-terminal Trp^{27} residue almost completely restored full affinity (84.1%) (Büllesbach and Schwabe, 2004). Therefore, Trp^{27} toward the C-terminus of the B-chain appears to be crucial for binding of INSL3 to the RXFP2 receptor. On the other hand, shortening the A-chain by six residues from the N-terminus had no effect on either receptor binding or cAMP production, although further truncation resulted in tightly bound ligands with no signaling activity; that is, they were INSL3 antagonists (Büllesbach and Schwabe, 2005). A recent structure–activity study has revealed that truncation of the INSL3 B-chain up to residue 8 causes a 60% reduction in receptor binding affinity that cannot be reversed by N-terminal acetylation (Büllesbach and Schwabe, 2007). These authors have also shown that the truncation of the B-chain also has an effect on the signaling activity of INSL3. Truncation of the B-chain up to six residues from the N-terminus has no effect on signaling and cAMP production, but truncation up to residues 7 and 8 leads to a significant drop in cAMP production, by 65% and 20%, respectively (Büllesbach and Schwabe, 2007). In a recent structure-activity study, the C-terminus of the INSL3 A-chain was chemically linked to the side chain of the residue at position 26 of the B-chain to keep Trp^{27} in a fixed position relative to the terminal region (Büllesbach and Schwabe, 2004). The analogs with a short distance between the α-carbons of the residue at the A-chain C-terminus and B26 had significantly reduced receptor-binding activity, whereas those analogs where the distance between the α-carbons of A26 and B26 were extended to 10–11 Å were highly active (Büllesbach and Schwabe, 2004).

Based on the primary binding motif (Arg^{12}–X–X–X–Arg^{16}–X–X–Ile^{19}) of relaxin-2, it was thought that the binding of INSL3 to the RXFP2 receptor is dependent on residues in the helical region of the B-chain. The residues at equivalent positions in INSL3 are His^{12}, Arg^{16}, and Val^{19}. Recent structure–activity studies by our group using single Ala substitution has shown that substituting Arg^{16} and Val^{19} with Ala significantly reduced the affinity of these analogs for the RXFP2 receptor (Rosengren et al., 2006). On the other hand, multi-Ala-substitution showed that His^{12} and Arg^{20} have a strong synergistic effect with Arg^{16}, suggesting that His^{12} and Arg^{20} may be involved in the initial step of receptor recognition that drive the electrostatic interaction between the basic residues of the peptide and acidic residues on the receptor (Rosengren et al., 2006).

The mechanism of receptor activation by INSL3 has recently been reported to be independent of the amino-acid side chains and is a function of certain peptide bonds at the N-terminus of the A-chain (Büllesbach and Schwabe, 2007). In this study, the authors have shown that progressive truncation of the A-chain from the N-terminus diminishes the signaling activity of LGR8, but a single-residue mutation of Arg^{A8} and Tyr^{A9} to alanine does not perturb the signaling activity (Büllesbach and Schwabe, 2007). In order to further explore the involvement of the backbone amide bond in the signaling activity of INSL3, the Tyr^{A9} has been replaced with D-Pro and had been shown to have no impact on receptor binding while severely retarding signaling activity. Further replacement of Tyr^{A9} with N-methyl alanine has also been shown to have a similar effect as that of D-Pro replacement (Büllesbach and Schwabe, 2007).

7.7 CLINICAL APPLICATION—INSL3 β-CHAIN PEPTIDES AS LGR8 ANTAGONISTS

As mentioned earlier, INSL3 binds to its receptor via residues primarily located on the helical region of the B-chain; however, the mode of interaction is not yet clearly understood. Because INSL3 causes maturation of germ cells in adults, any mimetics of the INSL3 B-chain that can prevent INSL3 binding and signaling might have potential as clinically useful fertility-control drugs. With identification of the important residues for binding of INSL3 on the B-chain (including Trp^{27}), our group recently designed and synthesized shortened analogs of the INSL3 B-chain. One of these cyclic peptide analogs of the INSL3 B-chain (Figure 7.4a) was shown to be able to completely displace[33] P-relaxin from LGR8 with a pKi ± SEM value of 5.99 ± 0.07 (Del Borgo et al., 2005). This compound appeared to be devoid of INSL3-like agonist activity but at a concentration of 10 μM, it showed a 70 to 80% inhibition of cAMP accumulation in HEK 293T cells with 1 nM INSL3 stimulation, that is, it acts as an INSL3 antagonist. *In vivo* administration of this peptide into the testes of rats resulted in a substantial decrease in testis weight, probably due to the inhibition of germ cell survival (Del Borgo et al., 2006).

In an attempt to improve the receptor binding affinity of this antagonist, the INSL3 B-chain was further modified by placing the disulfide bond at a position that was thought to be more structurally appropriate. This led to the development of two new analogs that were shorter than the previous analog and that possessed slightly improved receptor binding affinity (pKi ± SEM, 6.65 ± 0.09, $n = 3$ (Figure 7.4b); 6.34 ± 0.07, $n = 3$ (Figure 7.4c) (Shabanpoor et al., 2007). The presence of a secondary structure in these peptides was determined using circular dichroism (CD) spectroscopy. Using this technique, it was shown that these disulfide-constrained cyclic peptides are unstructured

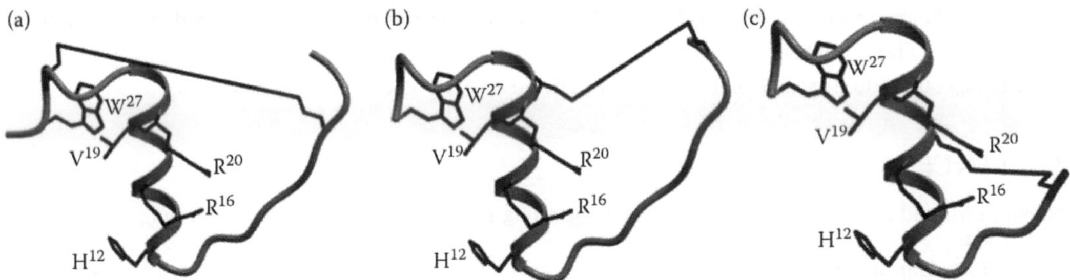

FIGURE 7.4 Three INSL3 B-chain analogs. (a) The first disulfide constraint analog of the INSL3 B-chain, which was found to be antagonist. (b) and (c) INSL3 B-chain analogs with improved receptor binding affinity; both are shorter than the analog in (a).

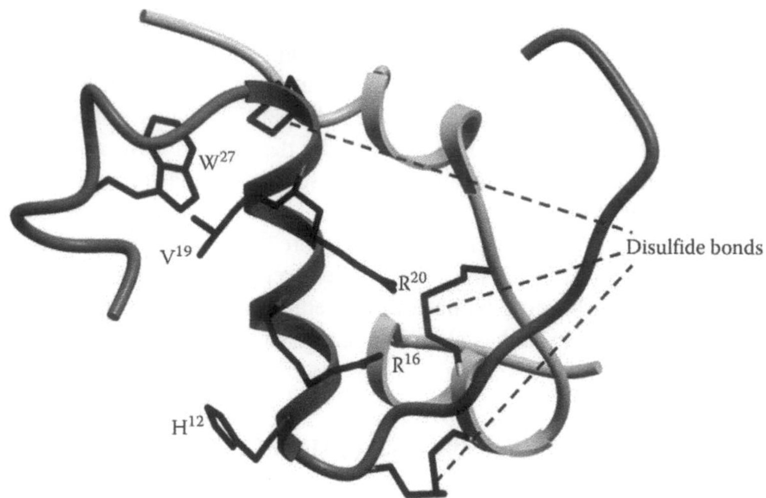

FIGURE 7.5 NMR solution structure of human INSL3 (A-chain in silver and B-chain in gray). Inter- and intradisulfide bonds are shown in black. The key receptor binding residues (His^{12}, Arg^{16}, Val^{19}, Arg^{20}, and Trp^{27}) located on the B-chain α-helix are highlighted in black.

in phosphate buffer and adopt an α-helical structure only in the presence of a cosolvent such as trifluoroethanol (TFE) (Shabanpoor et al., 2007). The low receptor-binding affinity of these peptides was thought to be, in part, due to the lack of native INSL3-like α-helical structure of the B-chain, which contains all the important residues for high receptor-binding affinity. These three INSL3 B-chain analogs, which act as antagonists, were designed based on a homology model using the X-ray crystal structure of human relaxin-2 as template because at the time of designing these analogs there was no tertiary structure of INSL3 available. With the recent determination of the solution NMR structure of human INSL3 (Figure 7.5), our analysis revealed that we would have likely proposed identical cyclic disulfide-constrained B-chain mimetics using the homology model as a template.

7.8 FUTURE DEVELOPMENTS AND CONCLUDING REMARKS

In the seven years since the INSL3 gene knockout experiments were reported, burgeoning physiochemical, biochemical, and physiological studies have demonstrated its key importance in fetal development and in fertility regulation. They have shown that both agonists and antagonists of the peptide will have significant clinical promise for use in the reproductive system in both male and female fertility management. In particular, the potential for a universal contraceptive based on such antagonists is enormous given that the population growth in many countries exceeds the capacity to provide care. The emergence of this most interesting member of the insulin superfamily promises that the next decade of study on this peptide will be equally productive and exciting.

REFERENCES

Adham, I.M. and Agoulnik, A.I., Insulin-like 3 signalling in testicular descent, *Int. J. Androl.*, 27, 257–265, 2004.

Adham, I.M., Burkhardt, E., Benahmed, M., and Engel, W., Cloning of a cDNA for a novel insulin-like peptide of the testicular Leydig cells, *J. Biol. Chem.*, 268, 26668–26672, 1993.

Adham, I.M. et al., The overexpression of the INSL3 in female mice causes descent of the ovaries, *Mol. Endocrinol.*, 16, 244–252, 2002.

Amory, J.K. et al., Elevated end-of-treatment serum INSL3 is associated with failure to completely suppress spermatogenesis in men receiving male hormonal contraception, *J. Androl.*, 28, 548–554, 2007.

Bathgate, R.A., Ivell, R., Sanborn, B.M., Sherwood, O.D., and Summers, R.J., International Union of Pharmacology LVII: Recommendations for the nomenclature of receptors for relaxin family peptides, *Pharmacol. Rev.*, 58, 7–31, 2006.

Bathgate, R.A. et al., Human relaxin gene 3 (H3) and the equivalent mouse relaxin (M3) gene. Novel members of the relaxin peptide family, *J. Biol. Chem.*, 277, 1148–1157, 2002.

Bogatcheva, N.V. et al., GREAT/LGR8 is the only receptor for insulin-like 3 peptide, *Mol. Endocrinol.*, 17, 2639–2646, 2003.

Büllesbach, E.E. and Schwabe, C., On the receptor binding site of relaxins, *Int. J. Pept. Protein Res.*, 32, 361–367, 1988.

Büllesbach, E.E. and Schwabe, C., A novel Leydig cell cDNA-derived protein is a relaxin-like factor, *J. Biol. Chem.*, 270, 16011–16015, 1995.

Büllesbach, E.E. and Schwabe, C., Tryptophan B27 in the relaxin-like factor (RLF) is crucial for RLF receptor-binding, *Biochemistry*, 38, 3073–3078, 1999.

Büllesbach, E.E. and Schwabe, C., The primary structure and the disulfide links of the bovine relaxin-like factor (RLF), *Biochemistry*, 41, 274–281, 2002.

Büllesbach, E.E. and Schwabe, C., Synthetic cross-links arrest the C-terminal region of the relaxin-like factor in an active conformation, *Biochemistry*, 43, 8021–8028, 2004.

Büllesbach, E.E. and Schwabe, C., LGR8 signal activation by the relaxin-like factor, *J. Biol. Chem.*, 280, 14586–14590, 2005.

Büllesbach, E.E. and Schwabe, C., Structure of the transmembrane signal initiation site of the relaxin-like factor (RLF/INSL3), *Biochemistry*, 46, 9722–9727, 2007.

Burkhardt, E. et al., A human cDNA coding for the Leydig insulin-like peptide (Ley I-L), *Hum. Genet.*, 94, 91–94, 1994.

Canto, P. et al., A novel mutation of the insulin-like 3 gene in patients with cryptorchidism, *J. Hum. Genet.*, 48, 86–90, 2003.

Claasz, A.A. et al., Relaxin-like bioactivity of ovine insulin 3 (INSL3) analogues, *Eur. J. Biochem.*, 269, 6287–6293, 2002.

Conklin, D. et al., Identification of INSL5, a new member of the insulin superfamily, *Genomics*, 60, 50–56, 1999.

Del Borgo, M.P. et al., Analogs of insulin-like peptide 3 (INSL3) B-chain are LGR8 antagonists *in vitro* and *in vivo*, *J. Biol. Chem.*, 281, 13068–13074, 2006.

Del Borgo, M.P., Hughes, R.A., and Wade, J.D., Conformationally constrained single-chain peptide mimics of relaxin B-chain secondary structure, *J. Pept. Sci.*, 11, 564–571, 2005.

Eigenbrot, C. et al., X-ray structure of human relaxin at 1.5 Å. Comparison to insulin and implications for receptor binding determinants, *J. Mol. Biol.*, 221, 15–21, 1991.

Feng, S. et al., Mutation analysis of INSL3 and GREAT/LGR8 genes in familial cryptorchidism, *Urology*, 64, 1032–1036, 2004.

Foresta, C. and Ferlin, A., Role of INSL3 and LGR8 in cryptorchidism and testicular functions, *Reprod. Biomed. Online*, 9, 294–298, 2004.

Fu, P. et al., Synthesis, conformation, receptor binding, and biological activities of monobiotinylated human insulin-like peptide 3, *J. Pept. Res.*, 63, 91–98, 2004.

Gorlov, I.P. et al., Mutations of the GREAT gene cause cryptorchidism, *Hum. Mol. Genet.*, 11, 2309–2318, 2002.

Halls, M.L., Bathgate, R.A., Roche, P.J., and Summers, R.J., Signaling pathways of the LGR7 and LGR8 receptors determined by reporter genes, *Ann. NY Acad. Sci.*, 1041, 292–295, 2005.

Hopkins, E.J., Layfield, S., Ferraro, T., Bathgate, R.A., and Gooley, P.R., The NMR solution structure of the relaxin (RXFP1) receptor lipoprotein receptor class A module and identification of key residues in the N-terminal region of the module that mediate receptor activation, *J. Biol. Chem.*, 282, 4172–4184, 2007.

Hsu, S.Y., New insights into the evolution of the relaxin–LGR signaling system, *Trends Endocrinol. Metab.*, 14, 303–309, 2003.

Hsu, S.Y. et al., Relaxin signaling in reproductive tissues, *Mol. Cell Endocrinol.*, 202, 165–170, 2003.

Hsu, S.Y. et al., Activation of orphan receptors by the hormone relaxin, *Science*, 295, 671–674, 2002.

Hudson, P., Haley, J., Cronk, M., Shine, J., and Niall, H., Molecular cloning and characterization of cDNA sequences coding for rat relaxin, *Nature*, 291, 127–131, 1981.

Ivell, R. and Bathgate, R.A., Reproductive biology of the relaxin-like factor (RLF/INSL3). *Biol. Reprod.*, 67, 699–705, 2002a.

Ivell, R. and Einspanier, A., Relaxin peptides are new global players, *Trends Endocrinol. Metab.*, 13, 343–348, 2002b.

Ivell, R., Hartung, S., and Anand-Ivell, R., Insulin-like factor 3: Where are we now? *Ann. NY Acad. Sci.*, 1041, 486–496, 2005.

James, R., Niall, H., Kwok, S., and Bryand-Greenwood, G., Primary structure of porcine relaxin: Homology with insulin and related growth factors, *Nature*, 267, 544–546, 1977.

Kawamura, K. et al., Paracrine regulation of mammalian oocyte maturation and male germ cell survival, *Proc. Natl Acad. Sci. USA*, 101, 7323–7328, 2004.

Klonisch, T. et al., Human medullary thyroid carcinoma: A source and potential target for relaxin-like hormones, *Ann. NY Acad. Sci.*, 1041, 449–461, 2005.

Kumagai, J. et al., INSL3/Leydig insulin-like peptide activates the LGR8 receptor important in testis descent, *J. Biol. Chem.*, 277, 31283–31286, 2002.

Lim, H.N., Raipert-de Meyts, E., Skakkebaek, N.E., Hawkins, J.R., and Hughes, I.A., Genetic analysis of the INSL3 gene in patients with maldescent of the testis, *Eur. J. Endocrinol.*, 144, 129–137, 2001.

Lok, S. et al., Identification of INSL6, a new member of the insulin family that is expressed in the testis of the human and rat, *Biol. Reprod.*, 62, 1593–1599, 2000.

Marin, P. et al., Novel insulin-like 3 (INSL3) gene mutation associated with human cryptorchidism, *Am. J. Med. Genet.*, 103, 348–349, 2001.

Nef, S. and Parada, L.F., Cryptorchidism in mice mutant for Insl3, *Nat. Genet.*, 22, 295–299, 1999.

Overbeek, P.A. et al., A transgenic insertion causing cryptorchidism in mice, *Genesis*, 30, 26–35, 2001.

Pusch, W., Balvers, M., and Ivell, R., Molecular cloning and expression of the relaxin-like factor from the mouse testis, *Endocrinology*, 137, 3009–3013, 1996.

Roche, P.J., Butkus, A., Wintour, E.M., and Tregear, G., Structure and expression of Leydig insulin-like peptide mRNA in the sheep, *Mol. Cell Endocrinol.*, 121, 171–177, 1996.

Rosengren, K.J. et al., Solution structure and characterization of the LGR8 receptor binding surface of insulin-like peptide 3, *J. Biol. Chem.*, 281, 28287–28295, 2006.

Scott, D.J. et al., Defining the LGR8 residues involved in binding insulin-like peptide 3, *Mol. Endocrinol.*, 21, 1699–1712, 2007.

Shabanpoor, F. et al., Design, synthesis, and pharmacological evaluation of cyclic mimetics of the insulin-like peptide 3 (INSL3) B-chain, *J. Pept. Sci.*, 13, 113–120, 2007.

Sherwood, O.D., Relaxin's physiological roles and other diverse actions, *Endocrinol. Rev.*, 25, 205–234, 2004.

Spiess, A.N. et al., Structure and expression of the rat relaxin-like factor (RLF) gene, *Mol. Reprod. Dev.*, 54, 319–325, 1999.

Sudo, S. et al., H3 relaxin is a specific ligand for LGR7 and activates the receptor by interacting with both the ectodomain and the exoloop 2, *J. Biol. Chem.*, 278, 7855–7862, 2003.

Tashima, L.S., Hieber, A.D., Greenwood, F.C., and Bryant-Greenwood, G.D., The human Leydig insulin-like (hLEY I-L) gene is expressed in the corpus luteum and trophoblast, *J. Clin. Endocrinol. Metab.*, 80, 707–710, 1995.

Zimmermann, S., Schottler, P., Engel, W., and Adham, I.M., Mouse Leydig insulin-like (Ley I-L) gene: Structure and expression during testis and ovary development, *Mol. Reprod. Dev.*, 47, 30–38, 1997.

Zimmermann, S. et al., Targeted disruption of the INSL3 gene causes bilateral cryptorchidism, *Mol. Endocrinol.*, 13, 681–691, 1999.

8 Somatostatin Analogs

Kjell Öberg

CONTENTS

8.1 INTRODUCTION

In 1968 Krülich and colleagues isolated from rat hypothalamus a substance with an inhibiting action on the release of pituitary growth hormone and called it GH-RIF (Krulich et al., 1968). In the same year Hellman and Lernmark reported on the presence of a potent inhibitor of insulin secretion in extracts of pancreatic islets (Hellman and Lernmark, 1969). In 1973 Guillemain and Brazeau at the Salk Institute isolated a tetradecapeptide finally called somatostatin-14 from extracts of more than a thousand sheep hypothalami (Brazeau et al., 1973). Using radioimmunoassay and immunohistochemistry methods it was shown that somatostatin-like immunoreactivity was heterogeneously distributed and present in tissues of many animal species, vertebrates as well as invertebrates (Yamada et al., 1977). The first somatostatin-producing neoplasm of the endocrine pancreas was described in 1977 by Larsson and colleagues and the clinical presentation included diabetes mellitus, gall-bladder disease, anemia and delayed gastric emptying and loss of weight (Larsson et al., 1977). Guillemin and Schally together received the Nobel Prize in 1977 for the isolation of somatostatin. The molecular form originally isolated in the hypothalamus was the peptide somatostatin (SS-14) containing 14 amino acids, with a molecular weight of 1640.

Another important biologically active product is somatostatin 28 (SS-28), an analog from the same precursor molecule that presents an N-terminal extended form (Gluschankof et al., 1984)

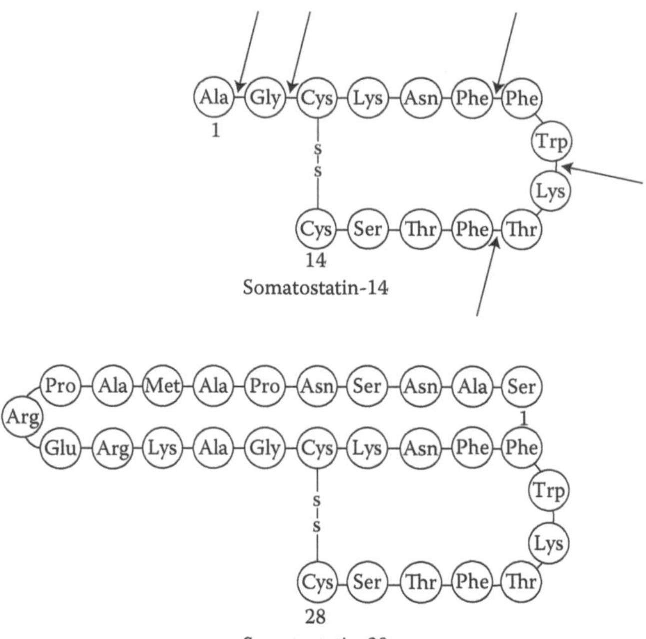

FIGURE 8.1 Primary structure of somatostatin-14 and somatostatin-28. Key sites of enzymatic segradation are marked with arrows for somatostatin-14. (Adapted from Susini, C. and Buscail, L., *Ann. Oncol.*, 17, 1733–1742, 2006.)

(Figure 8.1). The extremely short plasma half-life of somatostatin (~2 min) was a serious drawback to successful exploitation of its therapeutical potential. It became evident that stable and potent synthetic analogs would be required to answer some fundamental questions before one could think of an effective application in human medicine. In 1978 Vale and Rivier reported on an octapeptide analog

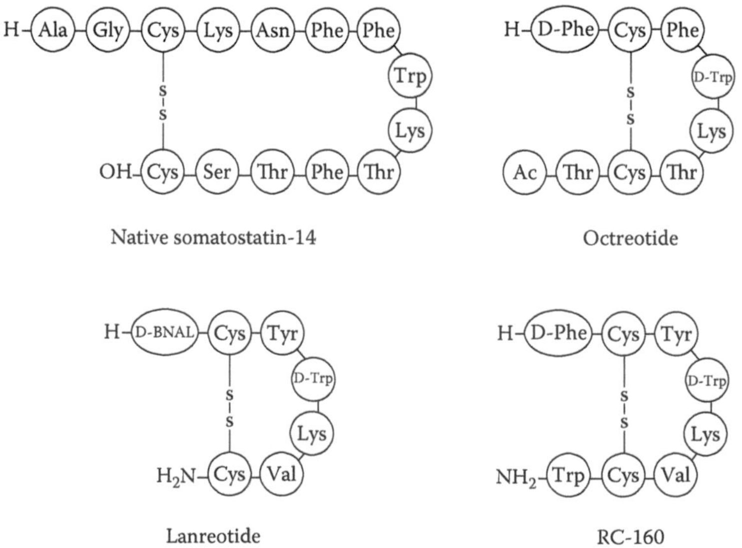

FIGURE 8.2 Chemical structure of native somatostatin-14 and the synthetic analogs octreotide, lanreotide, and RC-150 (Vapreotide). (Adapted from Susini, C. and Buscail, L., *Ann. Oncol.*, 17, 1733–1742, 2006.)

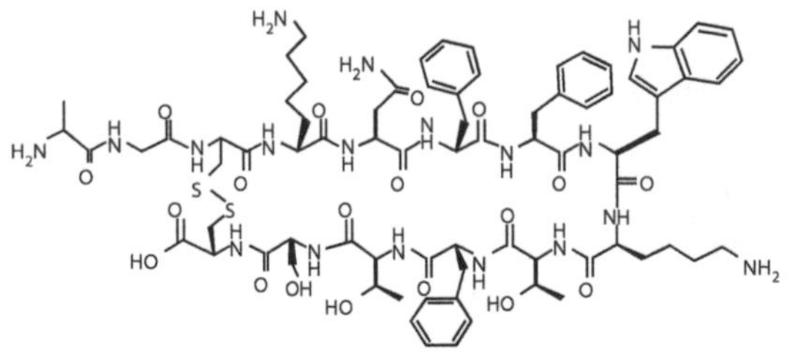

SRIF-14

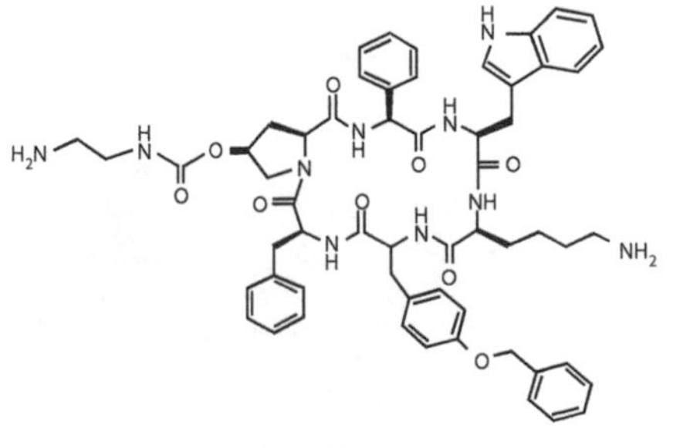

SMS 201-995

SOM230

Novel cyclohexapeptide

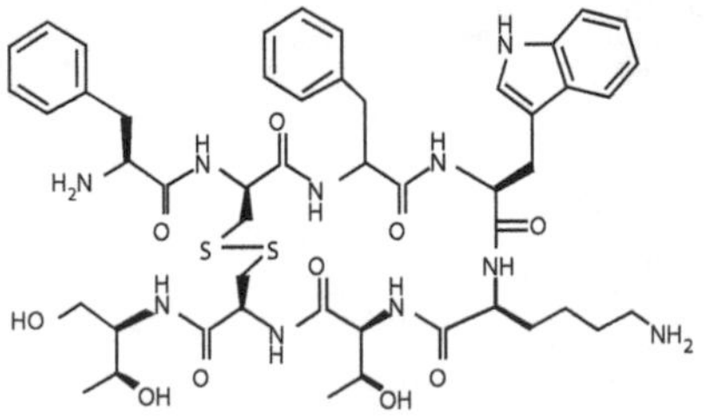

FIGURE 8.3 Chemical structure of somatostatin-14 octreotide (SMS201-995) and the new analog SOM230 (Pasireotide).

that displayed the full biological activity of somatostatin (Vale et al., 1978). In 1980 Bauer and colleagues at Sandoz synthesized an analog that resulted in the code of SMS201-995, later taking the name octreotide (Bauer et al., 1982). Its structure contains three unnatural amino acids and the molecule has therefore become resistant to metabolic degradation and has a longer duration of action.

Other companies and research institutes became interested in the field of somatostatin analogs and started to produce analogs that were chemically and biologically very similar to octreotide, being octapeptides and displaying high affinity to somatostatin receptor types 2 and 5. These analogs were lanreotide and RC-160 (Schally, 1988; Weckbecker et al., 1993; Bruns et al., 2002) (Figure 8.2). The most recent analog produced by Novartis is SOM230 (pasireotide), which binds to somatostatin receptors 1, 2, 3, and 5. (Figure 8.3) Both octreotide and lanreotide are available as long-acting formulations given once monthly. These two analogs have been the gold standard in the treatment of acromegaly and different functioning gastro-entero-pancreatic tumors (see Section 8.7).

8.2 SOMATOSTATIN RECEPTORS

Five somatostatin receptor subtypes (SSTR1–SSTR5) have been identified by gene-cloning techniques and one of these SSTR2 is expressed in two alternatively spliced forms (Figure 8.4) (O'Carroll et al., 1992; Yamada et al., 1992). These receptor subtypes are encoded by separate genes located on different chromosomes, are expressed in unique or partially overlapping distribution in multiple target organs, and differ in their coupling to second messengers signaling molecules (Table 8.1). These subtypes also differ in their binding affinity to specific somatostatin analogs, of importance for the use of somatostatin analogs in therapy and diagnostic imaging (Table 8.2). All somatostatin receptor subtypes are G protein coupled receptors and bind SS-14 and SS-28 with high affinity in the low nanomolar range, although SS-28 has a uniquely high affinity for SSTR-5 (Patel, 1999). SSTR-1 and SSTR-2 are the two most abundant subtypes in brain and probably functioning as presynaptic autoreceptors in the hypothalamus and the limbic forebrain, respectively, in addition to their postsynaptic actions (Blake et al., 2004). SSTR4 is most prominent in hippocampus. All subtypes are expressed in the pituitary. However, SSTR2 and SSTR5 are the most abundant receptors on somatotrophs (Thoss

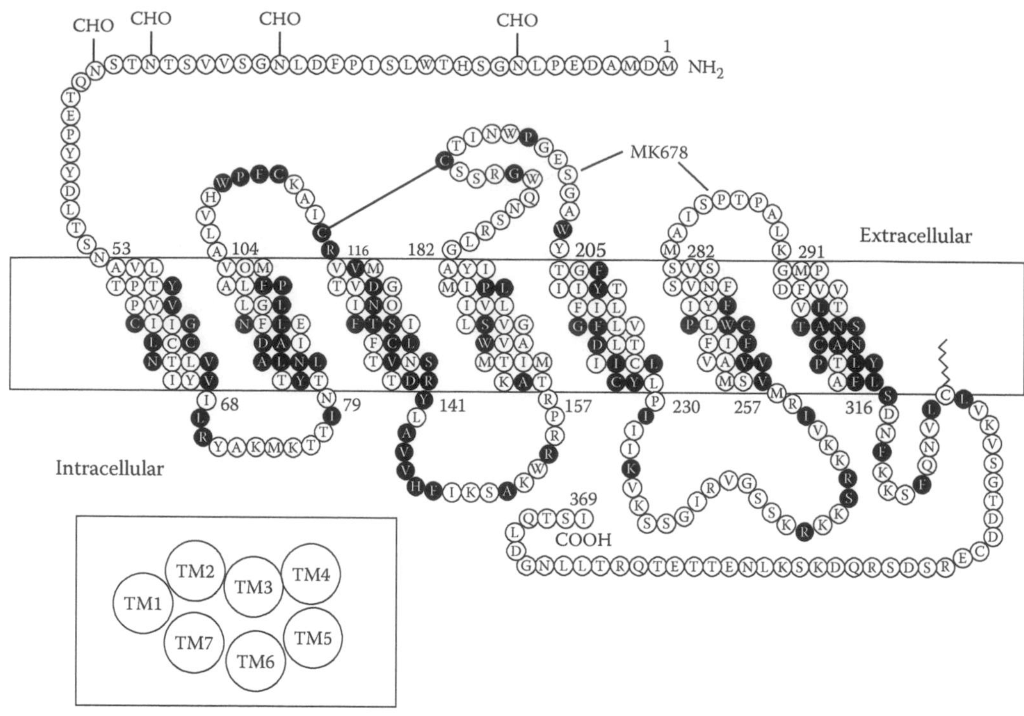

FIGURE 8.4 The biochemical structure of somatostatin receptor type 2A.

et al., 1996). SSTR2 and SSTR5 are also the most physiologically important receptors in pancreatic islets, regulating insulin and glucagon secretion (Strowski et al., 2000). Moreover, somatostatin receptor expressing cells are found in a number of other organs in the body, for example, the gastrointestinal tract, lungs, spleen, thymus, and adrenals. Lymphocytes and macrophages express SSTR2.

8.3 MOLECULAR AND BIOLOGICAL ACTIONS OF SOMATOSTATIN ANALOGS

Somatostatin and its analogs have diverse biological effects in many cells and organs throughout the body. The effects of somatostatin and somatostatin analogs are broadly inhibitory on the secretion of hormones as well as on proliferation and survival of different cells (Reichlin, 1983a, 1983b). Their action includes inhibition of endocrine secretion, GH, TSH, TRH, CRH, gastrin insulin, glucagon, cholecystokinin, vasoactive intestinal peptide, and secretin. They also inhibit exocrine secretion, gastric-, pancreatic-, and intestinal-fluid secretion, inhibit intestinal motility, absorption of nutrients and ions, and vascular contractility (Bloom et al., 1974; Dollinger et al., 1976; Krejs et al., 1980; Reichlin, 1983a, 1983b; Dueno et al., 1987). In addition to its modulatory role in neurotransmission and cognitive functions the peptides (Bissette and Myers, 1992) control the proliferation of normal and tumor cells (Pages et al., 1999; Schally et al., 2004). The biological effects of somatostatin analogs and somatostatin are mediated through five high-affinity plasma membrane receptors, SSTR1–5, which were cloned and identified in the early 1990s (O'Carroll et al., 1992; Yamada et al., 1992). It is well known that many tumors express somatostatin receptors, each tumor expressing more than one subtype (Bissette and Myers, 1992). Immunohistochemistry and autoradiography reveal that somatostatin receptors are highly expressed in gastrinomas, carcinoid tumors, and some other neuroendocrine tumors (Janson et al., 1998; Kulaksiz et al., 2002; Papotti et al., 2002; Fjallskog et al., 2003; Reubi and Waser, 2003) (Table 8.3, Figure 8.5). However,

TABLE 8.1
Summary of the Characteristics of Human Somatostatin Receptor Subtypes

Subtype	SST_1	SST_2	SST_3	SST_4	SST_5
Alternate names	$SRIF_2$, $SSTR_1$, $SFIF_{2A}$	$SRIF_1$, $SSTR_2$, $SFIF_{1A}$	$SSTR_3$ $SFIF_{1C}$	$SSTR_4$, $SSTR_5$, $SFIF_{2B}$	$SSTR_5$, $SSTR_4$, $SFIF_{1B}$
G-protein coupling	Yes	Yes	Yes	Yes	Yes
Adenylyl cyclase activity	↓	↓	↓	↓	↓
Tyrosine phosphatase activity	↑	↑	↑	↑	↑
MAP kinase activity	↑	↓	↑↓	↑	↓
C channels (GIRK)		↑	↑	↑	↑
Ca^{2+} channels	↓	↓			
Na^+/ H^+ exchanger	↑				
AMPA/kainate glutamate channels	↑	↓			
Phospholipase C/IP_3 activity	↑	↑	↑	↑	↑↓
Phospholipase A_2 activity				↑	
Gene/chromosome	SSR_1/14q13	$SSR_{2A,2B}$/17q24	SSR_3/22q13.1	SSR_4/20p11.2	SSR_5/16p13.3
Amino acids; Swiss prot. #	391; P30872	369/356; P30874	418; P32745	388; P31391	364; P35346
Molecular weight (kDa)	42.7	41.3	45.9	41.9	39.2
Transcript size (kb)	4.8	8.5	5.0	4.0	4.0
Glycosylation sites	3	4	2	1	3
Phosphorylation sites	6	8	5	3	5
Tissue distribution	Brain, pituitary, islets, stomach, liver, kidneys	Brain, pituitary, islets, stomach, kidneys	Brain, pituitary, islets, stomach	Brain, islets, stomach, lungs, placenta	Pituitary, islets, stomach

TABLE 8.2
Binding Affinity for Somatostatin 14 and Analogs for the Receptor Subtypes

Compound	SST_1	SST_2	SST_3	SST_4	SST_5
Somatostatin	0.93 ± 0.12	0.15 ± 0.02	0.56 ± 0.17	1.35 ± 0.4	0.29 ± 0.04
Octreotide	280 ± 80	0.38 ± 0.08	7.10 ± 1.4	>1000	6.3 ± 1
Lanreotide	180 ± 20	0.54 ± 0.08	140 ± 9	230 ± 40	17 ± 5
SOM230	9.3 ± 0.1	1.0 ± 0.1	1.5 ± 0.3	>100	0.16 ± 0.01

Source: Adapted from Bruns, C. et al., *Eur. J. Endocrinol.*, 146, 707–716, 2002.
Note: Data are mean IC_{50} ± SEM values (nmol/L).

the frequency and expression pattern of each subtype vary greatly in different tumor types, but also in each patient. Undifferentiated gastro-entero-pancreatic endocrine tumors express somatostatin receptors in lower density than well-differentiated tumors (Reubi et al., 2001). The majority of tumors express SSTR2 followed by less frequent expression of SSTR1, 5, and 3, while SSTR4 is expressed in only a minority of tumors (Reubi et al., 2001; Fjallskog et al., 2003). The two common analogs in clinical use, octreotide and lanreotide, bind preferentially to SSTR2 and 5, with moderate affinity for SSTR3 and low affinities for SSTR1 and 4 (Oberg, 2001). The multiple somatostatin receptor ligand SOM230 (pasireotide) exhibits high-affinity binding to SSTR2, 3, and 5 and moderate affinity for SSTR1 (Bruns et al., 2002). This ligand is now in early clinical trials.

The binding of somatostatin receptors triggers a wide variety of pertussis-toxin-sensitive G protein-dependent and -independent intracellular signals. Each receptor subtype is coupled to multiple intracellular transduction pathways.

8.4 ANTISECRETORY EFFECT OF SOMATOSTATIN AND SOMATOSTATIN ANALOGS

The antisecretory effects of somatostatin analogs on growth hormone are regulated by SSTR2, 5, and 1, whereas prolactin secretion is inhibited by SSTR1 and SSTR5 (Shimon et al., 1997; Hofland and Lamberts, 2004). ACTH is inhibited by signaling through receptor types 2 and 5, whereas calcitonin is mainly inhibited by SSTR2 signaling. In the pancreas insulin secretion is inhibited by SSTR5 and 2 and glucagon secretion by SSTR2, which is also involved in the inhibition of interferon-γ and gastric acid secretion. SSTR5 inhibits glucagon-like peptide 1 (GLP-1) and amylase

TABLE 8.3
Somatostatin Receptor Subtypes mRNA in Neuroendocrine Tumors

Tumor	SST1 (%)	SST2 (%)	SST3 (%)	SST4 (%)	SST5 (%)
Gastrinoma	79[a]	93	36	61	93
Insulinoma	76	81	38	58	57
Nonfunctioning pancreatic tumor	58	88	42	48	50
Carcinoid tumor of the gut	76	80	43	68	77

Source: Adapted from Plöckinger, U. and Wiedenmann, B., *Clin. Endocrinol. Metabol.*, 21, 145–162, 2007.
Note: SST, somatostatin receptor.
[a] The percentage of positive tumors for each SST. mRNA expression may overestimate the number of receptors present, depending on the technique used (PR-polymerase chain reaction, Northern blot, *in situ* hybridization).

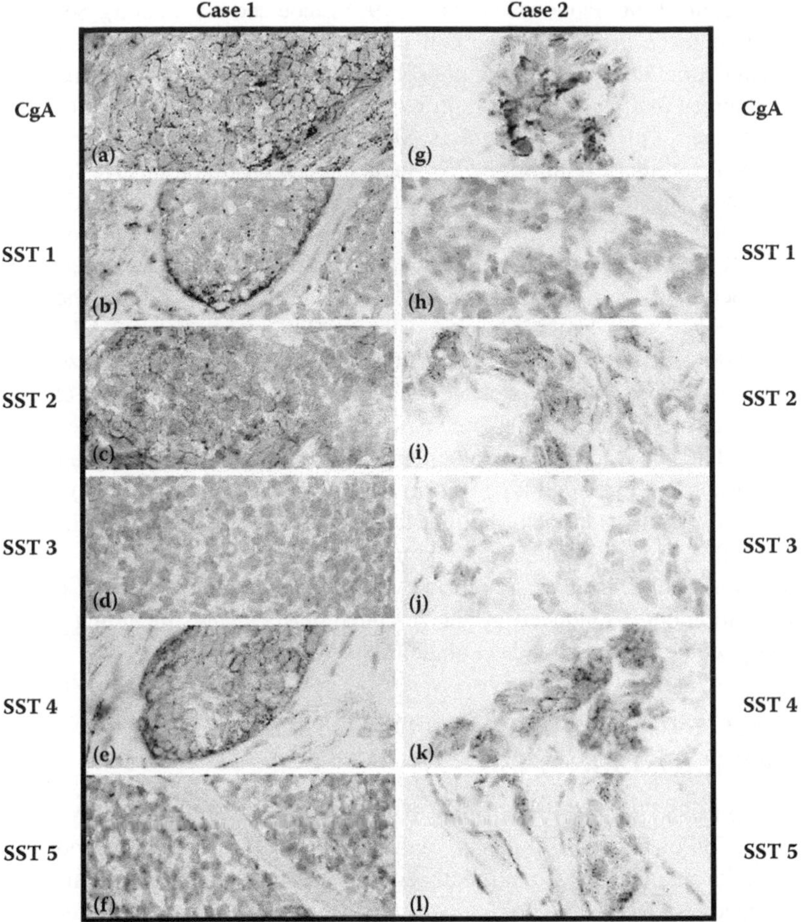

FIGURE 8.5 **(See color insert following page 176.)** Expression of different somatostatin receptors in two cases with endocrine pancreatic tumors. Content of somatostatin receptors are demonstrated with specific antibodies to different subtypes of somatostatin receptors. (CgA, chromogranin A; SST, somatostatin receptor). Case 1 demonstrates expression of receptor types 1, 2, 4, and 5, whereas Case 2 presents expression of receptor types 2, 4, and 5.

secretion (Dollinger et al., 1976; Reichlin, 1983a, 1983b; Strowski et al., 2000). Somatostatin-mediated inhibition of hormone secretion depends on inhibition of adenylate cyclase and/or regulation of ion channels. All five somatostatin receptors are functionally coupled to inhibition of adenylate cyclase via a pertussis-toxin-sensitive protein, $G_{i\alpha 1-3}$. However, somatostatin analog-induced inhibition of peptide secretion mainly results from a decrease in intracellular calcium (Ca^{2+}), which is achieved by either opening the K^+ channels or secondary inhibition of voltage-dependent Ca^{2+} currents. SSTR2, 3, 4, and 5 activate K^+ via $G_{i\alpha 2}$ or $G_{i\alpha 3}$, SSTR2 coupling being the most efficient (Reichlin, 1983a, 1983b). Somatostatin induced activation of Ca^{2+}-dependent phosphatase, and calcineurin may also participate in SSTR-mediated inhibition of exocytosis (Patel, 1999).

8.5 DIRECT ANTITUMOR EFFECTS OF SOMATOSTATIN AND SOMATOSTATIN ANALOGS

Direct antitumor activities mediated through SSTR1–5 expressed in tumor cells include inhibition of synthesis and production of autocrine and paracrine growth-promoting factors and inhibition of growth factors-mediated mitogenic signals and induction of apoptosis (Siegal et al., 1988; Brevini et al., 1993; Keri et al., 1996; Plonowski et al., 1999; Szende and Keri, 2003). Several mechanisms have been reported to be involved in the cell growth inhibition caused by somatostatin and its analogs, including stimulation of tyrosine phosphatases, regulation of MAP-kinases, inhibition of the Na^+–H^+ exchanger, and modulation of nitric oxide (NO) production. Somatostatin and its analogs activate a number of protein phosphatases, SHP-1 and SHP-2, which in turn leads to induction of the cyclin-dependent kinase inhibitor $p27^{Kip1}$ inducing cell proliferation arrest. The MAPK-ERK pathway is an important mediator of somatostatin-induced cell growth regulation. However, this pathway is differently regulated according to the SSTR subtypes and cell environment. SSTR1, 2, and 5 inhibit ERK activity stimulated by serum or growth factors, and this effect is related to the antiproliferative action of these peptides. Na^+–H^+ exchange regulates cell volume and cytoskeletal organization. SSTR1, 3, and 4 regulates this pathway (Sharma et al., 1996; Pages et al., 1999; Cattaneo et al., 2000; Florio et al., 2003; Lahlou et al., 2004). NO is now recognized as an important signaling molecule involved in the regulation of many physiological functions including cell proliferation. Inhibition of NO production may be involved in SSTR1- and SSTR3-mediated cell growth arrest, whereas SSTR2-mediated inhibition of cell proliferation involves inhibition or stimulation of NO production, according to the cell type (Florio et al., 2003; Arena et al., 2005). Besides blocking mitosis, somatostatin analogs are also thought to inhibit cell proliferation by inducing apoptosis (Sharma et al., 1996). This mechanism may be mediated through SSTR3. This is a p53-mediated apoptosis. However, SSTR2 induces apoptosis by a mechanism independent of p53. It also sensitizes tumor cells to apoptosis induced by the so-called death ligands including TNFα, FasL, and TRAIL (Rosskopf et al., 2003; Lattuada et al., 2002).

8.6 INDIRECT ANTITUMOR ACTION OF SOMATOSTATIN AND SOMATOSTATIN ANALOGS

Recent studies using SSTR2 gene transfer has demonstrated an inhibition of tumor angiogenesis, resulting in reduction of microvessel density and inhibition of intratumoral production of VEGF. Somatostatin and its analogs can also indirectly control tumor development and metastases by inhibition of angiogenesis. Over expression of peritumoral vascular somatostatin receptors with high affinity for somatostatin and SSTR2-binding analog octreotide has been reported in human primary colorectal carcinoma, small cell lung carcinoma, breast cancer, renal carcinoma, malignant lymphoma, and endocrine pancreatic tumors (Figure 8.6). This expression appears to be independent of receptor expression in the tumor and immunohistochemical staining has demonstrated expression on SSTR2 in proliferating angiogenic vessels in hypervascular tumors such as hepatocellular carcinoma. The inhibition of angiogenesis is thought to be the key pathway by which octreotide exerts its effect. Somatostatin and its analogs inhibit the proliferation of endothelial cell (HUVEC) in the chicken chorioallantoic membrane model (CAM). Somatostatin analogs also exert antiangiogenic action through a broad inhibition of both the release and the effect of angiogenic factors including VEGF, PDGF, IGF-1, and basic FGF. These growth factors are secreted by tumor cells as well as by infiltrating inflammatory cells, and stimulate endothelial and smooth muscle cell proliferation and migration (Danesi et al., 1997; Woltering et al., 1997; Lawnicka et al., 2000; Adams et al., 2004; Trieber et al., 2004). For review see Susini and Buscail (2006).

In patients with acromegaly, treatment with octreotide results in a tumor shrinkage in 70% of patients. Somatostatin analogs suppress the growth GH-IGF-1 axis by both central and peripheral mechanisms. They inhibit pituitary GH release, but they also inhibit hepatic GH-induced IGF-1 production via SSTR2 (Bevan et al., 2002; Bevan, 2005).

8.7 CLINICAL APPLICATIONS OF SOMATOSTATIN ANALOGS

The pharmacologic development and clinical application of somatostatin analog heralds a new era for peptidomergic therapy for a wide range of neuroendocrine- and nonendocrine-related disorders.

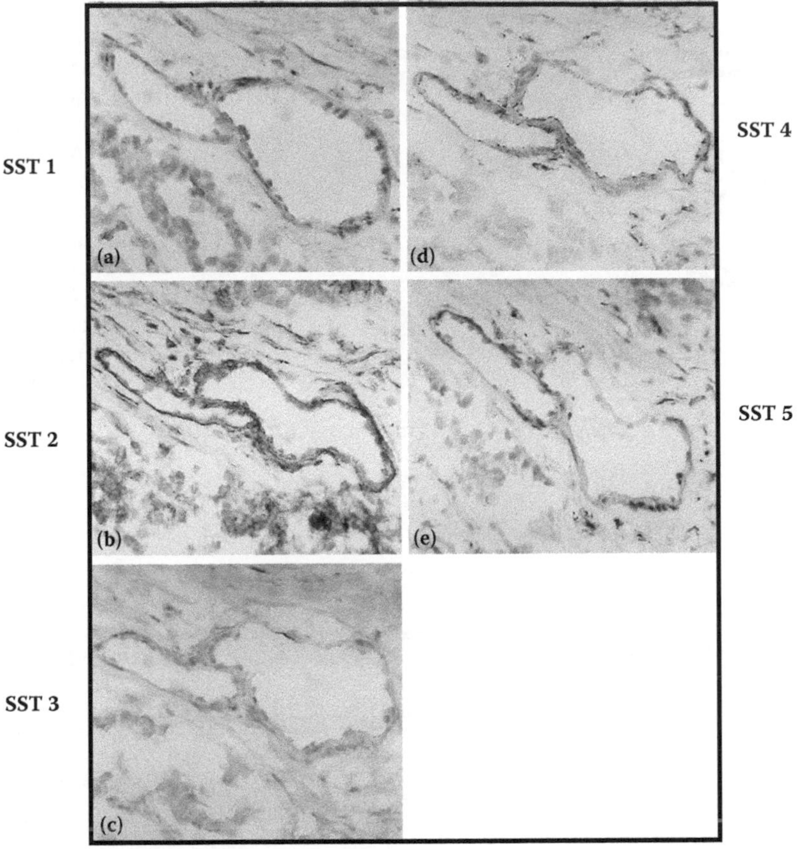

FIGURE 8.6 **(See color insert following page 176.)** Expression of somatostatin receptors type 2, 4, and 5 in newly formed intratumor blood vessels from an endocrine pancreatic tumor.

Clinicians have been provided with unique pharmacological tools that allow an effective peptide delivery and action in a safe therapeutic setting. In light of their unique properties, the prolonged circulation half-life, nontoxic side effects, and ease of injection, somatostatin analogs are clearly ideal therapeutic tools. An array of somatostatin receptor subtypes are ubiquitously expressed in endocrine-derived tissues as well is in other cell types. Somatostatin analogs inhibit pituitary function, gastrointestinal and exocrine pancreatic secretion, and reduce intestinal motility, but are also biologically active at other tissue sites expressing appropriate somatostatin receptor levels. Pharmacokinetic data on the metabolism and elimination half-life of somatostatin analogs (octreotide, lanreotide) show significant enhanced metabolic stability, a small volume of distribution, and low clearance, all resulting in a long duration of exposure and consequently a longer-lasting biological activity (Freda, 2002). Moreover, rebound hypersecretion of hormones does not occur, making these analogs highly useful for clinical application. The current clinically available synthetic somatostatin analogs (octreotide and lanreotide) display a high binding affinity for SSTR2 and lower affinity for SSTR3 and SSTR5, but no binding to SSTR1 and SSTR4 (Table 8.2). A significant number of novel somatostatin ligands have been developed. These analogs include SSTR-selective-, bispecific, universal, as well as chimeric dopamine and somatostatin ligands. Biomeasure generated many somatostatin-subtype specific analogs, which were predominantly directed towards SSTR2 (BIM-23190, BIM-23197) and SSTR5 (BIM-23268). Bispecific somatostatin analogs were also developed, such as BIM-23244. Apart from SSTR-selective and bispecific somatostatin analogs, somatostatin ligands with a more universal binding pattern (SOM230, KE108) have also been

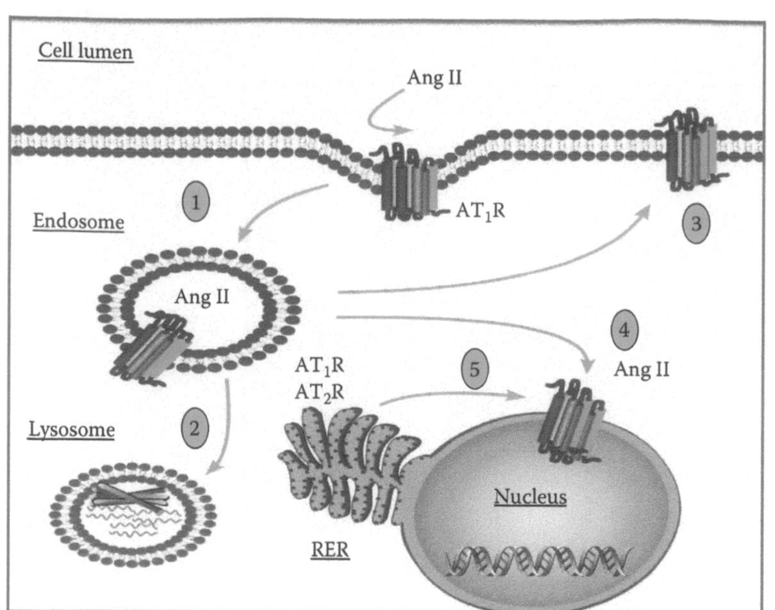

FIGURE 1.2 Scheme for the internalization of the Ang II–AT1 receptor complex in the kidney. Ang II binds to the AT1 receptor (AT1R) on the cell surface and the complex is internalized. Sequestration into endosomes may provide a stable intracellular pool of Ang II (1) while transport to lysosomes will facilitate peptide metabolism (2). The AT1R recycles to the plasma membrane (3) or the Ang II–AT1R may traffic to the nucleus (4). The direct trafficking of the AT1R or the AT2R from the endoplasmic reticulum (RER) or Golgi apparatus to the nucleus may also occur (5).

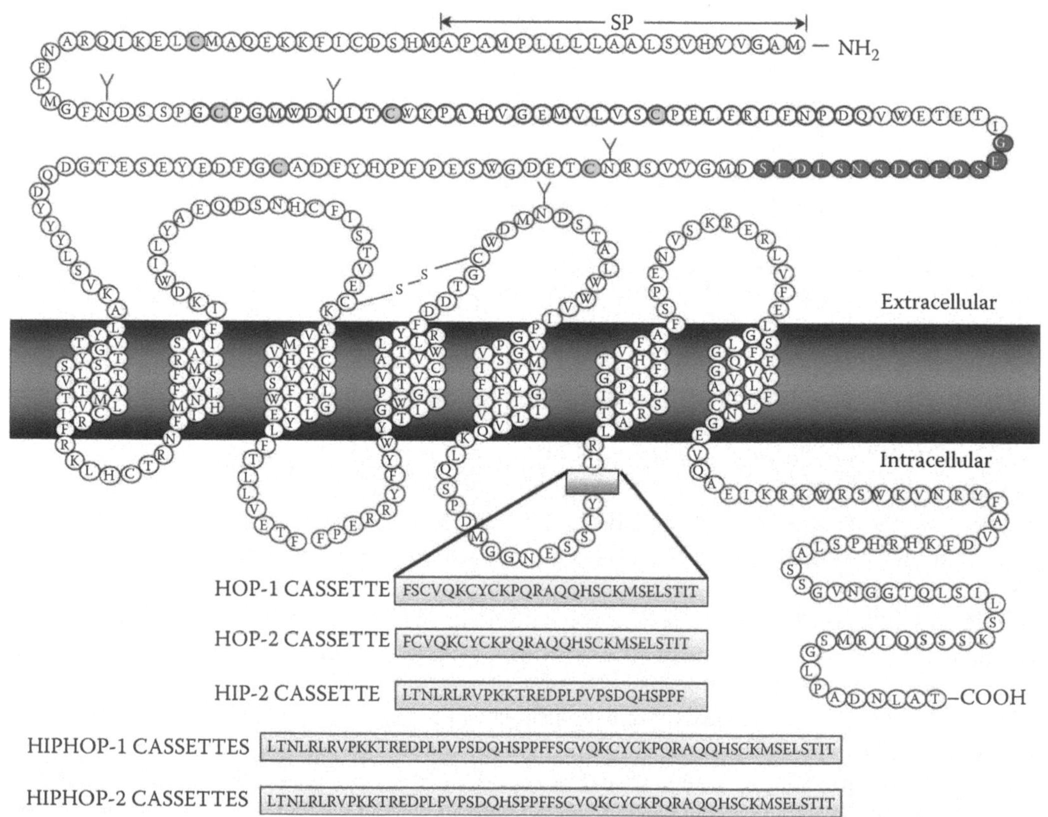

FIGURE 4.5 Schematic diagram of the human PAC_1 receptor. The predicted 468-amino-acid sequence of the null PAC_1 receptor isoform, $hPAC_1$-null, is shown along with the amino-acid cassettes that can be inserted into ic3 shown beneath. The signal peptide sequence (SP) is indicated by the arrows, and cleavage is predicted to occur between A^{20} and M^{21}. The potential N-glycosylation sites in the NT domain are indicated by the symbol Y. Amino acid sequences encoded by exon 4 are outlined in red, the additional 21-amino-acid sequences encoded by exons 5 and 6 are indicated in yellow and orange, respectively. The transmembrane spanning domains were predicted by hydrophobicity plot.

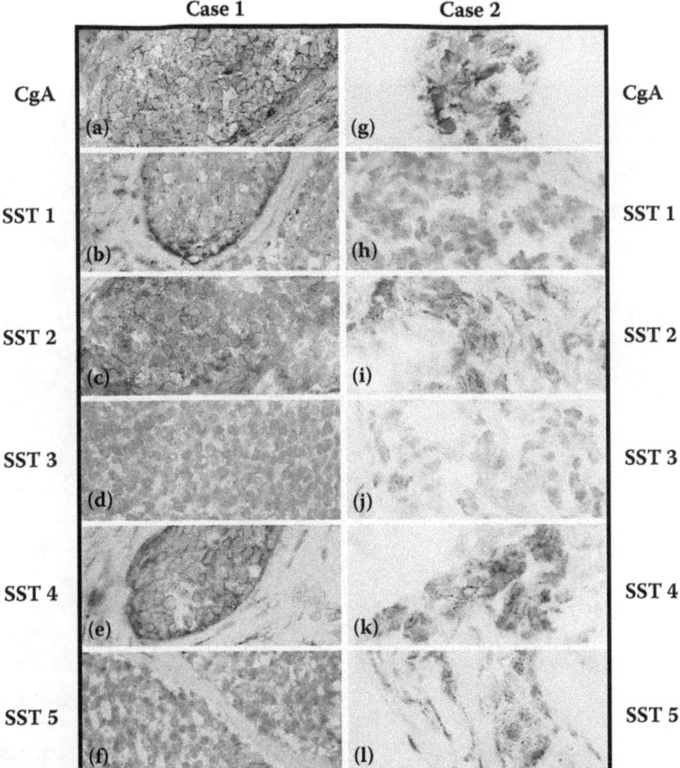

FIGURE 8.5 Expression of different somatostatin receptors in two cases with endocrine pancreatic tumors. Content of somatostatin receptors are demonstrated with specific antibodies to different subtypes of somatostatin receptors. (CgA, chromogranin A; SST, somatostatin receptor). Case 1 demonstrates expression of receptor types 1, 2, 4, and 5, whereas Case 2 presents expression of receptor types 2, 4, and 5.

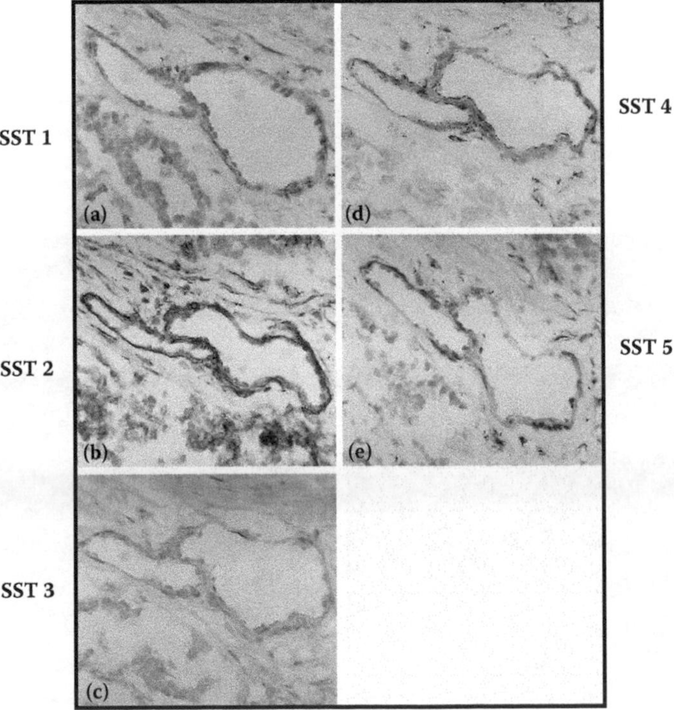

FIGURE 8.6 Expression of somatostatin receptors type 2, 4, and 5 in newly formed intratumor blood vessels from an endocrine pancreatic tumor.

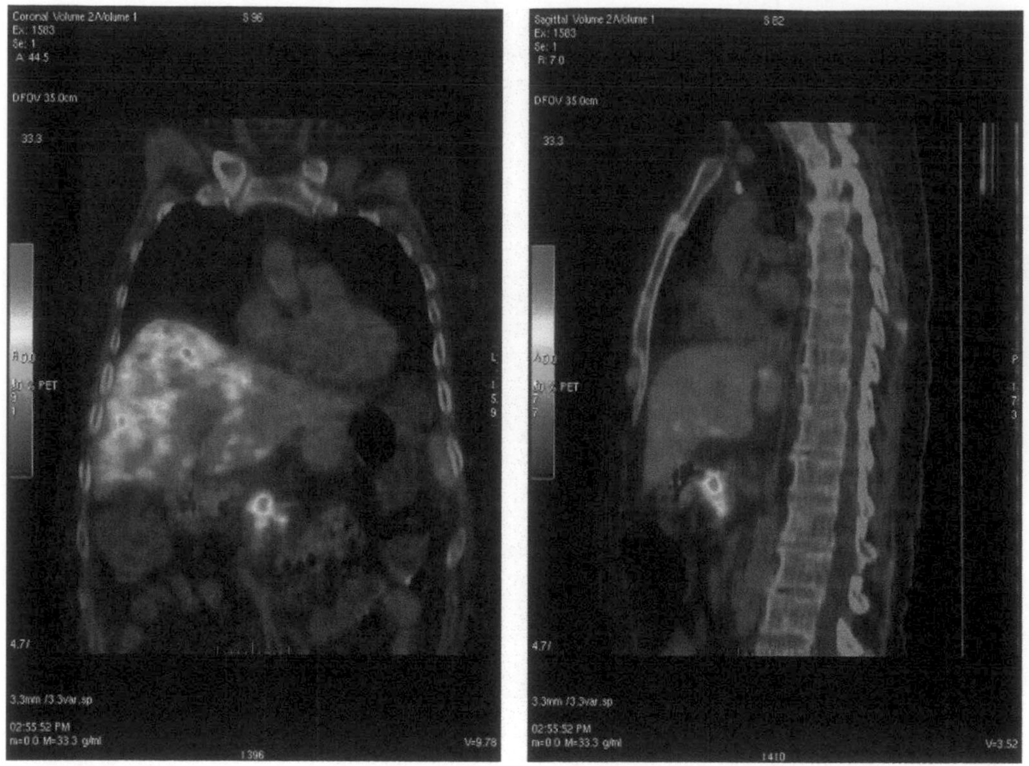

FIGURE 8.8 PET/CT scan with ^{68}Ga-DOTA-octreotide, a new tracer for neuroendocrine tumors, demonstrates lymph node metastases in the mesentery and multiple liver metastases.

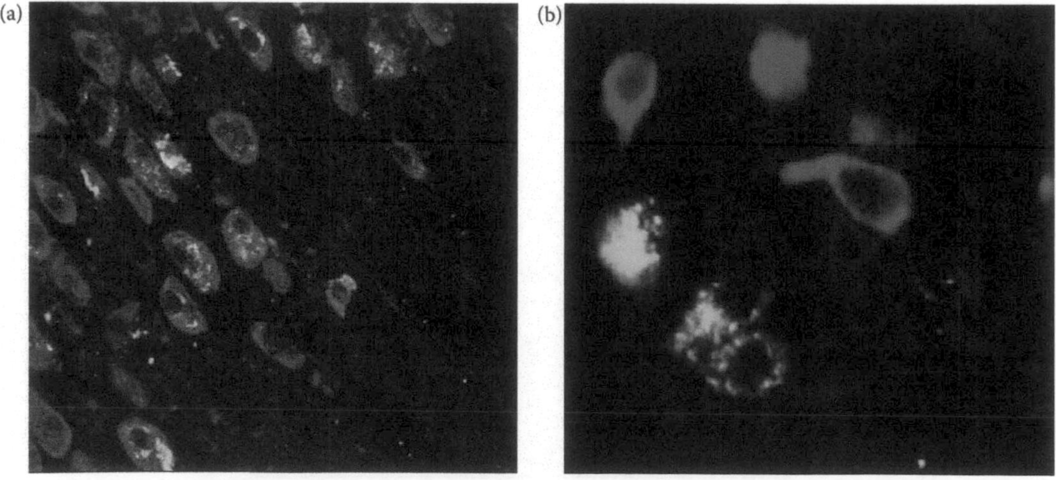

FIGURE 12.5 Immunohistochemistry of the human gastric antrum showing colocalization of orexin A with gastrin (a), and orexin A adjacent to somatostatin (b).

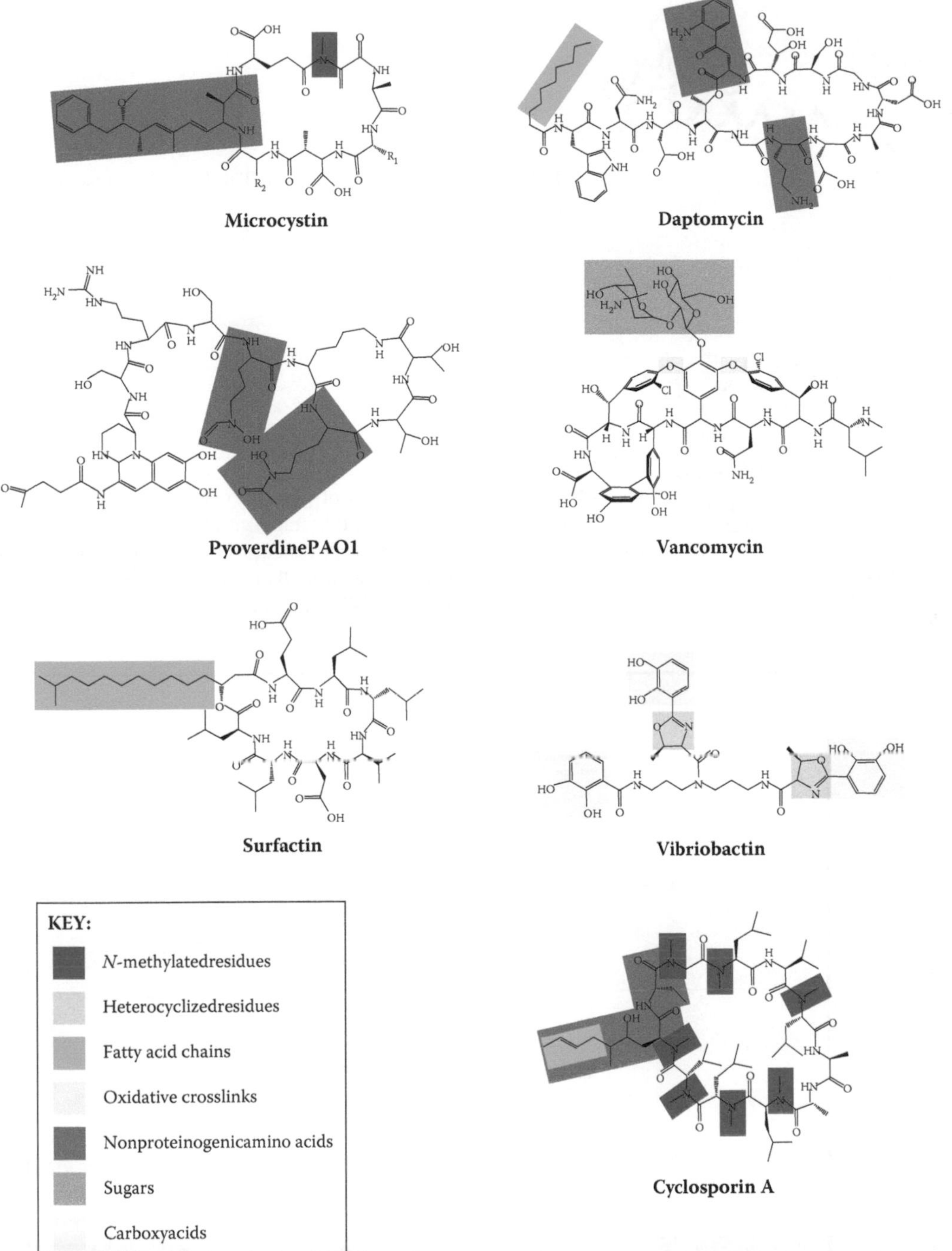

FIGURE 13.5 Examples of nonribosomal peptide structural diversity. Highlighted regions depict characteristic structural variations carried out by adenylation and modification domains. $R_{1,2}$, variable amino acids in the microcystin structure.

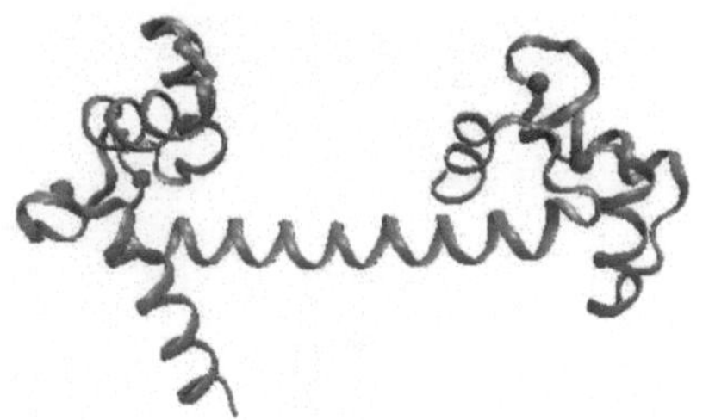

FIGURE 14.11 Ca^{2+} calmodulin. The four Ca^{2+} ions are indicated in red.

FIGURE 14.12 Representation of globular complex formed between Ca^{2+}CaM and the 26-residue peptide fragment of skeletal myosin light-chain kinase (shown in green) with residues 3–21 encompassed within Ca^{2+}CaM. (Redrawn and modified from Ikura, M., Kay, L.E., Krinks, M., and Bax, A., *Biochemistry*, 30, 5498–5504, 1991.)

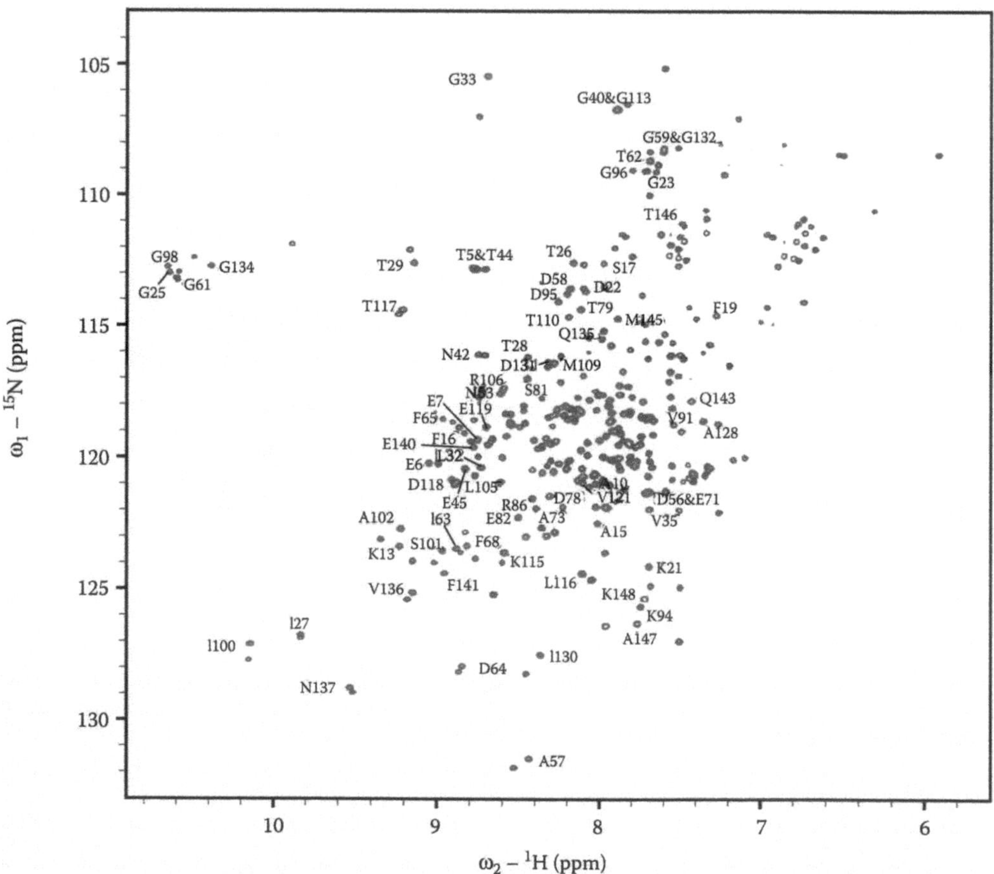

FIGURE 14.14 ^{15}N HSQC spectra of Ca^{2+} ^{15}N-CaM in the absence of peptide (red), and with the addition of caerin 1.8 in a 2:1 peptide:protein ratio (purple).

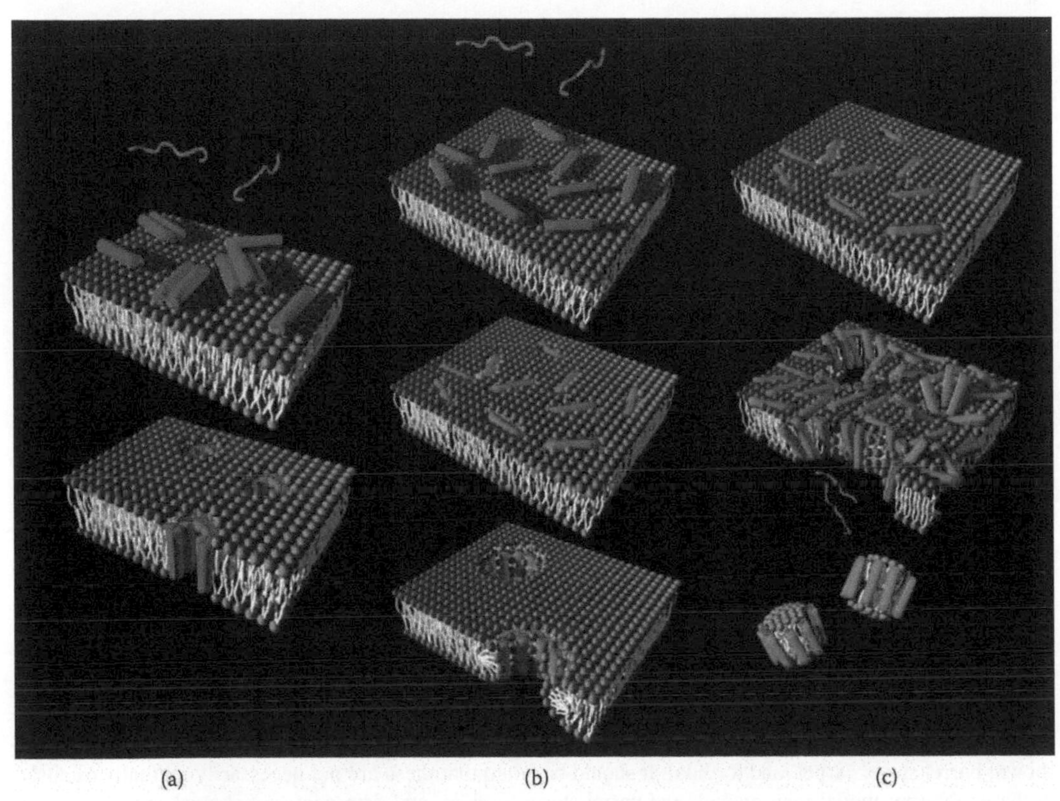

FIGURE 15.1 Bilayer disruption mechanisms. (a) Barrel stave mechanism. (b) Toroidal mechanism. (c) Carpet mechanism. Helical forms of peptides are represented by cylinders, while hydrophobic and hydrophilic regions are colored red and blue, respectively.

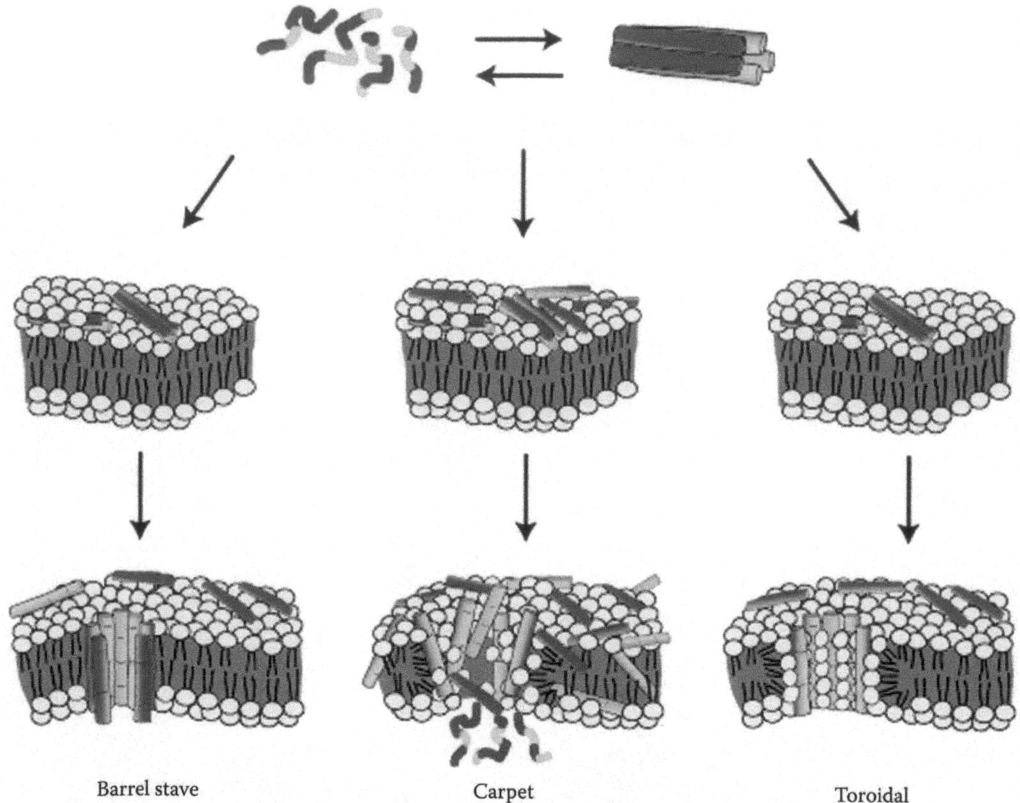

FIGURE 16.1 Proposed mechanisms of action of antimicrobial peptides. Three potential mechanisms are shown; barrel stave, carpet, and toroidal. It should be noted that these are not necessarily mutually exclusive in that the carpet model may precede the barrel stave, toroidal, and detergent (not shown) models, and the toroidal model may precede the detergent model (not shown). In the barrel-stave model peptides assemble on the surface of the membrane as monomers or oligomers then insert into the membrane. In the carpet model, once a threshold concentration of peptide is reached, the membrane is permeated and transient pores form. In the toroidal model the peptides cause the membrane to bend through the pore so the pore is lined with peptide and lipid head groups. Yellow represents the hydrophilic surface and magenta the hydrophobic surface. (Modified with permission from Huang et al., *Phys. Rev. Lett.*, 92, 198–304, 2004; copyright 2004 by the American Physical Society and from Shai et al., *Biopolymers*, 66, 236–248, 2002; copyright 2002 by John Wiley & Sons, Hoboken, NJ. With permission.)

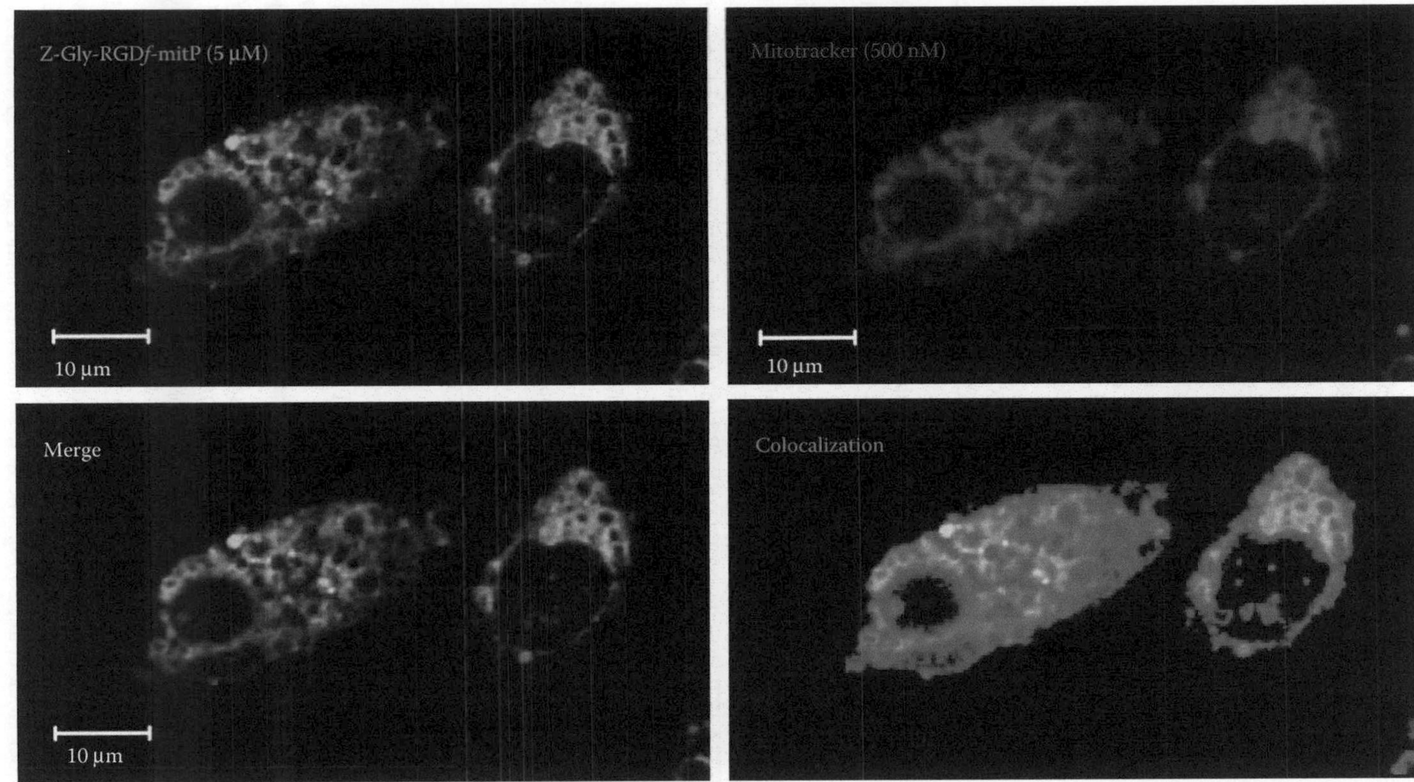

FIGURE 18.2 The chimeric mitP analog Z-Gly-RGD*f*-mitP demonstrates strong mitochondrial localizing properties. Live confocal cell imaging was used to visualize the colocalization of fluorescein-labeled Z-Gly-RGD*f*-mitP with mitochondria in ECV304 cells, costained with Mitotracker™ (Molecular Probes). Using Carl Zeiss imaging software, colocalized fluorophores are shown.

TABLE 8.4
Binding Potency of Somatostatin Analogs and Novel Chimeric Somatostatin–DA Compounds to SSTR and D2DR

	Receptor Subtype Binding Affinity IC50 (nM)					
Compound	**SSTR1**	**SSTR2**	**SSTR3**	**SSTR4**	**SSTR5**	**D2DR**
Somatostatin-14	2.3	0.2	1.4	1.8	0.9	nd
Octreotide	>1000	0.6	34.5	>1000	7	nd
Lanreotide	>1000	0.8	107	>1000	5.2	nd
SOM230	9.3	1	1.5	>1000	0.2	nd
KE-108	2.6	0.9	1.5	1.6	0.7	nd
BIM-23244	>1000	0.3	133	>1000	0.7	nd
BIM-23A387	293	0.1	77	>1000	26	22
BIM-23A761	462	0.06	52	>1000	3.7	27
BIM-23A765	>1000	0.2	76	>1000	7	9

Source: Adapted from Hofland, L.J. et al., *J. Endocrinol. Invest.* 28, Suppl. 11, 36–42, 2005.

synthesized. SOM230 (pasireotide) is a stable molecule with a relatively long elimination half-life in the circulation of 11.8 h, which is about five times longer than octreotide (2.3 h) (Bruns et al., 2002; Schmid and Schoeffter, 2004). K108 also seems to be a stable compound. Both SOM230 and KE108 show high binding affinity for the 5 SSTRs except for SSTR4. The binding potencies were comparable to native somatostatin-14 or somatostatin-28. Finally an interesting novel series of chimerics, somatostatin–dopamine compounds (dopastatins) with high binding affinity to SSTR2 and DAD2 (BIM-23A387) receptors or to SSTR2, SSTR5, and D2DR (BIM-23A761 and BIM 23A765) were developed. The dopamine moiety of these molecules also results in high binding affinity to D2DR (Saveanu et al., 2002; Jaquet et al., 2005) (Table 8.4). The development of the above-described novel somatostatin ligands has significantly improved our insight into the role of the individual SSTRs in the regulation of hormone secretion and other biological actions of SSTR signaling. Below is a review of the application of somatostatin analogs in different disease stages.

8.8 PITUITARY ADENOMAS

8.8.1 Acromegaly

The normal anterior pituitary gland expresses SSTR1, SSTR2, SSTR3, and SSTR5. GH-secreting pituitary adenomas express predominantly SSTR2 and SSTR5, showing a large variability between individual adenomas. Usually, SSTR5 is more highly expressed than SSTR2 (Jaquet, 2000; Park et al., 2004). The response of GH secretion to octreotide or native somatostatin has been shown to correlate positively to the expression of SSTR2 mRNA levels, but not with SSTR5 mRNA levels (Jaquet, 2000). Therefore, SSTR2 is a receptor predominantly involved in the inhibition of GH secretion by GH-secreting pituitary adenomas. Targeting both the SSTR2 and SSTR5 receptors by bispecific or universal somatostatin analogs results in a better inhibitory response of GH secretion. This is particularly noticed in adenomas expressing low SSTR2 level and high SSTR5 (Shimon et al., 1997; Saveanu et al., 2001b). This has also been demonstrated by a stronger inhibitory effect of the universal somatostatin ligand SOM230 in patients with acromegaly compared with treatment with octreotide alone (van der Hoek et al., 2004). The new somatostatin analog combined with DA moiety resulting in binding affinity to both SSTR2 and D2DR have demonstrated an enhanced potency in suppressing GH and prolactin release by cultured GH-secreting pituitary adenomas (Saveanu et al., 2002). Additionally, SSTR2- and SSTR5-D2DR specific compounds showed enhanced potency in suppressing hormone release by GH-secreting adenomas (Jaquet et al., 2005). Combination of somatostatin analogs with GH-receptor antagonists (Pegvisomant) can result in biochemical cure in ~95% of acromegalic patients (Feenstra et al., 2005). In a review of 921 patients with acromagely from 36 reports a tumor shrinkage was obtained in 42% of patients (Bevan et al., 2002).

8.8.2 Cushing's Disease

The clinically available somatostatin analogs octreotide and lanreotide seem not to be useful in the treatment of patients with corticotroph adenomas. That indicates that the SSTR2 expression is low in patients with Cushing's disease and may be suppressed by circulating cortisol levels. It has recently been shown that SSTR5 plays an important role in the regulation of basal and CRH-induced ACTH release in primary cultures of human corticotroph adenomas (Hofland et al., 2005; van der Hoek et al., 2005). SSTR5 expression is considered to be more resistant to the suppressive effect of glucocorticoids and therefore SOM230 could demonstrate a biological effect (Hofland et al., 2005). Somatostatin analogs targeting SSTR5 and SSTR2 by the recent generation of somatostatin analogs including SOM230 may be novel candidates for the medical treatment of patients with Cushing's disease.

8.8.3 Other Pituitary Adenomas

Octreotide and lanreotide has no suppressive effect on prolacting levels in patients with prolactinomas (Hofland and Lamberts, 2004). Recent data show that SSTR5 is the SSTR which predominantly expressed in prolactinomas and that SSTR5-preferring somatostatin analogs significantly suppress prolactin release (Jaquet et al., 1999; Hofland et al., 2005). SSTR5-preferring agonists can be an alternative to dopamine agonists in therapy in patients resistant to dopamine agonists. So called nonfunctioning pituitary adenomas are usually resistant to currently available somatostatin analogs, although exceptional cases have been published (Saveanu et al., 2001a). Usually, SSTR3 is more abundantly expressed in nonfunctioning pituitary adenomas, but the role of SSTR3 has yet to be further explored.

8.9 NEUROENDOCRINE GEP-TUMORS

SSTR2 is the receptor expressed most abundantly in GEP neuroendocrine tumors (Janson et al., 1998; Kulaksiz et al., 2002; Papotti et al., 2002; Fjallskog et al., 2003). Therefore, the successful clinical application of octreotide and lanreotide is related to the high binding affinity for SSTR2. The two analogs are mainly used for controlling symptoms related to hormonal hypersecretion by the tumor cells (Table 8.5), but there are also possibilities for visualization of SSTR-positive tumors using radiolabeled somatostatin analogs (Figures 8.7 and 8.8) as well as for SSTR targeted treatment

TABLE 8.5
Effects of Somatostatin Analogs on Hypersecretion Syndromes

Syndrome	Symptom	Hormone/ Neurotransmitter	Tumor Marker
Carcinoid syndrome	Flush > diarrhea	Serotonin	5-HIAA, CgA
Watery diarrhea syndrome	Diarrhea, dehydration, acidosis	VIP	VIP
Glucagonoma syndrome	Migratory necrolytic erythema	Glucagon	Glucagon
Zollinger–Ellison syndrome	Peptic ulceration, gastro-esophageal reflux disease	Gastrin	Gastrin

Source: Adapted from Plöckinger, U. and Wiedenmann, B., *Clin. Endocrinol. Metab.*, 21, 145–162, 2007.
Note: 5-HIAA, 5-hydroxyindol-acetic acid; CgA, chromogranin A; VIP, vasoactive intestinal peptide.

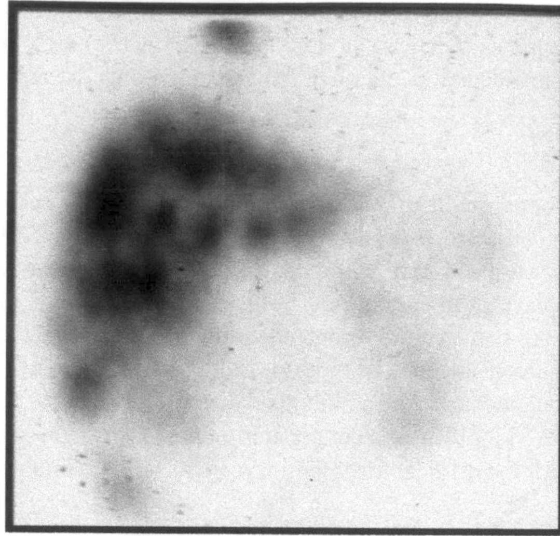

FIGURE 8.7 Somatostatin receptor scintigraphy in a patient with foregut carcinoid tumor, demonstrating the primary tumor in the lung and multiple liver metastases.

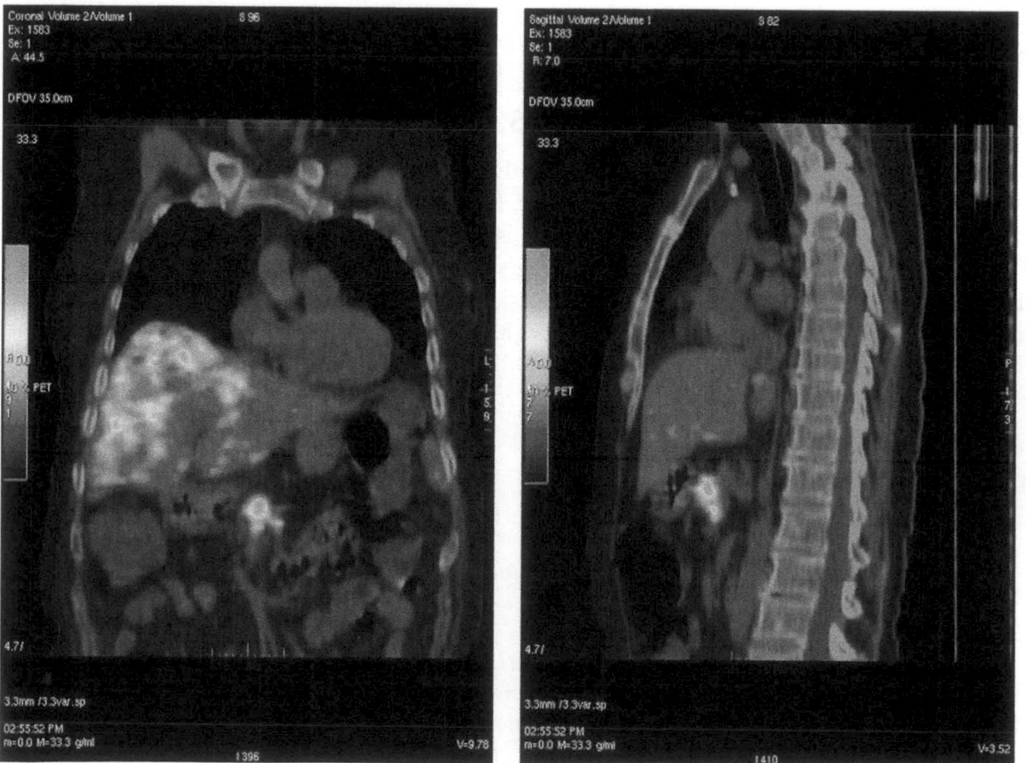

FIGURE 8.8 **(See color insert following page 176.)** PET/CT scan with ^{68}Ga-DOTA-octreotide, a new tracer for neuroendocrine tumors, demonstrates lymph node metastases in the mesentery and multiple liver metastases.

TABLE 8.6
Neuroendocrine Tumors. Somatostatin Analogue Therapy—Summary of Several Trials

Response		Standard Dose (100–1500 μg/d)	High Dose (>3000 μg/d)	Slow Release (20–30 mg/d/2–4w)
Symptomatic *n* (%)		146/228 (64)	11/26 (42)	76/119 (63)
Biochemical *n* (%)	CR	6/54 (11)	1/33 (3)	3/119 (3)
	PR	116/211 (55)	24/83 (72)	76/119 (64)
	SD	72/211 (34)	7/33 (21)	21/119 (18)
	PD	11/211 (11)	1/33 (3)	19/119 (15)
Tumor *n* (%)	CR	—	1/53 (2)	—
	PR	7/131 (5)	6/53 (11)	4/119 (3)
	SD	50/131 (38)	25/53 (47)	94/119 (79)
	PD	74/131 (56)	21/51 (39)	21/119 (18)

Note: CR, complete remission; PD, progressive disease; PR, partial remission; SD, stable disease.

using radiolabeled somatostatin analogs coupled to gamma- and beta-emitting isotopes. Using unlabeled somatostatin analogs as octreotide or lanreotide, unfortunately a considerable proportion of patients experience an escape from treatment within months to several years. This treatment escape is suggested to be due to several mechanisms including uncoupling of receptor activation or to outgrowth of receptor negative tumor clones (Hofland and Lamberts, 2003). Besides expression of SSTR2, neuroendocrine GEP tumors also express all the other subtypes of somatostatin receptors at different concentrations (Table 8.3). It has also recently been established that neuroendocrine GEP tumors also express dopamine receptors (Reubi and Waser, 2003). Therefore, the expression of other somatostatin receptors form novel additional targets in the treatment of these tumors, with for example universal somatostatin analogs, such as SOM230 or KE108 or also the new combined somatostatin and dopamine receptor analogs (dopastatin). May be in the future this will also reduce the problem of resistance to somatostatin analog treatment. The first clinical trials comparing octreotide with SOM230 are now just ongoing in GEP-NETs. The biochemical and symptomatic response rates for octreotide and lanreotide are 45 to 60% (Table 8.6). The antitumor effect is only 3–5% (Table 8.7) (Kvols et al., 1986; Maton, 1989; Vinik and Moattari, 1989; Saltz et al., 1993;

TABLE 8.7
Antiproliferative Effect of Somatostatin Analog in Patients with Progressive Disease

SSA	Dosage	*n*	CR	PR	SD	PD
Lanreotide	3,000 μg/day	22	0	1	7	14
Lanreotide	30 mg/2 weeks	35	0	1	20	14
Octreotide	600–1,500 μg/day	52	0	0	19	33
Octreotide	1,500–3,000 μg/day	58	0	2	27	29
Octreotide	600 μg/day	10	0	0	5	5
Lanreotide	15,000 μg/day	24	1	1	11	11
		201	1 (0.5%)	5 (3%)	89 (44%)	106 (53%)

Source: Adapted from Plöckinger, U. and Wiedenmann, B., *Clin. Endocrinol. Metab.*, 21, 145–162, 2007.
Note: CR, complete remission; PD, progressive disease; PR, partial remission; SD, stable disease.

Ruszniewski et al., 1996; Tomassetti et al., 1998; Eriksson and Oberg, 1999; Faiss et al., 1999; O'Toole et al., 2000; Tomassetti et al., 2000; Oberg, 2001; Welin et al., 2004; Panzuto et al., 2006). High-dose treatment might generate more antitumor regression (Welin et al., 2004; Faiss et al., 1999) (Tables 8.5 to 8.7).

8.10 OTHER GASTROINTESTINAL DISORDERS

As somatostatin analogs inhibit secretion of most gastrointestinal hormones, they reduce gut motility, inhibit pancreatic exocrine secretion, and also reduce the splancnic blood flow. These compounds have been widely applied in treating different gastrointestinal disorders. Functional diarrhea, hypersecreting syndromes, gastrointestinal fistulae, dumping syndrome, and variceal bleedings are areas where somatostatin analogs have been applied with varied success. It is currently used in patients with diarrhea secondary to chemotherapy. It is also used in patients undergoing pancreatic surgery to reduce the development of fistulaes and also for treating patients with portal hypertension and varices to reduce the frequency of bleedings (Mulvihill et al., 1986; Cascinu et al., 1994; Lamberts et al., 1996; Jenkins et al., 1997).

8.11 CANCER THERAPY

Experimental studies of the combination of octreotide with antimitotic drugs such as vincristine, methotrexate, fluouracil, and suramin resulted in a slightly additive action (Schally, 1988). Currently, the combination of chemotherapy with somatostatin analogs is considered for use in patients with gastroenteropancreatic neuroendocrine tumors. Furthermore, somatostatin analogs have also been used as carriers to deliver cytotoxic agents to cancer cells. Schally and coworkers synthesized novel target cytotoxic somatostatin octapeptide conjugates, such as AN-238, which contains RC-121 and octapeptides coupled to doxorubicin derivative, 2-pyrrolino-DOX (AN201). In human experimental cancer models AN238 was very effective (Kiaris et al., 2001). However, it has not yet been transferred to clinical trials.

8.12 OTHER APPLICATIONS

Radiolabeled somatostatin analogs are useful diagnostic tools for the detection and localization of small somatostatin receptor expressing tumors (Krenning et al., 1993; Hofmann et al., 2001; Maecke et al., 2005). In parallel with the development of these diagnostic tools, targeted radioactive treatment with the same compounds has been applied in a small series of patients, starting with ^{111}Indium-DTPA-Octreotide, moving to ^{90}Yttrium-DOTATOC and finally now ^{177}Lutetium-DOTA-Octreotate (Kwekkeboom et al., 2000; Forrer et al., 2007). The response rates of these new targeting treatments are ~30 to 40% objective responses. A new, yet developed strategy is somatostatin type 2 receptor gene-transfer, which has resulted in a significant inhibition of tumor growth *in vivo* in pancreatic cancer models (Vernejoul et al., 2002). The antitumor effect of *in vivo* gene transfer may also be enhanced by systemic chemotherapy with targeted cytotoxic therapy or with administration of radionuclide analogs (Rogers et al., 2002).

8.13 SIDE EFFECTS OF SOMATOSTATIN ANALOGS

Frequently occurring side effects, such as abdominal discomfort, bloating, and steatorrea, occur due to inhibition of pancreatic enzymes. They are mostly mild and subside spontaneously within the first week of treatment (Plockinger et al., 1990). Persistent steatorrea can be treated with supplementation of pancreatic enzymes. Cholestasis with subsequent cholecystolithiasis does occur in up to 60% of patients due to inhibition of cholecystokinin and the production of lithogenic bile (Trendle et al., 1997). Prophylactic therapy with chenodeoxycholic acid and ursodeoxycholic acid may be able to prevent the occurrence of gallstone disease in patients on long-term somatostatin analog treatment. Due to steatorrhea malabsorption, reduction of serum vitamin D concentration and subsequently reduced calcium absorption may occur. Serum vitamin B-12 concentration may decline, possibly due to direct inhibition of the intrinsic factor secretion of the parietal cells (Plockinger et al., 1998). Other rare side effects are cardiac bradycardia and diabetes mellitus.

8.14 SUMMARY

Octreotide was the first somatostatin analog coming into clinical use for more than two decades ago. It was really a breakthrough in the management of acromegaly and functioning GEP-NETs. Life-threatening conditions such as carcinoid crisis and WDHA-syndrome could be more easily managed. Other somatostatin analogs have further improved the therapeutic arsenal for different neuroendocrine tumors, particulary the development of long-acting formulations. The latter have significantly improved the quality of life for patients. The side effects are rather mild. In the future, new somatostatin analogs will come into clinical trials as well as high-dose treatment with currently available somatostatin analogs. The development of radiolabeled somatostatin analogs has significantly improved the diagnosis of neuroendocrine tumors. Most recently targeting treatment with ^{177}Lutetium and ^{90}Yttrium labeled somatostatin analogs has further increased the therapeutic range with significant antitumor effects. In the future, a combination with cytotoxic agents as well as gene transfer might form an additional therapeutic model.

REFERENCES

Adams, R.L., Adams, I.P., Lindow, S.W., and Atkin, S.L., Inhibition of endothelial proliferation by the somatostatin analogue SOM230, *Clin. Endocrinol. (Oxford)*, 61, 431–436, 2004.

Arena, S., Pattarozzi, A., Corsaro, A., Schettini, G., and Florio, T., Somatostatin receptor subtype-dependent regulation of nitric oxide release: Involvement of different intracellular pathways, *Mol. Endocrinol.*, 19, 255–267, 2005.

Bauer, W. et al., SMS 201–995: A very potent and selective octapeptide analogue of somatostatin with prolonged action, *Life Sci.*, 31, 1133–1140, 1982.

Bevan, J.S., Clinical review: The antitumoral effects of somatostatin analog therapy in acromegaly, *J. Clin. Endocrinol. Metab.*, 90, 1856–1863, 2005.

Bevan, J.S. et al., Primary medical therapy for acromegaly: An open, prospective, multicenter study of the effects of subcutaneous and intramuscular slow-release octreotide on growth hormone, insulin-like growth factor-I, and tumor size, *J. Clin. Endocrinol. Metab.*, 87, 4554–4563, 2002.

Bissette, G. and Myers, B., Somatostatin in Alzheimer's disease and depression, *Life Sci.*, 51, 1389–1410, 1992.

Blake, A.D., Badway, A.C., and Strowski, M.Z., Delineating somatostatin's neuronal actions, *Curr. Drug Targets CNS Neurol. Disord.*, 3, 153–160, 2004.

Bloom, S.R. et al., Inhibition of gastrin and gastric-acid secretion by growth-hormone release-inhibiting hormone, *Lancet*, 2, 1106–1109, 1974.

Brazeau, P. et al., Hypothalamic polypeptide that inhibits the secretion of immunoreactive pituitary growth hormone, *Science*, 179, 77–79, 1973.

Brevini, T.A., Bianchi, R., and Motta, M., Direct inhibitory effect of somatostatin on the growth of the human prostatic cancer cell line LNCaP: Possible mechanism of action, *J. Clin. Endocrinol. Metab.*, 77, 626–631, 1993.

Bruns, C., Lewis, I., Briner, U., Meno-Tetang, G., and Weckbecker, G., SOM230: A novel somatostatin peptidomimetic with broad somatotropin release inhibiting factor (SRIF) receptor binding and a unique antisecretory profile, *Eur. J. Endocrinol.*, 146, 707–716, 2002.

Cascinu, S., Fedeli, A., Fedeli, S.L., and Catalano, G., Control of chemotherapy-induced diarrhea with octreotide. A randomized trial with placebo in patients receiving cisplatin, *Oncology*, 51, 70–73, 1994.

Cattaneo, M.G. et al., Selective stimulation of somatostatin receptor subtypes: Differential effects on Ras/MAP kinase pathway and cell proliferation in human neuroblastoma cells, *FEBS Lett.*, 481, 271–276, 2000.

Danesi, R. et al., Inhibition of experimental angiogenesis by the somatostatin analogue octreotide acetate (SMS 201–995), *Clin. Cancer Res.*, 3, 265–272, 1997.

Dollinger, H.C., Raptis, S., and Pfeiffer, E.F., Effects of somatostatin on exocrine and endocrine pancreatic function stimulated by intestinal hormones in man, *Horm. Metab. Res.*, 8, 74–78, 1976.

Dueno, M.I., Bai, J.C., Santangelo, W.C., and Krejs, G.J., Effect of somatostatin analog on water and electrolyte transport and transit time in human small bowel, *Dig. Dis. Sci.*, 32, 1092–1096, 1987.

Eriksson, B. and Oberg, K., Summing up 15 years of somatostatin analog therapy in neuroendocrine tumors: Future outlook, *Ann. Oncol.*, 10, Suppl. 2, S31–S38, 1999.

Faiss, S. et al., Ultra-high-dose lanreotide treatment in patients with metastatic neuroendocrine gastroentero-pancreatic tumors, *Digestion*, 60, 469–476, 1999.

Feenstra, J. et al., Combined therapy with somatostatin analogues and weekly pegvisomant in active acromegaly, *Lancet*, 365, 1644–1646, 2005.

Fjallskog, M.L. et al., Expression of somatostatin receptor subtypes 1 to 5 in tumor tissue and intratumoral vessels in malignant endocrine pancreatic tumors, *Med. Oncol.*, 20, 59–67, 2003.

Florio, T. et al., Somatostatin inhibits tumor angiogenesis and growth via somatostatin receptor-3-mediated regulation of endothelial nitric oxide synthase and mitogen-activated protein kinase activities, *Endocrinology*, 144, 1574–1584, 2003.

Forrer, F. et al., Neuroendocrine tumors. Peptide receptor radionuclide therapy, *Best Pract. Res. Clin. Endocrinol. Metab.*, 21, 111–129, 2007.

Freda, P.U., Somatostatin analogs in acromegaly, *J. Clin. Endocrinol. Metab.*, 87, 3013–3018, 2002.

Gluschankof, P. et al., Enzymes processing somatostatin precursors: An Arg–Lys esteropeptidase from the rat brain cortex converting somatostatin-28 into somatostatin-14, *Proc. Natl. Acad. Sci. USA*, 81, 6662–6666, 1984.

Hellman, B. and Lernmark, A., Inhibition of the *in vitro* secretion of insulin by an extract of pancreatic alpha-1 cells, *Endocrinology*, 84, 1484–1488, 1969.

Hofland, L.J. and Lamberts, S.W., The pathophysiological consequences of somatostatin receptor internalization and resistance, *Endocr. Rev.*, 24, 28–47, 2003.

Hofland, L.J. and Lamberts, S.W., Somatostatin receptors in pituitary function, diagnosis, and therapy, *Front Horm. Res.*, 32, 235–252, 2004.

Hofland, L.J. et al., The multi-ligand somatostatin analogue SOM230 inhibits ACTH secretion by cultured human corticotroph adenomas via somatostatin receptor type 5, *Eur. J. Endocrinol.*, 152, 645–654, 2005.

Hofmann, M. et al., Biokinetics and imaging with the somatostatin receptor PET radioligand (68)Ga-DOTATOC: Preliminary data, *Eur. J. Nucl. Med.*, 28, 1751–1757, 2001.

Janson, E.T. et al., Determination of somatostatin receptor subtype 2 in carcinoid tumors by immunohistochemical investigation with somatostatin receptor subtype 2 antibodies, *Cancer Res.*, 58, 2375–2378, 1998.

Jaquet, P. et al., Quantitative and functional expression of somatostatin receptor subtypes in human prolactinomas, *J. Clin. Endocrinol. Metab.*, 84, 3268–3276, 1999.

Jaquet, P. et al., Human somatostatin receptor subtypes in acromegaly: Distinct patterns of messenger ribonucleic acid expression and hormone suppression identify different tumoral phenotypes, *J. Clin. Endocrinol. Metab.*, 85, 781–792, 2000.

Jaquet, P. et al., Efficacy of chimeric molecules directed towards multiple somatostatin and dopamine receptors on inhibition of GH and prolactin secretion from GH-secreting pituitary adenomas classified as partially responsive to somatostatin analog therapy, *Eur. J. Endocrinol.*, 153, 135–141, 2005.

Jenkins, S.A. et al., A multicentre randomised trial comparing octreotide and injection sclerotherapy in the management and outcome of acute variceal haemorrhage, *Gut*, 41, 526–533, 1997.

Keri, G. et al., A tumor-selective somatostatin analog (TT-232) with strong *in vitro* and *in vivo* antitumor activity, *Proc. Natl Acad. Sci. USA*, 93, 12513–12518, 1996.

Kiaris, H. et al., A targeted cytotoxic somatostatin (SST) analogue, AN-238, inhibits the growth of H-69 small-cell lung carcinoma (SCLC) and H-157 non-SCLC in nude mice, *Eur. J. Cancer*, 37, 620–628, 2001.

Krejs, G.J., Browne, R., and Raskin, P., Effect of intravenous somatostatin on jejunal absorption of glucose, amino acids, water, and electrolytes, *Gastroenterology*, 78, 26–31, 1980.

Krenning, E.P. et al., Somatostatin receptor scintigraphy with [111In-DTPA-D-Phe1]- and [123I-Tyr3]-octreotide: The Rotterdam experience with more than 1000 patients, *Eur. J. Nucl. Med.*, 20, 716–731, 1993.

Krulich, L., Dhariwal, A.P., and McCann, S.M., Stimulatory and inhibitory effects of purified hypothalamic extracts on growth hormone release from rat pituitary *in vitro*, *Endocrinology*, 83, 783–790, 1968.

Kulaksiz, H. et al., Identification of somatostatin receptor subtypes 1, 2A, 3, and 5 in neuroendocrine tumours with subtype specific antibodies, *Gut*, 50, 52–60, 2002.

Kvols, L.K. et al., Treatment of the malignant carcinoid syndrome. Evaluation of a long-acting somatostatin analogue, *New Engl. J. Med.*, 315, 663–666, 1986.

Kwekkeboom, D., Krenning, E.P., and de Jong, M., Peptide receptor imaging and therapy, *J. Nucl. Med.*, 41, 1704–1713, 2000.

Lahlou, H. et al., Molecular signaling of somatostatin receptors, *Ann. NY Acad. Sci.*, 1014, 121–131, 2004.

Lamberts, S.W., van der Lely, A.J., de Herder, W.W., and Hofland, L.J., Octreotide, *New Engl. J. Med.*, 334, 246–254, 1996.

Larsson, L.I. et al., Pancreatic somatostatinoma. Clinical features and physiological implications, *Lancet*, 1, 666–668, 1977.

Lattuada, D., Casnici, C., Venuto, A., and Marelli, O., The apoptotic effect of somatostatin analogue SMS 201–995 on human lymphocytes, *J. Neuroimmunol.*, 133, 211–216, 2002.

Lawnicka, H. et al., Effect of somatostatin and octreotide on proliferation and vascular endothelial growth factor secretion from murine endothelial cell line (HECa10) culture, *Biochem. Biophys. Res. Commun.*, 268, 567–571, 2000.

Maecke, H.R., Hofmann, M., and Haberkorn, U., (68)Ga-labeled peptides in tumor imaging, *J. Nucl. Med.*, 46, Suppl. 1, 172S–178S, 2005.

Maton, P.N., The use of the long-acting somatostatin analogue, octreotide acetate, in patients with islet cell tumors, *Gastroenterol. Clin. North Am.*, 18, 897–922, 1989.

Mulvihill, S., Pappas, T.N., Passaro, E. Jr, and Debas, H.T., The use of somatostatin and its analogs in the treatment of surgical disorders, *Surgery*, 100, 467–476, 1986.

Oberg, K., Established clinical use of octreotide and lanreotide in oncology, *Chemotherapy*, 47, Suppl. 2, 40–53, 2001.

O'Carroll, A.M., Lolait, S.J., Konig, M., and Mahan, L.C., Molecular cloning and expression of a pituitary somatostatin receptor with preferential affinity for somatostatin-28, *Mol. Pharmacol.*, 42, 939–946, 1992.

O'Toole, D. et al., Treatment of carcinoid syndrome: A prospective crossover evaluation of lanreotide versus octreotide in terms of efficacy, patient acceptability, and tolerance, *Cancer*, 88, 770–776, 2000.

Pages, P. et al., sst2 Somatostatin receptor mediates cell cycle arrest and induction of p27(Kip1). Evidence for the role of SHP-1, *J. Biol. Chem.*, 274, 15186–15193, 1999.

Panzuto, F. et al., Long-term clinical outcome of somatostatin analogues for treatment of progressive, metastatic, well-differentiated entero-pancreatic endocrine carcinoma, *Ann. Oncol.*, 17, 461–466, 2006.

Papotti, M. et al., Expression of somatostatin receptor types 1–5 in 81 cases of gastrointestinal and pancreatic endocrine tumors. A correlative immunohistochemical and reverse-transcriptase polymerase chain reaction analysis, *Virchows Arch*, 440, 461–475, 2002.

Park, C. et al., Somatostatin (SRIF) receptor subtype 2 and 5 gene expression in growth hormone-secreting pituitary adenomas: The relationship with endogenous SRIF activity and response to octreotide, *Endocrinol. J.*, 51, 227–236, 2004.

Patel, Y.C., Somatostatin and its receptor family, *Front Neuroendocrinol.*, 20, 157–198, 1999.

Plöckinger, U. and Wiedenmann, B., Best practice and research, *Clin. Endocrinol. Metabol.*, 21, 145–162, 2007.

Plockinger, U., Dienemann, D., and Quabbe, H.J., Gastrointestinal side-effects of octreotide during long-term treatment of acromegaly, *J. Clin. Endocrinol. Metab.*, 71, 1658–1662, 1990.

Plockinger, U. et al., Effect of the somatostatin analog octreotide on gastric mucosal function and histology during 3 months of preoperative treatment in patients with acromegaly, *Eur. J. Endocrinol.*, 139, 387–394, 1998.

Plonowski, A., Schally, A.V., Nagy, A., Sun, B., and Szepeshazi, K., Inhibition of PC-3 human androgen-independent prostate cancer and its metastases by cytotoxic somatostatin analogue AN-238, *Cancer Res.*, 59, 1947–1953, 1999.

Reichlin, S., Somatostatin, *New Engl. J. Med.*, 309, 1495–1501, 1983a.

Reichlin, S., Somatostatin (second of two parts), *New Engl. J. Med.*, 309, 1556–1563, 1983b.

Reubi, J.C. and Waser, B., Concomitant expression of several peptide receptors in neuroendocrine tumours: Molecular basis for *in vivo* multireceptor tumour targeting, *Eur. J. Nucl. Med. Mol. Imaging*, 30, 781–793, 2003.

Reubi, J.C., Waser, B., Schaer, J.C., and Laissue, J.A., Somatostatin receptor sst1–sst5 expression in normal and neoplastic human tissues using receptor autoradiography with subtype-selective ligands, *Eur. J. Nucl. Med.*, 28, 836–846, 2001.

Rogers, B.E. et al., Targeted radiotherapy with [(90)Y]-SMT 487 in mice bearing human nonsmall cell lung tumor xenografts induced to express human somatostatin receptor subtype 2 with an adenoviral vector, *Cancer*, 94, Suppl., 1298–1305, 2002.

Rosskopf, D. et al., Signal transduction of somatostatin in human B lymphoblasts, *Am. J. Physiol. Cell Physiol.*, 284, C179–C190, 2003.

Ruszniewski, P. et al., Treatment of the carcinoid syndrome with the long-acting somatostatin analogue lanreotide: A prospective study in 39 patients, *Gut*, 39, 279–283, 1996.

Saltz, L. et al., Octreotide as an antineoplastic agent in the treatment of functional and nonfunctional neuroendocrine tumors, *Cancer*, 72, 244–248, 1993.

Saveanu, A. et al., A luteinizing hormone-, alpha-subunit- and prolactin-secreting pituitary adenoma responsive to somatostatin analogs: *In vivo* and *in vitro* studies, *Eur. J. Endocrinol.*, 145, 35–41, 2001a.

Saveanu, A. et al., Bim-23244, a somatostatin receptor subtype 2- and 5-selective analog with enhanced efficacy in suppressing growth hormone (GH) from octreotide-resistant human GH-secreting adenomas, *J. Clin. Endocrinol. Metab.*, 86, 140–145, 2001b.

Saveanu, A. et al., Demonstration of enhanced potency of a chimeric somatostatin–dopamine molecule, BIM-23A387, in suppressing growth hormone and prolactin secretion from human pituitary somatotroph adenoma cells, *J. Clin. Endocrinol. Metab.*, 87, 5545–5552, 2002.

Schally, A.V., Oncological applications of somatostatin analogues, *Cancer Res.*, 48, Pt. 1, 6977–6985, 1988.

Schally, A.V. et al., New approaches to therapy of cancers of the stomach, colon, and pancreas based on peptide analogs, *Cell Mol. Life Sci.*, 61, 1042–1068, 2004.

Schmid, H.A. and Schoeffter, P., Functional activity of the multiligand analog SOM230 at human recombinant somatostatin receptor subtypes supports its usefulness in neuroendocrine tumors, *Neuroendocrinology*, 80, Suppl. 1, 47–50, 2004.

Sharma, K., Patel, Y.C., and Srikant, C.B., Subtype-selective induction of wild-type p53 and apoptosis, but not cell cycle arrest, by human somatostatin receptor 3, *Mol. Endocrinol.*, 10, 1688–1696, 1996.

Shimon, I. et al., Somatostatin receptor (SSTR) subtype-selective analogues differentially suppress *in vitro* growth hormone and prolactin in human pituitary adenomas. Novel potential therapy for functional pituitary tumors, *J. Clin. Invest.*, 100, 2386–2392, 1997.

Siegel, R.A., Tolcsvai, L., and Rudin, M., Partial inhibition of the growth of transplanted dunning rat prostate tumors with the long-acting somatostatin analogue sandostatin (SMS 201–995), *Cancer Res.*, 48, 4651–4655, 1988.

Strowski, M.Z., Parmar, R.M., Blake, A.D., and Schaeffer, J.M., Somatostatin inhibits insulin and glucagon secretion via two receptors subtypes: An *in vitro* study of pancreatic islets from somatostatin receptor 2 knockout mice, *Endocrinology*, 141, 111–117, 2000.

Susini, C. and Buscail, L., Rationale for the use of somatostatin analogs as antitumor agents, *Ann. Oncol.*, 17, 1733–1742, 2006.

Szende, B. and Keri, G., TT-232: A somatostatin structural derivative as a potent antitumor drug candidate, *Anticancer Drugs*, 14, 585–588, 2003.

Thoss, V.S., Perez, J., Probst, A., and Hoyer, D., Expression of five somatostatin receptor mRNAs in the human brain and pituitary, *Naunyn Schmiedebergs Arch Pharmacol.*, 354, 411–419, 1996.

Tomassetti, P., Migliori, M., and Gullo, L., Slow-release lanreotide treatment in endocrine gastrointestinal tumors, *Am. J. Gastroenterol.*, 93, 1468–1471, 1998.

Tomassetti, P., Migliori, M., Corinaldesi, R., and Gullo, L., Treatment of gastroenteropancreatic neuroendocrine tumours with octreotide LAR, *Aliment. Pharmacol. Ther.*, 14, 557–560, 2000.

Trendle, M.C., Moertel, C.G., and Kvols, L.K., Incidence and morbidity of cholelithiasis in patients receiving chronic octreotide for metastatic carcinoid and malignant islet cell tumors, *Cancer*, 79, 830–834, 1997.

Trieber, G., Wex, T., and Malfertheiner, P., Inhibition of angiogenesis rather than growth hormones is a key factor for octreotide treatment response in HCC patients, in *ASCO*, Abstract 91, 2004.

Vale, W., Rivier, J., Ling, N., and Brown, M., Biologic and immunologic activities and applications of somatostatin analogs, *Metabolism*, 27, Suppl. 1, 1391–1401, 1978.

van der Hoek, J. et al., A single-dose comparison of the acute effects between the new somatostatin analog SOM230 and octreotide in acromegalic patients, *J. Clin. Endocrinol. Metab.*, 89, 638–645, 2004.

van der Hoek, J. et al., Distinct functional properties of native somatostatin receptor subtype 5 compared with subtype 2 in the regulation of ACTH release by corticotroph tumor cells, *Am. J. Physiol. Endocrinol. Metab.*, 289, E278–E287, 2005.

Vernejoul, F. et al., Antitumor effect of *in vivo* somatostatin receptor subtype 2 gene transfer in primary and metastatic pancreatic cancer models, *Cancer Res.*, 62, 6124–6131, 2002.

Vinik, A. and Moattari, A.R., Use of somatostatin analog in management of carcinoid syndrome, *Dig. Dis. Sci.*, 34, Suppl., 14S–27S, 1989.

Weckbecker, G., Raulf, F., Stolz, B., and Bruns, C., Somatostatin analogs for diagnosis and treatment of cancer, *Pharmacol. Ther.*, 60, 245–264, 1993.

Welin, S.V. et al., High-dose treatment with a long-acting somatostatin analogue in patients with advanced midgut carcinoid tumours, *Eur. J. Endocrinol.*, 151, 107–112, 2004.

Woltering, E.A. et al., Somatostatin analogs: Angiogenesis inhibitors with novel mechanisms of action, *Invest. New Drugs*, 15, 77–86, 1997.

Yamada, Y. et al., Cloning and functional characterization of a family of human and mouse somatostatin receptors expressed in brain, gastrointestinal tract, and kidney, *Proc. Natl Acad. Sci. USA*, 89, 251–255, 1992.

Yamada, Y., Ito, S., Matsubara, Y., and Kobayashi, S., Immunohistochemical demonstration of somatostatin-containing cells in the human, dog, and rat thyroids, *Tohoku J. Exp. Med.*, 122, 87–92, 1977.

9 Tachykinins

Nigel M. Page, Parvathy Subramaniam, and Suzanne E. Newton

CONTENTS

The tachykinins represent the largest known peptide family and are responsible for a range of pleiotropic functions in both vertebrates and invertebrates. They are characterized by possessing the C-terminal motif FXGLM-NH_2, which is evolutionarily conserved, and elicits their biological functions through interaction with the tachykinin receptors, which in mammals are NK_1, NK_2, and NK_3. Until recently, these peptides were believed to be confined to the nervous system; however, it has become apparent that some of the most predominant sites of expression are actually in the periphery. A plethora of biological responses are associated with the tachykinins; here we review their role in the brain, adrenal gland, and reproductive systems. A number of promising NK antagonists have been developed, of which aprepitant is licenced for use in the treatment of chemotherapy-induced emesis.

9.1 INTRODUCTION

The tachykinins form the largest known peptide family and are responsible for a range of pleiotropic functions in both vertebrates and invertebrates (Severini et al., 2002). Their name derives from their ability to cause fast (*tachy*) contractions (*kinin*) on smooth muscle preparations. For many years, the tachykinins were considered to be almost exclusively peptides of neuronal origin and were subsequently often termed neurokinins (Patacchini et al., 2004). Recently, however, this dogma has been challenged and abundant expression has been found in a range of peripheral cells and tissues, including immune and inflammatory cells (Maggi, 1997) and even in the placenta, a tissue totally devoid of nerves (Page et al., 2000, 2003). They are involved in a wide range of biological processes such as immunomodulation, neurogenic inflammation and nociception, smooth muscle contraction, vasodilatation, hematopoiesis, and stimulation of endocrine gland secretion.

The mammalian members of the tachykinins include substance P (SP) (von Euler and Gaddum, 1931; Chang and Leeman, 1970), neurokinin A (NKA) (Kimura et al., 1983; Nawa et al., 1983; Minamino et al., 1984), and neurokinin B (NKB) (Kangawa et al., 1983; Kimura et al., 1983), as well as the recently discovered hemokinin-1 (HK-1) (Zhang et al., 2000) and the endokinins (A and B)

(EK) (Page et al., 2003; Page, 2004). In total, so far, over 50 tachykinins have been described from mammals, submammals, and invertebrates (Severini et al., 2002). Most are short peptides ~10 amino-acid residues in length; however, biologically active N-terminally extended forms of NKA, neuropeptide K (NPK) (Tatemoto et al., 1985), and neuropeptide gamma (NPγ) (Kage et al., 1988) have been reported to exist. The classical tachykinin peptide presents an amphipathic structure; a hydrophobic, methionine-amidated conserved C-terminal motif, FXGLM-NH_2, and a hydrophilic N-terminal of divergent sequence (Page, 2004), although some minor exceptions to this rule have been found in some nonmammalian tachykinins (Severini et al., 2002). The amidated C-terminal motif is essential for receptor interaction (Olham et al., 1997), whereas the unique N-terminal motif dictates receptor specificity (Page, 2004). Mammalian tachykinins can be classified into either aromatic or aliphatic tachykinins, based upon the residue in position X of the C-terminal motif. Aromatic tachykinins usually have a phenylalanine or tyrosine at position X, and include SP, HK-1, and EKA/B (Page, 2004). Aliphatic tachykinins include neurokinin A and B, which have the aliphatic residues of either valine or isoleucine at position X (Page, 2004). Tachykinins in mammals elicit their biological responses by interaction with three G-protein coupled receptors, namely NK_1, NK_2, and NK_3 (Section 9.3).

The voyage of discovery to the current number of known mammalian tachykinins spans 73 years from 1931 to 2004 (Page, 2004). In 1931, von Euler and Gaddum first reported an unidentified substance found in alcoholic extracts of equine brain and intestine that elicited pharmacological actions such as hypotension, vasodilatation, and smooth muscle contraction in rabbit tissue preparations. In 1934, Gaddum and Schild named this substance "substance P." However, attempts to produce a pure preparation of SP were unsuccessful until 36 years later, when Leeman and Hammerschlag (1967), whilst attempting to isolate corticotrophin-releasing factor stumbled upon a peptide, sialogen, which stimulated salivary secretion in anesthetized rats. It was Lembeck and Starke (1968) who realized that sialogen was indeed the same as SP, and this was further confirmed in 1970 by comparison of their biological properties (Chang and Leeman, 1970). The next year, Chang et al. (1971) established the structure and determined the sequence of SP from bovine hypothalamus. The early 1980s saw the confirmation of the existence of two additional members, NKA and NKB. By the 1990s, it could have been concluded that there were no more mammalian tachykinins remaining to be discovered, as what appeared to be a complete picture of three tachykinins each with their own preferred receptor had become apparent (Page, 2004). Nevertheless, the turn of the new century was to reveal a completely new and diverse group of tachykinins in mammals and, more unexpectedly, tachykinin gene-related peptides (Zhang et al., 2000; Page et al., 2003; Page, 2004). A further unexpected find was the release of a biologically active tachykinin, virokinin, from the furin-mediated cleavage of bovine respiratory syncytial virus (Zimmer et al., 2003). One of the most challenging future aspects of tachykinin research will be dissecting distinct roles for each of the tachykinins.

A parallel tale of the discovery of the nonmammalian tachykinins also developed, starting in 1949, when Erspamer identified eledoisin from the salivary glands of the Mediterranean octopus, *Eledone moschata*. By 1964, another eledoisin-like peptide had been discovered from the skin of the South American frog *Physalaemus biligonigerus* and was named physalaemin (Anastasi et al., 1964). This soon lead to a plethora of other related peptides being characterized from amphibians, mollusks, and fish, followed by the confirmation of authentic invertebrate tachykinins in the mosquito and migratory locust (Severini et al., 2002). Studies on these nonmammalian tachykinins have proven indispensable in understanding the roles of the mammalian tachykinins. For instance, they provided hints to the existence of NKA and NKB long before they were even isolated (Erspamer et al., 1980). It should also be noted that peptides with some sequence homology (30 to 45%) to the classical tachykinins already described have been identified from a range of invertebrates such as insects, echiuroid worms, and crustaceans (Nässel, 1999). These are characterized by a conserved C-terminal hexapeptide motif –GFX_1GX_2R–NH_2 (where X_1 is Phe, His, Leu, Met, Gln, or Tyr and X_2 is Met, Thr, or Val) (Nässel, 1999; Severini et al., 2002) and for this reason have been named tachykinin-related peptides.

As tachykinins and their receptors are implicated in various centrally and peripherally mediated processes, the application of tachykinin receptor antagonists has become a main focus of research. Extensive research has been undertaken to determine the role of the tachykinins, by blocking and inactivating the receptors to prevent inflammatory associated diseases such as preeclampsia, rheumatoid arthritis, asthma, allergic rhinitis, chronic obstructive pulmonary disease, and irritable bowel syndrome (reviewed in Lecci and Maggi, 2003). Furthermore, application of tachykinin receptor antagonists may represent a mode of treatment for centrally mediated and neurological disorders such as nociception, anxiety, depression, schizophrenia, bipolar mood, and neurodegenerative disorders (e.g., Parkinson's and Alzheimer's disease; Page, 2006b). To date, aprepitant (MK-869), an NK_1 receptor antagonist, is the only antagonist registered for patient use for the prevention of acute and delayed chemotherapy-induced nausea and vomiting (Section 9.6) (de Wit et al., 2004).

9.2 SOURCES OF PEPTIDES

The regional distribution of the tachykinins is ultimately affected by the alternative splicing of the tachykinin (*TAC*) mRNA transcripts and the manner in which the TAC precursors are processed. There are three different mammalian tachykinin genes: *TAC*1, *TAC*3, and *TAC*4 (Page, 2004). The transcription of *TAC*1 produces four different transcripts, α, β, γ, and δ, due to alternative splicing (Krause et al., 1987; Harmar et al., 1990). This gives rise to the possibility of producing four different peptides, namely SP, NKA, NPK, and NPγ. Substance P is encoded by all four *TAC*1 transcripts and therefore wherever *TAC*1 is expressed it is expected SP will be present. In the case of the other encoded tachykinins, NKA is encoded by β and γ*TAC*1, NPK by β*TAC*1, and NPγ by γ*TAC*1 (Krause et al., 1987; Harmar et al., 1990). In most locations, it has been shown that β– and γ*TAC*1 are the most abundantly expressed, and therefore SP and NKA are normally expected to be cosynthesized and released together.

Three alternative splice variants of *TAC*3 are known (α, β, and γ); nevertheless, these are found to encode only the single tachykinin, NKB, and no further information on the differential expression of the *TAC*3 transcripts is currently available (Page, 2004). *TAC*4 was first cloned in hematopoietic cells in mice, and its expression was believed to be limited to immune tissues (Zhang et al., 2000). In mice and rats, a single *TAC*4 transcript is found to encode only the single tachykinin HK-1 (Zhang et al., 2000; Page, 2004). In humans, four alternatively spliced variants are found, α, β, γ, and δ, which all encoded EKB. An additional, albeit rare, splice variant at the end of exon 1 encodes EKA; also, α*TAC*4 encodes EKC and β*TAC*4 encodes EKD (Page et al., 2003), of which the latter are classified as tachykinin gene-related peptides as they have a transition from the classical tachykinin motif (i.e., FXGLM–NH_2 to FQGLL–NH_2) (Page et al., 2003). Tissue-specific expression of *TAC*4 is seen, with α- and γTAC4 most abundantly expressed (Page et al., 2003; Page, 2004).

Tachykinins have been found widely distributed in both the central and peripheral nervous systems (Page, 2006a). Substance P is predominantly found in the brain (Figure 9.1), is distributed throughout the central nervous system (CNS), and is also located in primary afferent sensory neurons innervating a significant number of peripheral tissues (Holzer, 1988; Maggi and Meli, 1988; Lundberg, 1996). Neurokinin B, until recently, was believed to be present only in the CNS and the spinal cord (Kangawa et al., 1983; Moussaoui et al., 1992). In the brain, there are different patterns of distribution of *TAC*1 and *TAC*3 (reviewed in Page, 2006b). *TAC*4, although predominantly found in the periphery, has also been detected in some regions of the brain with a similar expression pattern to *TAC*1; however, it is present at consistently lower expression levels (Figure 9.1; Duffy et al., 2003; Page et al., 2003).

At the peripheral level, the primary source of tachykinins is considered to be the capsaicin-sensitive sensory nerves (Janscó et al., 1977); however, it has become apparent in recent years that tachykinins exist in other locations within the periphery. Within the nervous system these locations include expression by capsaicin-resistant large neurons bearing Aβ fibers present in sensory nerve

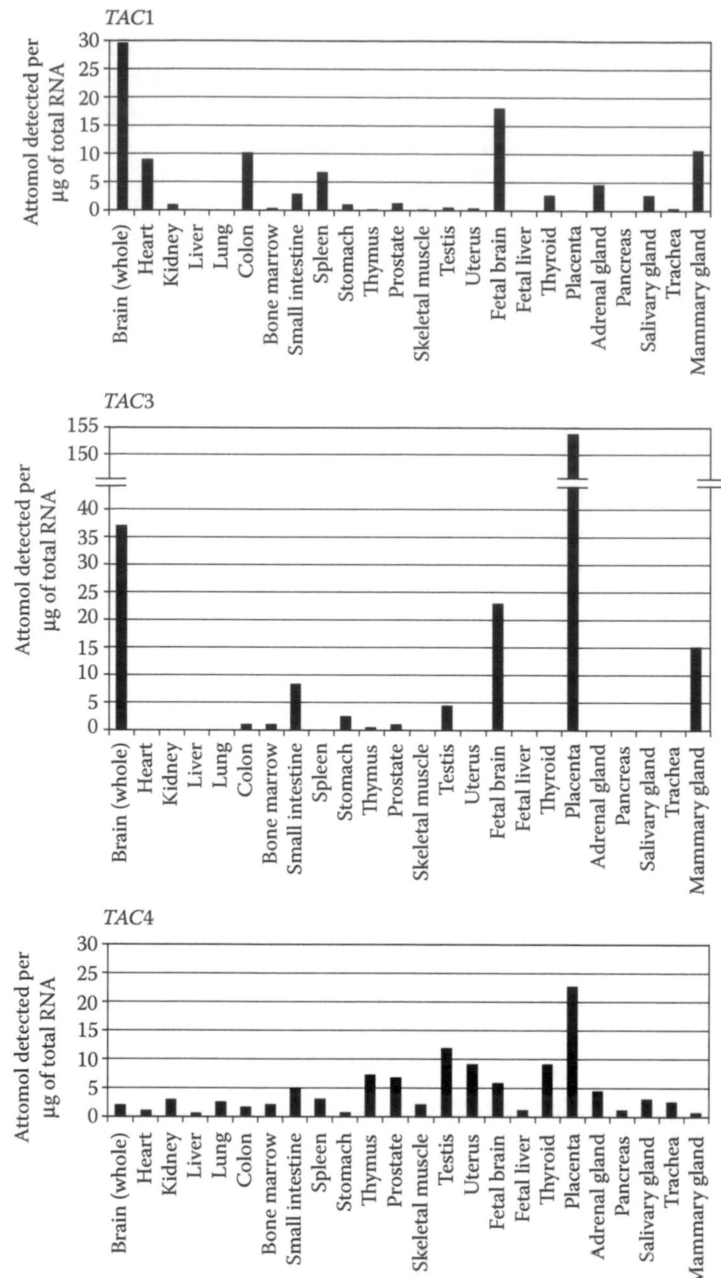

FIGURE 9.1 Real-time quantitative PCR analysis of the tissue distribution of tachykinin *TAC*1, *TAC*3, and *TAC*4 gene expression in 24 human tissues.

fibers following inflammatory stimuli (Neumann et al., 1996), and also in the enteric nervous system (Holzer and Holzer-Petsche, 1997). Notwithstanding, recent advances have challenged the notion that the tachykinins are exclusively of neuronal origin. For example, the most abundant site of expression of NKB has been found to be the placenta, a tissue that has no nervous innervation (Page et al., 2006; Figure 9.1). In the same way, *TAC*4 is also found to be most abundantly expressed in the placenta and in a wide range of peripheral cells and tissues (Page et al., 2003; Page, 2004; Figure 9.1). This has led to the name endokinin being given to the EK peptides to indicate that they may have endocrine rather than neuronal roles. Overall, tachykinins have been found in inflammatory cells such as dendritic cells, eosinophils, macrophages, and lymphocytes (Joos et al., 2000), in Leydig cells (Chiwakata et al., 1991), in the luteal cells and interstitial cells of the ovary (Debeljuk, 2006), in bone marrow cells (Murthy et al., 2007), in the syncytiocytotrophoblasts of the placenta (Page et al., 2000), in platelets (Graham et al., 2004), and in enterochromaffin cells of the intestinal mucosa and in fibroblasts (Bae et al., 2002).

Just as important as the locations of the tachykinins is the colocalization of the systems that can process and degrade them. Tachykinins are produced from large inactive preprotachykinin (TAC) precursor proteins and are secreted through the classical secretory pathway. TACs have an N-terminal hydrophobic signal peptide sequence, which directs the protein to the endoplasmic reticulum where the signal peptide sequence is removed. Further cleavage of the precursor occurs at designated cleavage sites (normally Lys and/or Arg) by specific enzymes that include the prohormone convertases. The products are further post-translationally modified by the action of carboxypeptidases that remove the C-terminal dibasic residues, allowing the action of the peptidylglycine α-amidating enzyme that converts the exposed C-terminal glycine residue into an amide (Page, 2004).

After the tachykinins have been released, they may be cleaved and inactivated by various proteolytic enzymes (Severeni et al., 2002). Substance P is more susceptible to proteolytic attack, whereas tachykinins with a *pyro*-Glu residue at their N-terminus (e.g., physalaemin) are more resistant to enzymatic degradation. Various enzymes are known to participate in the degradation of SP, including neutral endopeptidase (NEP), SP-degrading enzyme (SPDE), angiotensin-converting enzyme (ACE), dipeptidyl aminopeptide IV (DPIV), postproline endopeptidase (PEP), cathepsin-D, and cathepsin-E (Severini et al., 2002). In the brain, spinal cord, and periphery, NEP degrades SP, whereas in plasma, cerebrospinal fluid and the *substantia nigra* ACE catalyses the hydrolysis of SP (Harrison and Geppetti, 2001). ACE cleaves SP at two sites, between Phe^8 and Gly^9, and Gly^9 and Leu^{10}, whereas NEP cleaves between Gly^6 and Phe^7, and Gly^9 and Leu^{10} (Table 9.1) (Khawaja and Rogers, 1996).

TABLE 9.1
Peptide Sequences of Some Key Example Endogenous Tachykinins, Along with Those of the Tachykinin Gene-Related Peptides and the Tachykinin Related Peptides the Underlined Amino-Acid Residues Represent the Defining Motif

Tachykinins	Peptide Sequence	Accession Number
SP	RPKPQQFFGLM-NH_2	P67933
NKA	HKTDSFVGLM-NH_2	P19851
NPK	DADSSIEKQVALLKALYGHGQISHKRHKTDSFVGLM-NH_2	1105207A
NPγ	DAGHGQISHKRHKTDSFVGLM-NH_2	NP_054703
NKB	DMHDFFVGLM-NH_2	P67934
EKA/B	TGKASQFFGLM-NH_2	AAP30874
HK-1	SRTRQFYGLM-NH_2	NP_758831
Eledoisin	QPSKDAFIGLM-NH_2	P62934
Physalaemin	QADPNKFYGLM-NH_2	P08615
Tachykinin Gene-Related Peptides		
EKC	KKAYQLEHTFQGLL-NH_2	AAP30874
EKD	VGAYQLEHTFQGLL-NH_2	AAP30875
Tachykinin-Related Peptides		
LomTK1	GPSGFYGVR-NH_2	P16223
UruTK1	LRQSQFVGSR-NH_2	P40751

9.3 MOLECULAR MECHANISM OF ACTION

There are three tachykinin receptors in mammals, termed NK_1, NK_2, and NK_3 (Page et al., 2003). Receptors equivalent to the mammalian NK_1 receptor have been isolated from chicken, bull frog (Simmons et al., 1997), the cane toad (Liu et al., 2004), as well as from an invertebrate, the echiuroid worm (Kawada et al., 2002). All the mammalian tachykinin receptors share a similar overall structure comprising an extracellular amino terminal and an intracellular carboxyterminal bridged by seven domains spanning the cell membrane coupled to a G-protein (Strader et al., 1994). The NK_1 receptor is 47% identical to the NK_2 receptor; the NK_2 is 41% identical to the NK_3 receptor; and the NK_1 receptor is 51% identical to the NK_3 receptor (Almeida et al., 2004). The greatest identity (70%) is found between the receptors in the transmembrane domains (Gerard et al., 1993).

Each of the three traditional tachykinins (SP, NKA, and NKB) will ligand bind with each of the receptors, but with varying affinities. Substance P will preferentially bind with NK_1, NKA with NK_2, and NKB with NK_3. The EKs and HK-1 have a SP-like binding profile and so preferentially ligand bind with NK_1 (Page et al., 2003; Berger and Paige, 2005). Nevertheless, there are cases where, even in the presence of SP, NKA seems to act as the endogenous ligand of NK_1 receptors, for example in the rabbit iris and the rat spinal cord (Maggi, 2000; Trafton et al., 2001) or where SP appears more potent in stimulating NK_2 than NK_1 receptors, which is the case in hematopoietic progenitor cells (Bandari et al., 2003). In general, activation by tachykinins elicits several cellular responses. The most noted response is the accumulation of inositol phosphate induced by coupling of the receptors to $G_{q/11}$ (Mochizuki-Oda et al., 1994). However, activation of alternative G proteins has also been observed, including G_s, which increases cAMP, and G_o, which mobilizes arachidonic acid (Torrens et al., 1998).

There is evidence to suggest that binding sites on the mammalian NK_1 receptor for SP, NKA, and NKB are not distinct, but overlap and may involve different residues that are close in three dimensions (Strader et al., 1994). The presence of conserved groups of amino acids on the extracellular amino terminal (residues 23, 24, and 25) and first intracellular loop (residues 96 and 108) are thought to interact with the common carboxy terminals of the tachykinins, responsible for their common biological activity (Watling and Krause, 1993). A group of residues on the third extracellular loop is also required for high-affinity binding of the endogenous peptides (Fong et al., 1992b). As exemplified by the interaction of SP at the NK_1 receptor, the C-terminal sequence of each tachykinin is essential for activity, and the minimum length of a fragment with reasonable affinity for the tachykinin NK_1 receptor is the carboxy-terminal hexapeptide. Cloned human and rat tachykinin NK_1 receptors show 95% homology, with the majority of the differing residues localised at the C- and N-termini. Only six amino-acid residues differ between the human and rat tachykinin NK_1 receptors in the transmembrane segments (Sundelin et al., 1992). With the one exception of residue 226 in transmembrane segment six (valine in rat and isoleucine in mouse), the mouse and rat tachykinin NK_1 receptors have identical transmembrane regions (Sundelin et al., 1992). An important point here is that species-related differences are not evident when testing SP or other natural tachykinins. The variable affinities of antagonists at NK_1 receptors from different species are linked to species-dependent variations in the amino-acid sequence of the NK_1 receptor protein. The binding sites of some antagonists on NK_1 receptor proteins are at least partially nonoverlapping or partially distinct from binding sites of the agonists (Quartara and Maggi, 1997).

Also worthy of note is that a truncated form of the NK_1 receptor has also been reported, which is missing the C-terminal cytosolic region after the seventh transmembrane domain (Fong et al., 1992a; Mantyh et al., 1996). Such an alteration has been shown to change its effector systems (Fong et al., 1992a). The C-terminal tail also contains a series of sites that mediate the phosphorylation and internalization of the receptor directly following agonist binding and activation (Bohm et al., 1997; Vigna, 1999). Recently it has been shown that the transfection of the truncated but not the standard form of NK_1 receptors in breast cells can induce a transformed phenotype (Patel et al., 2005). This would suggest an alteration in G-protein interaction; however a distinct role for this subtype has yet to be determined, although its apparent predominance in the periphery may suggest it plays a role here (Page, 2006a).

The binding site for the NK_2 receptor appears to be after the fifth transmembrane domain, as NK_2 substitution at this segment increases the affinity of the NK_1 receptor for NKA (Gerard et al., 1993). The gene encoding the NK_2 receptor has been defined as *TACR*2 and yields two splice variants: the alpha isoform is the classical NK_2 receptor, whereas the beta-isoform lacks the amino acid sequence that forms the second intracellular loop to the third extracellular loop that includes the fourth transmembrane domain. Therefore, this latter deletion causes the tachykinins to neither bind to the beta isoform nor activate intracellular signals. Although it can be excluded as a functional tachykinin receptor, there is a possibility that it could be involved in other cellular functions (Bellucci et al., 2004). Neurokinin B interaction with the NK_3 receptor appears to be towards the carboxy terminal, within the seventh transmembrane domain and the third extracellular loop (Fong et al., 1992b; Gether et al., 1993). As a note, substitution of valine with glutamate in residue 97 (the first extracellular loop) of the NK_1 receptor confers on it a higher affinity for NKB. Furthermore, the NK_3 receptor has a glutamate residue at this position, indicating an interaction point for NKB in the first extracellular loop (Fong et al., 1992b).

9.4 BIOLOGICAL AND PATHOLOGICAL MODES OF ACTION

The widespread distribution of the tachykinins has led them to be implicated in a diverse range of biological actions, including, but by no means limited to, smooth muscle contraction, vasodilatation, nociception, immunomodulation, neuroprotection, neurogenic inflammation, mitogenesis, reproduction, gastrointestinal regulation, and hematopoiesis. Here, we will review a few of the aspects that are pertinent to some of our own interests. Perhaps the most notable is their role in the brain where they are believed to play very important roles as neurotransmitter, neuromodulatory, and neurotrophic agents. They occur particularly in large quantities in the areas involved in the control of several autonomic and endocrine functions, of affective and emotion responses, and of higher cerebral functions. Several reports have suggested that SP evokes the release of several other important modulators including dopamine, Met-enkephalin, and acetylcholine (Page, 2006b). Activation of these has been shown to increase locomotor activity and exert effects on memory-promoting activities. In addition, a number of tachykinin actions in the brain are demonstrated by direct intracerebroventricular injection. These include cardiovascular responses promoted by SP and NKA typical of defence reactions with associated increases in blood pressure, heart rate, visceral vasoconstriction, sympathetic efferent activity, and increased states of anxiety (Ebner and Singewald, 2006). Furthermore, the effects of noxious stimuli at the central and peripheral nervous system are accompanied by the release of tachykinins from the capsaicin-sensitive neurons of both the spinal cord and peripheral nervous systems causing pain, vasodilatation, and edema formation, observations defined under the umbrella of neurogenic inflammation (Bradesi et al., 2003).

The tachykinins have been implicated in a number of centrally mediated disease states. These have included, among others, the neurodegenerative disorders of Alzheimer's (Rösler et al., 2001) and Parkinson's disease (Chen et al., 2004), and several affective disorders (Rosenkranz, 2007). In the case of Alzheimer's disease, SP function has been related to memory and learning (Herpfer et al., 2007), where it is highly expressed in several areas of the brain from the forebrain to hindbrain. In this regard, the basal forebrain, particularly the nucleus of Meynert, has received attention, as it is these SP-containing nuclei that have been shown to degenerate in the senile dementia of Alzheimer's disease. Substance P has also been found to interact with the cholinergic ascending system of the nucleus of Meynert, and patients with Alzheimer's disease show a marked loss of cholinergic neurons from the nucleus of Meynert to the cortex (Nakajima et al., 1985) and diminished brain SP immunoreactivity. In Parkinson's disease, which is a serious motor disorder, degeneration of the dopamine neurons in the *substantia nigra* occurs (Wolters, 2006). Lines of evidence indicate a role for the tachykinins in its pathogenesis, as seen by the decrease of *TAC*1 and

SP-immunoreactivity found in the nigral and striatal regions of animal and postmortem Parkinson's disease patients (Tenovuo et al., 1984) and the neuroprotective effects exerted on neurons by tachykinins (Chen et al., 2004). In the case of the affective disorders, SP antagonists have been shown to suppress anxiety-related behaviors and defensive cardiovascular changes in rodents. However, research to ascertain the effectiveness of aprepitant (MK-869) to treat depression (Section 9.6) in phase III trials showed it to be no more effective than placebo. The debate as to whether tachykinins are involved in and could prove potential therapeutic agents in a number of psychiatric illnesses remains open. A list of potential target illnesses includes schizophrenia, panic disorder, generalized anxiety disorder, posttraumatic stress disorder, and obsessive-compulsive disorder. Additional work in the brain is also being performed to determine the role of the tachykinins in migraine (May and Goadsby, 2001), malignant brain tumor therapies (Lazarczyk et al., 2007), and traumatic brain injury (Donkin et al., 2007).

In addition to the important role of SP as a central agent in controlling the transition between acute and chronic stress, there is evidence that the tachykinins take part in the functional control of the hypothalamo–pituitary–adrenal (HPA) axis, where they are found both in the nerve fibers and secretory cells (Nussdorfer and Malendowicz, 1998). All three main mammalian tachykinins, SP, NKA, and NKB, have been found in the bovine adrenal medulla (Cheung et al., 1993), whereas in the rat and in humans SP-like immunoreactivity has been detected in the nerve fibers that penetrate the cortex and medulla (Linnoila et al., 1980). Intravenous infusions of SP have been shown to produce an increase in plasma cortisol levels in humans (Coiro et al., 1992), but *in vivo* administration of SP, NKA, and NKB incites substantial increases in corticosterone and aldosterone secretion in the rat (Nussdorfer et al., 1988). Furthermore, NK_1 receptor agonists have been shown to exert an inhibitory effect on the hypothalamo–pituitary corticotrophin releasing hormone/adrenocorticotropic hormone system, whereas NK_2 receptor agonists are shown to stimulate it, thereby controlling the secretion and growth of the adrenal cortex (Nussdorfer and Malendowicz, 1998). Overall, there are indications that SP is involved in the maintenance of normal growth and steroidogenic capacity of the rat *zona glomerulosa*, and that SP and NKA play an important role in the stimulation of adrenal growth during fetal life (Nussdorfer and Malendowicz, 1998). More recently, we have found one of the primary sites of expression of the EKs to be the adrenal gland (Page et al., 2003), and it is plausible that many of the functions previously assigned to SP, in reality, could be performed by the EKs. We have found the rat adrenal gland to express all three of the mammalian tachykinin genes and their respective receptor genes (Figure 9.2). In an experiment to determine the effects of fasting on the expression of the tachykinin genes and their receptors in the adrenal gland we have found significant expression differences in the upregulation of *TAC*4 and *TACR*2 and the downregulation of *TACR*1 and *TACR*3 (Figure 9.2). The effects of fasting clearly have defined effects on the tachykinin system in the adrenal gland. However, to date, there has been conflicting evidence in different species for the effects of fasting on the HPA axis. In humans it has been shown that during a period of fasting the stress response is reduced, but in rats the converse is most often observed (Kirschbaum et al., 1997). It will be interesting to see how this area of research develops.

A growing and exciting area of interest for the biological and pathological roles of the tachykinins is their role during reproduction. For a class of peptides that had been classified solely as neurotransmitters and believed confined to the nervous system, the recent discovery of them in the placenta, a tissue totally devoid of nervous innervations, has challenged their dogmatic roles (Page et al., 2001). Indeed, many of the sites of tachykinin gene expression are actually in the periphery tissues (Figure 9.1) and cells (Page et al., 2003). Excessive secretion of NKB into the maternal circulation during the third trimester of pregnancy was first reported during preeclampsia (Page et al., 2000). The source of this expression is believed to be the placenta, where elevated levels of *TAC*3 mRNA expression are observed in the preeclamptic placenta (Page et al., 2006). The precise mechanism of the increased NKB production remains unknown, but it is possible they it may represent a sustained synthesis because of the early failure of trophoblast invasion in the uterus. This theory is supported by the observation that female rats treated with the NK_3 receptor antagonist SR142801

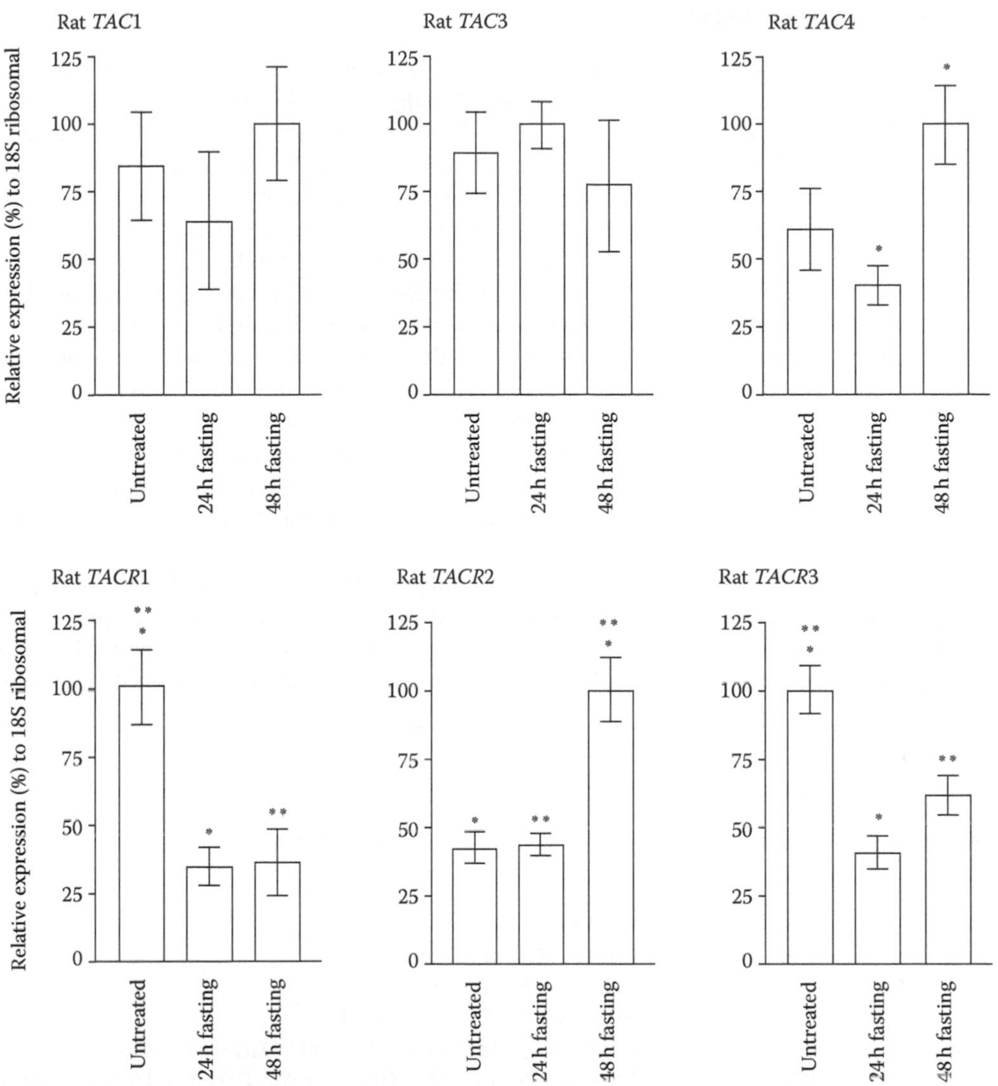

FIGURE 9.2 The effects of fasting for 24 and 48 h on rat adrenal gland mRNA expression levels of the tachykinin genes *TAC*1 (SP, NKA), *TAC*3 (NKB), and *TAC*4 (HK-1) and the tachykinin receptor genes *TACR*1 (NK_1), *TACR*2 (NK_2), and *TACR*3 (NK_3). The untreated columns are the fed rat controls ($n = 4$). Fasted rats were caged individually and deprived of food but not water for either 24 or 48 h ($n = 4$ for each). At the end of the stated periods the animals were sacrificed and the adrenals removed. Each bar represents the mean of four different rats with SEM shown by vertical lines. *,**$P < 0.05$, significant difference versus mRNA levels from corresponding treatment group.

before mating exhibit a tendency towards decreased fertility with a significant reduction in their litter sizes (Pintado et al., 2003). Diseases such as preeclampsia have been described as a maternal inflammatory response to pregnancy (Redman et al., 1999) and are yet another example of the roles of the tachykinins in stress and inflammation. Tachykinin immunoreactive nerves also supply the uteri of several species, and it is believed that the association of these nerves with smooth muscle may influence uterine contractility (Patak et al., 2003). This has led to suggestions that the tachykinins may possibly play physiological and pathological roles in normal and premature labor, stress-induced abortion, and menstrual disorders (Pennefather et al., 2004).

9.5 STRUCTURAL ANALOGS

Analogs for SP (Table 9.2) were the first to be developed, in an attempt to improve longevity and solubility issues. Some of the first analogs were based on the recognition that the minimum selective fragment at the NK_1 receptor was the hexapeptide SP^{6-11}. An example of a good selective agonist compound based on SP^{6-11} is [Arg^6, Sar^9]SP^{6-11}, presenting with an Arg at position 6 instead of Gln and a sarcosine at position 9 (Drapeau et al., 1987). Septide ([$pGlu^6$, Pro^9]SP^{6-11}) is another derivative of this short recognition sequence and was developed by Laufer and colleagues in 1986 to increase SP's selectivity. They also produced a water-soluble (WS) septide (Ac-[Arg^6, Pro^9]SP^{6-11}), and both versions showed high affinity and selectivity for the NK_1 receptor (Laufer et al., 1986). From this design many alterations were made, some of which produced analogs, and some of which did not. Perhaps the most notable modification to this design is the N-backbone cyclization, which, whilst preserving the side chains of the WS-septide, resulted in a more rigid conformation of the peptide analog. The size of this ring was found to be critical for affinity, as demonstrated by cycloseptide (EC_{50} of 5 nM on guinea pig ileum), which is an NK_1 receptor agonist that also shows greater stability towards degradation from peptidases (Byk et al., 1996). Cycloseptide is overall a 20--member ring that has a linkage between its N-terminal amine group and the nitrogen of the peptide amide Phe^8–Gly^9.

Tachykinins and their receptors mediate a wide range of physiological and potentially pathological processes, and are the basis of a wealth of research into identifying and developing specific, selective, and long-lasting antagonists, not only to understand their mechanism of action but also as potential new drug treatments. Much of the work performed in designing antagonists for the NK_1

TABLE 9.2
Examples of Synthetic Analogs[a]

Classification	Name	Structural Sequence
NK_1		
Agonists	SP^{6-11}	Gln–Phe–Phe–Gly–Leu–Met–NH_2
	[Arg6, Sar9]SP_{6-11}	Arg–Phe–Phe–Sar–Leu–Met–NH_2
	Septide	pGlu–Phe–Phe–Pro–Leu–Met–NH_2
	WS-septide	acArg–Phe–Phe–Pro–Leu–Met–NH_2
Antagonists	Spantide I	DArg–Pro–Lys–Pro–Gln–Gln–DTrp–Phe–DTrp–Leu–Leu–NH_2
	Spantide II	DLys(Nic)–Pro–Pal(3)–Pro–DPhe(Cl_2)–Asn–DTrp–Phe–DTrp–Leu–Nle–NH_2
	[DPro^9,Pro^{10},Trp^{11}]SP	Arg–Pro–Lys–Pro–Gln–Gln–Phe–Phe–DPro–Pro(Nic)–Trp–NH_2
	CP96345	*cis*-3-(z-methoxybenzyl)-amino-2-benzhydrylquinuclidine
	RP67580	perhydroisoindolone 7,7-diphenyl-2[1-imino-2(2-methoxy-phenyl)-ethyl] perhydroisoindol-4-one
NK_2		
Agonists	NKA^{4-10}	Asp–Ser–Phe–Val–Gly–Leu–Met–NH_2
	[βAla^8]NKA^{4-10}	Asp–Ser–Phe–Val–βAla–Leu–Met–NH_2
	[DPhe^6]NKA^{4-10}	Asp–Ser–DPhe–Val–Gly–Leu–Met–NH_2
Antagonists	GR94800	PhCO–Ala–Ala–DTrp–Phe–DPro–Pro–Nle–NH_2
NK_3		
Agonists	NKB^{4-10}	Asp–Phe–Phe–Val–Gly–Leu–Met–NH_2
	[MePhe^7]NKB	Asp–Met–His–Asp–Phe–Phe–MePhe–Gly–Leu–Met–NH_2
	Senktide	Suc–Asp–Phe–MePhe–Gly–Leu–Met–NH_2
Antagonists	[Gly^6]NKB^{3-10}	His–Asp–Phe–Gly–Val–Gly–Leu–Met–NH_2

[a] The underlined amino-acid residues represent the derivations from the endogenous sequence.

receptor has revolved around the introduction of D-amino acids. Spantide I (DArg–Pro–Lys–Pro–Gln–Gln–DTrp–Phe–DTrp–Leu–Leu–NH_2) was the first such pharmacologically useful antagonist developed. It is obtained by substituting positions 7 and 9 of SP with D-Trp and introducing D-Arg in the N-terminal position to enhance metabolic stability. Spantide I, however, produced neurotoxicity and potent histamine release (Folkers et al., 1984). Folkers and colleagues (1990) further developed Spantide II to produce a linear undecapeptide with seven residues different from SP (DLys (Nic)–Pro–Pal(3)–Pro–DPhe(Cl_2)–Asn–DTrp–Phe–DTrp–Leu–Nle–NH_2) and this was shown to be devoid of the neurotoxicity associated with Spantide I. Other attempts to produce NK_1 antagonists have involved the modification of the last three amino-acid positions of SP, an example being [DPro^9, *Pro^{10}, Trp^{11}]SP (Lavielle et al., 1994). There has also been a series of cyclic compounds produced including the cyclohexapeptide FK224. Each of these structures has highly constrained rings aimed at increasing selectivity (Hashimoto et al., 1992).

Nonpeptide antagonists for the tachykinin receptors are a growing research area, providing a rationale approach to future drug designs. Among the classes involved are steroids, perhydrosoindolones, benzylamino and benzylether quinuclidines, benzylamine piperidine, other piperidine-based structures, and tryptophan-based structures. Quartara and Maggi (1997) have characterized these classes in detail. However, some of the first nonpeptide NK_1 compounds presented unfavorable characteristics, including those of nonspecific effects and species-related differences (Maggi et al., 1993). For example, affinity of the quinuclidine derivative NK_1 receptor antagonist *cis*-3-(z-methoxybenzyl)-amino-2-benzhydrylquinuclidine (CP96345) is ~100-fold higher for human and guinea-pig NK_1 receptor than that for the rat and mouse NK_1 receptor. Conversely, perhydroisoindolone-7,7-diphenyl-2[1-imino-2(2-methoxy-phenyl)-ethyl]perhydroisoindol-4-one (RP67580) has a higher affinity for the rat and mouse NK_1 than for the human and guinea pig NK_1 receptor. The variable affinities of antagonists at NK_1 receptors from different species are linked to species-dependent variation in the amino-acid sequence of the NK_1 receptor protein. These positions are amino-acid positions 116 and 290, which contain Val and Ile in the human receptor and Leu and Ser, respectively, in the rat and mouse NK_1 receptor. Replacing the respective amino acids of the human receptor with the corresponding ones of the rat receptor, and vice versa, switched affinities of CP96345 and RP67580. In mutant receptors, the affinity of SP remained unchanged, indicating the varying amino-acid sequences across species (Fong et al., 1992b; Sachais et al., 1993). To an extent some of the unfavorable characteristics remain, with most of the nonpeptide antagonists still displaying some form of species variation. These include L-732,183 and L-733,060, which are the tryptophan and piperidine derivatives of CP-96345, respectively (Rupniak et al., 1997). To date, the quaternary ammonium SR-140333 stands out as the most potent and selective nonpeptide NK_1 antagonist, showing only a minimal species variation (Emonds-Alt et al., 1993).

Research into tachykinin NK_2 and NK_3 receptors has led to a more targeted approach in agonist and antagonist development. In many ways the story of NK_2-selective agonists has followed a similar story to that of SP. Again, the region that gave the minimum selectivity at the NK_2 receptor was determined. This was found to be NKA$^{4-10}$, which has given way to analogs such as [βAla^8]NKA$^{4-10}$, one of the most potent NK_2 receptor agonists (Rovero et al., 1989). Changes in amino-acid chirality have also proved beneficial in the modification of NKA$^{4-10}$ (e.g., Asp–Ser–DPhe–Val–Gly–Leu–Met–NH_2) and such changes have been found to increase not only metabolic stability but also affect potency and binding affinity (Warner et al., 2001). In the case of NKB, several modifications have been made to either its primary structure or its shortened version, NKB$^{4-10}$. MePhe^7 substitution of Val^7 is a classic example (Berthiaume et al., 1995). Senktide is another interesting NK_3 receptor analog being based on the SP$^{6-11}$ backbone (suc-[Asp^6, MePhe^8]SP$^{6-11}$); however, it does possess an Asp at position 6 and a methylated Phe at position 8. The behavior of NK_3 agonist analogs can be variable and depends not only on the species being examined but also on the tissue being tested (Nguyen et al., 1994).

Many of the first NK_2 receptor antagonists were based on substitution with D-amino acids, particularly D-Trp. Compounds such as GR94800 (PhCO–Ala–Ala–DTrp–Phe–DPro–Pro–Nle–NH_2)

which has an N-terminus protected by benzoate, has proved an extremely potent and selective NK_2 receptor antagonist (McElroy et al., 1992). Alternative strategies to develop NK_2 antagonist compounds have seen the creation of small cyclic peptides (Sakurada et al., 1994). Napadutant (MEN11420), a bicyclic hexapeptide with a sugar moiety is one such compound that has been shown to possess an interesting profile in humans with altered intestinal motility and perception (Lordal et al., 2001). However, common problems with such antagonists are the limitations of oral bioavailability, which have so far hindered any potential irritable bowel treatment drug candidates (Cialdai et al., 2006). Nonpeptide antagonists aim to keep the specificity and the affinity for the targeted receptor whilst allowing for it to stand up to the rigors of oral consumption. So far, targeted methods for the development of antagonistic drugs has produced aprepitant, a nonpeptide NK_1 antagonist (a molecule comprising of a morpholine core with two substitutes attached to an adjacent carbon ring) that is used in the treatment of chemotherapy-induced emesis (Hesketh et al., 2003a).

9.6 CLINICAL APPLICATIONS

The tachykinin peptides have been implicated in many disorders. However, the development of antagonists to treat aspects of these disorders has proved difficult and somewhat unfruitful. The primary hindrance has been the transition of drugs from animal models to clinical trials. Many promising tachykinin NK_1 receptor antagonists in animal studies have failed to have an effect on the desired disorder in larger human trials. Of particular merit is the suspension of NK_1 antagonist trials in the treatment of depression (Chahl, 2006). Tachykinin NK_1 receptor antagonists, however, have found a role in controlling emesis (Warr, 2008). Less is known of the potential of the NK_2 and NK_3 receptor antagonists, although the orally active NK_3 antagonist SB-223412 (talnetant) had completed phase II trials by 2004 for the potential treatment of urinary incontinence, irritable bowel syndrome, and schizophrenia (Evangelista, 2005).

There has been much speculation on the role of SP in the stress responses of animals. It is thought that a lack of noxious stimuli or trauma in a stressful situation may cause an excess of SP release in the brain's emotional centre, which could lead to depression and anxiety-like symptoms (Keller et al., 2006). It has been proposed that antagonism of the NK_1 receptor could provide a novel class of antidepressants. The story of the development of this hypothesis highlights the difficulty in developing responses seen in animal models into human drugs. Emotional pain response was first shown to be reduced with NK_1 antagonism when young mice were removed from their mothers (Boyce et al., 2001). In addition, direct injection of an SP agonist produced a range of defensive behavioral and cardiovascular changes in animals (Elliot, 1988; Krase et al., 1994; Aguiar and Brandão, 1996). These observations led to phase II clinical trials to determine if aprepitant, an NK_1 antagonist, could be prescribed as an antidepressant. Kramer and colleagues (1998) recruited ~200 patients for a pilot six-week double blind, placebo-controlled trial. The results from this looked promising, with patients who received aprepitant exhibiting a statistically significant improvement in their Hamilton Rating Scale for Depression (HAM-D); this level of improvement was comparable to the already established antidepressant paroxetine. An additional phase II clinical study using the NK_1 receptor antagonist L-759274 was also performed, with significant results confirming those seen for aprepitant (Kramer et al., 2004).

Five double-blind phase III clinical trials involving over 2500 patients were set up, with results pooled and collated by Keller and colleagues (2006). Unfortunately, this time, these trials demonstrated that aprepitant was no more effective than the placebo. This result was compounded by the fact that positron emission tomography (PET) used to establish the relationship between brain NK_1 occupancy and aprepitant plasma levels showed that the doses used were likely to achieve very high occupancies that would have effectively blocked the NK_1 receptor system throughout the study (Bergström et al., 2004). The observations from this last, large-scale study indicated that aprepitant would not be a suitable antidepressant, despite observations from previous trials. There are many reasons for the apparent false positives seen in the phase II trials. Perhaps keeping the identity of those patients receiving placebo or the active compound in the double-blind placebo-controlled trials proved harder in the smaller trials than in the larger phase III study (Keller et al., 2006).

A more successful story is that of the application of the NK_1 antagonist aprepitant to emesis. Chemically induced emesis is a major problem for chemotherapy patients. Since the 1950s, SP has been implicated in controlling the vomiting reflex where high concentrations of SP are detected in the brainstem emetic nuclei, nucleus tractus, and the area postrema (Armstrong et al., 1981). Tattersall and colleagues showed that NK_1 receptor antagonists exhibited broad-spectrum antiemetic activity in ferrets (Tattersall et al., 1994). Importantly, NK_1 receptor antagonists were able to attenuate markedly both acute and delayed emesis induced by the chemotherapy agent cisplatin in ferrets (Rudd et al., 1996), suggesting they could have important therapeutic use in patients undergoing chemotherapy. These studies led to clinical trials with the NK_1 receptor antagonists CP-122721 (Kris et al., 1997), CJ-11974 (Hesketh et al., 1999), and MK-869 (Navari et al., 1999), which confirmed the effectiveness of NK_1 receptor antagonists in preventing delayed emesis after cisplatin chemotherapy. Aprepitant (MK-869) became the first of a new class of drugs that blocks NK_1 receptors in the brainstem emetic centre and gastrointestinal tract. Published studies have shown that the addition of aprepitant to standard antiemetic therapy appears to have a significant effect on controlling cisplatin-induced acute as well as delayed emesis, with the benefit more pronounced in the delayed phase (Hesketh et al., 2003b; Poli-Bigelli et al., 2003; Schmoll et al., 2006). Aprepitant is recommended for use in highly emetogenic chemotherapy and, in part, moderately emetogenic chemotherapy. A study by Warr and colleagues is the basis for these recommendations; the triple combination of ondansetron, dexamethasone, and aprepitant was used in the first 24 h followed by aprepitant monotherapy for another two days, and proved to be superior to other antiemetic regimes (Warr et al., 2005).

For the moment at least, aprepitant, the tachykinin NK_1 receptor antagonist, remains the only tachykinin antagonist, licenced for use in humans. The trials will no doubt continue in the search for further clinical uses for tachykinin receptor agonists and antagonists.

9.7 COMMON METHODOLOGIES AND ASSAY SYSTEMS

The discovery of SP, the first tachykinin to be isolated, was performed using alcoholic extracts from whole rabbit brain and intestinal tissue, the extracts subsequently being used to monitor the contractile effects of longitudinal muscle of the rabbit's isolated intestine (von Euler and Gaddum, 1931). Today, acid extraction methods are used to isolate the tachykinins from tissues (Page et al., 2000). These techniques involve the washing of the tissues in phosphate buffered saline and homogenization in five to ten volumes of acid extraction buffer (1 M HCl containing 5% (vol/vol) formic acid, 1% (wt/vol) NaCl, and 1% (vol/vol) trifluoroacetic acid). After centrifugation at 3500*g* at 4°C the supernatant is loaded onto Sep-Pak C18 cartridges (Millipore Corporation) and washed with 0.1% (vol/vol) trifluoroacteic acid. The tachykinins are then eluted in between 30 and 50% acetonitrile in aqueous 0.1% (vol/vol) trifluoroacetic acid. Finally, they are then dried in a vacuum centrifuge with 1 mg mannitol before being resuspended in water. Samples can then be processed using additional peptide purification steps such as size exclusion chromatography, before immunological or sequence determination. Developments within the field of genomics have produced a vast array of data that are now stored in publicly available bio-informatic databases. More recently, these databases have been utilized in the discovery and characterization of novel tachykinins as exemplified by HK-1, CL14TKL-1, and EKA and B (Zhang et al., 2000; Jiang et al., 2003; Page et al., 2003). A very useful tool in the repertoire for hunting for new tachykinins is the basic local alignment search tool (BLAST) for comparing gene and protein sequences against others in the public databases.

Comparative analysis of expression levels of endogenous tachykinins has been extensively performed, first using immunohistochemical techniques. This is where tissues, after being fixed and sectioned onto slides, are probed with a specific tachykinin primary antibody. They are then further probed with a secondary antibody, which binds to the primary, and then a stain can be added. Stained areas represent areas of expression of the specific tachykinin (Shults et al., 1984; Hennig et al., 1995). However, the use of antibodies raised to even what is believed to be a specific tachykinin is not an exact science. The presence of the particularly antigenic amide group on each tachykinin has been problematic in raising specific antibodies. Moreover, previous extensive work using immunohistochemical and immunoassay techniques to determine the presence and quantity of the mammalian tachykinins in various tissues may be flawed. Strikingly, Page (2004) demonstrated that a competitive enzyme immunoassay for SP (Bachem, S-1180) was found to display 100% cross-reactivity with EKA/B and 70% cross-reactivity with HK-1. Moreover, all of the non-neuronal locations of the tachykinins have concerned mainly SP, the occurrence of which has been established mainly by immunocytochemistry using selective anti-SP antibodies (Severini et al., 2002). This calls into question whether these assays were actually measuring SP, HK-1, EKA, or EKB. A similar phenomenon is seen with the commercially available NKA antibody which displays 80% cross-reactivity with NKB (Bachem, S-2158) (Table 9.3).

More recently, and perhaps more reliably, quantitative analysis of mRNA expression has been used, where specific primers for the mRNA in question are developed and used in reverse transcription polymerase chain reactions (RT-PCR). These procedures are highly quantitative. In these methods, total RNA is isolated from selected tissues using methods such as Tri Reagent (Sigma-Aldrich, Poole, UK). Contaminating genomic DNA is removed by DNase I treatment and the RNA then repurified. First, strand cDNA synthesis is performed using 1 to 2 μg of total RNA using a reverse transcriptase in the presence of random hexameric primers. The resulting cDNA is diluted in nuclease free H_2O and 1 μL aliquots used in quantitative PCR. Specific Taqman® probes and primer sets are designed using Primer Express™ (Applied Biosystems) to span, where possible, the exon–exon junctions of a tachykinin precursor cDNA (examples in Page et al., 2006). Reactions are set up in triplicate in a total of 25 μL using ABsolute™ Quantitative PCR ROX master mix (ABgene, Epsom, UK) and PCR cycling performed on a PCR analyzer capable of light detection, for example, an ABI PRISM® 7700 sequence detector. Cycles are typically an initial denaturation/activation of the

TABLE 9.3
Radio-Ligands, Receptor Membrane Preparations, and Immunoassays Commonly Used in Tachykinin Research

Type of Assay Component	Details/Supplier of Assay Components
Radioligands	
NK_1	[^{3}H]SR140333 (Amersham Biosciences)
	[^{125}I]Substance P (Amersham Biosciences)
NK_2	[^{3}H]SR48968 (Amersham Biosciences)
	[^{125}I]Substance K (Amersham Biosciences)
Membrane Preparations	
NK_1	ChemiSCREEN™ Human NK_1 Tachykinin Receptor Membrane Preparation (Chemicon International)
NK_2	ChemiSCREEN™ Membrane Preparation Recombinant Human NK_2 Tachykinin Receptor (Chemicon International)
Immunoassays	
Substance P	Substance P–RIA Kit; Host: Guinea Pig (Bachem)
	Substance P–EIA Kit; Host: Rabbit High Sensitivity (Bachem)
Neurokinin A	Neurokinin A–EIA Kit; Host: Rabbit High Sensitivity (Bachem)
	Neurokinin A–RIA Kit; Host: Rabbit (Bachem)
Neurokinin B	Neurokinin B–RIA Kit; Host: Rabbit (Bachem)
	Neurokinin B–EIA Kit; Host: Rabbit High Sensitivity (Bachem)

Thermo-Start® DNA polymerase at 95°C for 15 min and then 40 cycles of 95°C for 15 sec and 60°C for 1 min. The optimal concentrations of the primers (between 50 and 900 nM) and probe (between 25 and 225 nM) are used to amplify each target gene, which are calculated using those combinations that give the lowest threshold cycle (C_T). Typical results from such an analysis are shown in Figure 9.1. Messenger RNA levels can give an indication of peptide expression levels as they are precursors to the final protein and are normally only synthesized when expression of the protein is required (Duffy et al., 2003; Berger and Paige, 2005).

Binding studies have been used to determine the affinity of the tachykinins to each of the tachykinin receptors. Radioligand binding is the most common technique used. This technique is also used to assess the agonist and antagonist binding affinities developed against the tachykinins and their receptors (Kurtz et al., 2002; Page et al., 2003; Cialdai et al., 2006). Membrane preparations for ligand binding can be made by either transfecting a cell line such as Chinese hamster ovary with each tachykinin receptor or by buying in ready prepared membranes. We have previously used bought prepared membranes for convenience and one current supplier is Chemicon International (Table 9.3). Competition assays with NK_1 and NK_2 receptors were conducted with 10 µg of membrane protein and 0.3 nM [^{3}H] SR140333 and 0.5 nM [^{3}H] SR48968 (Amersham Biosciences), respectively. It is also important to perform nonspecific binding in the presence of an antagonist for each receptor. We performed our NK_1 receptor incubations in 40 mM Hepes pH 7.4, 5 mM $MgCl_2$, 0.5% (wt/vol) bovine serum albumin and NK_2 receptor incubations in 20 mM Hepes pH 7.4, 1 mM $MnCl_2$, 0.1% (wt/vol) bovine serum albumin, all containing 10 µM phosphoramidon with or without various concentrations of cold ligand for 90 min at 27°C in 1 mL triplicate assays. The incubations were stopped by rapid filtration through Whatman GF/C filters, presoaked for 15 min in 0.5% (wt/vol) bovine serum albumin, with a Brandel tissue harvester (Gaithersburg). Membranes were washed four times with 5 mL phosphate buffered saline before being placed in vials with 2 mL of optiphase "Hisafe" 3 (Fisher) and counted in a liquid scintillation counter. Inhibition constants (K_i) can then be determined from the concentration–response curves using GraphPad Prism (GraphPad Software).

Postreceptor signaling event analysis uses various methods, the most common being SDS PAGE western blotting. Other techniques include immunohistochemistry, [^{3}H] thymidine incorporation (for DNA synthesis analysis), and radiolabeling of possible signaling products and inhibitors. Western blotting and immunohistochemistry both use antibodies raised against specific signaling proteins, for example, phospholipase C or mitogen activated protein kinase, which can be used to assess whether there is a poststimulation increase or decrease. Inhibitors can be used in conjunction with western blotting or immunohistochemistry to block a specific point, verifying their effect (Luo et al., 1996, Lallemend et al., 2003).

In addition, various animal studies have been performed to assess the effects of the tachykinins on whole body systems. Recently, the most common method being used is a direct administration of tachykinins into a specific brain region or spinal cord by intrathecal injection and using visual clues to assess the effects (Kart et al., 2004; Fu et al., 2005; Yoshioka et al., 2006). Potential agonist and antagonist drug therapies are also being tested in animal models, including those targeted at the periphery. These are usually given as oral preparations of the agonist or antagonist (Gannon and Millan, 2005; Cialdai et al., 2006).

9.8 FUTURE DEVELOPMENTS

We feel that, perhaps more so than ever, there has never been a more exciting time to be working in the field of tachykinin research. The recent explosion in the number of known tachykinins and their diverse distribution across the species and within different tissue types can only be testament to the growing interest in this field. The current list of known tachykinins at the last count was over 50 and we expect this number to double over the next few years. Open to debate is the relevance of the tachykinin-related peptides with the conserved C-terminal hexapeptide motif $-GFX_1GX_2R-NH_2$. These peptides have a sequence homology of 30 to 45% to the classical tachykinins, and it will be interesting to determine whether these, in evolutionary terms, represent the ancestral gene progenitors of the vertebrate tachykinins. It is proposed that today's diverse range of tachykinins is the result of a series of gene duplications, and from such predictions Conlon and Larhammar (2005) have suggested that there is at least one additional mammalian tachykinin gene still to be discovered. Evolution also appears to have moulded the selective events on the classical tachykinin motif, and we see examples of the tachykinin gene-related peptides represented in humans by the motif FQGLL–NH_2 (Page, 2004). The relevance of this divergent motif is not fully understood, and it has been shown to have little affinity at the known tachykinin receptors. (Page et al., 2003). Additional diversity in the tachykinins is also shown at the level of transcription, and for each tachykinin gene there are several known splice variants. For example, this leads in the case of *TAC*1 to the potential translation and processing of four distinct tachykinin peptides, namely SP, NKA, NPK, and NPγ. Each of these tachykinins has its own distinct affinities, potencies, and tissue selectivities at each of the three mammalian tachykinin receptors. Nonetheless, the potential diversity of the tachykinins does not stop at the genomic and transcriptomic level, and at the level of the translated tachykinin precursor there is the opportunity for post-translational modification. These can be seen, for example, in the differential proteolytic cleavage of the tachykinin precursors in the disease state of preeclampsia (Page, 2006a).

There remains a wealth of information to be unlocked from the complex control mechanisms and diversification of each individual tachykinin gene. At the current time, the permutations within the mammalian tachykinins alone seem to outstrip the number of known tachykinin receptors. Are there any more tachykinin receptors to be discovered? This is not an easy question to answer. It is apparent that the known receptors can display a multitude of configurations and functions. For instance, at least four alternative receptor conformations of the NK_1 receptor have been revealed, including the "classic," "septide-sensitive," "new NK_1-sensitive," and a C-terminal truncated form (Page, 2005). Therefore, one receptor subtype can display a range of profiles. It could also be the case that the site of receptor expression may also play just as an important role and an elegant study by Page (2006a) shows very defined expression patterns in the expression of the NK_1 receptor and its C-terminal truncated counterpart between the nervous and periphery tissues, respectively. This pattern also holds true in the expression patterns of SP and EKA/B, where SP is found predominantly in neuronal tissue and EKA/B in peripheral tissue. This has led to the proposal that, although SP and EKA/B would appear to have the same binding patterns and potencies at each of the three tachykinin receptors, their expression is independently controlled. SP appears to play a role predominantly as a neurotransmitter and EKA/B as an endocrine factor.

Certainly, there is plenty of scope for many challenges in the years to come, and although the drug industry appears to have had a setback in the application of the NK_1 antagonist aprepitant in the treatment of depression, there remain plenty of opportunities for the application of NK antagonists in an ever increasing list of pathologies associated with the tachykinins.

REFERENCES

Aguiar, M.S. and Brandão, M.L., Effects of microinjections of the neuropeptide substance P in the dorsal periaqueductal gray on the behaviour of rats in the plus-maze test, *Physiol. Behav.*, 60, 1183–1186, 1996.

Almeida, T.A. et al., Tachykinins and tachykinin receptors: Structure and activity relationships, *Curr. Med. Chem.*, 11, 2045–2081, 2004.

Anastasi, A., Erspamer, V., and Cei, J.M., Isolation and amino acid sequence of physalaemin, the main active polypeptide of the skin of *Physalaemus fuscumaculatus*, *Arch. Biochem. Biophys.*, 108, 341–348, 1964.

Armstrong, D.M. et al., Immuno-cytochemical localisation of catecholamine synthesising enzymes and neuropeptides in area protrema and medial nucleus tractus solitaries of rat brain, *J. Comp. Neurol.*, 196, 505–517, 1981.

Bae, S.J. et al., Substance P induced preprotachykinin-A mRNA, neutral endopeptidase mRNA, and substance P in cultured normal fibroblasts, *Inter. Arch. Allergy Immunol.*, 127, 316–321, 2002.

Bandari, P.S. et al., Crosstalk between neurokinin receptors is relevant to hematopoietic regulation: Cloning and characterization of neurokinin 2 promoter, *J. Neuroimmunol.*, 138, 65–75, 2003.

Bellucci, F. et al., Pharmacological evaluation of alpha and beta human tachykinin NK(2) receptor splice variants expressed in CHO cells, *Eur. J. Pharmacol.*, 499, 229–238, 2004.

Berger, A. and Paige, C.J., Hemokinin-1 has substance P like behaviour in U-251 MG astrocytoma cells. A pharmacological and functional study, *J. Neuroimmunol.*, 164, 48–56, 2005.

Bergström, M. et al., Human positron emission tomography studies of brain neurokinin 1 receptor occupancy by aprepitant, *Biol. Psychiatry*, 55, 1007–1012, 2004.

Berthiaume, N. et al., Characterization of receptors for kinins and neurokinins in the arterial and venous mesenteric vasculatures of the guinea-pig, *Br. J. Pharmacol.*, 115, 1319–1325, 1995.

Bohm, S.K. et al., Identification of potential tyrosine-containing endocytic motifs in the carboxyl-tail and seventh transmembrane domain of the neurokinin 1 receptor, *J. Biol. Chem.*, 272, 2363–2372, 1997.

Boyce, S. et al., Intra-amygdala injection of the substance P [NK (1) receptor] antagonist L-760735 inhibits neonatal vocalisation in guinea-pigs, *Neuropharmacology*, 41, 130–137, 2001.

Bradesi, S. et al., Stress-induced visceral hypersensitivity in female rats is estrogen-dependent and involves tachykinin NK1 receptors, *Pain*, 102, 227–234, 2003.

Byk, G. et al., Synthesis and biological activity of NK1 selective, N-backbone cyclic analogues of the C-terminal hexapeptide of substance P, *J. Med. Chem.*, 39, 3174–3178, 1996.

Chahl, L.A., Tachykinins and neuropsychiatric disorders, *Curr. Drug Targets*, 7, 993–1003, 2006.

Chang, M.M. and Leeman, S.E., Isolation of a sialogogic peptide from bovine hypothalamic tissue and its characterization as substance-P, *J. Biol. Chem.*, 245, 4784–4790, 1970.

Chang, M.M., Leeman, S.E., and Niall, H.D., Amino-acid sequence of substance P, *Nat New Biol.* 232, 86–87, 1971.

Chen, L.W., Yung, K.K., and Chan, Y.S., Neurokinin peptides and neurokinin receptors as potential therapeutic intervention targets of basal ganglia in the prevention and treatment of Parkinson's disease, *Curr. Drug Targets*, 5, 197–206, 2004.

Cheung, N.S., Basil, S., and Livett, B.G., Identification of multiple tachykinins in bovine adrenal medulla using an improved chromatographic procedure, *Neuropeptides*, 24, 91–97, 1993.

Chiwakata, C. et al., Tachykinin (substance P) gene-expression in Leydig-cells of the human and mouse testis, *Endocrinology*, 128, 2441–2448, 1991.

Cialdai, C. et al., MEN15596, a novel nonpeptide tachykinin NK2 receptor antagonist, *Eur. J. Pharmacol.*, 549, 140–148, 2006.

Coiro, V. et al., Stimulation of ACTH/cortisol by intravenously infused substance P in normal men: Inhibition by sodium valproate, *Neuroendocrinology*, 56, 459–463, 1992.

Conlon, J.M. and Larhammar, D., The evolution of neuropeptides, *Gen. Comp. Endo.*, 142, 53–59, 2005.

Debeljuk, L., Tachykinins and ovarian function in mammals, *Peptides*, 27, 736–742, 2006.

Donkin, J.J. et al., Substance P in traumatic brain injury, *Prog. Brain Res.*, 161, 97–109, 2007.

Drapeau, G. et al., Selective agonists for substance P and neurokinin receptors, *Neuropeptides*, 10, 43–54, 1987.

Duffy, R.A. et al., Centrally administered hemokinin-1 (HK-1), a neurokinin NK1 receptor agonist, produces substance P-like behavioral effects in mice and gerbils, *Neuropharmacology*, 45, 242–250, 2003.

Ebner, K. and Singewald, N., The role of substance P in stress and anxiety responses, *Amino Acids*, 31, 251–272, 2006.

Elliot, P.J., Place aversion induced by the substance P analogue, dimethyl-c7, is not state dependent: Implication of substance P in aversion, *Exp. Brain Res.*, 73, 354–356, 1988.

Emonds-Alt, X. et al., In vitro and in vivo biological activities of SR140333, a novel potent non-peptide tachykinin NK1 receptor antagonist, *Eur J Pharmacol.* 250, 403–413, 1993.

Erspamer, V., Ricerche preliminary sulla moschatina, *Experientia*, 5, 79–81, 1949.

Erspamer, V. et al., Sauvagine, a new polypeptide from phyllomedusa-sauvagei skin-occurance in various Phyllomedusa species and pharmacological actions on rat blood pressure and diuresis, *Naunyn Schmiedebergs Arch. Pharmacol.*, 312, 265–270, 1980.

Evangelista, S., Talnetant GlaxoSmithKline, *Curr. Opin. Investig. Drugs*, 6, 717–721, 2005.

Folkers, K. et al., Biological evaluation of substance P antagonists, *Br. J. Pharmacol.*, 83, 449–456, 1984.

Folkers, K. et al., Spantide II, an effective tachykinin antagonist having high potency and negligible neurotoxicity, *Proc. Natl Acad. Sci. USA*, 87, 4833–4835, 1990.

Fong, T.M. et al., Differential activation of the intracellular effector by two isoforms of the human neurokinin-1 receptor, *Mol. Pharmacol.*, 41, 24–30, 1992a.

Fong, T.M., Yu, H., and Strader, C.D., Molecular basis for species selectivity of neurokinin-1 receptor antagonists CP-96345 and RP67580, *J. Biol. Chem.*, 267, 25668–25671, 1992b.

Fu, C.Y. et al., Effects and mechanisms of supraspinal administration of rat/mouse hemokinin-1, a mammalian tachykinin peptide, on nociception in mice, *Brain Res.*, 1056, 51–58, 2005.

Gaddum, J.H. and Schild, H., Depressor substances in extracts of intestine, *J. Physiol.*, 83, 1–14, 1934.

Gannon, R.L. and Millan, M.J., The selective tachykinin neurokinin 1 (NK1) receptor antagonist GR 205, 171, stereospecifically inhibits light-induced phase advances of hamster circadian activity rhythms, *Eur. J. Pharmacol.*, 527, 86–93, 2005.

Gerard, N.P. et al., Molecular aspects of the tachykinin receptors, *Regul. Pept.*, 43, 21–35, 1993.

Gether, U., Johansen, T.E., and Schwartz, T.W., Chimeric NK1 (substance P)/NK3 (neurokinin B) receptors. Identification of domains determining the binding specificity of tachykinin agonists, *J. Biol. Chem.*, 268, 7893–7898, 1993.

Graham, G.J. et al., Tachykinins regulate the function of platelets, *Blood*, 104, 1058–1065, 2004.

Harmar, A.J., Hyde, V., and Chapman, K., Identification and cDNA sequence of delta preprotachykinin, a 4th splicing variant of the rat substance-P receptor, *FEBS Lett.*, 275, 22–24, 1990.

Harrison, S. and Geppetti, P., Substance P, *Int. J. Biochem. Cell Biol.*, 33, 555–576, 2001.

Hashimoto, M. et al., WS9326A, a novel tachykinin antagonist isolated from *Streptomyces violaceusniger* no. 9326. II. Biological and pharmacological properties of WS9326A and tetrahydro-WS9326A (FK224), *J. Antibiot.* (*Tokyo*), 45, 1064–1070, 1992.

Hennig, I.M. et al., Substance-P receptors in human primary neoplasms—tumoral and vascular localization, *Int. J. Cancer*, 61, 786–792, 1995.

Herpfer, I. et al., Effects of substance P on memory and mood in healthy male subjects, *Hum. Psychopharmacol.*, 22, 567–573, 2007.

Hesketh, P.J. et al., Randomised phase II study of the neurokinin 1 receptor antagonist CJ-11,974 in the control of cisplatin-induced emesis, *J. Clin. Oncol.*, 17, 338–343, 1999.

Hesketh, P.J. et al., The oral neurokinin-1 antagonist aprepitant for the prevention of chemotherapy-induced nausea and vomiting: A multinational, randomised, double-blind, placebo-controlled trial in patients receiving high-dose cisplatin—the Aprepitant Protocol 052 Study Group, *J. Clin. Oncol.*, 21, 4112–4119, 2003a.

Hesketh, P.J. et al., Differential involvement of neurotransmitters though the time course of cisplatin-induced emesis as revealed by therapy, *Eur. J. Cancer.*, 39, 1074–1080, 2003b.

Holzer, P., Local effector functions of capsaicin-sensitive sensory nerve endings: Involvement of tachykinins, calcitonin gene-related peptide, and other neuropeptides, *Neuroscience*, 24, 739–768, 1988.

Holzer, P. and Holzer-Petsche, U., Tachykinins in the gut. Part II. Roles in neural excitation, secretion, and inflammation, *Pharmacol. Ther.*, 73, 219–263, 1997.

Jancsó, G., Kiraly, E., and Jancsó-Gábor, A., Pharmacologically induced selective degeneration of chemosensitive primary sensory neurones, *Nature*, 270, 741–743, 1977.

Jiang, Y. et al., PepPat, a pattern-based oligopeptide homology search method and the identification of a novel tachykinin-like peptide, *Mamm. Genome*, 14, 341–349, 2003.

Joos, G.F., Germonpre, P.F., and Pauwels, R.A., Role of tachykinins in asthma, *Allergy*, 55, 321–337, 2000.

Kage, R. et al., Neuropeptide-gamma—a peptide isolated from rabbit intestine that is derived from gamma-preprotachykinin, *J. Neurochem.*, 50, 1412–1417, 1988.

Kangawa, K. et al., Neuromedin-K—a novel mammalian tachykinin identified in porcine spinal-cord, *Biochem. Biophys. Res. Comm.*, 114, 533–540, 1983.

Kart, E. et al., Neurokinin-1 receptor antagonism by SR140333: Enhanced *in vivo* ACh in the hippocampus and promnestic post-trial effects, *Peptides*, 25, 1959–1969, 2004.

Kawada, T. et al., A novel tachykinin-related peptide receptor—sequence, genomic organization, and functional analysis, *Eur. J. Biochem.*, 269, 4238–4246, 2002.

Keller, M. et al., Lack of efficacy of the substance P (neurokinin1 receptor) antagonist aprepitant in the treatment of major depressive disorder, *Biol. Psychiatry*, 59, 216–223, 2006.

Khawaja, A.M. and Rogers, D.F., Tachykinins: Receptor to effector, *Int. J. Biochem. Cell Biol.*, 28, 721–738, 1996.

Kimura, S. et al., Novel neuropeptides, neurokinin-alpha and neurokinin-beta isolated from porcine spinal-cord, *Proc. Jpn Acad. Series B Phys. Biol. Sci.*, 59, 101–104, 1983.

Kirschbaum, C. et al., Effects of fasting and glucose load on free cortisol responses to stress and nicotine, *J. Clin. Endocrinol. Metab.*, 82, 1101–1105, 1997.

Kramer, M.S. et al., Distinct mechanism for antidepressant activity by blockade of central substance P receptors, *Science*, 281, 1640–1645, 1998.

Kramer, M.S. et al., Demonstration of the efficacy and safety of a novel substance P (NK1) receptor antagonist in major depression, *Neuropsychopharmacotherapy*, 29, 385–392, 2004.

Krase, W., Koch, M., and Schnitzter, H.V., Substance P is involved in the sensitisation of the acoustic startle response by footshocks in rats, *Behav. Brain Res.*, 63, 81–88, 1994.

Krause, J.E. et al., 3 rat preprotachykinin messenger-RNAs encode the neuropeptides substance-P and neurokinin-A, *Proc. Natl Acad. Sci. USA*, 84, 881–885, 1987.

Kris, M.G. et al., Use of an NK1 receptor antagonist to prevent delayed emesis after cisplatin, *J. Natl Cancer Inst.*, 89, 817–818, 1997.

Kurtz, M.M. et al., Identification, localization, and receptor characterization of novel mammalian substance P-like peptides, *Gene*, 296, 205–212, 2002.

Lallemend, F. et al., Substance P protects ganglion neurons from apoptosis via PKC-Ca^{2+}-MAPK/ERK pathways, *J. Neurochem.*, 87, 508–521, 2003.

Laufer, R. et al., pGlu6, Pro9 SP(6-11), a selective agonist for the substance P receptor subtype, *J. Med. Chem.*, 29, 1284–1288, 1986.

Lavielle, S. et al., Highly potent substance P antagonists substituted with beta-phenyl- or beta-benzyl-proline at position 10, *Eur. J. Pharmacol.*, 258, 273–276, 1994.

Łazarczyk, M., Matyja, E., and Lipkowski, A., Substance P and its receptors—a potential target for novel medicines in malignant brain tumour therapies (mini-review), *Folia Neuropathol.*, 45, 99–107, 2007

Lecci, A. and Maggi, C.A., Peripheral tachykinin receptors as potential therapeutic targets in visceral diseases, *Expert Opin. Therap. Targets*, 7, 343–362, 2003.

Leemann, S.E. and Hammerschlag, R., Stimulation of salivary secretion by a factor extracted from hypothalamic tissue, *Endocrinology*, 81, 803–810, 1967.

Lembeck, F. and Starke, K., Substance P and salivary secretion, *Naunyn Schmiedebergs Arch. Exp. Pathol. Pharmakol.*, 259, 204–205, 1968.

Linnoila, R.I. et al., Distribution of [Met^5]- and [Leu^5]-enkephalin-, vasoactive intestinal polypeptide- and substance P-like immunoreactivities in human adrenal glands, *Neuroscience*, 5, 2247–2259, 1980.

Liu, L. et al., Molecular identification and characterization of three isoforms of tachykinin NK1-like receptors in the cane toad, *Bufo marinus*, *Am. J. Physiol. Regul. Integr. Comp. Physiol.*, 285, R575–R585, 2004.

Lordal, M. et al., A novel NK2 receptor antagonist prevents motility-stimulating effects of neurokinin A in small intestine, *Br. J. Pharmacol.*, 134, 215–223, 2001.

Lundberg, J.M., Pharmacology of cotransmission in the autonomic nervous system: Integrative aspects on amines, neuropeptides, adenosine triphosphate, amino acids, and nitric oxide, *Pharmacol. Rev.*, 48, 113–178, 1996.

Luo, W.H., Sharif, T.R., and Sharif, M., Substance P-induced mitogenesis in human astrocytoma cells correlates with activation of the mitogen-activated protein kinase signaling pathway, *Cancer Res.*, 56, 4983–4991, 1996.

Maggi, C.A. and Meli, A., The sensory-efferent function of capsaicin-sensitive sensory neurons, *Gen. Pharmacol.*, 19, 1–43, 1988.

Maggi, C.A. et al., Tachykinin receptors and tachykinin receptor antagonists, *J. Auton. Pharmacol.*, 13, 23–93, 1993.

Maggi, C.A., The effects of tachykinins on inflammatory and immune cells, *Regul. Pept.*, 70, 75–90, 1997.

Maggi, C.A., Principles of tachykininergic cotransmission in the peripheral and enteric nervous system, *Regul. Pept.*, 93, 53–64, 2000.

Mantyh, P.W. et al., Differential expression of two isoforms of the neurokinin-1 (substance P) receptor *in vivo*, *Brain Res.*, 719, 8–13, 1996.

May, A. and Goadsby, P.J., Pharmacological opportunities and pitfalls in the therapy of migraine, *Curr. Opin. Neurol.*, 14, 341–345, 2001.

McElroy, A.B. et al. Highly potent and selective heptapeptide antagonists of the neurokinin NK-2 receptor, *J. Med. Chem.*, 35, 2582–2591, 1992.

Minamino, N. et al., A novel mammalian tachykinin identified in porcine spinal cord, *Neuropeptides*, 4, 57–166, 1984.

Mochizuki-Oda, N. et al., Characterization of the substance P receptor-mediated calcium influx in cDNA transfected Chinese hamster ovary cells. A possible role of inositol 1,4,5-trisphosphate in calcium influx, *J. Biol. Chem.*, 269, 9651–9658, 1994.

Moussaoui, S.M. et al., Distribution of neurokinin B in rat spinal cord and peripheral tissues: Comparison with neurokinin A and substance P and effects of neonatal capsaicin treatment, *Neuroscience*, 48, 969–978, 1992.

Murthy, R.G. et al., Tachykinins and haematopoietic stem cell functions: Implications in clinical disorders and tissue regeneration, *Front Biosci.*, 12, 4779–4787, 2007.

Nakajima, Y. et al., Dissociated cell culture of cholinergic neurons from nucleus basalis of Meynert and other basal forebrain nuclei, *Proc. Natl Acad. Sci. USA*, 82, 6325–6329, 1985.

Nässel, D.R., Tachykinin-related peptides in invertebrates: A review, *Peptides*, 20, 141–158, 1999.

Navari, R.M. et al., Reduction of cisplatin-induced emesis by a selective neurokinin-1 receptor antagonist L-754,030 Antiemetic Trials Group, *New Engl. J. Med.*, 340, 190–195, 1999.

Nawa, H. et al., Nucleotide sequences of cloned cDNAs for two types of bovine brain substance P precursor, *Nature*, 306, 32–36, 1983.

Neumann, S. et al., Inflammatory pain hypersensitivity mediated by phenotypic switch in myelinated primary sensitive neurons, *Nature*, 384, 360–364, 1996.

Nguyen, Q.T. et al., Two NK-3 receptor subtypes: Demonstration by biological and binding assays, *Neuropeptides*, 27, 157–161, 1994.

Nussdorfer, G.G. et al., Effects of substance P on the rat adrenal zona glomerulosa *in vivo*, *Peptides*, 9, 1145–1149, 1988.

Nussdorfer, G.G. and Malendowicz, L.K., Role of tachykinins in the regulation of the hypothalamo–pituitary–adrenal axis, *Peptides*, 19, 949–968, 1998.

Oldham, C.D. et al., Amidative peptide processing and vascular function, *Am. J. Physiol.*, 273, C1908–C1914, 1997.

Page, N.M. et al., Excessive placental secretion of neurokinin B during the third trimester causes pre-eclampsia, *Nature*, 405, 797–800, 2000.

Page, N.M., Woods, R.J., and Lowry, P.J., A regulatory role for neurokinin B in placental physiology and pre-eclampsia, *Regul. Pep.*, 98, 97–104, 2001.

Page, N.M. et al., Characterisation of the endokinins: Human tachykinins with cardiovascular activity, *Proc. Natl Acad. Sci. USA*, 100, 6245–6250, 2003.

Page, N.M., Hemokinins and endokinins, *Cell Mol. Life Sci.*, 61, 1652–1663, 2004.

Page, N.M., New challenges in the study of the mammalian tachykinins, *Peptides*, 26, 1356–1368, 2005.

Page, N.M., Characterisation of the gene structures, precursor processing, and pharmacology of the endokinin peptides, *Vascul. Pharmacol.*, 45, 200–208, 2006a.

Page, N.M., Brain tachykinins, in *The Handbook of Biologically Active Peptides*, Kastin, A., Ed., Academic Press, Chapter 105, 2006b, p. 763–769.

Page, N.M., Dakour, J., and Morrish, D.W., Gene regulation of neurokinin B and its receptor NK3 in late pregnancy and pre-eclampsia, *Mol. Hum. Repro.*, 12, 427–433, 2006.

Patacchini, R. et al., Newly discovered tachykinins raise new questions about their peripheral roles and the tachykinin nomenclature, *Trends Pharmacol. Sci.*, 25, 1–3, 2004.

Patak, E. et al., Tachykinins and tachykinin receptors in human uterus, *Br. J. Pharmacol.*, 139, 523–532, 2003.

Patel, H.J. et al., Transformation of breast cells by truncated NK1 secondary to activation with preprotachykinin-A peptides, *Proc. Natl Acad. Sci. USA*, 102, 17436–17441, 2005.

Pennefather, J.N. et al., Mammalian tachykinins and uterine smooth muscle: The challenge escalates, *Eur. J. Pharmacol.*, 500, 15–26, 2004.

Pintado, C.O. et al., A role for tachykinins in female mouse and rat reproductive function, *Biol. Reproduction*, 69, 940–946, 2003.

Poli-Brigelli, S. et al., Addition of the neurokinin-1 receptor antagonist aprepitant to standard antiemetic therapy improves control of chemotherapy-induced nausea and vomiting. Results from a randomised, double-blind, placebo-controlled trial in Latin America, *Cancer*, 97, 3090–3098, 2003.

Quartara, L. and Maggi, C.A., The tachykinin NK1 receptor: Part I. Ligands and mechanisms of cellular activation, *Neuropeptides*, 31, 537–563, 1997.

Redman, C.W., Sacks, G.P., and Sargent, I.L., Preeclampsia: An excessive maternal inflammatory response to pregnancy, *Am. J. Obstet. Gynecol.*, 180, 499–506, 1999.

Rosenkranz, M.A., Substance P at the nexus of mind and body in chronic inflammation and affective disorders, *Psychol. Bull.*, 133, 1007–1037, 2007.

Rösler, N., Wichart, I., and Jellinger, K.A., *Ex vivo* lumbar and post mortem ventricular cerebrospinal fluid substance P-immunoreactivity in Alzheimer's disease patients, *Neurosci. Lett.*, 299, 117–120, 2001.

Rovero, P. et al., A potent and selective agonist for NK-2 tachykinin receptor, *Peptides*, 10, 593–595, 1989.

Rudd, J.A., Jordan, C.C., and Naylor, R.J., The action of the NK1 tachykinin receptor antagonist, CP99994, in antagonising the acute and delayed emesis induced by cisplatin in the ferret, *Br. J. Pharmacol.*, 119, 931–936, 1996.

Rupniak, N.M. et al., *In vitro* and *in vivo* predictors of the anti-emetic activity of tachykinin NK1 receptor antagonists, *Eur. J. Pharmacol.*, 326, 201–209, 1997.

Sachais, B.S. et al., Molecular basis for the species selectivity of the substance P antagonist CP96345, *J. Biol. Chem.*, 268, 2319–2323, 1993.

Sakurada, T. et al., Pharmacological characterisation of NK1 receptor antagonist, [D-Trp7]sendide, on behaviour elicited by substance P in the mouse, *Naunyn Schmiedebergs Arch. Pharmacol.*, 350, 387–392, 1994.

Schmoll, H.J. et al., Comparison of an aprepitant regime with a multiple-day ondonsetron regimen, both with dexamethasone, for antiemetic efficacy in high dose cisplatin treatment, *Ann. Oncol.*, 17, 1000–1006, 2006.

Severini, C. et al., The tachykinin peptide family, *Pharmacol. Rev.*, 54, 285–322, 2002.

Shults, C.W. et al., A comparison of the anatomical distribution of substance P and substance P receptors in the rat central nervous system, *Peptides*, 5, 1097–1128, 1984.

Simmons, M.A. et al., Molecular characteristics and functional expression of a substance P receptor from sympathetic ganglion of *Rona catesbeiana*, *Neuroscience*, 79, 1219–1229, 1997.

Strader, C.D. et al., Structure and function of G-protein coupled receptors, *Ann. Rev. Biochem.*, 63, 101–132, 1994.

Sundelin, J.D. et al., Molecular cloning of the murine substance K and substance P receptors, *Eur. J. Biochem.*, 203, 625–631, 1992.

Tatemoto, K. et al., Neuropeptide K: Isolation, structure, and biological activities of a novel brain tachykinin, *Biochem. Biophys. Res. Commun.*, 128, 947–953, 1985.

Tattersall, F.D. et al., Enantioselective inhibition of apomorphine-induced emesis in the ferret by the neurokinin (1) receptor antagonist CP99994, *Neuropharmacology*, 33, 259–260, 1994.

Tenovuo, O., Rinne, U.K., and Viljanen, M.K., Substance P immunoreactivity in the post-mortem parkinsonian brain, *Brain Res*, 303, 113–116, 1984.

Torrens, Y. et al., Functional coupling of the NK1 tachykinin receptor to phospholipase D in chinese hamster ovary cells and astrocytoma cells, *J. Neurochem.*, 70, 2091–2098, 1998.

Trafton, J.A., Abbadie, C., and Basbaum, A.I., Differential contribution of substance P and neurokinin A to spinal cord neurokinin-1 receptor signaling in the rat, *J. Neurosci.*, 21, 3656–3664, 2001.

Van Euler, U.S. and Gaddum, J.H., An unidentified depressor substance in certain tissue extracts, *J. Physiol.*, 72, 74–87, 1931.

Vigna, S.R., Phosphorylation and desensitization of neurokinin-1 receptor expressed in epithelial cells, *J. Neurochem.*, 73, 1925–1932, 1999.

Warner, F.J. et al., Structure–activity relationships of neurokinin A (4–10) at the human tachykinin NK(2) receptor: The role of natural residues and their chirality, *Biochem. Pharmacol.*, 61, 55–60, 2001.

Warr, D.G. et al., Efficacy and tolerability of aprepitant for the prevention of chemotherapy-induced nausea and vomiting in patients with breast cancer after moderately emetogenic chemotherapy, *J. Clin. Oncol.*, 23, 2822–2830, 2005.

Warr D.G., Chemotherapy and cancer-related nausea and vomiting, *Curr. Oncol.*, 15, S4–S9, 2008.

Watling, K.J. and Krause, J.E., The rising sun shines on substance P and related peptides, *Trends Pharmacol. Sci.*, 14, 81–84, 1993.

de Wit, R. et al., The oral NK(1) antagonist, aprepitant, given with standard antiemetics provides protection against nausea and vomiting over multiple cycles of cisplatin-based chemotherapy: A combined analysis of two randomised, placebo-controlled phase III clinical trials, *Eur. J. Cancer*, 40, 403–410, 2004.

Wolters, E.Ch., PD-related psychosis: Pathophysiology with therapeutical strategies, *J. Neural Transm. Suppl.*, 71, 31–37, 2006.

Yoshioka, D. et al., Intrathecal administration of the common carboxyl-terminal decapeptide in endokinin A and endokinin B evokes scratching behaviour and thermal hyperalgesia in the rat, *Neurosci. Lett.*, 410, 193–197, 2006.

Zhang, Y. et al., Hemokinin is a hematopoietic-specific tachykinin that regulates B lymphopoiesis, *Nat. Immunol.*, 1, 392–397, 2000.

Zimmer, G. et al., Virokinin, a bioactive peptide of the tachykinin family, is released from the fusion protein of bovine respiratory syncytial virus, *J. Biol. Chem.*, 278, 46854–46861, 2003.

10 Neurotensin

Paul R. Dobner and Robert E. Carraway

CONTENTS

10.1 INTRODUCTION

Susan Leeman, an Assistant Professor at Brandeis University in 1967, directed several graduate students in an attempt to isolate corticotrophin releasing hormone (CRH). With limited facilities, they extracted 1 kg batches of hypothalami and fractionated the material on columns of Sephadex. Assaying for CRH by i.v. injection into anesthetized rats, Leeman astutely noted that specific fractions induced unexpected visible effects that were protease-sensitive. Hypothesizing that the active substances were yet-to-be discovered peptides, Leeman assigned Michael Chang to isolate the factor that simulated salivary secretion, which turned out to be substance P. Bob Carraway, another graduate student, pursued the identification of the cyanosis-inducing factor, and he eventually isolated (Carraway and Leeman, 1973), sequenced (Carraway and Leeman, 1975b), and synthesized (Carraway and Leeman, 1975a) the active peptide, which was named neurotensin (NT) based on its hypotensive property and neural origin.

Interestingly, all of the phylogenetic variants of NT characterized to date share C-terminal homology with NT (Carraway et al., 1982), as is also true for the intraspecies variants, LANT-6 and neuromedin N (NMN) (Carraway and Ferris, 1983; Minamino et al., 1984). Convergent evolution appears to explain the C-terminal similarity of NT and xenopsin, a 9-amino-acid peptide first discovered in toxin glands in the skin of *Xenopus laevis* (Araki et al., 1973) and later shown to exist in mammals as the 25-amino-acid counterpart xenin, which is part of the vesicle coat protein coatomer α (Feurle, 1998). Other peptides sharing C-terminal homology with NT include NT-related peptide and histamine releasing peptide, which are fragments of albumin that can stimulate inflammatory mast cells (Carraway, 1987; Carraway et al., 1989).

Although NT was first isolated based on its inflammatory properties, many other biologic effects have been characterized over the years—effects on temperature regulation, pain perception, addictive behaviors, reproductive function, appetite, digestion, growth of the intestinal mucosa, and cancer cell growth—some of which involve complex interactions with monoamine transmitters and hormones. This chapter focuses on the biosynthesis of NT, its molecular signaling and mode of action, its structural analogs, and their potential clinical applications.

10.2 SOURCE OF PEPTIDES

10.2.1 Structure of the NT Precursor

The structure of the NT precursor protein was initially inferred from DNA sequencing of cDNA clones from a library derived from canine intestinal endocrine cells that express NT (Dobner et al., 1987). The basic structure of the precursor is evolutionarily conserved from fish to humans (Figure 10.1) with NT lying near the C-terminus preceded by the related 6-amino-acid peptide NMN. The peptide coding domains and associated processing sites are highly conserved from fish to human; however, the rest of the precursor sequence has diverged considerably, except for a collection of conserved basic (light shading), acidic (dark shading), and scattered uncharged residues. The NT and NMN coding domains are bounded and separated by conserved paired basic amino acid residues that function as proteolytic processing sites. The precursor ranges in length from 164 amino acids in the zebrafish to 170 in humans. Subsequent cloning and sequence analysis of the rat and human genes revealed that the coding region is divided into 4 exons, with exon 1 encoding the predicted signal peptide and exon 4 encoding both NT and NMN (Kislauskis et al., 1988; Bean et al., 1992).

10.2.2 Precursor Processing

The NT precursor is differentially processed by distinct prohormone convertases (PCs), likely accounting for the production of different precursor fragments and peptides in the brain and gastrointestinal tract (Carraway and Mitra, 1990, 1991; Shaw et al., 1990; Kitabgi et al., 1991). Transfection experiments have provided evidence that PC2 completely processes the precursor to yield both NT and NMN (Rovere et al., 1996), whereas PC1 and PC5A give rise to N-terminally extended NMN and NT (Barbero et al., 1998). NT is extensively colocalized with PC1, PC2, and PC5A in the

```
                  10        20        30        40         50        60        70           80        90
Human      MMAGMKIQLVCMLLLAFSSWSLCSDSEEEMKALEADFLTNMHTSKIS--KAHVPSWKMTLLNVCSLVNNLNSPAEET---GEVHEEELVARRKLP
Chimp       MAGMKIQLVCMLLLAFSSWSLCSDSEEEMKALEADFLTNMHTSKLKV-KAHVPSWKMTLLNVCSLVNNLNSPAEET---GEVHEEELVARRKLP
Macaque    MMAGMKIQLVCMLLLAFSSWSLCSDSEEEMKALEADLLTSMHTSKIS--KAHVPSWKMTLLNVCSLVNNLNSPAEET---GEVHEEQLITRRKLP
Dog        MMAGMKIQLVCMILLAFSSWSLCSDSEEEMKALEADLLTNMHTSKIS--KASVSSWKMTLLNVCSFVNNLNSQAEET---GEFREEELITRRKFP
Cow        MMAGMKIQLVCMILLAFSSWSLCSDSEEEMKALETDLLTNMHTSKIS--KASVPSWKMSLLNVCSLINNLNSQAEET---GEFHEEELITRRKFP
Rat         MIGMNLQLVCLTLLAFSSWSLCSDSEEDVKVLEADLLTNMHASKVS--KGSPPSWKMTLLNVCSLINNLNSAAEEA---GEMRDDDLVAKRKLP
Mouse       MRGMNLQLVCLTLLAFSSWSLCSDSEEDVKALEADLLTNMHTSKIS--KASPPSWKMTLLNVCSLINNVNSPAEEA---GDMHDDDLVGKRKLP
Possum     MIARMKIQLICLMFLTFTSQSLCSDSEEEMKALEADLLTNMYTSKIS--KARFPHWKMTLINVCNLVNNLNSQVEET---GETGDDELVMRRQFS
Platypus    MTTMKIQLACMMLLAFTSWSLCSDSEEEMKALEADLLTNMYTSKIS--KARLPYWKMTLLNVCSLINNLNSQAEET---GETGEEELITRKQFP
Chicken        MRAQLVCVVLLALASCSLCSDSEEEMKALEADLLTNMYTSKMN--RAKLSYWKVTLLNVCNLINNMNNQVGET---VEVDDEDLISGRQFP
Xenopus     MTWTRFQLVCLMLLAFTCSAMCSDSEEEMKALETDVLSNIYSSKVN--KARLPYWKMTLLNVCGIINNLNNQAEEP---EETGEDEFLLRRQYP
Zebrafish      MQMQLTSFLLLFLLCNGLCSDIDQGKRAIEDEVLRSLLTSKVKASRHIAPLWQLPLQDVCRMVNGLGDSWLEAWANEEAAEDTEVHADYEQ
                                    ^                  ^

            100       110       120       130       140       150       160       170
Human       TALDGFSLEAMLTIYQLHKICHSRAFQHWELIQE-DILDTGNDKNGKEEVIKRKIPYILKRQLYENKPRRPYILKRDSYYY
Chimp       TALDGFSLEAMLTIYQLHKICHSRAFQHWELIQE-DILDTGNDKNEKEEVIKRKIPYILKRQLYENKPRRPYILKRDSYYY
Macaque     TALDGFSLEAMLTIYQLHKICHSRAFQHWELIQE-DILDTGNDKNEKEEVIKRKIPYILKRQLYENKPRRPYILKRDSYYY
Dog         TALDGFSLEAMLTIYQLQKICHSRAFQQWELIQE-DVLDAGNDKNEKEEVIKRKIPYILKRQLYENKPRRPYILKRGSYYY
Cow         AALDGFSLEAMLTIYQLQKICHSRAFQHWELIQE-DILDAGNDKNEKEEVIKRKIPYILKRQLYENKPRRPYILKRGSYYY
Rat         LVLDDFSLEALLTVFQLQKICRSRAFQHWEIIQE-DILDHGNEKTEKEEVIKRKIPYILKRQLYENKPRRPYILKRASYYY
Mouse       LVLDGFSLEAMLTIFQLQKICRSRAFQHWEIIQE-DILDNVNDKNEKEEVIKRKIPYILKRQLYENKPRRPYILKRGSYYY
Possum      T-LDGFNLEAMLTIYQLQKICQSRVFQHWELLQD-DILETGNLNREKEEVMKRKTPYILKRQIHVNKARRPYILKRSSSYY
Platypus    TALDGFSLEAMLTVYQLQKICHSRAFQHWELLQE-DVLDPGNSNHEKEDIVKRKSPYILKRQLHANKARRPYILKRSSYY
Chicken     AALDGFSLEAMLTVYQLQKVCHSRAFQHWELLQQ-DAFDLENSSQEKE-IMKRKNPYILKRQLHVNKARRPYILKRSSYY
Xenopus     M--DGLNLEAMLTIVQLQKICQSRGLQHWEFMQH-DYLEPNSPNSEKDEPIKRKTPYILKRQTYVSKARRPYILKRGSYY
Zebrafish   RVSGTLLQ--MLEEMHDIQNL-CRVLQPRELQDEQEYLELEQNS---DSPLKRKSPYILKRQLRTNKSRRPYILKRSVIY
                                          ^
```

FIGURE 10.1 NT/N precursor sequence conservation. Precursor sequences from the indicated species were aligned using BLASTP. Residues that are identical in all species examined are indicated in bold, and conserved basic (K or R) and acidic (D or E) residues are indicated with either dark or light shading, respectively. Arrowheads below the sequences indicate intron positions. The human sequence is numbered and sequence gaps are indicated by hyphens. The peptide coding domains and associated processing sites (residues 141–170 in human) are 73% identical from human to fish. The human (*Homo sapiens*, 5453816), chimp (*Pan troglodytes*, 114646098), macaque (*Macaca mulatta*, 109097971), dog (*Canis familiaris*, 73978173), cow (*Bos taurus*, 85719294), rat (*Rattus norvegicus*, 156139120), mouse (*Mus musculus*, 13277358), chicken (*Gallus gallus*, 50728400), and Xenopus (*Xenopus laevis*, 147901221) are from Genbank (*Genbank identifier in parentheses*). The possum (*Monodelphis domestica*, ENSMODP00000025645), platypus (*Ornithorhynchus anatinus*, ENSOANP00000010598), and zebrafish (*Danio rerio*, ENSDARP00000075084) sequences are from Ensembl (Ensembl identifier in parentheses).

central nervous system (CNS), consistent with a role for these PCs in NT precursor processing *in vivo* (Villeneuve et al., 2000a, 2000b), and evidence from PC2-deficient mice suggests that this convertase is particularly important for NT precursor processing in the hypothalamus (Villeneuve et al., 2002).

10.2.3 Neuronal and Endocrine Sources of NT and NMN

NT is widely distributed in the limbic regions of the rodent brain (Uhl et al., 1977; Kahn et al., 1980; Hara et al., 1982; Jennes et al., 1982) and in a variety of other mammals, including humans (Mai et al., 1987). *In situ* hybridization experiments have provided more detail regarding the distribution of NT neurons (Alexander et al., 1989c; Williams et al., 1990; Sato et al., 1991). The reader is referred back to the original references for detailed descriptions of the distribution of NT in the brain and spinal cord, because the discussion here will be limited for the most part to regions where physiological functions have been ascribed to the peptide.

NT-immunoreactive (IR) perikarya and fibers are distributed throughout several hypothalamic regions, consistent with other evidence indicating that NT modulates certain hypothalamic functions. Concentrations of NT-IR perikarya and fibers have been found in the medial preoptic area (mPOA), where NT has been implicated in gonadotropin releasing hormone (GnRH) release (see Section 10.4.2.1), Numerous NT-IR neurons have also been detected in the paraventricular nucleus; however, NT expression in this region appeared to be largely due to colchicine treatment (Kiyama and Emson, 1991; Alexander and Leeman, 1994). Moderate numbers of NT-IR cell bodies were also detected in the periventricular nucleus and the lateral hypothalamus.

NT-IR neurons have also been localized to various structures in the basal ganglia where the peptide has been shown to influence dopamine (DA) signaling. NT expression displays a high degree of plasticity in the striatum, where both DA agonist and antagonist treatment results in increased NT content and gene expression, particularly in the caudate-putamen (CPu) (reviewed in Dobner, 2005). NT-IR has been colocalized to a subpopulation of DA neurons in the ventral tegmental area (VTA) in the rat (Studler et al., 1988; Febvret et al., 1991), but not human (Gaspar et al., 1990) or mouse (Smits et al., 2004); however, NT mRNA has been colocalized to melanin-containing, presumably DA neurons, in the human VTA (Bean et al., 1992), and is expressed throughout the VTA and the substantia nigra (SN) in DA transporter-deficient mice (Roubert et al., 2004). Thus, NT expression appears to be driven by imbalances in DA signaling in several interconnected basal ganglia structures. However, there appears to be considerable variation in expression between species in midbrain DA neurons.

Scattered NT-IR neurons have been reported in the NAc and in the dorsomedial and ventrolateral CPu in rats. Treatment of rats and mice with haloperidol, which blocks DA D_2 receptors, resulted in a robust increase in both NT (Zahm, 1992; Zahm et al., 1998) and NT mRNA (Merchant et al., 1992; Merchant and Dorsa, 1993) expression in the dorsolateral CPu and a more modest increase in the NAc, and these increases were maintained following chronic haloperidol administration (Merchant et al., 1994). These long-term changes in NT expression have been postulated to be involved in the therapeutic and motor side effects of haloperidol and other antipsychotic drugs (APDs) (see Section 10.4.1.1).

NT is prominently expressed in enteroendocrine N cells located in the mucosal layer of the small intestine in man and many other species (reviewed in Sundler et al., 1982). In fact, >85% of the NT in the adult rat is found in the gastrointestinal tract (Carraway and Leeman, 1976). The chicken counterpart of NMN, LANT-6, was first isolated from extracts of chicken intestine (Carraway and Ferris, 1983) and immunohistochemical analysis indicated that LANT-6 and NT were colocalized in enteroendocrine cells in birds, reptiles, and bony fish (Reinecke, 1985). Ingestion of a high fat diet in humans (reviewed in Rosell, 1982) and instillation of fats into the rat intestine (Ferris et al., 1985) stimulated NT release, and NT has been implicated in fat uptake (Armstrong et al., 1986) and the recycling of bile acid (Gui et al., 2001).

10.3 MOLECULAR MECHANISMS OF ACTION

10.3.1 Neurotensin Receptors

Two structurally related G protein-coupled NT receptors, NTR-1 (also NTS-1, NTSR-1) and NTR-2 (also NTS-2, NTSR-2), have been cloned and characterized. NTR-1 corresponds to the high-affinity NT receptor that appears to mediate many of the pharmacological and physiological effects of the peptide (Tanaka et al., 1990). NTR-2 binds NT with somewhat lower affinity and in contrast to NTR-1 is blocked by the antihistamine levocabastine (Chalon et al., 1996; Mazella et al., 1996). There is evidence supporting a role for NTR-2 in mediating, at least in part, the analgesic effects of NT (Dubuc et al., 1999). The development of two nonpeptide NT receptor antagonists, the relatively selective NTR-1 antagonist, SR 48692, and the NTR-1/NTR-2 antagonist, SR 142948A, greatly facilitated the investigation of the physiological functions of the endogenous peptide (Gully et al., 1993, 1997). A third receptor, NTR-3 (also NTS-3, NTSR-3), corresponds to the primarily intracellular sortilin protein, which has been implicated in intracellular vesicle trafficking (Mazella et al., 1998). A structurally related intracellular sorting protein, SorLA/LR11 (Jacobsen et al., 2001), also binds NT with moderate affinity and thus may be considered a fourth NT receptor (NTR-4). Of these receptors, NTR-1 has been studied in the most detail and there is now considerable information regarding critical amino-acid residues and regions involved in both NT and SR 48692 binding, receptor internalization, and G-protein coupling.

10.3.2 The High-Affinity NT Receptor NTR-1

10.3.2.1 NT and SR 48692 Binding to NTR-1

Mutational analysis has been used quite effectively to gain insights into NT receptor binding and the results are illustrated schematically in Figure 10.2. These analyses revealed that the major determinants for the binding of NT (dark shading), SR 48692 (light shading), or both ligands (dark shading, italic font) lie within transmembrane (TM) spanning domains 6 and 7, and, in the case of NT, in the third extracellular (EC) loop of rat NTR-1 (Botto et al., 1997a; Labbe-Jullie et al., 1998; Barroso et al., 2000). Consistent with the hydrophobic composition of SR 48692, mutation of a number of hydrophobic amino-acid residues within TM 6 and 7 had a severe impact on SR 48692 binding in transfected

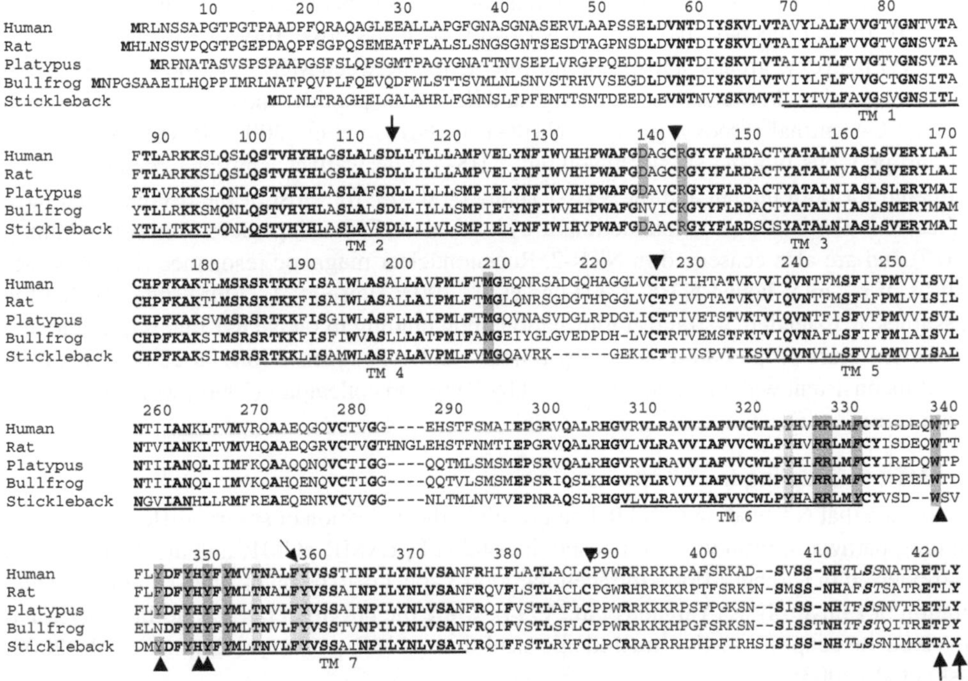

FIGURE 10.2 Sequence conservation and binding pocket of NTR-1. The NTR-1 sequences from several species from fish to human are depicted (amino acid numbers for rat are indicated). Conserved residues implicated as playing a major functional role are indicated by bold single-letter amino acid abbreviations. Regions corresponding to the predicted transmembrane spanning domains (TM1–7) are labeled and underlined. Residues that have been shown to play a major role in either NT or SR 48692 binding, are indicated by either dark or light shading, respectively. Residues indicated by italics with dark shading have been implicated in the binding of both ligands. Cysteine residues that are predicted to form a disulfide bond (C142 and C225 in rat) or to be palmitoylated (C388 in rat) are indicated by downward arrows. An aspartate residue that has been implicated in sodium sensitivity is indicated by a downward arrow. Serine and threonine residues that are required for interactions with β-arrestin are indicated in italics. Two residues at the C-terminus that have been implicated in receptor internalization are indicated by upward arrows. A phenylalanine residue, which when mutated to alanine results in constitutive activation of PI turnover, is indicated by a slanted arrow. Residues that have been implicated in G protein coupling are indicated by upward arrowheads. The human (*Homo sapiens*, 110611243), rat (*Rattus norvegicus*, 157822691), and bullfrog (*Rana catesbeiana*, 52078094) sequences are from Genbank and the platypus (*Ornithorhynchus anatinus*, ENSOANP00000007892) and stickleback (*Gastreosteus aculeatus*, ENSGACP00000005851) sequences are from Ensembl. Genus and species names as well as the Genbank and Ensembl identifiers are given in parentheses after the common name.

COS cells, including Tyr^{324}, Phe^{331}, Tyr^{351}, Phe^{358}, and Tyr^{359} (Labbe-Jullie et al., 1998). Two of these mutations, affecting residues predicted to be near the EC junctions of TM 6 and 7 (Phe^{331} to Ala and Tyr^{351} to Ala) also compromised NT binding (Figure 10.2). In addition, mutation of either Met^{208} in TM 4 or Arg^{327} (or a combination of Arg^{327} and Arg^{328}) in TM 6 markedly decreased both NT and SR 48692 binding (Botto et al., 1997a; Labbe-Jullie et al., 1998). Mutations in several other aromatic residues in EC loop 3 compromised NT binding without affecting SR 48692 binding, including Trp^{339}, Phe^{344}, Tyr^{347}, and Tyr^{349} (Labbe-Jullie et al., 1998; Barroso et al., 2000). These results indicate that the SR 48692 binding pocket is embedded in the TM domains 6 and 7, while the NT binding pocket lies more toward the EC surface, including at least a portion of EC loop 3, with some clear overlap between the two that most likely accounts for the competitive binding exhibited by the antagonist and NT.

A number of side chain interactions have been postulated based on receptor binding studies with various synthetic derivatives of NT(8–13) (Barroso et al., 2000). Although hypothetical, a number of interactions are supported by the data, including π–π contacts between Trp^{339}, Phe^{344}, and Tyr^{347} within EC 3 of the rat receptor and Tyr^{11} of NT(8–13) with the possibility of hydrogen bonding between the Tyr hydroxyl groups, a hydrophobic interaction between Met^{208} and Ile^{12}, cation-π and hydrophobic interactions between Phe^{331} and Arg^{9} and Ile^{12} of the peptide, and an ionic bond between Arg^{327} and the C-terminal carboxyl group of NT(8–13) (Barroso et al., 2000). Based on computer modeling and previous binding studies involving NT(8–13) derivatives, similar interactions have been postulated for both human and rat NTR-1 (Pang et al., 1996). Most of the residues that have been implicated in these side chain interactions are identical in NTR-1 sequences from fish to human (Figure 10.2) and are also conserved in NTR-2. Recent nuclear magnetic resonance (NMR) structural analysis of receptor bound NT(8–13) (Luca et al., 2003; Heise et al., 2005) suggests that the peptide assumes an extended β-strand conformation upon binding to rat NTR-1 that is qualitatively similar to the structure proposed previously by Barroso and colleagues (2000) based on mutational analysis, but inconsistent with the model proposed by Pang and colleagues (1996), which includes a proline type I turn.

10.3.2.2 NTR-1 Coupling to G Proteins

There is evidence that NT binding to NTR-1 can result in the activation of several different intracellular signaling pathways, including the phosphoinositol (PI), cAMP, cGMP, and arachadonic acid pathways (reviewed in Kitabgi, 2006). Intracellular loop 3 appears to be involved in G_q coupling and the simulation of PI hydrolysis (Yamada et al., 1994), while the proximal portion of the C-terminal tail is involved in coupling to the arachidonic acid and cAMP pathways (Najimi et al., 2002; Skrzydelski et al., 2003).

Ligand binding to G protein-coupled receptors is thought to result in a change in receptor conformation that results in G protein activation and dissociation from the receptor. Several mutations in NTR-1 EC loop 3 have been shown to affect G protein coupling to a greater extent than NT binding (indicated by upward arrowheads in Figure 10.2), suggesting that these mutations interfere in some way with the conformational changes required for G protein activation (Richard et al., 2001b). Interestingly, another point mutation, Phe^{358} to Ala in TM 7 (arrow in Figure 10.2), resulted in constitutive PI turnover, but not cAMP, production in transfected cells (Barroso et al., 2002). There is also evidence that substitution of Asp^{113} in TM 2 (downward arrow in Figure 10.2) is involved in both sodium sensitivity and G protein activation (Martin et al., 1999). Collectively these results suggest that Phe^{358} is most likely involved in side chain interactions that keep NTR-1 in an inactive state and that NT binding results in a change in conformation most likely involving EC loop 3 residues and Asp^{113} that results in G protein activation. In the future, further mechanistic insights might be gleaned from the identification of mutations that suppress the constitutive activity of the Phe^{358} mutant.

10.3.2.3 Sequence Conservation and Additional Functional Domains of NTR-1

Sequence comparisons also revealed the presence of conserved cysteine residues (Figure 10.2, upward arrows) at positions that are likely to be involved in the formation of a disulfide bond between

Cys^{142} near the top of TM 3 and Cys^{225} in EC loop 2 of rat NTR-1 (Barroso et al., 2000). In addition, by analogy with a variety of other G protein-coupled receptors (Qanbar and Bouvier, 2003), Cys^{388} (in rat NTR-1) in the membrane proximal portion of the C-terminal tail may be palmitoylated and could play a role in coupling to the cAMP pathway. These comparisons also revealed a high degree of sequence conservation in the region of NTR-1 encompassing TM 2, EC loop 1, and TM 3, which, together with the observation that mutation of Asp^{139} greatly decreases NT binding, may indicate that this region is involved in NT binding and the structural transitions required for G protein activation and signal transduction.

Several residues in the C-terminal tail of NTR-1 have also been implicated in receptor internalization and interactions with β-arrestin. In CHO cells, the membrane proximal portion of the C-terminal tail was required for internalization (Hermans et al., 1996; Najimi et al., 2002); however, in COS cells, conserved C-terminal Thr^{422} and Tyr^{424} residues were essential (Chabry et al., 1995). Interaction with β-arrestin, but not internalization, was disrupted by mutation of the contiguous potential phosphorylation sites, Ser^{415}, Thr^{416}, Ser^{417} residues in rat NTR-1, but not by mutation of several Ser residues upstream of these positions (Oakley et al., 2001).

10.3.3 NTR-2 Receptor Signaling

NTR-2 cDNA clones have been isolated from mouse, rat, and human, and encoded protein corresponds to the biochemically and pharmacologically defined low-affinity, levocabastine-sensitive NT receptor (Chalon et al., 1996; Mazella et al., 1996; Vita et al., 1998). The expression of mouse NTR-2 in frog oocytes revealed that NT could activate an inward Ca^{2+}-dependent chloride current, suggesting that the receptor was coupled to intracellular signaling pathways; however, this effect was also produced by the "antagonists" levocabastine and SR 48692 (Mazella et al., 1996; Botto et al., 1997b). The agonist effects of these antagonists were confirmed in COS cells transfected with either the human (Vita et al., 1998) or rat (Yamada et al., 1998; Richard et al., 2001a) NTR-2 receptor. A subsequent report by Gendron and colleagues indicated that the ability of SR 48692 to increase intracellular Ca^{2+} levels did not require NTR-2 expression (Gendron et al., 2004), contradicting earlier reports (Vita et al., 1998; Yamada et al., 1998). Although NT has been reported to antagonize NTR-2 activation in transfected cells, NT (and levocabastine) stimulated MAPK activity in cultured cerebellar granule cells, which express NTR-2 and NTR-3, but not NTR-1, which the authors interpreted as evidence that NT is capable of signaling through NTR-2 (Sarret et al., 2002). However, in view of recent evidence that NT can stimulate MAPK activity through NTR-3 in microglial cells (Martin et al., 2003; Dicou et al., 2004), it is difficult to ascribe the effects in cerebellar granule cells to NTR-2 with certainty.

10.3.4 NTR-3/Sortilin and NTR-4/SorLA

NTR-3/sortilin and NTR-4/SorLA/LR11 bind NT with relatively high affinity through a multifunctional binding domain containing multiple disulfide bridges, which was first identified in the yeast vacuolar sorting protein Vps10p (Petersen et al., 1997, 1999; Jacobsen et al., 2001), and although the receptors appear to be mainly intracellular they have also been shown to interact with certain cell surface receptors. These NT binding domains also bind a number of other proteins and protein fragments, including the propeptide fragments generated by furin cleavage of full-length sortilin and SorLA/LR11, and proneurotrophins (Petersen et al., 1999; Jacobsen et al., 2001; Nykjaer et al., 2004). In fact, NTR-3/sortilin appears to be a coreceptor that together with the $p75^{NTR}$ neurotrophin receptor binds unprocessed proneurotrophins, triggering programmed neuronal cell death, and NT can counteract this effect *in vitro* (Nykjaer et al., 2004). NTR-3/sortilin-deficient mice exhibit reductions in neuronal cell death in the developing retina, the aging superior cervical ganglia, and following neuronal injury (Jansen et al., 2007), offering the possibility that NT might be useful therapeutically in the prevention of injury-related neuronal cell death. NT also appears to either stimulate (Dal Farra et al., 2001) or inhibit (Martin et al., 2002) cell growth through NTR-3 or NTR-3/NTR-1 receptor complexes, respectively.

10.4 BIOLOGICAL MODE OF ACTION

NT has been implicated in a diverse array of functions in the both the CNS and the periphery and a detailed discussion of all of these could easily fill a volume. NT is closely associated with DA circuits, and in Sections 10.4.1.1 and 10.4.1.2 we discuss recent evidence implicating endogenous NT in certain APD responses and amphetamine sensitization. In Sections 10.4.2.1 to 10.4.2.3, the potential reproductive functions of endogenous NT are discussed, including the potential role of NT in feedback inhibition of PRL secretion and in the feedforward effects of estrogen on GnRH release. Possible NT involvement in gastrointestinal functions, mast cell stimulation, appetite control, stress responses, analgesia, and other functions are barely touched upon; however, the reader is referred to an excellent compendium of reviews touching on many of these functions recently published in *Peptides* (*Peptides* 27, 2006) for further information.

10.4.1 NT Balancing Act in the Basal Ganglia

NT expression in the striatum, particularly the dorsal striatum, is incredibly plastic and is markedly increased in response to both APDs and psychostimulants, apparently due to D_2 receptor blockade or D_1 receptor stimulation, respectively (reviewed in Dobner, 2005). This robust plasticity suggests that the induction of NT is an adaptive response to alterations in DA signaling that may normally serve to restore balance in the DA circuits.

10.4.1.1 NT and Antipsychotic Drugs

Central NT administration was found to produce a variety of effects that are similar to those of APDs, leading to the hypothesis that the peptide may be an endogenous neuroleptic (Nemeroff, 1980). Similar to APDs, NT attenuated amphetamine- and DA-stimulated locomotor activity following microinjection into the NAc, suggesting that NT might serve to limit mesolimbic DA signaling (Ervin et al., 1981; Kalivas et al., 1984). All clinically effective APDs are DA D_2 receptor antagonists, supporting the idea that increases in DA signaling underlie schizophrenic symptoms, particularly psychotic symptoms (Creese et al., 1976; Meyer-Lindenberg et al., 2002). NT displays no appreciable D_2 receptor affinity, but likely acts to influence either D_2 receptor affinity through receptor crosstalk (Agnati et al., 1983; von Euler et al., 1991; Li et al., 1995; Diaz-Cabiale et al., 2002) or transmitter release from striatal afferents expressing NTR-1 (Pickel et al., 2001; Ferraro et al., 2007).

Several recent studies have implicated endogenous NT in responses to certain APDs, but not others (Binder et al., 2001, 2002; Kinkead et al., 2005, 2008). These studies have relied on animal models of sensorimotor gating defects, including prepulse inhibition (PPI) of the acoustic startle reflex (Binder et al., 2001; Kinkead et al., 2005, 2008) and latent inhibition (LI) (Binder et al., 2001, 2002). PPI measures the ability of a mild acoustic stimulus delivered 30 to 500 msec prior to a more intense startle-evoking acoustic stimulus to inhibit the startle reflex, and APDs restore PPI following disruption by either isolation-rearing or pharmacological manipulations. Pretreatment with the NT antagonist SR 142948A blocked the ability of both haloperidol and quetiapine to restore defects in PPI in isolation-reared rats, suggesting that NT is required for certain APD effects (Binder et al., 2001). Furthermore, quetiapine treatment normalized NT mRNA levels in the NAc, which were significantly lower in isolation- compared to socially-reared animals, suggesting that the PPI deficit in isolation-reared animals was due to decreased NT signaling in the NAc (Binder et al., 2001). SR 142948A also attenuated both basal and haloperidol-induced LI in rats (Binder et al., 2001, 2002) and the relatively selective NTR-1 antagonist SR 48692 had similar effects on basal LI (Binder et al., 2002). Collectively, these results suggest that NT plays a role in both basal PPI and the enhancement of PPI by certain APDs.

The analysis of PPI in NT-deficient mice has confirmed and extended these findings (Kinkead et al., 2005). Interestingly, these studies revealed that not all APDs require NT for the enhancment of PPI. Although haloperidol and quetiapine augmentation of PPI was blocked in NT-deficient mice, the response to clozapine was spared (Kinkead et al., 2005). Basal PPI was also modestly reduced in these mice. These results indicate that there are both NT-dependent and NT-independent APDs and that the NT-dependent APDs include both typical (haloperidol) and atypical (quetiapine) APDs. A key future objective will be to delineate the anatomical sites and the mechanisms underlying NT action in these responses.

The analysis of c-Fos expression has provided an important approach toward the identification of neuronal circuits involved in APD actions (Deutch et al., 1992; Robertson and Fibiger, 1992) and has been used to uncover APD neuronal responses that require NT signaling (Dobner et al., 2001; Fadel et al., 2001; Binder et al., 2004). All APDs that have been examined induce Fos expression in the NAc, consistent with other evidence that the antipsychotic effect of these drugs involves the attenuation of mesolimbic DA signaling. In contrast, typical and atypical APDs have divergent effects on Fos in either the dorsolateral CPu or prefrontal cortex (PFC), respectively. Typical and atypical APDs differ in both their therapeutic and extrapyramidal side effect (EPS) profiles, and the regional differences in Fos activation have been hypothesized to reflect these differences, with Fos expression in the dorsolateral CPu and PFC perhaps accounting for the increased EPS liability of typical APDs and the improved therapeutic profile of atypical APDs, respectively (Deutch et al., 1992; Robertson and Fibiger, 1992). The analysis of Fos expression in rats pretreated with SR 48692 and in NT-deficient mice presaged the PPI results discussed above in that the clozapine response was unaffected, but haloperidol-evoked Fos expression was markedly attenuated in the central and dorsolateral CPu (Dobner et al., 2001; Fadel et al., 2001; Binder et al., 2004). Surprisingly, NT-deficient mice and rats pretreated with SR 142948A displayed normal haloperidol-induced catalepsy, suggesting that the population of affected neurons in the CPu is not involved in the production EPS, but some other function such as PPI (Fadel et al., 2001; Binder et al., 2004).

The mechanism through which NT influences neuronal activation in the dorsolateral CPu remains speculative; however, acute administration of the D_2 antagonist eticlopride resulted in a reduction in extracelluar NT levels in the NAc, and medial, but not lateral, CPu, suggesting that phasic NT release is not involved (Wagstaff et al., 1996). NTR-1 expression is mainly confined to DA and excitatory afferents in the striatum (Pickel et al., 2001), and intrastriatal NT administration increases extracellular DA and glutamate levels (reviewed in Ferraro et al., 2007). These observations coupled with evidence that haloperidol-evoked Fos expression requires glutamate receptor signaling (Boegman and Vincent, 1996; Hussain et al., 2001) suggest that NT may augment this response through the stimulation of corticostriatal afferents (Dobner et al., 2003). In fact, there is evidence that NT influences neuronal activity through such a mechanism in the striatum (Matsuyama, 2003) and elsewhere (Saleh et al., 1997; Chen et al., 2006); however, a recent report suggests that NT may also inhibit glutamate release in the dorsolateral striatum under certain circumstances (Yin et al., 2007).

10.4.1.2 NT Involvement in Psychostimulant Sensitization

Site-specific microinjection of NT into the VTA stimulates locomotor activity and increases DA turnover in the NAc, similar to psychostimulants (reviewed in Dobner, 2005). Furthermore, daily NT administration in this region resulted in the augmentation or sensitization of NT, but not amphetamine, stimulation of locomotor activity (Elliott and Nemeroff, 1986; Kalivas and Duffy, 1990). Repeated i.c.v. NT administration does, however, lead to cross-sensitization of amphetamine and cocaine locomotor responses (Rompre, 1997; Rompre and Bauco, 2006).

There is increasing evidence that endogenous NT may play a key role in amphetamine sensitization and also influence certain cocaine responses (Table 10.1). Daily pretreatment with SR 48692 for five days prior to initiating cocaine treatments delayed the development of cocaine sensitization by several days (Horger et al., 1994); however, coadministration of SR 48692 and cocaine was ineffective (Horger et al., 1994; Betancur et al., 1998). In contrast, SR 48692 attenuated or blocked

TABLE 10.1
Neurotensin Anatagonist Effects on Amphetamine and Cocaine Sensitization

Antagonist	Dose[a]	Drug	Dose[a]	Protocol	Antagonist Effect(s)	Reference
SR 48692	80 µg/kg, i.p., p.o.	Cocaine	15	SR 48692[b] alone 5× daily, 7 days drug-free, cocaine every other day 6×	Delay in sensitization, no effect on acute response	Horger et al., 1994
SR 48692	80 µg/kg	Cocaine	15	Cocaine 3× every other day, SR 48692[b] 60 min prior to cocaine first 2× only	No effect	
SR 48692	1	Cocaine	15	SR 48692[b] 1 h prior to acute cocaine	No effect on horizontal locomotor activity, marked decrease in rearing	Betancur et al., 1998
SR 48692	0.1, 1	Cocaine	15	SR 48692[b] alone 5× daily, SR 48692[b] 1 h prior to cocaine daily 10× , SR 48692[b] 7× daily (no cocaine), SR 48692[b] 1 h prior to cocaine challenge	Decreased acute locomotor activition, decreased rearing 1st, 2nd, and challenge doses only, no effect on stereotyped behavior	
SR 48692	40, 80, 160 µg/kg	Amph.	0.75, 1.5	SR 48692[b] 30 min prior to amph.[b] (1.5 mg/kg) 4× every other day, drug-free week, amph. challenge (0.75 mg/kg)	Attenuated sensitization at two higher SR 48692 doses; lowest SR 48692 dose accelerated sensitization, prolonged some responses	Rompre and Perron, 2000
SR 48692	0.3	Amph.	2	Saline 30 min prior to amph.[b] every other day 7× , SR 48692[b] 15 min prior to amph. challenge 9 days later (Swiss mice)	No effect on acute locomotor activation, blocked expression of sensitization	Costa et al., 2001
SR 48692	0.1, 1	Amph.	0.5, 1.0	SR 48692[b] daily for 14 or 21 (1 mg/kg only) days, amph.[b] every other day 4×, amph.[b] challenge 1 or 2 weeks later	No effect on acute locomotor activation, attenuated sensitization at both amph. doses; no effect on stereotyped behavior; no effect at lower SR 48692 dose	Panayi et al., 2002
SR 142948A	5 pmole per side	Amph.	1, s.c.	Bilateral intra-VTA[c] SR 142948A[b] 1 min prior to amph.[b] 1st day, drug-free week, SR 142948A[b] 1 min prior to amph.[b] challenge	No effect on acute locomotor activation, attenuated sensitization, except when administered only prior to amph. challenge	Panayi et al., 2005
SR 142948A	0.03, 0.1, 0.3	Amph.	1, s.c.	SR 143948A[b] prior to sensitizing dose only	No effect on acute or sensitized response	
SR 48692	0.3	Amph.	2	Amph.[b] every other day 6× followed by SR 48692[b] 7× daily and subsequent amph. challenge (next day) (Swiss mice)	High and low responders to novelty analyzed separately; attenuated sensitized response in low responder group only	Costa et al., 2007

continued

TABLE 10.1 (continued)

Antagonist	Dose[a]	Drug	Dose[a]	Protocol	Antagonist Effect(s)	Reference
SR 48692	1	Cocaine	15	Cocaine[b] 4× every other day, drug-free week, cocaine challenge, SR48692[b] 14× daily, additional cocaine challenges after 1 or 2 weeks SR 48692	No effect after 1 week, attenuated sensitized locomotor and rearing response to cocaine after two weeks; no effect on acute responses	Felszeghy et al., 2007
SR 48692	1	Cocaine	15	Cocaine[b] 4× every other day paired with compartment, SR 48692[b] daily for 10 days, cocaine in home cage 30 min after final SR 48692, CPP[d] measured 24 h later	Significantly reduced time spent in cocaine-paired chamber	

[a] i.p., mg/kg, except where indicated; [b] or vehicle; [c] ventral tegmental area; [d] conditioned place preference.

both the initiation and expression of amphetamine sensitization (Rompre and Perron, 2000; Costa et al., 2001; Panayi et al., 2002). In a single-dose sensitization protocol, intra-VTA administration of SR 142948A just prior to the sensitizing dose of amphetamine blocked the initiation of sensitization; however, similar treatment just prior to amphetamine challenge had no effect, suggesting that NT signaling in the VTA is important for the initiation of, but not the expression of, amphetamine sensitization (Panayi et al., 2005), although systemic SR 48692 treatment also blocked the expression of amphetamine sensitization in mice (Costa et al., 2001). An important caveat in the interpretation of these studies is that both SR 48692 and SR 142948A have been reported to act as agonists at NTR-2 (Vita et al., 1998; Yamada et al., 1998; Richard et al., 2001a), and the possibility that these agents influence amphetamine sensitization through the stimulation of NTR-2 cannot be discounted. In fact, Rompre and Perron reported that lower doses of SR 48692 actually accelerated sensitization and prolonged certain sensitized behavioral responses (Rompre and Perron, 2000). Because SR 48692 has an ~16-fold higher affinity for NTR-1 than NTR-2 in rats (Gully et al., 1993), this result could be interpreted to suggest that both NT stimulation of NTR-1 and SR 48692 stimulation of NTR-2 act to limit amphetamine sensitization. The observation that NT agonist treatment blocks nicotine sensitization is also consistent with the view that NT signaling through NTR-1 could serve to reign in sensitization (Fredrickson et al., 2003a, 2003b). Experiments in NT-, and NT receptor-deficient mice (Dobner et al., 2001; Pettibone et al., 2002; Leonetti et al., 2004; Maeno et al., 2004) will undoubtedly shed further light on the precise role that NT plays in psychostimulant sensitization.

If NT is involved in amphetamine sensitization, the mechanism most likely involves NT activation of VTA DA neurons, which are enmeshed in NT fibers. Recent retrograde tracer experiments have revealed that the majority (~70%) of the NT-expressing neurons that contribute to this dense fiber plexus are located in the hypothalamus, including a continuum extending from the lateral POA into the rostral lateral hypothalamus and a collection of neurons centered in the mPOA (Zahm et al., 2001; Geisler and Zahm, 2006), and disinhibition of this region stimulated locomotor activity through a mechanism involving NT signaling in the VTA (Reynolds et al., 2006). Acute methamphetamine treatment revealed another collection of inducible NT afferents in the NAc shell, which could potentially be involved in amphetamine sensitization (Geisler and Zahm, 2006).

The analysis of c-Fos expression has provided insights into the neuronal populations that require NT to respond to psychostimulants (Alonso et al., 1999; Fadel et al., 2006; Costa et al.,

2007). Both amphetamine and cocaine induction of Fos expression in the dorsal striatum was NT-dependent to some degree (Alonso et al., 1999; Fadel et al., 2006). In both NT-deficient mice and in rats pretreated with SR 48692, amphetamine-evoked Fos was selectively attenuated in the medial CPu (Fadel et al., 2006), an area that has been implicated in cognitive functions and behavioral flexibility (Ragozzino and Choi, 2004). More recently, daily SR 48692 administration for seven days following amphetamine sensitization was reported to suppress both behavioral sensitization and Fos activation in the NAc following amphetamine challenge in a subpopulation of mice that displayed a low locomotor response to novelty, perhaps indicating that NT signaling in the NAc may be important for the expression of amphetamine sensitization (Costa et al., 2007). Intriguingly, SR 48692 administration following cocaine sensitization also attenuated cocaine sensitization (Felszeghy et al., 2007). Together these results suggest that NT signaling may be important for the maintenance of sensitization, and that NT plays an important role in psychostimulant-induced neuronal activation in the medial striatum and perhaps also the NAc following sensitization.

10.4.2 Potential Involvement of NT in Reproductive Functions

10.4.2.1 NT Modulation of GnRH Release

NT is expressed in hypothalamic regions that are involved in the control of reproductive function and there is considerable evidence that NT may mediate, at least in part, the feedforward effects of estrogen on gonadotropin releasing hormone (GnRH) release, which in turn produces a surge in luteinizing hormone (LH) in rats (reviewed in Rostene and Alexander, 1997). NT administration in the mPOA stimulated LH release and increased the magnitude of an artificially induced LH surge. Similar administration of an α-NT antiserum produced the opposite effect (Ferris et al., 1984; Akema et al., 1987; Alexander et al., 1989b), and increased NT release from the median eminence was correlated with increases in LH and PRL release (Watanobe and Takebe, 1993). GnRH neurons are closely apposed to NT-IR fibers (Hoffman, 1985) and express NTR-1 (Smith and Wise, 2001), but not estrogen receptor (ER) α (Hrabovszky et al., 2000; Kallo et al., 2001), which is however expressed in NT neurons (Axelson et al., 1992; Herbison and Theodosis, 1992), suggesting that NT neurons may mediate, at least in part, the feedforward effects of estrogen on GnRH release.

10.4.2.2 Estrogen Regulation of NT Expression

NT expression in several hypothalamic regions was found to be exquisitely sensitive to circulating estrogen levels (Alexander et al., 1989a; Alexander, 1993; Alexander and Leeman, 1994), resulting in cyclical expression during the estrous cycle (Smith and Wise, 2001) and sexually dimorphic expression in rats (Alexander et al., 1991). Estrogen had similar effects on NT expression in mice (Watters and Dorsa, 1998), and these effects were mediated through estrogen effects on cAMP signaling rather than conventional ER transcriptional activation (Watters and Dorsa, 1998). These results suggest that cyclical increases in NT expression driven by increases in circulating estrogen levels may be involved in conveying the feedforward effects of estrogen on GnRH release.

10.4.2.3 NT Regulation of PRL Secretion

NT has also been implicated in the modulation of PRL release at both the hypothalamic and pituitary level (reviewed in Rostene and Alexander, 1997). Intraventricular NT administration resulted in the inhibition of PRL release, suggesting that NT is involved in negative feedback regulation in the hypothalamus; however, peripheral NT administration increased plasma PRL levels (reviewed in Rostene and Alexander, 1997). NT most likely directly activates tuberoinfundibular DA (TIDA) neurons in the hypothalamus leading to increased DA inhibition of pituitary lactotropes (Hentschel et al., 1998). The stimulatory effects of NT on PRL release following peripheral administration may be due to either antagonism of D_2 receptor-mediated DA inhibition or direct stimulation of pituitary lactotropes. Interestingly, estrogen increases NT expression in TIDA neurons and the corelease of NT with DA could attenuate DA inhibition of PRL secretion during the PRL surge at estrous (Alexander, 1993, 1999).

Despite the evidence supporting the involvement of NT in the control of reproductive function, NT-deficient mice can reproduce and successfully nurse their pups (Dobner et al., 2001); however, detailed assessments of the LH and PRL surge and other aspects of reproductive physiology may yet reveal either defects in reproductive hormonal regulation or compensatory mechanisms that allow for relatively normal function in the absence of NT.

10.5 STRUCTURAL ANALOGS

Many NT analogs have been synthesized and tested for biological activity and these analogs generally involve modifications of NT(8–13), designed to protect the peptides from protease digestion to extend their bioavailability. To understand the reasoning behind these modifications, we first discuss NT degradation to illustrate which peptide bonds are particularly protease sensitive, and discuss stabilized NT analogs that in some cases appear to penetrate the blood–brain barrier following peripheral administration.

10.5.1 NT is Rapidly Degraded by Several Endopeptidases

NT and NMN are both rapidly degraded by metallo-peptidases, which are present in brain membrane preparations, and all the NT analogs that have central effects following peripheral administration contain modifications that are designed to circumvent this degradation. NT is principally cleaved at the Arg^8–Arg^9, the Pro^{10}–Tyr^{11}, and the Tyr^{11}–Ile^{12} peptide bonds when incubated with brain membrane preparations (Checler et al., 1983, 1984, 1985). Thimet oligopeptidase and enkephalinase (neprilysin) have been implicated in the cleavage of the Arg^8–Arg^9 and Tyr^{11}–Ile^{12} bonds, respectively (Checler et al., 1983, 1984, 1985). Enkephalinase was also shown to be partially responsible for cleavage at the Pro^{10}–Tyr^{11} bond, with the remainder of the activity contributed by the novel endopeptidase, neurolysin (Checler et al., 1986a). Cleavage at any of these positions results in the inactivation of NT. NMN is degraded much more rapidly than NT due to the action of aminopeptidases at the unprotected N-terminus of the peptide, with the first Lys–Ile peptide bond being the most susceptible to cleavage (Checler et al., 1986b). There is also evidence that NMN is cleaved by endopeptidases (Dubuc et al., 1988), perhaps at the Pro–Tyr and/or Tyr–Ile peptide bonds, which are the preferred cleavage sites in NT.

10.5.2 Stabilized NT Analogs

In an attempt to identify specific agonists and antagonists, particularly those displaying an enhanced biological stability and/or the ability to cross the bloodbrain barrier, numerous NT analogs have been synthesized. The stability of NT(8–13) was greatly enhanced by methylation of the N-terminus and substitution of Tle for Ile^{12} (Tokumura et al., 1990). This analog was more than 40 times more potent than NT or NT(8–13) in inhibiting locomotion when injected i.p. into mice, suggesting that it crossed the blood–brain barrier (Heyl et al., 1994). Systematically replacing the five peptide bonds in NT(8–13) with CH_2NH (ψ, reduced) bonds, Kitabgi's group found that placement of the ψ-bond at all positions except at Arg^8–Arg^9 resulted in a profound loss of affinity for NT receptors (Lugrin et al., 1991). Further work led them to identify JMV449 (Table 10.2), a ψ-bonded peptide at the 8–9 position, which is more stable and more potent than NT in receptor binding and biologic activity (Dubuc et al., 1992). Richelson's group also studied variants of NT(8–13) and identified Arg^8 as a position that could be modified with D-Lys or D-Orn to promote stability without a major loss of receptor binding activity (Cusack et al., 1995). Further work led them to introduce

Table 10.2
Structures of Promising NT Analogs

Name	Structures	Reference
NT	<Glu–Leu–Tyr–Glu–Asn–Lys–Pro–Arg–Arg–Pro–Tyr–Ile–Leu	Carraway and Leeman, 1975b
JMV449	Lysψ(CH_2NH)Lys–Pro–Tyr–Ile–Leu	Dubuc et al., 1992
NT1	N–Me–Arg–Lys–Pro–Trp–Tle–Leu	Machida et al., 1993; Heyl et al., 1994
NT66L	D-Lys–Arg–Pro–*neo*-Trp–Tle–Leu	Tyler et al., 1999
NT69L	N–Me–Arg–Lys–Pro–*neo*-Trp–Tle–Leu	Tyler-McMahon et al., 2000
KK13	N_3–HLys–Arg–Pro–Tyr–Tle–Leu	Kokko et al., 2005
KK28	N–Me–Lys–Arg–Pro–Tyr–Tle–Leu	Hadden et al., 2005
JMV2012	cyclic–[Lys–Lys–Pro–Tyr–Ile–Leu]$_2$	Bredeloux et al., 2008
#17	DTPA–D-Lys–Pro–Gly(PipAm)–Arg–Pro–Tyr–Tle–Cha	Achilefu et al., 2003
NT-VIII	(NαHis)AcN–Me–Arg–Lys–Pro–Tyr–Tle–Leu	Garcia-Garayoa et al., 2002
NT-XI	(NαHis)AcLys–(ψCH2-NH)–Arg–Pro–Tyr–Tle–Leu	Buchegger et al., 2003

L-*neo*-Trp in the Tyr^{11} position, producing compounds NT66L and NT69L (Table 10.2), which potently induced hypothermia and analgesia when given i.p. to rats (Table 10.3) (Tyler-McMahon et al., 2000). Dix's group then developed two analogs (KK13 and KH28, Table 10.2) that featured substitution of azido-7-hexanoic acid for Arg^8, which elicited hypothermia and inhibited amphetamine-induced hyperlocomotion when given i.p. or orally to rats (Table 10.3) (Hadden et al., 2005; Kokko et al., 2005). Another strategy to enhance lipophilicity led to the cyclic dimer of (Lys^8, Lys^9)-NT8-13 called JMV2012 (Table 10.2), which potently produced analgesic and hypothermic effects when given i.v., s.c., or orally to mice (Table 10.3) (Bredeloux et al., 2008). These and other studies have shown that the structural features of NT(8–13) cannot be modified greatly without compromising receptor binding affinity. Whereas the N-terminus can be blocked and positively charged substituents can be placed at Arg^8 and Arg^9, only minor changes to Pro^{10}, Ile^{12}, and Leu^{13} are tolerated. Although the Tyr^{11} position can accept extra steric bulk, an aromatic side chain with H-bonding ability is essential.

10.6 CLINICAL APPLICATIONS

10.6.1 Analgesic Effects of Peripherally Administered NT Analogs

Intracisternal NT administration was first found to result in a μ opioid-independent analgesia using the hot plate and acetic acid induced writhing assays (Clineschmidt and McGuffin, 1977; Clineschmidt et al., 1979). Every stable NT(8–13) analog that has been tested produces potent analgesic effects following either i.p., i.v., s.c., or p.o. administration in rats, and in some cases mice and monkeys (Table 10.3). These results suggest that certain NT analogs may find a use clinically as μ opioid-independent analgesics with low abuse potential; however, several lines of evidence suggest that NT is capable of producing central reward-like effects (Glimcher et al., 1984, 1987; Heidbreder et al., 1992; Rompre et al., 1992). In a recent preclinical study in rhesus monkeys, i.v. administration of NT69L produced fairly potent analgesic effects, but had no effect on cocaine self-administration and was not self-administered, suggesting that there is little risk of NT69L abuse (Fantegrossi et al., 2005). However, this study also revealed that several side effects might limit the utility of NT69L if they cannot be circumvented, including a fairly dramatic acute hypotensive response, and increased defecation at higher doses (Fantegrossi et al., 2005).

TABLE 10.3
Biological Effects of NT Analogs

Analog	Route	Dose Range (mg/kg)	Species	Effects	Reference
NT1	s.c.	0.01–1	OF-1 mouse	Analgesia, attenuated locomotor activity	Sarhan et al., 1997
NT1	i.p.	ED_{50} 0.3, 0.4	SD rat	Hypothermia, analgesia	Tyler et al., 1999
NT66L	i.p.	ED_{50} 0.5, 0.07	SD rat	Hypothermia, analgesia	Tyler et al., 1999
NT67L	i.p.	1	SD rat	Hypothermia, analgesia	Tyler et al., 1999
NT69L	i.p.	ED_{50} 0.3, 0.3; 0.001–1	SD rat	Hypothermia, analgesia, attenuated stimulant-induced locomotor activity, suppressed appetite and reward responding, attenuated sensitization, blocks catalepsy, blocks apomorphine-induced climbing	Boules et al., 2000; Cusack et al., 2000; Tyler-McMahon et al., 2000; Boules et al., 2001; Fredrickson et al., 2003b; Fredrickson et al., 2003a; Boules et al., 2007
NT69L	s.c.	0.08–2	SD rat	Restored PPI	Shilling et al., 2003; Shilling et al., 2004
PD149163	s.c.	0.01–1	SD rat	Restored PPI	Feifel et al., 1999; Feifel et al., 2003; Shilling et al., 2004
PD149163	s.c.	0.01–3	LE/BB rat	Restored PPI	Feifel et al., 2004; Feifel et al., 2007
JMV 1193	i.v.	5	CD-1 mouse	Hypothermia, analgesia	Van Kemmel et al., 1996
JMV 2012	i.v., s.c., p.o.	0.1–30	CD-1 mouse	Hypothermia, analgesia	Bredeloux et al., 2008
KK13	i.p.	ED_{50} 0.91; 1–5	SD rat	Hypothermia, attenuated locomotor activity	Kokko et al., 2005
KK14	i.p.	5	SD rat	Hypothermia	Kokko et al., 2005
KH28	i.p., p.o.	ED_{50} 1; 0.5–30	SD rat	Hypothermia, attenuated locomotor activity	Hadden et al., 2005

Sequences of NT analogs: NT1, (Me)Arg–Lys–Pro–Trp–*tert*-Leu–Leu–OEt; NT66L, D-Lys–Arg–Pro–L–*neo*-Trp–Ile–Leu; NT67L, D-Lys–Arg–Pro–L–*neo*-Trp–*tert*-Leu–Leu; NT69L, (Me)Arg–Lys–Pro–L–*neo*-Trp–*tert*-Leu–Leu; PD149163, Lysψ(CH_2NH)–Lys–Pro–Trp–*tert*-Leu–Leu–OEt; JMV 1193, cyclic–Lys–Lys–Pro–Tyr–Ile–Leu; JMV 2012, cyclic–Lys–Lys–Pro–Tyr–Ile–Leu–Lys–Lys–Pro–Tyr–Ile–Leu; KK13, N_3–L-Hlys–Arg–Pro–Tyr–*tert*-Leu–Leu; KK14, N_3–L-Hlys–Arg–Pro–Trp–*tert*-Leu–Leu; KH28, L-Hlys–Arg–Pro–Tyr–*tert*-Leu–Leu.

10.6.2 Potential of NT Analogs as Antipsychotic Drugs

As outlined in Section 10.4.1.1, there is substantial preclinical evidence that NT mediates at least some of the effects of APDs, and stable NT analogs produce a variety of effects that are similar to those of APDs after peripheral administration (Table 10.3). Measurements of cerebrospinal fluid NT levels in schizophrenic patients suggest that decreased NT levels are associated with more severe symptoms, and are normalized following APD treatment (Widerlov et al., 1982; Sharma et al., 1997). In preclinical studies, NT analogs exhibit characteristics of atypical APDs in that they restore PPI following disruption through the pharmacological manipulation of several neurotransmitter systems (Table 10.3) without associated EPS (Feifel et al., 1999, 2003; Shilling et al., 2003, 2004). In addition, at least some of these compounds display APD-like activity following oral administration

(Hadden et al., 2005). Perhaps not surprisingly in view of this evidence, the NT antagonist SR 48692 was shown to be ineffective for the treatment of schizophrenia in clinical trials (Meltzer et al., 2004).

10.6.3 NT Analogs and NT Antagonists as Potential Addiction Treatments

The evidence presented in Section 10.4.1.2 suggests that both NT agonists and antagonists counteract amphetamine sensitization, although the evidence regarding NT antagonists is much more extensive. Intriguingly, recent reports indicate that daily SR 48692 administration following cocaine and amphetamine sensitization can abolish sensitized locomotor responses and attenuate conditioned place preference, suggesting that NT is required to maintain sensitized responses, and that NT antagonists might by analogy reverse the neuronal changes that underlie drug addiction (Costa et al., 2007; Felszeghy et al., 2007). The ability of SR 48692 to reverse amphetamine sensitization once it has been established suggests that NT antagonists may hold the greatest promise as potential treatments for addiction.

Paradoxically, NT agonists have also been shown to inhibit sensitization. Thus, peripheral NT69L administration prior to nicotine blocked both the development and expression of nicotine sensitization and the authors of this study mention that similar results were also obtained for amphetamine sensitization (Fredrickson et al., 2003a, 2003b, 2005), although, unlike SR 48692, NT69L also blocked acute locomotor activation resulting from amphetamine, cocaine, or nicotine administration (Boules et al., 2001; Fredrickson et al., 2003a).

10.6.4 NT Antagonists in the Treatment of Cancer

With the emerging interest in using the NT receptor to image cancer has come the development of a series of stabilized NT analogs conjugated to heavy metal chelators (Achilefu et al., 2003). Table 10.2 shows three compounds (#17, NT-VIII, and NT-XI) in which the chelator moiety is placed at the N-terminus and stabilizing substitutions are incorporated at Arg^{8}, Arg^{9}, and/or Ile^{10}. Studies with NT-VIII in mice bearing colon and prostate cancer xenografts demonstrated that the tumors were easily visualized in scintigraphic images after injection of this ^{99m}Tc-labeled analog (Garcia-Garayoa et al., 2002). The first human studies involved the use of ^{99m}Tc-labeled NT-XI (Table 10.2) to image ductal pancreatic cancer in four patients (Buchegger et al., 2003). In an attempt to use NT analogs as vectors for targeting and killing cancer, Falciani and colleagues used a branched tetramer of NT(8–13) (4 units of NT(8–13) attached via their C-termini to a branch consisting of 3 lysines) that was coupled to methothrexate to treat mice bearing colon cancer xenografts (Falciani et al., 2007). The tetrabranched form of NT(8–13) was more than 10-fold more stable than NT in plasma, and when the methothrexate conjugate was tested in mice, it produced a 60% reduction in tumor weight, compared to 10% for the free drug. These studies illustrate that NT analogs with enhanced stability show promise as tumor targeting agents.

10.7 COMMON METHODOLOGIES AND ASSAY SYSTEMS

The levels of NT-related peptides in tissues and blood are generally measured by radioimmunoassay (Carraway et al., 1980), and selective antisera have also been developed to measure other peptides from the NT precursor (Carraway et al., 1992). NT receptor binding assays have been devised that use membranes from rat brain (Mazella et al., 1985) and from cells transfected with NT receptors (Hermans et al., 1994). The most distinguishing feature of NT-related peptides is their ability to induce cyanosis in anesthetized rats, and this can be the basis of a simple and selective visual assay (Carraway and Leeman, 1973). Injection of rats with NT rapidly elevates blood glucose levels (Carraway et al., 1976) and leukotriene levels (Carraway et al., 1991). Infusion of NT into rats inhibits gastric acid secretion (Andersson et al., 1977) and enhances intestinal uptake of bile acids

(Gui et al., 2001). Although these effects are not specific for NT, they could be used to assay bioactivity. The contractile effects of NT on smooth muscles are also not unique but they provide simple, reliable, and reproducible assays (Kitabgi and Freychet, 1978). NT is the most potent peptide known to stimulate histamine secretion from isolated mast cells (Miller et al., 1995). Other nonspecific assays include antinociception (Clineschmidt et al., 1982), hypothermia (Bissette et al., 1982), effects on brain dopaminergic systems (Nemeroff et al., 1982), and growth of the intestinal mucosa (Wood et al., 1988).

Because NT receptors can couple to $G_{\alpha}q$, cells that over-express NTR-1 can be used to assay for NT agonist and antagonist activity by measuring the generation of inositol phosphates (Amar et al., 1986) or the elevation of intracellular [Ca^{2+}] (Gailly, 1998). Downstream effects provide other assays based on the release of ^{3}H-arachidonic acid (Hassan and Carraway, 2006), the activation of PKC (Carraway et al., 2008), the phosphorylation of MAP kinases and the stimulation of DNA synthesis (Hassan et al., 2004).

10.8 FUTURE DEVELOPMENTS

There are several areas where the availability of both NT antagonists and mouse genetic approaches should lead to key insights into both the physiological functions of the peptide and a much clearer understanding of NT's role in APD and psychostimulant responses. In the case of APDs, both pharmacological and genetic approaches have provided evidence that NT is required for certain behavioral responses to a subset of these drugs. These findings set the stage for future experiments involving site-specific microinjection of NT agonists and antagonists, and targeted NT or NT receptor gene disruption to map the underlying neuroanatomy and neuronal signaling events underlying the requirement for NT in these responses, including the identification of NT release sites and targets, and the neurotransmitters involved.

There is also reason for optimism that NT analogs and antagonists may prove effective clinically in the treatment of several disorders, most notably as novel antipsychotic and antiaddictive drugs. Advances in our understanding of NT receptor binding and the structural alterations involved in signal transduction will more than likely lead to the development of more potent and selective agonists and antagonists, including nonpeptide agonists. These may in turn provide new therapeutic approaches toward the treatment of a variety of disorders ranging from cancer to schizophrenia. Furthermore, studies indicating that there may be NT-dependent and NT-independent APDs may provide a rationale for polypharmacy in the treatment of schizophrenia.

NT is an inflammatory peptide that stimulates mast cell degranulation (Miller et al., 1995). Mast cells, by degrading NT, also appear to limit the apparently deleterious effects of NT in a mouse model of sepsis involving caecal ligation puncture (CLP) (Piliponsky et al., 2008). Survival time following severe CLP was enhanced in NT knockout mice, NTR-1 knockout mice, and in mice given the NT antagonist SR 142948A. Severe CLP increased peritoneal and plasma levels of NT, and in mast cell-deficient mice, NT levels were further enhanced and NT-induced hypotension was exaccerbated. Isolated mast cells degraded NT by a mechanism involving NTR-1 and neurolysin. These results, coupled with preliminary data indicating that NT levels were increased in patients presenting with septic shock (Piliponsky et al., 2008), suggest that NT antagonism could prove beneficial in the clinical setting.

Finally, sequence comparisons have made clear that the NT precursor protein and NTR-1 are highly conserved from fish to human, making it possible that the unique features of one organism may be exploited to further our understanding of NT function in all vertebrates.

REFERENCES

Achilefu, S. et al., Novel bioactive and stable neurotensin peptide analogues capable of delivering radiopharmaceuticals and molecular beacons to tumors, *J. Med. Chem.*, 46, 3403–3411, 2003.

Agnati, L.G. et al., Neurotensin *in vitro* markedly reduces the affinity in subcortical limbic ^{3}H-*N*-propylapomorphine binding sites, *Acta Physiol. Scand.*, 119, 459–461, 1983.

Akema, T., Praputpittaya, C., and Kimura, F., Effects of preoptic microinjection of neurotensin on luteinizing hormone secretion in unanesthetized ovariectomized rats with or without estrogen priming, *Neuroendocrinology*, 46, 345–349, 1987.

Alexander, M.J., Estrogen-regulated synthesis of neurotensin in neurosecretory cells of the hypothalamic arcuate nucleus in the female rat, *Endocrinology*, 133, 1809–1816, 1993.

Alexander, M.J., Colocalization of neurotensin messengenr ribonucleic acid (mRNA) and progesterone receptor mRNA in rat arcuate neurons under estrogen-stimulated conditions, *Endocrinology*, 140, 4995–5003, 1999.

Alexander, M.J. and Leeman, S.E., Estrogen-inducible neurotensin immunoreactivity in the preoptic area of the female rat, *J. Comp. Neurol.*, 345, 496–509, 1994.

Alexander, M.J. et al., Estrogen induces neurotensin/neuromedin N messenger ribonucleic acid in a preoptic nucleus essential for the preovulatory surge of luteinizing hormone in the rat, *Endocrinology*, 125, 2111–2117, 1989a.

Alexander, M.J. et al., Evidence that neurotensin participates in the central regulation of the preovulatory surge of luteinizing hormone in the rat, *Endocrinology*, 124, 783–788, 1989b.

Alexander, M.J. et al., Distribution of neurotensin/neuromedin N mRNA in rat forebrain: Unexpected abundance in hippocampus and subiculum, *Proc. Natl Acad. Sci. USA*, 86, 5202–5206, 1989c.

Alexander, M.J., Kiraly, Z.J., and Leeman, S.E., Sexually dimorphic distribution of neurotensin/neuromedin N mRNA in the rat preoptic area, *J. Comp. Neurol.*, 311, 84–96, 1991.

Alonso, R. et al., Blockade of neurotensin receptors suppresses the dopamine D1/D2 synergism on immediate early gene expression in the rat brain, *Eur. J. Neurosci.*, 11, 967–974, 1999.

Amar, S., Kitabgi, P., and Vincent, J.-P., Activation of phosphatidylinositol turnover by neurotensin receptors in the human colonic adenocarcinoma cell line HT29, *FEBS Lett.*, 201, 31–36, 1986.

Andersson, S. et al., Inhibition of gastric and intestinal motor activity in dogs by (Gln4) neurotensin, *Acta Physiol. Scand.*, 100, 231–235, 1977.

Araki, K. et al., Isolation and structure of a new active peptide "Xenopsin" on the smooth muscle, especially on a strip of fundus from a rat stomach, from the skin of *Xenopus laevis*, *Chem. Pharm. Bull.*, 21, 2801–2804, 1973.

Armstrong, M.J. et al., Neurotensin stimulates [3H]oleic acid translocation across rat small intestine, *Am. J. Physiol.*, 251, G823–G829, 1986.

Axelson, J.F., Shannon, W., and Van Leeuwen, F.W., Immunocytochemical localization of estrogen receptors within neurotensin cells in the rostral peroptic area of the hypothalamus, *Neurosci. Lett.*, 136, 5–9, 1992.

Barbero, P. et al., PC5-A-mediated processing of pro-neurotensin in early compartments of the regulated secretory pathway of PC5-transfected PC12 cells, *J. Biol. Chem.*, 273, 25339–25346, 1998.

Barroso, S. et al., Identification of residues involved in neurotensin binding and modeling of the agonist binding site in neurotensin receptor 1, *J. Biol. Chem.*, 275, 328–336, 2000.

Barroso, S. et al., Constitutive activation of the neurotensin receptor 1 by mutation of Phe^{358} in helix seven, *Br. J. Pharmacol.*, 135, 997–1002, 2002.

Bean, A.J. et al., Cloning of human neurotensin/neuromedin N genomic sequences and expression in the ventral mesencephalon of schizophrenics and age/sex matched controls, *Neuroscience*, 50, 259–268, 1992.

Betancur, C. et al., Role of endogenous neurotensin in the behavioral and neuroendocrine effects of cocaine, *Neuropsychopharmacology*, 19, 322–332, 1998.

Binder, E. et al., Enhanced neurotensin neurotransmission is involved in the clinically relevant behavioral effects of antipsychotic drugs: Evidence from animal models of sensorimotor gating, *J. Neurosci.*, 21, 601–608, 2001.

Binder, E.B. et al., Effects of neurotensin receptor antagonism on latent inhibition in Sprague–Dawley rats, *Psychopharmacology*, 161, 288–295, 2002.

Binder, E.B. et al., Neurotensin receptor antagonist SR 142948A alters Fos expression and extrapyramidal side effect profile of typical and atypical antipsychotic drugs, *Neuropsychopharmacology*, 29, 2200–2207, 2004.

Bissette, G. et al., Neurotensin and thermoregulation, *Ann. NY Acad. Sci.*, 400, 268–282, 1982.

Boegman, R.J. and Vincent, S.R., Involvement of adenosine and glutamate receptors in the induction of c-fos in the striatum by haloperidol, *Synapse*, 22, 70–77, 1996.

Botto, J.-M. et al., Identification in the rat neurotensin receptor of amino acid residues critical for the binding of neurotensin, *Mol. Brain Res.*, 46, 311–317, 1997a.

Botto, J.-M. et al., Effects of SR48692 on neurotensin-induced calcium-activated chloride currents in the Xenopus oocyte expression system: Agonist-like activity on the levocabastine-sensitive neurotensin receptor and absence of antagonist effect on the levocabastine insensitive neurotensin receptor, *Neurosci. Lett.*, 223, 193–196, 1997b.

Boules, M. et al., A novel neurotensin peptide analog given extracranially decreases food intake and weight in rodents, *Brain Res.*, 865, 35–44, 2000.

Boules, M. et al., A novel neurotensin analog blocks cocaine- and D-amphetamine-induced hyperactivity, *Eur. J. Pharmacol.*, 426, 73–76, 2001.

Boules, M. et al., The neurotensin receptor agonist NT69L suppresses sucrose-reinforced operant behavior in the rat, *Brain Res.*, 1127, 90–98, 2007.

Bredeloux, P. et al., Synthesis and biological effects of c(Lys–Lys–Pro–Tyr–Ile–Leu–Lys–Lys–Pro–Tyr–Ile–Leu) (JMV 2012), a new analogue of neurotensin that crosses the blood–brain barrier, *J. Med. Chem.*, 51, 1610–1616, 2008.

Buchegger, F. et al., Radiolabeled neurotensin analog, ^{99m}Tc-NT-XI evaluated in ductal pancreatic adenocarcinoma patients, *J. Nucl. Med.*, 44, 1649–1654, 2003.

Carraway, R., Isolation, structures, and biological activities of neurotensin-related peptides generated in extracts of avain tissue, *J. Biol. Chem.*, 262, 15886–15889, 1987.

Carraway, R., Hammer, R.A., and Leeman, S.E., Neurotensin in plasma: Immunochemical and chromatographic character of acid/acetone-soluble material, *Endocrinology*, 107, 400–406, 1980.

Carraway, R. and Leeman, S.E., The synthesis of neurotensin, *J. Biol. Chem.*, 250, 1912–1918, 1975a.

Carraway, R. and Leeman, S.E., Characterization of radioimmunoassayable neurotensin in the rat: Its differential distribution in the central nervous system, small intestine, and stomach, *J. Biol. Chem.*, 251, 7045–7052, 1976.

Carraway, R., Ruane, S.E., and Kim, H.R., Distribution and immunochemical character of neurotensin-like material in representative vertebrates and invertebrates: Apparent conservation of the COOH-terminal region during evolution, *Peptides*, 3, 115–123, 1982.

Carraway, R.E. et al., Structures of histamine-releasing peptides formed by the action of acid proteases on mammalian albumin(s), *J. Immunol.*, 143, 1680–1684, 1989.

Carraway, R.E. et al., Neurotensin elevates hematocrit and plasma levels of the leukotrienes, LTB4, LTC4, LTD4, and LTE4, in anesthetized rats, *Peptides*, 12, 1105–1111, 1991.

Carraway, R.E., Demers, L.M., and Leeman, S.E., Hyperglycemic effect of neurotensin, a hypothalamic peptide, *Endocrinology*, 99, 1452–1462, 1976.

Carraway, R.E., Hassan, S., and Dobner, P.R., Protein kinase C inhibitors alter neurotensin receptor binding and function in prostate cancer PC3 cells, *Regul. Pept.*, 147, 96–109, 2008.

Carraway, R.E., Mitra, S.P., and Salmonsen, R., Isolation and quantitation of several new peptides from the canine neurotensin/neuromedin N precursor, *Peptides*, 13, 1039–1047, 1992.

Carraway, R.E. and Ferris, C.F., Isolation, biological and chemical characterization, and synthesis of a neurotensin-related hexapeptide from chicken intestine, *J. Biol. Chem.*, 258, 2475–2479, 1983.

Carraway, R.E. and Leeman, S.E., The isolation of a new hypotensive peptide, neurotensin, from bovine hypothalami, *J. Biol. Chem.*, 248, 6854–6861, 1973.

Carraway, R.E. and Leeman, S.E., The amino acid sequence of a hypothalamic peptide, neurotensin, *J. Biol. Chem.*, 250, 1907–1911, 1975b.

Carraway, R.E. and Mitra, S.P., Differential processing of neurotensin/neuromedin N precursor(s) in canine brain and intestine, *J. Biol. Chem.*, 265, 8627–8631, 1990.

Carraway, R.E. and Mitra, S.P., Purification of large neuromedin N (NMN) from canine intestine and its identification as NMN-125, *Biochem. Biophys. Res. Comm.*, 179, 301–308, 1991.

Chabry, J. et al., Thr-422 and Tyr-424 residues in the carboxyl terminus are critical for the internalization of the rat neurotensin receptor, *J. Biol. Chem.*, 270, 2439–2442, 1995.

Chalon, P. et al., Molecular cloning of a levocabastine-sensitive neurotensin binding site, *FEBS Letts.*, 386, 91–94, 1996.

Checler, F. et al., Inactivation of neurotensin by rat brain synaptic membranes: Cleavage at the Pro^{10}–Tyr^{11} bond by endopeptidase 24.11 (enkephalinase) and a peptidase different from proline-endopeptidase, *J. Neurochem.*, 43, 1295–1301, 1984.

Checler, F., Vincent, J.-P., and Kitabgi, P., Degradation of neurotensin by rat brain synaptic membranes: Involvement of a thermolysin-like metalloendopeptidase (enkephalinase), angiotensin-converting enzyme and other unidentified peptidases, *J. Neurochem.*, 41, 375–384, 1983.

Checler, F., Vincent, J.-P., and Kitabgi, P., Inactivation of neurotensin by rat brain synaptic membranes partly occurs through cleavage at the Arg^8–Arg^9 peptide bond by a metalloendopeptidase, *J. Neurochem.*, 45, 1509–1513, 1985.

Checler, F., Vincent, J.-P., and Kitabgi, P., Purification and characterization of a novel neurotensin degrading peptidase from rat brain synaptic membranes, *J. Biol. Chem.*, 261, 11274–11281, 1986a.

Checler, G., Vincent, J.-P., and Kitabgi, P., Neuromedin N: High-affinity interaction with brain neurotensin receptors and rapid inactivation by brain synaptic peptidases, *Eur. J. Pharmacol.*, 126, 239–244, 1986b.

Chen, L., Yung, K.K.L., and Yung, W.H., Neurotensin selectively facilitates glutamatergic transmission in globus pallidus, *Neuroscience*, 141, 1871–1878, 2006.

Clineschmidt, B.V., Martin, G.E., and Veber, D.F., Antinocisponsive effects of neurotensin and neurotensin-related peptides, *Ann. NY Acad. Sci.*, 400, 283–306, 1982.

Clineschmidt, B.V. and McGuffin, J.C., Neurotensin administered intracisternally inhibits responsiveness of mice to noxious stimuli, *Eur. J. Pharmacol.*, 46, 395–396, 1977.

Clineschmidt, B.V., McGuffin, J.C., and Bunting, P.B., Neurotensin: Antinociceptive action in rodents, *Eur. J. Pharmacol.*, 54, 129–139, 1979.

Costa, F.G., Frussa-Filho, R., and Felicio, L.F., The neurotensin receptor antagonist, SR48692, attenuates the expression of amphetamine-induced behavioral sensitisation in mice, *Eur. J. Pharmacol.*, 428, 97–103, 2001.

Costa, F.G. et al., Blockade of neurotensin receptors during amphetamine discontinuation indicates individual variability, *Neuropeptides*, 41, 83–91, 2007.

Creese, I., Burt, D.R., and Snyder, S.H., Dopamine receptor binding predicts clinical and pharmacological potencies of antischizophrenic drugs, *Science*, 192, 481–483, 1976.

Cusack, B. et al., Pharmacological and biochemical profiles of unique neurotensin 8–13 analogs exhibiting species selectivity, stereoselectivity, and superagonism, *J. Biol. Chem.*, 270, 18359–18366, 1995.

Cusack, B. et al., Effects of a novel neurotensin peptide analog given extracranially on CNS behaviors mediated by apomorphine and haloperdiol, *Brain Res.*, 856, 48–54, 2000.

Dal Farra, C. et al., Involvement of the neurotensin receptor subtype NTR3 in the growth effect of neurotensin on cancer cell lines, *Int. J. Cancer*, 92, 503–509, 2001.

Deutch, A.Y., Lee, M.C., and Iadorola, M.J., Regionally specific effects of atypical and antipsychotic drugs on striatal Fos expression: The nucleus accumbens shell as a locus of antipsychotic action, *Mol. Cell Neurosci.*, 3, 332–341, 1992.

Diaz-Cabiale, Z. et al., Neurotensin-induced modulation of dopamine D2 receptors and their function in rat striatum: Counteraction by a NTR1-like receptor antagonist, *NeuroReport*, 13, 763–766, 2002.

Dicou, E., Vincent, J.-P., and Mazella, J., Neurotensin receptor-3/sortilin mediates neurotensin-induced cytokine/chemokine expression in a murine microglial cell line, *J. Neurosci. Res.*, 78, 92–99, 2004.

Dobner, P.R., Multitasking with neurotensin in the central nervous system, *Cell Mol. Life Sci.*, 62, 1946–1963, 2005.

Dobner, P.R., Deutch, A.Y., and Fadel, J., Neurotensin: Dual roles in psychostimulant and antipsychotic drug responses, *Life Sci.*, 73, 801–811, 2003.

Dobner, P.R. et al., Cloning and sequence analysis of cDNA for the canine neurotensin/neuromedin N precursor, *Proc. Natl Acad. Sci. USA*, 84, 3516–3520, 1987.

Dobner, P.R. et al., Neurotensin-deficient mice show altered responses to antipsychotic drugs, *Proc. Natl Acad. Sci. USA*, 98, 8048–8053, 2001.

Dubuc, I. et al., Hypothermic effect of neuromedin N in mice and its potentiation by peptidase inhibitors, *Eur. J. Pharmacol.*, 151, 117–121, 1988.

Dubuc, I. et al., JMV 449: A pseudopeptide analogue of neurotensin-(8–13) with highly potent and long-lasting hypothermic and analgesic effects in the mouse, *Eur. J. Pharmacol.*, 219, 327–329, 1992.

Dubuc, I. et al., Identification of the receptor subtype involved in the analgesic effect of neurotensin, *J. Neurosci.*, 19, 503–510, 1999.

Elliott, P.J. and Nemeroff, C.B., Repeated neurotensin administration in the ventral tegmental area: Effects on baseline and D-amphetamine-induced locomotor activity, *Neurosci. Lett.*, 68, 239–244, 1986.

Ervin, G.N. et al., Neurotensin blocks certain amphetamine-induced behaviours, *Nature*, 291, 73–76, 1981.

Fadel, J., Dobner, P.R., and Deutch, A.Y., The neurotensin antagonist SR 48692 attenuates haloperidol-induced striatal Fos expression in the rat, *Neurosci. Lett.*, 303, 17–20, 2001.

Fadel, J., Dobner, P.R., and Deutch, A.Y., Amphetamine-elicited striatal Fos expression is attenuated in neurotensin null mutant mice, *Neurosci. Lett.*, 402, 97–101, 2006.

Falciani, C. et al., Synthesis and biological activity of stable branched neurotensin peptides for tumour targeting, *Mol. Cancer Ther.*, 6, 2441–2448, 2007.

Fantegrossi, W.E. et al., Antinociceptive, hypothermic, hypotensive, and reinforcing effects of a novel neurotensin receptor agonist, NT69L, in rhesus monkeys, *Pharmacol. Biochem. Behav.*, 80, 341–349, 2005.

Febvret, A. et al., Further indication that distinct dopaminergic subsets project to the rat cerebral cortex: Lack of colocalization with neurotensin in the superficial dopaminergic fields of the anterior cingulate, motor, retrosplenial, and visual cortices, *Brain Res.*, 547, 37–52, 1991.

Feifel, D. et al., Novel antipsychotic-like effects on prepulse inhibition of startle produced by a neurotensin agonist, *J. Pharmacol. Exp. Ther.*, 288, 710–713, 1999.

Feifel, D. et al., The effects of chronic administration of established and putative antipsychotics on natural prepulse inhibition deficits in Brattleboro rats, *Behav. Brain Res.*, 181, 278–286, 2007.

Feifel, D., Melendez, G., and Shilling, P.D., A systemically administered neurotensin agonist blocks disruption of prepulse inhibition produced by a serotonin-2A agonist, *Neuropsychopharmacology*, 28, 651–653, 2003.

Feifel, D., Melendez, G., and Shilling, P.D., Reversal of sensorimotor gating deficits in Brattleboro rats by acute administration of clozapine and a neurotensin agonist, but not haloperidol: A potential predictive model for novel antipsychotic effects. *Neuropsychopharmacology*, 29, 731–738, 2004.

Felszeghy, K. et al., Neurotensin receptor antagonist administered during cocaine withdrawal decreases locomotor sensitization and conditioned place preference, *Neuropsychopharmacology*, 32, 2601–2610, 2007.

Ferraro, L. et al., Mesolimbic dopamine and cortico-accumbens glutamate afferents as major targets for the regulation of the ventral striato-pallidal GABA pathways by neurotensin peptides, *Brain Res. Rev.*, 55, 144–154, 2007.

Ferris, C.F. et al., Stimulation of luteinizing hormone release after stereotaxic microinjection of neurotensin into the medial preoptic area of rats, *Neuroendocrinology*, 38, 145–151, 1984.

Ferris, C.F. et al., Release and degradation of neurotensin during perfusion of rat small intestine with lipid, *Regul. Pept.*, 12, 101–111, 1985.

Feurle, G.E., Xenin—a review, *Peptides*, 19, 609–615, 1998.

Fredrickson, P. et al., Blockade of nicotine-induced locomotor sensitization by a novel neurotensin analog in rats, *Eur. J. Pharmacol.*, 458, 111–118, 2003a.

Fredrickson, P. et al., Novel neurotensin analog blocks the initiation and expression of nicotine-induced locomotor sensitization, *Brain Res.*, 979, 245–248, 2003b.

Fredrickson, P. et al., Neurobiologic basis of nicotine addiction and psychostimulant abuse: A role for neurotensin? *Psychiatr. Clin. N. Am.*, 28, 737–751, 2005.

Gailly, P., Ca^{2+} entry in CHO cells, after Ca^{2+} stores depletion, is mediated by arachidonic acid, *Cell Calcium*, 24, 293–304, 1998.

Garcia-Garayoa, E. et al., Preclinical evaluation of a new, stabilized neurotensin(8–13) pseudopeptide radiolabelled with ^{99m}Tc, *J. Nucl. Med.*, 43, 374–383, 2002.

Gaspar, P., Berger, B., and Febvret, A., Neurotensin innervation of the human cerebral cortex: Lack of colocalization with catecholamines, *Brain Res.*, 530, 181–195, 1990.

Geisler, S. and Zahm, D.S., Neurotensin afferents of the ventral tegmental area in the rat: (1) re-examination of their origins and (2) responses to acute psychostimulant and antipsychotic drug administration, *Eur. J. Neurosci.*, 24, 116–134, 2006.

Gendron, L. et al., Low-affinity neurotensin receptor (NTS2) signaling: Internalization-dependent activation of ERK1/2, *Mol. Pharmacol.*, 66, 1421–1430, 2004.

Glimcher, P.W. et al., Neurotensin: A new "reward peptide," *Brain Res.*, 291, 119–124, 1984.

Glimcher, P.W., Giovino, A.A., and Hoebel, B.G., Neurotensin self-injection in the ventral tegmental area, *Brain Res.*, 403, 147–150, 1987.

Gui, X., Dobner, P.R., and Carraway, R.E., Endogenous neurotensin facilitates enterohepatic bile acid circulation by enhancing intestinal uptake in rats, *Am. J. Gastrointest. Liver Physiol.*, 281, G1413–G1422, 2001.

Gully, D. et al., Biochemical and pharmacological profile of a potent and selective nonpeptide antagonist of the neurotensin receptor, *Proc. Natl Acad. Sci. USA*, 90, 65–69, 1993.

Gully, D. et al., Biochemical and pharmacological activities of SR 142948A, a new potent neurotensin receptor antagonist, *J. Pharmacol. Exp. Ther.*, 280, 802–812, 1997.

Hadden, M.K. et al., Design, synthesis, and evaluation of the antipsychotic potential of orally bioavailable neurotensin (8–13) analogues containing non-natural arginine and lysine residues, *Neuropharmacology*, 49, 1149–1159, 2005.

Hara, Y. et al., Ontogeny of the neurotensin-containing neuron system of the rat: Immunohistochemical analysis. I. Forebrain and diencephalon, *J. Comp. Neurol.*, 208, 177–195, 1982.

Hassan, S. and Carraway, R.E., Involvement of arachidonic acid metabolism and EGF receptor in neurotensin-induced prostate cancer PC3 cell growth, *Regul. Pept.*, 133, 105–114, 2006.

Hassan, S., Dobner, P.R., and Carraway, R., Involvement of MAP-kinase, PI3-kinase, and EGF-receptor in the stimulatory effect of neurotensin on DNA synthesis in PC3 cells, *Regul. Pept.*, 120, 155–166, 2004.

Heidbreder, C. et al., Balance of glutamate and dopamine in the nucleus accumbens modulates self-stimulation behavior after injection of cholecystokinin and neurotensin in the rat brain, *Peptides*, 13, 441–449, 1992.

Heise, H. et al., Probing conformational disorder in neurotensin by two-dimensional solid-state NMR and comparison to molecular dynamics simulations, *Biophys. J.*, 89, 2113–2120, 2005.

Hentschel, K. et al., Pharmacological evidence that neurotensin mediates prolactin-induced activation of tuberoinfundibular dopamine neurons, *Neuroendocrinology*, 68, 71–76, 1998.

Herbison, A.E. and Theodosis, D.T., Localization of oestrogen receptors in preoptic neurons containing neurotensin but not tyrosine hydroxylase, cholecystokinin, or luteinizing hormone-releasing hormone in the male and female rat, *Neuroscience*, 50, 283–298, 1992.

Hermans, E. et al., Rapid desensitization of agonist-induced calcium mobilization in transfected PC12 cells expressing the rat neurotensin receptor, *Biochem. Biophys. Res. Comm.*, 198, 400–407, 1994.

Hermans, E., Octave, J.N., and Maloteaux, J.-M., Interaction of the COOH-terminal domain of the neurotensin receptor with a G protein does not control the phopholipase C activation but is involved in the agonist-induced internalization, *Mol. Pharmacol.*, 49, 365–372, 1996.

Heyl, D.L. et al., Structure–activity and conformational studies of a series of modified C-terminal hexapeptide neurotensin analogues, *Int. J. Pept. Protein Res.*, 44, 233–238, 1994.

Hoffman, G.E., Organization of LHRH cells: Differential apposition of neurotensin, substance P and catecholamine axons, *Peptides*, 6, 439–461, 1985.

Horger, B.A. et al., Preexposure to, but not cotreatment with, the neurotensin antagonist SR 48692 delays the development of cocaine sensitization, *Neuropsychopharmacology*, 11, 215–222, 1994.

Hrabovszky, E. et al., Detection of estrogen receptor-beta messenger ribonucleic acid and ^{125}I-estrogen binding sites in luteinizing hormone-releasing hormone neurons of the rat brain, *Endocrinology*, 141, 3506–3509, 2000.

Hussain, N., Flumerfelt, B.A., and Rajakumar, N., Glutamatergic regulation of haloperidol-induced c-fos expression in the rat striatum and nucleus accumbens, *Neuroscience*, 102, 391–399, 2001.

Jacobsen, L. et al., Activation and functional characterization of the mosaic receptor SorLA/LR11, *J. Biol. Chem.*, 276, 22788–22796, 2001.

Jansen, P. et al., Roles for the pro-neurotrophin receptor sortilin in neuronal development, aging and brain injury, *Nature Neurosci.*, 10, 1449–1457, 2007.

Jennes, L., Stumpf, W.E., and Kalivas, P.W., Neurotensin: Topographical distribution in rat brain by immunohistochemistry, *J. Comp. Neurol.*, 210, 211–224, 1982.

Kahn, D. et al., Neurotensin neurons in the rat hypothalamus: An immunocytochemical study, *Endocrinology*, 107, 47–54, 1980.

Kalivas, P.W. and Duffy, P., Effect of acute and daily neurotensin and enkephalin treatments on extracellular dopamine in the nucleus accumbens, *J. Neurosci.*, 10, 2940–2949, 1990.

Kalivas, P.W., Nemeroff, C.B., and Prange, A.J.J., Neurotensin microinjection into the nucleus accumbens antagonizes dopamine-induced increase in locomotion and rearing, *Neuroscience*, 11, 919–930, 1984.

Kallo, I. et al., Oestrogen receptor β-immunreactivity in gonadotropin releasing hormone-expressin neurones: Regulation by oestrogen, *J. Neuroendocrinol.*, 13, 741–748, 2001.

Kinkead, B. et al., Neurotensin-deficient mice have deficits in prepulse inhibition: Restoration by clozapine but not haloperidol, olanzapine, or quetiapine, *J. Pharmacol. Exp. Ther.*, 215, 256–264, 2005.

Kinkead, B. et al., Endogenous neurotensin is involved in estrous cycle related alterations in prepulse inhibition of the acoustic startle reflex in female rats, *Psychoneuroendocrinology*, 33, 178–187, 2008.

Kislauskis, E. et al., The rat gene encoding neurotensin and neuromedin N: Structure, tissue-specific expression, and evolution of exon sequences, *J. Biol. Chem.*, 263, 4963–4968, 1988.

Kitabgi, P., Functional domains of the subtype 1 neurotensin receptor (NTS1), *Peptides*, 27, 2461–2468, 2006.

Kitabgi, P. and Freychet, P., Effects of neurotensin on isolated intestinal smooth muscles, *Eur. J. Pharmacol.*, 50, 349–357, 1978.

Kitabgi, P. et al., Marked variations of the relative distributions of neurotensin and neuromedin N in micropunched rat brain areas suggest differential processing of their common precursor, *Neurosci. Lett.*, 124, 9–12, 1991.

Kiyama, H. and Emson, P.C., Colchicine-induced expression of proneurotensin mRNA in rat striatum and hypothalamus, *Mol. Brain Res.*, 9, 353–358, 1991.

Kokko, K.P. et al., *In vivo* behavioral effects of stable, receptor-selective neurotensin[8–13] analogues that cross the blood–brain barrier, *Neuropharmacology*, 48, 417–425, 2005.

Labbe-Jullie, C. et al., Mutagenesis and modeling of the neurotensin receptor NTR1, *J. Biol. Chem.*, 273, 16351–16357, 1998.

Leonetti, M. et al., Specific involvement of neurotensin type 1 receptor in the neurotensin-mediated *in vivo* dopamine efflux using knock-out mice, *J. Neurochem.*, 89, 1–6, 2004.

Li, X.-M. et al., Neurotensin peptides antagonistically regulate postsynaptic dopamine D2 receptors in rat nucleus accumbens: A receptor binding and microdialysis study, *J. Neural. Transm.*, 102, 125–137, 1995.

Luca, S. et al., The conformation of neurotensin bound to its G protein-coupled receptor, *Proc. Natl Acad. Sci. USA*, 100, 10706–10711, 2003.

Lugrin, D. et al., Reduced peptide bond pseudopeptide analogues of neurotensin: Binding and biological activities, and *in vitro* metabolic stability, *Eur. J. Pharmacol.*, 205, 191–198, 1991.

Machida, R. et al., Pharmacokinetics of novel hexapeptides with neurotensin activity in rats, *Biol. Pharm. Bull.*, 16, 43–47, 1993.

Maeno, H. et al., Comparison of mice deficient in the high- or low-affinity neurotensin receptors, Ntsr1 or Ntsr2, reveals a novel function for Ntsr2 in thermal nociception, *Brain Res.*, 998, 122–129, 2004.

Mai, J.K., Triepel, J., and Metz, J., Neurotensin in the human brain, *Neuroscience*, 22, 499–524, 1987.

Martin, S. et al., Pivotal role of an aspartate residue in sodium sensitivity and coupling to G proteins of neurotensin receptors, *Mol. Pharmacol.*, 55, 210–255, 1999.

Martin, S. et al., Neurotensin receptor-1 and -3 complex modulates the cellular signaling of neurotensin in the HT29 cell line, *Gastroenterology*, 123, 1135–1143, 2002.

Martin, S., Vincent, J.-P., and Mazella, J., Involvement of the neurotensin receptor-3 in the neurotensin-induced migration of human microglia, *J. Neurosci.*, 23, 1198–1205, 2003.

Matsuyama, S. et al., Regulation of DARPP-32 Thr75 phosphorylation by neurotensin in neostriatal neurons: Involvement of glutamate signaling, *Eur. J. Neurosci.*, 18, 1247–1253, 2003.

Mazella, J. et al., Structure, functional expression, and cerebral localization of the levocabastine-sensitive neurotensin/neuromedin N receptor from mouse brain, *J. Neurosci.*, 16, 5613–5620, 1996.

Mazella, J. et al., The 100 kDa neurotensin receptor is gp95/sortilin, a non-G-protein-coupled receptor, *J. Biol. Chem.*, 273, 26273–26276, 1998.

Mazella, J., Kitabgi, P., and Vincent, J.-P., Molecular properties of neurotensin receptors in rat brain. Identification of subunits by covalent labeling, *J. Biol. Chem.*, 260, 508–514, 1985.

Meltzer, H.Y. et al., Placebo-controlled evaluation of four novel compounds for the treatment of schizophrenia and schizoaffective disorder, *Am. J. Psychiatry*, 161, 975–984, 2004.

Merchant, K.M., Dobner, P.R., and Dorsa, D.M., Differential effects of haloperidol and clozapine on neurotensin gene transcription in the rat neostriatum, *J. Neurosci.*, 12, 652–663, 1992.

Merchant, K.M. and Dorsa, D.M., Differential induction of neurotensin and c-*fos* gene expression by typical versus atypical antipsychotics, *Proc. Natl Acad. Sci. USA*, 90, 3447–3451, 1993.

Merchant, K.M. et al., Effects of chronic haloperidol and clozapine treatment on neurotensin and c-fos mRNA in rat neostriatal subregions, *J. Pharmacol. Exp. Ther.*, 271, 460–471, 1994.

Meyer-Lindenberg, A. et al., Reduced prefrontal activity predicts exaggerated striatal dopaminergic function in schizophrenia, *Nature Neurosci.*, 5, 267–271, 2002.

Miller, L.A. et al., Blockade of mast cell histamine secretion in response to neurotensin by SR 48692, a non-peptide antagonist of the neurotensin brain receptor, *Br. J. Pharmacol.*, 114, 1466–1470, 1995.

Minamino, N., Kangawa, K., and Matsuo, H., Neuromedin N: A novel neurotensin-like peptide indentified in porcine spinal cord, *Biochem. Biophys. Res. Comm.*, 122, 542–549, 1984.

Najimi, M. et al., Distinct regions of C-terminus of the high affinity neurotensin receptor mediate the functional coupling with pertussis toxin sensitive and insensitive G-proteins, *FEBS Lett.*, 512, 329–333, 2002.

Nemeroff, C.B., Neurotensin: Perchance an endogenous neuroleptic? *Biol. Psychiatry*, 15, 283–302, 1980.

Nemeroff, C.B. et al., Interactions of neurotensin with brain dopamine systems, *Ann. NY Acad. Sci.*, 400, 330–344, 1982.

Nykjaer, A. et al., Sortilin is essential for proNGF-induced neuronal cell death, *Nature*, 427, 843–848, 2004.

Oakley, R.H. et al., Molecular determinants underlying the formation of stable intracellular G protein-coupled receptor-β-arrestin complexes after receptor endocytosis, *J. Biol. Chem.*, 276, 19452–19460, 2001.

Panayi, F. et al., Chronic blockade of neurotensin receptors strongly reduces sensitized, but not acute, behavioral response to D-amphetamine, *Neuropsychopharmacology*, 26, 64–74, 2002.

Panayi, F. et al., Endogenous neurotensin in the ventral tegmental area contributes to amphetamine behavioral sensitization, *Neuropsychopharmacology*, 30, 871–879, 2005.

Pang, Y.-P. et al., Proposed ligand binding site of the transmembrane receptor for neurotensin(8–13), *J. Biol. Chem.*, 271, 15060–15068, 1996.

Petersen, C.M. et al., Molecular identification of a novel candidate sorting receptor purified from human brain by receptor-associated protein affinity chromatography, *J. Biol. Chem.*, 272, 3599–3605, 1997.

Petersen, C.M. et al., Propeptide cleavage conditions sortilin/neurotensin receptor-3 for ligand binding, *EMBO J.*, 18, 595–604, 1999.

Pettibone, D.J. et al., The effects of deleting the mouse neurotensin receptor NTR1 on central and peripheral responses to neurotensin, *J. Pharmacol. Exp. Ther.*, 300, 305–313, 2002.

Pickel, V.M. et al., High-affinity neurotensin receptors in the rat nucleus accumbens: Subcellular targeting and relation to endogenous ligand, *J. Comp. Neurol.*, 435, 142–155, 2001.

Piliponsky, A.M. et al., Neurotensin increases mortality and mast cells reduce neurotensin levels in a mouse model of sepsis, *Nature Med.*, 14, 392–398, 2008.

Qanbar, R. and Bouvier, M., Role of palmitoylation/depalmitoylation reactions in G-protein-coupled receptor function, *Pharmacol. Therapeut.*, 97, 1–33, 2003.

Ragozzino, M.E. and Choi, D., Dynamic changes in acetylcholine output in the medial striatum during place reversal learning, *Learn Mem.*, 11, 70–77, 2004.

Reinecke, M., Neurotensin: Immunohistochemical localization in central and peripheral nervous system and in endocrine cells and its functional role as neurotransmitter and endocrine hormone, *Progr. Histochem. Cytochem.*, 16, 1–175, 1985.

Reynolds, S.M. et al., Neurotensin antagonist acutely and robustly attenuates locomotion that accompanies stimulation of a neurotensin-containing pathway from rostrobasal forebrain to the ventral tegmental area, *Eur. J. Neurosci.*, 24, 188–196, 2006.

Richard, F. et al., Agonism, inverse agonism, and neutral antagonism at the constitutively active human neurotensin receptor 2, *Mol. Pharmacol.*, 60, 1392–1398, 2001a.

Richard, F. et al., Impaired G protein coupling of the neurotensin receptor 1 by mutations in extracellular loop 3, *Eur. J. Pharmacol.*, 433, 63–71, 2001b.

Robertson, G.S. and Fibiger, H.C., Neuroleptics increase c-Fos expression in the forebrain: Contrasting effects of haloperidol and clozapine, *Neuroscience*, 46, 315–328, 1992.

Rompre, P.-P., Repeated activation of neurotensin receptors sensitizes to the stimulant effect of amphetamine, *Eur. J. Pharmacol.*, 328, 131–134, 1997.

Rompre, P.-P. and Bauco, P., Neurotensin receptor activation sensitizes to the locomotor stimulant effect of cocaine: A role for NMDA receptors, *Brain Res.*, 1085, 77–86, 2006.

Rompre, P.P., Bauco, P., and Gratton, A., Facilitation of brain stimulation reward by mesencephalic injections of neurotensin-(1–13), *Eur. J. Pharmacol.*, 211, 295–303,1992.

Rompre, P. and Perron, S., Evidence for a role of endogenous neurotensin in the initiation of amphetamine sensitization, *Neuropharmacology*, 39, 1880–1892, 2000.

Rosell, S., The role of neurotensin in the uptake and distribution of fat, *Ann. NY Acad. Sci.*, 400, 183–195, 1982.

Rostene, W.H. and Alexander, M.J., Neurotensin and neuroendocrine regulation, *Front Neuroendocrinol.*, 18, 115–173, 1997.

Roubert, C. et al., Altered neurotensin mRNA expression in mice lacking the dopamine transporter, *Neuroscience*, 123, 537–546, 2004.

Rovere, C., Barbero, P., and Kitabgi, P., Evidence that PC2 is the endogenous pro-neurotensin convertase in rMTC 6-23 cells and that PC1- and PC2-transfected PC12 cells differentially process pro-neurotensin, *J. Biol. Chem.*, 271, 11368–11375, 1996.

Saleh, T.M. et al., Cholecystokinin and neurotensin inversely modulate excitatory synaptic transmission in the parabrachial nucleus *in vitro*, *Neuroscience*, 77, 23–35, 1997.

Sarhan, S. et al., Comparative antipsychotic profiles of neurotensin and a related systemically active peptide agonist, *Peptides*, 18, 1223–1227, 1997.

Sarret, P. et al., Pharmacology and functional properties of NTS2 neurotensin receptors in cerebellar granule cells, *J. Biol. Chem.*, 277, 36233–36243, 2002.

Sato, M. et al., Postnatal ontongeny of cells expressing prepro-neurotensin/neuromedin N mRNA in the rat forebrain and midbrain: A hybridization histochemical study involving isotope-labeled and enzyme-labeled probes, *J. Comp. Neurol.*, 310, 300–315, 1991.

Sharma, R.P. et al., CSF neurotensin concentrations and antipsychotic treatment in schizophrenia and schizoaffective disorder, *Am. J. Psychiatry*, 154, 1019–1021, 1997.

Shaw, C. et al., Differential processing of the neurotensin/neuromedin N precursor in the mouse, *Peptides*, 11, 227–235, 1990.

Shilling, P.D. et al., Neurotensin agonists block the prepulse inhibition deficits produced by a 5-HT_{2A} and an α_1 agonist, *Psychopharmacology*, 175, 353–359, 2004.

Shilling, P.D., Richelson, E., and Feifel, D., The effects of systemic NT69L, a neurotensin agonist, on baseline and drug-disrupted prepulse inhibition, *Behav. Brain Res.*, 143, 7–14, 2003.

Skrzydelski, D. et al., Differential involvement of intracellular domains of the rat NTS1 neurotensin receptor in coupling to G proteins: A molecular basis for agonist-directed trafficking of receptor stimulus, *Mol. Pharmacol.*, 64, 421–429, 2003.

Smith, M.J. and Wise, P.M., Neurotensin gene expression increases during proestrus in the rostral medial preoptic nucleus: Potential for direct communication with gonadotropin-releasing hormone neurons, *Endocrinology*, 142, 3006–3013, 2001.

Smits, S.M. et al., Species differences in brain pre-pro-neurotensin/neuromedin N mRNA distribution: The expression pattern in mice resembles more closely that of primates than rats, *Mol. Brain Res.* 125, 22–28, 2004.

Studler, J.M. et al., Extensive co-localization of neurotensin with dopamine in rat meso-cortico-frontal dopaminergic neurons, *Neuropeptides*, 11, 95–100, 1988.

Sundler, F. et al., Light and electron microscopic localization of neurotensin in the gastrointestinal tract, *Ann. NY Acad. Sci.*, 400, 94–103, 1982.

Tanaka, K., Masu, M., and Nakanishi, S., Structure and functional expression of the cloned rat neurotensin receptor, *Neuron*, 4, 847–854, 1990.

Tokumura, T. et al., Stability of a novel hexapeptide, (Me)Arg–Lys–Pro–tert-Leu–Leu–OEt, with neurotensin activity, in aqueous solution and in the solid state, *Chem. Pharm. Bull.*, 38, 3094–3098, 1990.

Tyler, B.M. et al., *In vitro* binding and CNS effects of novel neurotensin angonists that cross the blood–brain barrier, *Neuropharmacology*, 38, 1027–1034, 1999.

Tyler-McMahon, B.M. et al., Highly potent neurotensin analog that causes hypothermia and antinociception, *Eur. J. Pharmacol.*, 390, 107–111, 2000.

Uhl, G.R., Kuhar, M.J., and Snyder, S.H., Neurotensin: Immunohistochemical localization in rat central nervous system, *Proc. Natl Acad. Sci. USA*, 74, 4059–4063, 1977.

Van Kemmel, F.-M. et al., A C-terminal cyclic 8–13 neurotensin fragment analog appears less exposed to neprilysin when it crosses the blood–brain barrier than the cerebrospinal fluid–brain barrier in mice, *Neurosci. Lett.*, 217, 58–60, 1996.

Villeneuve, P. et al., Immunohistochemical evidence for the involvement of protein convertases 5A and 2 in the processing of pro-neurotensin in rat brain, *J. Comp. Neurol.*, 424, 461–475, 2000a.

Villeneuve, P., Seidah, N.G., and Beaudet, A., Immunohistochemical evidence for the implication of PC1 in the processing of proneurotensin in rat brain, *Neuroreport*, 11, 3443–3447, 2000b.

Villeneuve, P. et al., Altered processing of the neurotensin/neuromedin N precursor in PC2 knock down mice: A biochemical and immunohistochemical study, *J. Neurochem.*, 82, 783–793, 2002.

Vita, N. et al., Neurotensin is an antagonist of the human neurotensin NT2 receptor expressed in Chinese hamster ovary cells, *Eur. J. Pharmacol.*, 360, 265–272, 1998.

von Euler, G. et al., Neurotensin decreases the affinity of dopamine D2 agonist binding by a G protein-independent mechanism, *J. Neurochem.*, 56, 178–183, 1991.

Wagstaff, J.D., Gibb, J.W., and Hanson, G.R., Dopamine D2-receptors regulate neurotensin release from nucleus accumbens and striatum as measured by *in vivo* microdialysis, *Brain Res.*, 721, 196–203, 1996.

Watanobe, H. and Takebe, K., *In vivo* release of neurotensin from the median eminence of ovariectomized estrogen-primed rats as estimated by push–pull perfusion: Correlation with luteinizing hormone and prolactin surges, *Neuroendocrinology*, 57, 760–764, 1993.

Watters, J. and Dorsa, D.M., Transcriptional effects of estrogen on neuronal neurotensin gene expression involve cAMP/protein kinase A-dependent signaling mechanisms, *J. Neurosci.*, 19, 6672–6680, 1998.

Widerlov, E. et al., Subnormal CSF levels of neurotensin in a subgroup of schizophrenic patients: Normalization after neuroleptic treatment, *Am. J. Psychiatry*, 139, 1122–1126, 1982.

Williams, F.G., Murtaugh, M.P., and Beitz, A.J., The effect of acute haloperidol treatment on brain proneurotensin mRNA: *In situ* hybridization analyses using a novel fluorescence detection procedure, *Mol. Brain Res.*, 7, 347–358, 1990.

Wood, J.G. et al., Neurotensin stimulates growth of small intestine in rats, *Am. J. Physiol.*, 255, G813–G817, 1988.

Yamada, M. et al., Deletion mutation in the putative third intracellular loop of the rat neurotensin receptor abolishes polyphosphoinositide hydrolysis but not cylcic AMP formation in CHO-K1 cells, *Mol. Pharmacol.*, 46, 470–476, 1994.

Yamada, M. et al., Distinct functional characteristics of levocabastine sensitive rat neurotensin NT2 receptor expressed in Chinese hamster ovary cells, *Life Sci.*, 62, 375–380, 1998.

Yin, H.H., Adermark, L., and Lovinger, D.M., Neurotensin reduces glutamateric transmission in the dorsolateral striatum via retrograde endocannabinoid signaling, *Neuropharmacology*, 54, 79–86, 2007.

Zahm, D.S., Subsets of neurotensin-immunoreactive neurons revealed following antagonism of the dopamine-mediated suppression of neurotensin immunoreactivity in the rat striatum, *Neuroscience*, 46, 335–350, 1992.

Zahm, D.S. et al., Distinct and interactive effects of d-amphetamine and haloperidol on levels of neurotensin and its mRNA in subterritories in the dorsal and ventral striatum of the rat, *J. Comp. Neurol.*, 400, 487–503, 1998.

Zahm, D.S. et al., Neurons of origin of the neurotensinergic plexus enmeshing the ventral tegmental area in rat: Retrograde labeling and *in situ* hybridization combined, *Neuroscience*, 104, 841–851, 2001.

11 Twenty-Five Years of Galanin Research

Johan Runesson, John K. Robinson, Ulla Eriksson Sollenberg, and Ülo Langel

CONTENTS

Since the discovery of galanin 25 years ago, the importance of galanin has been shown in numerous physiological functions in the central nervous system (CNS), peripheral nervous system (PNS), and various organs. The relatively recent discovery of other members of the galanin peptide family has prompted a reevaluation and injection of energy into the galanin field. To date, we have only started to explore the domain of clinical galaninergic therapeutics.

11.1 INTRODUCTION

Galanin was first discovered by Professor Viktor Mutt and colleagues in 1983 (Tatemoto et al., 1983) at the Karolinska Institute in Stockholm using a chemical method for identification of amidated peptides followed by a random search for peptides with this characteristic in porcine intestine. Using this approach, they found a 29-amino-acid-long sequence which was named galanin after its N-terminal glycine and its C-terminal alanine. Galanin is a C-terminally amidated peptide in all species identified so far, with the exception of humans, where galanin is 30 amino acids long and displays a free acid at its C-terminal end. The N-terminal end of galanin is crucial for its biological activity and the first 14 amino acids are conserved between species (Table 11.1). In a membrane-mimicking environment, the structure of galanin is organized into a horseshoe-like shape, with both the C- and N-terminal ends adopting helical conformations bent with a β-bend around the proline in position 13 (Kulinski et al., 1997). The preprogalanin gene contains 6 exons, giving rise to a 123 amino acid precursor peptide which is proteolytically processed producing two peptides, galanin and message-associated peptide (GMAP). The galanin-peptide family is now known to consist of galanin, GMAP, galanin-like peptide (GALP), and alarin. GALP is a 60-amino-acid, nonamidated peptide and the amino acid sequence of GALP-(9–21) is completely identical to that of galanin(1–13). The newest member of the galanin-peptide family, alarin, is a 25-amino-acid-long peptide originating as a splice variant of the GALP mRNA (Santic et al., 2007).

Since its initial discovery, galanin has been shown to be expressed in both the central and peripheral nervous systems (CNS, PNS), as well as in the endocrine system and is now thought to be involved in a variety of physiological and pathological processes. Galanin-like immunoreactivity has primarily been investigated in rat brain, with the peptide predominantly identified in the hypothalamus, thalamus, ventral hippocampus, medulla oblongata, locus coeruleus, mid-brain, basal forebrain, pituitary, and spinal cord (review by Merchenthaler et al., 1993). Galanin has also been identified in glial cells (for a review see Ubink et al., 2003).

The galanin-peptide family has been the subject of 25 years of research, and above 3000 research publications. The goal of this chapter will be to provide an overview of the molecular and physiological functions of galanin peptides and receptors, their potential roles in neuropathologies, and to review available pharmacological tools.

TABLE 11.1
Sequences of the Galanin Peptide in Various Species

Species	Galanin Sequence
Human	GWTLNSAGYLLGPHAVGNHRSFSDKNGLT S
Rat	GWTLNSAGYLLGPHAIDNHRSFSDKHGLT amide
Mouse	GWTLNSAGYLLGPHAIDNHRSFSDKHGLT amide
Pig	GWTLNSAGYLLGPHAIDSHRSFHDKYGLA amide
Bovine	GWTLNSAGYLLGPHALDNHRSFQDKHGLA amide
Dog	GWTLNSAGYLLGPHAIDNHRSFHEKPGLT amide
Sheep	GWTLNSAGYLLGPHAIDNHRSFHDKHGLA amide
Chicken	GWTLNSAGYLLGPHAVDNHRSFNDKHGFT amide
Alligator	GWTLNSAGYLLGPHAIDNHRSFNEKHGIA amide
Trout	GWTLNSAGYLLGPHGIDGHRTLSDKHGLA amide
Frog	GWTLNSAGYLLGPHAIDNHRSFNDKHGLA amide
Bowfin	GWTLNSAGYLLGPHAVDNHRSLNDKHGLA amide

11.2 MOLECULAR MECHANISMS OF ACTION

Galanin has a fairly long half-life, 100 min in rat hypothalamus, 110 min in spinal cord, and 120 min in freshly isolated cerebrospinal fluid (Land et al., 1991a; Bedecs et al., 1995). This bioactive peptide signals via three receptor subtypes, galanin receptor types 1, 2, and 3 (GalR1–3), which are all seven-transmembrane spanning receptors signaling via G proteins.

GalR1 has been cloned from human Bowes melanoma cells (Habert-Ortoli et al., 1994), Rin14B cells (Parker et al., 1995), rat brain (Burgevin et al., 1995), and mouse (Jacoby et al., 1997; Wang et al., 1997c). The homology between species is high, with 92% of the residues in rGalR1 being identical to those of hGalR1. Important residues for ligand binding and receptor activity have been identified in transmembrane region VI (TMVI) and extracellular region III (ECIII) of GalR1 (Kask et al., 1996). In the model presented by Kask et al. (1996) Trp^{2} in galanin binds to either one or both of His^{264} and His^{267} in GalR1 and Tyr^{9} of galanin interacts with Phe^{282}, which was later confirmed and identified to be of an aromatic–aromatic nature (Berthold et al., 1997). Phe^{115} in TMIII was identified as the residue important for interaction with the N-terminal of galanin (Berthold et al., 1997), although this was questioned by Church et al. (2002), who suggested that Phe^{115} was not involved in direct ligand interaction, but rather in the stability of the GalR1. It has also been suggested that Glu^{271} in hGalR1, corresponding to Glu^{269} in rGalR1, is another residue important for interaction with the N-terminus of galanin (Kask et al., 1996; Church et al., 2002). Phe^{186} in ECII is thought to form hydrophobic interactions with Ala^{7} and Leu^{11} (Church et al., 2002). The extracellular N-terminus of GalR1 has been shown to have no impact on the binding affinity of galanin (Kask et al., 1996). The third intracellular loop, and especially its N-terminal, has been recognized as the domain that defines the receptor's signal transduction function (Razani et al., 2000).

GalR1 expression is regulated by cyclic AMP through a CREB-dependent mechanism (Zachariou et al., 2001; Hawes et al., 2005). Activation of GalR1 results in pertussis toxin-sensitive and forskolin-insensitive inhibition of adenylyl cyclase through interaction with $G\alpha_i/\alpha_o$ G proteins (Habert-Ortoli et al., 1994; Wang et al., 1997c). It can also increase intracellular Ca^{2+} via $\beta\gamma$ subunit activation of phospholipase C $\beta 2$ isoforms. Human GalR1 mRNA was initially reported to be found in fetal brain, small intestine tissue (Rossowski et al., 1993; Harbert-Ortoli et al., 1994), and in the gastrointestinal tract (Lorimer et al., 1996). However, a later study identified the GalR1 expression to be exclusively in the CNS and PNS (Waters and Krause, 2000), where it was detected in hypothalamus, amygdala, ventral hippocampus, thalamus, brainstem (medulla oblongata, locus coeruleus, and lateral parabrachial nucleus), and dorsal horn of the spinal cord (Gustafson et al., 1996), although broader central and peripheral tissue distribution has also been reported (Sullivan et al., 1997). GalR1 expression does not fluctuate during the development of the brain (Burazin et al., 2000).

GalR2 has been cloned from rat hypothalamus (Howard et al., 1997; Smith et al., 1997; Ahmad et al., 1998), mouse (Pang et al., 1998), and human (Bloomquist et al., 1998; Borowsky et al., 1998). The hGalR2 protein has a high sequence identity (87%) and similarity (92%) to rGalR2, although there is one notable difference—the 15-amino-acid extension of the C-terminal end in human GalR2 (Kolakowski et al., 1998). Some important residues known in GalR1 are also conserved in GalR2, in particular His^{264}, Ile^{256}, Trp^{260}, and Tyr^{271} (Bloomquist et al., 1998). A site-directed mutagenesis study identified His^{252} located in TMVI to be necessary for galanin binding to GalR2. Also of importance were the aromatic residues Phe^{264} and Tyr^{271} in ECIII, which probably mediate aromatic interactions between ECIII and galanin (Lundström et al., 2007), an observation that is also reported for GalR1 (Kask et al., 1998; Church et al., 2002). Changing the N-terminal tail dramatically changed the affinity of galanin for GalR2, which is likely explained by the disruption of a disulfide bridge between the N-terminal and ECIII, a cysteine residue that is missing in GalR1. The residue Ile^{256} in GalR2, corresponding to His^{267} in GalR1, is vital for the binding of the fragment Gal(2–11), but it is only partly responsible for the binding of galanin (Lundström et al., 2007). There is no amino acid corresponding to Glu^{269} in rGalR2, the residue in GalR1 proposed to interact with the N-terminal of galanin, which may account for the lack of importance for residue Gly^{1} in the binding affinity of galanin to rGalR2 (Wang et al., 1997a). Human GalR2 mRNA is more widely distributed than GalR1. It is found in several peripheral tissues, including the pituitary gland, gastrointestinal tract, skeletal muscle, heart, kidney, uterus, ovary, and testis, as well in regions in the nervous system such as the dentate gyrus, the hypothalamus, and in the olfactory and cortical regions, cerebellum, and in the brainstem and spinal cord (Smith et al., 1997; Ahmad et al., 1998; Bloomquist et al., 1998; Waters and Krause, 2000; Xia et al., 2005). Unlike GalR1, GalR2 expression levels vary during the development of the rat brain, with a broader distribution and a peak in expression before postnatal day 7, particularly in the cortex and thalamus, and much reduced levels after postnatal day 14 (Burazin et al., 2000). GalR2 is able to activate the stimulatory pathway of $G\alpha_{q/11}$ G proteins. This triggers intracellular phosphatidylinositol turnover, increasing intracellular Ca^{2+}, and probably activates phospholipase C as well; GalR2 is only weakly coupled to adenylate cyclase inhibition (Smith et al., 1997; Kolakowski et al., 1998).

GalR3 was first isolated from rat hypothalamic cDNA libraries (Wang et al., 1997b) and later also from human cDNA (Kolakowski et al., 1998; Smith et al., 1998). The sequence of the hGalR3 shares 36% homology with hGalR1 and 58% with hGalR2 and ~90% with rGalR3 (Kolakowski et al., 1998). There is some general acceptance of a weak and narrow GalR3 expression pattern relative to GalR1 and GalR2 transcript levels that is most prominent in the hypothalamus (Wang et al., 1997b; Smith et al., 1998; Mennicken et al., 2002), although some studies report a wider distribution of GalR3 throughout the central and peripheral tissues (Kolakowski et al., 1998; Waters and Krause, 2000). GalR3 actions are also mediated by $G\alpha_i/G\alpha_o$ G proteins, which are strongly coupled to the inhibition of adenylate cyclase (Kolakowski et al., 1998), which results in an inward K^+ current through GIRK channels (Smith et al., 1998).

There is evidence of the existence of additional and as of yet unidentified galanin receptors. Putative receptor(s) with affinity for N-terminally truncated galanin fragments such as galanin(3–29), which show low affinity for cloned receptors *in vitro*, have been suggested in rat pituitary (Wynick et al., 1993) and rat jejunal preparations (Rossowski et al., 1990). The discrepancy between *in vivo* and *in vitro* cloned receptors to galanin(1–15), led to the postulation of a potential galanin receptor in hippocampal neurons (Hedlund et al., 1992, 1994; Xu et al., 1999). In line with these hypotheses, a novel receptor, named GalRL, was cloned from both human and mouse with homology to the established GalRs and low affinity for the galanin peptide (Ignatov et al., 2004).

11.3 BIOLOGICAL FUNCTIONS

11.3.1 Stress Responses

High levels of galanin expression are evident in regions implicated in the regulation of mood and stress such as the hypothalamus, amygadala, locus coeruleus (LC) dorsal raphe nucleus (DRN) and ventral tegmental area (Skofitsch et al., 1986; Lu et al., 2005b). However, a consistent role for galanin in mediating stress responses has remained elusive. The effects of galanin seem to depend on factors such as the route and site of peptide administration and the nature of the behavioral response (Khoshbouei et al., 2002b). Some evidence suggest that galanin upregulates in response to stressful experiences. Repeated restraint stress increases preprogalanin mRNA expression in hypothalamus and amygdala (Makino et al., 1999; Sweerts et al., 1999). Also, chronic exposure to stressors such as social stress and repeated treadmill exercise increases mRNA expression in the LC (Holmes et al., 1995; O'Neal et al., 2001). However, less stressful exposures, such as swim stress and wheel running does not alter galanin gene expression (Austin et al., 1990; Soares et al., 1999). Consistent with an endogenous role in stress regulation, administration of a galanin antagonist attenuates anxiogenic-like effects of immobilization stress (Khoshbouei et al., 2002b). The selective α_2-antagonist yohimbine in combination with acute stress induces a galanin-mediated anxiolytic effect, which a galanin antagonist blocks completely. Neither stress nor yohimbine administration by themselves evoke the galanin-induced response, which has led to the hypotheses that galanin may represent a form of negative feedback regulation of noradrenalin modulation of a more severe stress response (Khoshbouei et al., 2002a).

Galanin-overexpressing (Gal OE) mice generally show few differences when compared to wild-type mice under baseline conditions, but do show a heightened anxiogenic response to yohimbine (Holmes et al., 2002; Kuteeva et al., 2005). GalR1 null mutant mice show heightened anxiety-like behavior in the elevated plus maze (Holmes et al., 2003), while GalR2 null mutant mice show no difference in phenotype compared to wild-type except an anxiogenic-like phenotype in the elevated plus-maze model (Krasnow et al., 2004; Gottsch et al., 2005). In summary, the degree that galanin modulates stress responses appears to depend upon the severity of the stressful situation (Holmes et al., 2003; Yoshitake et al., 2004; Kuteeva et al., 2005; Pirondi et al., 2005; Bailey et al., 2007).

11.3.2 Feeding and Metabolism

Central administration of galanin and GALP induces a strong feeding response in rats (Kyrkouli et al., 1986; Crawley et al., 1990; Lawrence et al., 2002). This induction by galanin was for the preferential consumption of fat, occurring most strongly at the end of the nocturnal (active) cycle, and in a manner relatively insensitive to acute need (Tempel et al., 1988; Kyrkouli et al., 1990; Akabayashi et al., 1994). In addition, hypothalamic levels of galanin are positively related to the amount of consumed fat (Tempel et al., 1988) and galanin causes a decrease in energy expenditure after administration into the paraventriculalar nucleus of the hypothalamus, suggesting a propensity towards promoting obesity (Menendez et al., 1992). Indeed, galanin gene expression in the hypothalamus is increased in genetically obese rats (Beck et al., 1993; Beck, 2007) and galanin knockout mice show decreased fat consumption (Adams et al., 2008). One explanation for this is that food consumption produces positive feedback onto galanin-containing neurons in the hypothalamus (Leibowitz et al., 1998; Leibowitz, 2007), although galanin may also simply enhance existing macronutrient preference (Smith et al., 1996; Crawley, 1999; Kyrkouli et al., 2006). Two studies have shown that galanin gene expression and galanin-induced feeding are reduced by central injection of leptin (Sahu, 1998a, 1998b), and central administration of insulin significantly decreases galanin mRNA levels in the paraventricular nucleus of the hypothalamus (Wang et al., 1997a), suggesting that galanin regulation of feeding is not entirely independent of hormonal signals.

It is likely that galanin and GALP are part of an extremely complex system of motivational regulation, involving many neurotransmitters and neurohormones in the brain and periphery, and it is not a simple task to separate specific roles for each, given the shared receptors. Although galanin OE and GalR1 knockout mice show no altered body weight regulation, body weight regulation (Hohmann et al., 2003; Wrenn et al., 2004; Zorrilla et al., 2007), pharmacological evidence points to a role for GalR1 (Lundström et al., 2005c) but not GalR2 or GalR3 in mediating galanin and GALP feeding induction (Man and Lawrence, 2008).

11.3.3 Learning and Memory

Central administration of galanin to rats results in impairment of performance in several different memory tasks: emotional memory (Kinney et al., 2002), spatial memory (Sundström et al., 1988; Ögren et al., 1992, 1996; Gleason et al., 1999) and working memory (e.g., Robinson and Crawley, 1993a, 1993b, 1994). Galanin OE mice have been shown to be similarly impaired in the performance of several learning and memory tasks such as emotional memory (Kinney et al., 2002), olfactory memory (Wrenn et al., 2002, 2003), and spatial location (Steiner et al., 2001; Wrenn et al., 2002). Studies with GalR1 knockout mice suggest that some, but not all memory task impairments resulting from excess galanin may involve signaling through GalR1 (Wrenn et al., 2004), an observation that is supported by the report that GalR2 knockout mice display no significant differences from wild-type littermates in learning and memory tasks (Bailey et al., 2007).

The deficits induced by galanin have been suggested to be the result of the inhibitory action of galanin on hippocampal cholinergic transmission and are based upon early *in vitro* and *in vivo* studies showing galanin inhibition of stimulated acetylcholine (ACh) release (reviewed in Crawley, 1996). Galanin produces a similar pattern of impairment to cholinergic muscarinic receptor blockade (Robinson and Crawley, 1994) and Gal OE mice also show reduced basal release of ACh in the ventral hippocampus (Laplante et al., 2004). However, Elvander and colleagues (2004) demonstrated that galanin infused into the medial septum facilitates spatial learning and enhances hippocampal ACh release, and that galanin stimulates ACh release in striatum (Ögren and Pramanik, 1991; Amoroso et al., 1992), dorsal hippocampus (Ögren et al., 1999), and from basal forebrain cultures (Jhamandas et al., 2002). These results refute the simpler theory that a decrease in hippocampal ACh is sufficient to produce deficits in learning and memory and demand more complex theories, such as the suggestion that galanin selectively impairs the more demanding components of cognitive tasks (McDonald et al., 1998) or that behavioral deficits may be the result of the convergence of galanin effects on multiple systems (see Robinson, 2004 for discussion).

11.3.4 Neurogenesis

Galanin mRNA has been detected in immature progenitor cells in proliferative zones of the CNS (Shen et al., 2003, 2005) and in the olfactory bulb, where it acts as an inhibitory factor on olfactory ensheathing cell proliferation (Xia et al., 2005). Galanin and all three subtypes of the receptor have been shown to be abundant in embryonic stem cells, indicating a putative role in regulating proliferation, migration, and/or differentiation of progenitor cells (Anisimov et al., 2002a, 2002b). The most abundant receptor subtype was GalR2 (Tarasov et al., 2002), which confirmed earlier results, implicating the role of GalR2 during nervous system development (Burazin et al., 2000). Galanin is essential for a third of the cholinergic neurons of the basal forebrain during development (O'Meara et al., 2000) which may, together with other data, indicate that nerve growth factor (NGF) and galanin function together as neurotrophines (Murphy et al., 1999; Holmes et al., 2000; O'Meara et al., 2000). A subset of sensory neurons in the dorsal root ganglion (DRG) also require galanin for their postnatal developmental survival (Holmes et al., 2000). Galanin also attenuates glutamate-, staurosporine-, and kainite-induced hippocampal cell death, primarily via GalR2 (Elliott-Hunt et al., 2004). Mice homozygous for a target mutation in the galanin gene have significantly fewer DRG neurons and an impaired regeneration after nerve crush injury, and cultured DRG mutant neurites are both fewer in number and shorter (Holmes et al., 2000). Gal knockout mice are deficient in neurite outgrowth, which can be attenuated by exogenous galanin (Mahoney et al., 2003). Such effects are mediated via GalR2 in a PKC-dependent manner. GalR1 knockout mice display no differences when compared to wild-type littermates in peripheral nerve regeneration (Jacoby et al., 2002; Holmes et al., 2003). Addition of galanin increases the length and the number of branches in cultured DRG neurons (Mahoney et al., 2003).

11.3.5 Nociception

Galanin has important but complex actions in mediating nociception. Galanin is expressed at high levels in DRG during development (Marti et al., 1987; Villar et al., 1989). Although a biphasic, dose-dependent effect can be shown electrophysiologically (e.g., Flatters et al., 2003), some consensus has emerged that galanin is primarily inhibitory of substance P and calcitonin gene-related peptide and facilitory of opiate-induced analgesia (reviewed by Wiesenfeld-Hallin et al., 2005). Gal OE animals show moderate inhibition of basic responses in pain sensation tests (Grass et al., 2003a; Holmes et al., 2003; Hygge-Blakeman et al., 2004). Gal knockout mice have nociceptive deficits following injury (Kerr et al., 2000, 2001) and it has been proposed that this finding is consistent with the role of galanin in promoting the survival of unmyelinated nociceptive c-fibers during development, which would therefore be absent in adult Gal knockout mice (Mahoney et al., 2003; Holmes et al., 2005).

Both GalR1 and GalR2 are expressed in DRG neurons, and it has been proposed that GalR1 mediates the inhibitory effects of high doses of galanin and GalR2 the excitatory effects of low doses of galanin on nociceptive tone and pain-like responses (Wiesenfeld-Hallin et al., 2005). However, the specific roles of these two receptor subtypes are not entirely clear. Studies using GalR1 knockout mice suggest a relatively minor role of GalR1 in baseline nociception, as there was little potentiation of responses in these animals (Grass et al., 2003b). However, GalR1 downregulation using antisense peptide nucleic acid (PNA) attenuated a galanin-induced inhibition of c-fiber flexor reflex (Pooga et al., 1998b) and a GalR1 agonist inhibited the development of neuropathic pain (Bartfai et al., 2004). In contrast, GalR2 receptor knockout animals showed reduced neuropathic pain (Hobson et al., 2006), and pharmacological stimulation of GalR2 enhanced pain-like reaction to cold and mechanical stimulation (Liu et al., 2001). Separating neurotrophic roles (both developmental and neuroprotective) from roles in neurotransmission is one major focus of current work. Additionally, understanding the central effects of galanin is of interest. Galanin injected into the arcuate nucleus of the hypothalamus has nociceptive effects, apparently by enhancing the actions of β-endorphin in the periaqueductal gray neurons that provide descending inhibition of spinal cord nociception (Sun et al., 2007).

11.3.6 Osmotic Regulation and Water Intake

Studies from as early as the late 1980s suggest that galanin might play a role in fluid intake regulation after galanin and vasopressin were shown to coexist in the hypothalamus (Melander et al., 1986; Skofitsch et al., 1989). Various experimental manipulations, such as injections of hypertonic saline and hemorrhage were shown to upregulate galanin mRNA. Galanin was shown to inhibit the release of vasopressin (Rökaeus et al., 1988; Kondo et al., 1991; Ciosek and Cisowska, 2003). A recent report shows that galanin inhibited the actions of angiotensin II in neurons in the subfornical organ, a site at which circulating angiotensin II is thought to have effects ultimately leading to the stimulation of drinking, further suggesting that galanin is an endogenous modulator of fluid intake and fluid conservation signals (Kai et al., 2006). Studies that have examined the effects on drinking behavior following central administration of galanin have confirmed that galanin inhibits fluid consumption, but only under conditions of water deprivation (Kyrkouli et al., 1990; Brewer et al., 2005a; Schneider et al., 2007).

11.3.7 Reproductive Behavior

Galanin is well situated to modify reproductive hormones and reproductive behavior, in that galanin coexists with luteinizing hormone releasing hormone (LHRH) in the medial preoptic area and medial diagonal band neurons, especially in female rats, and is coreleased with LHRH from the anterior pituitary (Merchenthaler et al., 1990; López et al., 1991). Evidence implicates galanin in the regulation of luteinizing hormone and follicle stimulating hormone in a manner that is sensitive to estrogen and testosterone levels (Kaplan et al., 1988; Scheffen et al., 2003; cf. Baranowska-Bik et al., 2005). Galanin is also upregulated by estrogens (Kaplan et al., 1988) and later, three estrogen response elements (EREs) were found in the promoter region of the preprogalanin gene.

The studies of the effects of galanin on sexual behavior suggest that translation of these endocrinological actions into behavior is not straightforward. Intracerebroventricularly (i.c.v.) administered galanin and galanin antagonists, respectively, inhibit and stimulate sexual behavior in male rats (Poggioli et al., 1992; Benelli et al., 1994), although galanin injected into the medial preoptic nucleus stimulates both female and male sexual behavior (Bloch et al., 1993, 1996). It is also far from clear if these effects are mediated endogenously by GALP, galanin, or both. GALP induces LHRH release and i.c.v. administered GALP stimulates sexual behavior in male rats (Matsumoto et al., 2001; Fraley et al., 2004). However, GALP inhibits sexual behavior in male mice (Kaufmann et al., 2005). Given the close anatomical and functional relationship between the hormonal systems, it has been proposed that galanin/GALP may serve as an interface point for neurohormonal systems involved in integrating and dually regulating feeding and reproductive behavior (Gundlach, 2002; Cunningham et al., 2004b; Kageyama et al., 2005; Crown et al., 2007).

11.3.8 Instrumental Behavior

Recent evidence also points to a modulatory role for galanin in mesolimbic dopamine neurotransmission associated with instrumental behavior. Rats engaging in various tests in which they were required to emit instrumental responses in an operant chamber (lever presses in Skinner boxes) showed less persistence in these reinforcement schedules (summarized in Brewer et al., 2005b). This galanin-induced inhibition of instrumental bar pressing for food rewards occurs even when galanin stimulated consumption of the same freely available food, indicating that the inhibition of the bar pressing is not the result of reduced "hunger" (Brewer and Robinson, 2008). Lesioning mid-brain dopamine neurons or blockade of dopamine neurotransmission in the nucleus accumbens results in a similar decrease in operant bar pressing without reducing consumption of the same food when made freely available (e.g., Salamone et al., 2007; Kelley et al., 2005). Indeed, some neurochemical evidence points to a modulatory role of galanin in mesolimbic dopamine neurotransmission (e.g., Rada et al., 1998; Ericson and Ahlenius, 1999; Counts et al., 2002), although considerable further investigation will be needed to determine the nature and implications of this dopamine–galanin interaction.

11.4 STRUCTURAL ANALOGS

The galanin peptide is a linear peptide containing 29 or 30 amino-acid residues (in human), with or without a carboxy-terminal amide, derived from preprogalanin precursor proteins (123 or 124 amino acids). Endogenous galanin has a high affinity for all three galanin receptors, GalR1, GalR2, and GalR3 (Wang et al., 1997b) (Table 11.2). In all species studied except the tuna fish, galanin (Kakuyama et al., 1997) shares an absolute conserved N-terminal region (residues 1–14) and a variant COOH-terminal region. The galanin fragment Gal(1–16) illustrates the importance of the N-terminal for receptor binding, because despite lacking half the complete sequence, it retains high-affinity binding (Table 11.2) when compared to the striking reduction in receptor affinity observed for Gal(3–29). Gal(1–16) has frequently been utilized as a substitute ligand whose effects represent the effects of galanin, although there are reports of a somewhat different pharmacological profile for Gal(1–15) and Gal(1–16) fragments compared to galanin itself (Hedlund, 1994; Xu et al., 1999). Shortening of the N-terminal fragment to Gal(1–11) results in the loss of high-affinity binding (Land et al., 1991b). Identification of important pharmacophores shows Trp^2, Asn^5, Tyr^9, and Gly^{12} to be of significant importance for binding of Gal(1–16) to unspecified subtypes of the galanin receptors (Land et al., 1991b), and Trp^2, Asn^5, Gly^8 and Tyr^9 of Gal(2–11) to be important for binding to GalR2 (Lundström et al., 2005a). Fragments of galanin are not ideal subtype selective ligands, although the galanin fragments Gal(2–11) and Gal(2–29) have moderate selectivity for GalR2 (Wang et al., 1997b; Lu et al., 2005b) (Table 11.2).

GALP was first isolated and cloned from porcine hypothalamus (Ohtaki, 1999). GALP has 60-amino-acid residues and is processed from a 115–120 amino-acid precursor, depending on the species (see Cunningham, 2004a for review). It has a nonamidated C-terminal and the sequence GALP(9–21) is identical to that of galanin(1–13). GALP binds all three galanin subtypes, with GalR3 exhibiting the highest affinity (Lang et al., 2005) (Table 11.2) and the presence of a putative receptor has also been suggested (Cunningham et al., 2002). The 25-amino-acid peptide alarin, is a splice variant of GALP, isolated from murine brain, thymus, skin (Santic et al., 2007), and human neuroblastic tumors (Santic et al., 2006).

TABLE 11.2
Affinities of Different Ligands of Galanin Receptor Subtypes, Determined as K_i

Ligand	K_i (nM) R1	R2	R3	Reference
rGal(1–29)	1.0	1.5	1.5	Wang et al., 1997b
	0.3 (h)	1.6 (h)	12 (h)	Borowsky et al., 1998
hGal(1–30)	0.4 (h)	2.3 (h)	69 (h)	Borowsky et al., 1998
Gal(1–16)	4.8	5.7	50	Wang et al., 1997b
Gal (2–11)	>5000 (h)	88	271	Lu et al., 2005b.
	879[a] (h)	1.8[a]	—	Liu et al., 2001
Gal(2–29)	85	1.9	13	Wang et al., 1997b
Gal(3–29)	>1000	>1000	>1000	Wang et al., 1997b
Gal(1–29)–D-Trp[2]	407	28	>1000	Smith et al., 1998
(p)GALP	4.3	0.24	—	Ohtaki et al., 1999
hGALP	77 (h)	28 (h)	10 (h)	Lang et al., 2005
M15	0.65	1.0	10	Smith et al., 1998
M35	4.8	8.2	4.7	Lu et al., 2005
M40	1.8	5.1	63	Lu et al., 2005
M617	0.23 (h)	5.71 (h)	—	Lundström et al., 2005b
M871	420 (h)	13 (h)	—	Sollenberg et al., 2006
Galnon	11,700 (h)	34,100	—	Bartfai et al., 2004
Galmic	34,200 (h)	>100,000	—	Bartfai et al., 2004
Sch 202596	1700[a] (h)	—	—	Chu et al., 1997
Dithiepine-1,1,4,4-tetroxide	190[a] (h)	>30,000[a] (h)	—	Scott et al., 2000
SNAP 37889	>10,000 (h)	>10,000 (h)	17.4 (h)	Swanson et al., 2005
SNAP 398299	>1000 (h)	>1000 (h)	5.3 (h)	Swanson et al., 2005
GalR3ant	>10,000 (h)	>10,000 (h)	15 (h)	Konkel et al., 2004
				Barr et al., 2006

Note: Displacement is performed on the rat galanin receptor unless indicated otherwise. p, porcine; h, human; —, not determined.

[a] Presented as IC_{50} values.

11.4.1 Chimeric Ligands

Several chimeric ligands have been synthesized, conjugating mammalian galanin (1–13) to other bioactive molecules (Table 11.3), M15 (also called galantide) (Bartfai et al., 1991), M32 (Wiesenfeld-Hallin et al., 1992), M35 (Ögren et al., 1992; Wiesenfeld-Hallin et al., 1992; Kask et al., 1995), C7 (Crawley et al., 1993), and M40 (Crawley et al., 1993; Bartfai et al., 1993). Although they all maintain antagonistic properties *in vivo* at doses between 0.1 and 10nM when delivered i.c.v. or intrathecally (i.t.) (Parker et al., 1995; Lu et al., 2005a), they all have a partial agonistic nature *in vitro* (Kask et al., 1995). Several other peptides have been synthesized and compared with the above-mentioned ligands by various groups (Yanaihara et al., 1993; Pooga et al., 1998a; Saar et al., 2001).

The first introduced chimeric peptide with antagonist effect was M15 (Bartfai et al., 1991). Here the N-terminal (1–13) of galanin, which is crucial for receptor recognition, was coupled to a C-terminal fragment in substance P(5–11), which acts as an agonist of the substance P receptor. M15 showed an approximately 10-fold higher affinity than endogenous galanin. Later, M35 was synthesized (Ögren et al., 1992) with increased *in vivo* stability (Wiesenfeld-Hallin et al., 1992). M15, M32, M35, and M40 have similar affinities to galanin and have been valuable tools in galanin research, but are

TABLE 11.3
Selection of Chimeric Galanin Receptor Peptide Ligands

Peptide	Sequence
rGalanin	GWTLNSAGYLLGPHAIDNHRSFSDKHGLT amide
Galanin(2–11)	WTLNSAGYLL amide
M15	Galanin(1–13)–substance P(5–11) amide GWTLNSAGYLLGPQQFFGLM amide
M35	Galanin(1–13)–bradykinin(2–9) amide GWTLNSAGYLLGPPPGFSPFR amide
M617	Galanin(1–13)–Gln^{14}–bradykinin(3–9) amide GWTLNSAGYLLGPQPGFSPFR amide
M40	Galanin(1–13)–$(Pro)_2$–$(Ala–Leu)_2$–Ala amide GWTLNSAGYLLGPPPALALA amide
M32	Galanin(1–13)–NPY(25–36) amide GWTLNSAGYLLGPRHYINLITRQRY amide
M871	Gal(2–13)–EHPPPALALA amide WTLNSAGYLLGPEHPPPALALA amide
C7	GWTLNSAGYLLGP(dR)PKPQQ(dW)F(dW)LL
pGALP	APVHRGRGGWTLNSAGYLLGPVLHPPSRAEGGGKG KTALGILD(LW/HY)KAIDGLPYPQSQLAS

Note: r = rat, p = porcine.

limited by their relative nonspecificity towards the different galanin receptors (Borowsky et al., 1998) and by their interaction with receptors other than galanin receptors (Berglund et al., 2001).

M617 (Table 11.3) resembles the M35 peptide, with a substitution of proline at position 14 to a glutamine. This monosubstitution results in a 25-fold selectivity for GalR1 over GalR2 *in vitro* (Table 11.2), a finding that is also confirmed in several other distinct *in vivo* models (Lundström et al., 2005b; Blackshear et al., 2007). M617 has been shown to produce antinociceptive effects (Jimenez-Andrade et al., 2006) and to delay the development of seizures in animal models (Mazarati et al., 2006).

The design of M871 (Sollenberg et al., 2006) (Table 11.3) was based on M40. M871 functions as a partial agonist and selectivity for GalR2 was later confirmed *in vivo* (Jimenez-Andrade et al., 2005; Alier et al., 2007; Kuteeva et al., 2008). The importance of the development of M617 and M871 and other subtype selective agonists and antagonists cannot be overestimated and holds the key for successful development of therapeutics.

11.4.2 Nonpeptide Agonists

Galnon (Saar et al., 2002) was identified after screening a combinatorial peptidomimetic library. It acts as an agonist in functional studies both *in vitro* and *in vivo* and has a K_i of 11.7 μM for GalR1 and K_i of 34.1 μM for GalR2 (Table 11.3) (Bartfai et al., 2004). It has been evaluated in models of anxiety and depression (Rajarao et al., 2007), feeding (Abramov et al., 2004), and pain (Wu et al., 2003). The downside of galnon is that it has multiple sites of interaction that produce unwanted physiological effects (Florén et al., 2005). Galmic is a nonpeptide agonist with higher affinity for GalR1 than GalR2, which under conditions of intrahippocampal administration is sixfold more potent than galnon in inhibiting self-sustaining status epilepticus, an *in vivo* model for epilepsy (Bartfai et al., 2004; Ceide et al., 2004). It is able to penetrate the blood–brain barrier, although its pharmacological importance is hampered by its low affinity (K_i of 34.2 μM for GalR1).

11.4.3 Nonpeptide Antagonists

The metabolite Sch 202596, isolated from an *Aspergillus* sp. culture found in an abandoned uranium mine in Tuolemone County California has a modest inhibitory action on GalR1 *in vitro* (IC_{50} 1.7 μM) (Chu et al., 1997). Sch 202596 was characterized as a molecule with a spirocoumaranone skeleton and so far has only been partly synthesized (Katoh et al., 2002). Several 1,4-dithiins and dithiipine-1,1,4,4-tetroxides with binding affinity to GalR1 have been identified by Scott and colleagues (2000). The research paper concludes that the compound 2,3-dihydro-2-(4-methylphenyl)-1,4-dithiepine-1,1,4,4-tetroxide is a submicromolar antagonist. It has an IC_{50} of 190 nM for GalR1 and an IC_{50} above 30 μM (the highest concentration tested) for GalR2. However, its reactive nature and low solubility make it unattractive from a drug discovery point of view. It has been used and evaluated in several studies (Mahoney et al., 2003; Kozoriz et al., 2006).

A series of 3-imonio-2-indolones were identified as specific GalR3 antagonists, with K_i values for GalR3 as low as 17 nM and above 10 μM for GalR1 and GalR2 (Konkel et al., 2006a). One of these was referred to as SNAP 37889 by Swanson et al. (2005). One drawback of these indolones, however, is their low aqueous solubility (<1 μg/mL). This feature motivated further studies, which identified a compound, 1,3-dihydro-1-[3-(2-pyrrolidinylethoxy)phenyl]-3-[[3-(trifluoromethyl) phenyl]imino]-2H-indol-2-one, with an even lower K_i at 5 nM and an increased water solubility (Swanson et al., 2005; Konkel et al., 2006b). Another of the synthesized indolones was evaluated *in vivo* by Barr and colleagues (2006), which together with other published works and several patent applications (Konkel et al., 2004) indicate that specific GalR3 ligands are in development.

11.5 CLINICAL APPLICATIONS

11.5.1 Alcoholism

One component of alcoholism is the alteration in systems controlling ingestive behavior, causing an abnormal intake of alcohol. This biologically oriented explanation implies that hypothalamic peptides implicated in feeding, such as galanin, may also be involved in alcoholism. Indeed, there is a growing body of evidence suggesting that galanin plays a modulatory role in acute alcohol consumption. Injection of galanin into the hypothalamus or the third ventricle in rats stimulates increased alcohol intake (Lewis et al., 2004; Rada et al., 2004; Schneider et al., 2007). Alcohol consumption also induces an increase in the expression of galanin mRNA in rats (Leibowitz et al., 2003). This suggests the possibility of a positive feedback loop between galanin and alcohol uptake very similar to the apparent positive feedback loop linking galanin and fat consumption (Thiele et al., 2004). The relevance of these findings to human alcoholism is implied by the finding that different galanin haplotypes have been found to be associated with alcoholism (Belfer et al., 2006). A recent genotyping study (Belfer et al., 2007) pinpoints GalR3 as mediating the alcoholism-related action of galanin. However, the complexity of the systems regulating food consumption and the clear importance of psychosocial factors in contributing to the development of alcoholism in humans suggests that galanin's role in this disorder will ultimately be quite limited.

11.5.2 Alzheimer's Disease (AD)

The Alzheimer's brain is characterized by neurofibrilliary tangles and neuritic plaques composed of astrocytes, glial cells, and neurites around an amyloid core (Walsh and Selkoe, 2004). A body of work has shown that markers for galanin do not decrease or even upregulate over the course of AD. For example, galanin binding sites are increased in the central nucleus and cortico-amygdaloid transition area of the amygdala in early stages of the disease, but not during the late stages of Alzheimer's disease (Perez et al., 2002). Galanin-positive fibers hyperinnervate the remaining cholinergic cell bodies in the cholinergic basal forebrain in late stages of the disease (Chan-Palay, 1988; Rodriguez-Puertas et al., 1997; Counts et al., 2006). Galanin is also overexpressed in other brain regions in AD, such as the entorhinal cortex (Deecher et al., 1998) and galanin immunoreativity (IR) is found within amyloid plaques of patients with AD (Kowall et al., 1989) and mice overexpressing the β-amyloid precursor protein (Diez et al., 2000, 2003). Recent evidence also implicates GALP as a putative gene involved in late-onset AD in a genomic-wide screening (Grupe et al., 2007).

These finding were initially interpreted as indicating that galanin was functioning in a pathological manner, increasing inhibitory actions on cholinergic neurotransmission at a time when those cells were being lost, thus contributing to the symptomatic memory loss of AD (Chan-Palay, 1988; Steiner et al., 2001; Rustay et al., 2005). However, the growing appreciation of the neurotrophic properties of galanin suggests that the neuronal damage caused by AD may be the trigger for increased expression of galanin (Counts et al., 2006). The hyperinnervation of galanin in brain regions with high neuronal damage during AD may therefore be an attempt to reestablish some equilibrium as the pathology progresses.

11.5.3 Cancer Growth and Pathology

Galanin has been found to be expressed in a variety of tumors, especially in neuroendocrine cells, found in small cell lung cancer, prostate carcinomas, breast cancer, and oral squamous cell (for review see Berger et al., 2003, 2005). GalR1 and GalR3 are significantly more highly expressed in neuroblastic tumor cells compared to their benign differentiated counterpart ganglioneuroma. This observation argues for a role of galanin in regulating neuroblastic tumor development (Perel et al., 2002). Galanin stimulates growth in lung cancer cells (Sethi and Rozengurt, 1991; Seufferlein and Rozengurt, 1996) but acts as an antiproliferative molecule in other studies (Iishi et al., 1995; Berger et al., 2002; Perel et al., 2002; El-Salhy, 2005). So although galanin interventions promise to have therapeutic utility in treating several kinds of cancers, much additional basic research and preclinical work will be required to fully understand the mechanisms and applications of galanin therapeutics for cancer.

11.5.4 Depression

Potential therapies for depression form one of the most active areas of research in which galaninergic drugs might serve. This interest has been stimulated by anatomical, physiological, neurochemical, and behavioral evidence. I.c.v. administration of galanin or infusion into the dorsal raphe nucleus (DRN), but not into hippocampus, reduces 5-HT release, an effect blocked by galanin receptor antagonists (Razani et al., 2000, 2001; Kehr et al., 2002). These, and several other studies, identify a link between galanin and 5-HT (for an excellent review, see Ögren et al., 2007). I.c.v. administration of galanin reduces the efficacy of antidepressant therapy (Yoshitake et al., 2003b) and desensitizes 5-HT_{1A} receptors (Yoshitake et al., 2003a), which taken together with the fact that all antidepressant drugs on the market decrease LC activity (see Weiss et al., 2005; LC hyperactivity produces depression-related changes via galanin) indicates that the modulation of galanin may be an important consideration when formulating antidepressant therapies. An important theoretical issue here is that galanin release is potentially regulated in a manner referred to as *frequency coded release of coexisting transmitters*, that is, when neurons only release low-weight transmitters like 5-HT, serotonin or NA at low frequencies of firing, but corelease neuropeptides and classical neurotransmitters at higher firing frequencies (e.g., 4 to 10 Hz; Consolo et al., 1994; Muschol and Salzberg, 2000). The hyperactivity of LC during depression means that both galanin and NA will be released onto 5-HT and GABAergic neurons of the DRN, but in the absence of depression, these inputs are exclusively mediated by NA signaling (Grenhoff et al., 1993; Pieribone et al., 1995; Weiss et al., 2005). A galanin receptor antagonist, in theory, should therefore have no effect in the absence of depression-mediated hyperactivity and a blocking effect when needed during LC hyperactivity.

Hippocampal NA and 5-HT levels are upregulated in Gal OE mice accompanied by increased depression-like behavior, which could be blocked by coadministration with a galanin receptor antagonist (Yoshitake et al., 2004; Pirondi et al., 2005). In an established genetic rat model of depression, the Flinders Sensitive Line, increased galanin binding sites in the DRN and attenuated levels of galanin-like immunoreactivity are associated with immobility in the forced swim model (Bellido et al., 2002; Husum et al., 2003). By studying one of several specific GalR3 antagonists, all triaminopurimidines listed in a patent application (Konkel et al., 2004), Swanson and colleagues (2005) showed that this antagonist induced both acute and chronic anxiolytic- and antidepressant-like behavior. Another GalR3 antagonist, from the same patent application, decreased immobility in the tail suspension test and increased activity in the forced swim test, providing evidence for an antidepressant-like effect of this compound (Barr et al., 2006). These triaminopurimidines (Konkel et al., 2006a, 2006b) and new discoveries may completely change antidepressant treatment in the future. Taken together, the current evidence suggests that galanin may play a prodepressive role through facilitation of NA following chronic stress, and that GalR3 antagonists may be useful in attenuating this prodepressive effect.

11.5.5 Diabetes

Increased plasma levels of galanin have been found to be higher among children with insulin-dependent diabetes mellitus compared to healthy children (Celi et al., 2005). Also, in patients in the fasting state with type II diabetes mellitus, a correlation between galanin plasma concentration and glucose concentration was observed in both sexes (Legakis et al., 2005). Insulin secretion from isolated pancreases of normal and diabetic rats is inhibited by galanin (Adeghate and Ponery, 2001). Systemic galanin inhibits glucose-stimulated insulin release in rodents, with a larger inhibition at the level of secretion than of proinsulin synthesis (Lindskog et al., 1995). GalR1 knockout mice have reduced levels of circulating insulin-like growth factor (Jacoby et al., 2002). When considered together, these data suggest that galanin controls glucose metabolism and feeding at several levels (Ahrén et al., 2004).

11.5.6 Neuropathic Pain

Early findings that demonstrated expression of galanin and its receptors in sensory neurons and the superficial spinal cord suggested that galanin was involved in nociceptive transmission (Skofitsch and Jacobowitz, 1985a, 1985b; Chńg et al., 1985). The first direct evidence that galanin might be involved in neuropathic pain was provided by an observed increase in galanin levels after peripheral nerve injury in primary DRG cells in rats (Hökfelt et al., 1987; Villar et al., 1989). Galanin was subsequently shown to be upregulated in a number of animal models of neuropathic pain, such as chronic nerve constriction (Nahin et al., 1994; Shi et al., 1999; Wilson-Gerwing and Verge, 2006), partial sciatic nerve ligation (PSNL) (Ma and Bisby, 1997; Shi et al., 1999), spinal nerve ligation (Fukuoka et al., 1998), photochemically induced ischemic nerve injury (Hao et al., 1999; Shi et al., 1999), spared nerve injury (Imbe et al., 2004), and single ligature nerve constriction (Coronel et al., 2008).

Increased levels of galanin mRNA and galanin have been observed in the DRG of monkeys after complete peripheral nerve injury (Wang et al., 2007) and in humans suffering from severe brachial plexus injury and herpes zoster virus infection (Landry et al., 2003). These studies, and several others, using endogenous and exogenous galanin, point towards an antinociceptive function of galanin (Shi et al., 1999; Hao et al., 1999; Flatters et al., 2003). In the PSNL model, galanin mRNA is upregulated in both injured and uninjured DRG neurons (Ma and Bisby, 1999). However, some evidence also points toward a pronociceptive function of galanin. A study with Gal knockout mice showed an inhibitory function of galanin under normal conditions; however, in damaged sensory neurons, galanin promoted the emergence of neuropathic pain through activation of excitatory galanin receptors (Kerr et al., 2000). Distinct roles for galanin receptor subtypes have not yet fully emerged, but some evidence points to the involvement of both GalR1 and GalR2. PNA antisense reagents coupled to a cellular transporter protein against GalR1 reduce the depressive effects of i.t. administration of galanin on spinal hyperexcitability, pointing to the involvement of GalR1. The initial excitatory effect of galanin was not reduced, but enhanced, suggesting that a different subtype of galanin receptor mediates initial excitatory events (Pooga et al., 1998b). A selective upregulation of GalR2 after facial nerve crush confirms the involvement of GalR2 (Burazin and Gundlach, 1998).

I.t. administration of galaninergic therapeutics may therefore have specific application to the treatment of neuropathic pain, a debilitating disorder particularly difficult to treat because of its chronic nature and complications that arise due to the addictive potential of opiate-based analgesics. Moreover, the clinical applicability of galanin is supported by studies using the galanin agonist galnon (Wiesenfeld-Hallin et al., 1992), the galanin antagonist M35 (Wu et al., 2003), galanin antisense oligonucleotides (Ji et al., 1994; Pooga et al., 1998b), and implants that overexpress galanin (Eaton et al., 1999).

11.5.7 Seizures and Epilepsy

Another active area of research for galanin with the promise of clinical applicability is the prevention and management of seizures. Galanin knockout mice have a lower threshold in three different test models, but galanin OE mice show increased resistance to induced seizures (Mazarati et al., 2000; Kokai et al., 2001). This resistance to seizures is also seen when over expressing galanin using adeno-associated viral transfection (Haberman et al., 2003; Lin et al., 2003). Galanin knockout mice do not have spontaneous seizures (Mazarati et al., 2000), but GalR1 knockout mice do (Jacoby et al., 2002; McColl et al., 2006). Knockdown of GalR2 with PNA antisense oligonucleotides increases the severity of seizures (Mazarati et al., 2004).

11.6 FUTURE DEVELOPMENTS

This review has provided an introduction and brief overview of a complex peptide, galanin. In the 25 years that galanin has been studied, much has been learned about its biochemical and physiological actions and its roles in both normal and pathological functions, particularly in the CNS and endocrine systems. At least three receptor subtypes have been isolated and studied, and some selective pharmacological tools are emerging. The relatively recent discovery of GALP has also prompted some reevaluation of the functions of galanin *per se* and galanin receptors, particularly in the regulation of food intake. New applications and functions, such as the potential involvement of galanin in regulating alcohol consumption, are still emerging. However, what is also clear is that while there appears to be a high degree of potential for galaninergic clinical therapeutics, no such therapeutic strategy has emerged past the animal model stage. Furthermore, many areas of basic research on galanin are relatively underdeveloped, encompassing a small body of studies that are more suggestive than conclusive. So one might say that galanin research is at a tipping point of sorts, whereby new developments are pivotal if interest in galanin research is to be revived and to translate these findings into the clinical setting. Otherwise, the utility of this promising bioactive peptide may be sadly forgotten.

ACKNOWLEDGMENTS

J.K.R. thanks Professors Sven Ove Ögren and Tomas Hökfelt for graciously hosting his sabbatical at the Karolinska Institutet during which this chapter was prepared.

REFERENCES

Abramov, U. et al., Regulation of feeding by galnon, *Neuropeptides*, 38, 55–61, 2004.

Adams, A.C., Clapham, J.C., Wynick, D., and Speakman, J.R., Feeding behaviour in galanin knockout mice supports a role of galanin in fat intake and preference, *J. Neuroendocrinol.*, 20, 199–206, 2008.

Adeghate, E. and Ponery, A.S., Large reduction in the number of galanin-immunoreactive cells in pancreatic islets of diabetic rats, *J. Neuroendocrinol.*, 13, 706–710, 2001.

Ahmad, S. et al., Cloning and evaluation of the role of rat GALR-2, a novel subtype of galanin receptor, in the control of pain perception, *Ann. NY Acad. Sci.*, 863, 108–119, 1998.

Ahrén, B. et al., Loss-of-function mutation of the galanin gene is associated with perturbed islet function in mice, *Endocrinology*, 145, 3190–3196, 2004.

Akabayashi, A. et al., Galanin-containing neurons in the paraventricular nucleus: A neurochemical marker for fat ingestion and body weight gain, *Proc. Natl Acad. Sci. USA*, 91, 10375–10379, 1994.

Alier, K.A. et al., Selective stimulation of GalR1 and GalR2 in rat substantia gelatinosa reveals a cellular basis for the anti- and pro-nociceptive actions of galanin, *Pain*, 137, 138–146, 2007.

Amoroso, D. et al., Mechanism of the galanin induced increase in acetylcholine release *in vivo* from striata of freely moving rats, *Brain Res.*, 589, 33–38, 1992.

Anisimov, S.V. et al., SAGE identification of gene transcripts with profiles unique to pluripotent mouse R1 embryonic stem cells, *Genomics*, 79, 169–176, 2002a.

Anisimov, S.V. et al., SAGE identification of differentiation responsive genes in P19 embryonic cells induced to form cardiomyocytes *in vitro*, *Mech. Devel.*, 117, 25–74, 2002b.

Austin, M.C., Cottingham, S.L., Paul, S.M., and Crawley, J.N., Tyrosine hydroxylase and galanin mRNA levels in locus coeruleus neurons are increased following reserpine administration, *Synapse*, 6, 351–357, 1990.

Bailey, K.R. et al., Galanin receptor subtype 2 (GalR2) null mutant mice display an anxiogenic-like phenotype specific to the elevated plus-maze, *Pharmacol. Biochem. Behav.*, 86, 8–20, 2007.

Baranowska-Bik, A. et al., Galanin modulates pituitary hormones release, *Neurol. Endocrinol. Lett.*, 26, 468–472, 2005.

Barr, A.M. et al., A novel, systemically active, selective galanin receptor type-3 ligand exhibits antidepressant-like activity in preclinical tests, *Neurosci. Lett.*, 405, 111–115, 2006.

Bartfai, T. et al., M-15: High-affinity chimeric peptide that blocks the neuronal actions of galanin in the hippocampus, locus coeruleus, and spinal cord, *Proc. Natl Acad. Sci. USA*, 88, 10961–10965, 1991.

Bartfai, T. et al., Galanin-receptor ligand M40 peptide distinguishes between putative galanin-receptor subtypes, *Proc. Natl Acad. Sci. USA*, 90, 11287–11291, 1993.

Bartfai, T. et al., Galmic, a nonpeptide galanin receptor agonist, affects behaviors in seizure, pain, and forced-swim tests, *Proc. Natl Acad. Sci. USA*, 101, 10470–10475, 2004.

Beck, B., Burlet, A., Nicolas, J.P., and Burlet, C., Galanin in the hypothalamus of fed and fasted lean and obese Zucker rats, *Brain Res.* 623, 124–130, 1993.

Beck, B., Hypothalamic galanin and early state of hyperphagia in obese Zucker rats, *Appetite*, 48, 206–210, 2007.

Bedecs, K., Langel, Ü., and Bartfai, T., Metabolism of galanin and galanin (1–16) in isolated cerebrospinal fluid and spinal cord membranes from rat, *Neuropeptides*, 29, 137–143, 1995.

Belfer, I. et al., Association of galanin haplotypes with alcoholism and anxiety in two ethnically distinct populations, *Mol. Psychiatry*, 11, 301–311, 2006.

Belfer, I. et al., Alcoholism is associated with GALR3 but not two other galanin receptor genes, *Genes Brain Behav.*, 6, 473–481, 2007.

Bellido, I. et al., Increased density of galanin binding sites in the dorsal raphe in a genetic rat model of depression, *Neurosci. Lett.*, 317, 101–105, 2002.

Benelli, A. et al., Galantide stimulates sexual behaviour in male rats, *Eur. J. Pharmacol.*, 260, 279–282, 1994.

Berger, A. et al., Galanin and galanin receptors in human gliomas, *Acta Neuropathol.*, 105, 555–560, 2003.

Berger, A. et al., Elevated expression of galanin receptors in childhood neuroblastic tumors, *Neuroendocrinology*, 75, 130–138, 2002.

Berger, A. et al., Galanin and galanin receptors in human cancers, *Neuropeptides*, 39, 353–359, 2005.

Berglund, M.M. et al., Binding of chimeric npy/galanin peptides M32 and M242 to cloned neuropeptide Y receptor subtypes Y1, Y2, Y4, and Y5, *Neuropeptides*, 35, 148–153, 2001.

Berthold, M. and Bartfai, T., Modes of peptide binding in G protein-coupled receptors, *Neurochem. Res.*, 22, 1023–1031, 1997.

Berthold, M. et al., Mutagenesis and ligand modification studies on galanin binding to its GTP-binding-protein-coupled receptor GalR1, *Eur. J. Biochem.*, 249, 601–606, 1997.

Blackshear, A. et al., Intracerebroventricular administration of galanin or galanin receptor subtype 1 agonist M617 induces c-Fos activation in central amygdala and dorsomedial hypothalamus, *Peptides*, 28, 1120–1124, 2007.

Bloomquist, B.T. et al., Cloning and expression of the human galanin receptor GalR2, *Biochem. Biophys. Res. Commun.*, 243, 474–479, 1998.

Bloch, G.J., Butler, P.C., Kohlert, J.G., and Bloch, D.A., Microinjection of galanin into the medial preoptic nucleus facilitates copulatory behavior in the male rat, *Physiol. Behav.*, 54, 615–624, 1993.

Bloch, G.J., Butler P.C., and Kohlert, J.G., Galanin microinjected into the medial preoptic nucleus facilitates female- and male-typical sexual behaviors in the female rat, *Physiol. Behav.*, 59, 1147–1154, 1996.

Borowsky, B. et al., Cloning and characterization of the human galanin GALR2 receptor, *Peptides*, 19, 1771–1781, 1998.

Brewer, A., Langel, Ü., and Robinson, J.K., Intracerebroventricular administration of galanin decreases free water intake and operant water reinforcer efficacy in water-restricted rats, *Neuropeptides*, 39, 117–124, 2005a.

Brewer, A., Echevarria, D.J., Langel, Ü., and Robinson, J.K., Assessment of new functional roles for galanin in the CNS, *Neuropeptides*, 39, 323–326, 2005b.

Brewer, A. and Robinson, J.K., Galanin stimulation of feeding is blocked by the addition of a response requirement, *Behav. Neurosci.*, 122, 949–953, 2008.

Burazin, T.C. and Gundlach, A.L., Inducible galanin and GalR2 receptor system in motor neuron injury and regeneration, *J. Neurochem.*, 71, 879–882, 1998.

Burazin, T.C., Larm, J.A., Ryan, M.C., and Gundlach, A.L., Galanin-R1 and -R2 receptor mRNA expression during the development of rat brain suggests differential subtype involvement in synaptic transmission and plasticity, *Eur. J. Neurosci.*, 12, 2901–2917, 2000.

Burgevin, M.C., Loquet, I., Quarteronet, D., and Habert-Ortoli, E., Cloning, pharmacological characterization, and anatomical distribution of a rat cDNA encoding for a galanin receptor, *J. Mol. Neurosci.*, 6, 33–41, 1995.

Ceide, S.C. et al., Synthesis of galmic: A nonpeptide galanin receptor agonist, *Proc. Natl Acad. Sci. USA*, 101, 16727–16732, 2004.

Celi, F. et al., Circulating acylated and total ghrelin and galanin in children with insulin-treated type 1 diabetes: Relationship to insulin therapy, metabolic control, and pubertal development, *Clin. Endocrinol.*, 63, 139–145, 2005.

Chan-Palay, V., Galanin hyperinnervates surviving neurons of the human basal nucleus of Meynert in dementias of Alzheimer's and Parkinson's disease: A hypothesis for the role of galanin in accentuating cholinergic dysfunction in dementia, *J. Comparative Neurol.*, 273, 543–557, 1988.

Chńg, J.L. et al., Distribution of galanin immunoreactivity in the central nervous system and the responses of galanin-containing neuronal pathways to injury, *Neuroscience*, 16, 343–354, 1985.

Ciosek, J. and Cisowska, A., Centrally administered galanin modifies vasopressin and oxytocin release from the hypothalamo-neurohypophysial system of euhydrated and dehydrated rats, *J. Physiol. Pharmacol.*, 54, 625–641, 2003.

Chu, M. et al., A new fungal metabolite, Sch 202596, with inhibitory activity in the galanin receptor GalR1 assay, *Tetrahedron Lett.*, 38, 6111–6114, 1997.

Church, W.B. et al., Molecular modelling and site-directed mutagenesis of human GALR1 galanin receptor defines determinants of receptor subtype specificity, *Protein Eng.*, 15, 313–323, 2002.

Consolo, S. et al., Impulse flow dependency of galanin release *in vivo* in the rat ventral hippocampus, *Proc. Natl Acad. Sci. USA*, 91, 8047–8051, 1994.

Counts, S.E. et al., Galanin inhibits tyrosine hydroxylase expression in midbrain dopaminergic neurons, *J. Neurochem.*, 83, 442–451, 2002.

Coronel, M.F., Brumovsky, P.R., Hökfelt, T., and Villar, M.J., Differential galanin upregulation in dorsal root ganglia and spinal cord after graded single ligature nerve constriction of the rat sciatic nerve, *J. Chem. Neuroanat.*, 35, 94–100, 2008.

Crown, A., Clifton, D.K., and Steiner, R.A., Neuropeptide signaling in the integration of metabolism and reproduction, *Neuroendocrinology*, 86, 175–182, 2007.

Counts, S.E. et al., Galanin inhibits tyrosine hydroxylase expression in midbrain dopaminergic neurons, *J. Neurochem.*, 83, 442–451, 2002.

Counts, S.E. et al., Galanin fiber hypertrophy within the cholinergic nucleus basalis during the progression of Alzheimer's disease, *Dementia Geriatric Cognitive Disorders*, 21, 205–214, 2006.

Crawley, J.N. et al., Activity of centrally administered galanin fragments on stimulation of feeding behavior and on galanin receptor binding in the rat hypothalamus, *J. Neurosci.*, 10, 3695–3700, 1990.

Crawley, J.N., Robinson, J.K., Langel, Ü., and Bartfai, T., Galanin receptor antagonists M40 and C7 block galanin-induced feeding, *Brain Res.*, 600, 268–272, 1993.

Crawley, J.N., Minireview. Galanin–acetylcholine interactions: Relevance to memory and Alzheimer's disease, *Life Sci.*, 58, 2185–2199, 1996.

Crawley, J.N., The role of galanin in feeding behavior, *Neuropeptides*, 33, 369–375, 1999.

Cunningham, M.J., Scarlett, J.M., and Steiner, R.A., Cloning and distribution of galanin-like peptide mRNA in the hypothalamus and pituitary of the macaque, *Endocrinology*, 143, 755–763, 2002.

Cunningham, M.J., Galanin-like peptide as a link between metabolism and reproduction, *J. Neuroendocrinol.*, 16, 717–723, 2004a.

Cunningham, M.J. et al., Galanin-like peptide as a possible link between metabolism and reproduction in the macaque, *J. Clin. Endocrinol. Metab.*, 89, 1760–1766, 2004b.

Deecher, D.C., Mash, D.C., Staley, J.K., and Mufson, E.J., Characterization and localization of galanin receptors in human entorhinal cortex, *Regul. Pept.*, 73, 149–159, 1998.

Diez, M. et al., Neuropeptides in hippocampus and cortex in transgenic mice overexpressing V717F beta-amyloid precursor protein—initial observations, *Neuroscience*, 100, 259–286, 2000.

Diez, M. et al., Neuropeptide alterations in the hippocampal formation and cortex of transgenic mice overexpressing beta-amyloid precursor protein (APP) with the Swedish double mutation (APP23), *Neurobiol. Dis.*, 14, 579–594, 2003.

Eaton, M.J. et al., Lumbar transplant of neurons genetically modified to secrete galanin reverse pain-like behaviors after partial sciatic nerve injury, *J. Periph. Nerv. Syst.*, 4, 245–257, 1999.

El-Salhy, M., Triple treatment with octreotide, galanin, and serotonin is a promising therapy for colorectal cancer, *Curr. Pharm. Des.*, 11, 2107–2117, 2005.

Elvander, E. et al., Intraseptal muscarinic ligands and galanin: Influence on hippocampal acetylcholine and cognition, *Neuroscience*, 126, 541–557, 2004.

Elliott-Hunt, C.R. et al., Galanin acts as a neuroprotective factor to the hippocampus, *Proc. Natl Acad. Sci. USA*, 101, 5105–5110, 2004.

Ericson, E. and Ahlenius, S., Suggestive evidence for inhibitory effects of galanin on mesolimbic dopaminergic neurotransmission, *Brain Res.*, 822, 200–209, 1999.

Flatters, S.J., Fox, A.J., and Dickenson, A.H., *In vivo* and *in vitro* effects of peripheral galanin on nociceptive transmission in naive and neuropathic states, *Neuroscience*, 116, 1005–1012, 2003.

Florén, A. et al., Multiple interaction sites of galnon trigger its biological effects, *Neuropeptides*, 39, 547–558, 2005.

Fraley, G.S. et al., Stimulation of sexual behavior in the male rat by galanin-like peptide, *Horm. Behav.*, 46, 551–557, 2004.

Fukuoka, T. et al., Change in mRNAs for neuropeptides and the GABA(A) receptor in dorsal root ganglion neurons in a rat experimental neuropathic pain model, *Pain*, 78, 13–26, 1998.

Gleason, T.C., Dreiling, J.L., and Crawley, J.N., Rat strain differences in response to galanin on the Morris water task, *Neuropeptides*, 33, 265–270, 1999.

Gottsch, M.L. et al., Phenotypic analysis of mice deficient in the type 2 galanin receptor (GALR2), *Mol. Cell. Biol.*, 25, 4804–4811, 2005.

Grass, S., Crawley, J.N., Xu, X-J., and Wiesenfeld-Hallin, Z., Reduced spinal cord sensitization to C-fibre stimulation in mice over-expressing galanin, *Eur. J. Neurosci.*, 17, 1829–1832, 2003a.

Grass, S. et al., Flexor reflex excitability in mice lacking galanin receptor galanin-R1, *Neurosci. Lett.*, 345, 153–156, 2003b.

Grenhoff, J. et al., Noradrenergic modulation of midbrain dopamine cell firing elicited by stimulation of the locus coeruleus in the rat, *J. Neural Transm.*, 93, 11–25, 1993.

Gundlach, A.L., Galanin/GALP and galanin receptors: Role in central control of feeding, body weight/obesity, and reproduction? *Eur. J. Pharmacol.*, 440, 255–268, 2002.

Grupe, A. et al., Evidence for novel susceptibility genes for late-onset Alzheimer's disease from a genome-wide association study of putative functional variants, *Hum. Mol. Genet.*, 16, 865–873, 2007.

Gustafson, E.L. et al., Distribution of a rat galanin receptor mRNA in rat brain, *Neuroreport*, 7, 953–957, 1996.

Haberman, R.P., Samulski, R.J., and McCown, T.J., Attenuation of seizures and neuronal death by adeno-associated virus vector galanin expression and secretion, *Nature Med.*, 9, 1076–1080, 2003.

Habert-Ortoli, E. et al., Molecular cloning of a functional human galanin receptor, *Proc. Natl Acad. Sci. USA*, 91, 9780–9783, 1994.

Hao, J.X. et al., Intrathecal galanin alleviates allodynia-like behaviour in rats after partial peripheral nerve injury, *Eur. J. Neurosci.*, 11, 427–432, 1999.

Hawes, J.J. et al., GalR1, but not GalR2 or GalR3, levels are regulated by galanin signaling in the locus coeruleus through a cyclic AMP-dependent mechanism, *J. Neurochem.*, 93, 1168–1176, 2005.

Hedlund, P.B., Yanaihara, N., and Fuxe, K., Evidence for specific N-terminal galanin fragment binding sites in the rat brain, *Eur. J. Pharmacol.*, 224, 203–205, 1992.

Hedlund, P.B., Finnman, U.B., Yanaihara, N., and Fuxe, K., Galanin-(1–15), but not galanin-(1–29), modulates 5-HT1A receptors in the dorsal hippocampus of the rat brain: Possible existence of galanin receptor subtypes, *Brain Res.*, 634, 163–167, 1994.

Hobson, S.-A. et al., Mice deficient for galanin receptor 2 have decreased neurite outgrowth from adult sensory neurons and impaired pain-like behavior, *J. Neurochem.*, 99, 1000–1010, 2006.

Hohmann, J.G. et al., Neuroendocrine profiles in galanin-overexpressing and knockout mice, *Neuroendocrinology*, 77, 354–366, 2003.

Hökfelt, T., Wiesenfeld-Hallin, Z., Villar, M., and Melander, T., Increase of galanin-like immunoreactivity in rat dorsal root ganglion cells after peripheral axotomy, *Neurosci. Lett.*, 83, 217–220, 1987.

Holmes, A., Yang, R.J., and Crawley, J.N., Evaluation of an anxiety-related phenotype in galanin overexpressing transgenic mice, *J. Mol. Neurosci.*, 18, 151–165, 2002.

Holmes, F.E. et al., Targeted disruption of the galanin gene reduces the number of sensory neurons and their regenerative capacity, *Proc. Natl Acad. Sci. USA*, 97, 11563–11568, 2000.

Holmes, F.E. et al., Transgenic overexpression of galanin in the dorsal root ganglia modulates pain-related behavior, *Proc. Natl Acad. Sci. USA*, 100, 6180–6185, 2003.

Holmes, F.E., Mahoney, S.A., and Wynick, D., Use of genetically engineered transgenic mice to investigate the role of galanin in the peripheral nervous system after injury, *Neuropeptides*, 39, 191–199, 2005.

Holmes, P.V. et al., Chronic social stress increases levels of preprogalanin mRNA in the rat locus coeruleus, *Pharmacol. Biochem. Behav.*, 50, 655–660, 1995.

Howard, A.D. et al., Molecular cloning and characterization of a new receptor for galanin, *FEBS Lett.*, 405, 285–290, 1997.

Husum, H. et al., Topiramate normalizes hippocampal NPY-LI in flinders sensitive line "depressed" rats and upregulates NPY, galanin, and CRH-LI in the hypothalamus: Implications for mood-stabilizing and weight loss-inducing effects, *Neuropsychopharmacology*, 28, 1292–1299, 2003.

Hygge-Blakeman, K. et al., Galanin over-expression decreases the development of neuropathic pain-like behaviors in mice after partial sciatic nerve injury, *Brain Res.*, 1025, 152–158, 2004.

Iishi, H. et al., Chemoprevention by galanin against colon carcinogenesis induced by azoxymethane in Wistar rats, *Int. J. Cancer*, 61, 861–863, 1995.

Ignatov, A., Hermans-Borgmeyer, I., and Schaller, H.C., Cloning and characterization of a novel G-protein-coupled receptor with homology to galanin receptors, *Neuropharmacology*, 46, 1114–1120, 2004.

Imbe, H. et al., Increase of galanin-like immunoreactivity in rat hypothalamic arcuate neurons after peripheral nerve injury, *Neurosci. Lett.*, 368, 102–106, 2004.

Jacoby, A.S. et al., Structural organization of the mouse and human GALR1 galanin receptor genes (Galnr and GALNR) and chromosomal localization of the mouse gene, *Genomics*, 45, 496–508, 1997.

Jacoby, A.S. et al., Critical role for GALR1 galanin receptor in galanin regulation of neuroendocrine function and seizure activity, *Brain Res. Mol. Brain Res.*, 107, 195–200, 2002.

Jhamandas, J.H., Harris, K.H., MacTavish, D., and Jassar, B.S., Novel excitatory actions of galanin on rat cholinergic basal forebrain neurons: Implications for its role in Alzheimer's disease, *J. Neurophysiol.*, 87, 696–704, 2002.

Ji, R.R. et al., Galanin antisense oligonucleotides reduce galanin levels in dorsal root ganglia and induce autotomy in rats after axotomy, *Proc. Natl Acad. Sci. USA*, 91, 12540–12543, 1994.

Jimenez-Andrade, J.M. et al., Mechanism by which peripheral galanin increases acute inflammatory pain, *Brain Res.*, 1056, 113–117, 2005.

Jimenez-Andrade, J.M. et al., Activation of peripheral galanin receptors: Differential effects on nociception, *Pharmacol. Biochem. Behav.* 85, 273–280, 2006.

Kai, A. et al., Galanin inhibits neural activity in the subfornical organ in rat slice preparation, *Neuroscience*, 143, 769–777, 2006.

Kageyama, H. et al., Galanin-like peptide in the brain: Effects on feeding, energy metabolism, and reproduction, *Regul. Pept.*, 126, 21–26, 2005.

Kakuyama, H. et al., Role of N-terminal active sites of galanin in neurally evoked circular muscle contractions in the guinea-pig ileum, *Eur. J. Pharmacol.*, 329, 85–91, 1997.

Kaplan, L.M. et al., Galanin is an estrogen-inducible, secretory product of the rat anterior pituitary, *Proc. Natl Acad. Sci. USA*, 85, 7408–7412, 1988.

Kask, K. et al., Binding and agonist/antagonist actions of M35, galanin(1–13)-bradykinin(2–9)amide chimeric peptide, in Rin m 5F insulinoma cells, *Regul. Pept.*, 59, 341–348, 1995.

Kask, K. et al., Delineation of the peptide binding site of the human galanin receptor, *EMBO J.*, 15, 236–244, 1996.

Kask, K. et al., Mutagenesis study on human galanin receptor GalR1 reveals domains involved in ligand binding, *Ann. NY Acad. Sci.*, 863, 78–85, 1998.

Kaufman, A.S., Buenzle, J., Fraley, G.S., and Rissman, E.F., Effects of galanin-like peptide (GALP) on locomotion, reproduction, and body weight in female and male mice, *Horm. Behav.*, 48, 141–151, 2005.

Katoh, T., Ohmori, O., Iwasaki, K., and Inoue, M., Synthetic studies on Sch 202596, an antagonist of the galanin receptor GalR1: An efficient synthesis of (±)-geodin, the spirocoumaranone part of Sch 202596, *Tetrahedron*, 58, 1289–1299, 2002.

Kehr, J. et al., Galanin is a potent *in vivo* modulator of mesencephalic serotonergic neurotransmission, *Neuropsychopharmacology*, 27, 341–356, 2002.

Kelley, A.E., Baldo, B.A., Pratt, W.E., and Will, M.J., Corticostriatal-hypothalamic circuitry and food motivation: Integration of energy, action, and instrumental behavior, *Physiol. Behav.*, 86, 773–795, 2005.

Kerr, B.J. et al., Galanin knockout mice reveal nociceptive deficits following peripheral nerve injury, *Eur. J. Neurosci.*, 12, 793–802, 2000.

Kerr, B.J. et al., Endogenous galanin potentiates spinal nociceptive processing following inflammation, *Pain*, 93, 267–277, 2001.

Khoshbouei, H. et al., Behavioral reactivity to stress: Amplification of stress-induced noradrenergic activation elicits a galanin-mediated anxiolytic effect in central amygdala, *Pharmacol. Biochem. Behav.*, 71, 407–417, 2002a.

Khoshbouei, H., Cecchi, M., and Morilak, D.A., Modulatory effects of galanin in the lateral bed nucleus of the stria terminalis on behavioral and neuroendocrine responses to acute stress, *Neuropsychopharmacol.*, 27, 25–34, 2002b.

Kinney, J.W. et al., Deficits in trace cued fear conditioning in galanin-treated rats and galanin-overexpressing transgenic mice, *Learn Mem.*, 9, 178–190, 2002.

Kokaia, M. et al., Suppressed kindling epileptogenesis in mice with ectopic overexpression of galanin, *Proc. Natl Acad. Sci. USA*, 98, 14006–14011, 2001.

Kolakowski, L.F., Jr, et al., Molecular characterization and expression of cloned human galanin receptors GALR2 and GALR3, *J. Neurochem.*, 71, 2239–2251, 1998.

Kondo, K. et al., Centrally administered galanin inhibits osmotically stimulated arginine vasopressin release in conscious rats, *Neurosci. Lett.*, 22, 245–248, 1991.

Konkel, M.J., Wetzel, L., and Talisman, I.J., 3-Imino-2-indolones for the treatment of depression and/or anxiety, U.S. patent 514/323, 2004.

Konkel, M.J. et al., 3-arylimino-2-indolones are potent and selective galanin GAL3 receptor antagonists, *J. Med. Chem.*, 49, 3757–3758, 2006a.

Konkel, M.J. et al., Amino substituted analogs of 1-phenyl-3-phenylimino-2-indolones with potent galanin Gal3 receptor binding affinity and improved solubility, *Bioorg. Med. Chem. Lett.*, 16, 3950–3954, 2006b.

Kowall, N.W. and Beal, M.F., Galanin-like immunoreactivity is present in human substantia innominata and in senile plaques in Alzheimer's disease, *Neurosci. Lett.*, 98, 118–123, 1989.

Kozoriz, M.G., Kuzmiski, J.B., Hirasawa, M., and Pittman, Q.J., Galanin modulates neuronal and synaptic properties in the rat supraoptic nucleus in a use and state dependent manner, *J. Neurophysiol.*, 96, 154–164, 2006.

Krasnow, S.M. et al., Analysis of the contribution of galanin receptors 1 and 2 to the central actions of galanin-like peptide, *Neuroendocrinology*, 79, 268–277, 2004.

Kulinski, T. et al., Conformational analysis of galanin using end-to-end distance distribution observed by Förster resonance energy transfer, *Eur. Biophys. J.*, 26, 145–154, 1997.

Kuteeva, E., Hökfelt, T., and Ögren, S.O., Behavioural characterisation of transgenic mice overexpressing galanin under the PDGF-B promoter, *Neuropeptides*, 39, 299–304, 2005.

Kuteeva, E. et al., Differential role of galanin receptors in the regulation of depression-like behavior and monoamine/stress-related genes at the cell body level, *Neuropsychopharmacology*, 33, 2790–2802, 2008.

Kyrkouli, S.E., Stanley, B.G., and Leibowitz, S.F., Galanin: Stimulation of feeding induced by medial hypothalamic injection of this novel peptide, *Eur. J. Pharmacol.*, 122, 159–160, 1986.

Kyrkouli, S.E., Stanley, B.G., Seirafi, R.D., and Leibowitz, S.F., Stimulation of feeding by galanin: Anatomical localization and behavioral specificity of this peptide's effects in the brain, *Peptides*, 11, 995–1001, 1990.

Kyrkouli, S.E., Strubbe, J.H., and Scheurink, A.J., Galanin in the PVN increases nutrient intake and changes peripheral hormone levels in the rat, *Physiol. Behav.*, 89, 103–109, 2006.

Land, T., Langel, Ü., and Bartfai, T., Hypothalamic degradation of galanin(1–29) and galanin(1–16): Identification and characterization of the peptidolytic products, *Brain Res.*, 558, 245–250, 1991a.

Land, T. et al., Linear and cyclic N-terminal galanin fragments and analogs as ligands at the hypothalamic galanin receptor, *Int. J. Pept. Protein Res.*, 38, 267–272, 1991b.

Landry, M. et al., Galanin expression in adult human dorsal root ganglion neurons: Initial observations, *Neuroscience*, 117, 795–809, 2003.

Lang, R. et al., Pharmacological and functional characterization of galanin-like peptide fragments as potent galanin receptor agonists, *Neuropeptides*, 39, 179–184, 2005.

Laplante, F., Crawley, J.N., and Quirion, R., Selective reduction in ventral hippocampal acetylcholine release in awake galanin-treated rats and galanin-overexpressing transgenic mice, *Regul. Pept.*, 122, 91–98, 2004.

Lawrence, C.B., Baudoin, F.M., and Luckman, S.M., Centrally administered galanin-like peptide modifies food intake in the rat: A comparison with galanin, *J. Neuroendocrinol.*, 14, 853–860, 2002.

Legakis, I., Mantzouridis, T., and Mountokalakis, T., Positive correlation of galanin with glucose in type 2 diabetes, *Diabetes Care*, 28, 759–760, 2005.

Leibowitz, S.F., Akabayashi, A., and Wang, J., Obesity on a high-fat diet: Role of hypothalamic galanin in neurons of the anterior paraventricular nucleus projecting to the median eminence, *J. Neurosci.*, 18, 2709–2719, 1998.

Leibowitz, S.F. et al., Ethanol intake increases galanin mRNA in the hypothalamus and withdrawal decreases it, *Physiol. Behav.*, 79, 103–111, 2003.

Leibowitz, S.F., Overconsumption of dietary fat and alcohol: Mechanisms involving lipids and hypothalamic peptides, *Physiol. Behav.*, 91, 513–521, 2007.

Leung, B. et al., Galanin in human pituitary adenomas: Frequency and clinical significance, *Clin. Endocrinol.*, 56, 397–403, 2002.

Lewis, M.J. et al., Galanin microinjection in the third ventricle increases voluntary ethanol intake, *Alcoholism Clin. Exp. Res.*, 28, 1822–1828, 2004.

Lin, E.J. et al., Recombinant AAV-mediated expression of galanin in rat hippocampus suppresses seizure development, *Eur. J. Neurosci.*, 18, 2087–2092, 2003.

Lindskog, S., Gregersen, S., Hermansen, K., and Ahrén, B., Effects of galanin on proinsulin mRNA and insulin biosynthesis in normal islets, *Regul. Pept.*, 58, 135–139, 1995.

Liu, H.X. et al., Receptor subtype-specific pronociceptive and analgesic actions of galanin in the spinal cord: Selective actions via GalR1 and GalR2 receptors, *Proc. Natl Acad. Sci. USA*, 98, 9960–9964, 2001.

López, F.J. et al., Galanin: A hypothalamic–hypophysiotropic hormone modulating reproductive functions, *Proc. Natl Acad. Sci. USA*, 88, 4508–4512, 1991.

Lorimer, D.D. and Benya, R.V., Cloning and quantification of galanin-1 receptor expression by mucosal cells lining the human gastrointestinal tract, *Biochem. Biophys. Res. Com.*, 222, 379–385, 1996.

Lu, X., Lundström, L., Langel, Ü., and Bartfai, T., Galanin receptor ligands, *Neuropeptides*, 39, 143–146, 2005a.

Lu, X., Lundström, L., and Bartfai, T., Galanin (2–11) binds to GalR3 in transfected cell lines: Limitations for pharmacological definition of receptor subtypes, *Neuropeptides*, 39, 165–167, 2005b.

Lundström, L., Lu, X., Langel, Ü., and Bartfai, T., Important pharmacophores for binding to galanin receptor 2, *Neuropeptides*, 39, 169–171, 2005a.

Lundström, L. et al., A galanin receptor subtype 1 specific agonist, *Int. J. Pept. Res. Ther.*, 11, 17–27, 2005b.

Lundström, L., Elmquist, A., Bartfai, T., and Langel, Ü., Galanin and its receptors in neurological disorders, *Neuromolecular Med.*, 7, 157–180, 2005c.

Lundström, L., Sollenberg, U.E., Bartfai, T., and Langel, Ü., Molecular characterization of the ligand binding site of the human galanin receptor type 2, identifying subtype selective interactions, *J. Neurochem.*, 103, 1774–1784, 2007.

Ma, W., and Bisby, M.A., Differential expression of galanin immunoreactivities in the primary sensory neurons following partial and complete sciatic nerve injuries, *Neuroscience*, 79, 1183–1195, 1997.

Ma, W. and Bisby, M.A., Increase of galanin mRNA in lumbar dorsal root ganglion neurons of adult rats after partial sciatic nerve ligation, *Neurosci. Lett.*, 262, 195–198, 1999.

Mahoney, S.A. et al., The second galanin receptor GalR2 plays a key role in neurite outgrowth from adult sensory neurons, *J. Neurosci.*, 23, 416–421, 2003.

Makino, S., Asaba, K., Nishiyama, M., and Hashimoto, K., Decreased type 2 corticotropin-releasing hormone receptor mRNA expression in the ventromedial hypothalamus during repeated immobilization stress, *Neuroendocrinology*, 70, 160–167, 1999.

Man, P.S. and Lawrence, C.B., The effects of galanin-like peptide on energy balance, body temperature, and brain activity in the mouse and rat are independent of the GALR2/3 receptor, *J. Neuroendocrinol.*, 20, 128–137, 2008.

Marti, E. et al., Ontogeny of peptide- and amine-containing neurones in motor, sensory, and autonomic regions of rat and human spinal cord, dorsal root ganglia, and rat skin, *J. Comp. Neurol.*, 266, 332–359, 1987.

Matsumoto, H. et al., Stimulation effect of galanin-like peptide (GALP) on luteinizing hormone-releasing hormone-mediated luteinizing hormone (LH) secretion in male rats, *Endocrinol.*, 142, 3693–3696, 2001.

Mazarati, A.M. et al., Modulation of hippocampal excitability and seizures by galanin, *J. Neurosci.*, 20, 6276–6281, 2000.

Mazarati, A. et al., Galanin type 2 receptors regulate neuronal survival, susceptibility to seizures and seizure-induced neurogenesis in the dentate gyrus, *Eur. J. Neurosci.*, 19, 3235–3244, 2004.

Mazarati, A. et al., Regulation of kindling epileptogenesis by hippocampal galanin type 1 and type 2 receptors: The effects of subtype-selective agonists and the role of G-protein-mediated signaling, *J. Pharmacol. Exp. Ther.*, 318, 700–708, 2006.

McColl, C.D. et al., Galanin receptor-1 knockout mice exhibit spontaneous epilepsy, abnormal EEGs and altered inhibition in the hippocampus, *Neuropharmacology*, 50, 209–218, 2006.

McDonald, M.P., Gleason, T.C., Robinson, J.K., and Crawley, J.N., Galanin inhibits performance on rodent memory tasks, *Ann. NY Acad. Sci.*, 863, 305–322, 1998.

Melander, T. et al., Coexistence of galanin-like immunoreactivity with catecholamines, 5-hydroxytryptamine, GABA, and neuropeptides in the rat CNS, *J. Neurosci.*, 6, 3640–3654, 1986.

Menendez, J.A., Atrens, D.M., and Leibowitz, S.F., Metabolic effects of galanin injections into the paraventricular nucleus of the hypothalamus, *Peptides*, 13, 323–327, 1992.

Mennicken, F. et al., Restricted distribution of galanin receptor 3 (GalR3) mRNA in the adult rat central nervous system, *J. Chem. Neuroanat.*, 24, 257–268, 2002.

Merchenthaler, I., Lopez, F.J., and Negro-Vilar, A., Colocalization of galanin and luteinizing hormone-releasing hormone in a subset of preoptic hypothalamic neurons: Anatomical and functional correlates, *Proc. Natl Acad. Sci. USA*, 87, 6326–6330, 1990.

Merchenthaler, I., Lopez, F.J., and Negro-Vilar, A., Anatomy and physiology of central galanin-containing pathways, *Prog. Neurobiol.*, 40, 711–769, 1993.

Murphy, P.G. et al., Endogenous interleukin-6 contributes to hypersensitivity to cutaneous stimuli and changes in neuropeptides associated with chronic nerve constriction in mice, *Eur. J. Neurosci.*, 11, 2243–2253, 1999.

Muschol, M. and Salzberg, B.M., Dependence of transient and residual calcium dynamics on action-potential patterning during neuropeptide secretion, *J. Neurosci.*, 20, 6773–6780, 2000.

Nahin, R.L., Ren, K., De Leon, M., and Ruda, M., Primary sensory neurons exhibit altered gene expression in a rat model of neuropathic pain, *Pain*, 58, 95–108, 1994.

Ögren, S.O. and Pramanik, A., Galanin stimulates acetylcholine release in the rat striatum, *Neurosci. Lett.*, 128, 253–256, 1991.

Ögren, S.O. et al., Evidence for a role of the neuropeptide galanin in spatial learning, *Neuroscience*, 51, 1–5, 1992.

Ögren, S.O., Kehr, J., and Schött, P.A., Effects of ventral hippocampal galanin on spatial learning and on *in vivo* acetylcholine release in the rat, *Neuroscience*, 75, 1127–1140, 1996.

Ögren, S.O. et al., Galanin and learning, *Brain Res.*, 848, 174–182, 1999.

Ögren, S.O., Razani, H., Elvander-Tottie, E., and Kehr, J., The neuropeptide galanin as an *in vivo* modulator of brain 5-HT1A receptors: Possible relevance for affective disorders, *Phys. Behav.*, 92, 172–179, 2007.

Ohtaki, T. et al., Isolation and cDNA cloning of a novel galanin-like peptide (GALP) from porcine hypothalamus, *J. Biol. Chem.*, 274, 37041–37045, 1999.

O'Meara, G. et al., Galanin regulates the postnatal survival of a subset of basal forebrain cholinergic neurons, *Proc. Natl Acad. Sci. USA*, 97, 11569–11574, 2000.

O'Neal, H.A., Van Hoomissen, J.D., Holmes, P.V., and Dishman, R.K., Prepro-galanin messenger RNA levels are increased in rat locus coeruleus after treadmill exercise training, *Neurosci. Lett.*, 299, 69–72, 2001.

Ozturk, G. and Tonge, D.A., Effects of leukemia inhibitory factor on galanin expression and on axonal growth in adult dorsal root ganglion neurons *in vitro*, *Exp. Neurol.*, 169, 376–385, 2001.

Pang, L. et al., The mouse GalR2 galanin receptor: Genomic organization, cDNA cloning, and functional characterization, *J. Neurochem.*, 71, 2252–2259, 1998.

Parker, E.M. et al., Cloning and characterization of the rat GALR1 galanin receptor from Rin14B insulinoma cells, *Brain Res. Mol. Brain Res.*, 34, 179–189, 1995.

Perel, Y. et al., Galanin and galanin receptor expression in neuroblastic tumours: Correlation with their differentiation status, *Br. J. Cancer*, 86, 117–122, 2002.

Perez, S., Basile, M., Mash, D.C., and Mufson, E.J., Galanin receptor over-expression within the amygdala in early Alzheimer's disease: An *in vitro* autoradiographic analysis, *J. Chem. Neuroanat.*, 24, 109–116, 2002.

Pieribone, V.A. et al., Galanin induces a hyperpolarization of norepinephrine-containing locus coeruleus neurons in the brainstem slice, *Neuroscience*, 64, 861–874, 1995.

Piroli, G.G. et al., Progestin regulation of galanin and prolactin gene expression in oestrogen-induced pituitary tumours, *J. Neuroendocrinol.*, 13, 302–309, 2001.

Pirondi, S. et al., The galanin-R2 agonist AR-M1896 reduces glutamate toxicity in primary neural hippocampal cells, *J. Neurochem.*, 95, 821–833, 2005.

Poggioli, R., Rasori, E., and Bertolini, A., Galanin inhibits sexual behavior in male rats, *Eur. J. Pharmacol.*, 213, 87–90, 1992.

Pooga, M. et al., Novel galanin receptor ligands, *J. Pept. Res.*, 51, 65–74, 1998a.

Pooga, M. et al., Cell penetrating PNA constructs regulate galanin receptor levels and modify pain transmission *in vivo*, *Nature Biotechnol.*, 16, 857–861, 1998b.

Pramanik, A. and Ögren, S.O., Galanin-evoked acetylcholine release in the rat striatum is blocked by the putative galanin antagonist M15, *Brain Res.*, 574, 317–319, 1992.

Rada, P., Mark, G.P., and Hoebel, B.G., Galanin in the hypothalamus raises dopamine and lowers acetylcholine release in the nucleus accumbens: A possible mechanism for hypothalamic initiation of feeding behavior, *Brain Res.*, 798, 1–6, 1998.

Rada, P., Avena, N.M., Leibowitz, S.F., and Hoebel, B.G., Ethanol intake is increased by injection of galanin in the paraventricular nucleus and reduced by a galanin antagonist, *Alcohol.*, 33, 91–97, 2004.

Rajarao, S.J. et al., Anxiolytic-like activity of the non-selective galanin receptor agonist, galnon, *Neuropeptides*, 41, 307–320, 2007.

Razani, H., Diaz-Cabiale, Z., Fuxe, K., and Ögren, S.O., Intraventricular galanin produces a time-dependent modulation of 5-HT1A receptors in the dorsal raphe of the rat, *Neuroreport*, 11, 3943–3948, 2000.

Razani, H. et al., Prolonged effects of intraventricular galanin on a 5-hydroxytryptamine(1A) receptor mediated function in the rat, *Neurosci. Lett.*, 299, 145–149, 2001.

Robinson, J.K., Galanin and cognition, *Behav. Cogn. Neurosci. Rev.*, 3, 222–242, 2004.

Robinson, J.K. and Crawley, J.N., Intraventricular galanin impairs delayed nonmatching-to-sample performance in rats, *Behav. Neurosci.*, 107, 458–467, 1993a.

Robinson, J.K. and Crawley, J.N., Intraseptal galanin potentiates scopolamine impairment of delayed nonmatching to sample, *J. Neurosci.*, 13, 5119–5125, 1993b.

Robinson, J.K. and Crawley, J.N., Analysis of anatomical sites at which galanin impairs delayed nonmatching to sample in rats, *Behav. Neurosci.*, 108, 941–950, 1994.

Rodriguez-Puertas, R. et al., 125I-galanin binding sites in Alzheimer's disease: Increases in hippocampal subfields and a decrease in the caudate nucleus, *J. Neurochem.*, 68, 1106–1113, 1997.

Rökaeus, A., Young, W.S. III, and Mezey, É., Galanin coexists with vasopressin in the normal ra hypothalamus and galanin's synthesis is increased in the Brattleboro (diabetes insipidus) rat, *Neurosci. Lett.*, 19, 45–50, 1988.

Rossowski, W.J. et al., Galanin binding sites in rat gastric and jejunal smooth muscle membrane preparations, *Peptides*, 11, 333–338, 1990.

Rossowski, W.J. et al., Galanin: Structure-dependent effect on pancreatic amylase secretion and jejunal strip contraction, *Eur. J. Pharmacol.*, 240, 259–267, 1993.

Rustay, N.R. et al., Galanin impairs performance on learning and memory tasks: Findings from galanin transgenic and GAL-R1 knockout mice. *Neuropeptides*, 39, 239–243, 2005.

Saar, K. et al., Characterisation of a new chimeric ligand for galanin receptors: Galanin(1–13)-[D-Trp(32)]-neuropeptide Y(25–36)amide, *Regul. Pept.*, 102, 15–19, 2001.

Saar, K. et al., Anticonvulsant activity of a nonpeptide galanin receptor agonist, *Proc. Natl Acad. Sci. USA*, 99, 7136–7141, 2002.

Sahu, A. et al., Effects of neuropeptide Y, NPY analog (norleucine4-NPY), galanin and neuropeptide K on LH release in ovariectomized (ovx) and ovx estrogen, progesterone-treated rats, *Peptides*, 8, 921–926, 1987.

Sahu, A., Evidence suggesting that galanin (GAL), melanin-concentrating hormone (MCH), neurotensin (NT), proopiomelanocortin (POMC), and neuropeptide Y (NPY) are targets of leptin signaling in the hypothalamus, *Endocrinology*, 139, 795–798, 1998a.

Sahu, A., Leptin decreases food intake induced by melanin-concentrating hormone (MCH), galanin (GAL), and neuropeptide Y (NPY) in the rat, *Endocrinology*, 11, 4739–4742, 1998b.

Salamone, J.D., Correa, M., Farrar, A., and Mingote, S.M., Effort-related functions of nucleus accumbens dopamine and associated forebrain circuits, *Psychopharmacology* (*Berlin*), 191, 461–482, 2007.

Santic, R. et al., Gangliocytes in neuroblastic tumors express alarin, a novel peptide derived by differential splicing of the galanin-like peptide gene, *J. Mol. Neurosci.*, 29, 145–152, 2006.

Santic, R. et al., Alarin is a vasoactive peptide, *Proc. Natl Acad. Sci. USA*, 104, 10217–10222, 2007.

Scheffen, J.R., Splett, C.L., Desotelle, J.A., and Bauer-Dantoin, A.C., Testosterone-dependent effects of galanin on pituitary luteinizing hormone secretion in male rats, *Biol. Reprod.*, 68, 363–369, 2003.

Schneider, E.R. et al., Orexigenic peptides and alcohol intake: Differential effects of orexin, galanin, and ghrelin, *Alcohol Clin. Exp. Res.*, 31, 1858–1865, 2007.

Scott, M.K. et al., 2,3-Dihydro-dithiin and -dithiepine-1,1,4,4-tetroxides: Small molecule non-peptide antagonists of the human galanin hGAL-1 receptor, *Bioorg. Med. Chem.*, 8, 1383–1391, 2000.

Sethi, T. and Rozengurt, E., Galanin stimulates Ca^{2+} mobilization, inositol phosphate accumulation, and clonal growth in small cell lung cancer cells, *Cancer Res.*, 51, 1674–1679, 1991.

Seufferlein, T. and Rozengurt, E., Galanin, neurotensin, and phorbol esters rapidly stimulate activation of mitogen-activated protein kinase in small cell lung cancer cells, *Cancer Res.*, 56, 5758–5764, 1996.

Shen, P.J., Larm, J.A., and Gundlach, A.L., Expression and plasticity of galanin systems in cortical neurons, oligodendrocyte progenitors and proliferative zones in normal brain and after spreading depression, *Eur. J. Neurosci.*, 18, 1362–1376, 2003.

Shen, P.J. et al., Galanin in neuro(glio)genesis: Expression of galanin and receptors by progenitor cells *in vivo* and *in vitro* and effects of galanin on neurosphere proliferation, *Neuropeptides*, 39, 201–205, 2005.

Shi, T.J. et al., Regulation of galanin and neuropeptide Y in dorsal root ganglia and dorsal horn in rat mononeuropathic models: Possible relation to tactile hypersensitivity, *Neuroscience*, 93, 741–757, 1999.

Skofitsch, G. and Jacobowitz, D.M., Galanin-like immunoreactivity in capsaicin sensitive sensory neurons and ganglia, *Brain Res. Bull.*, 15, 191–195, 1985a.

Skofitsch, G. and Jacobowitz, D.M., Immunohistochemical mapping of galanin-like neurons in the rat central nervous system, *Peptides*, 6, 509–546, 1985b.

Skofitsch, G., Sills, M.A., and Jacobowitz, D.M., Autoradiographic distribution of ^{125}I-galanin binding sites in the rat central nervous system, *Peptides*, 7, 1029–1042, 1986.

Skofitsch, G., Jacobowitz, D.M., Amann, R., and Lembeck, F., Galanin and vasopressin coexist in the rat hypothalamo-neurohypophyseal system, *Neuroendocrinology*, 49, 419–427, 1989.

Smith, B.K., York, D.A., and Bray, G.A., Effects of dietary preference and galanin administration in the paraventricular or amygdaloid nucleus on diet self-selection, *Brain Res. Bull.*, 39, 149–154, 1996.

Smith, K.E. et al., Expression cloning of a rat hypothalamic galanin receptor coupled to phosphoinositide turnover, *J. Biol. Chem.*, 272, 24612–24616, 1997.

Smith, K.E. et al., Cloned human and rat galanin GALR3 receptors. Pharmacology and activation of G-protein inwardly rectifying K^+ channels, *J. Biol. Chem.*, 273, 23321–23326, 1998.

Soares, J. et al., Brain noradrenergic responses to footshock after chronic activity-wheel running. *Behav. Neurosci.*, 113, 558–566, 1999.

Sollenberg, U.E., Lundström, L., Bartfai, T., and Langel, Ü., M871—A novel peptide antagonist selectively recognizing the galanin receptor type 2, *Int. J. Pept. Res. Ther.*, 12, 115–119, 2006.

Steiner, R.A. et al., Galanin transgenic mice display cognitive and neurochemical deficits characteristic of Alzheimer's disease, *Proc. Natl Acad. Sci. USA*, 98, 4184–4189, 2001.

Sullivan, K.A., Shiao, L.L., and Cascieri, M.A., Pharmacological characterization and tissue distribution of the human and rat GALR1 receptors, *Biochem. Biophys. Res. Comm.*, 233, 823–828, 1997.

Sun, Y.G., Gu, X.L., and Yu, L.C., The neural pathway of galanin in the hypothalamic arcuate nucleus of rats: Activation of beta-endorphinergic neurons projecting to periaqueductal gray matter, *J. Neurosci. Res.*, 85, 2400–2406, 2007.

Sundström, E., Archer, T., Melander, T., and Hökfelt, T., Galanin impairs acquisition but not retrieval of spatial memory in rats studied in the Morris swim maze, *Neurosci. Lett.*, 88, 331–335, 1988.

Swanson, C.J. et al., Anxiolytic- and antidepressant-like profiles of the galanin-3 receptor (Gal3) antagonists SNAP 37889 and SNAP 398299, *Proc. Natl Acad. Sci. USA*, 102, 17489–17494, 2005.

Sweerts, B.W., Jarrott, B., and Lawrence, A.J., Expression of preprogalanin mRNA following acute and chronic restraint stress in brains of normotensive and hypertensive rats, *Brain Res. Mol. Brain Res.*, 69, 113–123, 1999.

Tarasov, K.V. et al., Galanin and galanin receptors in embryonic stem cells: Accidental or essential? *Neuropeptides*, 36, 239–245, 2002.

Tatemoto, K. et al., Galanin—a novel biologically active peptide from porcine intestine, *FEBS Lett.*, 164, 124–128, 1983.

Tempel, D.L., Leibowitz, K.J., and Leibowitz, S.F., Effects of PVN galanin on macronutrient selection, *Peptides*, 9, 309–314, 1988.

Thiele, T.E. et al., Overlapping peptide control of alcohol self-administration and feeding, *Alcohol Clin. Exp. Res.*, 28, 288–294, 2004.

Ubink, R., Calza, L., and Hökfelt, T., 'Neuro'-peptides in glia: Focus on NPY and galanin, *Trends Neurosci.*, 26, 604–609, 2003.

Villar, M.J. et al., Neuropeptide expression in rat dorsal root ganglion cells and spinal cord after peripheral nerve injury with special reference to galanin, *Neuroscience*, 33, 587–604, 1989.

Walsh, D.M. and Selkoe, D.J., Deciphering the molecular basis of memory failure in Alzheimer's disease, *Neuron*, 44, 181–193, 2004.

Wang, S. et al., Genomic organization and functional characterization of the mouse GalR1 galanin receptor, *FEBS Lett.*, 411, 225–230, 1997a.

Wang, S., He, C., Hashemi, T., and Bayne, M., Cloning and expressional characterization of a novel galanin receptor. Identification of different pharmacophores within galanin for the three galanin receptor subtypes, *J. Biol. Chem.*, 272, 31949–31952, 1997b.

Wang, S. et al., Molecular cloning and pharmacological characterization of a new galanin receptor subtype, *Mol. Pharmacol.*, 52, 337–343, 1997c.

Wang, S. et al., Differential intracellular signaling of the GalR1 and GalR2 galanin receptor subtypes, *Biochemistry*, 37, 6711–6717, 1998.

Wang, L.H., Lu, Y.J., Bao, L., and Zhang, X., Peripheral nerve injury induces reorganization of galanin-containing afferents in the superficial dorsal horn of monkey spinal cord, *Eur. J. Neurosci.*, 25, 1087–1096, 2007.

Waters, S.M. and Krause, J.E., Distribution of galanin-1, -2, and -3 receptor messenger RNAs in central and peripheral rat tissues, *Neuroscience*, 95, 265–271, 2000.

Weiss, J.M. et al., Testing the hypothesis that locus coeruleus hyperactivity produces depression-related changes via galanin, *Neuropeptides*, 39, 281–287, 2005.

Wiesenfeld-Hallin, Z. et al., Galanin-mediated control of pain: Enhanced role after nerve injury, *Proc. Natl Acad. Sci. USA*, 89, 3334–3337, 1992.

Wiesenfeld-Hallin, Z., Xu, X.J., Crawley, J.N., and Hökfelt, T., Galanin and spinal nociceptive mechanisms: Recent results from transgenic and knock-out models, *Neuropeptides*, 39, 207–210, 2005.

Wilson-Gerwing, T.D. and Verge, V.M., Neurotrophin-3 attenuates galanin expression in the chronic constriction injury model of neuropathic pain, *Neuroscience*, 141, 2075–2085, 2006.

Wrenn, C.C. et al., Galanin peptide levels in hippocampus and cortex of galanin-overexpressing transgenic mice evaluated for cognitive performance, *Neuropeptides*, 36, 413–426, 2002.

Wrenn, C.C., Harris, A.P., Saavedra, M.C., and Crawley, J.N., Social transmission of food preference in mice: Methodology and application to galanin-overexpressing transgenic mice, *Behav. Neurosci.*, 117, 21–31, 2003.

Wrenn, C.C. et al., Learning and memory performance in mice lacking the GAL-R1 subtype of galanin receptor, *Eur. J. Neurosci.*, 19, 1384–1396, 2004.

Wu, W.P. et al., Systemic galnon, a low-molecular weight galanin receptor agonist, reduces heat hyperalgesia in rats with nerve injury, *Eur. J. Pharmacol.*, 482, 133–137, 2003.

Wynick, D. et al. Characterization of a high-affinity galanin receptor in the rat anterior pituitary: Absence of biological effect and reduced membrane binding of the antagonist M15 differentiate it from the brain/gut receptor, *Proc. Natl Acad. Sci. USA*, 90, 4231–4235, 1993.

Xia, C.Y., Yuan, C.X., and Yuan, C.G., Galanin inhibits the proliferation of glial olfactory ensheathing cells, *Neuropeptides*, 39, 453–459, 2005.

Xu, Z.Q. et al., Electrophysiological evidence for a hyperpolarizing, galanin (1–15)-selective receptor on hippocampal CA3 pyramidal neurons, *Proc. Natl Acad. Sci. USA*, 96, 14583–14587, 1999.

Yanaihara, N. et al., Galanin analogues: Agonist and antagonist, *Regul. Pept.*, 46, 93–101, 1993.

Yoshitake, T. et al., Galanin attenuates basal and antidepressant drug-induced increase of extracellular serotonin and noradrenaline levels in the rat hippocampus, *Neurosci. Lett.*, 339, 239–242, 2003a.

Yoshitake, T. et al., Activation of 5-HT(1A) autoreceptors enhances the inhibitory effect of galanin on hippocampal 5-HT release *in vivo*, *Neuropharmacology*, 44, 206–213, 2003b.

Yoshitake, T. et al., Enhanced hippocampal noradrenaline and serotonin release in galanin-overexpressing mice after repeated forced swimming test, *Proc. Natl Acad. Sci. USA*, 101, 354–359, 2004.

Zachariou, V. et al., Galanin receptor 1 gene expression is regulated by cyclic AMP through a CREB-dependent mechanism, *J. Neurochem.*, 76, 191–200, 2001.

Zorrilla, E.P. et al., Galanin type 1 receptor knockout mice show altered responses to high-fat diet and glucose challenge, *Physiol. Behav.*, 91, 479–485, 2007.

12 Bioactive Peptides in Gut–Brain Signaling

Erik Näslund, Per Grybäck, and Per M. Hellström

CONTENTS

Bioactive peptides are in focus as regulatory compounds for brain and gastrointestinal functions where appetite, hunger, and satiety as well as functional gastrointestinal diseases including functional dyspepsia and irritable bowel syndrome are of main interest. Bioactive peptide signaling molecules and their specific receptors are of interest as targets for pharmacological control where physiological regulatory mechanisms are not optimized. The glucagon-like peptides, GLP-1 and GLP-2, have profound roles in glucose homeostasis, appetite regulation, and intestinal proliferation, and are currently under review for actions in irritable bowel syndrome. The pancreatic polypeptide family, PP, PYY_{1-36}, and PYY_{3-36} are under investigation for a role in appetite regulation as "satiety hormones," making these two peptide families interesting targets for the treatment of overweight, obesity, and metabolic disease. In contrast, ghrelin, considered to be an appetite-stimulating "hunger hormone" with gastrokinetic properties, is in focus for the treatment of malnutrition as seen in diabetic gastroparesis, renal insufficiency, and malignant disease. Similarly, the orexins A and B are considered as potential orexigens. Over the years, research with bioactive peptides in the clinical setting has shown many possibilities for the regulation of bodily functions with few adverse reactions and only mild side effects, principally nausea. A drawback to its development is its limited bioavailability in clinical use. Today, advanced biotechnology can overcome this problem, making the nasal or pulmonary route of administration most applicable.

12.1 INTRODUCTION

Over the last 20 years we have been able to study the properties of a vast number of biogenic peptides released from the gastrointestinal (GI) tract. These studies have shown that biogenic peptides regulate an array of functions in the body ranging from regulation of appetite to GI motility and gastric emptying, with subsequent control of metabolic balance, monitored as blood sugar concentrations. The biogenic peptides mainly seem to act on gut function, with effects on appetite that follow through complex gut–brain interactions involving not only efferent and afferent vagal pathways, but also sympathetic as well as enteric pathways.

Based on histological and physiological findings some peptides have been examined more than others. The glucagon-like peptides [glucagon-like peptide-1 (GLP-1), glucagon-like peptide-2 (GLP-2), oxyntomodulin (OXM)], the pancreatic polypeptide family [pancreatic polypeptic (PP), peptide YY1-36 (PYY1-36) and peptide YY3-36 (PYY3-36)] as well as ghrelin and the orexins A and B have received most attention. As the basic mechanisms of these peptides are revealed by studies of interactions with their respective receptors and receptor panels, we have witnessed the development of peptide analogs that can specifically hit certain peptide receptors in order to stimulate or inhibit specific peptide actions. A fundamental observation from this research suggests that receptor blockade seems more readily accomplished than stimulation, which in addition to a high receptor affinity also requires the ability to stimulate intrinsic activity, most likely through a three-dimensional compatibility of the analog with the receptor complex. The search for such analogs has been focused on drug candidates for the treatment of diseases such as diabetes, gastroparesis, obesity, short bowel syndrome, and poor GI motility, as seen under postoperative conditions.

This chapter aims to review the present knowledge of these peptides and the mechanisms by which they should act to overcome certain symptoms and diseases. We will also expand on further possible roles for GI peptides in the treatment of different diseases.

12.2 THE GLUCAGON-LIKE PEPTIDES

The preproglucagon gene is processed differently depending on the tissue studied. In the brain and gut the three main biologically active products form the preproglucagon gene are GLP-1, GLP-2, and OXM, but in the pancreas it is glucagon that is produced (Holst, 1994).

12.2.1 Glucagon-Like Peptide-1

The major form of GLP-1 in the bloodstream is the 30-amino-acid peptide GLP-1(7–36), which after a meal is rapidly released from enteroendocrine L-cells in the gut into the circulation. Originally, GLP-1 was found to decrease blood glucose through stimulation of glucose-dependent insulin secretion and inhibition of glucagon secretion, but it was later also found to decrease gastric emptying (Drucker, 2002). Central administration of GLP-1 inhibits feeding in rodents, and peripheral administration of GLP-1 to both normal-weight and obese humans results in reduced appetite and decreased food intake (Turton et al., 1996; Flint et al., 1998; Näslund et al., 1999a). The combination of incretin and anorectic effects of GLP-1 has resulted in interest in GLP-1 as a potential treatment for diabetes mellitus type 2. Exendin-4 is a naturally occurring agonist of the GLP-1 receptor, isolated from the saliva of the lizard *Heloderma suspectum*, a Gila monster native to several Southwestern American states (Eng et al., 1992). Exendin-4 has a longer half-life than GLP-1 and is therefore more suitable as a pharmaceutical. It has recently been approved for the treatment of type 2 diabetes in humans (exenatide; marketed as Byetta) (Amori et al., 2007). Another strategy using GLP-1 in the treatment of type-2 diabetes is to increase the amount of endogenous GLP-1. GLP-1 has a short half-life of minutes, dependent on it being degraded by the proteolytic enzyme dipeptidyl peptidase-IV (DPP-IV). DPP-IV cleaves GLP-1 at the NH_2-terminal dipeptide where the penultimate amino-acid residues are proline and alanine. The truncated GLP-1 molecule retains no biological activity. DPP-IV inhibitors are now being introduced for the treatment of type 2 diabetes (vildagliptin, sitagliptin; marketed as Galvus and Januvia, respectively). In contrast to exenatide, which is administered subcutaneously, the DPP-IV inhibitors can be taken orally (Idris and Donnelly, 2007) and have a similar effect on blood sugar control in type 2 diabetes as exendin-4 (Amori et al., 2007).

The focus of the GLP-1 agonists has been on their incretin effects. However, under physiological circumstances GLP-1 is released within a short time after food intake, and brings about a rapid blood sugar control (within 10 to 30 min), long before insulin release is stimulated to maintain blood sugar levels (40 to 60 min). This is considered to be due to an initial powerful inhibition of gastric emptying, which is followed by an insulin release caused by GLP-1 as part of the entero-insular axis. After food intake GLP-1 evokes anorexic effects. Most likely this is due to the profound inhibitory impact of the hormone on gastric emptying (Figure 12.1).

GLP-1 is also believed to mediate its anorectic effects via the GLP-1 receptor in the central nervous system (CNS), which is present both in the nucleus of the solitary tract (NTS) and the arcuate nucleus (ARC) of the hypothalamus. The NTS is considered to be sensitive to neural signals mediated through the afferent vagus projecting to the hypothalamic satiety centers, whereas the ARC seems to be sensitive to endocrine input into other areas of the hypothalamus associated with the regulation of food intake. As a result of these observations, short-term studies with GLP-1 have demonstrated an effect of the peptide on body weight. Prandial subcutaneous injections of GLP-1 given 30 min before food intake in obese subjects resulted in a slower gastric emptying, reduced food intake, and less subjective sensations of hunger and increased satiety, resulting in a weight loss of 0.55 ± 0.2 kg over 5 days of treatment (Näslund et al., 1994). Similarly, chronic subcutaneous administration of GLP-1 to obese patients with type 2 diabetes for 6 weeks resulted in an average of 1.9 kg weight loss in conjunction with inhibited gastric emptying over the treatment period (Zander et al., 2002). Treatment with exendin-4 has been shown to result in a decrease in weight of 1.4 kg versus placebo. However, no effect on body weight was seen with the DPP-IV inhibitors (Amori et al., 2007). Thus, there may be a future role for GLP-1 and GLP-1 analogs in the treatment of obesity (Figure 12.2).

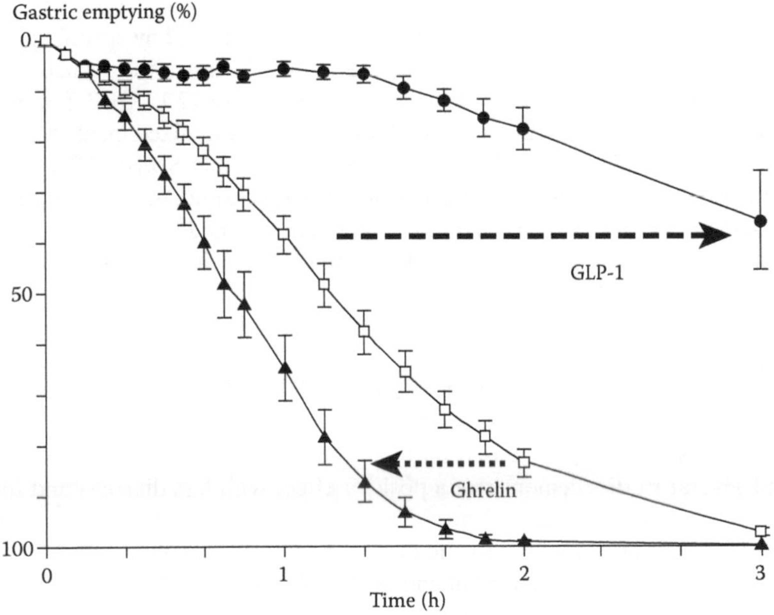

FIGURE 12.1 The "working-range" of the stomach in terms of gastric emptying rate when fully stimulated by ghrelin (half-emptying time, 47 ± 5 min) and inhibited with GLP-1 (half-emptying time estimate, 245 ± 58 min. No dumping symptoms were observed with ghrelin. Under pathophysiological conditions a gastric retention exceeding 60% at 2 h is considered to be "severe" gastroparesis.

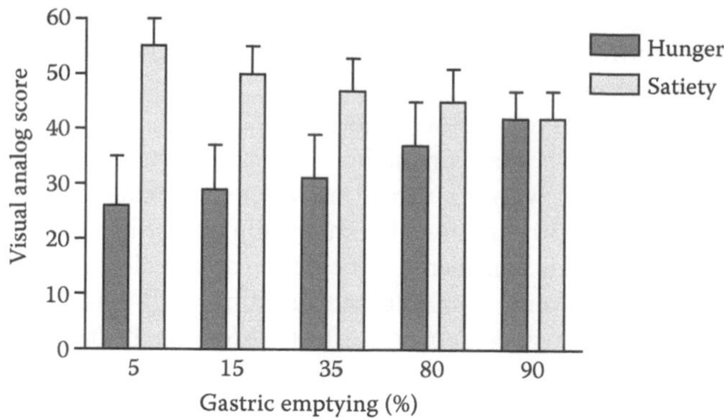

FIGURE 12.2 Hunger and satiety determined by visual analog scores at different grades of gastric emptying.

GLP-1 is widely recognized as a smooth muscle relaxing agent. Early studies showed that GLP-1 caused a marked inhibition of gastric emptying in man (Drucker, 2002). Later experimental studies have shown that this native peptide is capable of reducing fasting motility of the small intestine in healthy volunteers as well as in patients suffering from irritable bowel syndrome (IBS) (Figure 12.3) (Hellström et al., 2008a). This has led further to phase II studies in patients with IBS utilizing a GLP-1 analog, ROSE-010, which has a longer half-life than the natural hormone, and is maintained in the circulation during the acute phase of an attack of abdominal pain, resulting in a marked pain-relieving effect (Hellström et al., 2008b).

12.2.2 Glucagon-Like Peptide-2

GLP-2 is a biogenic peptide of 33-amino-acid residues that is created by specific post-translational proteolytic cleavage of proglucagon in a process that also liberates the related GLP-1 from the L-cells, which are mainly found in the ileum and colon (Figure 12.1). GLP-2 is released into the circulation in parallel with GLP-1 in an equimolar fashion after a mixed meal. A loss of more than two-thirds of the small intestine results in short bowel syndrome (SBS). SBS is characterized by insufficient absorption of nutrients, electrolytes, and fluids, resulting in severe diarrhea, nutritional deficiencies, and an ensuing dependence on parenteral nutrition and fluids as well as electrolyte and nutrient supplements, such as vitamins A, D, E, K, and B_{12}, calcium, magnesium, iron, folic acid, and zinc (Martin et al., 2006).

Experimental studies in animals and humans demonstrate that GLP-2 is a trophic hormone that plays and important role in intestinal growth adaptation (Drucker et al., 1996; Martin et al., 2006). Administration of GLP-2 has been shown to regulate cell proliferation, apoptosis, nutrient absorption, and epithelial and intestinal permeability (Scott et al., 1998; Martin et al., 2005), but has no major impact on gastric emptying. Based on this knowledge, GLP-2 has been studied in patients with SBS and several studies demonstrate a positive effect with less diarrhea and improved nutritional absorption after administration of the peptide (Jeppeson et al., 2001; Drucker, 2002). However, similar to GLP-1, GLP-2 has a short half-life of only minutes. Recently, a DPP-IV-resistant GLP-2 analog, teduglutide, has been studied in patients with SBS. In this study over 21 days it was demonstrated that teduglutide was safe, well tolerated, and resulted in a significant increase in small intestinal villus height, crypt depth, and mitotic index. There was also a clear clinical benefit in reducing fecal wet weight and fecal energy excretion (Jeppeson et al., 2005). Obviously, GLP-2 and pharmacologically developed GLP-2 analogs may have a future therapeutic role in treating SBS, especially in children where a congenital intestinal malformation has the utmost importance for their development and growth.

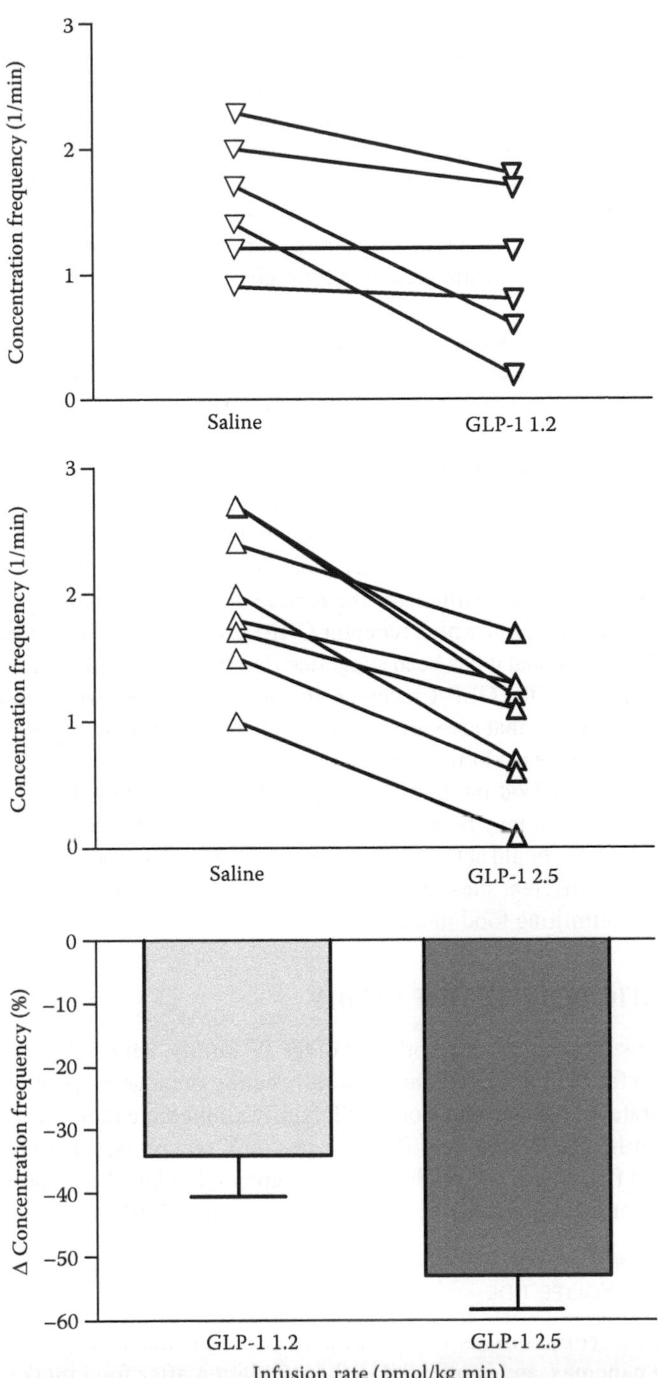

FIGURE 12.3 Contractile activity of the human small intestine inhibited by two different infusion rates of GLP-1. Values are given in absolute numbers and as percentual inhibition.

12.2.3 Oxyntomodulin

Oxyntomodulin (OXM) is a naturally occurring 37-amino-acid peptide in the gut. It is a proglucagon-derived peptide from amino-acid residues 33–72 of the molecule and is present in the brain and gut. OXM is also released after food intake from the L-cells in the intestine, into the circulation.

OXM has been shown to inhibit food intake in both rodents and man. In rats, intracerebroventricular administration and injection of OXM into the hypothalamus inhibits food intake after a fasting period. This effect seems to be dose-dependent and does not alter the behavior of the animal (Dakin et al., 2001). Peripheral administration of OXM twice daily to rats has also demonstrated a reduced weight gain. In addition, rats pair-fed the same amount of food as the OXM group gained more weight than the OXM-treated group, suggesting that the peptide not only influences appetite but also energy expenditure (Dakin et al., 2004). The mechanism of action of OXM is not well understood. It is known to bind both to the GLP-1 receptor and the glucagon receptor, but it is not known whether the effects of the hormone are mediated through these receptors or through an unidentified receptor.

In humans, OXM reduces food intake in both normal-weight and obese subjects. In healthy, normal-weight volunteers an infusion of OXM reduced subjective appetite visual analog scores and food intake during a meal (19%), and shortened the duration of a meal. This appetite-reducing effect persisted over 12 h, but with no report of nausea or any specific gustatory effects of OXM (Cohen et al., 2003). OXM has also been given subcutaneously to obese men over a four-week period. The OXM-treated group reduced their energy intake, which was followed by a reduced body weight of 2.3 kg as compared to 0.5 kg in the control group. Again, no report of nausea after peptide administration or change in palatability of the meal was made (Wynne et al., 2005).

Similarly to GLP-1, OXM also stimulates the release of insulin (Schjoldager et al., 1998); however, there is uncertainty about on which receptor OXM exerts its primary effects. Yet, there is no identified receptor for OXM, and it has been suggested that the effects of OXM on metabolic homeostasis should be mediated by the GLP-1 receptor. In spite of this, the two peptides should have the capability of promoting effects that are partly differential. Thus, OXM activates neurons in the hypothalamus, while GLP-1 acts in the brainstem (Baggio et al., 2004). Clearly, the findings of the effects of OXM on metabolism and food intake open the possibility of additional peptidergic influences in energy balance, and as such it may be a future candidate for drug development in the treatment for obesity. If GLP-1 and OXM should act through one and the same receptor, both resulting in decreased food intake but acting at different sites, activation of separate areas of the brain would speak in favor of synergistic effects delimiting food intake.

12.3 PANCREATIC POLYPEPTIDE FAMILY

The biogenic peptides PP and PYY are both part of the PP family, which also includes neuropeptide Y (NPY). NPY is so far the most potent and centrally acting orexogenic peptide that preferentially promotes carbohydrate intake. Members of the PP family all mediate their effects through the common Y receptor family. There are five different types of Y receptors, which differ with regard to their distribution and function. PYY binds to all Y-receptors, but has the highest affinity for Y_2 and Y_5, but PP has a higher affinity for the Y_4 receptor (Larhammar, 1996).

12.3.1 Pancreatic Polypeptide

Pancreatic polypeptide (PP) consists of 36 amino acids, is primarily expressed in the endocrine cells D-cells of the pancreas, and is released to the circulation after food intake. When PP is given peripherally to mice a reduction of food intake is seen, in conjunction with increased oxygen consumption and slowed gastric emptying, suggesting an increased energy expenditure (Asakawa et al., 2003). Transgenic mice overexpressing PP are lean, an effect that seems to be mediated by increased cholecystokinin (CCK) signaling. In such mice, plasma CCK seems to be adapted to a higher level, and administration of a CCK antagonist significantly increased their food intake (Ueno et al., 2007).

In humans, PP inhibits food intake in normal-weight subjects and in patients with Prader–Willi syndrome (Bernston et al., 1993; Batterham et al., 2003b). Our group has studied the effect of PP on gastric emptying, subjective appetite ratings, and insulin secretion in normal-weight subjects. We found a delayed gastric emptying of a solid meal and in consequence a delayed insulin peak after the meal. However, no effect was found on appetite ratings, which is at variance with previously published studies. One explanation may be the lower dose of PP used in our studies (0.75 and 2.25 pmol/kg/min intravenously) compared to others (10 pmol/kg/min intravenously), which may have elicited additional effects (Batterham et al., 2003b; Schmidt et al., 2005).

12.3.2 Peptide YY

Peptide YY (PYY), consisting of 36 amino-acid residues, is released from the enteroendocrine L-cells in the distal gut after food intake, most markedly after a fatty meal. PYY is found in two circulating forms, PYY_{1-36} and the truncated form PYY_{3-36}. PYY has come into focus as a possible treatment for obesity as it was shown that peripheral administration of PYY_{3-36} is able to decrease food intake in rodents for up to 12 h (Batterham et al., 2002). Similar results have been shown in man with a reduction of subjective appetite scores and food intake after peripheral administration of PYY_{3-36} (Batterham et al., 2003a).

It is believed that the effect of PYY on food intake is mediated primarily by Y_2 receptors in the ARC of the hypothalamus. PYY_{3-36} decreases NPY mRNA in the hypothalamus and peripheral PYY_{3-36} increases *c-fos* immunoreactivity in the ARC where Y_2 receptors are found (Batterham et al., 2002). Another interesting possible mechanism for PYY to mediate increased satiety is its effect on gut motility. PYY and GLP-1 are considered to be "ileal brake" hormones, which inhibit gut motility and slow down the rate of gastric emptying, which causes a prolonged period of fullness and satiety (Näslund et al., 1999b).

There are reports of new drugs targeting the PP/PYY pathway in the treatment of obesity. TM30338 is a Y_2 and Y_4 receptor agonist, mimicking both PP and PYY, and has been shown to reduce food intake in humans after subcutaneous administration. Similarly, PYY_{3-36} is in phase II trials as a nasal preparation for the treatment of obesity (Cooke et al., 2006; Gantz et al., 2007).

12.4 THE GHRELIN CONCEPT

The 28-amino-acid peptide ghrelin has a unique position amongst the gut peptide hormones, being the so far solely only known endocrine peptide that is connected to our circadian sensations of hunger, as featured by gastric hunger contractions, which promote food intake. The peptide exists in an endocrinological inactive pure peptide form, and an active octanoylated form. Ghrelin is mainly elaborated from the proghrelin gene by the P/D1 cells lining the fundus of the human stomach and the epsilon cells of the pancreas. In rodents X/A-cells in the stomach produce ghrelin. The ghrelin receptor (GR1a) is a G protein-coupled receptor, formerly known as the GHS receptor (growth hormone secretagogue receptor) (Kojima et al., 1999).

From a physiological point of view this peptide displays opposite functions to the previously "satiety-promoting" peptides described in this review. Ghrelin stimulates food intake in all species, including man (Tschöp et al., 2000; Wren et al., 2001). In the periphery ghrelin stimulates gastric emptying (Figure 12.1). As an interesting point it seems that ghrelin increases the emptying rate from the fundus as well as from the antrum. Hence, it seems that ghrelin selectively stimulates the motor function underlying the linear phase of gastric emptying, thus leaving the lag phase unaffected. Thus, our data speak in favor of ghrelin acting primarily on the propulsive phase of gastric emptying and leaving the adaptive relaxation of the stomach unaffected (Figure 12.4). Ghrelin also increases gut motility in rodents and man *in vivo*, and its plasma levels are suppressed by the ingestion of nutrients (Kojima et al., 1999; Cummings et al., 2001; Wren et al., 2001; Levin et al., 2006).

Ghrelin has important central nervous actions and influences neuronal activity in several areas of the brain involved in energy metabolism. Of central importance is the hormonal influence on

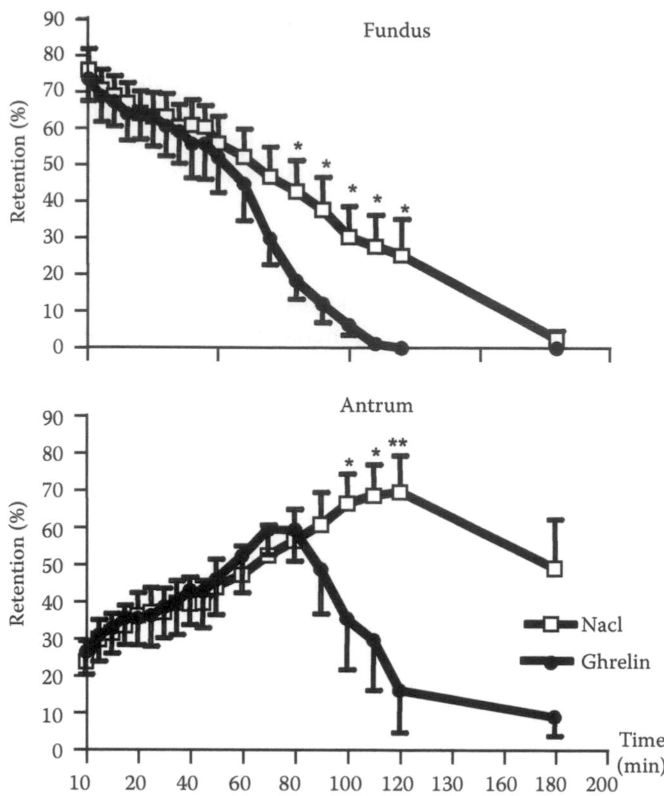

FIGURE 12.4 Gastric emptying profile as stimulated by ghrelin (15 pmol/kg min). Gastric emptying rate from the fundus (top) is compared to the antrum (bottom) using ghrelin as stimulant and physiological saline as control. * $P < 0.05$, ** $P < 0.01$.

functional ghrelin receptors of the ARC in the hypothalamus, the brainstem, and mesolimbic reward centers. The functional ghrelin receptors are also operative in afferent vagal nerves terminating in the NTS of the brainstem (as is the case for GLP-1 and different Y receptors) (Date et al., 2002).

From a pharmacodynamic point of view, ghrelin receptor blockade would diminish appetite and in consequence with that decrease food intake (Esler et al., 2007). Recently it was reported that vaccination of rats with ghrelin immunoconjugates decreased body weight due to the subsequent immune response against endogenous ghrelin. Thus, research scientists at Scripps are currently developing an antiobesity vaccine, directed against the hormone ghrelin (Zorrilla et al., 2006; Scripps.edu). The vaccine uses the immune system, specifically antibodies, to bind to selected targets, directing the body's own immune response against them. This prevents ghrelin from reaching the CNS, thus producing a desired reduction in weight gain. Whether this would work in humans is not yet clear, and one has to take into account the multitude of effects that ghrelin itself, or an inhibitor, would bring about not only by the direct effects of the peptide on the brain and GI tract, but also through the secondary effects on gastric emptying, which is of importance for the absorption of food stuffs in the small intestine, hence regulating insulin release and metabolic homeostasis.

A separate area of interest is the stimulatory action of ghrelin on GI motility. Ghrelin stimulates gastric emptying in normal humans (Figure 12.1) (Levin et al., 2006) and patients with diabetic gastroparesis (Murry et al., 2005). A ghrelin analog has also been shown to counteract postoperative ileus in rats (Poitras et al., 2005). Further investigations in healthy human volunteers using native ghrelin suggest that ghrelin preferentially stimulates gastric emptying, while having no effect on motility of the small intestine and colon as measured by transit studies employing the hydrogen breath test, radioopaque markers, or bowel movements (Häglund et al., unpublished data) during a 6-h infusion of ghrelin. As studies in rats have demonstrated that a penetration across the blood–brain barrier is necessary for the effect of ghrelin on colonic motility to take place (Shimizu et al., 2006), it is possible that natural ghrelin does not reach a sufficient penetration to the CNS to cause an effect on colonic motility. In line with this, a ghrelin analog, TZP-101, is currently under investigation as a possible stimulatory agent for shortening postoperative ileus in man (Venkova et al., 2007). A motility-stimulating effect should be of great interest in surgical care because not only the inhibition of motility itself, but also other postsurgical reactions such as nausea and even vomiting would be alleviated, with positive effects on quality of life in connection with abdominal surgery.

A second peptide has been isolated from the proghrelin gene. This peptide of 23 amino acids binds to the orphan G protein-coupled receptor GPR39 and has been named obestatin. Obestatin has been reported to suppress food intake, jejunal contractions *in vitro* and gastric emptying *in vivo* in mice (Zhang et al., 2005). However, as judged from work by our group and others, the role of obestatin as a regulator of motility and appetite through a GPR39 receptor-dependent mechanism can be questioned; it explains earlier findings, namely that removing the ghrelin gene (which includes obestatin) from mice did not significantly reduce their appetite (Bassil et al., 2007). Even though the GPR39 receptor is likely to be of importance for regulation of gastric emptying and food intake, it is still an orphan searching its physiological ligand.

12.5 THE OREXIN FAMILY

12.5.1 Orexin A and B

The two related peptides Orexin A (OXA, hypocretin-1) and Orexin B (OxB, or hypocretin-2) with approximately 50% sequence identity, are produced by cleavage of a single 130-amino-acid precursor protein, prepro-orexin, expressed by one and the same gene. OXA consists of 33-amino-acid residues and has two disulfide bonds, while OXB is a linear 28-amino-acid residue peptide. Studies suggest that OXA is of greater biological importance than OXB and research has therefore focused on OXA. The orexin peptides bind to the orexin receptor, which is a G protein coupled receptor. In the brain these peptides are found in a small population of cells in the lateral and posterior hypothalamus, which is involved with food intake, and send projections throughout the brain. In addition, it stimulates wakefulness and energy expenditure (Kirchgessner and Liu, 1999).

OXA and OXB both stimulate food consumption when administered intracerebroventricularly to rats. OXA and OXB gene expression in the brain is highly restricted to distinct populations of neurons located in specific hypothalamic regions, specifically the lateral hypothalamic area, a region implicated in hunger feeding behavior. Orexin-containing neurons excite various brain nuclei with important roles in wakefulness including the dopamine, norepinephrine, histamine, and acetylcholine systems and appear to play an important role in stabilizing wakefulness and sleep. The loss of orexins in the brain is associated with narcolepsy. Orexins are also found in the GI tract. Neurons in the GI submucosal and myenteric plexuses, and endocrine cells in the intestinal mucosa and pancreatic islets display OXA as well as orexin receptor immunoreactivity (Kirchgessner and Liu, 1999; Näslund et al., 2002). OXA may also be of importance for GI functions such as motility, acid secretion, and modulation of both insulin and glucagon release in the rat (Nowak et al., 2000; Ouedrago et al., 2003).

Plasma concentrations of OXA are increased during fasting (Komaki et al., 2001), and orexin-positive neurons in the gut, like those in the hypothalamus, are activated by fasting, indicating functional importance not only in the brain, but also in the GI tract (Figure 12.5) (Kirchgessner and Liu, 1999).

Orexin receptors are found on vagal afferent neurons in both rat and man, and may be of importance for an OXA-dependent inhibition of vagal afferent activity and satiety responses to CCK

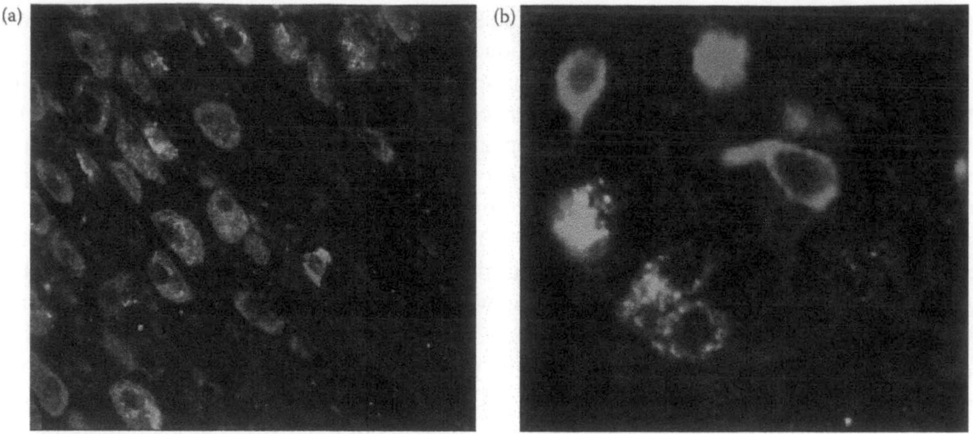

FIGURE 12.5 (See color insert following page 176.) Immunohistochemistry of the human gastric antrum showing colocalization of orexin A with gastrin (a), and orexin A adjacent to somatostatin (b).

(Burdyga et al., 2003). Intravenous administration of OXA to man does not influence appetite ratings; however, plasma concentrations of leptin decrease during a 3-h infusion of OXA, suggesting a gut-fat mass cross-talk that may influence long-term metabolic homeostasis (Ehrström et al., 2005). Laboratory studies in healthy volunteers have shown experimental sleep restriction to be associated with an adverse impact on glucose homeostasis. Insulin sensitivity decreases rapidly and markedly without adequate compensation in pancreatic beta cell function, resulting in an elevated risk of diabetes. The orexins are today claimed to be an important link between metabolic control and sleep regulation. Sleep curtailment is associated with a reduction of the satiety factor, leptin, and an increase in the hunger-promoting hormone, ghrelin. Thus, sleep loss may alter the ability of leptin and ghrelin to accurately signal our energy needs. The adverse impact of sleep deprivation on appetite regulation is likely to be driven by increased activity in neuronal populations expressing the excitatory orexin peptides that promote both waking and feeding. Many epidemiologic studies have shown an association between short sleep and higher body mass index after controlling for a variety of possible confounders, with severe consequences on a long-term basis (Näslund et al., 2007a). Interesting physiological studies have recently been made showing that OXA mediates the acid-induced bicarbonate secretion of the duodenum of the rat, dependent on a nutritional component for the effect to be accomplished (Flemström et al., 2003; Bengtsson et al., 2007). This may open a completely new concept for the action of orexin peptides in the complex digestive functions in the duodenal mixing segment where not only nutrients, but also hydrochloric acid and bicarbonate secretion are involved, as well as the release and effect of several gastrointestinal mediators, some of which reaching the brain.

12.6 CONCLUSIONS

Bioactive peptides are mainly involved with appetite and food intake, primarily in association with changes in GI motility and gastric emptying rate, and secondly blood sugar control and metabolic homeostasis. In terms of diabetes and glucose control, GLP-1 agonists and DPP-IV inhibitors are already in medical use. Obesity, a health issue increasing at an alarming rate, could be a target for treatment by many of the peptidergic mechanisms presented in this review. Short-term treatment with GLP-1 or OXM has resulted in weight loss (Näslund and Hellström, 2007b). There are, however, problems associated with the development of peptides as drugs. They have low bioavailability and need to be given by a parenteral route, and their half-life is also short. Some of these issues can be overcome and new means of drug delivery such as nasal or pulmonary formulations may alleviate some of these issues. An advantage in this setting is that the bioactive peptides represent specific physiological functions, which should make them less likely to cause side effects. Having carried out experimental research with a multitude of bioactive peptides in humans for more than two decades, the risk of side effects is surprisingly low, with nausea being dominant. However, from a pharmacokinetic point of view it might be more strategic to develop pharmacological "small molecules" that can hit a specific receptor. Furthermore, using this strategy it seems easier to find a ligand that can block a certain receptor than finding one that can also elicit a biological function, with the inherent problems such as receptor desensitization and multiple receptor subgroups. From a pharmacological standpoint the "receptor antagonist approach" also seems far more attractive, as the peptide itself is working under physiological conditions, which should reduce the risk of dose-dependent side effects.

In summary, bioactive gut peptides and their receptors may become useful targets for the treatment of functional disorders, such as obesity, functional dyspepsia, IBS, SBS, and malnutrition, in the same way as they now are recognized treatment options for type 2 diabetes.

ACKNOWLEDGMENTS

Supported by Karolinska Institutet and the Stockholm County Research Council.

REFERENCES

Amori, R.E., Lau, J., and Pittas, A.G., Efficacy and safety of incretin therapy in type 2 diabetes: Systematic review and meta-analysis, *JAMA*, 298, 194–206, 2007.

Asakawa, A. et al., Characterization of the effects of pancreatic polypeptide in the regulation of energy balance, *Gastroenterology*, 124, 1325–1336, 2003.

Baggio, L.L., Huang, Q., Brown, T.J., and Drucker, D.J., Oxyntomodulin and glucagon-like peptide-1 differentially regulate murine food intake and energy expenditure, *Gastroenterology*, 127, 546–558, 2004.

Bassil, A.K. et al., Little or no ability of obestatin to interact with ghrelin or modify motility in the rat gastrointestinal tract, *Br. J. Pharmacol.*, 150, 58–64, 2007.

Batterham, R.L. et al., Gut hormone PYY(3-36) physiologically inhibits food intake, *Nature*, 418, 650–654, 2002.

Batterham, R.L. et al., Inhibition of food intake in obese subjects by PYY3-36, *N. Engl. J. Med.*, 349, 941–948, 2003a.

Batterham, R.L. et al., Pancreatic polypeptide reduces appetite and food intake in humans, *J. Clin. Endocrinol. Metabol.*, 88, 3989–3992, 2003b.

Bengtsson, M.W. et al., Food-induced expression of orexin receptors in rat duodenal mucosa regulates the bicarbonate secretory response to orexin-A, *Am. J. Physiol.*, 293, G501–G509, 2007.

Berntson, G.G. et al., Pancreatic polypeptide infusions reduce food intake in Prader–Willi syndrome, *Peptides*, 14, 497–503, 1993.

Burdyga, G. et al., Localization of orexin-1 receptors to vagal afferent neurons in the rat and humans, *Gastroenterology*, 124, 129–139, 2003.

Cohen, N.M. et al., Oxyntomodulin suppresses appetite and reduces food intake in humans, *J. Clin. Endocrinol. Metabol.*, 88, 4696–4701, 2003.

Cooke, D. and Bloom, S., The obesity pipeline: Current strategies in the development of anti-obesity drugs, *Nat. Rev. Drug Discov.*, 5, 919–931, 2006.

Cummings, D.E. et al., A preprandial rise in plasma ghrelin levels suggests a role in meal initiation in humans, *Diabetes*, 50, 1714–1719, 2001.

Dakin, C.L. et al., Oxyntomodulin inhibits food intake in the rat, *Endocrinology*, 142, 4244–4250, 2001.

Dakin, C.L. et al., Peripheral oxyntomodulin reduces food intake and body weight gain in rats, *Endocrinology*, 145, 2687–2695, 2004.

Date, Y. et al., The role of the gastric vagal nerve in ghrelin-induced feeding and growth hormone secretion in rats, *Gastroenterology*, 123, 1120–1128, 2002.

Drucker, D.J., Ehrlich, P., Asa, S.L., and Brubaker, P.L., Induction of intestinal epithelial proliferation by glucagon-like peptide 2, *Proc. Natl Acad. Sci. USA*, 93, 7911–7916, 1996.

Drucker, D.J., Biological actions and therapeutic potential of the glucagon-like peptides, *Gastroenterology*, 122, 531–544, 2002.

Ehrström, M. et al., Inhibitory effect of exogenous orexin A on gastric emptying, plasma leptin, and the distribution of orexin and orexin receptors in the gut and pancreas in man, *J. Clin. Endocrinol. Metab.*, 90, 2370–2377, 2005.

Eng, J. et al., Isolation and characterization of exendin-4, an exendin-3 analogue, from *Heloderma suspectum* venom. Further evidence for an exendin receptor on dispersed acini from guinea pig pancreas, *J. Biol. Chem.*, 267, 7402–7405, 1992.

Esler, W.P. et al., Small-molecule ghrelin receptor antagonists improve glucose tolerance, suppress appetite, and promote weight loss, *Endocrinology*, 148, 5175–5185, 2007.

Flemström, G., Sjöblom, M., Jedstedt, G., and Akerman, K.E., Short fasting dramatically decreases rat duodenal secretory responsiveness to orexin A but not to VIP or melatonin, *Am. J. Physiol.*, 285, G1091–G1096, 2003.

Flint, A., Raben, A., Astrup, A., and Holst, J.J., Glucagon-like peptide 1 promotes satiety and suppresses energy intake in humans, *J. Clin. Invest.*, 101, 515–520, 1998.

Gantz, I. et al., Efficacy and safety of intranasal peptide YY3–36 for weight reduction in obese adults, *J. Clin. Endocrinol. Metabol.*, 92, 1754–1757, 2007.

Hellström, P.M. et al., GLP-1 suppresses gastrointestinal motility and inhibits the migrating motor complex in healthy subjects and patients with irritable bowel syndrome, *Neurogastroenterol. Motil.*, 20, 649–659, 2008a.

Hellström, P.M. et al., Clinical trial: The GLP-1 analogue ROSE-010 for management of acute pain in patients with irritable bowel syndrome: A randomized, placebo-controlled, double-blind study, *Aliment. Pharmacol. Ther.,* 2008 Oct 10 (Epub ahead of print), 2008b.

Holst, J.J., Glucagonlike peptide 1: A newly discovered gastrointestinal hormone, *Gastroenterology*, 107, 1848–1855, 1994.

Idris, I. and Donnelly, R., Dipeptidyl peptidase-IV inhibitors: A major new class of oral antidiabetic drug, *Diabetes Obes. Metab.*, 9, 153–165, 2007.

Jeppesen, P.B. et al., Glucagon-like peptide 2 improves nutrient absorption and nutritional status in short-bowel patients with no colon, *Gastroenterology*, 120, 1720–1728, 2001.

Jeppesen, P.B. et al., Teduglutide (ALX-0600), a dipeptidyl peptidase IV resistant glucagon-like peptide 2 analogue improves intestinal function in short bowel syndrome patients, *Gut*, 54, 1224–1231, 2005.

Kirchgessner, A.L. and Liu, M., Orexin synthesis and response in the gut, *Neuron*, 24, 941–951, 1999.

Kojima, M. et al., Ghrelin is a growth-hormone-releasing acylated peptide from stomach, *Nature*, 402, 656–660, 1999.

Komaki, G. et al., Orexin-A and leptin change inversely in fasting non-obese subjects, *Eur. J. Endocrinol.*, 144, 645–651, 2001.

Larhammar, D., Structural diversity of receptors for neuropeptide Y, peptide YY, and pancreatic polypeptide, *Regul. Pept.*, 65, 165–174, 1996.

Levin, F. et al., Ghrelin stimulates gastric emptying and hunger in normal-weight humans, *J. Clin. Endocrinol. Metabol.*, 91, 3296–3302, 2006.

Martin, G.R., Beck, P.L., and Sigalet, D.L., Gut hormones and short bowel syndrome: The enigmatic role of glucagon-like peptide-2 in the regulation of intestinal adaption, *World J. Gastroenterol.*, 12, 4117–4129, 2006.

Martin, G.R. et al., Nutrient-stimulated GLP-2 release and crypt cell proliferation in experimental short bowel syndrome, *Am. J. Physiol.*, 288, G431–G438, 2005.

Murry, C.D. et al., Ghrelin enhances gastric emptying in diabetic gastroparesis: A double blind, placebo controlled, crossover study, *Gut*, 54, 1693–1698, 2005.

Näslund, E. and Hellström, P.M., Drug targets modulating the gut–appetite–metabolism axis, *Drug Discov. Today: Therap. Strat.*, 4, 189–193, 2007a.

Näslund, E. and Hellström, P.M., Appetite signaling: From gut peptides and enteric nerves to brain, *Physiol. Behav.*, 92, 256–262, 2007b.

Näslund, E. et al., Prandial subcutaneous injections of glucagon-like peptide-1 cause weight loss in obese human subjects, *Br. J. Nutr.*, 91, 439–446, 1994.

Näslund, E. et al., Energy intake and appetite are suppressed by glucagon-like peptide-1 (GLP-1) in obese men, *Int. J. Obes. Relat. Metab. Disord.*, 23, 304–311, 1999a.

Näslund, E. et al., GLP-1 slows solid gastric emptying and inhibits insulin, glucagon, and PYY release in humans, *Am. J. Physiol.*, 277, R910–R916, 1999b.

Näslund, E. et al., Localization and effects of orexin on fasting motility in the rat duodenum, *Am. J. Physiol.*, 282, G470–G479, 2002.

Nowak, K.W. et al., Acute orexin effects on insulin secretion in the rat: *In vivo* and *in vitro* studies, *Life Sci.*, 66, 449–454, 2000.

Ouedrago, R., Näslund, E., and Kirchgessner, A.L., Glucose regulates the release of orexin-a from the endocrine pancreas, *Diabetes*, 52, 111–117, 2003.

Poitras, P., Polvino, W.J., and Rocheleau, B., Gastrokinetic effect of ghrelin analog RC-1139 in the rat. Effect on post-operative and on morphine induced ileus, *Peptides*, 26, 1598–1601, 2005.

Schjoldager, B.T., Baldissera, F.G., Mortensen, P.E., and Holst, J.J., Oxyntomodulin: A potential hormone from the distal gut. Pharmacokinetics and effecs on gastric acid and insulin secretion in man, *Eur. J. Clin. Invest.*, 18, 499–503, 1998.

Schmidt, P.T. et al., A role for pancreatic polypeptide in the regulation of gastric emptying and short-term metabolic control, *J. Clin. Endocrinol. Metabol.*, 90, 5241–5246, 2005.

Scott, R.B., Kirk, D., MacNaughton, W.K., and Meddings, J.B., GLP-2 augments the adaptive response to massive intestinal resection in rat, *Am. J. Clin. Nutr.*, 275, G911–G921, 1998.

Scripps.edu, "Scripps Research Scientists Successfully Test New Anti-Obesity Vaccine" at The Scripps Research Institute, available at www.scripps.edu.

Shimizu, Y. et al., Evidence that stimulation of ghrelin receptors in the spinal cord initiates propulsive activity in the colon of the rat, *J. Physiol.*, 576, 329–338, 2006.

Tschöp, M., Smiley, D.L., and Heiman, M.L., Ghrelin induces adiposity in rodents, *Nature*, 407, 908–913, 2000.

Turton, M.D. et al., A role for glucagon-like peptide-1 in the central regulation of feeding, *Nature*, 379, 69–72, 1996.

Ueno, N., Asakawa, A., Satoh, Y., and Inui, A., Increased circulating cholecystokinin contributes to anorexia and anxiety behaviour in mice overexpressing pancreatic polypeptide, *Regul. Pept.*, 141, 8–11, 2007.

Venkova, K., Fraser, G., Hoveyda, H.R., and Greenwood-Van Meerv, B., Prokinetic effects of a new ghrelin receptor agonist TZP-101 in a rat model of postoperative ileus, *Dig. Dis. Sci.*, 52, 2241–2248, 2007.

Wren, A.M. et al., Ghrelin enhances appetite and increases food intake in humans, *J. Clin. Endocrinol. Metab.*, 86, 5992–5995, 2001.

Wynne, K. et al., Subcutaneous oxyntomodulin reduces body weight in overweight and obese subjects: A double-blind, randomized, controlled trial, *Diabetes*, 54, 2390–2395, 2005.

Zander, M., Madsbad, S., Madsen, J.L., and Holst, J.J., Effect of 6-week course of glucagon-like peptide 1 on glycemic control, insulin sensitivity, and beta-cell function in type 2 diabetes: A parallel group study, *Lancet*, 359, 824–830, 2002.

Zhang, J.V. et al., Obestatin, a peptide encoded by the ghrelin gene, opposes ghrelin's effect on food intake, *Science*, 310, 996–999, 2005.

Zorrilla, E.P. et al., Vaccination against weight gain, *Proc. Natl Acad. Sci. USA*, 103, 13226–13231, 2006.

Part II

Nonribosomal Peptides

13 Nonribosomal Peptides

Alexandra A. Roberts, Leanne A. Pearson, and Brett A. Neilan

CONTENTS

Nonribosomal peptides are a diverse group of compounds synthesized via modular nonribosomal peptide synthetase (NRPS) enzyme complexes. They are produced by a variety of prokaryotes, including streptomycetes, bacillus, and cyanobacteria, and lower-order eukaryotes such as fungi and sponges. NRPS modules often have the ability to incorporate nonproteinogenic amino acids into the synthesized natural product, while structural complexity is achieved by various NRPS tailoring and modification domains. Therefore, a myriad of different peptides can be synthesized ribosomally, far more than those synthesized ribosomally. As such, these bioactive peptides have a varied range of attributes, including antibiotic, antiproliferative, immunomodulatory, metal-sequestering, and toxic properties. In this chapter, we describe the modular nature of NRPSs and the role each domain plays in nonribosomal peptide synthesis. The diverse range of nonribosomal peptide structures and activities are discussed in detail as well as the current field of combinatorial biosynthesis for creating novel bioactive peptides from these nonribosomal platforms.

13.1 INTRODUCTION

Among the vast array of bioactive peptides, there are some that are synthesized via ribosome-independent mechanisms. These compounds are produced by nonribosomal peptide synthetase (NRPS) enzymes in an assembly-line mechanism (Marahiel et al., 1997; Mootz et al., 2002b). Nonribosomal peptide synthesis was first confirmed by the ability of cell-free extracts to synthesize the antibiotic tyrocidine in the presence of ribosome inhibitors (Mach et al., 1963). Studies following this focused on the characterization of the synthetase enzymes from tyrocidine and gramicidin synthesis (Gevers et al., 1969; Kleinkauf et al., 1971; Lee and Lipmann, 1975). However, it has only been in the last two decades that the genes encoding NRPS enzymes, and the mechanisms for nonribosomal peptide synthesis, have finally been elucidated (Krause and Marahiel, 1988; Mittenhuber et al., 1989; Diez et al., 1990; Kleinkauf and Dohren, 1990). NRPS genes are usually arranged in large gene clusters approximately 20 to 80 kb in size. The recent explosion of genome sequencing projects has rapidly increased the rate of identification of NRPS gene clusters, compared to the traditional discovery methods of fermentation and bioactivity assays (Omura et al., 2001; Oliynyk et al., 2007; Udwary et al., 2007). Recent bioinformatic analysis has revealed that approximately half of the currently sequenced genomes harbor thiotemplate modular enzymes that may be involved in producing nonribosomal peptides or related polyketide compounds (Donadio et al., 2007).

The evolution of these nonribosomal expression systems has allowed for the production of peptide-based compounds with a relatively low ATP per mole product expenditure, with less dependence on RNA intermediates. While ribosomal synthesis requires ATP for aminoacyl-tRNA synthesis, proofreading, elongation, and translocation (Kallow et al., 1998), it is suggested that a sixfold lower consumption of ATP is required per peptide bond in nonribosomal synthesis (Kershaw et al., 1999). Another advantage of nonribosomal synthesis is the vast structural diversity of the compounds. This is due, in part, to the wide range of available substrates compared to the 20 amino acids available in ribosomal synthesis. In contrast to ribosomal peptides such as insulin, which incorporate proteinaceous amino acids, nonribosomal peptides can comprise unique and highly modified residues such as α-amino butyric acid (Abu) and (4*R*)-4-[(*E*)-2-butenyl]-4-methyl-threonine (Bmt) in the antibiotic cyclosporin. There are over 300 different amino-, hydroxy-, or carboxy-acid substrates that have been identified in nonribosomal peptide compounds (Kleinkauf and Dohren, 1990). Other unique features of NRPS compounds include fatty acid chains, macrocyclic and heterocyclic rings, *N*- and *O*-methylations, *N*-formylations, glycosylations, halogenations, and D-amino acids (Finking and Marahiel, 2004; Sieber and Marahiel, 2005). Nonribosomal peptides usually comprise between 2 and 20 residues, although the longest nonribosomal peptide is thought to be the linear toxin polytheonamide B from the marine sponge *Theonella swinhoei*, which contains 48 amino acids in alternating D- and L-forms (Hamada et al., 2005).

The unique amino acids and structures of nonribosomal peptides confer useful and interesting properties such as antimicrobial, toxic, and immunosuppressive activities. For this reason nonribosomal peptides have been of particular interest to pharmaceutical and agrichemical companies and are the focus of many drug discovery initiatives (Salomon et al., 2004). Well-known examples of nonribosomal peptides include antibiotics such as bacitracin and immunosuppressants such as cyclosporin (Lawen and Zocher, 1990; Konz et al., 1997). The modular nature of NRPSs and the interesting properties of nonribosomal peptides have driven current research into the development of novel compounds using combinatorial biosynthesis. Combinatorial peptide libraries have resulted in a practically limitless array of novel compounds, some of which have increased or new properties of value to industrial or medical applications (Baltz et al., 2006).

13.2 SOURCES OF NONRIBOSOMAL PEPTIDES

NRPS systems are usually found in marine and soil bacteria, or in lower-order eukaryotes such as filamentous fungi and marine sponges. The majority are produced by actinobacteria, ascomycetes, bacillus, cyanobacteria, and myxobacteria. However, similar nonribosomal mechanisms have also been discovered in higher eukaryotes such as mice and the fruit fly *Drosophila melanogaster*, where single NRPS modules are involved in more complex pathways (Kasahara and Kato, 2003; Richardt et al., 2003).

Many bioactive compounds of nonribosomal origin are also isolated from marine macroorganisms including ascidians, sponges, and molluscs (Moore, 2006). However, it is becoming increasingly apparent, with the development of culture-independent sequencing and cloning, that some of these natural products may actually be produced by symbiotic or dietary prokaryotes. For example, it is thought that the dolastatin group of compounds are produced by the cyanobacteria *Lyngbya* and *Symploca* rather than their symbiotic sea hare host, *Dolabella auricularia* (Luesch et al., 2002). Some nonribosomal peptide-producing microbes are also parasites or epiphytes on plants; however, production of these compounds has not yet been confirmed in plants themselves (Bender et al., 1999; Audenaert et al., 2002).

13.3 THE MODULAR STRUCTURE OF NRPSs

NRPS megasynthetases are the largest known enzymes and can be over 1000 kDa in size (Wenzel et al., 2006). Traditional NRPS assembly lines are made up of repeated modules where each module is responsible for the incorporation of a specific amino acid into the nascent peptide chain. There are three types of NRPSs (Mootz et al., 2002b). Type A NRPSs are linear in structure, where the organization of the modules corresponds to the primary structure of the resulting peptide. This type is the most common in bacterial systems. Type B NRPSs are used iteratively so that each module contributes more than one activated amino acid to the final product. This is the case for the biosynthesis of the siderophore enterobactin in which each of the two modules are used three times (Shaw-Reid et al., 1999; Schwarzer et al., 2003). Type C or nonlinear NRPSs often do not have enzymes in a linear module arrangement. For example, in vibriobactin synthesis the intermediates are not bound as assembly-line thiosters, but are free soluble molecules (Keating et al., 2000). These systems may also have enzymes associated with the NRPS that are not covalently attached to the enzyme complex. Often these are involved in forming cyclizations or branch-points on the peptide structure (Mootz et al., 2002b). This complicates the characterization of NRPSs and combinatorial biochemistry attempts as modules often cannot be deduced from the enzyme or compound structure (Finking and Marahiel, 2004).

Traditional NRPS modules are made up of a minimum of three domains: the adenylation domain, which selects and activates the cognate amino acid; the peptidyl carrier protein domain, which tethers the growing peptide chain and transports it to the following module; and the condensation domain, which is responsible for peptide bond formation between adjacent amino acids. There are also modification and tailoring domains, which add to the diversity of nonribosomal peptide structure and function.

13.3.1 ADENYLATION DOMAINS

Adenylation domains (A-domains) are responsible for recognition of the cognate amino acid and subsequent activation as an aminoacyl adenylate (Figure 13.1). ATP is consumed in this Mg^{2+} dependent reaction. Despite their functional homology to tRNA-synthetases in ribosomal peptide

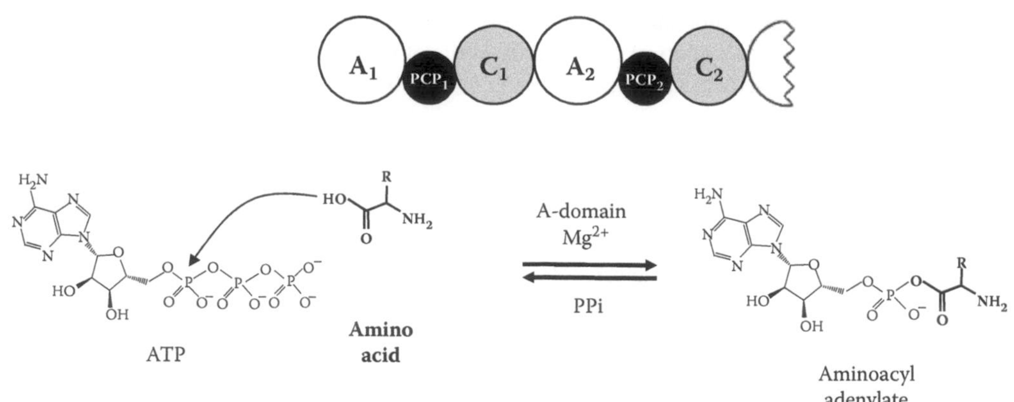

FIGURE 13.1 Recognition and activation of a cognate amino acid (bold) by the adenylation (A)-domain (white). ATP is consumed in this Mg^{2+}-dependent reaction. (From Sieber, S.A. and Marahiel, M.A., *Chem. Rev.*, 105, 715–738, 2005. With permission.)

synthesis, these enzymes do not share structural or sequence similarity (Weber and Marahiel, 2001). A-domains are approximately 550 amino acids in size and are made up of a large N-terminal subunit and a smaller C-terminal subunit. The N-terminal subunit comprises two domains whose interface forms the substrate binding pocket, and the C-terminal subunit is thought to form a lid over this binding pocket during substrate binding (Conti et al., 1997; May et al., 2002). Based on the crystal structure of the Phe-activating A-domain of gramicidin synthetase A, ten highly conserved core motifs (A1–10) were suggested to be involved in ATP-binding, hydrolysis, and substrate adenylation (Marahiel et al., 1997). Ten residues surrounding the A4 and A5 motifs, however, were less conserved and were shown to be responsible for substrate binding specificity (Stachelhaus et al., 1999). This signature sequence, which is found within the cleft between the two N-terminal domains, is known as the *nonribosomal code* and has been successfully used to predict the substrate specificity of uncharacterized A-domains.

There are some limitations of this specificity inferring code, such as the error rate in predicting amino acid specificity for sequences that do not exactly match characterized sequences. However, recent advances in bioinformatic analysis have been used to complement Stachelhaus' nonribosomal code with increased accuracy (Rausch et al., 2005). The nonribosomal code is also limited in its ability to effectively predict A-domains with relaxed specificity. These relaxed A-domains encode a signature sequence for a specific substrate but they are able to activate multiple amino acids. For example, it is thought that two relaxed specificity A-domains may result in the vast array of isoforms of the cyanobacterial toxin, microcystin (Tillett et al., 2000; Mikalsen et al., 2003). However, this is not reflected in their signature sequences.

13.3.2 Peptidyl Carrier Protein Domains

After activation by the A-domains, the aminoacyl adenylate intermediate is transferred to the next module via the relatively small peptidyl carrier protein domain (PCP-domain). PCP-domains, also known as thiolation domains (T-domains), are approximately 80 amino acids in size and are inactive when first translated. They require activation by a 4′-phosphopantetheinyl transferase (PPTase), which covalently attaches a 20-Å-long 4′-phosphopantetheine (4′PP) arm from a coenzyme A substrate (Figure 13.2) (Lambalot et al., 1996). The 4′PP arm is tethered to the invariant serine residue within the conserved motif ([I/L]GG[D/H]SL) on the loop between the first and second helices of the four-helix PCP-domain structure (Stein et al., 1994; Weber et al., 2000). The holo-PCP-domain then covalently binds the activated aminoacyl adenylate from the upstream A-domain as a thioester

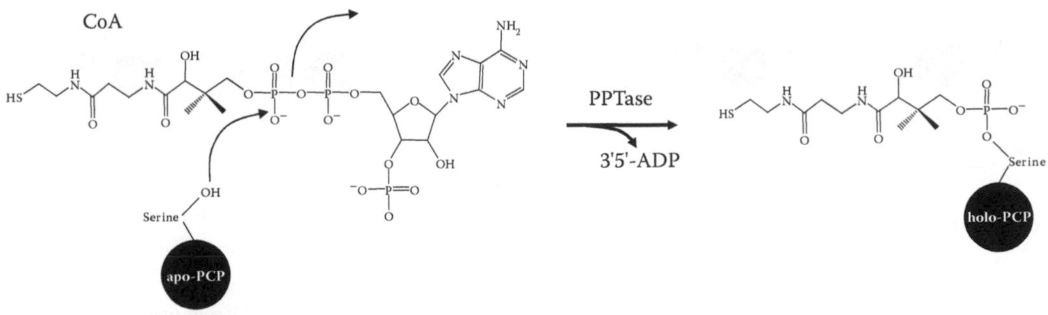

FIGURE 13.2 Modification of the apo-peptidyl carrier protein (PCP)-domain (black) via the action of a phosphopantetheinyl transferase (PPTase). A 4′-phosphopantetheinyl arm from coenzyme A (CoA) is covalently attached to activate the PCP-domain. (Adapted from Lambalot, R.H. et al., *Chem. Biol.*, 3, 923–936, 1996.)

for transfer to the next module in the NRPS complex (Figure 13.3) (Kleinkauf et al., 1971; Lee and Lipmann, 1975; Stachelhaus et al., 1996). This system is analogous to pantetheine-bound fatty acid chains in fatty acid synthesis.

13.3.3 Condensation Domains

Condensation (C) domains are responsible for peptide bond formation between the growing peptide chain on the upstream PCP-domain and the activated amino acid tethered to the downstream PCP-domain. They are approximately 450 amino acids in size and comprise a donor (acyl group of the growing peptide chain) and an acceptor (activated amino acid) binding site (Mootz et al., 2002b). Peptide bond formation occurs via nucleophilic attack of the donor acyl group by the amino group of the acceptor aminoacyl adenylate (Figure 13.4).

C-domains are characterized by seven conserved motifs with the His-motif, C3 [HHxxxDG], critical for the catalytic function and structure of the domain (Marahiel et al., 1997). The His-motif is found in all C-domains except cyclization (Cy)-domains, which instead harbor an aspartate motif [DxxxxD] for heterocyclization of the residues (Keating et al., 2002). Mutational analysis has revealed that the second His^{147} residue and the Asp^{151} residue of the C3 motif, along with an upstream

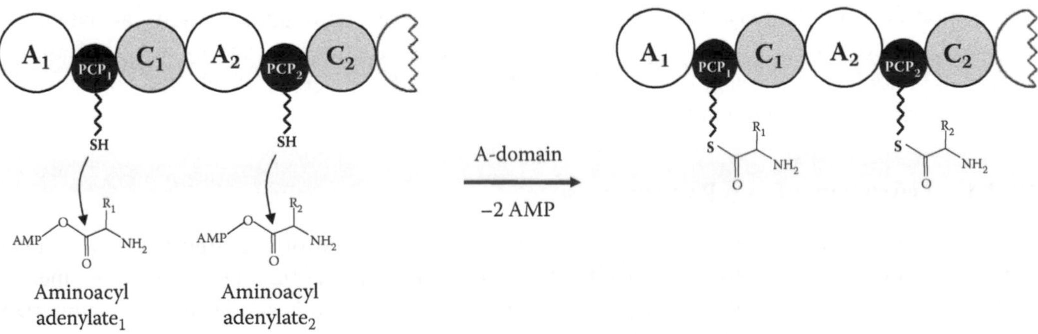

FIGURE 13.3 Covalent attachment of the activated aminoacyl adenylate from the A-domain onto the free thiol group (bold) of the 4′-phosphopantetheinyl-PCP-domain (black). (From Sieber, S.A. and Marahiel, M.A., *Chem. Rev.*, 105, 715–738, 2005. With permission.)

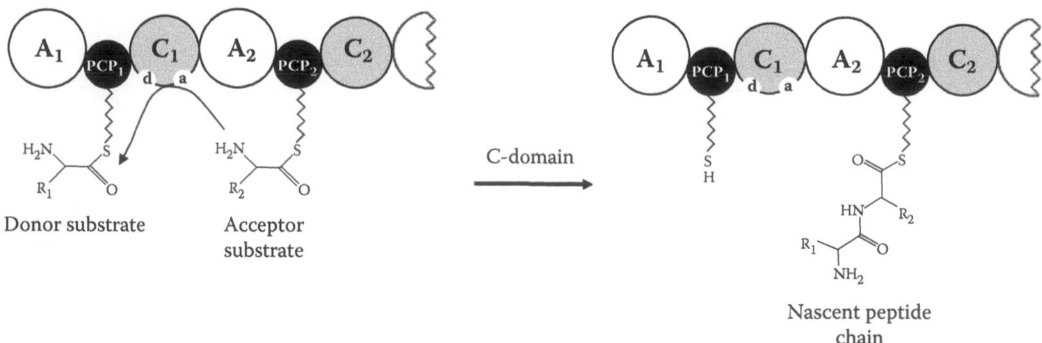

FIGURE 13.4 Peptide elongation by the condensation (C)-domain (gray), which catalyzes an attack of the nucleophilic amine of the acceptor (a) substrate onto the electrophilic thioester of the donor (d) substrate. (From Sieber, S.A. and Marahiel, M.A., *Chem. Rev.*, 105, 715–738, 2005. With permission.)

Arg^{62} residue were critical in peptide bond formation *in vitro*. This confirmed the function of C-domains in the condensation of nonribosomal peptides (Stachelhaus et al., 1998; Bergendahl et al., 2002).

Although A-domains are responsible for the activation of the correct amino-acid precursor, recent research has revealed the role of C-domains in the selectivity and specificity of nonribosomal peptide synthesis. The acceptor sites of C-domains have high selectivity and stereospecificity for the following amino acid and are therefore responsible for the selection of the correct substrate for elongation (Doekel and Marahiel, 2000). Although the donor sites of C-domains still possess strong stereoselectivity, there seems to be less selectivity towards the specific amino acid donated (Belshaw et al., 1999).

There are five phylogenetically distinct subtypes of C-domains: Starter, ${}^{L}C_L$, ${}^{D}C_L$, Dual E/C, and Cy-domains (Rausch et al., 2007). Although ${}^{L}C_L$ C-domains function in forming peptide bonds between two L-amino acids; the other C-domains are involved in tailoring and modification of the peptides.

13.4 NONRIBOSOMAL PEPTIDE STRUCTURAL DIVERSITY—MODIFICATION AND TAILORING DOMAINS

The modular NRPS enzymes can include modification domains that not only allow for the incorporation of nonproteinogenic amino acids, but the modification of the peptide via associated enzymes. This can occur *in cis* with NRPS-associated tailoring domains or *in trans* via distinct enzymes. These modifications increase the structural diversity of the compounds, which can be linear, branched, cyclic, double cyclic, or partially cyclic. These structural variations also increase the variety of biological function of the nonribosomal peptides. The diversity of selected nonribosomal peptide structures is depicted in Figure 13.5.

13.4.1 Epimerization and Racemization

Many nonribosomal peptides, especially antibiotics, are comprised of D-amino acids, which can function to inhibit proteolysis by L-specific proteases (Luo et al., 2002). There are three mechanisms by which D-amino acids are integrated into nonribosomal peptide synthesis. First, epimerization domains (E-domains) can convert the C-terminal L-amino acid into its D-counterpart *in cis* on the nascent peptide chain. The ~450-amino-acid E-domains are located downstream of the PCP-domain in D-amino acid incorporating modules. The equilibrium between L- and D-amino acyl substrates is selected for by the following ${}^{D}C_L$ C-domain, which ensures the correct amino acid form

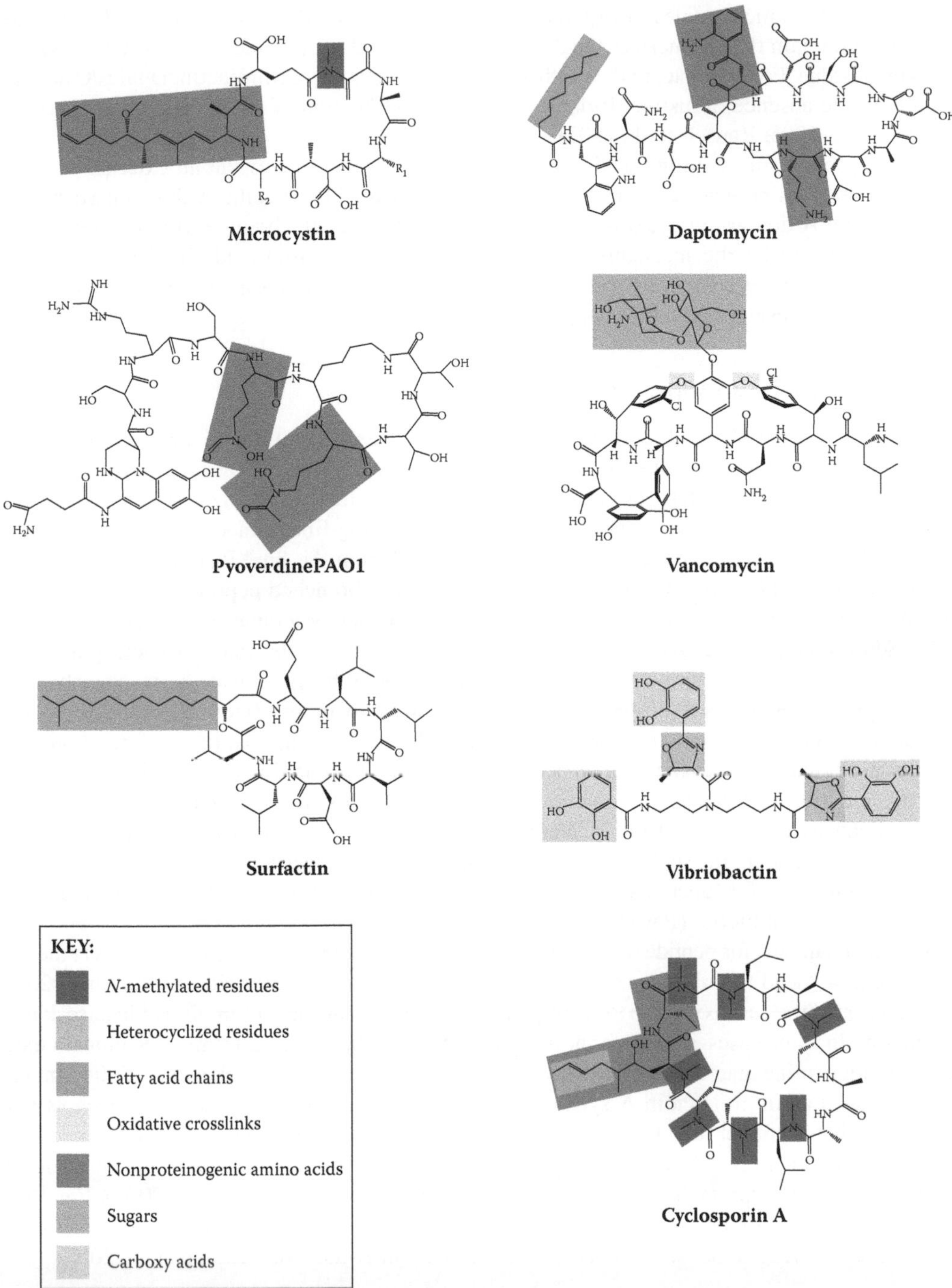

FIGURE 13.5 **(See color insert following page 176.)** Examples of nonribosomal peptide structural diversity. Highlighted regions depict characteristic structural variations carried out by adenylation and modification domains. $R_{1,2}$, variable amino acids in the microcystin structure.

is selected (Ehmann et al., 2000). For example, the SnbC A-domain from pristinamycin I biosynthesis activates L- but not D-aminobutyric acid, even though only D-aminobutyric acid is incorporated into the final compound (Thibaut et al., 1997). The L-aminobutyric acyl adenylate is then epimerized by the associated E-domain.

Alternatively, some peptides incorporating a D-amino acid are the result of a Dual E/C domain, a combined domain that epimerizes the donated L-amino acid for peptide bond formation with the following residue. For example, in the synthesis of the biosurfactant arthrofactin, Dual E/C domains function in the absence of distinct E-domains for the modification of amino acids to their D-form (Balibar et al., 2005; Rausch et al., 2007).

Naturally occurring L-amino acids can also be epimerized *in trans* via an external racemase enzyme. The D-amino acids can then be selected and activated by specific A-domains such as the D-Ala specific A-domain in cyclosporin A synthesis (Zocher et al., 1986; Hoffmann et al., 1994). This is thought to be the mechanism for incorporation of D-amino acids in *Pseudomonas* sp. nonribosomal peptides such as the syringomycin synthetase, which has no encoded E-domains yet comprises two D-residues (Guenzi et al., 1998).

13.4.2 Peptide Termination and Release

Following peptide synthesis, the final product must be released from the NRPS for enzyme turnover. Thioesterase (Te) domains are one such mechanism for peptide termination and release. Te-domains are approximately 280 amino acids in size and their specialization for each product results in the release of a diverse range of structures including linear, macrocyclic, and branched products. The completed peptide is covalently transferred from the final PCP to the Te-domain at the C-terminus of the NRPS. Formation of a linear or cyclic/branched peptide depends on the disruption of this intermediate by attack of either a water molecule or an internal nucleophile, respectively (Schwarzer et al., 2001). For cyclic or branched structures, the internal nucleophile is the N-terminus or an internal amine group such as in tyrocidine or bacitracin synthesis. Nonribosomal products are often cyclized, probably due to their ability to more effectively resist proteolytic degradation than linear compounds (Jegorov et al., 2006). It is thought that the conformation of the Te-domain regulates the cyclization of products. A closed Te-domain conformation may restrict water from acting as a nucleophile, which subsequently results in a cyclized, rather than linear, product (Schwarzer et al., 2001). Other interactions can include side-chain hydroxyl groups or β-OH groups in fatty-acid chains such as those in daptomycin and surfactin (Walsh, 2004). These lipid groups are important for functional interaction with hydrophobic targets, such as cell membranes in the case of many antibiotics (Powell et al., 2007).

Other mechanisms for peptide release are less common, but can include condensation via the final C-domain in a "head to tail" manner as in cyclosporin synthesis (Grünewald and Marahiel, 2006). The reduction of the carboxy group to corresponding aldehydes can also occur via a reductase domain (R-domain). In some cases, such as for nostocyclopeptide synthesis, the C-terminal R-domain mediates the unique imine macrocyclization between the N- and C-terminus (Becker et al., 2004). In other examples, such as in gramicidin A synthesis, the aldehydes are further reduced to alcohols via a mechanism that is not yet fully understood (Kessler et al., 2004). This is also the case for the release of the BT antibiotic in *Brevibacillus texasporus* where the alcoholic C-terminal residue provides the compound with superprotease resistance despite its linear structure (Wu et al., 2005). Finally, in the unique case of vibriobactin, the acceptor amines are soluble for each C-domain and therefore the final product is released into solution rather than via a Te- or R-domain (Keating et al., 2000).

13.4.3 Cyclization and Heterocyclization

The heterocyclization of cysteine and serine/threonine to five-membered thiazoline and oxazoline rings is carried out by cyclization (Cy)-domains. In these NRPS modules, the C-domains are replaced by the Cy-domains such as in bacitracin synthesis (Konz et al., 1997). The role of the Cy-domain in heterocyclization was confirmed by inactivation of the domain *in vitro*, which resulted in the expression of a noncyclic dipeptide (Duerfahrt et al., 2004). The cyclization reaction involves an independent amide bond condensation reaction followed by a cyclization reaction of the amide side chain and subsequent dehydration to the heterocyclic ring (Walsh et al., 2001). Further modifications can occur after heterocyclization including reduction of the thiazoline ring to a tetrahydro thiazolidine ring by R-domains, as in the synthesis of pyochelin (Patel and Walsh, 2001). Alternatively, oxidation of the dehydro heterocycles by oxidase (Ox)-domains can result in heteroaromatic thiazole or oxazole rings, such as those found in epothilone and bleomycin (Schneider et al., 2003).

Cyclization or heterocyclization of nonribosomal peptides increases structural rigidity for compounds involved in receptor or molecule binding. They are also important for biological activities such as metal chelation or for promoting protein and nucleic acid interactions (Roy et al., 1999). For example, vibriobactin's iron-sequestering activity is a result of the heterocyclization of two threonines to oxazoline rings, which imposes constraints on the branched structure (Grünewald and Marahiel, 2006). An oxazoline ring has also been identified in other siderophores such as agrobactin from *Agrobacterium tumefaciens*, parabactin from *Paracoccus denitrificans*, and mycobactin from *Mycobacterium tuberculosis* (Griffiths et al., 1984).

13.4.4 Other Nonribosomal Peptide Structural Modifications

Other modifications, occurring after synthesis, may include oxidoreductase reactions, which have a dramatic effect on the stereochemistry and bioactivity of natural products. They can generate or remove chiral centers, introduce highly reactive functional groups, interconvert H-bond donor/acceptor sites, and change overall solubility (Rix et al., 2002). Tailoring enzymes involved in these reactions include oxidases, peroxidases, reductases, dehydrogenases, and oxygenases. For example, oxidative crosslinking between aryl side chains of nonribosomal peptides can occur by cytochrome P450-dependent oxygenases (Bischoff et al., 2005). Glycopeptide antibiotics such as vancomycin, teicoplanin, and balhimycin have these highly constrained crosslinked structures, which form the binding site for bacterial peptidoglycan strands (Walsh, 2004). These antibiotics are also glycosylated via the action of glycosyltransferases, such as in the addition of the disaccharide D-glucose-vancosamine on the fourth residue of vancomycin (Baltz, 2002). A similar glycopeptide antibiotic, chloeremomycin, which has a variation in the sugar decorations has increased activity as an antibiotic against vancomycin-resistant enterococci (Cooper et al., 1996).

N-methylation can also occur via *N*-methyltransferase (NMT)-domains, which are located between the A- and T-domains of the corresponding modules (Marahiel et al., 1997). In this instance, nonmethylated amino acids are activated by the A-domains and are subsequently methylated before peptide bond formation (Billich and Zocher, 1987). The highly methylated cyclosporin compound comprises seven out of eleven *N*-methylated amino acids, and correspondingly, seven of the eleven domains encode NMTs (Weber et al., 1994). Methylated amino acids may be incorporated to help the peptide resist proteolytic breakdown (Sieber and Marahiel, 2005). Furthermore, the importance of methylations on activity was shown by Kohli and colleagues, who revealed that cyclization and *N*-methylation increased the potency of peptides for the inhibition of ligand binding by integrin receptors (Kohli et al., 2002). A methylated residue has also been shown to increase the potency of microcystin toxicity in mouse bioassays (Sivonen and Jones, 1999). Further modification of this *N*-methylated residue in microcystin also occurs *in trans*, via dehydration, to incorporate *N*-methyl-dehydroalanine from the L-Ser substrate. The microcystin synthetase complex also encodes for another form of methylation by the *O*-methyltransferase (OMT)-domain. This is thought to be involved in the transfer of a methyl group to the hydroxyl group on the unusual amino acid 3-amino-9-methoxy-2,6,8-trimethyl-10-phenyl-4,6-decadienoic acid (Adda) (Tillett et al., 2000; Christiansen et al., 2003; Rouhiainen et al., 2004). Other modifications in nonribosomal peptides can include halogenation, such as in the antibiotic complestatin; *N*-hydroxylation and acetylation, in the siderophore amphibactin; and *N*-formylation, in linear gramicidin; which all provide structural conformations and increased biological activity (Chui et al., 2001; Martinez et al., 2003; Kessler et al., 2004).

Finally, nonribosomal peptides can be modified via *N*-acylation of the first amino acid by the N-terminal C-domain, such as in the biosynthesis of the siderophore yersiniabactin (Suo et al., 2001; Tang et al., 2007). This initiation domain, known as the starter C-domain, acylates the first amino acid with a β-hydroxy-carboxylic fatty acid instead of an amino acid (Rausch et al., 2007). This first module architecture has only recently been described; however, many well-known metabolites, including surfactants such as surfactin, fengycin, and lichenysin are also synthesized in this manner.

13.4.5 Hybrid Polyketide/Nonribosomal Peptides

Polyketide synthases (PKSs) increase the diversity of NRPS products when combined in hybrid NRPS/PKS systems, such as those found in the myxobacterial anticancer compound epothilone and in the cyanobacterial toxins microcystin and nodularin (Tang et al., 2000; Tillett et al., 2000; Moffitt and Neilan, 2004). PKSs have a similar modular structure to NRPSs, although they produce secondary metabolites from organic acid precursors such as coenzyme A (CoA) or malonyl CoA instead of amino acids (Cane and Walsh, 1999; McDaniel et al., 2005; Weissman and Leadlay, 2005). Their modular structure includes a ketosynthase (KS)-domain, which catalyses C–C bond formation for compound elongation; an acetyltransferase (AT)-domain which selects the activated cognate substrate; and an acyl carrier protein (ACP)-domain. This ACP-domain is homologous to the PCP-domain in NRPS synthesis and also requires the addition of the 4'-PP arm for activation. Many PKS products are pharmaceutically important drugs such as erythromycin or rapamycin (Schwecke et al., 1995; Oliynyk et al., 2007).

There are several mechanisms by which PKS and NRPS enzymes can communicate to produce hybrid compounds (Du et al., 2001). One mechanism is via functional hybridizations between NRPS and PKS modules where the enzymes are linked on the same protein in a linear arrangement, such as in TA compound production in *Myxococcus xanthus* (Paitan et al., 1999). Alternatively, the NRPS and PKS enzymes can interact from locations on separate proteins, as in nostopeptolide synthesis (Hoffmann et al., 2003). A combination of both linked and separate interactions can also be involved in the synthesis of a compound, such as in the production of yersiniabactin (Pelludat et al., 1998).

Hybrid polyketide/nonribosomal peptides can also be synthesized without communication between PKS and NRPS enzymes. This can occur via the conversion of a polyketide intermediate into an amino-acid precursor for activation by the A-domain in the NRPS assembly line, such as in cyclosporin biosynthesis in *Tolypocladium niveum* (Offenzeller et al., 1996). The *Pseudomonas syringae* hybrid toxin, coronatine, is synthesized via a unique mechanism where the polyketide and nonribosomal peptide moieties are synthesized separately and are later ligated to form the final compound (Liyanage et al., 1995). Hybrid PKS/NRPS lipopeptides can also incorporate short carboxylic acids into the peptidyl compound by transfer and subsequent condensation of the β-hydroxy acyl group to the NRPS-bound amino acid as in syringomycin and fengycin synthesis (Du et al., 2001). Alternatively, a β-amino fatty acid can be converted to an ACP-bound β-amino acid via an aminotransferase (AMT)-domain as in mycosubtilin and microcystin synthesis (Duitman et al., 1999; Tillett et al., 2000).

Although this chapter focuses on nonribosomal peptides, the activities of some hybrid polyketide/nonribosomal peptides are also presented as examples of structural and functional diversity. The majority of NRPS genes for the nonribosomal peptides described here have been elucidated. However, some putative nonribosomal compounds are included based on their similar structures or activities when compared with confirmed NRPS peptides.

13.5 ACTIVITIES OF NONRIBOSOMAL PEPTIDES

13.5.1 Antibiotics

The first nonribosomal peptides to be discovered were antibiotics, including penicillin and gramicidin in the early twentieth century. However, the last few decades have seen increasing resistance of bacterial strains to the antibiotics in medical use (Coates et al., 2002). This has driven research into the development of novel antibiotics, which has been facilitated by the elucidation of antibiotic synthetase genes. The current focus is on novel drug discovery or to use novel scaffolds to rationally design new combinatorial antibiotics (Walsh, 2004).

Most antibiotics are produced by a variety of fungal and bacterial species, particularly *Streptomyces* sp., which produce approximately two-thirds of the known natural antibiotics (Bentley et al., 2002). Antibiotics have a wide variety of molecular targets, including cell wall synthesis, cell membranes, protein synthesis and DNA replication. This means that they can target a variety of organisms including Gram-positive and Gram-negative bacteria, fungi, insects, nematodes, and viruses.

Perhaps the most well-known antibiotics are the β-lactam compounds, which comprise a β-lactam ring that contains three carbons and one nitrogen atom. Antibiotics such as penicillin and cephalosporin contain this β-lactam ring attached to a sulfur-containing ring. This ring structure mimics the D-alanyl–D-alanine substrate for the transpeptidase enzyme involved in peptidoglycan crosslinking (Tipper and Strominger, 1965; Yocum et al., 1979). The transpeptidase active site is irreversibly acylated by the β-lactams, which results in inhibition of cell wall biosynthesis and, consequently, bacteriostatic activity towards Gram-positive organisms.

The precursor to the synthesis of penicillin and cephalosporin is the tripeptide δ-aminoadipoyl-cysteinyl-D-valine (ACV). This linear tripeptide precursor is synthesized nonribosomally by ACV synthetase. ACV has no known antibiotic activity and requires further modification to penicillins in fungi, or cephalosporins and cephamycins in fungi and bacteria (Aharonowitz et al., 1992). A range of different penicillin isoforms with varying antibiotic activities can be produced by feeding fermentation cultures different precursors and substrates (Behrens et al., 1948). Semisynthetic penicillin derivatives have also been developed that have broad-spectrum activities, resistance to acid for use in oral treatments, or resistance to the β-lactamases that break down natural penicillins (Doyle et al., 1961; O'Callaghan et al., 1976).

Another well-known group of cell wall synthesis inhibitors is the glycopeptide group I antibiotics, such as vancomycin, which is produced by actinobacteria. This heptapeptide is characterized by five aromatic residues with electron-rich side chains that allow oxidative crosslinking, resulting in a highly constricted architecture (Kahne et al., 2005). Other vancomycin-like antibiotics include balhimycin and chloroeremomycin, which differ in activity due to their different side chain glycosylations (Trauger and Walsh, 2000; Recktenwald et al., 2002). Glycopeptide group I antibiotics work by binding to the D-Ala–D-Ala precursor in peptidoglycan synthesis, which inhibits the transpeptidase involved in cell wall crosslinking (Chatterjee and Perkins, 1966; Nieto and Perkins, 1971). This means that D-Ala–D-Ala analogs, such as penicillin, can affect the activity of vancomycin (Perkins, 1969). These glycopeptide group I antibiotics are used as a last resort in the treatment of methicillin-resistant *Staphylococcus aureus* infections or for patients with β-lactam antibiotic allergies (Boger, 2001).

The emerging resistance of vancomycin-resistant strains has driven research into developing new drugs with novel mechanisms of action. Ramoplanin, another nonribosomal glycolipid, produced by *Actinoplanes* sp., is one such compound (Walker et al., 2005). It is currently in phase III clinical trials for use against vancomycin-resistant enterococci (Montecalvo, 2003). Ramoplanin blocks cell wall biosynthesis by inhibiting bacterial transglycosylases, which polymerize lipid intermediate II into peptidoglycan (Lo et al., 2000). Other cell wall synthesis inhibitors include metal-dependent antibiotics such as bacitracin, produced by some *Bacillus* species. The free amino group on this partially cyclic compound is important for binding the C_{55}-isoprenyl pyrophosphatase (IPP) involved in cell wall polymer formation and a divalent metal cation that prevents IPP recycling (Stone and Strominger, 1971; Storm and Strominger, 1973).

Daptomycin is another metal-dependent antibiotic, which has been used to treat skin infections from Gram-positive pathogens, such as methicillin and vancomycin-resistant *Staphylococcus aureus* (Arbeit et al., 2004). Its unique mechanism of action is via the Ca^{2+}-dependent binding to lipoteichoic acid in the cell wall, as well as permeation of the cell membrane, which disrupts the membrane potential (Canepari et al., 1990; Alborn et al., 1991). It is thought that the two aspartic acid residues within the cyclic structure are crucial for activity and for Ca^{2+} binding. Other antibiotics within this structural group include A54145 and calcium-dependent antibiotic (CDA). CDA is a partially cyclic lipopeptide produced by various streptomycetes and fungi including *Streptomyces*, *Tolypocladium*, and *Alternaria* species. It comprises 11 amino acids with an N-terminal hydroxylated fatty acid that allows it to permeate the cell membrane (Lakey et al., 1983; Kempter et al., 1997; Hojati et al., 2002). From there CDA can form channels that affect the transmembrane ion potentials by transporting monovalent cations across the lipid bilayer.

Cell membrane permeating antibiotics also include the alamethicins, which are linear peptides produced by the mould *Trichoderma viride*. One of the nonproteinagenic amino acids, 2-aminoisobutyric acid (Aib), stabilizes the α-helical structure of the compound, which allows the formation of the voltage-gated ion channels (Mueller and Rudin, 1968; Mak and Webb, 1995; Johansson et al., 2004). This mechanism provides alamethicins a broad-spectrum bactericidal, inhibitory, or toxic activity against a wide variety of organisms including Gram-positive bacteria, fungi, plants, arthropods, and mammals (Beven and Wroblewski, 1997; Leitgeb et al., 2007). A similar mode of action is used by linear gramicidins, which are produced by *Bacillus brevis*. Their structure comprises hydrophobic amino acids including 13 alternating L- and D-amino acids, resulting in a dimeric β-helix structure for channel formation in bacterial membranes (Cotten et al., 1997; Wallace, 2000). The gramicidins are active against topical infections of Gram-positive cocci and some Gram-negative bacteria including *Neisseria*, and are often used in conjunction with other antibiotics such as the similar membrane-permeating tyrocidines (Govaerts et al., 2001). Some broad-range membrane-permeating antibiotics also possess antiviral properties, including the potassium ionophore valinomycin, which has antibiotic activity against insects and nematodes and antiviral activity against the causative agent of severe acute respiratory syndrome (SARS) (Perkins et al., 1990; Wu et al., 2004). However, the toxicity of valinomycin limits its use in clinical treatments, and therefore the development of structural analogs is of interest for use against the virus (Cheng, 2006).

Other nonribosomal peptide antibiotics can act by inhibiting protein synthesis, such as the streptogramin B group, which includes pristinamycin I and virginiamycin SI. There are two types of streptogramin B antibiotics: type A and type B. Type A are polyunsaturated peptolides that contain a substituted aminodecanoic acid and oxazole ring that inhibit peptide elongation on the ribosome. Type B are higher-molecular-weight cyclic heterodetic peptides that usually contain a pipecolic acid derivative and that stimulate the release of the peptidyl-tRNA (Bycroft, 1977; Pulsawat et al., 2007). The combined use of both type A and B streptogramin B compounds results in a synergistic bactericidal effect on Gram-positive organisms. This synergy may be explained by an increased affinity of the B molecules for the ribosome when in the presence of A molecules (Cocito, 1979; Thibaut et al., 1997).

Nonribosomal antibiotics can also function by inhibiting DNA replication, as in the hybrid toxin albicidin. This compound is produced by the sugarcane pathogen *Xanthamonas albilineans* and has antibiotic activity against prokaryotes (Royer et al., 2004). However, albicidin also has phytotoxic properties as it inhibits chloroplastic DNA gyrase, which blocks chloroplast differentiation and results in leaf chlorosis (Hashimi et al., 2007). This demonstrates the wide spectrum of activity that many nonribosomal antibiotics possess, including antitumor, toxic, or biosurfactant properties, which will be discussed in the following sections.

13.5.2 Antitumor Agents

A vast array of nonribosomal peptides, especially from the marine environment, have been isolated in the process of screening for antitumor compounds (Proksch et al., 2002; Simmons et al., 2005; Dunlap et al., 2007). These antineoplastics inhibit the growth of tumors via several mechanisms, including the inhibition of protein synthesis, the alteration of the lysosomal membrane, noncovalent binding of DNA, and inhibition of microtubule assembly to disrupt the cell cycle. The majority of discovered compounds are not suitable for clinical use due to toxicity issues; however, improvements in our understanding of nonribosomal biosynthesis may enable the rational design of more efficacious drugs (Burja et al., 2001; Proksch et al., 2002; Dunlap et al., 2007).

The didemnin family of linear and partially cyclic compounds isolated from ascidians were the first marine natural products to be tested in human clinical trials. Although its antiproliferative mechanism is still not fully understood, it is thought that didemnin B inhibits protein synthesis through binding of elongation factor 1α (EF-1α), which prevents EF-2 binding to the ribosome (Crews et al., 1994). Specifically, it causes apoptosis in proliferating cells via a tyrosine kinase- and capsase-dependent pathway (Baker et al., 2002). Preliminary clinical results showed promising activities against lung, metastatic breast, and cervical cancer; however, serious side effects were identified such as neuromuscular toxicity (Vera and Joullié, 2002). This led to the characterization of other naturally produced didemnin derivatives such as dehydrodidemnin B (aplidine) from *Aplidium albicans*, which has a pyruvate moiety instead of the lactic acid side chain (Sakai et al., 1996). Dehydrodidemnin B is ten times as potent as didemnin B, and showed less nonspecific toxicity in *in vitro* trials (Vera and Joullié, 2002). The mechanism for dehydrodidemnin B activity is via inhibition of ornithine decarboxylase biosynthesis, which is involved in cell proliferation and transformation (Lobo et al., 1997). Initial trials have shown promising results in shrinking lung and colorectal tumors (Maroun et al., 2006).

Another antitumor agent isolated from the marine environment is kahalalide F, a partially cyclic compound produced by the marine mollusk *Elysia rufescens* and the green algae *Bryopsis pennata*. Although the mechanism of action is still unknown, it is thought that kahalalide F works by altering the function of the lysosomal membrane and subsequently affecting the cell architecture (García-Rocha et al., 1996; Suarez et al., 2003). This unique mode of action distinguishes it from all other known antitumor agents.

A significant number of antitumor agents have mechanisms that involve cleavage of DNA. One example is the glycopeptide antibiotic bleomycin, which is produced by *Streptomyces verticillus*, and which is used in the treatment of testicular cancer and lymphoma. Its mechanism of action is via metal- and oxygen-dependent oxidative cleavage of DNA (Aso et al., 1999). At low doses, bleomycin causes cell cycle arrest in the mitotic stage, but at high doses, the accumulation of DNA breakages results in cell apoptosis (Chen and Stubbe, 2005). Bleomycin has been used in chemotherapy treatment for lymphomas and testicular cancer; however, drug resistance and toxicity issues are barriers to its widespread use (Einhorn, 2002). Therefore, the current focus lies in developing improved bleomycin-derived compounds via combinatorial biochemistry. The bleomycin synthetase gene cluster encodes for nine NRPS modules, one PKS module and five sugar biosynthesis genes among others, reflecting the complex branched structure of these compounds and the subsequent difficulty in manipulating the enzymes (Du et al., 2000; Shen et al., 2002).

The bicyclic actinomycin, produced by *Steptomyces* species, are another complex group of DNA-binding antineoplastics. Actinomycin are characterized by a chromophoric phenoxazinone dicarboxylic acid moiety to which two pentapeptide lactone rings are attached via an amide linkage (Pfennig et al., 1999). Their mechanism of action is to intercalate DNA between adjacent guanine–cytosine or guanine–guanine base pairs to block transcription, which causes cell apoptosis (Bailey et al., 1994). Consequently, they have been used for over 40 years in the treatment of human neoplasms (Reich, 1963). Actinomycin are also used in conjunction with other anticancer therapeutics to treat soft tissue cancers such as testicular cancer, tumors in the uterus or womb, and Wilm's tumor, which affects the kidneys (Watt, 1968; Miyazaki et al., 2003; Covens et al., 2006).

Other modes of action for antitumor agents include the inhibition of microtubule depolymerization. An example of this is the mixed nonribosomal/polyketide anticancer drug epothilone, which is produced by the myxobacterium *Sorangium cellulosum* So ce90. There is only one NRPS module, among nine PKS modules, which is involved in the heterocyclization and oxidation of the cysteine to a thiazole ring to form the 2-methylthiazol starter unit (Molnár et al., 2000). Epothilone has cytostatic activity due to its ability to bind to β-tubulin, which results in cell cycle arrest in mitosis and the formation of microtubules in nonmitotic cells (Kowalski et al., 1997). It is currently undergoing human clinical trials in the treatment of a variety of cancers including breast, lung, and prostate cancer (Fumoleau et al., 2007).

The cryptophycins, produced by *Nostoc* species of cyanobacteria, are another potent group of tubulin-destabilizing compounds (Panda et al., 1998). Their characterization has led to the production of a synthetic analog, cryptophycin 52, which is currently in phase II clinical trials for the treatment of ovarian cancer (D'Agostino et al., 2006). However, the recent determination of the cryptophycin gene cluster may enable more efficient production yields and more promising derivatives with less dose-limiting toxicities (Magarvey et al., 2006). Another group of cyanobacterial tubulin-destabilizing agents are the dolastatins, including the analogs symplostatin and lyngbyastatin. These peptides were originally thought to be produced by the herbivorous sea hare *Dolabella auricularia*; however, more recently these compounds have been isolated in higher abundances from its dietary cyanobacterial species, *Symploca* and *Lyngbya* (Harrigan et al., 1998; Luesch et al., 1999, 2001). Although the NRPS genes for these linear and cyclic peptides have not been elucidated, it is generally presumed that they are nonribosomal in origin (Salomon et al., 2004). This is supported by the sequence analysis of the similar antiproliferative and antimitotic nonribosomal peptide, curacin A, from *Lynbgya majuscula* (Chang et al., 2004). Curacin A was originally tested in preclinical trials for use in the treatment of breast cancer. However, the trials were terminated due to curacin A's irreversible and nonspecific binding activity (Verdier-Pinard et al., 1999; Burja et al., 2001).

Dolastatins are effective microtubule inhibitors that also induce cell apoptosis through a mechanism involving the oncoprotein bcl-2, which is an apoptosis inhibitor that is overexpressed in some cancers (Kalemkerian et al., 1999). They possess *in vivo* antitumor activity against several types of murine tumors (Pettit et al., 1987; Bates et al., 1997). The most widely researched of these compounds is dolastatin 10, a linear peptide comprising three unique amino acids and a complex primary amine, dolaphenine (Simmons et al., 2005). It has undergone clinical trials for the treatment of several cancers including lung and breast cancer (Krug et al., 2000; Perez et al., 2005). However, the disappointing results and toxicities of dolastatin 10 led to the development of synthetic analogs including TZT-1027, which contains a benzylamine group instead of a thiazole ring (Miyazaki et al., 1995). This drug exhibits antitumor activity with greater efficacy than its parent compound in mouse models (Kobayashi et al., 1997). However, recent poor results in phase II clinical trials for advanced lung cancer shows the complexity and cyclical nature of cancer drug development (Riely et al., 2007).

13.5.3 Immunomodulators

Immunomodulatory activity has been identified for several cyclic or partially cyclic bioactive peptides that contain between 3 and 13 monomers. However, the NRPS genes have only been determined for the cyclic compound, cyclosporin A, which is produced by the fungus *Tolypocladium niveum*. Cyclosporin A is an immunomodulatory compound with anti-inflammatory, antifungal, and antiparasitic activity (Lawen and Zocher, 1990). It is synthesized by a single large synthetase of 15,281 amino acids comprising eleven modules (Weber et al., 1994). This corresponds to cyclosporin's peptide structure, which contains eleven amino acids, of which seven are methylated and two are unusual residues (Offenzeller et al., 1996). Natural variations on nine of the amino acids result in cyclosporins with different activities including those of unknown function. Cyclosporin A has been used extensively in the medical field to prevent organ rejection after transplantation (Duncan and Craddock, 2006). Its mechanism of immunomodulation involves inhibition of T-lymphocyte activation via suppression of cytokine gene transcription including interferon γ, tumor necrosis factor (TNF), and interleukin-2 (Cebrat et al., 1996).

The majority of putative nonribosomal peptides with immunomodulatory properties have been isolated from Pacific Ocean sponges. Marine sponges are a rich source of bioactive peptides with antibiotic, toxic, antitumor, and immunomodulatory activities (Proksch et al., 2002). Hymenistatin I is one such compound, produced by the sponge species *Hymeniacidon*, which possesses immunosuppressant activity comparable to cyclosporin A in *in vivo* experiments (Cebrat et al., 1996). The mechanisms of immunomodulation, however, are different to cyclosporin A, as hymenistatin I stimulates interleukin-1 production, which may subsequently affect T-cell activation as well as the inflammation response. Similarly, the partially cyclic compound keramamide, produced by *Theonella* sp., has immunosuppressant activity due to its ability to inhibit an intracellular signal transduction pathway in human neutrophils (Kobayashi et al., 1991). As culture-independent sequencing techniques improve, the status of many of these putative immunomodulatory compounds may be confirmed. This may facilitate chemical synthesis of these compounds for novel and improved drug development.

13.5.4 Protease Inhibitors

The role of peptides as protease inhibitors has been of interest to the medical and pharmaceutical fields for the development of novel drugs with reduced side effects. Many of the nonribosomal peptides with protease inhibitory activity are produced by cyanobacterial species, such as the angiotensin converting enzyme (ACE) inhibitor, microginin, from the cyanobacterial species *Microcystis* spp. (Yamaguchi et al., 1989). ACE is a peptidase involved in blood pressure regulation in humans, which makes ACE inhibitors important compounds in the development of novel antihypertensive agents. Over 30 structural variants of microginin with varying activities have been identified. Each isoform comprises four to six amino acids including a characteristic unusual decanoic acid derivative, 3-amino-2-hydroxy-decanoic acid (Ahda). The genes encoding the microginin synthetase enzyme complex have recently been elucidated and consequently, combinatorial biosynthesis utilizing this NRPS gene cluster is the subject of interest for the development of hypertensive agents with increased efficacy (Kramer, 2007).

Anabaenopeptins, also produced by cyanobacterial species, including *Anabaena*, *Oscillatoria*, and *Aphanizomenon*, are carboxypeptidase inhibitors (CPIs) with potential applications in the treatment of thrombotic disorders or as antitumor compounds (Vendrell et al., 2000; Arolas et al., 2005). Anabaenopeptins are partially cyclic compounds with an amino acid linked via a urea moiety to the D-Lys on the cyclic scaffold (Kodani et al., 1999). Their mode of action is likely to be via this linked amino acid because anabaenopeptin G with Tyr in this position is a vastly more potent carboxypeptidase A inhibitor than anabaenopeptin H with an Arg substitution (Itou et al., 1999). This structure may be similar to the C-terminal tail of characterized potato and leach CPIs which mimic the substrate to bind and inactivate the enzyme active site (Arolas et al., 2005).

Serine protease inhibitor (serpin) activity is exhibited by members of the cyanobacterial aeruginosin family of peptides, including oscillarin, aeruginosin, and dysinosin. This is due to their linear peptide chain, which mimics the structure of serine protease substrates (Murakami et al., 1995; Rios Steiner et al., 1998; Carroll et al., 2002; Hanessian et al., 2004). X-ray crystallography has shown that this occurs via antiparallel β-strand binding of the aeruginosin to the three active sites of the enzyme (Sandler et al., 1998). The genes for aeruginosin production have yet to be discovered; however, structure-based studies identified a (2*S*,3a*S*,6*R*-hydroxy,7*aS*) octahydroindole 2-carboxylic acid (L-Choi) core, which is most likely produced nonribosomally. Serine proteases are crucial in blood coagulation in the body and, uncontrolled, can cause blood clotting, resulting in stroke and myocardial infarctions (Kadono et al., 2005). Therefore serpins, such as aeruginosins, may be useful targets for developing new anticoagulants.

13.5.5 Siderophores

Siderophores are low-molecular-weight compounds secreted by bacteria, in iron-limiting environments, to scavenge and bind iron for the cell's requirements. Once bound, the ferrisiderophore complex is taken up by the bacterium via specific outer membrane receptors and released in the cell for use as a cofactor in redox-dependent reactions (Neilands, 1981; Bagg and Neilands, 1987). This reaction must be tightly regulated to prevent a toxic accumulation of iron in the cell. Siderophore production is repressed under high iron concentrations and is usually indirectly controlled by the global ferric uptake regulator protein, Fur, so that they are only produced under iron-limiting conditions (Visca et al., 1992; Xu et al., 2002; Agnoli et al., 2006). As siderophores are potent iron chelators, they have potential uses in treating acute and chronic metal toxicity (Griffiths et al., 1984).

Many of the peptidic siderophores are synthesized nonribosomally, including well-characterized compounds such as enterobactin and bacillibactin produced by *Escherichia coli* and *Bacillus subtilis*, respectively (Gehring et al., 1998; May et al., 2001). There are a wide variety of siderophore architectures including linear, branched, and partially cyclic structures with most containing two to thirteen monomers. Many siderophores are secreted by invasive pathogenic bacteria and are thought to be virulence factors associated with the ability to cause disease (Payne and Finkelstein, 1978). The mechanism for pathogenicity is via the bound ferrisiderophore complex, as iron is usually stored intracellularly or bound to heme proteins in higher organisms (Weinberg, 1984). One such example is the linear siderophore anguibactin, which comprises only two monomers and is produced by *Vibrio anguillarium.* This species is the causative agent of vibriosis in salmonid fish, which is characterized by fatal septicemia. Disruption of anguibactin synthesis results in a dramatic reduction in the pathogen's virulence (Di Lorenzo et al., 2004). Similarly, the hybrid NRPS/PKS compound yersiniabactin is a siderophore and virulence factor produced by the plague-causing bacterium *Yersinia pestis.* The gene cluster for yersiniabactin synthesis is located on a high pathogenicity island, which supports the hypothesized role of this siderophore in *Y. pestis* pathogenicity (Arnold et al., 2003). The hypothesized role of siderophores in virulence and disease reveals their potential as therapeutic targets against opportunistic pathogens.

Several siderophores are produced by pathogenic species of *Pseudomonas.* These include the pyoverdines, or pseudobactins, which are a diverse group of over 70 peptides that vary in structure and length. These fluorescent pigments are characterized by three domains: a quinoline-derived chromophore, and an acyl and peptidyl chain, which are attached to this chromophore at the C- and N-terminus, respectively (Lamont and Martin, 2003). The peptide chain plays a role in iron-binding with the chromophore and may also be involved in binding the iron complex to its cell membrane receptor. In *P. aeruginosa*, pyoverdine controls the production of virulence factors that are involved in the pathogenicity of this bacterium in immunocompromised patients (Lamont et al., 2002). Pyoverdine may also act in a quorum-sensing like manner to increase virulence when cell numbers reach a density for effective infection (Stintzi et al., 1998; Lamont et al., 2002).

P. aeruginosa also produces pyochelin, another nonribosomal siderophore, which is derived from salicylic acid. Pyochelin has a lower affinity for iron than pyoverdine, but it may be able to acquire additional metals for the cell, such as Co(II) and Mo(VI). This was demonstrated by the binding affinity of these metals with the siderophore and the repression of pyochelin synthesis at high Co(II) and Mo(VI) concentrations (Visca et al., 1992). It is also thought that pyochelin, as well as pyoverdine, may have a role in inducing resistance of certain plant roots and foliage to fungal pathogens (Lemanceau et al., 1992; Audenaert et al., 2002). Pseudomonads can produce a wide variety of other siderophores including the 8-hydroxy-4-methoxy-2-quinoline thiocarboxylic acid, quinolobactin, which is produced by *P. fluorescens* ATCC 17400 (Cornelis and Matthijs, 2002), and the amphiphilic linear compounds corrugatin and ornibactin, produced by *P. corrugata* and *Burkholderia cepacia*, respectively (Agnoli et al., 2006).

Among the vast range of nonribosomally produced siderophores discovered, very few others possess an amphiphilic structure. Those that do include marinobactins, aquachelins, and amphibactins, which are usually characterized by a unique peptide domain, to coordinate Fe(III), and a variable fatty acid moiety. Up to ten isoforms of each compound are produced, which differ on this fatty acid tail. The amphiphilic nature of these siderophores enables both water solubility as well as cell membrane association to minimize diffusion of the compound into the surrounding environment (Xu et al., 2002; Martinez et al., 2003). This characteristic is important for the producing organisms that inhabit iron-limited marine environments. Marinobactins, aquachelins, and amphibactins, produced by *Marinobacter* sp., *Halomonas aquamarina*, and *Vibrio* sp. R10, also have biosurfactant properties that may enable the cell/surface interfaces to change hydrophobicity, potentially for biofilm production (Neu, 1996).

13.5.6 Biosurfactants

Most nonribosomal biosurfactants are amphiphilic in structure, with a fatty acid tail attached to a cyclic peptide backbone, which enables the compounds to aggregrate at interfaces and form micelles. Biosurfactants also have the ability to lower the surface tension of liquids and consequently increase the solubility of otherwise poorly soluble compounds (Rosenberg and Ron, 1999). Applications resulting from this include increasing the bioavailability of hydrophobic pollutants by increasing the access of the interface to remediation bacteria (Kuiper et al., 2004). Biosurfactants may also exhibit include antifungal properties. Therefore, these compounds have stimulated the development of new pharmaceuticals to combat the increasing problem of resistance to known antifungals (Sorensen et al., 1996).

The majority of nonribosomal biosurfactants are produced by *Bacillus* and *Pseudomonas* species. The best characterized is surfactin, a potent surfactant secreted by *B. subtilis* under conditions of nutrient limitation (Takahashi et al., 2006). First isolated in 1968, surfactin was shown to prevent fibrin clot formation by inhibiting the formation of the fibrin polymer from its monomers (Arima et al., 1968). It also exhibits antibiotic activity via the formation of cationic pores in cell membranes, which leads to cell lysis (Sheppard et al., 1991). Surfactin comprises seven amino acids linked in a cyclic arrangement by a β-hydroxy fatty acid comprising 13 to 15 carbon atoms. The NRPS modules follow the traditional type A architecture, which reflects the order of the amino acids in the final product. The amino acids in positions 2, 4, and 7 can vary depending on the nitrogen source in the culture media, which results in various surfactants with different activities (Baumgart et al., 1991; Peypoux et al., 1994; Grangemard et al., 1997). The natural variation in surfactants, as well as the relative ease of genetically manipulating *B. subtilis* and type A NRPSs, has led to a multitude of approaches for the combinatorial biosynthesis of these compounds.

Natural isoforms of surfactin include fengycins and plipstatins produced by *B. subtilis* sp. and *B. cereus* sp.. However, these lipopeptides only possess certain antifungal properties of surfactins and do not have antibacterial or antiyeast activities. Fengycins and plipstatins affect the cytoplasmic membranes of filamentous fungi, which can result in morphological changes in the cells (Vanittanakom et al., 1986). A structurally similar biosurfactant is the cyclic antibiotic lichenysin produced by *B. licheniformis* BAS50. Lichenysin possesses higher surfactant activities than surfactin, although it is produced in much lower quantities in the host (Yakimov et al., 2000). *B. licheniformis* BAS50 also has relatively high halotolerance, which may enable applications for lichenysin as a biosurfactant in high salt environments (Yakimov et al., 1995).

In contrast to the other *Bacillus* surfactants, iturins, including bacillomycin and mycosubtilin, possess a β-amino fatty acid rather than a β-hydroxy fatty acid modification (Duitman et al., 1999). These heptapeptides comprise D- and L-amino acids that vary on four of the residues that affect the hemolytic and antifungal activity (Peypoux et al., 1985; Maget-Dana and Peypoux, 1994). *B. amyloliquefaciens* strain FZB42 produces several nonribosomal peptide surfactants, including bacillomycin and fengycin, which act in synergy to compete with organisms in its rhizosphere habitat (Koumoutsi et al., 2004). Consequently, this strain is able to stimulate plant growth and inhibit plant pathogens.

Many *Pseudomonas* species that produce biosurfactants also associate with plants and are often the causative agents for soft-rot diseases including brown blotch in rice and corn (Ballio et al., 1996). Nonribosomal biosurfactants facilitate bacterial dispersal across surface-aqueous interfaces and disrupt plant membranes due to their amphipathic nature (Lindlow and Brandl, 2003). One group of pseudomonad biosurfactants are the syringomycins, which include cormycin, pseudomycin, syringostatin, and syringotoxin. Syringomycins are partially cyclic nonapeptides, produced by *P. corrugata* sp. and *P. syringae* sp., and have potent antifungal and antiyeast activities due to their formation of ion channels in cell membranes (Scaloni et al., 2004). Syringomycins are characterized by a macrocycle comprising nine amino acids including unusual residues such as diaminobutanoic acid (Dab) and an N-terminal 3-hydroxy fatty acid chain of 10 to 14 carbon atoms (Sorensen et al., 1996). The syringomycin synthetase genes reveal that the NRPS modules are not in the traditional colinear organization and are more similar to fungal NRPSs (Guenzi et al., 1998).

Similar *Pseudomonas* biosurfactants are the tolaasins, including corpeptin, fuscopeptin, and syringopeptin, produced by *P. corrugata*, *P. fuscovaginae*, *P. syringae*, and *P. tolaasii*. However, these compounds have a longer and more hydrophobic peptide group than the syringomycins (Emanuele et al., 1998). Tolaasins are partially cyclic compounds of between 18 and 26 amino acids, with the last five residues forming a polar macrocycle. They have antimicrobial activity, particularly towards Gram-positive organisms, although the different isoforms have varying activities depending on structural modifications (Bassarello et al., 2004). Syringopeptin also possesses phytotoxic activity due to its ability to form pores in plasma membranes and to cause stomatal closure in plants (Bender et al., 1999).

Other biosurfactants produced by *Pseudomonas* species include the antifungal amphicins, such as arthrofactin, lokisin, pholipeptin A, and tensin. Amphicins are characterized by an N-terminal helix structure and a β-hydroxydecanoic acid that is attached to a C-terminal macrocyclized peptide chain (Sørensen et al., 2001). The genes for arthofactin show a traditional colinear arrangement. One of the genes in the cluster encodes a putative ATP-binding cassette (ABC) transporter, which is thought to be involved in the secretion of arthofactin (Roongsawang et al., 2003). Similar ABC transporters have also been identified in syringomycin and microcystin synthetases (Guenzi et al., 1998; Tillett et al., 2000). Putisolvin, produced by *Pseudomonas putida* strain PCL1445, can inhibit the formation of *Pseudomonas* biofilms on polyvinyl chloride and can break down preexisting biofilms by altering the outer membrane hydrophobicity (Kuiper et al., 2004). This interesting property has potential for use in the medical field, where biofilm formation on artificial surfaces and equipment is a major problem (Stewart and William Costerton, 2001). Other *Pseudomonas* biosurfactants with interesting activities include viscosin, which causes broccoli head rot by allowing the bacteria to penetrate the waxy plant, and massetolide, which has *in vitro* activity against *Mycobacterium tuberculosis*, the causative agent for tuberculosis (Gerard et al., 1997; Braun et al., 2001).

Finally, a group of biosurfactants produced by *Serratia marcescens* called serrawettins are surface active exolipids with variable structures. The surfactant properties of serrawettins allow them to promote spreading growth of the bacterium on hard surfaces (Matsuyama et al., 1992). The nonribosomal biosynthesis of serrawettin W1 was confirmed via the mutational analysis of a PPTase, which resulted in the loss of serrawettin WI synthesis and also of another nonribosomally produced exolipid, prodigiosin, in the same strain (Sunaga et al., 2004).

13.5.7 Toxins

Although many antibiotics and antitumor agents can also be classified as toxins, the following compounds are distinguished by their ability to cause disease. Most of the nonribosomal toxins are produced by fungi or cyanobacteria with toxicities against plants or animals, including humans.

The plant toxins can be classified into two groups: host-specific toxins (HSTs) and nonspecific plant toxins. HSTs are the pathogenicity determinants of host specificity for the producing organism (Walton, 1996). These mycotoxins allow pathogenic fungi to infect more plant tissue and can also cause growth inhibition, chlorosis, necrosis, and electrolyte leakage (Pringle and Scheffer, 1964). HSTs can be polyketides, terpenoids, saccharides, and peptides. All of the peptidic HSTs are produced nonribosomally except Ptr toxin, which is produced by *Pyrenophora tritici-repentis* (Tuori et al., 2000).

A well-characterized group of nonribosomal peptide HSTs are AM-toxins, which are cyclic tetrapeptides containing the nonproteinogenic α-hydroxy-isovaleric acid (Hyv) and dehydroalanine (Dha) (Ueno et al., 1977). AM-toxins are produced by *Alternaria alternate*, which is pathogenic to specific varieties of apple and pear trees and causes Alternaria blotch disease. The HST is produced in the spore-germinated fluids and causes veinal necrosis in leaves by acting on chloroplasts and the cell wall/plasma membrane interface (Nishimura and Kohmoto, 1983). Similar to tentoxin, its activity may be due to the ring structure and the unsaturated bond in the Dha residue.

The cyclic peptide lactone, destruxin, is another HST produced by several fungal species including *Oospora destructor*, *Alternaria brassicae*, and *Trichothecium* sp. Although it has a wide range of hosts and activities, the toxin is primarily linked with pathogenicity and in facilitating fungal infection (Samuels et al., 1988a). In addition to its phytotoxicity towards *Brassica* plants such as canola and cabbage, destruxin is also an insecticidal neurotoxic compound affecting tobacco hornworm, desert locust, and vine weevil (Bains and Tewari, 1987; Samuels et al., 1988b). This neurotoxicity occurs via tetanic muscular paralysis and flaccidity due to opening of the membrane Ca^{2+} channels, which results in muscle depolarization. The compound also inhibits vacuolar-type ATPases involved in maintaining acidic homeostasis in organelles (Muroi et al., 1994). Destruxin contains five amino acids, one α-hydroxy acid, and an ester bond that is essential for activity (Cavelier et al., 1996). There are 23 isoforms of the compound, with destruxin E being the most potent due to its epoxide moiety (Kershaw et al., 1999). Other HSTs include the histone deacetylase inhibitor, HC-toxin, which is required for pathogenicity of the filamentous fungus *Cochliobolus carbonum* on maize (Brosch et al., 1995), and the mitochondrial glycine decarboxylase, victorin, produced by *C. victoriae*, which causes victoria blight in oats (Markham and Hille, 2001).

In contrast to HSTs, there are also nonspecific phytotoxins that affect a broader spectrum of organisms than the producing organism infects (Yoder, 1980). These nonribosomal toxins are usually virulence factors such as phaseolotoxin and coronatine as well as the surfactant antibiotic syringomycin. Phaseolotoxin, produced by *Pseudomonas syringae* pv. *phaseolicola*, is a linear tripeptide with the structure N^{δ}(N′-sulfodiaminophosphinyl)-ornithyl-alanyl-homoarginine (Ferguson and Johnston, 1980). It causes ornithine accumulation in legumous plants, which is characterized by a chlorotic zone around the site of infection due to the diffusion of the toxin (Mitchell and Bieleski, 1977). Its toxicity is due to the binding and reversible inhibition of the ornithine carbamoyltransferase involved in arginine biosynthesis (Ferguson et al., 1980). Recently, the identification of the phaseolotoxin gene cluster showed its location near the genes for the biosythesis of the precursors ornithine and homoarginine and the gene for phaseolotoxin resistance (Aguilera et al., 2007).

Coronatine is a linear virulence factor and nonspecific plant toxin produced by several pathovars of *Pseudomonas syringae*. It causes chlorosis, induction of hypertrophy, inhibition of root elongation, and stimulation of ethylene production in a wide variety of plant species such as Italian ryegrass and soybean (Mitchell and Bieleski, 1977). Coronatine is a highly modified NRPS/PKS hybrid that comprises a polyketide-derived unit, coronafacic acid, and an *allo*-isoleucine-derived residue, coronamic acid (Parry and Mafoti, 1986). The coronatine synthesis enzymes include a discrete A- and PCP-domain that binds the allo-isoleucine precursor (Couch et al., 2004). This bound thioester is subsequently cyclized and hydrolyzed to produce free coronamic acid for attachment to the polyketide moiety via a ligase-catalyzed amide bond (Liyanage et al., 1995).

Finally, the *Alternaria tenuis* cyclic tetrapeptide tentoxin is another virulence factor and nonspecific phytotoxin characterized by the highly modified *N*-methyldehydrophenylalanine residue (Meyer et al., 1974; Mitchell, 1984). Tentoxin causes chlorosis in lettuce, potato, and spinach plants by altering chloroplast structure and disrupting chloroplast function (Halloin and Hagedorn, 1975). The chloroplast coupling factor (CF_1) is the receptor site for tentoxin in sensitive species, resulting in the inactivation of CF_1 ATPase and electron transport phosphorylation (Steele et al., 1976). CF_1 from resistant species, such as radish, have a lower binding affinity for the toxin and therefore have no disruption of chloroplast function.

Nonribosomal peptide toxins are also known to affect animals, including humans. One such compound is cerulide, a heat-stable cyclic depsipeptide. Cerulide is produced by *Bacillus cereus* and can cause emetic gastroenteritis when ingested (Agata et al., 1994). Its mechanism of action is via the formation of vacuoles, which results in swelling of the mitochondria (Agata et al., 1995; Paananen et al., 2002). Cerulide comprises three repeated tetrapeptides, D-O-Leu-D-Ala-O-Val-Val, in a structure similar to the antibiotic valinomycin. Other toxins that affect animals include microcystins and nodularins, which are hepta- and pentapeptides produced by bloom-forming cyanobacterial species such as *Microcystis*, *Aphanizomnenon*, *Planktothrix*, *Anabaena*, and *Nodularia*. Microcystin and nodularin are hepatotoxic to organisms such as humans and livestock that ingest contaminated water (Carmichael, 1992; Carmichael et al., 2001). Symptoms include gastroenteritis, liver damage, liver cancer, and even death in extreme cases. The unusual polyketide-derived Adda residue and the stereochemistry of the double conjugated bond is critical for toxicity (An and Carmichael, 1994). Their mode of action involves the uptake of the molecule into hepatocytes and inhibition of the serine/threonine protein phosphatases, which results in uncontrolled phosphorylation and liver cytoskeleton damage (Dawson, 1998; Gehringer, 2004).

Similarly, lyngbyatoxins are hybrid polyketides/nonribosomal peptides produced by the marine cyanobacterium *Lyngbya majuscula*. They are powerful activators of protein kinase C, which causes the skin irritation "swimmer's itch" (Basu et al., 1991). The indolactam V core structure of lyngbyatoxin is synthesized by an NRPS that comprises two modules, the first encoding an N-methyltransferase domain, and the second encoding a C-terminal reductase for peptide release (Edwards and Gerwick, 2004). This precursor is then modified by a P450 mono-oxygenase/cyclase, a prenyl transferase, and a reductase/oxidase to result in the final toxin structure. Other cyanobacterial hybrid toxins include barbamide and jamaicamide. Although they are not known to cause disease, they are toxic to mollusc and fish species (Chang et al., 2002; Edwards et al., 2004). Barbamides are chlorinated lipopeptides while jamaicamides are neurotoxins that block voltage-gated sodium channels. These complex hybrid polyketide/nonribosomal peptides are elegant examples of natural combinatorial biosynthesis, which have evolved as a competitive advantage for the cyanobacteria (Welker and von Dohren, 2006).

13.6 FUTURE DEVELOPMENTS FOR NONRIBOSOMAL PEPTIDES—COMBINATORIAL BIOSYNTHESIS

One of the most promising sources for new drugs is via combinatorial biosynthesis of the modular NRPS/PKS enzymes, which potentially allows for the production of a limitless number of novel compounds (Keller and Schauwecker, 2003; Reeves, 2003). The structural complexity and unique substrates of nonribosomal synthesis allow the production of otherwise chemically inaccessible compounds. As new and improved techniques for bioactive compound mining and culture-independent sequencing emerge, many more NRPS/PKS genes and compounds will undoubtedly be elucidated (Wilkinson and Micklefield, 2007). This will facilitate the understanding of nonribosomal peptide structural and functional diversity for use in the rational design of bioactive compounds (Van Lanen and Shen, 2006). This section describes the recent advances in combinatorial biosynthesis as a platform for the future production of a vast range of novel compounds.

13.6.1 Biosynthesis of Natural and Unnatural Nonribosomal Peptides *in Vitro*

Initial work on the combinatorial biosynthesis of novel compounds focused on the *in vitro* reconstruction of assembly lines such as those for enterobactin from *E. coli*, pyochelin from *Pseudomonas* sp., and yersiniabactin from *Yersinia pestis* (Gehring et al., 1998; Patel and Walsh, 2001; Miller et al., 2002). In these experiments, the NRPS and PKS proteins were expressed separately and the final compounds were synthesized *in vitro* from precursors. Another example of *in vitro* expression involved the reassociation of separate surfactin modules, which broke the colinearity rule and yet still resulted in surfactin formation (Menkhaus et al., 1993).

The next approach undertaken for the development of novel compounds was the site-directed mutagenesis of A-domains to alter the amino acids that were activated by the modules. In this way, two residues involved in Phe specificity from the Phe-activating A-domain from gramicidin S synthesis were altered, resulting in Leu activation at comparable catalytic efficiencies (Stachelhaus et al., 1999). Similarly, one residue of an Asp-activating A-domain from surfactin synthesis in *B. subtilis* was altered to code for Asn, although this time with a reduction in catalytic activity (Eppelmann et al., 2002).

Unnatural dimodule fusions have also been attempted *in vitro* in order to explore the complex mechanisms involved in combinatorial biosynthesis. In these experiments, the A-domain retained catalytic activity with fusions at the hinge region of the A-domain and the binding motif in the PCP-domain (Symmank et al., 1999). However, peptide bond formation was only successful with C-domain fusions in the conserved C3 motif, indicating that there are important interactions between both the C- and A-domains and PCP- and C-domains (Symmank et al., 2002). Although some instances of novel *in vitro* dipeptide production with intradomain fusions were successful, such as the conversion of an internal elongation module into an initiation or termination module (Doekel and Marahiel, 2000) or swapping a Val-incorporating A-domain with an *N*-methyl Val domain from the actinomycin NRPS (Schauwecker et al., 2000); it was Marahiel and colleagues who suggested that CA-domains were an inseparable unit and that interactions may be hindered if incorrect fusion domains are created. Therefore they used whole module fusions to create tripeptides from the tyrocidine synthetase NRPS with a Te-domain for peptide release (Mootz et al., 2000). Using this technique, novel tripeptides were synthesized at rates comparable to native NRPSs.

Other attempts at synthesizing unnatural peptides from the tyrocidine synthetase system revealed important intermodule interactions that were controlled by communication-mediating (COM) domains (Hahn and Stachelhaus, 2004). These small regions of 15 to 30 amino acids are present between modules that prevent the incorrect production of compounds. Sequence similarity is not usually conserved between different COM domains, which ensures the integrity of natural production formation. However, this complicates combinatorial biosynthesis as compatible COM-domains are critical for effective compound production. This was shown by the shift in selectivity of interacting tyrocidine synthetase modules by point mutations in the COM-domains (Hahn and Stachelhaus, 2006). Due to sequence similarity between selected COM-domains, novel dipeptides were produced *in vitro* from interacting tyrocidine and gramicidin S synthetase modules (Stachelhaus et al., 1998), and tyrocidine and surfactin synthetase modules (Hahn and Stachelhaus, 2004). Furthermore, an artificial, *in vitro* synthetase was created using modules from the tyrocidine, bacitracin, and surfactin synthetases (Hahn and Stachelhaus, 2006). By using identical COM-domains for each module, multiple interactions were permitted, which resulted in the formation of both the expected novel tripeptide L-Phe–D-Orn–L-Leu as well as the unnatural dipeptide L-Phe–L-Leu.

The first example of NRPS tailoring domain bioengineering was achieved by Cy-domain replacements (Duerfahrt et al., 2004). In this study, an *in vitro* NRPS dimodule was formed by combining the first two modules of bacitracin A synthesis in *Bacillus licheniformis* ATCC10716 with the Te-domain of tyrocidine synthesis for peptide release. This dipeptide contained a Cy-domain for cyclization of the second amino acid to a thiazole ring. Module exchange with other Cy-modules was carried out to produce novel dipeptides *in vitro*, but with yields at 80-fold less than expressed by the native dimodule.

13.6.2 Biosynthesis of Natural and Unnatural Nonribosomal Peptides *in Vivo*

While manipulation of nonribosomal peptides *in vitro* has incorporated approaches such as site-directed mutagenesis, and domain and module swapping, addition, and deletions, *in vivo* expression relies on the genetic amenability of the host. Consequently, this has limited *in vivo* combinatorial biosynthesis efforts to a few species. Therefore, research has focused on producing the natural products in heterologous hosts, such as engineering *E. coli* and *P. putida* strains to synthesize the hybrid PKS/NRPS compounds yersiniabactin and myxothiazol, respectively (Eppelmann et al., 2001; Pfeifer et al., 2003; Wenzel et al., 2005).

Initial attempts at *in vivo* combinatorial biosynthesis were limited to novel surfactin biosynthesis in *Bacillus subtilis*. Later work has included the modification of NRPS domains and modules from pyoverdine synthesis in *Pseudomonas aeruginosa* and most recently for the production of daptomycin-derived compounds in *Steptomyces* species. However, as techniques improve for the genetic manipulation of a wider range of hosts, other NRPS systems will undoubtedly be targeted for rational design. A summary of successful *in vivo* combinatorial biosynthesis approaches is depicted in Figure 13.6.

13.6.2.1 Novel Surfactin-Derived Compounds

Much of the pioneering work on nonribosomal peptide combinatorial biosynthesis was attempted on the surfactin NRPS gene cluster. Surfactin synthetase contains seven NRPS modules in a colinear arrangement on three synthetase enzymes. A C-terminal Te-domain is responsible for termination and cyclization of the product. One approach for genetic manipulation of the surfactin synthetase was to swap and delete minimal modules (A- and PCP-domains). The first attempts involved the deletion of the Leu^7 minimal module from surfactin synthetase, which resulted in a truncated, linear,

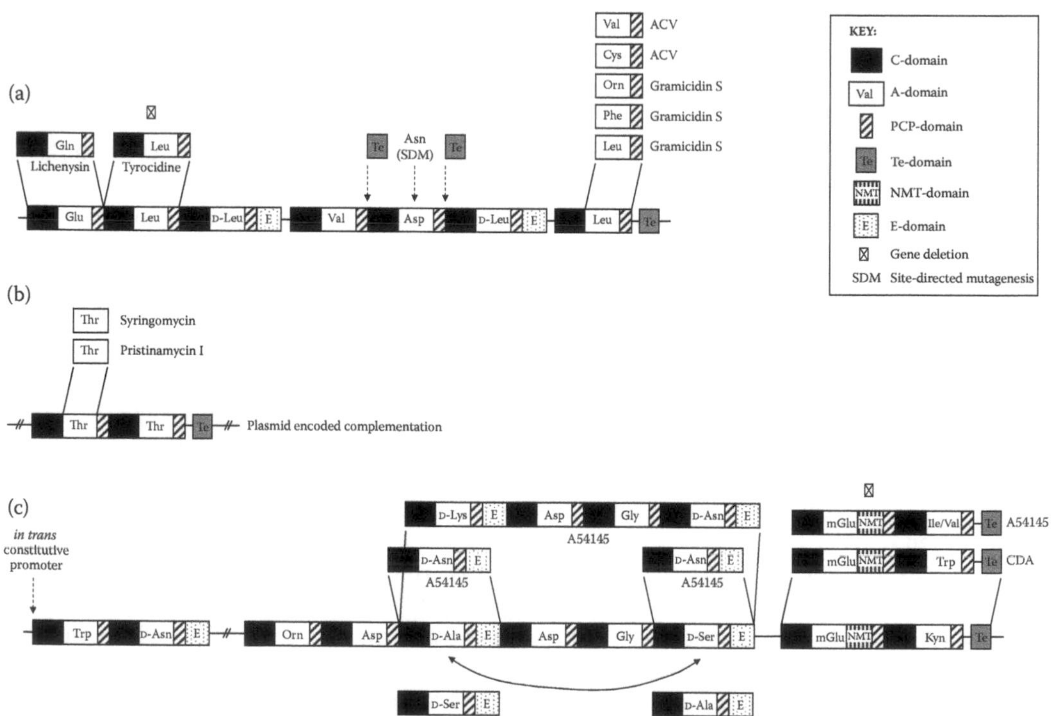

FIGURE 13.6 Summary of successful *in vivo* combinatorial biosynthesis approaches. (a) Manipulation of surfactin synthetase from *B. subtilis*. (b) Manipulation of pyoverdine synthetase from *P. aeruginosa* PAO1. The genomic *pvdD* gene, encoding two Thr-encoding modules and one Te-domain, was inactivated and subsequently complemented by plasmid encoded dimodules. (c) Manipulation of daptomycin synthetase from *S. roseosporus*. Replacement domains and modules originate from the gene clusters of the compounds indicated. A-domains from modules containing E-domains are depicted with their D-amino acid specificities despite actually activating L-residues before subsequent epimerization.

but inactive, product (Stachelhaus et al., 1995). However, substitutions of the minimal modules into Leu^7 modules, via homologous recombination, resulted in surfactins with different amino acids incorporated into the corresponding position. These included a heterologous Leu-activating module, noncognate Phe and Orn didomains from gramicidin S synthetase, and Cys- and Val-activation domains from *Penicillium chrysogenum* ACV synthetase. These modified surfactins were still active, although at lower levels than the wild type, due to the disruption of intermodular linkers that may have affected modular interactions. Later, the same group attempted to swap the Leu^2 minimal module from the surfactin synthetase. However, this time the resulting peptides were prematurely cyclized, resulting in weak activity or inactive derivates (Schneider et al., 1998). This may reflect the conserved Leu^2 residue in surfactin compared to the natural variability at the Leu^7 site, which can also be Val^7 (Baumgart et al., 1991; Peypoux et al., 1991).

An alternative strategy to produce novel surfactins was to move the C-terminal Te-domain to different module positions. This resulted in the production of truncated linear tetra- and pentapeptides at 23% and 9% yields compared to the cyclic heptapeptide parent compound (de Ferra et al., 1997). During this experiment, linker regions were identified in which additional modules could be inserted to potentially create novel compounds.

Whole module replacements, which were the most successful in *in vitro* studies, were also attempted between C-domain active sites *in vivo*. Using the template of the structurally similar lichenysin, Yakimov and colleagues swapped the Glu^1 residue to Gln^1, which resulted in a surfactin that retained high yields but possessed increased bioactivity similar to lichenysin (Yakimov et al., 2000). Similarly, Symmank et al. showed that fusion of modules in the conserved C-domain C3 motif allowed successful production of novel surfactin-derived compounds, however, yield and bioactivity were greatly reduced (Symmank et al., 2002). Other attempts included the successful exchange of the Leu^2 domain with another Leu activating domain from tyrocidine biosynthesis (Mootz et al., 2002a). This is in contrast to the dimodule swapping of the same domain, which did not restore surfactin synthesis effectively. The same group also deleted the entire module encoding for the incorporation of Leu^2 into the cyclic surfactin compound, resulting in the production of an active surfactin with reduced ring size. Although the Te-domain was able to macrocyclize and release this smaller unnatural product, there was a higher rate of hydrolysis and a dramatic reduction in production efficiency (Mootz et al., 2002a). These experiments showed that whole module exchanges were more effective than minimal module replacements. However, they also revealed the complexities of combinatorial biosynthesis as they were limited by either conservative module replacements or by reduced yield and bioactivity.

The most promising attempt at the combinatorial biosynthesis of surfactin used site-directed mutagenesis to create a *B. subtilis* mutant with altered amino acid specificity of the strictly conserved Asp^5 A-domain. This resulted in a novel surfactin with an Asn in the fifth residue position. Although this change is relatively conservative, naturally produced [Asn^5]surfactin has not yet been identified (Eppelmann et al., 2002). The limitless potential for combinatorial biosynthesis shows the value of this approach over biomining in the identification of novel compounds.

13.6.2.2 Novel Pyoverdine-Derived Compounds

Ackerley and Lamont (2004) were the first to use a system outside of surfactin biosynthesis to explore combinatorial biosynthesis. They engineered a *P. aeruginosa* pyoverdine PAO1 mutant lacking the NRPS gene, *pvdD*, which encodes for two Thr-incorporating modules and a Te-domain (Ackerley and Lamont, 2004). Heterologous plasmid-encoded Thr-incorporating A-domains from pristinamycin I and syringomycin synthetases were able to restore pyoverdine production (Figure 13.6b). However, noncognate A-domains, such as the Cys- and Val-incorporating A-domains from ACV synthetase and the Ser-incorporating domain from the pyoverdine gene cluster, were not able to restore pyoverdine synthesis. This was repeated with CA-didomain replacements; however, none were able to complement pyoverdine synthesis. This suggested that there are C-domain specificities for the donor amino acid as well as the acceptor residue in some instances.

13.6.2.3 Novel Daptomycin-Derived Compounds

The daptomycin group of antibiotics include daptomycin, calcium-dependent antibiotic (CDA), and A54145, which are all produced on NRPS enzymes with similar architecture (Figure 13.6c). Initial studies on developing novel daptomycin-related compounds focused on directed fermentations where the cultures were fed analogous substrates to yield novel CDA peptides with modified arylglycine residues, such as 4-fluorophenylglycine instead of 4-hydroxyphenylglycine (Hojati et al., 2002). More recently, novel CDA-derived compounds were developed via the mutation of the conserved serine of the NRPS PCP-domain, which prevented phosphopantetheinylation and resulted in the inactivation of this domain (Powell et al., 2007). Novel CDA compounds were then synthesized by feeding a variety of different substrates for the C-domain downstream of this mutated PCP-domain.

A multitude of approaches have also been underaken by Baltz and colleagues to develop novel daptomycin-related products with comparable yields and antimicrobial activity. These involved the replacement of the daptomycin synthetase genes with similar genes from related A54145 and CDA gene clusters from *Streptomyces* species. One method used *in trans* chromosomal complementation of an inactivated NRPS gene, *dptD*, which encodes two modules for the incorporation of mGlu12 and the unusual amino acid kynurenine13 (Kyn13), as well as a Te-domain. The first module encoded by *dptD* is responsible for the incorporation of mGlu in all three NRPSs, while the second module incorporates Ile or Val in A54145 and Trp in CDA (Miao et al., 2006). The yields for novel daptomycin production were 25% and 50% that of the wild-type strain, and both compounds retained antimicrobial activity. Similarly, the same group also inactivated the first daptomycin NRPS gene, *dptA*, and replaced it with its own gene placed *in trans* under the control of a strong constitutive expression promoter. This increased yields for the previous mutants to 40% and 69% (Coeffet-Le Gal et al., 2006).

Another method used to develop novel daptomycin-related compounds was the rearrangement of entire modules within the NRPS gene cluster. The λ-Red-mediated recombination approach was used whereby the target module, encoded on a plasmid, was replaced with an antibiotic resistance cassette. The cassette was exchanged with other modules and was then used to replace the original NRPS gene cluster *in vivo*, via homologous recombination of the flanking regions. (Datsenko and Wanner, 2000). This technique was used to swap the D-Ala^{8} residue for D-Ser^{11} and vice versa. Compound expression was approximately 60% and 20% of the original expression and both displayed antimicrobial activity (Nguyen et al., 2006). The same technique was also used to incorporate heterologous D-Asn modules from A54145 synthesis, which showed that the native E-domains downstream of the D-Ala^{8} and D-Ser^{11} positions were able to catalyse the incorporation of D-Asn into those sites.

Other combinatorial biosynthesis approaches included the deletion of the NMT-domain responsible for methylation of the mGlu^{12} residue, which resulted in an unmethylated residue at that position (Nguyen et al., 2006). Although this novel daptomycin derivative had the lowest bioactivity of those developed, it also had a significantly lower LD_{50} in mouse bioassays. Therefore, it might be of use in clinical settings if coupled with the manipulation of another module to restore activity. Finally, the same group replaced the genes for four of the modules from the daptomcyin NRPS gene, *dptBC*, with the homologous genes in A54145 synthesis. This resulted in the production of a daptomycin analogue with D-Ala/D-Lys and D-Ser/D-Asn substitutions. Although this novel compound was produced at low yields, its bioactivity was comparable to wildtype daptomycin (Nguyen et al., 2006). The Baltz group have developed over 100 novel daptomycin-derived compounds by using a variety of combinatorial biosynthesis approaches combined with the naturally occurring derivatives from directed fermentation (Baltz et al., 2006; Nguyen et al., 2006). This research has provided the platform for the development of future compounds for use in clinical settings.

13.7 CONCLUSIONS

The quest so far for the development of novel nonribosomal peptides via combinatorial biosynthesis highlights the complexities involved in this approach. This reflects the diverse and complex structure and function of nonribosomal peptides, despite their synthesis by repeated modular enzymes. The most successful combinatorial methods altered A-domain specificities or replaced modules with relatively conserved amino acid changes. However, in order to harness the structural and functional diversity of these compounds, further work is required to fully understand the mechanisms involved in NRPS domain and module interactions. The ever increasing numbers of characterized NRPS gene clusters will facilitate this aim and will enable more effective rational design of novel compounds with increased yields and activities.

ACKNOWLEDGMENT

The authors thank David Roberts for editorial comments on this manuscript.

REFERENCES

Ackerley, D.F. and Lamont, I.L., Characterization and genetic manipulation of peptide synthetases in *Pseudomonas aeruginosa* PAO1 in order to generate novel pyoverdines, *Chem. Biol.*, 11, 971–980, 2004.

Agata, N. et al., A novel dodecadepsipeptide, cereulide, isolated from *Bacillus cereus* causes vacuole formation in HEp-2 cells, *FEMS Microbiol. Lett.*, 121, 31–34, 1994.

Agata, N., Ohta, M., Mori, M., and Isobe, M., A novel dodecadepsipeptide, cereulide, is an emetic toxin of *Bacillus cereus*, *FEMS Microbiol. Lett.*, 129, 17–19, 1995.

Agnoli, K. et al., The ornibactin biosynthesis and transport genes of *Burkholderia cenocepacia* are regulated by an extracytoplasmic function σ factor which is a part of the Fur regulon, *J. Bacteriol.*, 188, 3631–3644, 2006.

Aguilera, S. et al., Functional characterization of the gene cluster from *Pseudomonas syringae* pv. *phaseolicola* NPS3121 involved in synthesis of phaseolotoxin, *J. Bacteriol.*, 189, 2834–2843, 2007.

Aharonowitz, Y., Cohen, G., and Martin, J.F., Penicillin and cephalosporin biosynthetic genes: Structure, organization, regulation, and evolution, *Annu. Rev. Microbiol.*, 46, 461–495, 1992.

Alborn, W.E., Jr, Allen, N.E., Preston, D.A., Daptomycin disrupts membrane potential in growing *Staphylococcus aureus*, *Antimicrob. Agents Chemother.*, 35, 2282–2287, 1991.

An, J. and Carmichael, W.W., Use of a colorimetric protein phosphatase inhibition assay and enzyme linked immunosorbent assay for the study of microcystins and nodularins, *Toxicon* 32, 1495–1507, 1994.

Arbeit, R.D. et al., The safety and efficacy of daptomycin for the treatment of complicated skin and skin-structure infections, *Clin. Infect. Dis.*, 38, 1673–1681, 2004.

Arima, K., Kakinuma, A., and Tamura, G., Surfactin, a crystalline peptidelipid surfactant produced by *Bacillus subtilis*: Isolation, characterization and its inhibition of fibrin clot formation, *Biochem. Biophys. Res. Commun.*, 31, 488–494, 1968.

Arnold, D.L., Pitman, A., and Jackson, R., Pathogenicity and other genomic islands in plant pathogenic bacteria, *Mol. Plant Pathol.*, 4, 407–420, 2003.

Arolas, J.L. et al., The three-dimensional structures of tick carboxypeptidase inhibitor in complex with A/B carboxypeptidases reveal a novel double-headed binding mode, *J. Mol. Biol.*, 350, 489–498, 2005.

Aso, M., Kondo, M., Suemune, H., and Hecht, S.M., Chemistry of the bleomycin-induced alkali-labile DNA lesion, *J. Am. Chem. Soc.*, 121, 9023–9033, 1999.

Audenaert, K., Pattery, T., Cornelis, P., and Hofte, M., Induction of systemic resistance to *Botrytis cinerea* in tomato by *Pseudomonas aeruginosa* 7NSK2: role of salicylic acid, pyochelin, and pyocyanin, *Mol. Plant–Microbe Interact.*, 15, 1147–1156, 2002.

Bagg, A. and Neilands, J.B. Molecular mechanisms of regulation of siderophore-mediated iron assimilation, *Microbiol. Rev.*, 51, 509–518, 1987.

Bailey, S.A., Graves, D.E., and Rill, R., Binding of actinomycin D to the T(G)nT motif of double-stranded DNA: Determination of the guanine requirement in nonclassical, non-GpC binding sites, *Biochemistry*, 33, 11493–11500, 1994.

Bains, P.S. and Tewari, J.P., Purification, chemical characterization and host-specificity of the toxin produced by *Alternaria brassicae*, *Physiol. Mol. Plant Pathol.*, 30, 259–271, 1987.

Baker, M.A., Grubb, D.R., and Lawen, A., Didemnin B induces apoptosis in proliferating but not resting peripheral blood mononuclear cells, *Apoptosis*, 7, 407–412, 2002.

Balibar, C.J., Vaillancourt, F.H., and Walsh, C.T., Generation of D amino acid residues in assembly of arthrofactin by dual condensation/epimerization domains, *Chem. Biol.*, 12, 1189–1200, 2005.

Ballio, A. et al., Structure of fuscopeptins, phytotoxic metabolites of *Pseudomonas fuscovaginae*, *FEBS Lett.*, 381, 213–216, 1996.

Baltz, R., Combinatorial glycosylation of glycopeptide antibiotics, *Chem. Biol.*, 9, 1268–1270, 2002.

Baltz, R., Brian, P., Miao, V., and Wrigley, S., Combinatorial biosynthesis of lipopeptide antibiotics in *Streptomyces roseosporus*, *J. Ind. Microbiol. Biotechnol.*, 33, 66–74, 2006.

Bassarello, C. et al., Tolaasins A–E, five new lipodepsipeptides produced by *Pseudomonas tolaasii*, *J. Nat. Prod.*, 67, 811–816, 2004.

Basu, A., Kozikowski, A.P., Sato, K., and Lazo, J.S., Cellular sensitization to *cis*-diaminedichloroplatinum(II) by novel analogues of the protein kinase C activator lyngbyatoxin A, *Cancer Res.*, 51, 2511–2514, 1991.

Bates, R.B. et al., Dolastatins. 26. Synthesis and stereochemistry of dolastatin 11, *J. Am. Chem. Soc.*, 119, 2111–2113, 1997.

Baumgart, F. et al., Identification of amino acid substitutions in the lipopeptide surfactin using 2D NMR spectroscopy, *Biochem. Biophys. Res. Commun.*, 177, 998–1005, 1991.

Becker, J.E., Moore, R.E., and Moore, B.S., Cloning, sequencing, and biochemical characterization of the nostocyclopeptide biosynthetic gene cluster: Molecular basis for imine macrocyclization, *Gene*, 325, 35–42, 2004.

Behrens, O.K. et al., Biosynthesis of penicillins: Biological precursors for benzylpenicillin (penicillin G), *J. Biol. Chem.*, 175, 751–764, 1948.

Belshaw, P.J., Walsh, C.T., and Stachelhaus, T., Aminoacyl-CoAs as probes of condensation domain selectivity in nonribosomal peptide synthesis, *Science*, 284, 486–489, 1999.

Bender, C.L., Alarcon-Chaidez, F., and Gross, D.C., *Pseudomonas syringae* phytotoxins: Mode of action, regulation, and biosynthesis by peptide and polyketide synthetases, *Microbiol. Mol. Biol. Rev.*, 63, 266–292, 1999.

Bentley, S.D. et al., Complete genome sequence of the model actinomycete *Streptomyces coelicolor* A3(2), *Nature*, 417, 141–147, 2002.

Bergendahl, V., Linne, U., and Marahiel, M.A., Mutational analysis of the C-domain in nonribosomal peptide synthesis, *Eur. J. Biochem.*, 269, 620–629, 2002.

Beven, L. and Wroblewski, H., Effect of natural amphipathic peptides on viability, membrane potential, cell shape and motility of mollicutes, *Res. Microbiol.*, 148, 163–175, 1997.

Billich, A. and Zocher, R., N-Methyltransferase function of the multifunctional enzyme enniatin synthetase, *Biochemistry*, 26, 8417–8423, 1987.

Bischoff, D. et al., The biosynthesis of vancomycin-type glycopeptide antibiotics—a model for oxidative side-chain cross-linking by oxygenases coupled to the action of peptide synthetases, *ChemBioChem*, 6, 267–272, 2005.

Boger, D.L., Vancomycin, teicoplanin, and ramoplanin: Synthetic and mechanistic studies, *Med. Res. Rev.*, 21, 356–381, 2001.

Braun, P.G., Hildebrand, P.D., Ells, T.C., and Kobayashi, D.Y., Evidence and characterization of a gene cluster required for the production of viscosin, a lipopeptide biosurfactant, by a strain of *Pseudomonas fluorescens*, *Can. J. Microbiol.*, 47, 294–301, 2001.

Brosch, G. et al., Inhibition of maize histone deacetylases by HC toxin, the host-selective toxin of *Cochliobolus carbonum*, *Plant Cell*, 7, 1941–1950, 1995.

Burja, A.M. et al., Marine cyanobacteria—a prolific source of natural products, *Tetrahedron*, 57, 9347–9377, 2001.

Bycroft, B.W., Configurational and conformational studies on the group A peptide antibiotics of the mikamycin (streptogramin, virginiamycin) family, *J. Chem. Soc.*, 1, 2464–2470, 1977.

Cane, D.E. and Walsh, C.T., The parallel and convergent universes of polyketide synthases and nonribosomal peptide synthetases, *Chem. Biol.*, 6, 319–325, 1999.

Canepari, P., Boaretti, M., Lleo, M.M., and Satta, G., Lipoteichoic acid as a new target for activity of antibiotics: Mode of action of daptomycin (LY146032), *Antimicrob. Agents Chemother.*, 34, 1220–1226, 1990.

Carmichael, W.W., Cyanobacteria secondary metabolites—the cyanotoxins, *J. Appl. Bacteriol.*, 72, 445–459, 1992.

Carmichael, W.W. et al., Human fatalities from cyanobacteria: Chemical and biological evidence for cyanotoxins, *Environ. Health. Perspect.*, 109, 663–668, 2001.

Carroll, A.R. et al., Dysinosin A: A novel inhibitor of factor VIIa and thrombin from a new genus and species of Australian sponge of the family Dysideidae, *J. Am. Chem. Soc.*, 124, 13340–13341, 2002.

Cavelier F., Jacquier, R., Mercadier, J.-L., and Verducci, J., Destruxin analogues: Depsi peptidic bond replacement by amide bond, *Tetrahedron*, 52, 6173–6186, 1996.

Cebrat, M., Wieczorek, Z., and Siemion, I.Z., Immunosuppressive activity of hymenistatin I, *Peptides*, 17, 191–196, 1996.

Chang, Z. et al., The barbamide biosynthetic gene cluster: A novel marine cyanobacterial system of mixed polyketide synthase (PKS)-non-ribosomal peptide synthetase (NRPS) origin involving an unusual trichloroleucyl starter unit, *Gene*, 296, 235–247, 2002.

Chang, Z. et al., Biosynthetic pathway and gene cluster analysis of curacin A, an antitubulin natural product from the tropical marine cyanobacterium *Lyngbya majuscula*, *J. Nat. Prod.*, 67, 1356–1367, 2004.

Chatterjee, A.N. and Perkins, H.R., Compounds formed between nucleotides related to the biosynthesis of bacterial cell wall and vancomycin, *Biochem. Biophys. Res. Commun.*, 24, 489–494, 1966.

Chen, J. and Stubbe, J., Bleomycins: Towards better therapeutics, *Nat. Rev. Cancer*, 5, 102–112, 2005.

Cheng, Y.-Q., Deciphering the biosynthetic codes for the potent anti-SARS-CoV cyclodepsipeptide valinomycin in *Streptomyces tsusimaensis* ATCC 15141, *ChemBioChem*, 7, 471–477, 2006.

Christiansen, G. et al., Microcystin biosynthesis in *Planktothrix*: Genes, evolution and manipulation, *J. Bacteriol.*, 185, 564–572, 2003.

Chui, H.-T. et al., Molecular cloning and sequence analysis of the complestatin biosynthetic gene cluster, *Proc. Natl Acad. Sci. USA*, 98, 8548–8553, 2001.

Coates, A., Hu, Y., Bax, R., and Page, C., The future challenges facing the development of new antimicrobial drugs, *Nat. Rev. Drug Discov.*, 1, 895–910, 2002.

Cocito, C., Antibiotics of the virginiamycin family, inhibitors which contain synergistic components, *Microbiol. Rev.*, 43, 145–198, 1979.

Coeffet-Le Gal, M.-F. et al., Complementation of daptomycin *dptA* and *dptD* deletion mutations *in trans* and production of hybrid lipopeptide antibiotics, *Microbiology*, 152, 2993–3001, 2006.

Conti, E., Stachelhaus, T., Marahiel, M.A., and Brick, P., Structural basis for the activation of phenylalanine in the non-ribosomal biosynthesis of gramicidin S, *EMBO J.*, 16, 4174–4183, 1997.

Cooper, R.D.G. et al., Reductive alkylation of glycopeptide antibiotics: Synthesis and antibacterial activity, *J. Antibiot.*, 49, 575–581, 1996.

Cornelis, P. and Matthijs, S., Diversity of siderophore-mediated iron uptake systems in fluorescent pseudomonads: Not only pyoverdines, *Environ. Microbiol.*, 4, 787–798, 2002.

Cotton, M., Xu, F., and Cross, T.A., Protein stability and conformational rearrangements in lipid bilayers: Linear gramicidin, a model system, *Biophys. J.*, 73, 614–623, 1997.

Couch, R. et al., Characterization of CmaA, an adenylation-thiolation didomain enzyme involved in the biosynthesis of coronatine, *J. Bacteriol.*, 186, 35–42, 2004.

Covens, A. et al., Phase II trial of pulse dactinomycin as salvage therapy for failed low-risk gestational trophoblastic neoplasia, *Cancer*, 107, 1280–1286, 2006.

Crews, C.M. et al., GTP-dependent binding of the antiproliferative agent didemnin to elongation factor 1 alpha, *J. Biol. Chem.*, 269, 15411–15414, 1994.

D'Agostino, G. et al., A multicenter phase II study of the cryptophycin analog LY355703 in patients with platinum-resistant ovarian cancer, *Int. J. Gynecol. Cancer*, 16, 71–76, 2006.

Datsenko, K.A. and Wanner, B.L., One-step inactivation of chromosomal genes in *Escherichia coli* K-12 using PCR products, *Proc. Natl Acad. Sci. USA*, 97, 6640–6645, 2000.

Dawson, R.M., The toxicology of microcystins, *Toxicon*, 36, 953–962, 1998.

de Ferra F. et al., Key role of the thioesterase-like domain for the efficient production of recombinant peptides, *J. Biol. Chem.*, 272, 25304–25309, 1997.

Di Lorenzo, M.S. et al., A nonribosomal peptide synthetase with a novel domain organization is essential for siderophore biosynthesis in *Vibrio anguillarum*, *J. Bacteriol.*, 186, 7327–7336, 2004.

Diez, B. et al., The cluster of penicillin biosynthetic genes. Identification and characterization of the *pcbAB* gene encoding the alpha-aminoadipyl-cysteinyl-valine synthetase and linkage to the *pcbC* and *penDE* genes, *J. Biol. Chem.*, 265, 16358–16365, 1990.

Doekel, S. and Marahiel, M.A., Dipeptide formation on engineered hybrid peptide synthetases, *Chem. Biol.*, 7, 373–384, 2000.

Donadio, S., Monciardini, P., and Sosio, M., Polyketide synthases and nonribosomal peptide synthetases: The emerging view from bacterial genomics, *Nat. Prod. Rep.*, 24, 1073–1109, 2007.

Doyle, F.P., Nayler, J.H.C., Smith, H., and Stove, E.R., Some novel acid-stable penicillins, *Nature*, 191, 1091–1092, 1961.

Du, L. et al., The biosynthetic gene cluster for the antitumor drug bleomycin from *Streptomyces verticillus* ATCC15003 supporting functional interactions between nonribosomal peptide synthetases and a polyketide synthase, *Chem. Biol.*, 7, 623–642, 2000.

Du, L., Sanchez, C., and Shen, B., Hybrid peptide–polyketide natural products: Biosynthesis and prospects toward engineering novel molecules, *Metab. Eng.*, 3, 78–95, 2001.

Duerfahrt, T., Eppelmann, K., Müller, R., and Marahiel, M.A., Rational design of a bimodular model system for the investigation of heterocyclisation in nonribosomal peptide biosynthesis, *Chem. Biol.*, 11, 261–271, 2004.

Duitman, E.H. et al., The mycosubtilin synthetase of *Bacillus subtilis* ATCC6633: A multifunctional hybrid between a peptide synthetase, an amino transferase, and a fatty acid synthase, *Proc. Natl Acad. Sci. USA*, 96, 13294–13299, 1999.

Duncan, N. and Craddock, C., Optimizing the use of cyclosporin in allogeneic stem cell transplantation, *Bone Marrow Transplant*, 38, 169–174, 2006.

Dunlap, W.C. et al., Biomedicinals from the phytosymbionts of marine invertebrates: A molecular approach, *Methods*, 42, 358–376, 2007.

Edwards, D.J. and Gerwick, W.H., Lyngbyatoxin biosynthesis: Sequence of biosynthetic gene cluster and identification of a novel aromatic prenyltransferase, *J. Am. Chem. Soc.*, 126, 11432–11433, 2004.

Edwards, D.J. et al., Structure and biosynthesis of the jamaicamides, new mixed polyketide–peptide neurotoxins from the marine cyanobacterium *Lyngbya majuscula*, *Chem. Biol.*, 11, 817–833, 2004.

Ehmann, D.E., Trauger, J.W., Stachelhaus, T., and Walsh, C.T., Aminoacyl-SNACs as small-molecule substrates for the condensation domains of nonribosomal peptide synthetases, *Chem. Biol.*, 7, 765–772, 2000.

Einhorn, L.H., Curing metastatic testicular cancer, *Proc. Natl Acad. Sci. USA*, 99, 4592–4595, 2002.

Emanuele, M.C. et al., Corceptins, new bioactive lipodepsipeptides from cultures of *Pseudomonas corrugata*, *FEBS Lett.*, 433, 317–320, 1998.

Eppelmann, K., Doekel, S., and Marahiel, M.A., Engineered biosynthesis of the peptide antibiotic bacitracin in the surrogate host *Bacillus subtilis*, *J. Biol. Chem.*, 276, 34824–34831, 2001.

Eppelmann, K., Stachelhaus, T., and Marahiel, M.A., Exploitation of the selectivity-conferring code of nonribosomal peptide synthetases for the rational design of novel peptide antibiotics, *Biochemistry*, 41, 9718–9726, 2002.

Ferguson, A.R. and Johnston, J.S., Phaseolotoxin: Chlorosis, ornithine accumulation and inhibition of ornithine carbamoyltransferase in different plants, *Physiol. Plant Pathol.*, 16, 269–275, 1980.

Ferguson, A.R., Johnston, J.S., and Mitchell, R.E., Resistance of *Pseudomonas syringae* pv, *phaseolicola* to its own toxin, phaseolotoxin, *FEMS Microbiol. Lett.*, 7, 123–125, 1980.

Finking, R. and Marahiel, M., Biosynthesis of nonribosomal peptides, *Annu. Rev. Microbiol.*, 58, 453–488, 2004.

Fumoleau, P., Coudert, B., Isambert, N., and Ferrant, E., Novel tubulin-targeting agents: Anticancer activity and pharmacologic profile of epothilones and related analogues, *Ann. Oncol.*, 18, 9–15, 2007.

García-Rocha, M., Bonay, P., and Avila, J., The antitumoral compound kahalalide F acts on cell lysosomes, *Cancer Lett.*, 99, 43–50, 1996.

Gehring, A., M., Mori, I., and Walsh, C.T., Reconstitution and characterization of the *Escherichia coli* enterobactin synthetase from EntB, EntE, and EntF, *Biochemistry*, 37, 2648–2659, 1998.

Gehringer, M.M., Microcystin-LR and okadaic acid-induced cellular effects: A dualistic response, *FEBS Lett.*, 557, 1–8, 2004.

Gerard, J., et al., Massetolides A-H, antimycobacterial cyclic depsipeptides produced by two Pseudomonads isolated from marine habitats, *J. Nat. Prod.*, 60, 223–229, 1997.

Gevers, W., Kleinkauf, H., and Lipmann, F., Peptidyl transfers in gramicidin S biosynthesis from enzyme-bound thioester intermediates, *Proc. Nat Acad. Sci. USA*, 63,1335–1342, 1969.

Govaerts, C. et al., Structure elucidation of four related substances in gramicidin with liquid chromatography/mass spectrometry, *Rapid Commun. Mass Spectrom.*, 15, 128–134, 2001.

Grangemard, I. et al., Lipopeptides with improved properties: Structure by NMR, purification by HPLC and structure–activity relationships of new isoleucyl-rich surfactins, *J. Pept. Sci.*, 3, 145–154, 1997.

Griffiths, G.L., Sigel, S.P., Payne, S.M., and Neilands, J.B., Vibriobactin, a siderophore from *Vibrio cholerae*. *J. Biol. Chem.*, 259, 383–385, 1984.

Grünewald, J. and Marahiel, M.A., Chemoenzymatic and template-directed synthesis of bioactive macrocyclic peptides, *Microbiol. Mol. Biol. Rev.*, 70, 121–146, 2006.

Guenzi, E. et al., Characterization of the syringomycin synthetase gene cluster. A link between prokaryotic and eukaryotic peptide synthetases, *J. Biol. Chem.*, 273, 32857–32863, 1998.

Hahn, M. and Stachelhaus, T., Selective interaction between nonribosomal peptide synthetases is facilitated by short communication-mediating domains, *Proc. Natl Acad. Sci. USA*, 101, 15585–15590, 2004.

Hahn, M. and Stachelhaus, T., Harnessing the potential of communication-mediating domains for the biocombinatorial synthesis of nonribosomal peptides, *Proc. Natl Acad. Sci. USA*, 103, 275–280, 2006.

Halloin, J.M. and Hagedorn, D.J., Effects of tentoxin on enzymic activities in cucumber and cabbage cotyledons, *Mycopathologia*, 55, 159–162, 1975.

Hamada, T., Matsunaga, S., Yano, G., and Fusetani, N., Polytheonamides A and B, highly cytotoxic, linear polypeptides with unprecedented structural features, from the marine sponge, *Theonella swinhoei*, *J. Am. Chem. Soc.*, 127, 110–118, 2005.

Hanessian, S., Tremblay, M., and Petersen, J.F.W., The *N*-acyloxyiminium ion aza-Prins route to octahydroindoles: Total synthesis and structural confirmation of the antithrombotic marine natural product oscillarin, *J. Am. Chem. Soc.*, 126, 6064–6071, 2004.

Harrigan, G.G. et al., Isolation, structure determination, and biological activity of dolastatin 12 and lyngbyastatin 1 from *Lyngbya majuscula*/*Schizothrix calcicola* cyanobacterial assemblages, *J. Nat. Prod.*, 61, 1221–1225, 1998.

Hashimi, S.M. et al., The phytotoxin albicidin is a novel inhibitor of DNA gyrase, *Antimicrob. Agents Chemother.*, 51, 181–187, 2007.

Hoffmann, D., Hevel, J.M., Moore, R.E., and Moore, B.S., Sequence analysis and biochemical characterization of the nostopeptolide A biosynthetic gene cluster from *Nostoc* sp. GSV224, *Gene*, 311, 171–180, 2003.

Hoffmann, K., Schneider-Scherzer, E., Kleinkauf, H., and Zocher, R., Purification and characterization of eucaryotic alanine racemase acting as key enzyme in cyclosporin biosynthesis, *J. Biol. Chem.*, 269, 12710–12714, 1994.

Hojati, Z. et al., Structure, biosynthetic origin, and engineered biosynthesis of calcium-dependent antibiotics from *Streptomyces coelicolor*, *Chem. Biol.*, 9, 1175–1187, 2002.

Itou, Y., Suzuki, S., Ishida, K., and Murakami, M., Anabaenopeptins G and H, potent carboxypeptidase A inhibitors from the cyanobacterium *Oscillatoria agardhii* (NIES-595), *Bioorg. Med. Chem. Lett.*, 9, 1243–1246, 1999.

Jegorov, A., Hajduch, M., Sulc, M., and Havlicek, V., Nonribosomal cyclic peptides: Specific markers of fungal infections, *J. Mass Spectrom.*, 41, 563–576, 2006.

Johansson, F.I., Michalecka, A.M., Møller, I.M., and Rasmusson, A.G., Oxidation and reduction of pyridine nucleotides in alamethicin-permeabilized plant mitochondria, *Biochem. J.*, 380, 193–202, 2004.

Kadono, S. et al., Structure of human factor VIIa/tissue factor in complex with a peptide-mimetic inhibitor: High selectivity against thrombin by introducing two charged groups in P2 and P4, *Acta Crystallogr. Sect. F*, 61, 169–173, 2005.

Kahne, D., Leimkuhler, C., Lu, W., and Walsh, C., Glycopeptide and lipoglycopeptide antibiotics, *Chem. Rev.*, 105, 425–448, 2005.

Kalemkerian, G.P. et al., Activity of dolastatin 10 against small-cell lung cancer *in vitro* and *in vivo*: Induction of apoptosis and bcl-2 modification, *Cancer Chemother. Pharmacol.*, 43, 507–515, 1999.

Kallow, W., von Dohren, H., and Kleinkauf, H., Penicillin biosynthesis: Energy requirement for tripeptide precursor formation by δ-(L-α-aminoadipyl)-L-cysteinyl-D-valine synthetase from *Acremonium chrysogenum*, *Biochemistry*, 37, 5947–5952, 1998.

Kasahara, T. and Kato, T., A new redox-cofactor vitamin for mammals, *Nature*, 422, 832, 2003.

Keating, T.A., Marshall, C.G., and Walsh, C.T., Reconstitution and characterization of the *Vibrio cholerae* vibriobactin synthetase from VibB, VibE, VibF, and VibH, *Biochemistry*, 39, 15522–15530, 2000.

Keating, T.A., Marshall, C.G., Walsh, C.T., and Keating, A.E., The structure of VibH represents nonribosomal peptide synthetase condensation, cyclization and epimerization domains, *Nature Struct. Biol.*, 9, 522–526, 2002.

Keller, U. and Schauwecker, F., Combinatorial biosynthesis of nonribosomal peptides, *Comb. Chem. High Throughput Screen*, 6, 527–540, 2003.

Kempter, C. et al., CDA: Calcium-dependent peptide antibiotics from *Streptomyces coelicolor* A3(2) containing unusual residues, *Angew. Chem. Int. Ed.*, 36, 498–501, 1997.

Kershaw, M.J. et al., The role of destruxins in the pathogenicity of *Metarhizium anisopliae* for three species of insect, *J. Invertebr. Pathol.*, 74, 213–223, 1999.

Kessler, N. et al., The linear pentadecapeptide gramicidin is assembled by four multimodular nonribosomal peptide synthetases that comprise 16 modules with 56 catalytic domains, *J. Biol. Chem.*, 279, 7413–7419, 2004.

Kleinkauf, H. and Dohren, H., Nonribosomal biosynthesis of peptide antibiotics, *Eur. J. Biochem.*, 192, 1–15, 1990.

Kleinkauf, H., Roskoski, R., and Lipmann, F., Pantetheine-linked peptide intermediates in gramicidin S and tyrocidine biosynthesis, *Proc. Natl Acad. Sci. USA*, 68, 2069–2072, 1971.

Kobayashi, J.I. et al., Keramamides B–D: Novel peptides from the Okinawan marine sponge *Theonella* sp, *J. Am. Chem. Soc.*, 113, 7812–7813, 1991.

Kobayashi, M. et al., Antitumor activity of TZT-1027, a novel dolastatin 10 derivative, *Cancer Sci.*, 88, 316–327, 1997.

Kodani, S., Suzuki, S., Ishida, K., and Murakami, M., Five new cyanobacterial peptides from water bloom materials of lake Teganuma (Japan), *FEMS Microbiol. Lett.*, 178, 343–348, 1999.

Kohli, R.M., Takagi, J., and Walsh, C.T., The thioesterase domain from a nonribosomal peptide synthetase as a cyclization catalyst for integrin binding peptides, *Proc. Natl Acad. Sci. USA*, 99, 1247–1252, 2002.

Konz, D., Klens, A., Schörgendorfer, K., and Marahiel, M.A., The bacitracin biosynthesis operon of *Bacillus licheniformis* ATCC 10716: Molecular characterization of three multi-modular peptide synthetases, *Chem. Biol.*, 4, 927–937, 1997.

Koumoutsi, A. et al., Structural and functional characterization of gene clusters directing nonribosomal synthesis of bioactive cyclic lipopeptides in *Bacillus amyloliquefaciens* strain FZB42, *J. Bacteriol.*, 186, 1084–1096, 2004.

Kowalski, R.J., Giannakakou, P., and Hamel, E., Activities of the microtubule-stabilizing agents epothilones A and B with purified tubulin and in cells resistant to paclitaxel (Taxol®), *J. Biol. Chem.*, 272, 2534–2541, 1997.

Kramer, D., *Microginin Producing Proteins and Nucleic Acids Encoding a Microginin Gene Cluster as well as Methods for Creating Microginins*, W.I.P. Organization, Germany, 2007.

Krause, M. and Marahiel, M.A., Organization of the biosynthesis genes for the peptide antibiotic gramicidin S, *J. Bacteriol.*, 170, 4669–4674, 1988.

Krug, L.M. et al., Phase II study of dolastatin-10 in patients with advanced non-small-cell lung cancer, *Ann. Oncol.*, 11, 227–228, 2000.

Kuiper, I. et al., Characterization of two *Pseudomonas putida* lipopeptide biosurfactants, putisolvin I and II, which inhibit biofilm formation and break down existing biofilms, *Mol. Microbiol.*, 51, 97–113, 2004.

Lakey, J.H. et al., A new channel-forming antibiotic from *Streptomyces coelicolor* A3(2) which requires calcium for its activity, *J. Gen. Microbiol.*, 129, 3565–3573, 1983.

Lambalot, R.H. et al., A new enzyme superfamily—the phosphopantetheinyl transferases, *Chem. Biol.*, 3, 923–936, 1996.

Lamont, I.L. et al., Siderophore-mediated signaling regulates virulence factor production in *Pseudomonas aeruginosa*, *Proc. Natl Acad. Sci. USA*, 99, 7072–7077, 2002.

Lamont, I.L. and Martin, L.W., Identification and characterization of novel pyoverdine synthesis genes in *Pseudomonas aeruginosa*, *Microbiology*, 149, 833–842, 2003.

Lawen, A. and Zocher, R., Cyclosporin synthetase. The most complex peptide synthesizing multienzyme polypeptide so far described, *J. Biol. Chem.*, 265, 11355–11360, 1990.

Lee, S.G. and Lipmann, F., The tyrocidine synthetase system, *Methods Enzymol.*, 43, 585–602, 1975.

Leitgeb, B. et al., The history of alamethicin: A review of the most extensively studied peptaibol, *Chem. Biodivers.*, 4, 1027–1051, 2007.

Lemanceau, P. et al., Effect of pseudobactin 358 production by *Pseudomonas putida* WCS358 on suppression of fusarium wilt of carnations by nonpathogenic *Fusarium oxysporum* Fo47, *Appl. Environ. Microbiol.*, 58, 2978–2982, 1992.

Lindlow, S.E. and Brandl, M.T., Microbiology of the phyllosphere, *Appl. Environ. Microbiol.*, 69, 1875–1883, 2003.

Liyanage, H., Penfold, C., Turner, J., and Bender, C.L., Sequence, expression and transcriptional analysis of the coronafacate ligase-encoding gene required for coronatine biosynthesis by *Pseudomonas syringae*, *Gene*, 153, 17–23, 1995.

Lo, M.C. et al., A new mechanism of action proposed for ramoplanin, *J. Am. Chem. Soc.*, 122, 3540–3541, 2000.

Lobo, C., García-Pozo, S.G., Núñez de Castro, I., and Alonso, F.J., Effect of dehydrodidemnin B on human colon carcinoma cell lines, *Anticancer Res.*, 17, 333–336, 1997.

Luesch, H., Harrigan, G.G., Goetz, G., and Horgen, F.D., The cyanobacterial origin of potent anticancer agents originally isolated from sea hares, *Curr. Med. Chem.*, 9, 1791–1806, 2002.

Luesch, H. et al., Isolation of dolastatin 10 from the marine cyanobacterium *Symploca* species VP642 and total stereochemistry and biological evaluation of its analogue symplostatin 1, *J. Nat. Prod.*, 64, 907–910, 2001.

Luesch, H., Yoshida, W.Y., Moore, R.E., and Paul, V.J., Lyngbyastatin 2 and norlyngbyastatin 2, analogues of dolastatin G and nordolastatin G from the marine cyanobacterium *Lyngbya majuscula*, *J. Nat. Prod.*, 62, 1702–1706, 1999.

Luo, L. et al., Timing of epimerization and condensation reactions in nonribosomal peptide assembly lines: Kinetic analysis of phenylalanine activating elongation modules of tyrocidine synthetase B, *Biochemistry*, 41, 9184–9196, 2002.

Mach, B., Reich, E., and Tatum, E.L., Separation of the biosynthesis of the antibiotic polypeptide tyrocidine from protein biosynthesis, *Proc. Natl Acad. Sci. USA*, 50, 175–181, 1963.

Magarvey, N.A. et al., Biosynthetic characterization and chemoenzymatic assembly of the cryptophycins. Potent anticancer agents from *Nostoc* cyanobionts, *ACS Chem. Biol.*, 1, 766–779, 2006.

Maget-Dana, R. and Peypoux, F., Iturins, a special class of pore-forming lipopeptides: Biological and physicochemical properties, *Toxicology*, 87, 151–174, 1994.

Mak, D.O. and Webb, W.W., Two classes of alamethicin transmembrane channels: Molecular models from single-channel properties, *Biophys. J.*, 69, 2323–2336, 1995.

Marahiel, M.A., Stachelhaus, T., and Mootz, H.D., Modular peptide synthetases involved in nonribosomal peptide synthesis, *Chem. Rev.*, 97, 2651–2673, 1997.

Markham, J.E. and Hille, J., Host-selective toxins as agents of cell death in plant–fungus interactions, *Mol. Plant Pathol.*, 2, 229–239, 2001.

Maroun, J.A. et al., Phase I study of Aplidine in a dailyx5 one-hour infusion every 3 weeks in patients with solid tumors refractory to standard therapy. A National Cancer Institute of Canada Clinical Trials Group study: NCIC CTG IND 115, *Ann. Oncol.*, 17, 1371–1378, 2006.

Martinez, J.S. et al., Structure and membrane affinity of a suite of amphiphilic siderophores produced by a marine bacterium, *Proc. Natl Acad. Sci. USA*, 100, 3754–3759, 2003.

Matsuyama, T. et al., A novel extracellular cyclic lipopeptide which promotes flagellum-dependent and -independent spreading growth of *Serratia marcescens*, *J. Bacteriol.*, 174, 1769–1776, 1992.

May, J.J., Kessler, N., Marahiel, M.A., and Stubbs, M.T., Crystal structure of DhbE, an archetype for aryl acid activating domains of modular nonribosomal peptide synthetases, *Proc. Natl Acad. Sci. USA*, 99, 12120–12125, 2002.

May, J.J., Wendrich, T.M., and Marahiel, A., The *dhb* operon of *Bacillus subtilis* encodes the biosynthetic template for the catecholic siderophore 2,3-dihydroxybenzoate-glycine-threonine trimeric ester bacillibactin, *J. Biol. Chem.*, 276, 7209–7217, 2001.

McDaniel, R., Welch, M., and Hutchinson, C.R., Genetic approaches to polyketide antibiotics. 1, *Chem. Rev.*, 105, 543–558, 2005.

Menkhaus, M. et al., Structural and functional organization of the surfactin synthetase multienzyme system, *J. Biol. Chem.*, 268, 7678–7684, 1993.

Meyer, W.L. et al., The amino acid sequence and configuration of tentoxin, *Biochem. Biophys. Res. Commun.*, 56, 234–240, 1974.

Miao, V. et al., Genetic engineering in *Streptomyces roseosporus* to produce hybrid lipopeptide antibiotics, *Chem. Biol.*, 13, 269–276, 2006.

Mikalsen, B. et al., Natural variation in the microcystin synthetase operon *mcyABC* and impact on microcystin production in *Microcystis* strains, *J. Bacteriol.*, 185, 2774–2785, 2003.

Miller, D.A. et al., Yersiniabactin synthetase: A four-protein assembly line producing the nonribosomal peptide/polyketide hybrid siderophore of *Yersinia pestis*, *Chem. Biol.*, 9, 333–344, 2002.

Mitchell, R.E., The relevance of non-host specific toxins in the expression of virulence by pathogens, *Annu. Rev. Phytopathol.*, 22, 215–245, 1984.

Mitchell, R.E. and Bieleski, R.L., Involvement of phaseolotoxin in halo blight of beans: Transport and conversion to functional toxin, *Plant Physiol.*, 60, 723–729, 1977.

Mittenhuber, G., Weckermann, M. and Marahiel, M.A., Gene cluster containing the genes for tyrocidine synthetases 1 and 2 from *Bacillus brevis*: Evidence for an operon, *J. Bacteriol.*, 171, 4881–4887, 1989.

Miyazaki, J. et al., The limited efficacy of methotrexate, actinomycin D and cisplatin (MAP) for patients with advanced testicular cancer, *Jpn. J. Clin. Oncol.*, 33, 391–395, 2003.

Miyazaki, K. et al., Synthesis and antitumour activity of novel dolastatin 10 analogs, *Chem. Pharm. Bull.*, 43, 1706–1718, 1995.

Moffitt, M.C. and Neilan, B.A., Characterization of the nodularin synthetase gene cluster and proposed theory of the evolution of cyanobacterial hepatotoxins, *Appl. Environ. Microbiol.*, 70, 6353–6362, 2004.

Molnár, I. et al., The biosynthetic gene cluster for the microtubule-stabilizing agents epothilones A and B from *Sorangium cellulosum* So ce90, *Chem. Biol.*, 7, 97–109, 2000.

Montecalvo, M.A., Ramoplanin: A novel antimicrobial agent with the potential to prevent vancomycin-resistant enterococcal infection in high-risk patients, *J. Antimicrob. Chemother.*, 51, 31–35, 2003.

Moore, B.S., Biosynthesis of marine natural prodcts: Macroorganisms (Part B), *Nat. Prod. Rep.*, 23, 615–629, 2006.

Mootz, H.D. et al., Decreasing the ring size of a cyclic nonribosomal peptide antiobiotic by in-frame module deletion in the biosynthetic genes, *J. Am. Chem. Soc.*, 124, 10980–10981, 2002a.

Mootz, H.D., Schwarzer, D., and Marahiel, M.A., Construction of hybrid peptide synthetases by module and domain fusions, *Proc. Natl Acad. Sci. USA*, 97, 5848–5853, 2000.

Mootz, H.D., Schwarzer, D., and Marahiel, M.A., Ways of assembling complex natural products on modular nonribosomal peptide synthetases, *ChemBioChem*, 3, 490–504, 2002b.

Mueller, P. and Rudin, D.O., Action potentials induced in biomolecular lipid membranes, *Nature*, 217, 713–719, 1968.

Murakami, M. et al., Aeruginosins 98-A and B, trypsin inhibitors from the blue-green alga *Microcystis aeruginosa* (NIES-98), *Tetrahedron Lett.*, 36, 2785–2788, 1995.

Muroi, M., Shiragami, N., and Takatsuki, A., Destruxin B, a specific and readily reversible inhibitor of vacuolar-type H^+-translocating ATPase, *Biochem. Biophys. Res. Commun.*, 205, 1358–1365, 1994.

Neilands, J.B., Microbial iron compounds, *Annu. Rev. Biochem.*, 50, 715–731, 1981.

Neu, T.R., Significance of bacterial surface-active compounds in interaction of bacteria with interfaces, *Microbiol. Rev.*, 60, 151–166, 1996.

Nguyen, K.T. et al., Combinatorial biosynthesis of novel antibiotics related to daptomycin, *Proc. Natl Acad. Sci. USA*, 103, 17462–17467, 2006.

Nieto, M. and Perkins, H.R., Modifications of the acyl-D-alanyl–D-alanine terminus affecting complex-formation with vancomycin, *Biochem. J.*, 123, 789–803, 1971.

Nishimura, S. and Kohmoto, K., Host-specific toxins and chemical structures from *Alternaria* species, *Annu. Rev. Phytopathol.*, 21, 87–116, 1983.

O'Callaghan, C.H., Sykes, R.B., Griffiths, A., and Thornton, J.E., Cefuroxime, a new cephalosporin antibiotic: Activity *in vitro*, *Antimicrob. Agents Chemother.*, 9, 511–519, 1976.

Offenzeller, M. et al., Biosynthesis of the unusual amino acid (4*R*)-4-[(*E*)-2-butenyl]-4-methyl-L-threonine of cyclosporin A: Enzymatic analysis of the reaction sequence including identification of the methylation precursor in a polyketide pathway, *Biochemistry*, 35, 8401–8412, 1996.

Oliynyk, M. et al., Complete genome sequence of the erythromycin-producing bacterium *Saccharopolyspora erythraea* NRRL23338, *Nature Biotechnol.*, 25, 447–453, 2007.

Omura, S. et al., Genome sequence of an industrial microorganism *Streptomyces avermitilis*: Deducing the ability of producing secondary metabolites, *Proc. Natl Acad. Sci. USA*, 98, 12215–12220, 2001.

Paananen, A. et al., Inhibition of human natural killer cell activity by cereulide, an emetic toxin from *Bacillus cereus*, *Clin. Exp. Immunol.*, 129, 420–428, 2002.

Paitan, Y. et al., The first gene in the biosynthesis of the polyketide antibiotic TA of *Myxococcus xanthus* codes for a unique PKS module coupled to a peptide synthetase, *J. Mol. Biol.*, 286, 465–474, 1999.

Panda, D. et al., Antiproliferative mechanism of action of cryptophycin-52: Kinetic stabilization of microtubule dynamics by high-affinity binding to microtubule ends, *Proc. Natl Acad. Sci. USA*, 95, 9313–9318, 1998.

Parry, R.J. and Mafoti, R., Biosynthesis of coronatine, a novel polyketide, *J. Am. Chem. Soc.*, 108, 4681–4682, 1986.

Patel, H.M. and Walsh, C.T., *In vitro* reconstitution of the *Pseudomonas aeruginosa* nonribosomal peptide synthesis of pyochelin: Characterization of backbone tailoring thiazoline reductase and N-methyltransferase activities, *Biochemistry*, 40, 9023–9031, 2001.

Payne, S.M. and Finkelstein, R.A., The critical role of iron in host-bacterial interactions, *J. Clin. Invest.*, 61, 1428–1440, 1978.

Pelludat, C. et al., The yersiniabactin biosynthetic gene cluster of *Yersinia enterocolitica*: Organization and siderophore-dependent regulation, *J. Bacteriol.*, 180, 538–546, 1998.

Perez, E.A. et al., Phase II trial of dolastatin-10 in patients with advanced breast cancer, *Invest. New Drugs*, 23, 257–261, 2005.

Perkins, H.R., Specificity of combination between mucopeptide precursors and vancomycin or ristocetin, *Biochem. J.*, 111, 195–205, 1969.

Perkins, J.B. et al., *Streptomyces* genes involved in biosynthesis of the peptide antibiotic valinomycin, *J. Bacteriol.*, 172, 3108–3116, 1990.

Pettit, G.R. et al., The isolation and structure of a remarkable marine animal antineoplastic constituent: Dolastatin 10, *J. Am. Chem. Soc.*, 109, 6883–6885, 1987.

Peypoux, F. et al., Isolation and characterization of a new variant of surfactin, the [Val7]surfactin, *Eur. J. Biochem.*, 202, 101–106, 1991.

Peypoux, F. et al., [Ala4]Surfactin, a novel isoform from *Bacillus subtilis* studied by mass and NMR spectroscopies, *Eur. J. Biochem.*, 224, 89–96, 1994.

Peypoux, F. et al., Structure of bacillomycin F, a new peptidolipid antibiotic of the iturin group, *Eur. J. Biochem.*, 153, 335–340, 1985.

Pfeifer, B.A., Wang, C.C.C., Walsh, C.T., and Khosla, C., Biosynthesis of yersiniabactin, a complex polyketide–nonribosomal peptide, using *Escherichia coli* as a heterologous host, *Appl. Environ. Microbiol.*, 69, 6698–6702, 2003.

Pfennig, F., Schauwecker, F., and Keller, U., Molecular characterization of the genes of actinomycin synthetase I and of a 4-methyl-3-hydroxyanthranilic acid carrier protein involved in the assembly of the acylpeptide chain of actinomycin in *Streptomyces*, *J. Biol. Chem.*, 274, 12508–12516, 1999.

Powell, A. et al., Engineered biosynthesis of nonribosomal lipopeptides with modified fatty acid side chains, *J. Am. Chem. Soc.*, 129, 15182–15191, 2007.

Pringle, R.B. and Scheffer, R.P., Host-specific plant toxins, *Annu. Rev. Phytopathol.*, 2, 133–156, 1964.

Proksch, P., Edrada, R.A., and Ebel, R., Drugs from the seas—current status and microbiological implications, *Appl. Microbiol. Biotechnol.*, 59, 125–134, 2002.

Pulsawat, N., Kitani, S., and Nihira, T., Characterization of biosynthetic gene cluster for the production of virginiamycin M, a streptogramin type A antibiotic, in *Streptomyces virginiae*, *Gene*, 393, 31–42, 2007.

Rausch, C. et al., Phylogenetic analysis of condensation domains in NRPS sheds light on their functional evolution, *BMC Evol. Biol.*, 7, 78, 2007.

Rausch, C. et al., Specificity prediction of adenylation domains in nonribosomal peptide synthetases (NRPS) using transductive support vector machines (TSVMs), *Nucl. Acids Res.*, 33, 5799–5808, 2005.

Recktenwald, J. et al., Nonribosomal biosynthesis of vancomycin-type antibiotics: A heptapeptide backbone and eight peptide synthetase modules, *Microbiology*, 148, 1105–1118, 2002.

Reeves, C.D., The enzymology of combinatorial biosynthesis, *Crit. Rev. Biotechnol.*, 23, 95–147, 2003.

Reich, E., Biochemistry of actinomycins, *Cancer Res.*, 23, 1428–1441, 1963.

Richardt, A. et al., Ebony, a novel nonribosomal peptide synthetase for β-alanine conjugation with biogenic amines in *Drosophila*, *J. Biol. Chem.*, 278, 41160–41166, 2003.

Riely, G.J. et al., A phase 2 study of TZT-1027, administered weekly to patients with advanced non-small cell lung cancer following treatment with platinum-based chemotherapy, *Lung Cancer*, 55, 181–185, 2007.

Rios Steiner, J.L., Murakami, M., and Tulinsky, A., Structure of thrombin inhibited by aeruginosin 298-A from a blue-green alga, *J. Am. Chem. Soc.*, 120, 597–598, 1998.

Rix, U., Fischer, C., Remsing, L.L., and Rohr, J., Modification of post-PKS tailoring steps through combinatorial biosynthesis, *Nat. Prod. Rep.*, 19, 542–580, 2002.

Roongsawang, N. et al., Cloning and characterization of the gene cluster encoding arthrofactin synthetase from *Pseudomonas* sp. MIS38, *Chem. Biol.*, 10, 869–880, 2003.

Rosenberg, E. and Ron, E.Z., High- and low-molecular-mass microbial surfactants, *Appl. Microbiol. Biotechnol.*, 52, 154–162, 1999.

Rouhiainen, L. et al., Genes coding for hepatotoxic heptapeptides (microcystins) in the cyanobacterium *Anabaena* strain 90, *Appl. Environ. Microbiol.*, 70, 686–692, 2004.

Roy, R.S. et al.,Thiazole and oxazole peptides: Biosynthesis and molecular machinery, *Nat. Prod. Rep.*, 16, 249–263, 1999.

Royer, M. et al., Albicidin pathotoxin produced by *Xanthomonas albilineans* is encoded by three large PKS and NRPS genes present in a gene cluster also containing several putative modifying, regulatory, and resistance genes, *Mol. Plant–Microbe Interact.*, 17, 414–427, 2004.

Sakai, R. et al., Structure–activity relationships of the didemnins, *J. Med. Chem.*, 39, 2819–2834, 1996.

Salomon, C.E., Magarvey, N.A., and Sherman, D.H., Merging the potential of microbial genetics with biological and chemical diversity: An even brighter future for marine natural product drug discovery, *Nat. Prod. Rep.*, 21, 105–121, 2004.

Samuels, R.I., Charnley, A.K., and Reynolds, S.E., The role of destruxins in the pathogenicity of 3 strains of *Metarhizium anisopliae* for the tobacco hornworm *Manduca sexta*, *Mycopathologia*, 104, 51–58, 1988a.

Samuels, R.I., Reynolds, S.E., and Charnley, A.K., Calcium channel activation of insect muscle by destruxins, insecticidal compounds produced by the entomopathogenic fungus *Metarhizium anisopliae*, *Comp. Biochem. Physiol. C, Comp. Pharmacol.*, 90, 403–412, 1988b.

Sandler, B., Murakami, M., and Clardy, J., Atomic structure of the trypsin-aeruginosin 98-B complex, *J. Am. Chem. Soc.*, 120, 595–596, 1998.

Scaloni, A. et al., Structure, conformation and biological activity of a novel lipodepsipeptide from *Pseudomonas corrugata*: Cormycin A, *Biochem. J.*, 384, 25–36, 2004.

Schauwecker, F., Pfennig, F., Grammel, N., and Keller, U., Construction and *in vitro* analysis of a new bi-modular polypeptide synthetase for synthesis of N-methylated acyl peptides, *Chem. Biol.*, 7, 287–297, 2000.

Schneider, A., Stachelhaus, T., and Marahiel, M.A., Targeted alteration of the substrate specificity of peptide synthetases by rational module swapping, *Mol. Gen. Genet.*, 257, 308–318, 1998.

Schneider, T.L., Shen, B., and Walsh, C.T., Oxidase domains in epothilone and bleomycin biosynthesis: Thiazoline to thiazole oxidation during chain elongation, *Biochemistry*, 42, 9722–9730, 2003.

Schwarzer, D., R. Finking, and M. A. Marahiel. Nonribosomal peptides: From genes to products, *Nat. Prod. Rep.*, 20, 275–287, 2003.

Schwarzer, D., Mootz, H.D., and Marahiel, M.A., Exploring the impact of different thioesterase domains for the design of hybrid peptide synthetases, *Chem. Biol.*, 8, 997–1010, 2001.

Schwecke, T. et al., The biosynthetic gene cluster for the polyketide immunosuppressant rapamycin, *Proc. Natl Acad. Sci. USA*, 92, 7839–7843, 1995.

Shaw-Reid, C.A. et al., Assembly line enzymology by multimodular nonribosomal peptide synthetases: The thioesterase domain of *E. coli* EntF catalyzes both elongation and cyclolactonization, *Chem. Biol.*, 6, 385–400, 1999.

Shen, B. et al., Cloning and characterization of the bleomycin biosynthetic gene cluster from *Streptomyces verticillus* ATCC15003, *J. Nat. Prod.*, 65, 422–431, 2002.

Sheppard, J.D., Jumarie, C., Cooper, D.G., and Laprade, R., Ionic channels induced by surfactin in planar lipid bilayer membranes, *Biochim. Biophys. Acta*, 1064, 13–23, 1991.

Sieber, S.A. and Marahiel, M.A., Molecular mechanisms underlying nonribosomal peptide synthesis: Approaches to new antibiotics, *Chem. Rev.*, 105, 715–738, 2005.

Simmons, T.L. et al., Marine natural products as anticancer drugs, *Mol. Cancer Ther.*, 4, 333–342, 2005.

Sivonen, K. and Jones, G., Cyanobacterial toxins, in *Toxic Cyanobacteria in Water*, E. and F.N Spon, London, 1999, p. 41–111.

Sørensen, D. et al., Cyclic lipoundecapeptide amphisin from *Pseudomonas* sp. strain DSS73, *Acta Cryst.*, C57, 1123–1124, 2001.

Sorensen, K.N., Kim, K.H., and Takemoto, J.Y., *In vitro* antifungal and fungicidal activities and erythrocyte toxicities of cyclic lipodepsinonapeptides produced by *Pseudomonas syringae* pv. *syringae*. *Antimicrob. Agents Chemother.*, 40, 2710–2713, 1996.

Stachelhaus, T., Hüser, A., and Marahiel, M.A., Biochemical characterization of peptidyl carrier protein (PCP), the thiolation domain of multifunctional peptide synthetases, *Chem. Biol.*, 3, 913–921, 1996.

Stachelhaus, T., Mootz, H.D., Bergendah, V., and Marahiel, M.A., Peptide bond formation in nonribosomal peptide biosynthesis: Catalytic role of the condensation domain, *J. Biol. Chem.*, 273, 22773–22781, 1998.

Stachelhaus, T., Mootz, H.D., and Marahiel, M.A., The specificity-conferring code of adenylation domains in nonribosomal peptide synthetases, *Chem. Biol.*, 6, 493–505, 1999.

Stachelhaus, T., Schneider, A., and Marahiel, M.A., Rational design of peptide antibiotics by targeted replacement of bacterial and fungal domains, *Science*, 269, 69–72, 1995.

Steele, J.A. et al., Chloroplast coupling factor 1: A species-specific receptor for tentoxin, *Proc. Natl Acad. Sci. USA*, 73, 2245–2248, 1976.

Stein, T. et al., Detection of 4'-phosphopantetheine at the thioester binding site for L-valine of gramicidin S synthetase 2, *FEBS Lett.*, 340, 39–44, 1994.

Stewart, P.S. and J. William Costerton. Antibiotic resistance of bacteria in biofilms, *Lancet*, 358, 135–138, 2001.

Stintzi, A., Evans, K., Meyer, J.M., and Poole, K., Quorum-sensing and siderophore biosynthesis in *Pseudomonas aeruginosa*: *LasR/lasI* mutants exhibit reduced pyoverdine biosynthesis, *FEMS Microbiol. Lett.*, 166, 341–345, 1998.

Stone, K.J. and Strominger, J.L., Mechanism of action of bacitracin: Complexation with metal ion and C_{55}-isoprenyl pyrophosphate, *Proc. Natl Acad. Sci. USA*, 68, 3223–3227, 1971.

Storm, D.R. and Strominger, J.L., Complex formation between bacitracin peptides and isoprenyl pyrophosphates. The specificity of lipid–peptide interactions, *J. Biol. Chem.*, 248, 3940–3945, 1973.

Suarez, Y. et al., Kahalalide F, a new marine-derived compound, induces oncosis in human prostate and breast cancer cells, *Mol. Cancer Ther.*, 2, 863–872, 2003.

Sunaga, S. et al., Identification and characterization of the *pswP* gene required for the parallel production of prodigiosin and serrawettin W1 in *Serratia marcescens*, *Microbiol. Immunol.*, 48, 723–728, 2004.

Suo Z., Tseng, CC., and Walsh, C.T., Purification, priming, and catalytic acylation of carrier protein domains in the polyketide synthase and nonribosomal peptidyl synthetase modules of the HMWP1 subunit of yersiniabactin synthetase, *Proc. Natl Acad. Sci. USA*, 98, 99–104, 2001.

Symmank, H., Franke, P., Saenger, W., and Bernhard, F., Modification of biologically active peptides: Production of a novel lipohexapeptide after engineering of *Bacillus subtilis* surfactin synthetase, *Protein Eng.*, 15, 913–921, 2002.

Symmank, H., Saenger, W., and Bernhard, F., Analysis of engineered multifunctional peptide synthetases. Enzymatic characterization of surfactin synthetase domains in hybrid bimodular systems, *J. Biol. Chem.*, 274, 21581–21588, 1999.

Takahashi, T. et al., Inhibition of lipopolysaccharide activity by a bacterial cyclic lipopeptide surfactin, *J. Antibiot.*, 59, 35–43, 2006.

Tang, G.-L., Cheng, Y.-Q., and Shen, B., Chain initiation in the leinamycin-producing hybrid nonribosomal peptide/polyketide synthetase from *Streptomyces atroolivaceus* S-140: Discrete, monofunctional adenylation enzyme and peptidyl carrier protein that directly load D-alanine, *J. Biol. Chem.*, 282, 20273–20282, 2007.

Tang, L. et al., Cloning and heterologous expression of the epothilone gene cluster, *Science*, 287, 640–642, 2000.

Thibaut, D. et al., Purification of peptide synthetases involved in pristinamycin I biosynthesis, *J. Bacteriol.*, 179, 697–704, 1997.

Tillett, D. et al., Structural organization of microcystin biosynthesis in *Microcystis aeruginosa* PCC7806: An integrated peptide–polyketide synthetase system, *Chem. Biol.*, 7, 753–764, 2000.

Tipper, D.J. and Strominger, J.L., Mechanism of action of penicillins: A proposal based on their structural similarity to acyl–D-alanyl–D-alanine, *Microbiology*, 54, 1133–1141, 1965.

Trauger, J.W. and Walsh, C.T., Heterologous expression in *Escherichia coli* of the first module of the nonribosomal peptide synthetase for chloroeremomycin, a vancomycin-type glycopeptide antibiotic, *Proc. Natl Acad. Sci. USA*, 97, 3112–3117, 2000.

Tuori, R.P., Wolpert, T.J., and Ciuffetti, L.M., Heterologous expression of functional Ptr ToxA, *Mol. Plant–Microbe Interact.*, 13, 456–464, 2000.

Udwary, D.W. et al., Genome sequencing reveals complex secondary metabolome in the marine actinomycete *Salinispora tropica*, *Proc. Natl Acad. Sci. USA*, 104, 10376–10381, 2007.

Ueno, T. et al., Mass spectrometry of *Alternaria mali* toxins and related cyclodepsipeptides, *Biol. Mass Spectrom.*, 4, 134–142, 1977.

Van Lanen, S.G. and Shen, B., Microbial genomics for the improvement of natural product discovery, *Curr. Opin. Microbiol.*, 9, 252–260, 2006.

Vanittanakom, N., Loeffler, W., and Koch, U., Fengycin—a novel antifungal lipopeptide antibiotic produced by *Bacillus subtilis* F-29-3, *J. Antibiot.*, 39, 888–901, 1986.

Vendrell, J., Querol, E., and Aviles, F.X., Metallocarboxypeptidases and their protein inhibitors: Structure, function and biomedical properties, *Biochim. Biophys. Acta*, 1477, 284–298, 2000.

Vera, M.D. and Joullié, M.M., Natural products as probes of cell biology: 20 years of didemnin research, *Med. Res. Rev.*, 22, 102–145, 2002.

Verdier-Pinard, P. et al., Biosynthesis of radiolabeled curacin A and Its rapid and apparently irreversible binding to the colchicine site of tubulin, *Arch. Biochem. Biophys.*, 370, 51–58, 1999.

Visca, P. et al., Metal regulation of siderophore synthesis in *Pseudomonas aeruginosa* and functional effects of siderophore-metal complexes, *Appl. Environ. Microbiol.*, 58, 2886–2893, 1992.

Walker, S. et al., Chemistry and biology of ramoplanin: A lipoglycodepsipeptide with potent antibiotic activity, *Chem. Rev.*, 105, 449–476, 2005.

Wallace, B.A., Common structural features in gramicidin and other ion channels, *BioEssays*, 22, 227–234, 2000.

Walsh, C.T., Polyketide and nonribosomal peptide antibiotics: Modularity and versatility, *Science*, 303, 1805–1810, 2004.

Walsh, C.T. et al., Tailoring enzymes that modify nonribosomal peptides during and after chain elongation on NRPS assembly lines, *Curr. Opin. Chem. Biol.*, 5, 525–534, 2001.

Walton, J.D., Host-selective toxins: Agents of compatibility, *Plant Cell*, 8, 1723–1733, 1996.

Watt, J., The use of cytotoxic drugs in the surgery of malignant disease, *J. Bone Joint Surg. Br.*, 50-B, 511–523, 1968.

Weber, G., Schörgendorfer, K., Schneider-Scherzer, E., and Leitner, E., The peptide synthetase catalyzing cyclosporine production in *Tolypocladium niveum* is encoded by a giant 45.8-kilobase open reading frame, *Curr. Genet.*, 26, 120–125, 1994.

Weber, T. et al., Solution structure of PCP, a prototype for the peptidyl carrier domains of modular peptide synthetases, *Structure*, 8, 407–418, 2000.

Weber, T. and Marahiel, M.A., Exploring the domain structure of modular nonribosomal peptide synthetases, *Structure*, 9, R3–R9, 2001.

Weinberg, E.D., Iron withholding: A defense against infection and neoplasia, *Physiol. Rev.*, 64, 65–102, 1984.

Weissman, K.J. and Leadlay, P.F., Combinatorial biosynthesis of reduced polyketides, *Nat. Rev. Micro.*, 3, 925–936, 2005.

Welker, M. and von Dohren, H., Cyanobacterial peptides—Nature's own combinatorial biosynthesis, *FEMS Microbiol. Rev.*, 30, 530–563, 2006.

Wenzel, S.C. et al., Heterologous expression of a myxobacterial natural products assembly line in pseudomonads via Red/ET recombineering, *Chem. Biol.*, 12, 349–356, 2005.

Wenzel, S.C. et al., Nonribosomal peptide biosynthesis: Point mutations and module skipping lead to chemical diversity, *Angew. Chem. Int. Ed.*, 45, 2296–2301, 2006.

Wilkinson, B. and Micklefield, J., Mining and engineering natural-product biosynthetic pathways, *Nature Chem. Biol.*, 3, 379–386, 2007.

Wu, C.-Y. et al., Small molecules targeting severe acute respiratory syndrome human coronavirus, *Proc. Natl. Acad. Sci. USA*, 101, 10012–10017, 2004.

Wu, X., Ballard, J., and Jiang, Y.W., Structure and biosynthesis of the BT peptide antibiotic from *Brevibacillus texasporus*, *Appl. Environ. Microbiol.*, 71, 8519–8530, 2005.

Xu, G., Martinez, J.S., Groves, J.T., and Butler, A., Membrane affinity of the amphiphilic marinobactin siderophores, *J. Am. Chem. Soc.*, 124, 13408–13415, 2002.

Yakimov, M.M., Giuliano, L., Timmis, K.N., and Golyshin, P.N., Recombinant acylheptapeptide lichenysin: High level of production by *Bacillus subtilis* cells, *J. Mol. Microbiol. Biotechnol.*, 2, 217–224, 2000.

Yakimov, M.M., Timmis, K.N., Wray, V., and Fredrickson, H.L., Characterization of a new lipopeptide surfactant produced by thermotolerant and halotolerant subsurface *Bacillus licheniformis* BAS50, *Appl. Environ. Microbiol.*, 61, 1706–1713, 1995.

Yamaguchi, K., Murakami, M., and Okino, T., Screening of angiotensin-converting enzyme inhibitory activities in microalgae, *J. Appl. Phycol.*, 1, 271–275, 1989.

Yocum, R.R., Waxman, D.J., Rasmussen, J.R., and Strominger, J.L., Mechanism of penicillin action: Penicillin and substrate bind covalently to the same active site serine in two bacterial D-alanine carboxypeptidases, *Proc. Natl Acad. Sci. USA*, 76, 2730–2734, 1979.

Yoder, O.C., Toxins in pathogenesis, *Annu. Rev. Phytopathol.*, 18, 103–129, 1980.

Zocher, R. et al., Biosynthesis of cyclosporin A: Partial purification and properties of a multifunctional enzyme from *Tolypocladium inflatum*, *Biochemistry*, 25, 550–553, 1986.

Part III

Host-Defense Peptides and Antimicrobials

14 Anuran Host-Defense Peptides That Complex with Ca^{2+} Calmodulin and Inhibit the Synthesis of the Cell Signaling Agent Nitric Oxide by Neuronal Nitric Oxide Synthase

Jason R. Doyle, John H. Bowie, Rebecca J. Jackway, Lyndon E. Llewellyn, Tara L. Pukala, Margit A. Apponyi, and Grant W. Booker

CONTENTS

14.1 INTRODUCTION

The skin glands of an amphibian contain a potent chemical arsenal, which is secreted from specialized glands onto the dorsal surface and into the gut of the animal in response to a variety of stimuli. Bioactive peptides comprise a significant component of the skin secretion, playing a role in host defense and regulation of dermal physiological function. There are several thousand species of frogs and toad worldwide and extensive research has shown that many of these animals contain a variety of active peptides in skin glands, including neuropeptides, hormones, and antimicrobial peptides, which together protect the anurans from predators both large and small (Bevins and Zasloff, 1990; Lazarus and Attila, 1993; Erspamer, 1994; Apponyi et al., 2004; Pukala et al., 2006b).

Amphibians evolved from freshwater fish some 400 to 350 million years ago. There are more than 200 species of anuran (frogs and toads) in Australia: many of these are indigenous to Australia, presumably evolving following the separation of Australia from Gondwanaland between 100 and 80 million years ago. We have been working on the skin peptide profiles of Australian anurans for the past 15 years, having investigated the skin peptides of some 40 species, mainly frogs, and have isolated and identified more than 200 active peptides (Apponyi et al., 2004; Pukala et al., 2006b). We obtain these peptides using the noninvasive electrical stimulation method (Tyler et al., 1992). Species investigated include tree frogs of the genus *Litoria*, many of which produce the potent smooth muscle contractant caerulein (Anastasi et al., 1968; Erspamer, 1994) and the hinged antimicrobial caerin 1 peptides (Figure 14.2), isolated from *Litoria splendida* and *L. caerulea* (Stone et al., 1992; Erspamer, 1994) (Figures 14.1 and 14.2).

pEQDY(SO_3)TGWMDF(NH_2) Caerulein
GLLSVLGSVAKHVLPHVVPVIAEHL(NH_2) Caerin 1.1

FIGURE 14.1 *Litoria splendida*, geographical location, caerulein and caerin 1.1 (www.2docstock.com/NT-animals.html).

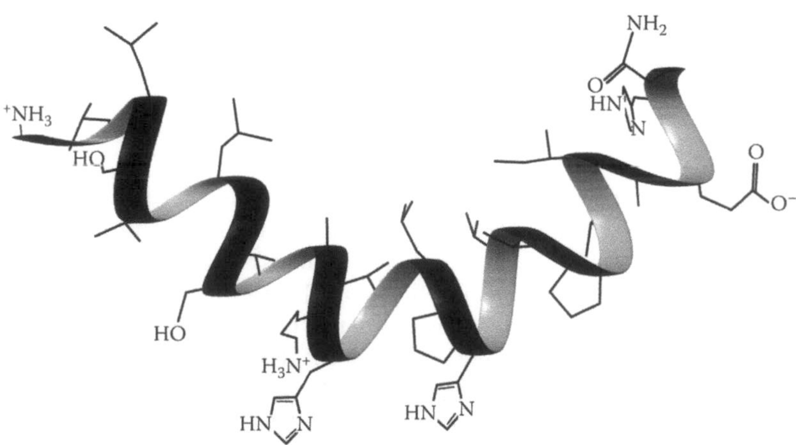

FIGURE 14.2 The solution structure of the wide-spectrum antibiotic caerin 1.1 determined by 2D NMR spectroscopy in the solvent system TFE/H_2O (1:1). (From Pukala et al., *Nat. Prod. Rep.*, 23, 368–393, 2006b. With permission.)

14.2 THE ISOLATION AND IDENTIFICATION OF nNOS ACTIVE PEPTIDES

The largest tree frog in Australia, the white lipped tree frog, *Litoria infrafrenata*, is unusual because it was the first frog we studied where none of the skin peptides exhibited the classical smooth muscle contracting and/or antimicrobial activities described above. The skin peptides of *L. infrafrenata* must be defense peptides. The animal would not expend such an amount of energy to biosynthesize a peptide that was inactive. The major skin peptide from *L. infrafrenata* is frenatin 3 (Figure 14.3) (Waugh et al., 1996; Zhou et al., 2005). The solution structure of frenatin 3 [determined using 2D nuclear magnetic resonance spectroscopy (NMR) in the membrane-mimicking solvent trifluoroethanol/water (TFE/H_2O)] is most unusual, comprising an N-terminal α-helical section to Leu14 and a random section from Gly15 onwards (Figure 14.4). What defense role does frenatin 3 effect in *L. infrafrenata*?

At the same time we were studying *L. infrafrenata*, the Australian Institute of Marine Science (Townsville MC, Queensland, Australia) was carrying out a bioactivity screening program of material obtained from aquatic organisms. As part of this survey it was discovered that frenatin 3, at μM concentrations, stopped the synthesis of the cell signaling molecule nitric oxide (NO) by the protein neuronal nitric oxide synthase (nNOS) (Doyle et al., 2002). Since that time it has been shown that some

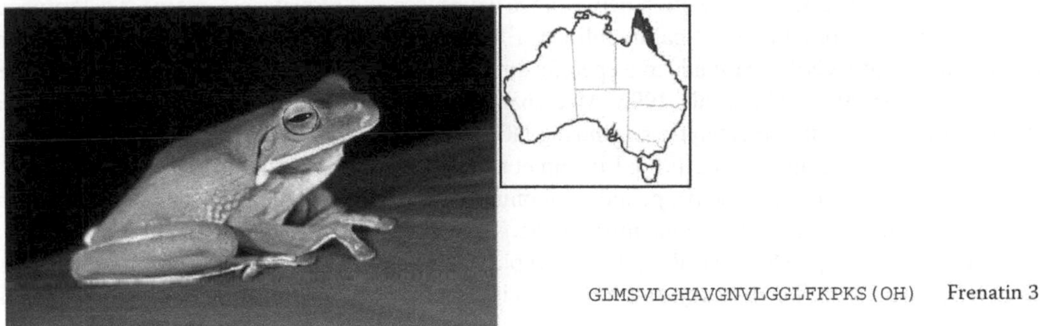

FIGURE 14.3 *Litoria infrafrenata*, geographical location and frenatin 3 (www.jcu.edu/school/tbiol/zoology/herp/fullsize/ksb-litoria-infrafrenata-tully.jpg).

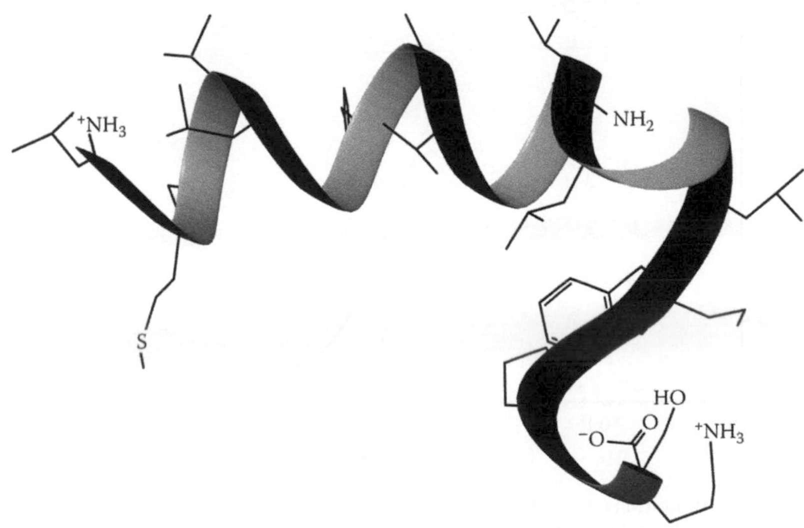

FIGURE 14.4 The solution structure of frenatin 3, determined by 2D NMR spectroscopy in the solvent system TFE/H_2O (1:1). (From Pukala et al., *Nat. Prod. Rep.*, 23, 368–393, 2006b. With permission.)

(but not all) frogs and toadlets of the genera *Crinia*, *Litoria*, and *Uperoleia* have at least one skin peptide that deactivates nNOS (Doyle et al., 2003; Apponyi et al., 2004; Pukala et al., 2006b). In complete contrast, no frog of the *Limnodynastes* genus that we have investigated, namely, *Limnodynastes dumerilii* (Raftery et al., 1993), *L. fletcheri* (Bradford et al., 1993), *L. interioris* (Raftery et al., 1993), *L. salmini* (Bradford et al., 1993), and *L. terraereginae* (Raftery et al., 1993) exudes either nNOS inhibitor peptides or antimicrobial peptides from its skin glands (Pukala et al., 2006b).

NO is unique among biological signals for its rapid diffusion, ability to permeate cell membranes and intrinsic instability, properties that eliminate the need for extracellular cell receptors or targeted NO degradation. NO differs from other neurotransmitters and hormones in that its synthesis is regulated by three nitric oxide synthase isoforms: neuronal (nNOS), endothelial (eNOS), and inducible (iNOS). They are among the most complex enzymes known (e.g., for nNOS see Bredt et al., 1991; Marletta, 1993), and oxidize L-arginine (L-Arg) to NO and citrulline, thereby controlling NO distribution and concentration (Marletta, 1994; Kerwin, 1995; Andrew and Mayer, 1999; Alderton et al., 2001). All isoforms are composed of two domains (Watanabe et al., 1996; Lincoln et al., 1997; Andrew and Mayer, 1999; Steuhr, 1999): (i) an N-terminal catalytic oxygenase domain that binds heme (iron protoporphyrin IX), tetrahydrobiopterin (BH4), and the substrate L-Arg, and (ii) a C-terminal reductase domain that binds flavin mononucleotide (FMN), flavin adenine dinucleotide (FAD), and NADPH. Communication between the oxygenase and reductase domains is determined by the regulatory proteins Ca^{2+} calmodulin (Ca^{2+}CaM) (for nNOS and eNOS), and calmodulin (CaM) (for iNOS). Ca^{2+}CaM binds to a specific amphipathic α-helical region at the centre of NOS (Marletta, 1993; Watanabe et al., 1996; Matsubara et al., 1997), and facilitates electron transfer through the enzyme to the catalytic site during NO production (Abu-Soud and Stuehr, 1993; Abu-Soud et al., 1994; Gachhui et al., 1996; Lincoln et al., 1997; Matsubara et al., 1997).

Inhibition of nNOS by an active peptide is monitored by measuring the change in the conversion of [^{3}H]arginine to [^{3}H]citrulline plus nitric oxide. The [^{3}H] citrulline is isolated and monitored by scintillation counting (Doyle et al., 2002). A typical test result for a peptide-inhibiting nNOS is shown in Figure 14.5. The potency of inhibition is indicated as an IC_{50} value in this chapter, but it is

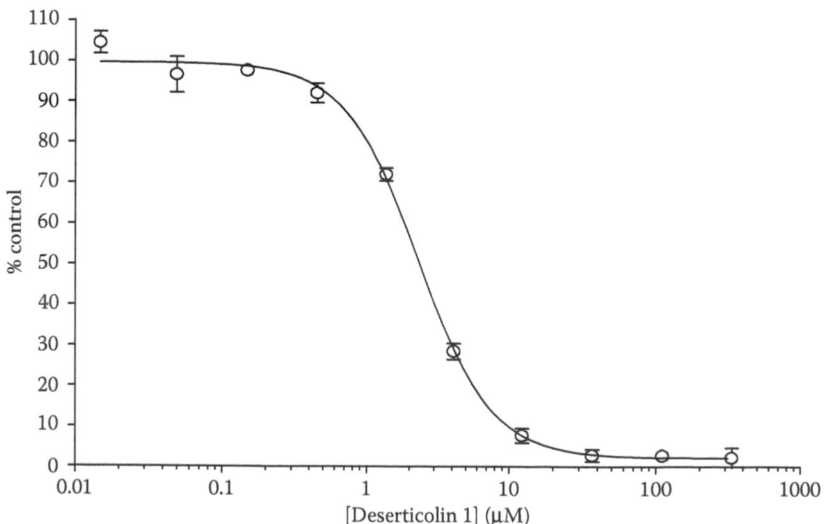

FIGURE 14.5 nNOS inhibition curve (decrease in counts of [^{3}H]-citrulline) with increasing concentration of deserticolin 1 (GLADFLNKAVGKVVDFVKS–NH_2) (from *Crinia deserticola*). IC_{50} = 2.4 μM; Hill slope = 1.6; K_i = 11.9 nM if inhibition is by an uncompetitive mechanism. (Calculated as in Brandt, R.B., Laux, J.E., and Yates, S.W., *Biochem. Med. Metab. Biol.*, 37, 344–349, 1987.)

important to understand what IC_{50} means in this context. The method used to quantify nNOS inhibition is a classical enzymic approach that titrates the concentration of an inhibitor while keeping all other experimental conditions constant. This enables computation of the concentration of peptide required to inhibit enzyme activity by 50%, namely the IC_{50}, which can then be used to compare potency between peptides. Although the IC_{50} reflects affinity of the peptide for the enzyme, it is not however the K_i, or inhibition constant. The IC_{50} of a compound may change if there are significant changes in experimental conditions such as variation of substrate or co-factor concentration. Whereas the IC_{50} might vary, the K_i, however, should be invariant, as it is related to the more familiar dissociation constant. For those systems where the inhibitor directly competes for the same binding site that metabolizes the enzyme substrate, the K_i is calculated from the IC_{50} as follows (Cheng and Prusoff, 1973):

$$K_i = \frac{IC_{50}}{1 + \dfrac{\text{Substrate}}{K_M}}$$

where K_M is the Michaelis constant, the affinity of the enzyme for its substrate (Brandt et al., 1987). The effect of this equation is that the K_i is a fraction of the IC_{50}. These peptides, however, do not exert their activity by competing directly with the substrate (i.e., arginine), but rather through an allosteric effect via the CaM binding site. The relationship between IC_{50} and K_i is more complex for allosteric inhibitors. For uncompetitive inhibitors, the equation relating IC_{50} to K_M differs from the above Cheng–Prusoff equation by inverting the term (Substrate)/K_M to K_M/(Substrate). The change in the equation means that when the substrate concentration used experimentally is substantially less than K_M, then K_i is less than IC_{50} (Brandt et al., 1987). For noncompetitive inhibitors, the K_i is equivalent to the IC_{50} no matter the relationship between the concentrations of the substrate and the K_M (Brandt et al., 1987). The importance of these concepts is that the potency of these peptides may be greater than the reported IC_{50} values with their K_is potentially being nanomolar even if their IC_{50}s are micromolar.

The nNOS-inhibiting amphibian peptides, which are often major components of skin secretions, are usually active with IC_{50} values in the μM concentration range. These may play a regulatory role in the animal, as NO is known to be involved in sight, reproduction, and gastric modulation in anurans (Molero et al., 1998; Gobbetti and Zeran, 1999; Renteria and Constantine-Paton, 1999). In addition, these nNOS inhibitors almost certainly form part of the animal's primary defense, by interfering with the NO messenger capabilities of an invading predator or pathogen. In the latter context, even bacteria have been shown to produce NOS (Choi et al., 1997; Morita et al., 1997). A summary of all currently known natural nNOS inhibiting amphibian peptides is presented in Table 14.1. Here it can be seen that these active peptides fall into three well-defined groups:

1. The aurein/citropin group (Wegener et al., 1999; Rozek et al., 2000; Apponyi et al., 2004)
2. The caerin 1 peptides, particularly those with a Phe residue in position 3 (Wong et al., 1997; Bowie et al., 1998; Apponyi et al., 2004)
3. The frenatin 3/dahlein 5 type peptides, characterized by a C-terminal CO_2H group and a Lys–X–Lys motif near this terminus (Brinkworth et al., 2003; Apponyi et al., 2004)

There is also a miscellaneous group 4 that includes some disulfide peptides isolated from European anurans of the genus *Rana*, for example brevinin 1R (IC_{50} 4.0 μM) from *Rana ridibunda* (Figure 14.6) (Artemenko et al., 2007b). The three most active natural anuran nNOS inhibitors are shown in Figures 14.7 through 14.9, namely aurein 2.3 (IC_{50} 1.8 μM) from *Litoria aurea*, caerin 1.8 (IC_{50} 1.7 μM) from *Litoria chloris*, and dahlein 5.6 (IC_{50} 1.6 μM) from *Litoria dahlii* (Apponyi et al., 2004).

TABLE 14.1
nNOS Activities of Natural Amphibian Peptides in Order of Activities within a Group

Name	Species	Sequence	MW	IC_{50}		Hill Slope	Charge
				μg/mL	μM		
Group 1							
Citropin 1.1.2	*L. citropa*	DVIKKVASVIGGL-NH_2	1297	55.0	42.4	1.5	+3
Aurein 1.1	*L. aurea*	GLFDIIKKIAESI-NH_2	1444	49.0	33.9	2.0	+1
Citropin 1.2.3	*L. citropa*	GLFDIIKKVAS-NH_2	1071	24.2	22.6	2.2	+2
Lesueurin	*L. lesueuri*	GLLDILKKVGKVA-NH_2	1353	25.0	18.5	1.4	+3
Signiferin 2.1	*C. signifera*	IIGHLIKTALGMLGL-NH_2	1548	25.7	16.6	2.4	+2
Citropin 1.1	*L. citropa*	GLFDVIKKVASVIGGL-NH_2	1614	13.3	8.2	2.0	+2
Uperin 3.6	*U. mjobergii*	GVIDAAKKVVNVLKNLF-NH_2	1827	8.0	4.4	1.5	+3
Aurein 2.2	*L. aurea*	GLFDIVKKVVGALGSL-NH_2	1613	7.0	4.4	2.5	+2
Aurein 2.4	*L. aurea*	GLFDIVKKVVGTLAGL-NH_2	1630	3.9	2.4	2.4	+2
Aurein 2.3	*L. aurea*	GLFDIVKKVVGAIGSL-NH_2	1613	2.9	1.8	1.7	+2
Group 2							
Caerin 1.10	*L. chloris*	GLLSVLGSVAKHVLPHVVPVIAEKL-NH_2	2573	105.7	41.0	1.6	+2
Caerin 1.1	*L. caerulea*	GLLSVLGSVAKHVLPHVVPVIAEHL-NH_2	2582	94.4	36.6	1.4	+1
Caerin 1.20	Hybrid[a]	GLFGILGSVAKHVLPHVIPVVAEHL-NH_2	2600	70.9	27.2	1.5	+1
Caerin 1.6	*L. chloris*	GLFSVLGAVAKHVLPHVVPVIAEKL-NH_2	2591	22.0	8.4	1.7	+2
Caerin 1.9	*L. chloris*	GLFGVLGSIAKHVLPHVVPVIAEKL-NH_2	2591	16.5	6.4	1.6	+2
Caerin 1.19	*L. gracilenta*	GLFKVLGSVAKHLLPHVAPIIAEKL-NH_2	2649	10.3	3.9	4.7	+3
Caerin 1.19.3	*L. gracilenta*	GSVAKHLLPHVAPIIAEKL-NH_2	1991	—	—	—	+2
Caerin 1.8	*L. chloris*	GLFKVLGSVAKHLLPHVVPVIAEKL-NH_2	2662	4.5	1.7	3.6	+3
Caerin 1.8.1	*L. chloris*	GSVAKHLLPHVVPVIAEKL-NH_2	2006	—	—	—	+2

Group 3							
Signiferin 3.1	*C. signifera*	GIAEFLNYIKSKA-NH_2	1579	117.1	81.2	3.4	+2
Citropin 2.1	*L. citropa*	GLIGSIGKALGGLLVDVLKPKL-OH	2160	67.5	31.2	1.5	+3
Signiferin 4.3	*C. signifera*	GFADLFGKAVDFIKSRV-NH_2	1867	31.0	16.6	2.4	+2
Splendipherin	*L. splendida*	GLVSSIGKALGGLLADVVKSKGQPA-OH	2364	21.8	9.1	1.3	+2
Aurein 5.2	*L. aurea*	GLMSSIGKALGGLIVDVLKPKTPAS-OH	2450	19.0	7.7	1.4	+2
Frenatin	*L. infrafrenata*	GLMSVLGHAVGNVLGGLFKPKS-OH	2180	14.8	6.8	1.4	+2
Caerin 2.6	Hybrid[a]	GLVSSIGKVLGGLLADVVKSKGQPA-OH	2392	15.8	6.6	1.9	+2
Dahlein 5.2	*L. dahlii*	GLLGSIGNAIGAFIANKLKPK-OH	2080	5.4	2.6	4.6	+3
Dahlein 5.6	*L. dahlii*	GLLASLGKVFGGYLAEKLKPK-OH	2187	3.5	1.6	2.1	+3
Dahlein 5.3	*L. dahlii*	GLLASLGKVLGGYLAEKLKP-OH	2025	2.9	1.4	3.0	+2
Group 4							
Unnamed	*R. arvalis*[b]	FLPLLAASFACTVTKKC-OH	1809	131.8	72.8	1.1	+2
Unnamed	*R. arvalis*[b]	KNLLASALDKLKCKVTGC-OH	1902	61.7	32.1	1.8	+3
Brevinin 1R	*R. ridibunda*[c]	FFPAIFRLVAKVVPSIICSVIKKC-OH	2661	10.3	3.9	5.3	+4
Fallaxin 3	*L. fallax*[d]	GLLSFLPKVIGVIGHLIHPPS-NH_2	2193	33.7	15.4	4.0	+3
Dahlein 4.2	*L. dahlii*	GLWQFIKDKIKDAATGLVTGIQS-NH_2	2486	27.8	11.1	2.1	+2
Deserticolin 1	*C.deserticola*[e]	GLADFLNKAVGKVVDFVKS-NH_2	2005	4.8	2.4	1.6	+2
Cupiennin 1a[f]	*Cupiennius salei*	GFGALFKFLAKKVAKTVAKQAAKQGAKYVVNKQME-NH_2	3795	4.9	1.3	3.5	+8

Note: For individual references see (Pukala et al., 2006b) unless indicated to the contrary.

[a] Pukala et al., 2006a.

[b] Artemenko et al., 2007a.

[c] Artemenko et al., 2007b.

[d] Jackway, Bowie, and Tyler 2007.

[e] Maselli, Bowie, and Tyler 2007.

[f] Cupiennin 1a is both an antimicrobial and an nNOS inhibitor contained in the venom of the South American spider *Cupiennius salei*. It forms a 1:1 complex with Ca^{2+}CaM with the N-terminal region of cupriennin 1a held within the globular CaM complex (Pukala et al., 2007b).

FFPAIFRLVAKVVPSIICSVTKKC(OH) Brevinin 1R

FIGURE 14.6 *Rana ridibunda*, geographic location and brevinin 1R.

GLFDIVKKVVGAIGSL(NH_2) Aurein 2.3

FIGURE 14.7 *Litoria aurea* geographical location, and aurein 2.3. (From Barker, J., Grigg, G.C., and Tyler, M.J., *A Field Guide to Australian Frogs*, Surrey Beatty and Sons, Chipping Norton, New South Wales, Australia, p. 50, 1995, With permission.)

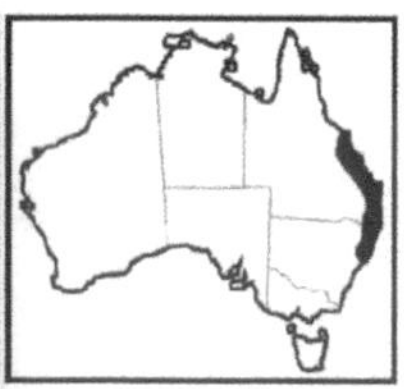

GLFKVLGSVAKHLLPHVVPVIAEKL(NH_2) Caerin 1.8

FIGURE 14.8 *Litoria chloris*, geographical location, and caerin 1.8 (www.fats.org.au/pr.html).

An interesting feature of three of the key peptides of groups 1 to 3, namely aurein 2.2 (group 1), caerin 1.1 (group 2), and frenatin 3 (group 3), is that although the sequences and solution structures of these active peptides are quite different, it is likely that they all evolved from the same precursor peptide. This can be seen from the results of cDNA cloning experiments summarized in Figure 14.10

GLLASLGKVFGGYLAEKLKPK(OH) Dahlein 5.6

FIGURE 14.9 *Litoria splendida* geographic location, and dahlein 5.6 (www.ryanphotographic.com/Litoria.com).

for aurein 2.2 (Chen et al., 2005), caerin 1.1 (from *Litoria splendid*, Pukala et al., 2006a; see also from *Litoria caerulea*, vanHoye et al., 2003), and frenatin 3 (Zhou et al., 2005). The signal regions of these three peptides are conserved, and the three spacer peptides also exhibit similarities. Antibiotic disulfide peptides from northern hemisphere anurans of the genus *Rana* have also been studied by cDNA cloning techniques (vanHoye et al., 2003). The example of the disulfide antibiotic brevinin 1E (from *Rana esculenta*) is shown, for comparison with the other peptides, in Figure 14.10. Brevinin 1E has a completely different sequence from those of the aureins, caerins, and frenatin 3 peptides: even so, its signal region is similar to the other examples shown in Figure 14.8. Perhaps all four active peptides shown in Figure 14.10 originated from the same ancestor peptide? If so, this evolutionary process was initiated before Gondwanaland dispersed more than 100 million years ago (compare also vanHoye et al., 2003).

The activities of peptides of groups 1 and 2 are multifaceted. Apart from the nNOS activities, peptides of group 1 are also membrane-active antimicrobial and anticancer compounds active at μM concentrations, and shown by 2D NMR studies to adopt amphipathic α-helical structures on the lipid bilayers of cell membranes (Apponyi et al., 2004). The caerins 1 (of group 2) are also membrane-active antimicrobial and anticancer active compounds, but in this case the peptides adopt a more complex structure on the lipid bilayer, with two α-helical sections separated by a central

Signal (Pre)

MAFLKKSLFLVLFLGLVSLSIC	Aurein 2.2
MASLKKSLFLVIFLGLVSLSIC	Caerin 1.1
MHFLKKSIFLVLFLGLVSLSIC	Frenatin 3
MFTLKKSNLLLPFLGTIMLSLC	Brevinin 1E

Spacer (Pro)

EKEKRQNEEDEDENEAANHEEGSEEKR	Aurein 2.2
EEEKRQEDEDEHEEEGESQEEGSEEKR	Caerin 1.1
EKEKREDQNEEEVDE---NEEESEEKR	Frenatin 3
EEERDADEEERRDNFDESEVEVEKR	Brevinin 1E

Active peptide

GLFDIVKKVVGALGSL(G)	Aurein 2.2
GLLSVLGSVAKHVLPHVVPVIAEHL(G)	Caerin 1.1
GLMSVLGHAVGNVLGGLFKPKS	Frenatin 3
FLPLLAGLAAMFLPKIRCKITRKC	Brevinin 1E

FIGURE 14.10 Prepropeptide sequences for aurein 2.2, caerin 1.1, and frenatin 3 from cloned cDNAs.

flexible hinge initiated by Pro15 (Apponyi et al., 2004) (Figure 14.2). In contrast, peptides of group 3 show no other activity (in our testing regime) except for nNOS activity. There is a specific exception to this: splendipherin is nNOS active and is also the aquatic male sex pheromone of *Litoria splendida* (Wabnitz et al., 1999, 2000).

Detailed studies of these amphibian peptides during *in vitro* testing show the observed Hill slopes (Fersht, 1987) to be greater than 1 (Table 14.1, see also Figure 14.5). A Hill slope equal to 1 indicates an interaction between a single active enzyme element and a single substrate, and would be expected if the peptide bound directly to the arginine substrate site (on heme) for nNOS. However, a Hill slope value of greater than unity is an indication of positive cooperativity, and suggests that some other noncompetitive interaction is causing the inhibition of nNOS (Fersht, 1987). We thought the most likely interaction of the nNOS active peptide would be to form a complex with the regulatory protein Ca^{2+}CaM, inducing a conformational change in the Ca^{2+}CaM that might prevent the binding of this cofactor to nNOS. Three peptides representative of the three active groups, namely citropin 1.1 (group 1), caerin 1.8 (group 2), and frenatin 3 (group 3) (Table 14.1) were chosen for particular study. The nNOS-deactivating experiments were set up such that addition of each active peptide stopped production of [^{3}H] citrulline. Excess Ca^{2+}CaM was then added to each test system. In each case, regeneration of nNOS activity commenced and proceeded slowly to about 50% of the [^{3}H] citrulline production recorded before the amphibian peptide was added. Thus it appears that the amphibian peptides do interact with Ca^{2+}CaM, either stopping it attaching to the CaM binding site on nNOS, or alternatively causing ineffectual binding at that site.

Ca^{2+}CaM is a dumbbell-shaped 148-residue-protein containing up to four Ca^{2+} units (Figure 14.11), which is required for the activation of nNOS, acting as an electron shuttle and calcium transporter. It also alters the conformation of the reductase domain of nNOS, allowing reactions to proceed at the heme site. Ca^{2+}CaM wraps itself around the CaM α-helical binding domain on eNOS, converting CaM into a globular shape (Aoyagi et al., 2003), analogous to that shown in Figure 14.12 for the 26-residue peptide fragment of skeletal myosin light-chain kinase with residues 3 to 21 encompassed within Ca^{2+}CaM (Ikura et al., 1991).

CaM is not only the regulatory protein for the NOS isoforms. It is also the regulatory protein for adenylate cyclase and for a range of phosphorylating kinases, including calcineurin (Klee and Vanaman, 1982). CaM is also involved in the regulation of the eukarytic cytoskeleton, and it is required by some protozoa for ciliate movement (Nakaoka et al., 1984). In addition to the experiments with citropin 1.1, caerin 1.8, and frenatin 3 with nNOS outlined above, we have carried out analogous experiments with calcineurin. These three peptides all inhibited the operation of calcineurin at μM concentrations, and the addition of excess Ca^{2+}CaM to each test system partially restored the operation of the calcineurin (Doyle et al., 2002). The likely scenario is therefore that the nNOS active peptides shown in Table 14.1 operate by complexation of Ca^{2+}CaM.

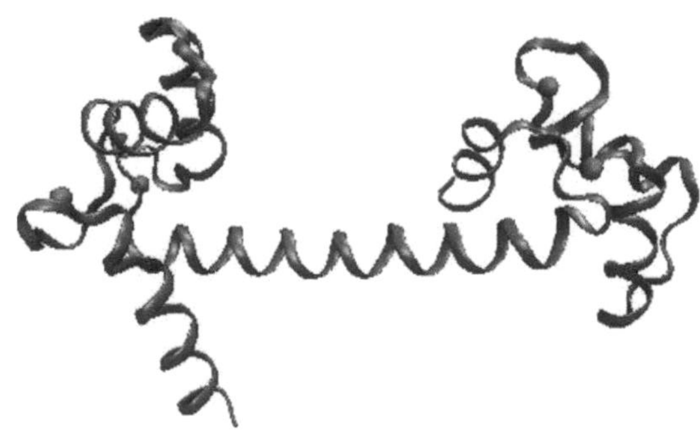

FIGURE 14.11 **(See color insert following page 176.)** Ca^{2+} calmodulin. The four Ca^{2+} ions are indicated in red.

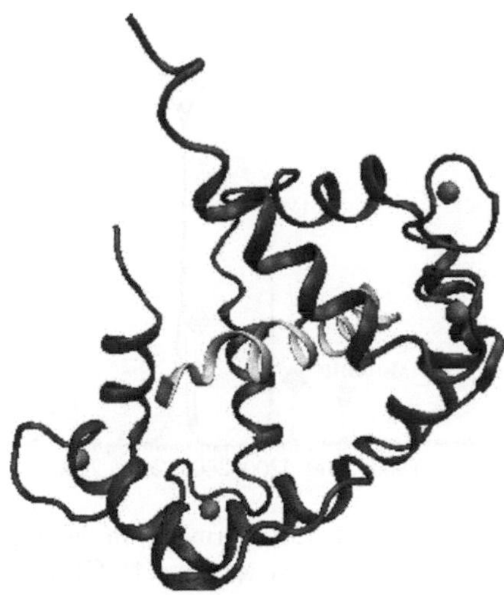

FIGURE 14.12 **(See color insert following page 176.)** Representation of globular complex formed between Ca^{2+}CaM and the 26-residue peptide fragment of skeletal myosin light-chain kinase (shown in green) with residues 3–21 encompassed within Ca^{2+}CaM. (Redrawn and modified from Ikura, M., Kay, L.E., Krinks, M., and Bax, A., *Biochemistry*, 30, 5498–5504, 1991.)

14.3 CONFIRMATION OF THE FORMATION OF ACTIVE PEPTIDE–Ca^{2+}CaM COMPLEXES

14.3.1 The Application of Electrospray Mass Spectrometry

Complexes formed between a peptide and a protein or between two proteins can often be observed using mass spectrometric methods, in particular electrospray mass spectrometry (ESMS), and the structures of these complexes often mirror those formed between the two interacting systems in aqueous conditions (Loo, 1997; Nemirovskiy et al., 1997).

The nNOS active peptides citropin 1.1, aurein 2.3 (group 1, Table 14.1), caerin 1.8 (group 2) and splendipherin, frenatin 3, dahlein 5.1, and dahlein 5.6 (group 3) were chosen for the mass spectrometric study. These peptides were tested for the formation of complexes in the gas phase with Ca^{2+}CaM using ESI negative ion mass spectrometry with a Micromass Q-TOF 2 mass spectrometer. Samples were directly injected in water containing ammonium acetate using a syringe pump at a flow rate of 10 mL/min. All samples showed the formation of Ca^{2+}–CaM–peptide complexes in the ratio 4:1:1. As an illustration, the ESI mass spectrum of the Ca^{2+}CaM complex with caerin 1.8 in the negative mode is recorded in Figure 14.13. Adducts are observed in multiply-charged states, usually $(M\text{-}6H)^{6-}$, $(M\text{-}7H)^{7-}$, $(M\text{-}8H)^{8-}$, and $(M\text{-}9H)^{9-}$ (Pukala et al., 2008; c.f. also Schnier et al., 1995).

Conformational changes can be readily monitored by mass spectrometry using H/D exchange experiments, because the rate of exchange of a given proton is dependent upon structurally related features such as solvent exposure and hydrogen bonding (Chattopadhyaya et al., 1992; Wintrode and Privalov, 1997). The extent of H/D exchange for Ca^{2+}CaM was monitored over a time course of 90 min both in the presence and absence of equimolar caerin 1.8 (the most active peptide listed in Table 14.1). During the 90 min time course, Ca^{2+}CaM exchanged a maximum of 117 amide protons for deuterons as determined by the increase in molecular mass. The extent of deuterium incorporation is therefore 80% of the total available amide hydrogens. In contrast, in the presence of caerin 1.8, only 107 (or 73%) of available amide protons are exchanged. The reduction in deuterium incorporation must result from new protection of exchangeable amide hydrogens, because the peptide bound CaM assumes a more compact conformation. These results indicate that a conformation change of Ca^{2+}CaM occurs upon addition of caerin 1.8, to a tighter, less solvent accessible form.

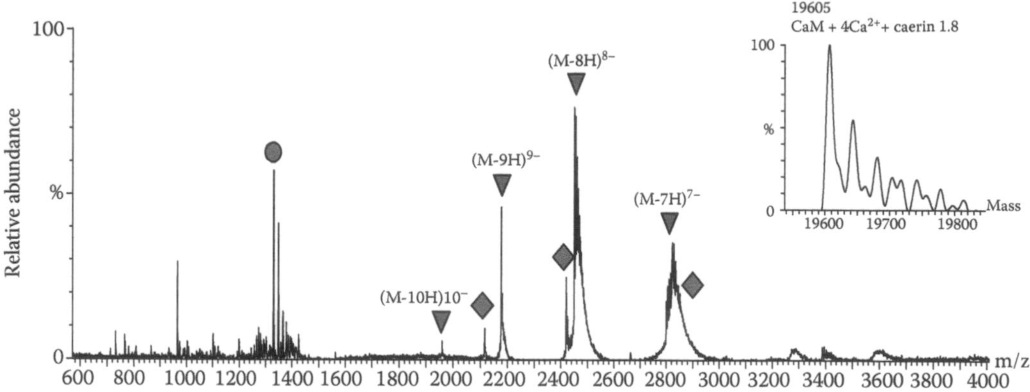

FIGURE 14.13 ESI-MS spectra of Ca^{2+}CaM with caerin 1.8, in the negative ion mode. Inset shows the transformed spectrum. Free peptide (●); free CaM (◆); peptide–Ca^{2+}CaM complexes (▼). Caerin 1.8 is GLFKVLGSVAKHLLPHVVPVIAEKL–NH_2.

14.3.2 The Application of Nuclear Magnetic Resonance Spectroscopy

The mass spectrometric investigations outlined above indicate that under ESI-MS conditions, 1:1:4 peptide–CaM–Ca^{2+} complexes are formed in which the whole of CaM is involved in the binding of the peptide. In principle, it should also be possible to identify the formation of a peptide–Ca^{2+}CaM complex by NMR titration. In such an experiment, different concentrations of the active peptide are titrated against ^{15}N-labeled Ca^{2+}CaM in high resolution 2D heteronuclear correlation spectroscopy (HSQC) experiments, with the chemical shift changes accompanying complex formation being monitored with increasing peptide concentration. The limiting factor here is that many of the peptides shown in Table 14.1 are quite hydrophobic and consequently not sufficiently soluble in water to enable such a titration to be carried out. Amongst the peptides in this category are the aureins 2 and citropins 1 (group 1, Table 14.1), and frenatin 3 and dahleins 5.2 and 5.3 (group 3). The three natural peptides chosen as suitable for such studies are caerin 1.8 (group 2, IC_{50} = 1.7 μM) together with dahlein 5.6 (group 3, IC_{50} = 1.6 μM) and splendipherin (group 3, IC_{50} = 9.1 μM). The results of the experiments are, in summary, that all three peptides form 1:1 complexes with Ca^{2+}CaM. The complex with splendipherin (the least active of the three peptides) shows weak binding, while the other two systems (the two most active natural nNOS inhibitors) show strong binding. NMR data for the 1:1 caerin 1.8–^{15}N-labeled Ca^{2+}CaM complex are shown in Figure 14.14 (Pukala et al., 2008, unpublished work).

Two peptide–CaM binding modes have been identified by previous NMR studies. In the first, Ca^{2+}CaM adopts a compact, globular shape, with the peptide engulfed in a hydrophobic channel formed by the two terminal domain (Ikura et al., 1991, 1992; Roth et al., 1991; Meador et al., 1992; Vetter and Leclerc, 2003). This is typified by the 26-residue peptide fragment of skeletal myosin light-chain kinase with residues 3 to 21 encompassed within Ca^{2+}CaM (Ikura et al., 1991, 1992) (Figure 14.12), and is also adopted by Ca^{2+}CaM when it binds to the CaM binding domain of eNOS (Aoyagi et al., 2003). The second binding example is when the C-terminal lobe of Ca^{2+}CaM binds with the 20-residue binding domain of the plasma membrane Ca^{2+} pump–Ca^{2+}CaM complex, with the first 12 residues of the peptide encompassed within the C-terminal end of Ca^{2+}CaM (Elshordt et al., 1999). The chemical shift data for the Ca^{2+}CaM caerin 1.8 and dahlein 5.6 complexes are consistent with a globular CaM structure in accordance with the data provided by ESI mass spectrometry (Pukala et al., 2008).

The complex formed between caerin 1.8 and Ca^{2+}CaM has been studied in more detail in order to determine how much of caerin 1.8 is contained within the Ca^{2+}CaM complex and how much of the

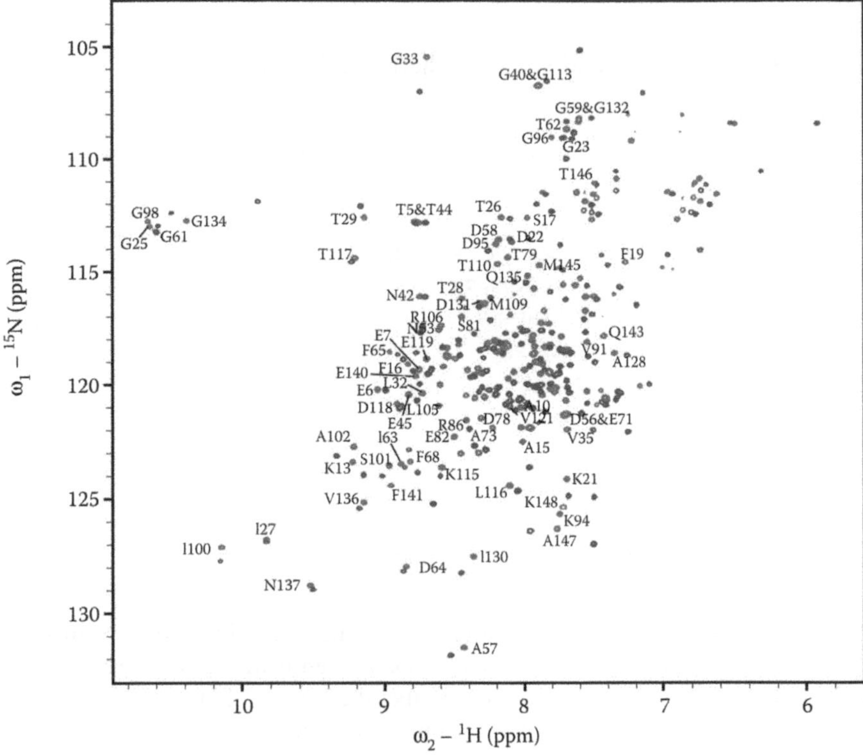

FIGURE 14.14 **(See color insert following page 176.)** ^{15}N HSQC spectra of Ca^{2+} ^{15}N-CaM in the absence of peptide (red), and with the addition of caerin 1.8 in a 2:1 peptide:protein ratio (purple).

peptide is required for nNOS activity. 2D NMR experiments show that caerin 1 peptides have random conformations in water but adopt helical–hinge–helical structures in membrane-mimicking solvents like trifluoroethanol (TFE)/water (Apponyi et al., 2004) (see Figure 14.2 for caerin 1.1). Caerin 1.8 will thus be conformationally random in aqueous conditions as it approaches the cofactor Ca^{2+}CaM:

GLFK**VL**GSV**A**KHL**L**PHV**V**PVI**AE**KL–NH_2 ^{15}N labeled caerin 1.8

^{15}N HSQC titrations of several ^{15}N labeled peptides (^{15}N labeled backbone nitrogens are identified in bold face above) with varying concentrations of Ca^{2+}CaM were carried out (Pukala et al., 2008). The chemical shift changes were then tracked by overlaying the spectra. Large chemical shift changes were obtained for Val^5, Leu^6, Ala^{10}, and Leu^{14}. A slight shift was noted for Val^{18}, but there were no changes in the chemical shifts of Ala^{22} and Glu^{23}. Thus the N-terminal region of caerin 1.8 is contained within the Ca^{2+}CaM complex, with that section of caerin 1.8 after Val^{18} lying outside the complex. The chemical shift data are consistent with the N-terminal region of caerin 1.8 up to Val^{18} being α-helical inside the globular CaM. This should be compared with the structure shown in Figure 14.12, which shows 18 residues of skeletal myosin light-chain kinase encompassed within Ca^{2+}CaM (Ikura et al., 1991).

A number of synthetic modifications of caerin 1.8 were synthesized in order to determine whether their activity data are in accordance with the structural data determined by NMR methods. The sequences of the modifications and their nNOS activity data are recorded in Table 14.2. The structural and activity data are consistent. Interestingly, the most active modification of caerin 1.8 is that where the two Pro residues are replaced by Ala (mod 1). Removing the N-terminal end of the

TABLE 14.2
nNOS Activities of Caerin 1.8 and Some Synthetic Modifications

Name	Species	Sequence	MW	IC_{50} mg/mL	IC_{50} mM	Hill Slope	Charge
Caerin 1.8	*L.chloris*	GLFKVLGSVAKHLLPHVVPVIAEKL-NH_2	2662	4.5	1.7	3.6	+3
Mods							
1		GLFKVLGSVAKHLLAHVVAVIAEKL-NH_2	2606	2.6	1.0	1.3	+3
2		GLFKVLGSVAKHLLPHVVP-NH_2	2010	3.0	1.5	1.4	+3
3		GLFKVLGSVAKHLLPHV-NH_2	1814	4.1	1.7	1.3	+3
4		GLFKVLGSVAKHLLP-NH_2	1577	4.9	3.1	1.5	+3
5		GLFKVLGSVAKHL-NH_2	1367	5.1	3.7	1.6	+3
6		GLFKVLGSVAK-NH_2	1117	3.7	3.3	1.4	+3
7		GLFKVLGSV-NH_2	918	48.0	52.3	0.9	+2
8		GLFKVLGS-NH_2	819		>850	—	+2
9		GSVAKHLLPHVVPVIAEKL-NH_2	2006	—	—	—	+2

peptide (mod 9) destroys the activity, whereas removal of the C-terminal region from Val18 (mod 3) has minimal effect on the activity. nNOS activity remains significant to modification 7 (11 residues, GLFKVLGSVAK–NH_2, $IC_{50} = 3.3\ \mu M$), but after that the activity drops away. So the NMR data show that up to 18 residues of caerin 1.8 lie within a globular CaM complex, while activity data indicate that the first 11 of these residues are the minimum prerequisite for significant nNOS inhibition (Pukala et al., 2008).

Finally, although we were not able to use NMR to determine a structure of the citropin 1.1–Ca^{2+}CaM complex (because of solubility problems with the hydrophobic citropin 1.1), ESI mass spectrometric data indicated the formation of a strong complex, which required the presence of all four Ca^{2+} ions. This again suggests the formation of a globular complex. A number of synthetic modifications of citropin 1.1 have been made; their structures and nNOS activity data are recorded in Table 14.3 (Doyle et al., 2003). The following major points follow from consideration of these data:

1. The activity of the synthetic all D-isomer is much less than that of the natural L-isomer, suggesting that the helical arrangement of citropin 1.1 within the complex is important [citropin 1.1 is an amphipathic α-helix in the membrane mimicking solvent system TFE/water (Figure 14.15)] (Doyle et al., 2003).
2. Removal of residues from either end of citropin 1.1 reduce the activity suggesting that most of the 16-residue citropin 1.1 is contained within the CaM complex.
3. The activity increases with increasing positive charge of the peptide.
4. At least one of Lys^7 and Lys^8 must be present for activity.
5. The hydrophilic residues Asp^4 and Ser^{11} may be replaced by Ala with minimal loss in activity.

14.4 CONCLUSIONS

Of the 40 species of Australian frogs that we have studied, 17 species exude at least one nNOS active peptide from the skin glands onto the skin when the animal is hurt, stressed, or attacked. These nNOS inhibitors almost certainly form part of the animal's primary defense, by interfering with the NO messenger capabilities of an invading predator or pathogen. nNOS active peptides have been identified in the Australian anuran genera *Litoria*, *Crinia*, and *Uperoleia*. Analogous peptides

TABLE 14.3
nNOS Activities of Citropin 1.1 and Some Synthetic Modifications

Name	Species	Sequence	MW	IC_{50} µg/mL	IC_{50} µM	Hill Slope	Charge
Citropin 1.1	*L. citropa*	GLFDVIKKVASVIGGL-NH_2	1614	13.3	8.2	2.0	+2
D-citropin 1.1		GlfdvikkvasviGGl-NH_2	1614	49.5	30.7	1.0	+2
Citropin 1.1.2	*L. citropa*	DVIKKVASVIGGL-NH_2	1297	55.0	42.4	1.5	+2
Citropin 1.2.3	*L. citropa*	GLFDIIKKVAS-NH_2	1071	24.2	22.6	2.2	+2
Retro citropin 1.1		LGGIVSAVKKIVDFLG-NH_2	1614	24.2	15.0	1.3	+2
1		GLFDVAAAVASVIGGL-NH_2	1550	29.6	19.0	0.7	0
2		GLFDVIKKVASVIGGA-NH_2	1572	12.4	7.9	1.7	+2
3		GLFKVIKKVASKIGGL-NH_2	1643	11.5	7.0	2.3	+3
4		GLFDVIKKVAAVIGGL-NH_2	1599	8.0	5.0	1.4	+2
5		GLFDVIKAVASVIGGL-NH_2	1557	7.0	4.5	2.1	+1
6		GLFEVIKKVASVIGGL-NH_2	1628	6.8	4.2	1.8	+2
7		GLFDVIKKVASVIKGL-NH_2	1683	6.8	4.0	3.0	+3
8		GLADVIKKVASVIGGL-NH_2	1537	5.1	3.3	1.6	+2
9		GLFDVIKKVASKIKGL-NH_2	1714	5.0	2.9	2.1	+4
10		GLFAVIKKVASVIGGL-NH_2	1570	4.3	2.7	2.1	+3
11		GLFDVIAKVASVIGGL-NH_2	1557	3.8	2.4	3.4	+1
12		GLFKVIKKVASVIGGL-NH_2	1627	3.4	2.1	2.1	+4
13		GLFDVIKKVASVIKKL-NH_2	1756	3.5	2.0	2.5	+4
14		KLFAVIKKVAAVIGGL-NH_2	1625	2.1	1.3	3.0	+3
15		GLFKVIKKVAKVIKKL-NH_2	1810	2.2	1.2	3.3	+7
16		GLFAVIKKVAKVIKKL-NH_2	1753	2.1	1.2	2.2	+6
17		GLFAVIKKVAAVIKKL-NH_2	1696	1.9	1.1	3.8	+5
18		GLFAVIKKVAAVIRRL-NH_2	1752	1.9	1.1	4.6	+5
19		GLFAVIKKVASVIKKL-NH_2	1712	1.6	1.0	2.3	+5
20		KLFAVIKKVAAVIRRL-NH_2	1823	1.9	1.0	4.4	+5
21		GLFDVIAKVASVIKKL-NH_2	1699	1.6	0.9	4.0	+3

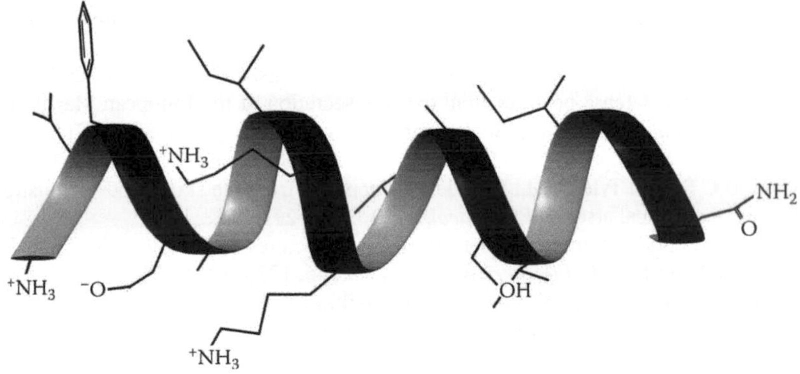

FIGURE 14.15 The solution structure of citropin 1.1 determined by 2D NMR in the solvent system TFE/H_2O (1:1). This is the expected structure of citropin 1.1 within a globular Ca^{2+}CaM–citropin 1.1 complex. (Adapted from Pukala, T.L. et al., *Nat. Prod. Rep.*, 23, 368–393, 2006b.)

are likely to be produced by anurans from other continents, because we have shown (in one case) that a disulfide antibiotic peptide from a European *Rana* species also inhibits the formation of NO by nNOS. In this context, we have also identified a 35-residue nNOS-inhibiting peptide in the venom of a South American spider (Pukala et al., 2007b). In one Australian tree frog, *Litoria infrafrenata*, the major skin peptide shows only nNOS activity. Most studied species, however, produce both nNOS inhibitors and antimicrobial peptides. The nNOS inhibiting peptide often has other activities, including wide-spectrum antimicrobial activity.

Peptides that are nNOS active are hydrophobic and cationic. They are normally conformationally unstructured in water. When the 3D structure is determined by 2D NMR in a membrane-mimicking solvent like TFE/water (1:1), some peptides adopt a simple α-helical structure, while others contain both α-helical and flexible hinge regions. These peptides adopt the more structured conformation when complexed with Ca^{2+}CaM. For those systems that have been studied in detail (by ESI-MS and 2D NMR), the active peptide is part of a globular Ca^{2+}CaM complex. The change in shape of Ca^{2+}CaM inhibits its binding to the α-helical Ca^{2+}CaM binding domain of nNOS. Some peptides, like the 16-residue citropin 1.1, are mostly contained within the globular CaM, whereas the hinged peptide caerin 1.8 has only the N-terminal portion of the peptide inside the complex. *In vitro* testing of synthetic modifications of caerin 1.8 confirm that it is the N-helical portion of caerin 1.8 that is necessary for nNOS inhibition.

ACKNOWLEDGMENT

J.H.B. thanks the Australian Research Council for funding the anuran peptide research.

REFERENCES

Abu-Soud, H. and Stuehr, D., Nitric oxide synthases reveal a role for calmodulin in controlling electron transfer, *Proc. Natl Acad. Sci. USA*, 90, 10769–10772, 1993.

Abu-Soud, H., Yoho, L., and Stuehr, D., Calmodulin controls neuronal nitric-oxide synthase by a dual mechanism. Activation of intra- and interdomain electron transfer, *J. Biol. Chem.*, 269, 32047–32050, 1994.

Alderton, W.K., Copper, C.E., and Knowles, R.G., Nitric oxide synthases: Structure, function and inhibition, *Biochem. J.*, 357, 593–615, 2001.

Anastasi, A., Erspamer, V., and Endean, R., Isolation and amino acid sequence of caerulein, the active decapeptide of the skin of *Hyla caerulea*, *Arch. Biochem. Biophys.*, 125, 57–68, 1968.

Andrew, P.J. and Mayer, B., Enzymatic function of nitric oxide synthases, *Cardiovascular Res.*, 43, 521–531, 1999.

Aoyagi, M., Arvai, A.S., Tainer, J.A., and Getzoff, E.D., Structural basis for endothelial nitric oxide synthase binding to calmodulin, *EMBO J.*, 22, 766–775, 2003.

Apponyi, M.A. et al., Host-defense peptides of Australian anurans: Structure, mechanism of action and evolutionary significance, *Peptides*, 25, 1035–1054, 2004.

Artemenko, K., Samgina,T., Lebedev, A.T., and Bowie, J.H., unpublished observations, 2007a.

Artemenko, K. et al., Host defense peptides from the skin secretion of the European Marsh Frog *Rana ridibunda*, *Mass Spektrometria* (Russian Society for Mass Spectrometry), 79–89, 2007b.

Bevins, C.L. and Zasloff, M., Peptides from frog skin, *Annu. Rev. Biochem.*, 59, 395–414, 1990.

Bowie, J.H., Chia, B.C.S., and Tyler, M.J., Host defense peptides from the skin glands of Australian amphibians: A powerful chemical arsenal, *Pharmacol. News*, 5, 16–21, 1998.

Bradford, A.M. et al., Peptides from Australian frogs. The structures of the dynastins from *Limnodynastes salmini* and fletcherin from *L. Fletcheri*, *Aust. J. Chem.*, 46, 1235–1244, 1993.

Brandt, R.B., Laux, J.E., and Yates, S.W., Calculation of inhibitor Ki and inhibitor type from the concentration of inhibitor for 50% inhibition for Michaelis–Menten enzymes, *Biochem. Med. Metab. Biol.*, 37, 344–349, 1987.

Bredt, D.S. et al., Cloned and expressed nitric oxide synthase structurally resembles cytochrome P-450 reductase, *Nature*, 351, 714–718, 1991.

Brinkworth, C.S. et al., The solution structure of frenatin 3, a neuronal nitric oxide synthase inhibitor from the giant tree frog, *Litoria infrafrenata*, *Biopolymers*, 70, 424–434, 2003.

Chattopadhyaya, R., Meador, W.E., Means, A.R., and Quiocho, F.A., Calmodulin structure refined at 1.7 Å resolution, *J. Mol. Biol.*, 228, 1177–1192, 1992.

Chen, T. et al., The structural organization of aurein presursor cDNAs from the skin secretion of the Australian green and golden bell frog, *Litoria aurea, Regul. Pept.*, 128, 75–83, 2005.

Cheng, Y. and Prusoff, W.H., Relationship between the inhibition constant (K_i) and the concentration of inhibitor which causes 50 per cent inhibition (IC_{50}) of an enzymatic reaction, *Biochem. Pharmacol.*, 22, 3099–3108, 1973.

Choi, W.S. et al., Identification of NOS in *Staphylococcus aureus*, *Biochem, Biophys. Res. Commun.*, 237, 554–558, 1997.

Doyle, J.R. et al., Amphibian peptides that inhibit nNOS: The isolation of lesueurin from the skin secretion of the Australian Stony Creek Frog *Litoria lesueuri*, *Eur. J. Biochem.*, 269, 100–109, 2002.

Doyle, J.R. et al., nNOS inhibition, antimicrobial and anticancer activity of the amphibian skin peptide, citropin 1.1 and synthetic modifications: The solution structure of a modified citropin 1.1, *Eur. J. Biochem.*, 270, 1141–1153, 2003.

Elshordt, B. et al., NMR solution structure of a complex of calmodulin with a binding peptide of the Ca^{2+} pump, *Biochemistry*, 38, 12320–12332, 1999.

Erspamer, V., Bioactive secretions in the amphibian integument, in *Amphibian Biology: The Integument*, Heatwole, H., Ed., Surrey, Beatty, and Sons, Chipping Norton, 1994, pp. 178–350.

Fersht, A., *Enzyme Structure and Mechanism*, W.H. Freeman, New York, NY, USA, 1987.

Gachhui, R. et al., Characterization of the reductase domain of rat neuronal nitric oxide synthase generated in the methylotrophic yeast *Pichia pastoris*, *J. Biol. Chem.*, 271, 20594–20602, 1996.

Gobbetti, A. and Zeran, M., Hormonal and cellular brain mechanisms regulating the amplexus of male and female water frog (*Rana esculenta*), *J. Neuroendocrinol.*, 11, 589–596, 1999.

Ikura, M., Kay, L.E., Krinks, M., and Bax, A., Triple-resonance multidimensional NMR study of calmodulin complexed with the binding domain of skeletal muscle myosin light-chain kinase: Indication of a conformational change in the central helix, *Biochemistry*, 30, 5498–5504, 1991.

Ikura, M. et al., Solution structure of a calmodulin–target peptide complex by multidimensional NMR, *Science*, 256, 632–638, 1992.

Jackway, R.J., Bowie, J.H., and Tyler M.J., unpublished observations, 2007.

Kerwin, J.F., Nitric oxide: A new paradigm for second messengers, *J. Med. Chem.*, 38, 4343–4362, 1995.

Klee, C.B. and Vanaman, T.C., Calmodulin, *Adv. Protein Chem.*, 35, 213–321, 1982.

Lazarus, L.H. and Attila, M., The toad, ugly and venomous, wears yet a precious jewel in his skin, *Prog. Neurobiol.*, 41, 473–507, 1993.

Lincoln, J., Hoyle, C.H.V. and Burnstock, G., Synthesis and properties of nitric oxide, in *Nitric Oxide in Health and Disease*, Cambridge University Press, Cambridge, 1997, p. 12–26.

Loo, J.A., Studying noncovalent protein complexes by electrospray ionization mass spectrometry, *Mass Spectrom. Rev.*, 16, 1–23, 1997.

Marletta, M.A., Nitric oxide synthase structure and mechanism, *J. Biol. Chem.*, 268, 12231–12234, 1993.

Marletta, M.A., Nitric oxide synthase: Aspects concerning structure and catalysis, *Cell*, 78, 729–730, 1994.

Maselli, V.M., Bowie, J.H., and Tyler, M.J., unpublished observations, 2007.

Matsubara, M., Hayashi, N., Titani, K., and Taniguchi, H., Circular dichroism and ^{1}H NMR studies on the structures of peptides derived from the calmodulin-binding domains of inducible and endothelial nitric-oxide synthase in solution and in complex with calmodulin, *J. Biol. Chem.*, 272, 23050–23056, 1997.

Meador, W.E., Means, A.R., and Quiocho, F.A., Target enzyme recognition by calmodulin: 2.4 Å structure of a calmodulin peptide complex, *Science*, 257, 1251–1255, 1992.

Molero, M. et al., Modulation by nitric oxide of gastric acid secretion in toads, *Acta Physiol. Scand.*, 164, 229–236, 1998.

Morita, H. et al., Synthesis of nitric oxide from the two equivalent guanidino nitrogens of L-arginine by *Lactobacillus fermentum*, *J. Bacteriol.*, 179, 7812–7815, 1997.

Nakaoka, Y., Tanaka, H., and Oosawa, F., Ca^{2+} dependent regulation of beat frequency of cilia in *Paramecium*, *J. Cell. Sci.*, 65, 223–231, 1984.

Nemirovskiy, O., Ramanthan, R., and Gross, M.L., Investigation of calcium-induced, noncovalent association of calmodulin with melittin by electrospray ionization mass spectrometry, *J. Am. Soc. Mass Spectrom.*, 8, 809–812, 1997.

Pukala, T.L. et al., Host-defense peptide profiles of the skin secretions of interspecific hybrid tree frogs and their parents, female *Litoria splendida* and male *Litoria caerulea*, *FEBS J.*, 273, 3511–3519, 2006a.

Pukala, T.L. et al., Host-defense peptides from the glandular secretions of amphibians: Structure and activity, *Nat. Prod. Rep.*, 23, 368–393, 2006b.

Pukala, T.L. et al., Solution structure and interaction of cupiennin 1a, a spider venom peptides, with phospholipid bilayers, *Biochemistry*, 46, 3576–3585, 2007a.

Pukala, T.L. et al., Cupiennin 1a, an antimicrobial peptide from the venom of the neotropical wandering spider *Cupiennius salei*, also inhibits the formation of nitric oxide by neuronal nitric oxide synthase, *FEBS J.*, 274, 1778–1784, 2007b.

Pukala, T.L. et al., Binding studies of nNOS active amphobian peptides with Ca^{2+} calmodulin, using negative ion electrospray mass spectrometry. *Rapid Commun. Mass Spectrom.*, 22, 3501–3509, 2008.

Pukala, T.L. et al., unpublished work, 2008.

Raftery, M.J. et al., Peptides from Australian frogs. The structures of the dynastins from the Banjo Frogs *Limnodynastes interioris*, *L.dumerilii* and *L.terraereginae*, *Aust. J. Chem.*, 46, 833–842, 1993.

Renteria, R.C. and Constantine-Paton, M., Nitric oxide in the retinotectal system: A signal but not a retrograde messenger during map refinement and segregation, *J. Neurosci.*, 19, 7066–7076, 1999.

Roth, S.M. et al., Structure of the smooth muscle myosin light-chain kinase calmodulin-binding domain peptide bound to calmodulin, *Biochemistry*, 30, 10078–10084, 1991.

Rozek, T. et al., The antibiotic and anticancer active aurein peptides from the Australian bell frogs *Litoria aurea* and *Litoria raniformis*. The solution structure of aurein 1.2, *Eur. J. Biochem.*, 267, 5330–5341, 2000.

Schnier, P.D., Gross, D.S., and Williams, E.R., On the maximum charge state and proton transfer reactivity of peptide and protein ions formed by electrospray ionization, *J. Am. Soc. Mass Spectrom.*, 6, 1086–1097, 1995.

Stone, D.J.M., Bowie, J.H., Tyler, M.J., and Wallace, J.C., The structure of caerin 1.1, a novel antibiotic peptide from Australian tree frogs, *Chem. Commun.*, 6, 400–402, 1992.

Stuehr, D.J., Mammalian nitric oxide synthases, *Biochim. Biophys. Acta*, 1411, 217–230, 1999.

Tyler, M.J., Stone, D.J.M., and Bowie, J.H., A novel method for the release and collection of dermal, glandular secretions from the skin of frogs, *J. Pharm. Toxicol. Methods*, 28, 199–200, 1992.

vanHoye, D., Bruston, F., Nicolas, P., and Amiche, M., Antimicrobial peptides from hylid and ranid frogs originated from a 150-million-year-old ancestral precursor with a conserved signal peptide but a hypermutable antimicrobial domain, *Eur. J. Biochem.*, 270, 2068–2081, 2003.

Vetter, S.W. and Leclerc, E., Novel aspects of calmodulin target recognition and activation, *Eur. J. Biochem.*, 270, 404–414, 2003.

Wabnitz, P.A. et al., Aquatic sex pheromone of a male tree frog, *Nature*, 401, 2249–2251, 1999.

Wabnitz, P.A. et al., Differences in the skin peptides of the male and female Australian Magnificent Tree Frog *Litoria splendid*—a three year monitoring survey: The discovery of the male sex pheromone splendipherin together with phe8 caerulein and the antibiotic caerin 1.10, *Eur. J. Biochem.*, 267, 269–275, 2000.

Watanabe, Y., Hu, Y., and Hidaka, H., Identification of a specific amino acid cluster in the calmodulin-binding domain of the neuronal nitric oxide synthase, *FEBS Lett.*, 403, 75–78, 1996.

Waugh, R.J. et al., The structures of the frenatin peptides from the skin secretion of the Giant Tree Frog *Litoria infrafrenata*, *J. Pept. Sci.*, 2, 117–124, 1996.

Wegener, K.L. et al., Host defense peptides from the skin glands of the Australian Blue Mountains tree frog *Litoria citropa*. Solution structure of the antibacterial peptide citropin 1.1, *Eur. J. Biochem.*, 265, 627–637, 1999.

Wintrode, P.L. and Privalov, P.L., Energetics of target peptide recognition by calmodulin: A calorimetric study, *J. Mol. Biol.*, 266, 1050–1062, 1997.

Wong, H., Bowie, J.H., and Carver, J.A., The solution structure and activity of caerin 1.1, an antimicrobial peptide from the Australian green tree frog, *Litoria splendida*, *Eur. J. Biochem.*, 247, 545–557, 1997.

Zhou, M., Chen, T., Walker B., and Shaw, C., Novel frenatins from the skin of the Australasian giant white-lipped tree frog. *Litoria infrafrenata*: Cloning of precursor cDNAs and identification in defensive secretion, *Regul. Pept.*, 26, 2445–2451, 2005.

15 Host-Defense Peptides from the Secretion of the Skin Glands of Frogs and Toads: Membrane-Active Peptides from the Genera *Litoria*, *Uperoleia*, and *Crinia*

John H. Bowie, Rebecca J. Jackway, Frances Separovic, John A. Carver, and Michael J. Tyler

CONTENTS

15.1 INTRODUCTION

Frogs evolved from freshwater fish some 300 to 100 million years ago in the Devonian period. They have survived to this day, due, at least in part, to the host-defense compounds that are contained in specialized glands on the dorsal surface and in the gut. There are different classes of compounds secreted from these glands, among which are alkaloids, amines, and peptides. Among the bioactive peptides are smooth muscle active compounds, hormones, opioids, antibiotic, anticancer, antiviral, and antifungal peptides, which protect the frogs from predators both large and small (for reviews see Bevins and Zasloff, 1990; Lazarus and Attila, 1993; Erspamer, 1994; Apponyi et al., 2004; Pukala et al., 2006). More recently, it has been demonstrated that some frogs also contain peptides that inhibit the formation of the cell signaling agent nitric oxide by neuronal nitric oxide synthase (nNOS). These peptides act by complexing with the regulatory protein Ca^{2+} calmodulin, changing the shape of calmodulin, which hinders its attachment to the calmodulin binding domain on nNOS (Pukala et al., 2006; see also Chapter 14 on nNOS active peptides).

Early methods used to obtain active compounds from frog skins involved the killing of large numbers of animals (Erspamer, 1994). Methods used today that do not kill frogs include injection of noradrenaline (see for example, Gibson et al., 1986) or the noninvasive electrical stimulation (ES) method (Tyler et al., 1992). It is normal with the sophisticated analytical instruments used today for one secretion from one animal (by ES) to provide sufficient material to enable the identification of the entire skin peptide profile. Active peptides are separated by electrophoresis, column chromatography, or more commonly by high-performance liquid chromatography. Sequence determination of peptides is carried out using mass spectrometry and/or Edman degradation methods, with the secondary structure determined by 2D and/or 3D NMR, or less likely by X-ray methods (Pukala et al., 2006).

Active peptides are first formed as a larger prepropeptide (signal-spacer-active peptide), with the peptide stored in the appropriate gland as the inactive propeptide (spacer-active peptide). When the animal is stressed, unwell, or attacked, the propeptide is cleaved by a proteolytic enzyme and the active peptide exuded from the gland in a gel onto the skin. RNA/cDNA encoding is now a routine method providing the structures of the initially formed prepropeptides (van Hoye et al., 2003; Zhou et al., 2005).

It has been assumed that these host-defense peptides are only produced by an anuran when the animal has formed the dorsal glands in the metamorph stage; this has been supported by DNA data (Clark et al., 1994). However, this raises the question as to how tadpoles protect themselves in a static aqueous environment that may abound with pathogens. In at least one frog species, *Litoria splendida* (where the fertilized egg to metamorph cycle takes some 90 days), it has been demonstrated that the tadpole produces precisely the same active host-defense peptides as does the adult, but how and where these native peptides are formed from the prepropeptide is not understood (Wabnitz et al., 1998).

This chapter focuses on frogs of the genera *Litoria*, *Uperoleia*, and *Crinia*; all are found in Australasia. There are thousands of species of anuran (frogs and toads) worldwide, but only some 200 in Australasia. Many of these are indigenous to Australia, some having evolved following or during that period from 100 to 80 million years ago when Australia separated from Gondwanaland. Many Australian frogs contain as major components in their skin peptides one or other of the following types of peptide, namely (i) smooth muscle active agents (normally active at nM concentrations), which sometimes also act as opioids, (ii) a cocktail of antimicrobial peptides (normally active at μM concentrations), and (iii) an nNOS-inhibiting peptide (active at μM concentrations).

This report deals with membrane-active antibiotic peptides, many of which have multifaceted activity, exhibiting antibiotic, anticancer, antifungal, and antiviral activity, and sometimes nNOS inhibition. The antimicrobial activity is a function of the penetration of the cell membrane of the organism, effecting lysis of the membrane of the bacterium or the envelope of the virus (these peptides are not active against viruses without envelopes). A number of scenarios have been proposed to explain membrane penetration and lysis. The first is the "barrel stave" or pore-forming mechanism, where α-helical peptides (containing at least 20 amino-acid residues) bind initially to the outside of the lipid bilayer, then penetrate the bilayer to produce pores that are perpendicular to the bilayer (Matsuzaki, 1998; Hara et al., 2001; Ambroggio et al., 2004; Huang et al., 2004). The barrel stave process is shown in Figure 15.1a. The second process is a modification of the pore mechanism and is called the "toroidal" model (Ludtke et al., 1996; Matsuzaki, 1998). Here, lipids are interspersed between the peptides of the pore such that negatively charged head groups are associated with positive charges of the peptides' side chains and the aqueous pore interior. The pores formed by this process are larger than those formed by the barrel stave mechanism. The toroidal model is shown in Figure 15.1b. The third process is called the "carpet" mechanism. Here, peptides remain bound to the membrane interface and disrupt the lipid bilayer with a carpet-like effect. Holes are formed due to strain on the bilayer, which degrades into micelle complexes (Shai and Oren, 2001).

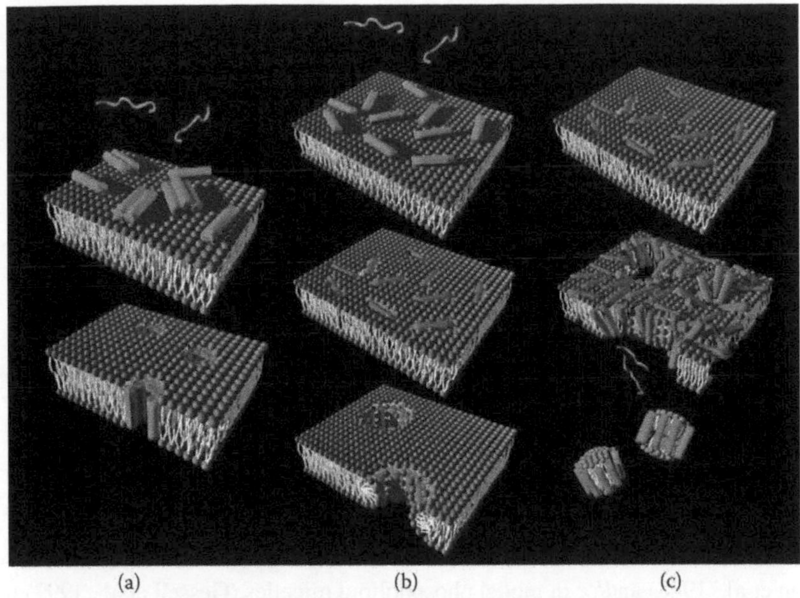

FIGURE 15.1 **(See color insert following page 176.)** Bilayer disruption mechanisms. (a) Barrel stave mechanism. (b) Toroidal mechanism. (c) Carpet mechanism. Helical forms of peptides are represented by cylinders, while hydrophobic and hydrophilic regions are colored red and blue, respectively.

A representation of the carpet mechanism is shown in Figure 15.1c. All three processes result in cell lysis.

Historically, the first antimicrobial peptides from an anuran that were studied in depth were the magainins, isolated from the African clawed frog *Xenopus laevis*, isolated independently by Williams (Gibson et al., 1986; Giovannini et al., 1986; Terry et al., 1988) and Zasloff (Zasloff, 1987; Cruciani et al., 1991). This and subsequent research on the magainins has been so important for the understanding of amphibian antimicrobial agents that it is summarized in brief here. The sequence of magainin 2 is shown in Figure 15.2. It is of interest to note that the magainin peptides are cytotoxic to *Xenopus laevis*, so after the magainins have been on the dorsal surface of the frog for a sufficient period, an endoprotease cleaves the magainins centrally, removing all activity (Resnick et al., 1991).

GIGKFHSAKKFGKAFVGEIMNS(OH) Magainin 2

FIGURE 15.2 *Xenopus laevis*, geographic location, magainin 2 (expasy.org/spotlight/back_issues/sptlt007.shtml).

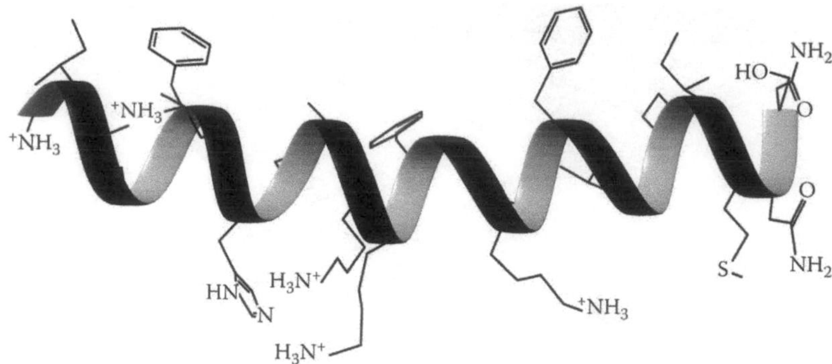

FIGURE 15.3 2D NMR structure (in TFE/H_2O) of magainin 2. (From Pukala, T.L. et al., *Nat. Prod. Rep.*, 23, 368–393, 2006b. With permission.)

Two-dimensional NMR studies in the membrane-mimicking solvent trifluoroethanol–water (TFE/H_2O) (Marion et al., 1988) and/or in model phospholipid micelles (Gessell et al., 1997) together with Fourier transform infrared (FTIR) investigations (Matsuzaki, 1998, 1999) show that the magainins adopt stable α-helical conformations in membrane environments. Magainin 2 is shown in Figure 15.3. The magainins are cationic, hydrophobic, and amphipathic, and exhibit activity against both Gram-negative and -positive organisms, generally at the minimum inhibitory concentration (MIC) range of 50 to 100 μg/mL. The magainins penetrate the bacterial membrane by means of a pore-type mechanism (Matsuzaki, 1998, 1999; Tachi et al., 2002). The magainins and some synthetic derivatives also exhibit anticancer (Ohsaki et al., 1992), antiviral (Aboudy et al., 1994), and antifungal activity (Lee et al., 2002) and also lyse protozoa (U.S. patent, 1992). The magainins show spermicidal activity (Wojak et al., 2000; Mystkowska et al., 2001; Zasloff and Anderson, 2001), and it has been suggested that the synthetic modification Ala(8,13,18)magainin 2 may show potential as an anti-implantation strategy for intercepting pregnancy (Dhawan et al., 2000).

A feature of some of the peptides that we will discuss is that, like the magainins, they often have other activities as well as their antibiotic activity. For example, the best studied membrane-active peptide from the genus *Litoria*, caerin 1.1 (GLLSVLGSVAKHVLPHVVPVIAEHL–NH_2), is a wide-spectrum antibiotic, an anticancer agent active at μM concentrations against most types of human tumor, is active against enveloped viruses at μM concentrations, is an antifungal agent, kills nematodes, and inhibits the formation of NO from nNOS (Pukala et al., 2006, and references cited later).

In the following discussion, only native (active) peptides isolated and identified from anuran secretions will be described: those peptides identified from DNA sequences but not detected in glandular secretions will not be included.

15.2 SIMPLE α-HELICAL AMPHIPATHIC PEPTIDES

Many of the membrane-active peptides listed in Table 15.1 are small cationic peptides that adopt amphipathic α-helical structures at the lipid bilayer of bacterial cells. Among these are the aureins (Rozek et al., 2000; Chen et al., 2005), citropins 1 (Wegener et al., 1999), dahleins 1 (Wegener et al., 2001), fallaxidin 2 (Jackway et al., 2008), signiferins 2 (Maselli et al., 2004), and the uperins 2–7 (Bradford et al., 1996a, 1996b). Figures 15.4 through 15.7 show the frogs *Litoria aurea* and *Litoria citropa*, and the toadlets *Crinia signifera* and *Uperoleia mjobergii*, together with their geographical locations and major amphipathic skin peptides. The antibiotic activities of selected peptides are listed in Table 15.2.

TABLE 15.1
Antibiotic Peptides from the Genera *Litoria, Uperoleia,* and *Crinia*

Name	Sequence	MW	Activity	Species	Reference
Aurein 1.1	GLFDIIKKIAESI-NH_2	1444 +	w	*L. raniformis*	Rozek et al., 2000
Aurein 1.2	GLFDIIKKIAESF-NH_2	1478 +/–	w	*L. raniformis*	Rozek et al., 2000
Aurein 2.1	GLLDIVKKVVGAFGSL-NH_2	1613 +	w	*L. aurea, L. raniformis*	Rozek et al., 2000
Aurein 2.2	GLFDIVKKVVGALGSL-NH_2	1613 +	w	*L. aurea*	Rozek et al., 2000
Aurein 2.3	GLFDIVKKVVGAIGSL-NH_2	1613 +	w	*L. aurea*	Rozek et al., 2000
Aurein 2.4	GLFDIVKKVVGTLAGL-NH_2	1627 +	w	*L. aurea*	Rozek et al., 2000
Aurein 2.5	GLFDIVKKVVGAFGSL-NH_2	1647 +	w	*L. aurea, L. raniformis*	Rozek et al., 2000
Aurein 2.6	GLFDIAKKVIGVIGSL-NH_2	1627 +	w	*L. raniformis*	Rozek et al., 2000
Aurein 3.1	GLFDIVKKIAGHIAGSI-NH_2	1736 +	w	*L. aurea, L. raniformis*	Rozek et al., 2000
Aurein 3.2	GLFDIVKKIAGHIASSI-NH_2	1766 +	w	*L. aurea, L. raniformis*	Rozek et al., 2000
Aurein 3.3	GLFDIVKKIAGHIVSSI-NH_2	1794 +	w	*L. raniformis*	Rozek et al., 2000
Aurein 5.2	GLMSSIGKALGGLIVDVLKPKTPAS-OH	2450 +	n	*L. aurea, L. raniformis*	Rozek et al., 2000; Chen et al., 2005
Caerin 1.1	GLLSVLGSVAKHVLPHVVPVIAEHL-NH_2	2582 +/–	w	*L. splendida, L. caerulea, L. gilleni*	Stone et al., 1993; Waugh et al., 1993; Wabnitz et al., 2000
Caerin 1.1.1	LSVLGSVAKHVLPHVVPVIAEHL-NH_2	2412	Inactive		
Caerin 1.1.2	SVLGSVAKHVLPHVVPVIAEHL-NH_2	2309	Inactive		
Caerin 1.1.3	VLPHVVPVIAEHL-NH_2	1321	Inactive		
Caerin 1.2	GLLGVLGSVAKHVLPHVVPVIAEHL-NH_2	2552 +	w	*L. caerulea*	Stone et al., 1993
Caerin 1.3	GLLSVLGSVAQHVLPHVVPVIAEHL-NH_2	2582 +	w	*L. caerulea*	Stone et al., 1993
Caerin 1.4	GLLSSLGSVAKHVLPHVVPVIAEHL-NH_2	2600 +/–	w	*L. caerulea, L. gilleni*	Stone et al., 1993; Waugh et al., 1993
Caerin 1.5	GLLSVLGSVVKHVLPHVVPVIAEHL-NH_2	2610 +/–	w	*L. caerulea*	Stone et al., 1993
Caerin 1.6	GLFSVLGAVAKHVLPHVVPVIAEKL-NH_2	2591 +/–	w	*L. splendida, L. xanthomera, L. chloris*	Steinborner et al., 1997a, 1998; Wabnitz et al., 2000
Caerin 1.6.1	FSVLGAVAKHVLPHVVPVIAEKL-NH_2	2421	Inactive		
Caerin 1.7	GLFKVLGSVAKHLLPHVAPVIAEKL-NH_2	2634 +/–	w	*L. xanthomera, L. chloris*	Steinborner et al., 1997a, 1998

continued

TABLE 15.1 (continued)

Name	Sequence	MW	Activity	Species	Reference
Caerin 1.8	GLFKVLGSVAKHLLPHVVPVIAEKL-NH_2	2662 +/–	w	*L. chloris*	Steinborner et al., 1998
Caerin 1.9	GLFGVLGSIAKHVLPHVVPVIAEKL-NH_2	2591 +/–	w	*L. chloris*	Steinborner et al., 1998
Caerin 1.10	GLLSVLGSVAKHVLPHVVPVIAEKL-NH_2	2573 +/–	w	*L. splendida*	Wabnitz et al., 2000
Caerin 1.11	GLLGAMFKVASKVLPHVVPAITEHF-NH_2	2659 +	w	*L. eucnemis*	Brinkworth et al., 2002
Caerin 1.17	GLFSVLGSVAKHLLPHVAPIIAEKL-NH_2	2606 +	w	*L. gracilenta*	Maclean et al., 2006
Caerin 1.18	GLFSVLGSVAKHLLPHVVPVIAEKL-NH_2	2620 +	w	*L. gracilenta*	Maclean et al., 2006
Caerin 1.19	GLFKVLGSVAKHLLPHVAPIIAEKL-NH_2	2600 +	w	*L. gracilenta*	Maclean et al., 2006
Caerin 1.19.1	FKVLGSVAKHLLPHVAPIIAEKL-NH_2	2590	Inactive	*L. gracilenta*	Maclean et al., 2006
Caerin 1.19.2	KVLGSVAKHLLPHVAPIIAEKL-NH_2	2477	Inactive	*L. gracilenta*	Maclean et al., 2006
Caerin 1.19.3	GSVAKHLLPHVAPIIAEKL-NH_2	1991	Inactive	*L. gracilenta*	Maclean et al., 2006
Caerin 1.20	GLFGILGSVAKHVLPHVIPVVAEHL-NH_2	2600 +	w	*L. caerulea/L. splendida* hybrid	Pukala et al., 2006
Caerin 2.1	GLVSSIGRALGGLLADVVKSKGQPA-OH	2392 –	n	*L. splendida*	Wabnitz et al., 2000
Caerin 2.2	GLVSSIGRALGGLLADVVKSKEQPA-OH	2464 +/–	n	*L. caerulea*	Stone et al., 1993
Caerin 2.2.1	ALGGLLADVVKSKEQPA-OH	1695	Inactive	*L. caerulea*	Stone et al., 1993
Caerin 2.5	GLVASIGRALGGLLADVVKSKEQPA-OH	2448 +	n	*L. gilleni*	Waugh et al., 1993
Caerin 2.6	GLVSSIGKVLGGLLADVVKSKGQPA-OH	2392 +	n	*L. caerulea/L. splendida* hybrid	Pukala et al., 2006
Caerin 2.7	GLVSSIGKALGGLLVDVVKSKGQPA-OH	2392 +	n	*L. caerulea/L. splendida* hybrid	Pukala et al., 2006
Caerin 3.1	GLWQKIKDKASELVSGIVEGVK-NH_2	2382 +	n	*L. splendida, L. caerulea*	Wabnitz et al., 2000
Caerin 3.2	GLWEKIKEKASELVSGIVEGVK-NH_2	2397 +	n	*L. caerulea*	Stone et al., 1993
Caerin 3.3	GLWEKIKEKANELVSGIVEGVK-NH_2	2424 +/–	n	*L. caerulea*	Stone et al., 1993
Caerin 3.4	GLWEKIREKANELVSGIVEGVK-NH_2	2452 +/–	n	*L. caerulea*	Stone et al., 1993
Caerin 3.5	GLWEKVKEKANELVSGIVEGVK-NH_2	2392 +	n	*L. gracilenta*	Maclean et al., 2006
Caerin 4.1	GLWQKIKSAAGDLASGIVEGIKS-NH_2	2326 +/–	n	*L. caerulea*	Stone et al., 1993
Caerin 4.2	GLWQKIKSAAGDLASGIVEAIKS-NH_2	2340 +/–	n	*L. caerulea*	Stone et al., 1993
Caerin 4.3	GLWKIKQAAGDLASGIVEGIKS-NH_2	2353 +/–	n	*L. caerulea*	Stone et al., 1993
Citropin 1.1	GLFDVIKKVASVIGGL-NH_2	1613 +	w	*L. citropa*	Wegener et al., 1999
Citropin 1.1.1	FDVIKKVASVIGGL-NH_2	1444	Inactive	*L. citropa*	Wegener et al., 1999
Citropin 1.1.2	DVIKKVASVIGGL-NH_2	1297	Inactive	*L. citropa*	Wegener et al., 1999
Citropin 1.1.3	GLFDVIKKVASVIGLASP-NH_2	1813 +	n	*L. citropa*	Wegener et al., 1999

Citropin 1.2	GLFDIIKKVASVVGGL-NH_2	1613 +	w	*L. citropa*	Wegener et al., 1999
				L. subglandulosa	Brinkworth et al., 2004
Citropin 1.3	GLFDIIKKVASVIGGL-NH2	1627 +	w	*L. citropa*	Wegener et al., 1999
Citropin 2.1	GLIGSIGKALGGLLVDVLKPKL-NH_2	2160 +	n	*L. citropa*	Wegener et al., 1999
Citropin 2.1.3	GLIGSIGKALGGLLVDVLKPKLQAAS-OH	2517 +	n	*L. citropa*	Wegener et al., 1999
Dahlein 1.1	GLFDIIKNIVSTL-NH_2	1430 +	w	*L. dahlii*	Wegener et al., 2001
Dahlein 1.2	GLFDIIKNIFSGL-NH_2	1434 +	w	*L. dahlii*	Wegener et al., 2001
Fallaxidin 2.1	GLLDLAKHVIGIASKL-NH_2	1645 +	n	*L. fallax*	Jackway et al., 2008
Fallaxidin 2.2	GLLFLAKHVIGIASKL-NH_2	1678 +	n	*L. fallax*	Jackway et al., 2008
Fallaxidin 3	GLLSFLPKVIGVIGHLIHPPS-NH_2	2193 +	w	*L. fallax*	Jackway et al., 2008
Maculatin 1.1	GLFGVLAKVAAHVVPAIAEHF-NH_2	2145 +/–	w	*L. genimaculata*	Brinkworth et al., 2002
Maculatin 1.2	GLFGVLAKVASHVVAAIAEHFQA-NH_2	2360 +	n	*L. genimaculata*	Brinkworth et al., 2002
Maculatin 1.3	GLLGLLGSVVSHVVPAIVGHF-NH_2	2068 +	w	*L. eucnemis*	Brinkworth et al., 2002
Maculatin 1.4	GLLGLLGSVVSHVLPAITQHL-NH_2	2121 +	w	*L. eucnemis*	Brinkworth et al., 2002
Maculatin 2.1	GFVDFLKKVAGTIANVVT-NH_2	1878 +	w	*L. genimaculata*	Brinkworth et al., 2002
Maculatin 3.1	GLLQTIKEKLESLESLAKGIVSGIQA-NH_2	2395 +	n	*L. genimaculata*	Brinkworth et al., 2002
Signiferin 2.1	IIGHLIKTALGMLGL-NH_2	1547 +	w	*C. signifera*	Maselli et al., 2004
Signiferin 2.2	IIGHLIKTALGFLGL-NH_2	1563 +	w	*C. signifera*	Maselli et al., 2004
Uperin 2.1	GIVDFAKKVVGGIRBALGI-NH_2	1925 +	n	*U. inundata*	Bradford et al., 1996a
Uperin 2.3	GFFDLAKKVVGGIRNALGI-NH_2	1973 +	n	*U. inundata*	Bradford et al., 1996a
Uperin 2.5	GIVDFAKGVLGKIKNVLGI-NH_2	1939 +	n	*U. inundata*	Bradford et al., 1996a
Uperin 2.8	GILDVAKTLVGKLRNVLGI-NH_2	1977 +	w	*U.mjobergii*	Bradford et al., 1996b
Uperin 3.1	GVLDAFRKIATVVKNVV-NH_2	1826 +	n	*U. inundata*	Bradford et al., 1996a
Uperin 3.5	GVGDLIRKAVSVIKNIV-NH_2	1778 +	w	*U. mjobergii*	Bradford et al., 1996b
Uperin 3.6	GVIDAAKKVVNVLKNLP-NH_2	1826 +	w	*U. mjobergii*	Bradford et al., 1996b
Uperin 4.1	GVGSFIHKVVSAIKNVA-NH_2	1723 +	n	*U. inundata*	Bradford et al., 1996a
Uperin 7.1	GWFDVVKHIASAV-NH_2	1427 +	n	*L. ewingi*	Steinborner et al., 1997
Uperin 7.1.1	FDVVKHIASAV-NH_2	1184	Inactive	*L. ewingi*	Steinborner et al., 1997

Note: +, Gram positive; –,Gram negative; w, wide spectrum; n, narrow spectrum.

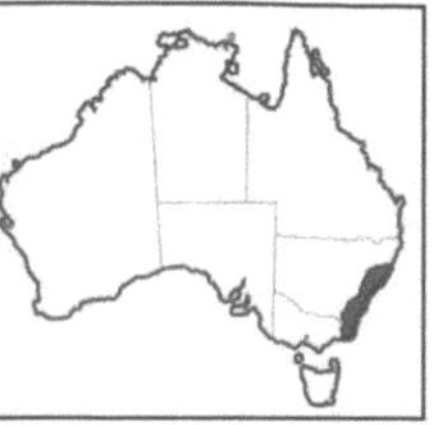

GLFDIIKKIAESF(NH$_2$) Aurein 1.2
GLFDIVKKVVGTLAGL(NH$_2$) Aurein 2.4

FIGURE 15.4 *Litoria aurea*; geographic location; aureins 1.2 and 2.4. (From Barker, J., Grigg, G.C., and Tyler, M.J., *A Field Guide to Australian Frogs*, Surrey Beatty and Sons, Chipping Norton, New South Wales, Australia, p. 100, 1995. With permission.)

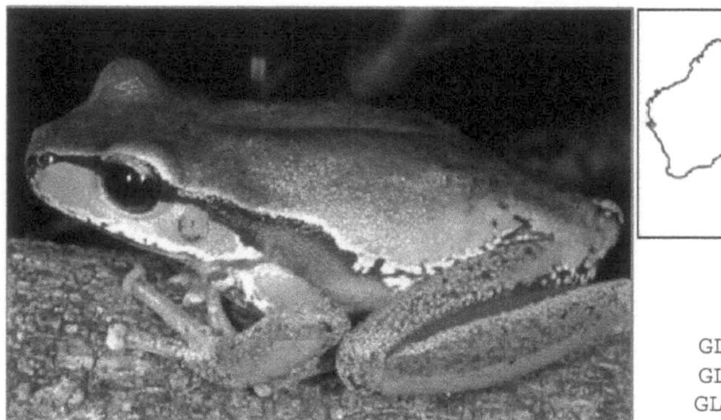

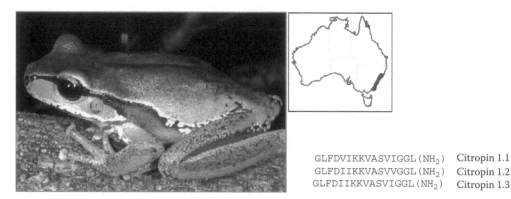

GLFDVIKKVASVIGGL(NH$_2$) Citropin 1.1
GLFDIIKKVASVVGGL(NH$_2$) Citropin 1.2
GLFDIIKKVASVIGGL(NH$_2$) Citropin 1.3

FIGURE 15.5 *Litoria citropa*; geographic location; citropins 1.1, 1.2, and 1.3 (frogs.org.au/frogs/show_image.php?image_id=398).

IIGHLIKTALGMLGL(NH$_2$) Signiferin 2.1
IIGHLIKTALGFLGL(NH$_2$) Signiferin 2.2

FIGURE 15.6 *Crinia signifera*; geographic location; signiferins 2.1 and 2.2 (www.answers.com/topic/crinia-signifera-jpg).

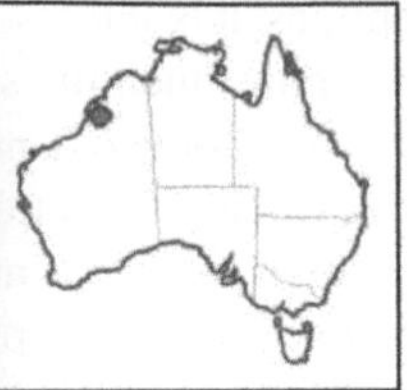

GVGDLIRKAVSVIKNIV(NH_2)	Uperin 3.5
GVIDAAKKVVNVLKNLP(NH_2)	Uperin 3.6

FIGURE 15.7 *Uperoleia mjobergii*; geographic location; uperins 3.5 and 3.6. (From Tyler, Smith, and Johnstone, *Frogs of Western Australia*, Lamb. Print. Perth, WA, 1994, plate 21. With permission.)

The peptides listed in Table 15.2 are wide-spectrum antibiotics. They exhibit activity against a range of Gram-positive organisms (such as *Staphylococcus aureus*), but less activity against Gram-negative organisms (such as *Escherichia coli*). Of the compounds listed in Table 15.2, the most active are uperins 3.5 and 3.6 from the toadlet *Uperoleia mjobergii*. Although there are some features

TABLE 15.2
Antibiotic Activities of Selected α-Helical Peptides[a,b]

	Sequence
Aurein 1.2	GLFDIIKKIAESI-NH_2
Aurein 2.4	GLFDIVKKVVGTLAGL-NH_2
Citropin 1.1	GLFDVIKKVASVIGGL-NH_2
Citropin 1.2	GLFDIIKKVASVVGGL-NH_2
Dahlein 1.2	GLFDIIKNIFSGL-NH_2
Signiferin 2.1	IIGHLIKTALGFLGL-NH_2
Uperin 3.5	GVGDLIRKAVSVIKNIV-NH_2
Uperin 3.6	GVIDAAKKVVNVLKNLP-NH_2

Bacterium[c]	A1.2	A2.4	Cit1.1	Cit1.2	D1.2	S2.1	U3.5	U3.6
Bacillus cereus	100	25	50	25	100	25	25	25
Enterococcus faecalis	—	—	—	—	100	100	—	—
Leuconostoc lactis	12	12	6	3	25	12	3	3
Listeria innocua	100	100	25	100	—	50	25	50
Micrococcus luteus	100	25	12	12	—	25	12	50
Staphylococcus aureus	50	12	25	25	100	12	50	25
S. epidermidis	50	25	12	25	100	25	12	12
Streptococcus uberis	50	25	25	12	100	25	12	12
Enterobacter clocae	—	—	—	—	—	—	—	—
Escherichia coli	—	—	—	—	—	—	25	25
Pasteurella multocida	100	25	—	—	—	—	25	25
Pseudomonas aeruginosa	—	—	—	—	—	—	—	—

[a] Minimum inhibitory concentration (MIC) values (μg/mL).
[b] —, no activity ≤100 μg/mL.
[c] Pathogens listed in the first group are Gram-positive organisms, those in the second group are Gram-negative organisms.

in common with these peptides, it is interesting that the overall sequences of some of these peptides show differences, yet they show much the same overall spectrum of antibiotic activities. The most likely reasons why each animal produces a number of active peptides is probably twofold: (i) because mixtures of active peptides often give better antibiotic cover that the single peptides, and (ii) this may be an evolutionary trend to produce a number of peptides so that an attacking organism cannot become resistant to every host-defense peptide. These peptides are cytotoxic to the host, so after they have been on the skin for a period (usually 10 to 30 min) a proteolytic enzyme cleaves residues from the N-terminal end of the peptide, destroying its activity (see, e.g., citropins 1.1.1 to 1.1.3, Table 15.1; Wegener et al., 1999). This same feature has been observed previously for the magainins, but in these cases, the enzyme cleaves towards the centre of the molecule (Resnick et al., 1991). This problem of toxicity to the host is probably why these animals secrete quite large amounts of antibiotic peptides that only show activity at μM concentrations.

The peptides listed in Figures 15.4 through 15.7 and Table 15.2 are amphipathic. They adopt α-helical structures on the bacterial lipid bilayer with well-defined hydrophilic and hydrophobic zones. This has been demonstrated by NMR studies in the membrane-mimicking solvent system TFE/H_2O and for model phospholipids for aurein 1.2 (Rozek et al., 2000), citropin 1.1 (Wegener et al., 1999), and uperin 3.6 (Chia et al., 1999). The helical structure of citropin 1.1 shown in Figure 15.8 is typical: a simple α-helix with hydrophilic and hydrophobic side chains localized on opposite sides of the helix. A prerequisite of a minimum of 20 amino-acid residues is required for a peptide in order to effect full penetration through the bacterial lipid bilayer in order to cause breaching of the membrane by a pore-forming mechanism (see Figure 15.1) (as adopted, e.g., by the magainins; Dathe and Weiprecht, 1999; Sitaram and Nagaraj, 2002; Tachi et al., 2002). There are also some small peptides that dimerize in order to breach the bilayer (Andreu et al., 1992). On the other hand, there are other antimicrobial, amphipathic peptides with more than 20 amino-acid residues that do not fully span the bilayer (Matsuzaki 1998; Hara et al., 2001; Ambroggio et al., 2004; Huang et al., 2004); these lyse bacterial cells by means of the carpet mechanism (Shai and Oren, 2001).

Solid-state NMR experiments with aurein 1.2 (13 residues) and citropin 1.1 (16 residues) show that these peptides insert into model bilayers at an angle of about 50° as shown for aurein 1.2 in Figure 15.9 (Balla et al., 2004; Marcotte et al., 2003). Aurein 1.2 and citropin 1.1 breach the bilayer by means of the carpet mechanism (Figure 15.2), and it is likely that the other small antimicrobial peptides listed in Table 15.2 operate by the same mechanism.

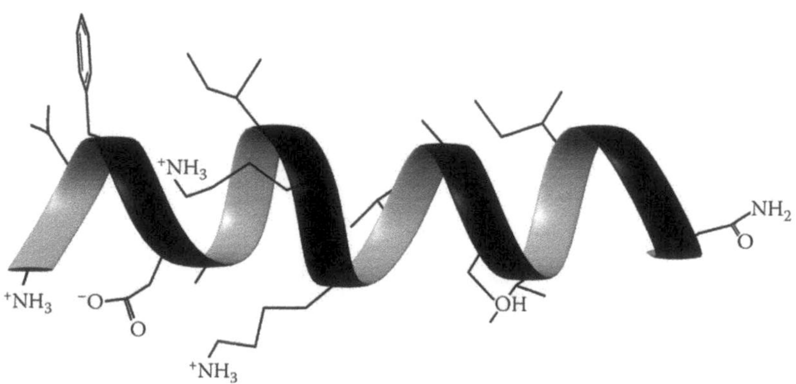

FIGURE 15.8 2D NMR structure of citropin 1.1 in TFE/H_2O (1:1). (From Pukala, T.L. et al., *Prod. Rep.*, 23, 368–393, 2006b. With permission.)

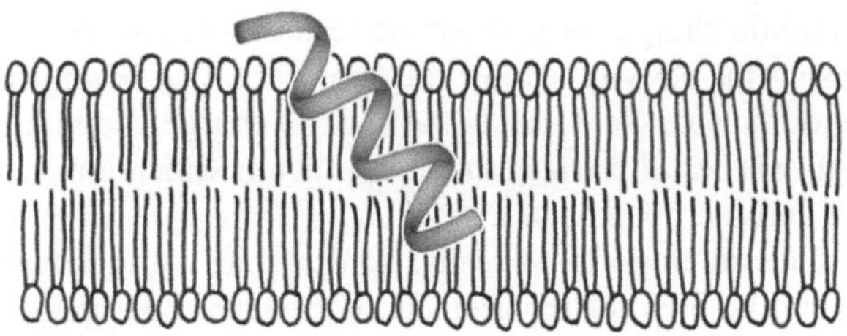

FIGURE 15.9 Small amphipathic peptide (aurein 1.2) penetrating the bacterial lipid bilayer. (From Marcotte, I. et al., *Chem. Phys. Lipids*, 122, 107–120, 2003. With permission.)

A large number of synthetic modifications of the cationic antibiotic citropin 1.1 have been prepared in order to study structure/activity relationships (Doyle et al., 2003). A selection of these, with antibiotic activities, is listed in Table 15.3. A number of features can be seen in Table 15.3:

1. Natural L-citropin 1.1 has, within experimental error, the same spectrum of antibiotic activities as the synthetic D-isomer. This is a feature of membrane active peptides (Barra and Simmaco, 1995; Wong et al., 1997).
2. Substituting Lys^7 and Lys^8 for Ala removes the cationic charge and destroys all antibiotic activity.

TABLE 15.3
Antibiotic Activities of Citropin 1.1 and of Some Synthetic Modifications[a,b]

	Sequence
Citropin 1.1	GLFDVIKKVASVIGGL-NH$_2$
Citropin 1.1D	G**lfdvikkvasvi**GG**l**-NH$_2$
Syn mod 2	GL**A**DVIKKVASVIGGL-NH$_2$
Syn mod 3	GLF**A**VIKKVASVIGGL-NH$_2$
Syn mod 4	GLFDVIKKVASVIGG**A**-NH$_2$
Syn mod 5	GLFDVI**AA**VASVIGGL-NH$_2$
Syn mod 6	GLFDVI**A**KVASVI**KK**L-NH$_2$
Syn mod 7	GLF**A**VIKKVASVI**KK**L-NH$_2$

Bacterium[c]	**Cit1.1**	**Cit1.1D**	**Mod2**	**Mod3**	**Mod4**	**Mod5**	**Mod6**	**Mod7**
Bacillus cereus	50	50	50	25	—	—	12	25
Enterococcus faecalis	—	—	—	100	—	—	100	100
Leuconostoc lactis	6	3	100	3	25	—	6	3
Listeria innocua	25	25	100	25	—	—	25	12
Micrococcus luteus	12	25	50	12	—	—	12	6
Staphylococcus aureus	25	25	100	25	—	—	6	12
S. epidermidis	12	12	100	12	—	—	6	6
Streptococcus uberis	25	12	100	25	100	—	12	25
Enterobacter clocae	—	—	—	—	—	—	—	—
Escherichia coli	—	—	—	100	—	—	50	50
Pasteurella multocida	—	—	—	—	—	—	100	100
Pseudomonas aeruginosa	—	—	—	—	—	—	—	—

[a] Minimum inhibitory concentration (MIC) values (μg/mL).

[b] —, no activity $\leq$100 mg/mL.

[c] Pathogens listed in the first group are Gram-positive organisms, those in the second group are Gram-negative organisms.

3. Increasing the cationic charge (of hydrophilic residues) enhances the overall activity, in particular increasing the activity against the Gram-negative organism *E. coli*. Replacing Asp^4 with Ala (modification 3) increases the positive charge and marginally increases the activity. This feature is best illustrated for citropin 1.1 modifications 6 and 7, which are Lys rich and effective antimicrobial agents.
4. Hydrophobic residues are important because they interact with the hydrocarbon portion of the bilayer. This can be seen for replacement of the large hydrophobic residue Phe^3 with Ala, which reduces the activity, whereas in contrast, replacement of the N-terminal Leu^{16} with Ala removes most of the activity.

Finally, and in the above context, it is of interest to consider the sequence of signiferin 2.1 (Tables 15.1 and 15.2). The signiferins 2 are the only antibiotic peptides listed in Table 15.1 that do not contain an N-terminal Gly. Antibiotic peptides commencing with Gly are common for anurans from some other genera, but in others, for example the genus *Rana*, other residues as well as Gly can be N-terminal (Barra and Simmaco, 1995; Pukala et al., 2006). Why the signiferins 2 have an N-terminal Ile is not clear, because if the Ile is replaced with Gly the activity (of GIGHLIKTALGFLGL–NH_2) is, within experimental error, the same as that of signiferin 2.1 (Table 15.2) (Maselli et al., 2007).

15.3 THE HINGED CAERIN 1 AND MACULATIN MEMBRANE-ACTIVE ANTIBIOTICS

A number of tree frogs of the genus *Litoria* produce the unique cationic caerin 1 wide-spectrum antimicrobial peptides. These include *Litoria caerulea* (Waugh et al., 1993a), *L. chloris* (Steinborner et al., 1998), *L. eucnemis* (Brinkworth et al., 2002), *L. ewingi* (Steinborner et al., 1997a), *L. gilleni* (Waugh et al., 1993b), *L. gracilenta* (Maclean et al., 2006), *L. splendida* (Stone et al., 1992a; Wabnitz et al., 2000), and *L. xanthomera* (Steinborner et al., 1997b). The related maculatin peptides are produced by *L. genimaculata* (Brinkworth et al., 2002) and *L. eucnemis* (Brinkworth et al., 2002). *L. eucnemis* is unique in that it is the only frog studied that produces both caerin 1 and maculatin 1 antibiotic peptides. The frogs *L. caerulea*, *L. xanthomera*, and *L. genimaculata*, their geographic locations, together with their wide-spectrum antibiotic peptides are illustrated in Figures 15.10, 15.11, and 15.12, respectively.

The activities of a selection of the caerin 1 antibiotics together with the related maculatin 1.1 are listed in Table 15.4 (against the standard pathogens that we use for our antibiotic testing regime). Fifteen natural caerins 1 have been isolated; they are all wide-spectrum antibiotics with activities

GLLSVLGSVAKHVLPHVVPVIAEHL(NH_2) Caerin 1.1
GLLGVLGSVAKHVLPHVVPVIAEHL(NH_2) Caerin 1.2
GLLSVLGSVAQHVLPHVVPVIAEHL(NH_2) Caerin 1.3
GLLSSLGSVAKHVLPHVVPVIAEHL(NH_2) Caerin 1.4
GLLSVLGSVVKHVLPHVVPVIAEKL(NH_2) Caerin 1.5

FIGURE 15.10 *Litoria caerulea*; geographic location; caerins 1.1, 1.2, 1.3, 1.4, and 1.5 (http://commons.wikimedia.org/wiki/Image:Litoria_caerulea.jpg).

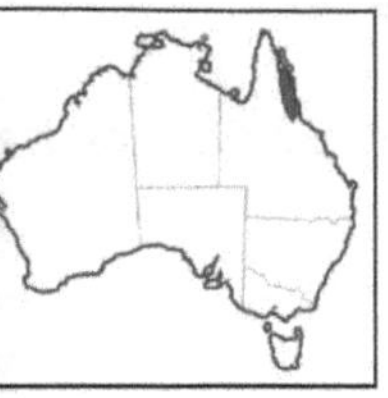

GLFSVLGAVAKHVLPHVVPVIAEKL(NH_2) Caerin 1.6
GLFKVLGSVAKHLLPHVAPVIAEKL(NH_2) Caerin 1.7

FIGURE 15.11 *Litoria xanthomera*; geographic location; caerins 1.6 and 1.7 (http://commonms.wikimedia.org/wiki/Image:Litoria xanthomera.jpg).

comparable to those of the α-helical uperins 3.5 and 3.6 (see Table 15.2). They are mainly active against Gram-positive organisms, but some show activity against Gram-negative organisms like *E. coli*. It is interesting that peptides having such different sequences (cf. Tables 15.2 and 15.4) and shapes (see below), have such similar ranges of antibiotic activities.

The caerin 1 and maculatin 1 peptides have helical regions at each end of the molecule separated by a flexible hinge initiated by Pro[15]. This is demonstrated in 2D NMR experiments in both TFE/H_2O and model lipids for caerin 1.1 (Wong et al., 1997; Pukala et al., 2004), caerin 1.4 (Wegener et al., 2003), and maculatin 1.1 (Chia et al., 2000a). The solution structure of caerin 1.1 is shown in Figure 15.13a.

Solid-state NMR studies suggest that caerin 1 peptides may operate via the carpet mechanism; penetration of these hinged peptides occurs as shown in Figure 15.14 (Balla et al., 2004; Marcotte et al., 2003). Solid-state NMR investigations have also been used to show that caerin 1.1 and maculatin 1.1 interact with the membranes of the live Gram-positive bacteria *Bacillus cereus* and *Staphylococcus epidermidis* (Chia et al., 2000b). FTIR spectroscopy (Chia et al., 2002) and confocal fluorescence spectroscopy (Ambroggio et al., 2005) suggest a pore-type mechanism for maculatin 1.1. It seems unusual that the related and hinged peptides caerin 1.1 and maculatin 1.1 should lyse bacteria by different mechanisms; further work needs to be carried out to confirm the mechanism of action of these membrane-active peptides.

GLFGVLAKVAAHVVPAIAEHF(NH2) Maculatin 1.1
GLLGLLGSVVSHVVPAIVGHF(NH2) Maculatin 1.3

FIGURE 15.12 *Litoria genimaculata*; geographic location; maculatins 1.1 and 1.3. (From Barker, J., Grigg, G.C., and Tyler, M.J., *A Field Guide to Australian Frogs*, Surrey Beatty and Sons, Chipping Norton, New South Wales, Australia, p. 50, 1995. With permission.)

TABLE 15.4
Antibiotic Activities of Selected Caerins 1 and Maculatin 1.1[a,b]

	Sequence
Caerin 1.1	GLLSVLGSVAKHVLPHVVPVIAEHL-NH$_2$
Caerin 1.3	GLLSVLGSVA**Q**HVLPHVVPVIAEHL-NH$_2$
Caerin 1.4	GLLS**S**LGSVAKHVLPHVVPVIAEHL-NH$_2$
Caerin 1.6	GL**F**SVLG**A**VAKHVLPHVVPVIAE**K**L-NH$_2$
Caerin 1.7	GL**FK**VLGSVAKHLLPHV**A**PVIAE**K**L-NH$_2$
Caerin 1.8	GL**FK**VLGSVAKHLLPHVVPVIAE**K**L-NH$_2$
Caerin 1.19	GL**FK**VLGSVAKHLLPHV**A**PVI**I**E**K**L-NH$_2$
Maculatin 1.1	GLFGVLAKVAAHVVPAIAEHF-NH$_2$

Bacterium[c]	C1.1	C1.3	C1.4	C1.6	C1.7	C1.8	C1.19	M1
Bacillus cereus	50	50	50	—	—	50	50	25
Enterococcus faecalis	25	—	—	—	—	100	—	—
Leuconostoc lactis	1.5	25	12	3	3	25	3	3
Listeria innocua	25	100	100	50	50	25	12	100
Micrococcus luteus	12	1.5	0.4	25	12	6	25	12
Staphylococcus aureus	3	100	100	6	12	6	12	6
S. epidermidis	12	100	25	12	50	12	12	12
Streptococcus uberis	12	100	100	25	50	25	25	3
Enterobacter clocae	—	—	—	—	—	—	—	—
Escherichia coli	—	50	—	—	100	—	—	—
Pasteurella multocida	25	25	25	25	25	50	50	50
Pseudomonas aeruginosa	—	—	—	—	—	—	—	—

[a] Minimum inhibitory concentration (MIC) values (μg/mL).
[b] —, no activity ≤100 μg/mL.
[c] Pathogens listed in the first group are Gram-positive organisms, those in the second group are Gram-negative organisms.

A number of synthetic modifications of caerin 1.1 have been made to investigate structure/activity relationships of this molecule. A selection of these data is recorded in Table 15.5. The following points can be made:

1. As was the case with citropin 1.1, the native L-form of caerin 1.1 and the synthetic all D-isomer show the same activity within experimental error, a feature characteristic of membrane-active peptides (Barra and Simmaco, 1995; Wong et al., 1997).
2. Modification of the hinge changes the activity of the peptide. If the two Pro residues are replaced by Gly (modification 2, Table 15.5) the activity is reduced. In contrast, if the two Pro are replaced by Ala (modification 3, Table 15.5) the activity is minimal. This feature can be explained by comparison of the 2D structures of caerin 1.1 and the Gly and Ala modifications shown in Figure 15.13. Both Pro and Gly are helix breakers, so replacement of Pro by Gly still leaves a modified hinge region in modification 2 (Figure 15.13b). This peptide is still active, but less so than caerin 1.1. Replacement of the two Pro residues by Ala removes the central hinge and gives a slightly bent α-helix (Figure 15.13c, modification 3; Table 15.5). This peptide has lost most of the activity of caerin 1.1.
3. It is possible to manipulate the activity of caerin 1.1 by increasing the cationic charge at hydrophilic residues throughout the peptide. This is shown dramatically for modification 6

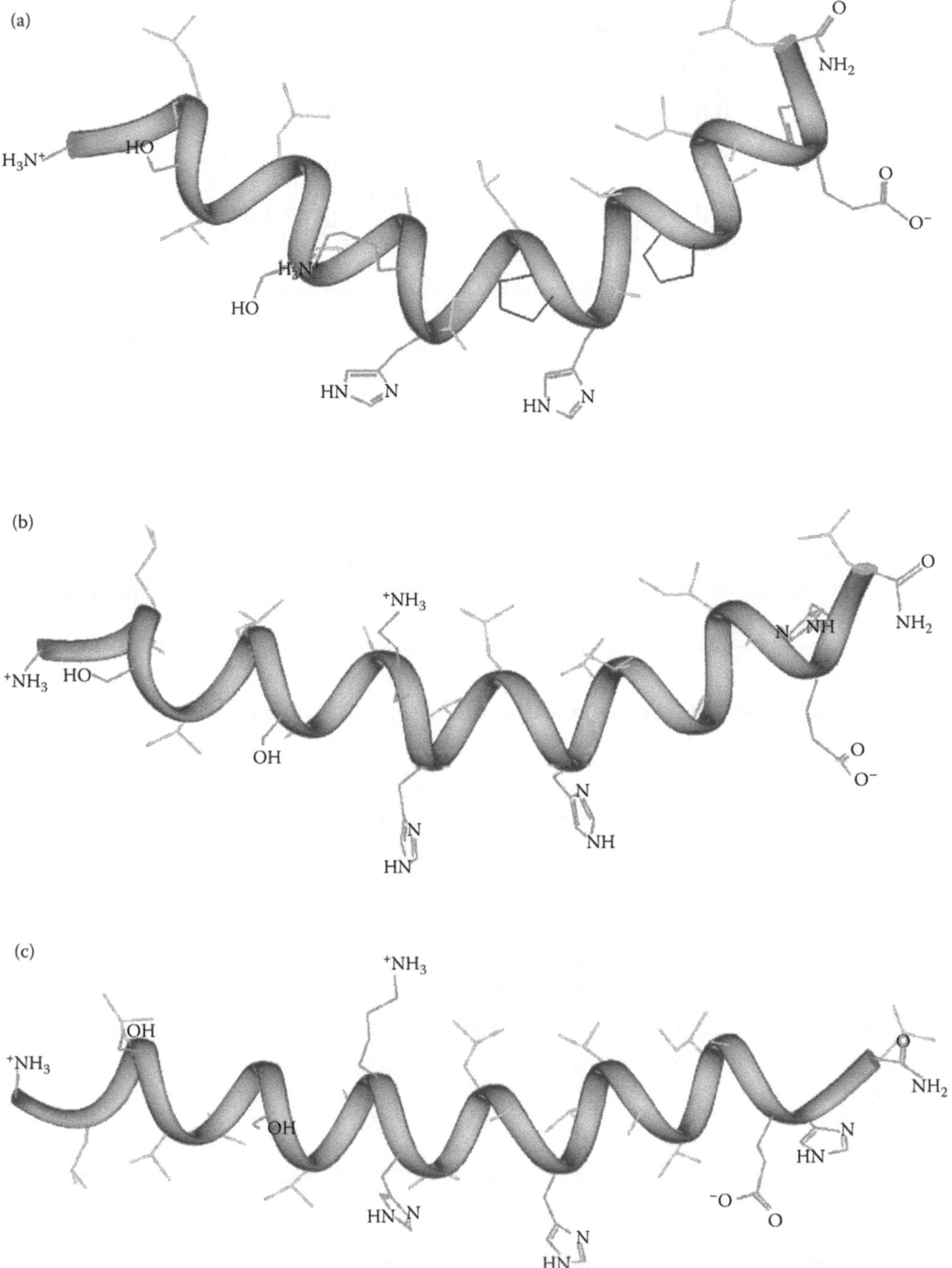

FIGURE 15.13 Solution structures of (a) caerin 1.1, (b) Gly^{15,19}-caerin 1.1, and (c) Ala^{15,19}-caerin 1.1 [2D NMR spectra measured in TFE/H_2O (1:1)]. (From Pukala, T.L., Brinkworth, C.S., Carver, J.A.., and Bowie, J.H., *Biochemistry*, 43, 937–944, 2004. With permission.)

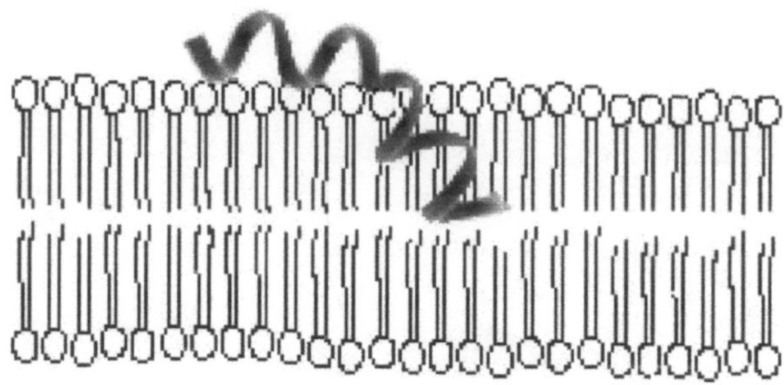

FIGURE 15.14 Caerin 1.1 penetrating bacterial bilayer by the carpet mechanism. (From Marcotte, I. et al., *Chem. Phys. Lipids*, 122, 107–120, 2003. With permission.)

(Table 15.5) which has a charge of +9. Here, the activity against Gram-positive organisms is lessened, but there is a significant increase in activity against Gram-negative pathogens. As an illustration, the MIC of caerin 1.1 syn. mod. 6 against *E. coli* is 6 μg/mL.

15.4 MISCELLANEOUS MEMBRANE-ACTIVE PEPTIDES

We have outlined above how a number of tree frogs of the genus *Litoria* contain a cocktail of hinged caerin 1 peptides that are mainly active against Gram-positive organisms. Some of these species, in

TABLE 15.5
Antibiotic Activities of Caerin 1.1 and Some Synthetic Modifications[a,b]

	Sequence
Caerin 1.1	GLLSVLGSVAKHVLPHVVPVIAEHL-NH_2
Caerin 1.1D	G**llsvl**G**svakhvlphvvpviaehl**-NH_2
Syn mod 1	GLLSVLGSVAKHVLPHVVPVIAEH**A**-NH_2
Syn mod 2	GLLSVLGSVAKHVL**G**HVV**G**VIAEHL-NH_2
Syn mod 3	GLLSVLGSVAKHVL**A**HVV**A**VIAEHL-NH_2
Syn mod 4	GLLSVLGSVAKHVLPHVVPVIA**AA**L-NH_2
Syn mod 5	GLLSVLGSVAKHVLPHVVPVIA**K**HL-NH_2
Syn mod 6	GLL**KK**LKKVAK**K**VLP**K**VVPVIAE**K**L-NH_2

Bacterium[c]	C1.1	C1.1D	Mod1	Mod2	Mod3	Mod4	Mod5	Mod6
Bacillus cereus	50	50	—	50	—	50	50	—
Enterococcus faecalis	25	25	—	100	—	50	—	—
Leuconostoc lactis	1.5	3	12	12	25	3	12	50
Listeria innocua	25	50	—	50	—	25	25	100
Micrococcus luteus	12	6	100	12	—	6	25	100
Staphylococcus aureus	3	6	100	25	—	12	25	100
S. epidermidis	12	12	—	100	—	25	12	25
Streptococcus uberis	12	12	50	12	—	25	25	12
Enterobacter clocae	—	—	—	—	—	—	—	100
Escherichia coli	—	—	—	50	—	—	100	6
Pasteurella multocida	25	25	—	100	—	25	100	12
Pseudomonas aeruginosa	—	—	—	—	—	—	—	50

[a] Minimum inhibitory concentration (MIC) values (μg/mL).
[b] —, no activity ≤100 μg/mL.
[c] Pathogens listed in the first group are Gram-positive organisms, those in the second group are Gram-negative organisms.

particular *L. caerulea* (Waugh et al., 1993a), *L. splendida* (Stone et al., 1992b; Wabnitz et al., 2000), *L. gillini* (Waugh et al., 1993b), and *L. gracilenta* (Maclean et al., 2006) contain, in addition, narrow-spectrum caerin antibiotics that are active against one or several pathogens. A selection of these peptides is listed in Table 15.6. Of particular relevance are those caerins 3 and 4 that are active against Gram-negative organisms like *E. coli.* A 2D NMR investigation in TFE/H_2O of caerin 4.1 indicates this peptide to be an amphipathic α-helix with hydrophobic and hydrophilic residues occupying specific zones (Chia et al., 2000c). These tree frogs therefore produce two types of antibiotic peptides, the wide-spectrum caerins 1, which are active mainly against Gram-positive organisms, together with caerins 3 and 4, which target one or several Gram-negative organisms.

We have recently isolated fallaxidin 3 (Tables 15.1 and 15.6) from *Litoria fallax* (Jackway et al., 2008) and the presence of Pro^7, Pro^{20}, and Pro^{21} residues might suggest that this molecule is unlikely to exhibit antimicrobial activity. Not only is this peptide a wide-spectrum antibiotic against Gram-positive organisms (Table 15.6), but it also inhibits the formation of nitric oxide from neuronal nitric oxide with an IC_{50} of 15.4 μM. Both activities suggest that fallaxidin 3 can adopt a helical conformation. The 2D NMR data for fallaxidin 3, determined in the membrane-mimicking solvent system TFE/H_2O, is shown in Figure 15.15. This is as complicated a 2D structure for a small antimicrobial peptide as we have seen. The first three residues are conformationally flexible: there is an α-helix from Ser^4 to I^{10}, a flexible hinge from Gly^{11} to Gly^{14}, a second helix from His^{15} to His^{18}, followed by a random section from Pro^{19} to Ser^{21}.

TABLE 15.6
Antibiotic Activities of Caerins 2, 3, and 4 and Fallaxidin 3[a,b]

	Sequence
Caerin 2.2	GLVSSIGRALGGLLADVVKSKQPA-OH
Caerin 2.5	GLVASIGRALGGLLADVVKSKEQPA-OH
Caerin 3.1	GLWQKIKDKASELVSGIVEGVK-NH_2
Caerin 3.3	GLWEKIKIKANELVSGIVEGVK-NH_2
Caerin 4.1	GLWQKIKSAAGDLASGIVEGIKS-NH_2
Caerin 4.2	GLWQKIKSAAGDLASGIVEAIKS-NH_2
Caerin 4.4	GLWQKIKNAAGDLASGIVEGIKS-NH_2
Fallaxidin 3	GLLSFLPKVIGVIGHLIHPPS-NH_2

Bacterium[c]	**C2.2**	**C2.5**	**C3.1**	**C3.3**	**C4.1**	**C4.2**	**C4.3**	**F3**
Bacillus cereus	—	—	—	—	—	—	—	25
Enterococcus faecalis	—	—	—	—	—	—	—	100
Leuconostoc lactis	—	—	—	—	—	—	—	3
Listeria innocua	—	—	—	—	—	—	—	50
Micrococcus luteus	50	<0.4	<0.4	3	12	6	25	12
Staphylococcus aureus	—	—	—	—	—	—	—	50
S. epidermidis	—	—	—	—	—	—	—	25
Streptococcus uberis	—	—	—	—	—	—	—	12
Enterobacter clocae	—	—	—	—	—	—	—	—
Escherichia coli	—	—	—	25	25	50	50	—
Pasteurella haemolytica	25	—	—	6	<0.4	<0.4	<0.4	—
Pseudomonas aeruginosa	—	—	—	—	—	—	—	—

[a] Minimum inhibitory concentration (MIC) values (μg/mL).
[b] —, no activity ≤100 μg/mL.
[c] Pathogens listed in the first group are Gram-positive organisms, those in the second group are Gram-negative organisms.

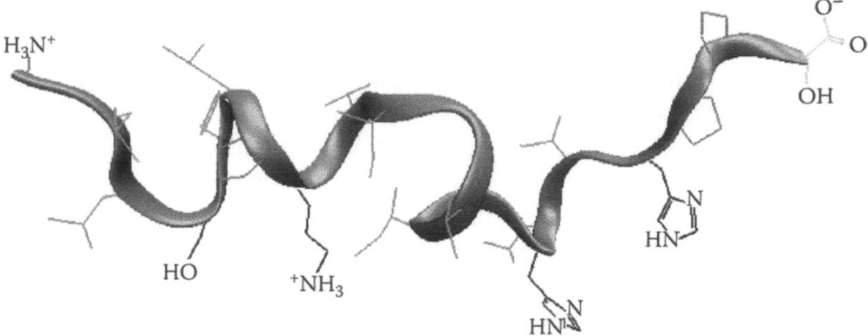

FIGURE 15.15 2D NMR (in TFE/H_2O) structure for fallaxidin 3.

15.5 ANTICANCER ACTIVE PEPTIDES

The wide-spectrum antibiotic peptides listed in Table 15.1 all exhibit anticancer activity in the 60 human tumor test regime of the National Cancer Institute in Washington DC (Monks et al., 1991). In contrast, the narrow-spectrum antibiotics (e.g., those listed in Table 15.5) show no such activity. The wide-spectrum antibiotic peptides are active against all human tumor cancer types tested, namely leukemia, lung, colon, CNS, melanoma, ovarian, renal, prostate, and breast cancers, at EC_{50} concentrations of 10^{-5} molar or better. The most active natural peptides together with some synthetic modifications are shown in Table 15.7, in which the average activity for all cancers within a particular type are listed, together with the total cancers (out of a maximum test total of 60) against

TABLE 15.7
Anticancer Activities of Some Membrane-Active Peptides

	Sequence
Citropin 1.1	GLFDVIKKVASVIGGL-NH_2
Syn mod3	GLF**A**VIKKVASVIGGL-NH_2
Syn mod8	GLF**A**VIKKVA**K**VI**KK**L-NH_2
Uperin 3.6	GVIDAAKKVVNVLKNLF-NH_2
Caerin 1.1	GLLSVLGSVAKHVLPHVVPVIAEHL-NH_2
Caerin 1.1D	G**llsvl**G**svakhvlphvvpviaehl**-NH_2
Syn mod4	GLLSVLGSVAKHVLPHVVPVIA**AA**L-NH_2
Caerin 1.9	GL**FG**VLGS**I**AKHVLPHVVPVIAE**K**L-NH_2

Cancer[a]	Cit1.1	Mod3	Mod8	U3.6	C1.1	C1.1D	Mod4	C1.9
Leukemia	5	6	5	6	5	5	6	6
Lung	5	6	5	5	6	6	6	5
Colon	5	6	5	5	5	5	6	5
CNS	5	6	6	6	6	6	6	5
Melanoma	5	6	6	5	6	6	6	6
Ovarian	5	6	5	5	5	5	6	6
Renal	5	6	5	5	6	6	5	5
Prostate	5	6	5	5	6	6	6	6
Breast	5	6	5	6	6	6	6	6
Total[b]	55	56	58	55	59	58	55	60

[a] EC_{50} values: 5 and 6 mean 10^{-5} and 10^{-6} molar. These are average values for all cancers tested within that particular group.
[b] Total is the number (out of 60 cancers tested) of individual cancers against which that particular peptide is active.

which the peptide is active. Citropin 1.1 and uperin 3.6 are typical of membrane-active peptides that adopt simple α-helical structures on a membrane: citropin shows activities in the 10^{-5} molar concentration, while uperin 3.6 is the most active of this group of peptides with some activities in the 10^{-6} molar range. The most active synthetic modification is citropin 1.1 mod. 3, which has a charge of +3. The hinged caerins 1 are the most active anticancer peptides of all of the peptides listed in Table 15.1. The activities of the natural L-caerin 1.1 and the synthetic all D-isomer are very similar, ruling out interaction with specific chiral receptors and suggesting that the mechanism of anticancer activity is very similar to that of the antibiotic activity of the peptides. The most active synthetic caerin 1 is modification 4, which has an overall charge of +2. In general terms the antibacterial and anticancer activity trends of these peptides are similar. There are some minor differences in the detail of the data, but not sufficient to alter the general trend. The concentrations at which citropin 1.1, caerin 1.1, and uperin 3.6 show antimicrobial and anticancer activities is less than the concentrations at which these three peptides lyse red blood cells (Apponyi et al., 2004).

15.6 MEMBRANE-ACTIVE PEPTIDES THAT ARE ACTIVE AGAINST ENVELOPED VIRUSES

The first report of antiviral activity among host-defense peptides of Australian frogs was for caerin 1.1, which was shown to be active against the enveloped viruses HIV MIC = 7.7 μM) and *Herpes simplex* 1 (MIC = 9.2 μM) (Apponyi et al., 2004). A recent survey of 14 antimicrobial peptides against HIV has demonstrated that the hinged peptides caerin 1.1, caerin 1.9, and maculatin 1.1 (see Table 15.1 for sequences) are active at MIC 7.8, 1.2, and 11.3 μM, respectively. Maculatin 1.1 and caerin 1.1 inhibit HIV infection of T cells within minutes of exposure to the virus at concentrations below those at which these peptides are toxic to target cells. The peptides also stop the transfer of HIV by dendridic cells to T cells. Other caerins 1 and maculatins 1 and some synthetic modifications are currently under investigation. The antimicrobial peptides magainin 2, dermaseptin, and a number of *Rana* peptides are inactive in this context (Van Compernolle et al., 2005).

15.7 FUNGICIDES

Many antimicrobial peptides isolated from African, American, and European anurans have been routinely tested against the fungus *Candida albicans* and often show activity at the μM concentration against this organism (Pukala et al., 2006). We have not tested peptides against this fungus, but there is evidence that some antimicrobial peptides from the genus *Litoria* are fungicides.

Amphibian populations are declining worldwide, a consequence not only of habitat destruction but also because of infections by viruses and fungicides (Stuart et al., 2004; Mendelson et al., 2006). Ranaviruses are responsible for amphibian decline in some areas of northern America and Europe (Collins and Storfer, 2003). Another major culprit is the zoosporic chytrid fungus (*Batrachochytrium dendrobatidis*), a fungus that is seriously affecting anuran populations in Central America and Australia (Mutschmann et al., 2000; Woodhams et al., 2006). Anecdotal evidence suggests that those Australian anurans that do not produce antimicrobial peptides in their skin secretion (e.g., frogs of the genus *Limnodynastes*) succumb more readily to the chytrid fungus than those species that do produce antimicrobials (e.g., from the genus *Litoria*). For example, caerin 1.1 and maculatin 1.1 destroy the chytrid fungus at μM concentrations (Woodhams et al., 2006). Unfortunately, even with the protection of these hinged peptides, *Litoria* species are still killed following infection by the zoospores of the chytrid fungus.

15.8 CONCLUSIONS

Anurans worldwide produce host-defense peptides that protect them against predators, and it may be argued that this, at least in part, has allowed frogs and toads to survive for the hundreds of millions of years since they first evolved from freshwater fish in the Devonian period. Typical of these host-defense peptides are (i) the membrane-active peptides that have the multifaceted activities outlined in this chapter, and (ii) those peptides that inhibit the synthesis of the cell signalling molecule NO by nitric oxide synthase isoforms (see Chapter 14). The situation in Australia is particularly interesting because when Gondwanaland dispersed some 100 million years ago, followed 20 million years later by the separation of Australia from Antarctica, Australia was left geographically isolated. Many life forms, including anurans, evolved to produce species indigenous to Australia. A consequence of this is that Australian anurans have developed host-defense peptides that often have different sequences/shapes to those produced by anurans in other continents. In spite of these peptide structural differences, the host-defense capabilities of anuran skin peptides worldwide are similar.

ACKNOWLEDGMENT

The authors thank the Australian Research Council for the financial support of our anuran research program.

REFERENCES

Aboudy, Y. et al., Activity of two synthetis amphiphiloic peptides and magainin-2 against *Herpes simplex* virus types 1 and 2, *Int. J. Pept. Protein Res.*, 43, 573–578, 1994.

Ambroggio, E.E., Separovic, F., Bowie, J.H., and Fidelio, G.D., Surface behaviour and peptide–lipid interactions of the antibiotic peptides maculatin 1.1 and citropin 1.1, *Biochem. Biophys. Acta—Biomembranes*, 1664, 31–37, 2004.

Ambroggio, E.E. et al., Direct visualisation of membrane leakage by antibiotic peptides maculatin 1.1, citropin 1.1 and aurein 1.1, *Biophys. J.*, 89, 1874–1881, 2005.

Andreu, D. et al., Shortened cecropin A–melittin hybrids—significant size reduction retains potent antibiotic activity, *FEBS Lett.*, 296, 190–194, 1992.

Apponyi, M.A. et al., Host-defense peptides of Australian anurans: Structure, mechanism of action and evolutionary significance, *Peptides*, 25, 1035–1054, 2004.

Balla, M.S., Bowie, J.H., and Separovic, F., Solid state NMR study of antimicrobial peptides from Australian frogs in phospholipid membranes, *Eur. J. Biophys.*, 33, 109–116, 2004.

Barra, D. and Simmaco, M., Amphibian skin: A promising resource for antimicrobial peptides, *TIBTECH*, 13, 205–209, 1995.

Bevins, C.L. and Zasloff, M., Peptides from frog skin, *Annu. Rev. Biochem.*, 59, 395–414, 1990.

Bradford, A.M. et al., Novel peptides from the dorsal glands of the Australian Flood Plain Toadlet *Uperoleia inundata*, *Aust. J. Chem.*, 49, 475–484, 1996a.

Bradford, A.M., Bowie, J.H., Tyler, M.J., and Wallace, J.C., New antibiotic aurein peptides from the dorsal glands of the Australian toadlet *Uperoleia mjobergii*, *Aust. J. Chem.*, 49, 1325–1331, 1996b.

Brinkworth, C.S., Bowie, J.H., Tyler, M.J., and Wallace, J.C., A comparison of the host defense peptides of the New Guinea Tree Frog *Litoria genimaculata* and the Fringed Tree Frog *Litoria eucnemis*, *Aust. J. Chem.*, 46, 1235–1244, 2002.

Chen, T. et al., The structural organization of aurein presursor cDNAs from the skin secretion of the Australian green and golden bell frog, *Litoria aurea*, *Regulatory Peptides*, 128, 75–83, 2005.

Chia, B.C.S., Carver, J.A., Mulkhern, T.D., and Bowie, J.H., The solution structure of uperin 3.6, an antibiotic peptide from the granular dorsal glands of the Australian toadlet *Uperoleia mjobergii*, *J. Peptide Res.*, 54, 137–145, 1999.

Chia, B.C.S., Carver, J.C., Mulhern, T.D., and Bowie, J.H., Maculatin 1.1, an antimicrobial peptide from the Australian tree frog *Litoria genimaculata*. Solution structure and biological activity, *Eur. J. Biochem.*, 267, 1894–1908, 2000a.

Chia, B.C.S. et al., The interaction of amphibian peptides with the membranes of live bacteria: An investigation using ^{31}P solid state NMR, *Lett. Pept. Sci.*, 7, 151–156, 2000b.

Chia, B.C.S. et al., Caerin 4.1, an antibiotic peptide from the Australian tree frog *Litoria caerulea*. The NMR-derived solution structure, *Aust. J. Chem.*, 53, 257–265, 2000c.

Chia, B.C.S. et al., The orientation of the antibiotic peptide maculatin 1.1 in DMPG and DMPC bilayers. Support for a pore forming mechanism, *FEBS Lett.*, 512, 47–51, 2002.

Clark, D.P., Durell, S., Maloy, W.L., and Zasloff, M., Ranalexin—a novel antimicrobial peptide from bullfrog (*Rana catesbeiana*) skin; structurally related to the bacterial antibiotic, polymyxin, *J. Biol. Chem.*, 269, 10849–10855, 1994.

Collins, J.P. and Stoprfer, A., Global amphibian declines: Sorting the hypotheses, *Diversity Distrib.*, 9, 89–93, 2003.

Cruciani, R.A. et al., Antibiotic magainins exert cytolytic activity against transformed cell lines through channel formation, *Proc. Natl Acad. Sci. USA*, 88, 3792–3796, 1991.

Dathe, M. and Wieprecht, T., Structural features of helical antimicrobial peptides: Their potential to modulate activity on model membranes and biological cells, *Biochim. Biophys. Acta–Biomembranes*, 1462, 71–87, 1999.

Dhawan, L. et al., Antinidatory effect of vaginally administered (Ala(8,13,8))-magainin 2 amide in the rhesus monkey, *Contraception*, 62, 39–43, 2000.

Doyle, J.R. et al., nNOS inhibition, antimicrobial and anticancer activity of the amphibian skin peptide, citropin 1.1 and synthetic modifications: The solution structure of a modified citropin 1.1, *Eur. J. Biochem.*, 270, 1141–1153, 2003.

Erspamer, V., Bioactive secretions in the amphibian integument, in *Amphibian Biology: The Integument*, Heatwole, H., Ed., Surrey Beatty and Sons, Chipping Norton, New South Wales, Australia, pp. 178–350, 1994.

Gessell, J., Zasloff, M., and Opella, S.J., Two dimensional ^{1}H NMR experiments show that the 23 residue magainin antibiotic peptide is an α-helix in docecylphosphocholine micelles, sodium dodecylsulfate micelles and trifluoroethanol/water solution, *J. Biomol. NMR*, 9, 137–141, 1997.

Gibson, B.W., Poulter, L., Williams, D.H., and Maggio, J.E., Novel peptide fragments originating from PGLa and the caerulein and xenopsin precursors from *Xenopus laevis*, *J. Biol. Chem.*, 261, 5341–5349, 1986.

Giovannini, M.G., Poulter, L., Gibson, B.W., and Williams, D.H., Biosynthesis and degradation of peptides derived from *Xenopus laevis* prohormones, *Biochem. J.*, 243, 113–120, 1986.

Hara, T. et al., Heterodimer formation between the antimicrobial peptides Magainin 2 and PGLa in lipid bilayers: A cross-linking study, *Biochemistry*, 40, 12395–12399, 2001.

van Hoye, D., Burston, F., Nicolas, P., and Amiche, M., Antimicrobial peptides from hylid and ranid frogs originated from a 150 million year old ancestral precursor with conserved signal peptide but a hypermutable antimicrobial domain. *Eur. J. Biochem.*, 270, 2062–2082, 2003.

Huang, H.W., Chen, F.Y., and Lee, M.T., Molecular mechanism of peptide-induced pores in membranes, *Phys. Rev. Lett.*, 92, 198–304, 2004.

Jackway, R.J., et al., The fallaxidin peptides from the skin secretion of the Eastern Dwarf Tree Frog *Litoria fallax*. Sequence determination by positive and negative ion electrospray mass spectrometry: Antimicrobial activity and cDNA cloning of the fallaxidins. *Rapid Commun. Mass Spectrom.*, 22, 3207–3216, 2008.

Lazarus, L.H. and Attila, M., The toad, ugly and venomous, wears yet a precious jewel in his skin, *Prog. Neurobiol.*, 41, 473–507, 1993.

Lee, D.G. et al., Influence on the plasma membrane of *Candida albicans* by HP (2–9)-magainin 2 (1–12) hybrid peptide, *Biochem. Biophys. Res. Commun.*, 297, 885–889, 2002.

Ludtke, S.J. et al., Membrane pores induced by magainin, *Biochemistry*, 35, 13723–13728, 1996.

Maclean, M.J. et al., New caerin antimicrobial peptides from the skin secretion of the Dainty Green Tree Frog *Litoria gracilenta*. Sequence determination using positive and negative ion electrospray mass spectrometry, *Toxicon*, 47, 664–675, 2006.

Marcotte, I. et al., Interaction of antimicrobial peptides from Australian amphibians with lipid membranes, *Chem. Phys. Lipids*, 122, 107–120, 2003.

Marion, D., Zasloff, M., and Bax, A., A two-dimensional study of the antimicrobial peptide magainin 2, *FEBS Lett.*, 227, 21–26, 1988.

Maselli, V.M., Brinkworth, C.S., Bowie, J.H., and Tyler, M.J., Host defense skin peptides of the Australian Common Froglet *Crinia signifera*. Sequence determination using positive and negative ion electrospray mass spectrometry, *Rapid Commun. Mass Spectrom.*, 18, 2155–2161, 2004.

Maselli, V.M., Bowie, J.H., and Tyler, M.J., unpublished observations, 2007.

Matsuzaki, K., Magainins as paradigm for the mode of action of pore forming polypeptides, *Biochim. Biophys. Acta*, 1376, 391–400, 1998.

Matsuzaki, K., Why and how are peptide–lipid interactions utilised for self defense? Magainins and tachyplesins as archetypes, *Biochim. Biophys. Acta—Biomembranes*, 1462, 1–10, 1999.

Mendelson, J.R. et al., Responding to amphibian loss, *Science*, 314, 1541–1542, 2006.

Monks, A. et al., Feasibility of a high-flux anticancer drug screen using a diverse panel of cultured human tumour cells, *J. Natl. Cancer Inst.*, 83, 757–766, 1991; see also http://dtp.nci.nih.gov.

Mutschmann, F., Berger, L., Zwart, P., and Gaedicke, C., Chrytridiomycosis on amphibians—first report from Europe, *Berl. Muench. Tieraerztl. Wochenschr.*, 113, 380–389: see also available at http://www.ncbi.nlm.nih.gov?entrez/query.fegi?cmd=Retrieve&dbd=Pub-Med&list_11084755&dopt=Abstract, 2000.

Mystkowska, E.T., Niermierko, A., Komar, A., and Sawicki, W., Embryotoxicity of magainin-2-amide and its enhancement by cyclodextrin, albumin, hydrogen peroxide and acidification, *Hum. Reprod.*, 16, 1457–1463, 2001.

Ohsaki, Y., Gazdar, A.F., Chen, H.C., and Johnson, B.E., Antitumor activity of magainin analogues against human lung cancer cell lines, *Cancer Res.*, 52, 3534–3451, 1992.

Pukala, T.L., Brinkworth, C.S., Carver, J.A.., and Bowie, J.H., Investigating the importance of the flexible hinge in caerin 1.1: The solution structures and activity of two synthetically modified caerin peptides, *Biochemistry*, 43, 937–944, 2004.

Pukala, T.L. et al., Host-defense peptides from the glandular secretions of amphibians: Structure and activity, *Nat. Prod. Rep.*, 23, 368–393, 2006.

Resnick, N.M., Maloy, W.I., Guy, H.R., and Zasloff, M., A novel endopeptidase from *Xenopus* that recognises α-helical secondary structure, *Cell*, 66, 541–554, 1991.

Rozek, T. et al., The antibiotic and anticancer active aurein peptides from the Australian Bell Frogs *Litoria aurea* and *Litoria raniformis*. The solution structure of aurein 1.2, *Eur. J. Biochem.*, 267, 5330–5341, 2000.

Shai, Y. and Oren, Z., From "carpet" mechanism to de-novo designated diastereomeric cell-selective antimicrobial peptides, *Peptides*, 22, 1629–1641, 2001.

Sitaram, N. and Nagaraj, R., The therapeutic potential of host-defense antimicrobial peptides, *Curr. Drug Targets*, 3, 259–267, 2002.

Steinborner, S.T., Bowie, J.H., Tyler, M.J., and Wallace, J.C., An unusual combination of peptides from the skin glands of Ewing's Tree Frog *Litoria ewingi*—sequence determination and antimicrobial activity, *Aust. J. Chem.*, 50, 889–894, 1997a.

Steinborner, S.T. et al., New caerin peptides from the skin glands of the Australian tree frog *Litoria xanthomera*. Sequence determination by mass spectrometry, *Rapid Commun. Mass Spectrom.*, 11, 997–1101, 1997b.

Steinborner, S.T. et al., New antibiotic caerin 1 peptides from the skin secretion of the Australian tree frog *Litoria chloris*, *J. Pept. Protein Res.*, 51, 121–126, 1998.

Stone, D.J.M., Bowie, J.H., Tyler, M.J., and Wallace, J.C., The structure of caerin 1.1, a novel antibiotic peptide from Australian tree frogs, *Chem. Commun.*, 6, 400–402, 1992a.

Stone, D.J.M. et al., The structures of the caerins and caeridin 1 from *Litoria splendida*, *J. Chem. Soc. Perkin Trans.*, 1, 3173–3179, 1992b.

Stuart, S.N. et al., Status and trends of amphibian declines and extinctions worldwide, *Science*, 306, 1783–1786, 2004.

Tachi, T., Epand, R.F., Epand, R.M., and Matsuzaki, K., Position-dependent hydrophobicity of the antimicrobial magainin peptide affects the mode of peptide–lipid interactions and selective toxicity, *Biochemistry*, 41, 10723–10731, 2002.

Terry, A.S. et al., The cDNA sequence coding for prepro-PGS(prepromagainins) and aspects of the processing of this prepro-polypeptide, *J. Biol. Chem.*, 263, 5745–5751, 1988.

Tyler, M.J., Stone, D.J.M., and Bowie, J.H., A novel method for the release and collection of dermal, glandular secretions from the skin of frogs, *J. Pharm. Toxicol. Methods*, 28, 199–200, 1992.

US Pat 07/963007, filed 19/10/1992; now US pat. 5,643,878.

Van Compernolle, S.E. et al., Amphibian antimicrobial skin peptides that potently inhibit HIV infection and dendridic cell mediated transfer of virus to T cells, *J. Virol.*, 79, 12088–12094, 2005.

Wabnitz, P.A. et al., First record of host defense peptides in tadpoles. The magnificent tree frog *Litoria splendid*, *J. Pept. Res.*, 52, 477– 481, 1998.

Wabnitz, P.A. et al., Differences in the skin peptides of the male and female Australian Magnificent Tree Frog *Litoria splendida*—a three year monitoring survey: The discovery of the male sex pheromone splendipherin together with phe8 caerulein and the antibiotic caerin 1.10, *Eur. J. Biochem.*, 267, 269–275, 2000.

Waugh, R.J. et al., Peptides from Australian Frogs. The structures of the caerins from *Litoria caerulea*, *J. Chem. Soc. Perkin Trans.*, 1, 573–576, 1993a.

Waugh, R.J. et al., Peptides from Australian frogs: The structures of the caerins and caeridins from *Litoria gillini*, *J. Chem. Res.*, 39, 937–961, 1993b.

Wegener, K.L. et al., Host defense peptides from the skin glands of the Australian Blue Mountains Tree Frog *Litoria citropa*. Solution structure of the antibacterial peptide citropin 1.1, *Eur. J. Biochem.*, 265, 627–637, 1999.

Wegener, K.L. et al., Bioactive dahlein peptides from the skin secretions of the Australian aquatic frog *Litoria dahlii*: Sequence determination by electrospray mass spectrometry, *Rapid Commun. Mass Spectrom.*, 15, 1626–1734, 2001.

Wegener, K.L., Carver, J.A., and Bowie, J.H., The solution and micelle structures and activity of caerin 1.1 and caerin 1.4 in aqueous trifluoroethanol and dodecylphosphocholine micelles, *Biopolymers*, 69, 42–59, 2003.

Wojak, C. et al., Cyclodextrin enhances spermicidal effects of magainin-2-amide, *Contraception*, 61, 99–103, 2000.

Wong, H., Bowie, J.H., and Carver, J.A., The solution structure and activity of caerin 1.1, an antimicrobial peptide from the Australian Green Tree Frog, *Litoria splendid*, *Eur. J. Biochem.*, 247, 545–557, 1997.

Woodhams, D.C. et al., Population trends associated with skin defenses against chytridiomycosis in Australian frogs, *Oecologia*, 146, 531–540, 2006.

Zasloff, M., Magainins, a class of antimicrobial peptides from *Xenopus* skin: Isolation and characterisation of two active forms and partial cDNA sequence of a precursor, *Proc. Natl Acad. Sci. USA*, 84, 5449–5453, 1987.

Zasloff, M. and Anderson, M., The development of antimicrobial peptides of animal origin as vaginal microbiocides: The challenges ahead and the potential, *AIDS*, 15, Suppl. 1, S54–S55, 2001.

Zhou, M., Chen, T., Walker, B., and Shaw, C., Novel frenatins from the skin of the Australasian giant white-lipped tree frog, *Litoria infrafrenata*: Cloning of the precursor cDNAs and identification in defensive skin secretion, *Regul. Pept.*, 26, 2445–2451, 2005.

16 Antimicrobial Peptides

Alison M. McDermott

CONTENTS

16.1 INTRODUCTION

Antimicrobial peptides (AMPs) are widely distributed effectors of innate immunity in animals and plants. They are typically small peptides, most less than 50 amino acids (although some antimicrobial proteins such as bactericidal/permeability increasing protein (BPI) are also mentioned in Sections 16.6.2 and 16.6.3), are amphipathic and typically carry an overall positive charge (+2 or greater) due to a relative excess of amino acids such as arginine and lysine (while most AMPs are cationic in nature a few anionic peptides are also mentioned). They have been found in every species examined to date and over 900 have been identified across species. The peptides show a broad spectrum of antimicrobial activity and many have additional effects on mammalian cell behaviors.

The existence of antimicrobial peptides can be traced back to the late 1800s when Ehrlich first identified that the granules of neutrophils (also referred to here as polymorphonuclear cells, PMNs) contained basic as well as acidic proteins (Ehrlich, 1891; Ehrlich and Lazarus, 1898). In 1905,

Petterson reported antimicrobial activity in aqueous extracts of pus from human patients with empyma and attributed this activity to basic proteins similar to the protamines of salmon sperm. Some fifty years passed until Hirsch (1956) reported "...another biochemical mechanism, which may kill certain bacteria in the cytoplasm of polymorphonuclear leucocytes." Hirsch referred to his substance, a protein, as "phagocytin," although he never specifically identified it, and showed it was distinct from lysozyme, which was the only previously known bactericidal component of PMNs. Hirsch's contribution was significant in that it revealed multiple PMN killing mechanisms and was instrumental in re-igniting interest in the field of innate immunity which had, with the exception of Sir Alexander Fleming's (later of penicillin fame) discovery of lysozyme in 1922 (Fleming, 1922), lain rather dormant since the turn of the century.

Spitznagel and Chi (1963) observed that the cytoplasmic granules of guinea pig PMNs stained strongly for cationic arginine-rich proteins and that during phagocytosis the granules collected around the ingested bacteria then seemed to disappear. They also observed that the organism then became histochemically positive for arginine-rich peptides, substances not normally present in bacteria. Then, within a short time, Zeya and Spitznagel (1963) provided convincing evidence of the existence of bactericidal cationic components of PMNs. They separated proteins from guinea pig neutrophils electrophoretically and observed several highly basic fractions (more so than the well-known basic proteins lysozyme and ribonuclease) with potent activity against Gram-positive and Gram-negative bacteria and *Candida albicans*. Notably a preparation of Hirsch's "phagocytin" generated comparable fractions. Neither Hirsch (1956) or Zeya and Spitznagel (1963) observed bacterial cell lysis (a known effect of AMPs, see Section 16.3.1.1) in response to treatment with their preparations. However, this may have been related to dilution effects on the AMPs in their preparation. Thus it does appear likely that both groups had isolated AMPs from neutrophils. Indeed, amino acid analysis by Zeya and Spitznagel showed the cationic proteins to have up to 35% arginine and 14% cysteine, features now considered characteristic of a group of AMPs called defensins. They also began to address the mechanism of antibacterial activity of these peptides, showing that they inhibited the respiratory activity of *Escherichia coli* and damaged bacterial permeability barriers (Zeya and Spitznagel, 1966, 1968, 1969).

In 1972, Garcia-Olmedo's group reported the identification of the first of several groups of plant antimicrobial peptides—the thionins (Fernandez de Caleya et al., 1972). Then in 1981, Steiner and colleagues reported the purification and characterization of two small basic proteins P9A and P9B, with potent antibacterial activity, from the hemolymph of the moth *Hyalophora cecropia*, which they named cecropins. The latter have been shown to be an important part of the immune system of insects (Bulet and Stocklin, 2005) and a mammalian homolog has been isolated from porcine intestine (Lee et al., 1989).

Also in the early 1980s, Lehrer and colleagues purified various antimicrobial peptides from rabbit PMNs (Selsted et al., 1984, 1985a) and in 1985, Ganz and colleagues reported the existence of related peptides in human PMNs and named them "defensins" based upon a presumed role in host defence. As discussed in more detail (Section 16.3) these neutrophil defensins are members of the α class of mammalian defensins and are present in multiple vertebrate and invertebrate species and plants. A short time later, Romeo and colleagues identified a group of peptides with antimicrobial activity in bovine PMNs, which they referred to as bactenecins (Romeo et al., 1988; Gennaro et al., 1989). Ultimately bactenecins were, along with several other peptides such as protegrins (porcine peptides identified in 1993 by Kokryakov et al.) and bovine indolicidin (Selsted et al., 1992), classified as members of the "cathelicidin" family by Zanetti and colleagues (1995), as all have a conserved proregion with sequence similarity to members of the cystatin superfamily of cysteine proteinase inhibitors and a variable C-terminal antimicrobial domain. The isolation of the only human cathelicidin, LL-37, was simultaneously reported in 1995 by Cowland and colleagues, Agerberth and colleagues, and Larrick and colleagues.

In 1993, Selsted and colleagues, studying bovine neutrophils, reported a second class of mammalian defensins, the β-defensins, which differed from the α-defensins previously identified in rabbit and human neutrophils (Selsted et al., 1984, 1985a, 1985b; Ganz et al., 1985, 1989) in the arrangement of their tridisulfide motifs. Notably, these β-defensins shared a consensus sequence with tracheal antimicrobial peptide (TAP) from bovine tracheal mucosa that was described by Diamond and colleagues in 1991. The search for mammalian epithelial AMPs, initially sparked by the discovery of a group of AMPs named magainins in amphibian skin (Zasloff et al., 1987), led to the identification of a group of mouse α-defensins called cryptins that are present in Paneth cells of the intestine (Ouellette et al., 1989). Similar α-defensin peptides, named human defensin (HD)-5 and -6, were identified in human intestinal Paneth cells, and thus were the first human epithelial AMPs to be identified (Jones and Bevins, 1992, 1993). Subsequently other human epithelial defensins, this time of the β class, namely human β-defensin (hBD)-1, -2, -3, and -4 were identified in plasma and a large variety of human epithelial tissues (Bensch et al., 1995; Harder et al., 1997, 2001; Garcia et al., 2001a, 2001b). Thus, by the mid-1990s it became obvious that not only were AMPs important components of neutrophils but that they were widely distributed throughout the body in areas commonly exposed to pathogens.

The 1990s and 2000s saw the discovery of numerous AMPs across a wide variety of species. There are too many to mention here, but each discovery is important to the specific species and to the field of immunology in general. Indeed, the observation that AMPs are produced by multiple species, from bacteria themselves, through plants, invertebrates to vertebrates, indicates that they are an "ancient" and important defense system. The numerous families of peptides that have now been identified across species can be (admittedly somewhat loosely) divided into groups based upon amino acid composition and structure. Some examples are presented in Table 16.1. Databases of both plant and animal AMPs are maintained at the following websites: http://www.bbcm.units.it/~antimic/home.html, http://aps.unmc.edu/AP/main.php, and http://defensins.bii.a-star.edu.sg/. In the remainder of this chapter, the focus will be on the mammalian (with an emphasis on human) defensins and cathelicidins, as these have become the most intensely studied AMPs. Other AMPs will be introduced and referred to as necessary to highlight specific concepts.

The identification of the various AMPs has been accompanied by intense study of their mechanism of antimicrobial action. As will be discussed in detail later (Section 16.3.1), with some 900 peptides having been identified, a single mechanism of action is not likely, indeed several modes have been proposed and many AMPs appear to have multiple actions leading to bacterial killing. Significant progress has been made by several research groups such as those of Huey Huang (Rice University, Houston, TX) and Yechiel Shai (The Weizmann Institute of Science, Rehovot, Israel) to name but two, but there is still much to learn about how AMPs act, particularly if we are to capitalize on their obvious potential as novel antimicrobial agents.

Furthermore, although first recognized for their antimicrobial effects, a large number of AMPs have now been shown to have a variety of effects on mammalian cell function. These, discussed in more detail in Section 16.3.2, include chemotactic effects on immune and inflammatory cells, stimulation of expression of multiple genes, activation of immune cells, and stimulation of epithelial cell proliferation. This has led to the designation of many AMPs as "multifunctional peptides" as they have not only direct antimicrobial activity but also immunomodulatory effects and modulate wound healing. Indeed, it has been suggested that for at least some AMPs, modulating mammalian cell behavior, rather than antimicrobial activity, is their primary function *in vivo*.

The discovery of AMPs as ubiquitous host-defense molecules expressed at the interface between pathogen and host has significantly broadened our understanding of innate immunity in humans and other species. Furthermore, it has revealed a possible new weapon in the ongoing war between microbes and man. As will be discussed (Section 16.6), it was not long after the identification and isolation of the first AMPs that the notion was put forth that they may represent a novel group of antimicrobial pharmaceuticals. Although none have yet been successfully brought to market, there is intense interest in developing AMP-based products for a variety of anti-infective indications. Furthermore, their more recently revealed immunomodulatory and wound-healing effects now further expands their possible clinical applications.

TABLE 16.1
Some Examples of Antimicrobial Peptides of Different Classes

Peptide	Source	Reference
Cationic Peptides		
Linear α-helical peptides		
Andropin	Fruit fly (*Drosophila melanogaster*)	Samakovlis et al., 1991
Cecropin A	Silk moth (*Hyalophora cecropia*)	Steiner et al., 1981
CRAMP	Mouse	Gallo et al., 1997
LL-37	Human	Cowland et al., Agerberth et al., and Larrick et al., 1995
Magainin I	Frog (*Xenopus laevis*)	Zasloff, 1987
Pleurocidin	Flounder (*Pleuronectes americanus*)	Cole et al., 1997
Peptides rich in certain amino acids (in parentheses)		
Bactenecin 5 (proline)	Ox	Frank et al., 1990
Histatins (histidine)	Human	Oppenheim et al., 1988
Indolicidin (tryptophan)	Ox	Selsted et al., 1992
PR-39 (proline)	Pig	Agerberth et al., 1991
Peptides with cysteines that form disulfide bonds		
Single disulfide peptides		
Brevinin-1	Frog (*Rana brevipoda porsa*)	Morikawa et al., 1992
Renalexin	Bullfrog (*Rana catesbeiana*)	Clark et al., 1994
Two disulfide bonds		
Polyphemusin 1	Horseshoe crab (*Limulus polyphemus*)	Miyata et al., 1989
Protegrin 1	Pig	Kokryakov et al., 1993
Three disulfide bonds		
Cryptdin 1	Mouse	Ouellette & Lualdi, 1990
Gallinacin 1	Chicken	Harwig et al., 1994
hBD-1	Human	Bensch et al., 1995
HD-5	Human	Jones & Bevins, 1992
HNP-1	Human	Ganz et al., 1985
NP-1	Rabbit	Ganz et al., 1989
Spheniscin 1	Penguin (*Aptenodytes patagonicus*)	Thouzeau et al., 2003
TAP	Ox	Diamond et al., 1991
Four or more disulfide bonds		
DmAMP1	Dahlia (*Dahlia merckii*)	Thevissen et al., 1996
NaD1	Ornamental tobacco (*Nicotiana alata*)	Lay et al., 2003
RsAFP2	Radish (*Raphanus sativus*)	Terras et al., 1995
Anionic Peptides		
Dermcidin	Human	Schittek et al., 2001
Maximin H5	Amphibians	Lai et al., 2002
Thymosin β4	Human	Tang et al., 2002

16.2 SOURCE OF PEPTIDES

The primary sources of human AMPs (Table 16.2) are immune and inflammatory cells, particularly neutrophils, and epithelial tissue. The two major categories are the defensins and cathelicidin. Human defensins are 29 to 45 amino acids in length and are characterized by the presence of six cysteine residues that interact to form three disulfide bonds and a β-sheet structure. As alluded to above, there are two classes of human defensin, referred to as α and β. This classification reflects the different length of peptide segments between the cysteines and different patterns of connectivity of the cysteines: the α-defensin arrangement is C1–C6, C2–C4, C3–C5, whereas that of β-defensins is C1–C5, C2–C4, and C3–C6 (Ganz, 2003). A third defensin class, referred to as theta (or minidefensins), has been identified in leukocytes of nonhuman primates (Tang et al., 1999; Leonova et al., 2001; Tran et al., 2002). These 18-amino-acid theta-defensins are unique in that they are formed by splicing of two truncated α-defensins to create a circular structure. Six genes for human theta-defensins have been identified, but premature stop codons prevent their translation (Nguyen et al., 2003).

Six human α-defensins and four β-defensins have been identified and characterized. Of the six α-defensins, four were discovered in neutrophils, hence they were named human neutrophil peptide (HNP)-1 through 4 (Ganz et al., 1985; Selsted et al., 1985b; Wilde et al., 1989). Mature HNP 1–4 are localized to a subset of azurophil (myeloperoxidase specific) granules (Faurschou et al., 2002) where their molar content greatly exceeds that of other granule peptides (together they represent 30% of total granule protein) and may be as high as 10 mg/mL (Ganz, 1987). HNP-1 and -3 are 30 amino acids in length and differ only in the first amino acid (see Table 16.2). HNP-2 is 29 amino acids long and appears to be a proteolytic product of HNP-1 and/or -3 (Selsted et al., 1985b). HNP-4 is 33 amino acids in length, of which 22 differ from the other HNPs (Wilde et al., 1989) and is a minor component of the neutrophil granules. HNPs are initially synthesized as preprosequences consisting

TABLE 16.2
Amino Acid Sequences of the Major Human Antimicrobial Peptides

Peptide	Sequence
α-Defensins	
HNP-1	ACYCRIPACIAGERRYGTCIYQGRLWAFCC
HNP-2	CYCRIPACIAGERRYGTCIYQGRLWAFCC
HNP-3	DCYCRIPACIAGERRYGTCIYQGRLWAFCC
HNP-4	VCSCRLVFCRRTELRVGNCLIGGVSFTYCCTRV
HNP-5	ATCYCRTGRCATRESLSGVCEISGRLYRLCCR
HD-6	AFTCHCRRSCYSTEYSYGTCTVMGINHRFCCL
β-Defensins	
hBD-1	GLGHRSDHYNCVSSGGQCLYSACPIFTKIQGTCYRGKAKCCK
hBD-2	GIGDPVTCLKSGAICHPVFCPRRYKQIGTCGLPGTKCCKKP
hBD-3	GIINTLQKYYCRVRGGRCAVLSCLPKEEQIGKCSTRGRKCCRRKK
hBD-4	EFELDRICGYGTARCRKKCRSQEYRIGRCPNTYACCLRKWDESLLNRTKP
LL-37	LLGDFFRKSKEKIGKEFKRIVQRIKDFLRNLVPRTES
Histatin 5	DSHAKRHHGYKRKFHEKHHSHRGY
LEAP family	
LEAP-1 (hepcidin)	DTHFPICIFCCGCCHRSKCGMCCKT
LEAP-2	MTPFWRGVSLRPIGASCRDDSECITRLCRKRRCSLSVAQE
Dermcidin	SSLLEKGLDGAKKAVGGLGKLGKDAVEDLESVGKGAVHDVKDVLDSV

of an amino terminal signal sequence (approximately 19 amino acids), an anionic propiece (approximately 45 amino acids), and a C-terminal mature peptide (Daher et al., 1988; Valore and Ganz, 1992). The propiece is required for correct subcellular trafficking and sorting of proHNPs and acts as an intramolecular inhibitor of HNP activity, presumably through charge neutralization (Liu and Ganz, 1995; Valore et al., 1996). Recent studies by Wu and colleagues (2003, 2007) indicate roles for the propiece in chaperoning the folding of HNPs, first by preventing aggregation of the C-terminal defensin domain and also by lowering the free energy of the transition state in the folding pathway by direct intramolecular interactions with the defensin domain.

HNPs are synthesized in neutrophil precursor cells in the bone marrow known as promyelocytes, but are not actively synthesized by mature circulating neutrophils (Harwig et al., 1992; Arnljots et al., 1998; Cowland and Borregaard, 1999). Removal of the signal sequence occurs rapidly, whereas processing to yield the mature AMP takes many hours, with the final steps occurring as the granules mature (Valore and Ganz, 1992). Although the major activity of HNPs is presumed to occur when the azurophil granules fuse with the phagocytic vacuole-containing ingested pathogen, HNPs are released extracellularly upon neutrophil activation (Ganz, 1987) and can be detected in plasma and body fluids. As would be expected their concentration rises with infection (Ihi et al., 1997). Neutrophils are the primary source of HNPs, but these α-defensins have been localized to monocytes and lymphocytes (Agerberth et al., 2000) and immature monocyte-derived dendritic cells (Rodriguez-Garcia et al., 2007). Notably in the latter, HNP 1–3 production was upregulated in response to proinflammatory cytokines such as interleukin (IL)-1β. Also, recently it was reported that macrophages can acquire HNPs, and other antimicrobial substances, through phagocytosis of apoptotic neutrophils (Tan et al., 2006). HNPs have also been detected in nonimmune/inflammatory cells and tissues in various disease states as discussed in Section 16.4.2.

The other two known α-defensins are human defensin (HD)-5 and -6, which were first isolated from intestinal Paneth cells (Jones and Bevins, 1992, 1993). The latter are specialized epithelial cells found at the base of the crypt's of Lieberkuhn in the duodenum and ileum and were named for the 19th-century Austrian physiologist Joseph Paneth. HD-5 has now also been localized to the female reproductive tract (Quayle et al., 1998), and is found "infrequently" in nasal and bronchial epithelial cells (Frye et al., 2000). Recent quantitative RT-PCR studies showed that HD-5 and -6 are the most abundant antimicrobial substances expressed in the small intestine, with the amount of HD-5 being sixfold higher than that of HD-6 (Wehkamp et al., 2006). These intestinal defensins are stored in the Paneth cells in proforms, which are then cleaved, during or perhaps after their release in to the intestinal lumen. Processing is carried out by the serine proteinase trypsin, the specific isoforms of which that are required are also produced by the Paneth cells as inactive zymogens (Ghosh et al., 2002). Extrapolating from findings with murine Paneth cell defensins, the cryptdins, human Paneth cell defensin synthesis is likely constitutive with secretion occurring in response to bacteria/bacterial products or cholinergic stimuli through as yet ill-defined pathways (Ayabe et al., 2000, 2002).

The genes for HNP-1, -3, and -4 and HD-5 and -6 all reside on chromosome 8p23 (Linzmeier et al., 1999). The genes for HNP-1 to -4 have two introns and three exons, with the last two encoding the prepropeptide (Linzmeier et al., 1993). In contrast, the genes for HD-5 and -6 have only two exons (Linzmeier et al., 1993; Mallow et al., 1996). Several studies have shown gene copy number polymorphisms for some α-defensin genes. Mars and colleagues (1995) first reported that individuals can inherit copies of chromosome 8 with two or three copies of the genes DEFA1 and DEFA3, which code for HNP-1 and HNP-3, respectively (and HNP-2 as it appears to be derived from either 1 or 3). The locus for these defensin genes is now referred to as DEFA1A3 and copy number polymorphism has been confirmed by Aldred and colleagues (2005), who reported between 4 and 11 copies per diploid genome, and Linzmeier and Ganz (2005), who reported 5 to 14 copies with a mode of 10. Notably, it has been reported that 10 to 37% of individuals, depending upon the populations tested, lack DEFA3, the gene for HNP-3 (Aldred et al., 2005; Linzmeier and Ganz, 2005; Ballana et al., 2007). Linzmeier and Ganz (2005) also observed that the amount of HNP-1 and

HNP-3 expressed in neutrophils was proportional to the copy number. Together, these observations raise the question of whether or not this variability in gene copy numbers may result in differences in individual resistance to infection. In contrast to HNP-1 and -3, the genes for HNP-4, HD-5, and HD-6 (DEFA4, DEFA5, DEFA6, respectively) are only found as two copies per diploid genome (Linzmeier and Ganz, 2005).

The four characterized human β-defensins are named human β-defensin (hBD) 1 through 4. They are primarily expressed in epithelial tissue but some (hBD-1 and -2) have also been found in immune cells (monocytes, macrophages, and dendritic cells) (Duits et al., 2002; Ryan et al., 2003). hBD-1, which was first isolated from plasma (Bensch et al., 1995), is constitutively expressed by a variety of epithelia including airway epithelia (McCray and Bentley, 1997), urogenital tissue (Valore et al., 1998), nasolacrimal duct (Paulsen et al., 2001), mammary gland (Jia et al., 2001), and corneal and conjunctival epithelia (Hattenbach et al., 1998; Haynes et al., 1999; Lehmann et al., 2000; McDermott et al., 2003; Narayanan et al., 2003). hBD-2 expression is inducible in many epithelia in response to bacterial products [often acting via toll-like receptor (TLR) activation] and cytokines (Harder et al., 1997; McNamara et al., 1999; O'Neil et al., 1999; Wada et al., 2001; Wang et al., 2003; McDermott et al., 2003). hBD-3 expression is also inducible in many epithelial tissues (Harder et al., 2001; Garcia et al., 2001a) whilst in some tissues such as the ocular surface epithelia its expression is constitutive (Narayanan et al., 2003). The expression of hBD-4 is more restricted than that of the other defensins, with testes and epididymis having the highest levels (Garcia et al., 2001b; Yamaguchi et al., 2002). Expression of hBD-4 is inducible, although not by the cytokines reported to upregulate hBD-2 and -3 (Garcia et al., 2001b).

Like α-defensins, the β-defensins also arise by cleavage of a larger precursor, this time consisting of a leucine-rich signal sequence, a short or absent propiece, and the mature peptide. In skin, hBD-2 is found stored in lamellar bodies, which are lipid-containing vesicles secreted in to the extracellular space (Oren et al., 2003). In most cases the β-defensin precursors are encoded in two separate exons separated by an intron of variable length, with one exon encoding the signal sequence and propiece, and the other encoding the mature peptide. The gene for hBD-2 has three binding sites for NF-κB (Liu et al., 1998), and binding sites for other transcription factors such as activator protein (AP)-1 and nuclear factor for IL-6 (Diamond et al., 2000; Harder et al., 2000), which is in keeping with its status as an inducible AMP.

The genes for hBD 1 to 4 (known as DEFB1, DEFB4, DEFB103, and DEFB104, respectively) all map to chromosome 8p22–23 close to the α-defensin genes (Liu et al., 1997). Using computational search strategies it has been established that the human genome contains some 39 β-defensin genes and pseudogenes that are distributed in five clusters, two each on chromosomes 8 and 20 and one on chromosome 6 (Schutte et al., 2002; Patil et al., 2005). Some of these novel genes have been investigated. Yamaguchi and colleagues (2002) studied DEFB5 and -6 and found that their products, named hBD-5 and -6, were specifically expressed in the epididymis. Cloning of DEFB25 to 29, which reside on chromosome 20, shows them also to be preferentially expressed in the male genital tract (Rodriguez-Jimenez et al., 2003). Premratanachai and colleagues (2004) observed that DEFB7, -9, -11, and -12 were constitutively expressed by gingival keratinocytes, whereas expression of DEFB8 and -14 was induced by cytokines and infection. hBD-5 and the product of DEFB9 have also been found to be expressed in the lung (Kao et al., 2003), and DEFB118 is present in epididymis (Yenugu et al., 2004).

As with the α-defensins HNP-1 and HNP-3, copy number variation has been reported for hBD-2, -3, and -4. Hollox and colleagues (2003) reported that a cluster containing DEFB4, DFEB103, and DEFB104 was present as 2 to 12 copies per diploid genome. Similarly, Linzmeier and Ganz (2005) reported copy numbers of 2 to 8 with a mode of 6 per diploid genome for DEFB4 and DEFB103. Again, gene copy number correlated with expression, suggesting that it may have significant consequences for the functioning of the immune system. DEFB1, the gene for hBD-1, only appears as a single copy (Linzmeier and Ganz, 2005) but as noted later (Section 16.4.2) exhibits a number of single nucleotide polymorphisms.

The second major group of mammalian AMPs is the cathelicidins, and while several species such as cow and pig express multiple cathelicidins only one, LL-37, is expressed in humans. The term cathelicidin was coined by Zanetti and colleagues (1995) to describe AMPs that have a "cathelin precursor" domain, the sequence of which is very similar to that of cathelin, a protein in porcine neutrophils that is an inhibitor of the cysteine protease cathepsin L (i.e., cathepsin-L-inhibitor) (Ritonja et al., 1989). The N-terminal cathelin domain is highly conserved among species and is 99 to 114 amino acids long (Zanetti et al., 1995). In contrast, the C-terminal antimicrobial domain varies considerably in sequence and length (12 to 100 amino acids). Identified by three independent groups in 1995 (Agerberth et al., 1995; Cowland et al., 1995; Larrick et al., 1995), the human cathelicidin LL-37 is so called because it begins with two leucine residues and is 37 amino acids long. Like most cathelicidins, the gene (CAMP) for LL-37, which maps to chromosome 3 (Gudmundsson et al., 1996) and exists as a single copy (Linzmeier and Ganz, 2006), has four exons, the first three of which code for a 29-amino-acid signal sequence and the cathelin domain, with the fourth coding for the mature peptide (Bals and Wilson, 2003). The full-length precursor is referred to as hCAP18 (human cationic antimicrobial protein 18) owing to its mass (18 kDa) and similarity to a rabbit antimicrobial protein CAP18 (Larrick et al., 1991; Cowland et al., 1995).

Initially, LL-37 was isolated from neutrophils where it is stored as hCAP18 in the secondary (or specific) granules. In contrast to the azurophil granules, the home of HNP 1–4, which are preferentially delivered to phagocytic vacuoles, the contents of the smaller secondary/specific granules are chiefly delivered to the extracellular milieu (Borregaard et al., 1993; Bainton, 1999). Processing of hCAP18 to yield LL-37 occurs during its secretion and is mediated by proteinase 3 (Sorensen et al., 2001). In addition to neutrophils, LL-37 is now known to be produced by immune cells (Agerberth et al., 2000), mast cells (DiNardo et al., 2003), and a variety of epithelia including lung (Bals et al., 1998), skin (Frohm et al., 1997), squamous epithelial of mouth, tongue, esophagus, cervix and vagina (Frohm-Nilsson et al., 1999), eccrine sweat glands (Murakami et al., 2002), and ocular surface (Gordon et al., 2005). LL-37 expression is constitutive in some tissues but inducible by pathogens, proinflammatory cytokines, and vitamin D3 in others (Wang et al., 2004a; Huang et al., 2006; Li et al., 2008; Ooi et al., 2007). Regulatory motifs involved in mediating LL-37 upregulation likely include NF-κB, γ-interferon response element, and NF-IL-6 (Gudmundsson et al., 1996). As noted for hBD-2 (Oren et al., 2003), LL-37 is also stored in lamellar bodies in the epidermis (Braff et al., 2005a). Notably, while hCAP18 has no known biological activity, in addition to LL-37, the cathelin domain has also been shown to be active and has antimicrobial activity and inhibits proteases (Zaiou et al., 2003). Post-secretory processing of LL-37 by serine proteases has been shown to occur at the skin surface, generating smaller peptides (referred to as KR-20, KS-30, and RK-31) with greater antimicrobial activity than the parent molecule (Murakami et al., 2004). The proteases responsible for this have since been identified as kallikrein 5 and 7 (Yamasaki et al., 2006).

Although defensins and cathelicidin represent the most well characterized human AMPs a number of others have been identified. Liver expressed antimicrobial peptide (LEAP)-1 was identified in blood ultrafiltrate by Krause and colleagues in 2000. This 25-amino-acid peptide has four disulfide bonds and is highly expressed in liver (hence its name). It was also isolated from urine by Park and colleagues (2001), who referred to it as hepcidin. A second AMP that is also synthesized by the liver, LEAP-2, has also been isolated from blood (Krause et al., 2003). Several circulating forms were identified that differed in length, but which all had two disulfide bonds. Messenger RNA for LEAP-1 and -2 has been reported to be present in ocular surface epithelial cells (McIntosh et al., 2005), although we were not able to confirm this (Huang et al., 2007a). Hepcidin (LEAP-1) is now recognized as a peptide hormone important for iron homeostasis, and through this action may further contribute to host defense by limiting iron availability to microbes (Ganz, 2006). Dermcidin is a 47-amino-acid AMP expressed in sweat glands and secreted into sweat (Schittek et al., 2001). Histatins are a group of histidine-rich AMPs produced by the parotid and submandibular salivary glands and, in contrast to most other AMPs which are bactericidal, these peptides are primarily antifungal (Kavanagh and Dowd, 2004). Thymosin β-4, a G-actin sequestering anionic peptide, has also been shown to have antimicrobial activity and is present in platelets and various tissues (Tang et al., 2002; Huang et al., 2007a). A significant number of chemokines, including MIP-3α, CCL1, CXCL1-3, and CXCL9-11, have also been found to have antimicrobial activity (Cole et al., 2001; Yang et al., 2003). However, for molecules such as Thymosin β-4 and chemokines, microbial killing is unlikely to be their major activity *in vivo*. A number of other small peptides with established functions but which have also been found to have antimicrobial activity, such as neuropeptide Y and α-melanocyte-stimulating hormone, are discussed in a recent review by Radek and Gallo (2007).

16.3 BIOLOGICAL AND MOLECULAR MECHANISMS OF ACTION

The major function of many AMPs is presumed to be the killing of microorganisms, particularly bacteria. However, some have additional, and perhaps in some cases primary, actions on mammalian cells. These two types of biological activity are, for the most part, completely independent. Here, the current thinking on the mechanism of antimicrobial action will be discussed first, followed by what is known of the pathways through which AMPs modulate mammalian cell function.

16.3.1 Antimicrobial Action

As a group, AMPs demonstrate direct broad-spectrum antimicrobial activity against a wide range of bacteria, fungi, and some viruses. Peak antibacterial activity typically occurs in the low μM range, but the various AMPs often differ significantly in their ability to kill specific bacteria. For example HNP-4 was up to 13-fold more potent against *E. coli* than the other neutrophil defensins, and HD-6 had minimal activity against this organism (Ericksen et al., 2005). *In vitro* antimicrobial activity is rapid (15 to 90 min), and maximal effects are most commonly observed under conditions of low ionic strength, with low concentrations of divalent cations and plasma proteins, although some AMPs, including hBD-3, are less salt sensitive than others, likely attributable to an increased positive surface charge (Schibli et al., 2002). Thus, the antimicrobial effectiveness of AMPs *in vivo* has been called into question. However, an interesting study by Dorschner and colleagues (2006) suggests that carbonate is the critical ionic factor that is present *in vivo* that allows AMPs to exert significant antimicrobial activity. Growth in carbonate was observed to cause significant changes in bacterial gene expression and a decreased cell wall thickness that enhanced susceptibility to AMPs.

That AMPs, particularly those produced by epithelial tissues, are active *in vivo* has also been questioned because the level of these peptides *in vivo*, even following bacterial/cytokine stimulation, is frequently much less than that required for *in vitro* killing. "Local concentrating" effects have been suggested as one way to circumvent this issue, and indeed, some epithelial AMPs are found stored in such ways as to raise their levels (e.g., hBD-2 and LL-37 are stored in lamellar bodies in skin; Oren et al., 2003; Braff et al., 2005a). It has also been found that many AMPs show synergistic/additive activity with other AMPs and with other known antimicrobial molecules such as lysozyme (Nagaoka et al., 2000; Singh et al., 2000; Yan and Hancock, 2001). Synergy between AMPs has also been suggested as a further mechanism to help overcome salt sensitivity (Nagaoka et al., 2000). In keeping with there being in-built mechanisms to circumvent issues of salt sensitivity and so on, studies using genetically modified animals do confirm that AMPs have significant antimicrobial activity *in vivo*. For example mice lacking the gene Cnlp, which codes for cathelin-related antimicrobial peptide (CRAMP), the mouse homolog of LL-37, show increased susceptibility to skin infection caused by group A streptococcus (Nizet et al., 2001) and corneal infection caused by *Pseudomonas aeruginosa* (Huang et al., 2007b), and transgenic mice expressing the gene for human intestinal defensin HD-5 were much more resistant to an oral challenge with *Salmonella typhimurium* than their wild-type counterparts (Salzman et al., 2003).

16.3.1.1 Antibacterial Action

16.3.1.1.1 Cationic Peptides

The primary site of AMP action is the microbial cell membrane, disruption of which leads to permeabilization, loss of essential intracellular components and death. The preference of AMPs for microbial versus mammalian cell membranes is the result of surface charge differences. Mammalian cells are typically neutral because of an asymmetric distribution of phospholipids, with the outer leaflet of the cell membrane having zwitterionic phosphatidylcholine and sphingomyelin and the inner, the negatively charged phosphatidlyserine (Verkleij et al., 1973). In addition, cholesterol, which is present in mammalian but not bacterial cell membranes, makes it more difficult for AMPs to disrupt the bilayer (Matsuzaki et al., 1995; Tytler et al., 1995). It should be noted, however, that although many AMPs show a preference for prokaryotic cells and do not affect mammalian cells, several, particularly at high concentrations, are cytotoxic to eukaryotic cells (see Section 16.4.1).

Antimicrobial effectiveness is influenced by a number of factors including amino acid sequence (content and length), net positive charge, amphipathicity, hydrophobicity, folding, and oligomerization. A number of different mechanisms have been proposed, but all have electrostatic interaction between the positively charged AMP and the negatively charged microbial membrane as the initial interaction. Negatively charged molecules on the outer membrane of Gram-negative bacteria include phosphate groups on lipopolysaccharide, whereas Gram-positive organisms have teichoic acids. Also, in contrast to the arrangement of the mammalian cell membrane described above, the membrane of both Gram-positive and -negative organisms has anionic phospholipids in the outer leaflet. Lehrer and colleagues were the first to document membrane permeabilization by an AMP in intact bacteria (Lehrer et al., 1989). They reported that the human neutrophil α-defensin HNP-1 sequentially permeabilized the outer and inner membrane of the Gram-negative organism *E. coli*, with permeabilization of the inner membrane being the lethal event. Huang (2000) suggested that AMPs bind to the membrane in two states dependent upon the lipid composition of the membrane. At low peptide/lipid ratios the peptides adsorb in the lipid headgroup region lying parallel to the plane of the bilayers in a functionally inactive state, which stretches the membrane (the surface or S state). Once a threshold peptide/lipid ratio is reached, there is then a transition such that the peptide is oriented perpendicular to the bilayer (the multiple-pore or I state), which mediates antimicrobial activity.

A number of models of membrane disruption have been proposed (Figure 16.1). In the carpet model (Shai, 2002), AMPs, either as single molecules or oligomers, first assemble as a covering on the surface of the membrane. Such "carpeting" may also precede some of the other putative modes of permeabilization discussed below. In general, AMPs are unstructured in solution then adopt a three-dimensional structure upon interaction with a membrane such that the molecule becomes amphiphilic. The AMP molecules orient themselves so that the hydrophobic portion is towards the membrane lipids and the hydrophilic region faces the polar head groups of the phospholipids. On reaching a threshold concentration the bilayer curvature is disrupted, there is thinning of the outer leaflet, and thus the membrane is permeated. Transient holes may also bc present before the collapse of the membrane. Numerous peptides have been reported to use the carpet model of action, including cecropin (from the silk moth; Gazit et al., 1995) and ovispirin (a derivative of bovine SMAP-29; Yamaguchi et al., 2001). Indeed, this model explains why numerous peptides have antimicrobial activity regardless of their length and amino acid composition (Shai, 2002).

In the barrel-stave model, peptides form traditional ion-channel pores, with the peptides forming the staves. In order to achieve this, peptides must be hydrophobic to enable penetration of the membrane, they have to be able to self-assemble while associated with the membrane, and if α-helical must be at least 22 amino acids in length or 8 amino acids if they have a β-sheet arrangement. The peptides aggregate in the membrane so that hydrophobic regions align with the lipid with the hydrophilic regions forming the interior of the pore. Recruitment of additional monomers leads to a steady increase in pore size and leakage of intracellular components subsequently leads to cell

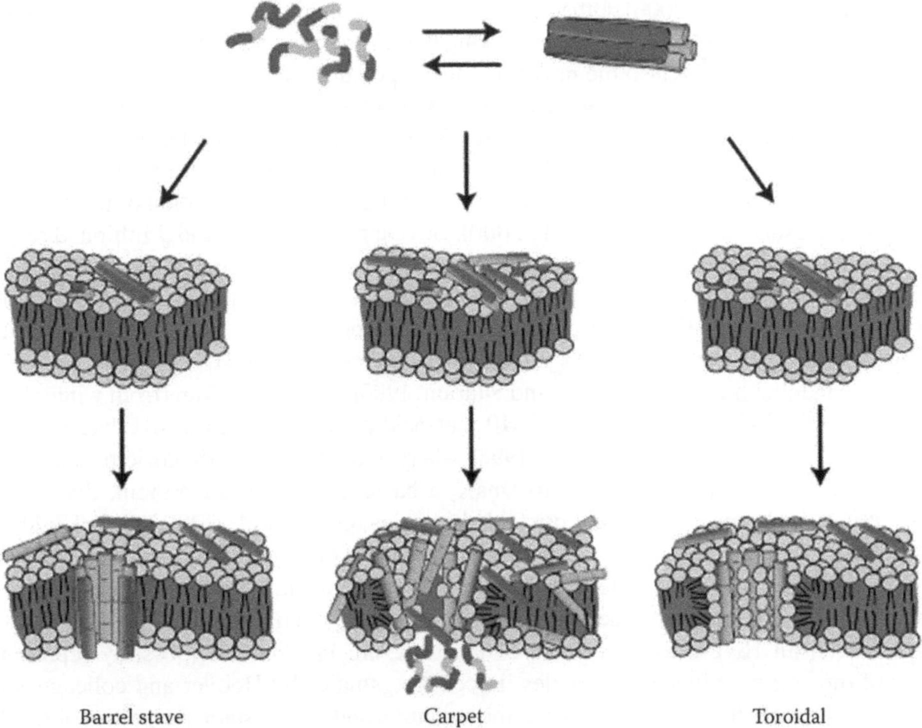

FIGURE 16.1 (See color insert following page 176.) Proposed mechanisms of action of antimicrobial peptides. Three potential mechanisms are shown; barrel stave, carpet, and toroidal. It should be noted that these are not necessarily mutually exclusive in that the carpet model may precede the barrel stave, toroidal, and detergent (not shown) models, and the toroidal model may precede the detergent model (not shown). In the barrel-stave model peptides assemble on the surface of the membrane as monomers or oligomers then insert into the membrane. In the carpet model, once a threshold concentration of peptide is reached, the membrane is permeated and transient pores form. In the toroidal model the peptides cause the membrane to bend through the pore so the pore is lined with peptide and lipid head groups. Yellow represents the hydrophilic surface and magenta the hydrophobic surface. (Modified with permission from Huang et al., *Phys. Rev. Lett.*, 92, 198–304, 2004; copyright 2004 by the American Physical Society and from Shai et al., *Biopolymers*, 66, 236–248, 2002; copyright 2002 by John Wiley & Sons, Hoboken, NJ. With permission.)

death (van't Hof et al., 2001). Peptides acting in this manner would be effective at relatively low concentrations as only a few pores would be required to disrupt the transmembrane potential (Shai, 2002). Only a few AMPs such as pardaxin (from the fish *Pardachirus marmoratus*) and alamethicin (a fungal peptide that forms pores 1.8 nm in diameter composed of eight monomers) appear to act via this mechanism (Rapaport and Shai, 1991; He et al., 1996).

In the toroidal pore (also referred to as wormhole) model the AMPs insert into the membrane and cause the monolayers to bend through the pore so that the interior is lined with both AMP and lipid head groups (Matsuzaki et al., 1996). Interactions between the positively charged peptides are minimized due to the negatively charged phospholipids lining the pore. This model differs from the barrel-stave model as the peptides are always associated with the lipid-head groups even when inserted into the membrane in a perpendicular fashion (Yang et al., 2001). One family of AMPs that uses this mechanism is the magainins (Matsuzaki, 1998). The latter were observed to form pores with an inner diameter of 3 to 5 nm and outer diameter of 7 to 8.4 nm and contained 4 to 7 magainin monomers (Yang et al., 2001).

In the detergent-type mechanism the AMPs first carpet the membrane, ultimately reaching a high enough local concentration to allow them to act as detergents and break the membrane into smaller fragments (Bechinger et al., 2006). Other models include the sinking raft model where the binding of AMPs creates a large membrane curvature, making the AMPs sink and form transient pores (Pkorny and Almeida, 2004). There is also the molecular electroporation model where AMPs create a difference in electrical potential in the membrane leading to pore formation (Miteva et al., 1999).

There is also evidence to suggest that several AMPs utilize intracellular targets in addition to (or sometimes instead of) membrane permeabilization to effect bacterial killing. For example, both magainins (from xenopus), and cecropins (from silk moth) induce transcription of the stress-related genes micF and osmY in *E. coli* (Oh et al., 2000). Bovine bactenecins 5 and 7 inhibited respiration and protein and RNA synthesis, as well as disrupting the inner and outer membranes of *E.coli* (Skerlavaj et al., 1990). In contrast, porcine PR-39 does not cause membrane permeabilization but kills *E.coli* by preventing DNA and protein synthesis (Boman et al., 1993). Other peptides also shown to inhibit bacterial protein and/or nucleic acid synthesis include HNP-1 and -2 (Lehrer et al., 1989), bovine indolicidin (Subbalakshmi and Sitaram, 1998), and pleurocidin (from winter flounder) (Patrzykat et al., 2002). The lantibiotic AMP mersacidin (from Gram-positive bacteria) inhibits peptidoglycan biosynthesis (Brotz et al., 1998) whereas drococin, pyrrhocoricin, and apidaecin (proline-rich AMPs from insects) bind to DnaK, a bacterial heat-shock protein, diminishing its ATPase activity and capacity to refold misfolded proteins (Otvos et al., 2000; Kragol et al., 2001). These modes of AMP "intracellular killing" are reviewed in more detail by Brogden (2005).

Studies are still ongoing to determine the precise mechanism of antimicrobial action of the human AMPs. Defensins form dimers in solution and Wimley and colleagues (1994) reported that neutrophil defensin HNP-2 formed pores of up to 2.5 nm in diameter (possibly representing a hexamer of dimers) in unilamellar vesicles. In contrast, studies by Hoover and colleagues (2000) indicated that whereas hBD-2 forms dimers, the structural and electrostatic properties of its oligomers were compatible with the carpet model rather than the formation of bilayer spanning pores. The details of defensin action have still to be elucidated, but it is known that disulfide bonding is not required for defensin antimicrobial activity (Wu et al., 2003; Kluver et al., 2005). Interestingly, the activity of hBD-3 against Gram-negative organisms appears to reside in the C-terminal region, whereas that against Gram-positives and fungi resides in the N-terminal region (Hoover et al., 2003). In contrast to most other linear AMPs, LL-37 assumes a secondary structure and forms oliogomers in aqueous solution (Johansson et al., 1998; Oren et al., 1999). Biophysical studies on its mechanism of antimicrobial action were recently reviewed by Durr and colleagues (2006). Although the barrel stave and detergent-like mechanisms have been ruled out, NMR studies provide supportive evidence for the toroidal-pore mechanism (Henzler Wildman et al., 2003), while others favor the carpet model (Oren et al., 1999; Neville et al., 2006). Interestingly, the results of a recent study indicate that LL-37 utilizes different mechanisms of membrane perturbation depending upon the headgroup charge and hydrocarbon chain length of the lipids (Sevcsik et al., 2007). The antimicrobial activity of LL-37 resides in the C-terminal region, with a 13-residue core peptide consisting of amino acids 17 to 29 having the most activity (Li et al., 2006).

16.3.1.1.2 Anionic Peptides

Although most AMPs are cationic in nature a number have been identified that are anionic, including the human peptide dermcidin (Schittek et al., 2001). Clearly the mechanism of antimicrobial action of these peptides must be rather different from that of their cationic cousins. Mammalian anionic peptides are known to require zinc for antimicrobial activity and complex with it. Thus, it has been speculated that zinc may form a salt bridge that allows the peptide to overcome the negative charge at the microbial surface. The peptide then penetrates the cell membrane and may attach to ribosomes and inhibit ribonuclease activity (Brogden et al., 2003).

16.3.1.2 Antifungal Action

The antifungal action of AMPs is the least well studied of their activities. As noted above, within human AMPs, the salivary histatins are primarily antifungal, although other AMPs such as defensins and LL-37 also have activity. Within the histatin family, histatin 5, a cleavage product of histatin 3, has the highest activity against *Candida albicans*, the fungal species most well studied to date, with regard to AMP activity (Kavanagh and Dowd, 2004). Early studies showed that lysis of the cell membrane was not a significant effect of histatin 5 (Baev et al., 2002). Now it is known that this AMP acts by first binding to specific cell wall proteins, the heat shock proteins Ssa1p and Ssa2p (Li et al., 2003). The peptide is then internalized, and the mitochondrion appears to be its major target. Having bound to the mitochondrion, histatin 5 causes release of ATP and other small nucleotides and ions into the cytoplasm and extracellular milieu (Koshlukova et al., 1999). Recently ATP efflux has been shown to be dependent on the Trk1p potassium transporter (Baev et al., 2004). Accumulation of extracellular ATP may activate receptors that trigger cell death or the loss of ATP in itself may contribute to death of the fungal cell (Kavanagh and Dowd, 2004). It has also been shown that histatin 5 induces the formation of reactive oxygen species, which in turn could damage organelles, DNA, and membranes, so further contributing to the demise of the cell (Helmerhorst et al., 2001).

The mechanisms of defensin and cathelicidin antifungal activity have also begun to be investigated. Although these appear to have similarity to histatin 5 in that there is leakage of intracellular contents, significant differences have been revealed. For example, HNP-1, hBD-2, and hBD-3 candidacidal activity does not involve Trk1p action (Vylkova et al., 2006). Additionally there are also differences in the mechanism of fungicidal activity of the various defensins as HNP-1 activity does not require interaction with Ssap1/2p, whereas as that of the β-defensins does. Vlykova and colleagues (2007) reported that the candidacidal activity of hBD-3 resides in the N-terminal 17 amino acids, and that in contrast to its antibacterial activity, its antifungal activity is salt-sensitive. They also observed differences in the antifungal effects of hBD-2 and hBD-3; for example, the activity of the former was cation selective. While Vlykova and colleagues (2007) observed that the β-defensins killed without causing membrane disruption, an earlier study by Feng and colleagues (2005) did show disruption of the fungal membrane by hBD-2.

Although histatin 5 has been documented to be internalized by fungal cells, a recent study with LL-37 showed that this AMP remained associated with the cell wall and cell membrane, presumably through interaction with specific protein components (den Hertog et al., 2005). Further, while both histatin 5 and LL-37 exposure was associated with leakage of ATP and other small components, treatment with LL-37 also permitted leakage of larger components (up to 40 kDa) and was consistent with "detergent-like properties." Membrane permeabilization of *C. albicans* by LL-37 and its murine homolog CRAMP was also reported by Lopez-Garcia and colleagues (2005).

There is much yet to be clarified regarding the details of defensin and cathelicidin antifungal action, but findings to date certainly imply that, as with antibacterial activity, multiple mechanisms of AMP antifungal activity exist.

16.3.1.3 Antiviral Action

A number of AMPs have potent activity against a variety of viruses. Most studies have been conducted with enveloped viruses, but activity against nonenveloped viruses such as adenovirus has been reported. Current evidence suggests that AMPs have both direct antiviral effects and indirect effects through interactions with host target cells.

The first human AMPs shown to have antiviral activity were the neutrophil defensins (Daher et al., 1986). HNP-1, -2, and -3 had potent neutralizing activity against herpes simplex virus (HSV)-1. HNP-1 also showed activity against HSV-2, vesicular stomatitis virus, influenza, and cytomegalovirus. Subsequently these α-defensins have also been shown to have activity against human immunodeficiency virus (HIV)-1 (Zhang et al., 2002), adenovirus (Bastian and Schafer, 2001), papillomavirus (Buck et al., 2006), and influenza virus (Salvatore et al., 2007). Also, the Paneth cell α-defensin HD-5 was shown to be inhibitory to adenovirus (Gropp et al., 1999) and papillomavirus (Buck et al., 2006).

hBD-1 and -2 showed little activity against human papillomavirus (Buck et al., 2006) and while hBD-1 blocked adenovirus (Gropp et al., 1999), hBD-2 did not (Bastian and Schafer, 2001). hBD-2 and -3 (but not hBD-1) inhibited HIV-1 replication (Quinones-Mateu et al., 2003) and hBD-3 also blocked influenza virus (Leikina et al., 2005) and HSV (Hazrati et al., 2006) infection. LL-37 has been shown to be inhibitory to vaccinia virus (Howell et al., 2004), adenovirus (Gordon et al., 2005), and HIV (Bergman et al., 2007), and although we observed activity against HSV-1 (Gordon et al., 2005), Yasin and colleagues (2000) did not.

Much remains to be understood regarding the mechanism of AMP antiviral activity and multiple modes of activity have been described (Klotman and Chang, 2006). Gropp and colleagues (1999) suggested that the ability of hBD-1 and HD-5 to block adenovirus was likely due to an effect before internalization. On a cautionary note, AMP antiadenovirus activity may pose a problem for adenovirus-mediated gene therapy strategies (Gropp et al., 1999; Bastian and Schfer, 2001). In the case of human papilloma virus, it appeared that HNPs and HD-5 prevented the virions escaping from endocytic vesicles and translocating to the nucleus (Buck et al., 2006). However, with influenza virus, hBD-3 formed a barricade of immobilized proteins that prevented hemagglutinin-mediated viral fusion (Leikina et al., 2005).

For obvious reasons, the effects of AMPs on HIV infection are currently the subject of intense interest. It has been shown that HNP-1 can, in the absence of serum, directly inactivate HIV-1 before it enters a cell. However in the presence of serum, HNP-1 acts on the target CD4+ T cell (likely through interfering with protein kinase C signaling, which is essential for viral replication) to block HIV-1 infection at the steps of nuclear import and transcription. Both actions could be operational *in vivo* (Chang et al., 2005). HNP-1 and -2 also inhibited HIV infection of macrophages, and this was attributed to the peptides stimulating production of CC-chemokines, which may contribute to viral inhibition through competition for receptors (Guo et al., 2004).

Direct binding of AMPs to the viral envelope via charge interactions, that is, similar to antibacterial activity, may account for some of their direct inhibition of HIV. In addition, HNP 1–3 have been reported to function as lectins that bind to the HIV envelope glycoprotein gp120 and T cell CD4, thus preventing viral entry (Wang et al., 2004; Furci et al., 2007). Interestingly, the binding of HNP-4, which has the most potent anti-HIV activity of the neutrophil defensins, to gp120 or CD4 is one to two orders of magnitude lower than that of HNP-1, indicating a different mechanism of anti-HIV activity (Wang et al., 2004b; Wu et al., 2005). Sun and colleagues (2005) attributed the anti-HIV activity of hBD-2 to direct inactivation of virions and the ability of the peptide to inhibit the formation of reverse-transcribed HIV DNA products. They did not observe any effect on viral fusion or expression of HIV coreceptors on CD4+ T cells. On the other hand Quinones-Mateu and colleagues (2003) observed that hBD-2 and -3 caused downregulation of HIV-1 coreceptor CXCR4 in peripheral blood mononuclear cells and T cells. They also showed that hBD-2 and -3 bind directly to HIV-1 virions. Some defensins are also capable of stimulating the release of cytokines, which may in turn contribute to inhibition of viral replication (Chang et al., 2005).

16.3.1.4 Microbial Resistance to AMPs

From an evolutionary stand point AMPs are ancient molecules, having been around some 100 million years. Given this time span it is not surprising that microorganisms have developed resistance mechanisms. The latter include the secretion of proteases that break down AMPs; for example, aureolysin and serin protease V8 from *Staphylococcus aureus* break down LL-37 (Sieprawska-Lupa et al., 2004). Some bacteria produce proteins that are located extracellularly and which bind AMPs, so preventing them from reaching the cytoplasmic membrane; for example, *S. aureus* produces staphylokinase, which binds defensins (Jin et al., 2004). Others modulate the electrostatic properties of their cell surface; for example, *Streptococcus agalactiae* resistance to defensins is related to alanylation of teichoic acids (Poyart et al., 2003). Bacteria that produce AMP-specific exporters have also been reported; for example, *Neisseria gonorrhoeae* has an energy-dependent efflux system referred to as MtrCDE, which confers resistance to protegrin-1 (a porcine AMP) and LL-37 (Shafer et al., 1998).

What is most striking, however, is how the ability of AMPs to inhibit microorganisms has remained so efficient during evolution. Despite the development of various mechanisms of resistance there are no organisms that are completely resistant to all AMPs. This has been attributed to the fact that organisms cannot completely change the composition of their cytoplasmic membrane and the high metabolic costs involved in acquiring the mechanisms that would confer resistance to all AMPs. This reduced likelihood of an organism developing AMP resistance is one of the features that make them attractive as novel antimicrobial agents (Section 16.6). For a detailed review of bacterial AMP resistance mechanisms (the most well studied to date) the reader is referred to a recent review by Kraus and Peschel (2006).

16.3.2 Modulation of Mammalian Cell Activity

In addition to antimicrobial activities, both defensins and cathelicidin have been shown to have a variety of effects on mammalian cells, ranging from stimulation of cytokine production, chemotaxis, and stimulation of cell proliferation. These effects, often (although not always) occur at AMP concentrations much lower than those required for antimicrobial activity and are not sensitive to the effects of serum or physiological salt concentrations. A number of studies have established that these effects are independent of antimicrobial action and are mediated via surface receptors on the target cells. Rather than having their own unique receptors, AMPs utilize receptors for cytokines and growth factors and are notoriously promiscuous. LL-37, for example, has been shown to utilize no fewer than three receptors. Immune, inflammatory, and epithelial cells appear to be the major targets, but a number of other activities have been associated with certain AMPs. For example, LL-37 is known to bind and neutralize LPS (Larrick et al., 1991, 1995) and thus may have a role in dampening endotoxin-mediated inflammation and damage. Indeed, LL-37 has been shown to protect rats from sepsis caused by Gram-negative bacteria (Cirioni et al., 2006). Some defensins have been observed to inhibit the classical and in some cases also the lectin pathway of complement activation (van den Berg et al., 1998; Bhat et al., 2007; Groeneveld et al., 2007), although there is also at least one report of human neutrophil defensins activating the classical pathway (Prohaszka et al., 1997). Also, while LL-37 appears to be proangiogenic, the neutrophil defensins are reported to be antiangiogenic (Chavakis et al., 2004; Economopoulou et al., 2005). Some of the known biological activities and examples of the AMPs that show these are presented in Figure 16.2 and the effects of AMPs on immune, inflammatory, and epithelial cell activities and the receptors and pathways that mediate them are described below.

16.3.2.1 AMP Effects on Immune and Inflammatory Cells

Immune and inflammatory cells are a major target of AMP action. The known effects of defensins on these cell types will first be described.

hBD-2, but not hBD-1 or -3, has been reported to be chemotactic for tumor necrosis factor (TNF)-α treated neutrophils (Garcia et al., 2001a; Niyonsaba et al., 2004). This effect was mediated by chemokine receptor CCR6. On the other hand, HNP-1 was found to suppress neutrophil migration and also oxygen radical production and phagocytic killing of *S. aureus* induced by various stimuli (Kaplan et al., 1999; Grutkoski et al., 2003). HNP-1, -2, but not -3, were reported by Territo and colleagues (1989) to be chemotactic for monocytes; however, this could not be repeated by Chertov and colleagues (1996). HNPs were also reported to stimulate cytokine production in monocytes activated by *S. aureus* or phorbol myristate acetate (Chaly et al., 2000). hBD-3, but not hBD-2, was chemotactic for monocytes (Yang et al., 1999; Garcia et al., 2001a) and induced expression of various costimulatory molecules in monocytes via activation of TLR 1 and 2 (Funderburg et al., 2007). hBD-1 through 4 (and also mBD-8) were found to stimulate macrophage migration via a Gi protein (not CCR6) and mitogen activated protein kinase (MAPK) pathways (Soruri et al., 2007). Similarly, HNP-1, -3, and HD-5 were chemotactic for macrophages via a Gi protein and MAPK

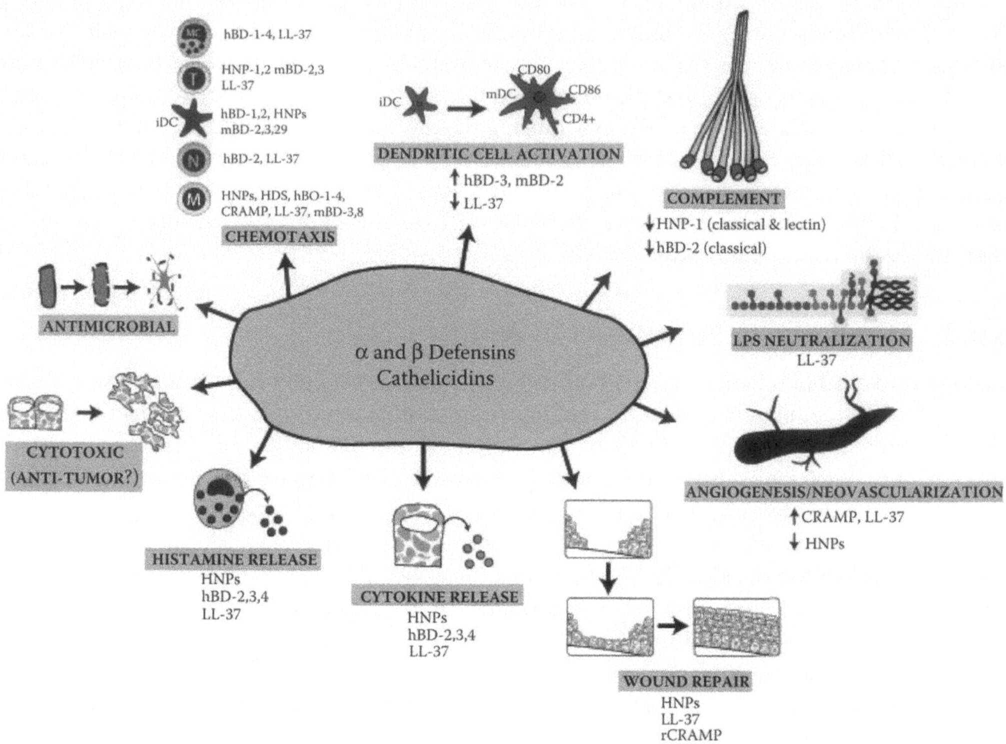

FIGURE 16.2 Summary of potential functions of antimicrobial peptides. This summary is largely based on *in vitro* data and focuses on the activities of human defensins and LL-37. The reader should be aware that there are discrepancies in the literature and where this exists the functionality being reported represents the majority view for that particular function. HNP, human neutrophil peptide; hBD, human β-defensin; HD, human defensin; CRAMP, cathelin-related antimicrobial peptide; rCRAMP, rat CRAMP; mBD, mouse β-defensin; MC, mast cell; T, T cell; N, neutrophil; M, monocyte/macrophage; iDC, immature dendritic cell; mDC, mature dendritic cell.

pathways (Grigat et al., 2007). hBD-1 through -3 also induced the expression of a variety of cytokines and chemokines in peripheral blood mononuclear cells (Boniotto et al., 2006).

HNP-1 and -2 were first reported to be chemotactic for T cells by Chertov and colleagues (1996). Subsequently, Yang and colleagues (2000) and Obata-Onai and colleagues (2002) showed that HNPs were chemotactic for naïve but not memory CD4 and CD8 positive T cells. The effects on CD4 T-cell migration were pertussis toxin sensitive, indicating involvement of a Gi linked receptor, but this has yet to be identified. More recently, Grigat and colleagues (2007) have reported that HNP-1, -3, and HD-5 mobilize both naïve and memory CD4-positive T cells. hBD-2 was chemoattractive for memory but not naïve CD4 positive T cells, an effect that was found to be mediated via CCR6 (Yang et al., 1999). However, a recent study contradicts this finding, indicating no effect of hBD-1 through -4 on memory T cells (Soruri et al., 2007).

Both human α and β defensins have been found to be chemoattractive for immature but not mature dendritic cells (Yang et al., 1999, 2000; Hubert et al., 2007). These effects are mediated via Gi coupled receptors, and for hBD-2 the receptor involved has been identified as CCR6 (Yang et al., 1999). mBD-2, -3, and -29 have also been shown to be chemoattractants for immature dendritic cells, with CCR6 also being involved (Biragyn et al., 2001, 2002; Conejo-Garcia et al., 2004). However, recent studies by Soruri and colleagues (2007) and Grigat and colleagues (2007) dispute these earlier findings, showing no effects of α or β defensins on immature dendritic cell migration. mBD-2 was observed to activate immature dendritic cells, which then exhibited a Th1, polarized response (Biragyn et al., 2002). This was shown to be mediated via TLR4. Recently Funderburg et al. (2007) reported that hBD-3, acting via TLR 1 and 2, was capable of activating dendritic cells.

HNPs and hBD-2 (but not hBD-1) induced histamine release from mast cells, which involved Gi coupled receptors and phospholipase C (for hBD-2) (Befus et al., 1999; Niyonsaba et al., 2001). HNP-1, -3, and HD-5 are chemotactic for mast cells but did not influence calcium fluxes in the cells (Grigat et al., 2007). hBD-2 was also found to be chemotactic for mast cells; again, this involved a Gi coupled receptor (not CCR6) and phospholipase C (Niyonsaba et al., 2002a). Soruri and colleagues (2007) recently reported that hBD-1 through -4 stimulated mast cell chemotaxis and calcium fluxes, although some of this data is contrary to that of Niyonsaba and colleagues (2002a), who did not observe any effect of hBD-1 on mast cells.

In sum, while some discrepancies in the literature exist, the majority of studies indicate that α and β defensins have chemotactic effects on macrophages, T cells, immature dendritic cells, and mast cells. Also, some defensins can stimulate gene expression in monocytes, activate dendritic cells and induce mast cell histamine release. Many of these effects appear to be mediated via a Gi-linked receptor, in some cases identified as CCR6, while involvement of TLRs is implicated in mediating defensin effects on dendritic cells. Overall, these findings strongly support a role for defensins in regulating immune and inflammatory responses.

In keeping with the concept that AMPs are multifunctional molecules, LL-37 (and its homologs) has also been shown to exert effects, often similar to those of the defensins, on immune and inflammatory cells. This cathelicidin has been shown to be chemotactic for human neutrophils, an effect that is mediated via formyl-peptide receptor-like-1 (FPRL-1) (Agerberth et al., 2000; De Yang et al., 2000; Tjabringa et al., 2006). Similarly, CRAMP, the murine homolog of LL-37, is chemotactic for leukocytes via mFRP2, the murine homolog of FPRL-1 (Kurosaka et al., 2005). LL-37 has also been shown to activate neutrophil NADPH oxidase, IL-8 and HNP secretion, and calcium mobilization (Zheng et al., 2007). Interestingly LL-37 also suppressed neutrophil apoptosis (Nagaoka et al., 2006). This involved activation of extracellular signal-related kinase (ERK), expression of Bcl-X1 and inhibition of caspase 3, and was mediated via both FPRL-1 and the purinergic receptor P2X7. LL-37 has been reported to be internalized by a number of cell types (Sandgren et al., 2004; Lau et al., 2005; Bandholtz et al., 2006; Lande et al., 2007) and thus would be capable of activating P2X7 via its cytoplasmic C-terminal domain.

LL-37 is also chemotactic for monocytes (again via FPRL-1), but not monocyte-derived immature dendritic cells (De Yang et al., 2000), and modulates the expression of multiple genes in macrophages (Scott et al., 2002). This AMP also suppresses the effects of toll-like receptor activation on monocyte cytokine production (Mookherjee et al., 2006) and stimulates monocyte IL-8 production through activation of p38 and ERK MAPKs (Bowdish et al., 2004). LL-37 has also been shown to promote maturation and release of IL-1β in LPS-primed monocytes via P2X7 (Elssner et al., 2004). LL-37 is also chemotactic for CD4-positive T cells (Agerberth et al., 2000; De Yang et al., 2000) and mast cells (Niyonsaba et al., 2002b). This AMP also induces mast cell histamine release and calcium mobilization (Niyonsaba et al., 2001). The effects on mast cells are mediated via a Gi coupled receptor (but not FRPL-1) and activation of phospholipase C.

The effects of LL-37 on dendritic cells, the primary antigen presenting cells of the immune response, appear complex. Davidson and colleagues (2004) reported that immature dendritic cells derived from monocytes in the presence of LL-37 (in addition to IL-4 and granulocyte monocyte-colony stimulating factor (GM-CSF)) had enhanced expression of certain activation markers (e.g., CD86) and increased endocytic activity compared to those derived in just IL-4 and GM-CSF. Furthermore, in addition to this ability to modulate differentiation, upon LPS stimulation to induce dendritic cell maturation, those cells derived in the presence of LL-37 preferentially produced IL-12 and IL-6 and thus would be capable of driving a Th1-mediated immune response. In a recent study Kandler and colleagues (2006) reported that immature dendritic cells were less able to respond to TLR ligands (e.g., LPS and flagellin) in the presence of LL-37 and so had a reduced ability to activate

T cells. Similar to Kandler and colleagues (2006), Di Nardo and colleagues (2007) reported that murine CRAMP and LL-37 prevented LPS-induced maturation of mouse and human dendritic cells, respectively. Bandholtz and colleagues (2006) reported that LL-37 was taken up into immature dendritic cells and that this was associated with increased expression of HLA-DR and CD86, indicative of maturation and activation. Together these studies suggest that LL-37 can have different effects on dendritic cells depending upon their maturation state and the presence of other modulating molecules. A recent study has also shown that LL-37 is able to interact with self DNA, creating a complex that is then internalized by plasmacytoid dendritic cells, which in turn become activated via TLR9 (Lande et al., 2007). This has implications for a role of LL-37 in autoimmune diseases such as psoriasis.

16.3.2.2 AMP Effects on Epithelial Cells

A range of epithelial cells and tissues have also been identified as targets for AMP activity. The primary effects of defensins relate to stimulation of cytokine/chemokine production and cell proliferation.

HNP1–3 has been shown to stimulate IL-8, monocyte chemotactic protein-1 (MCP-1), and granulocyte monocyte colony stimulating factor secretion from A549 lung epithelial cells and IL-8 and IL-6 secretion by primary bronchial epithelial cells (van Wetering et al., 1997, 2002). A subsequent study by Sakamoto and colleagues (2005) showed that in contrast to HNPs, hBD-2 did not modulate cytokine production by bronchial epithelial cells and suggested that the effects of the HNP-1 were mediated via activation of nuclear factor kappa B (NFκB). Most recently it has been shown that HNP-induced epithelial IL-8 release is mediated via the purinergic receptor P2Y6 signaling pathway and activation of ERK1/2 and phosphoinositide 3-kinase (PI3K)/Akt pathways (Khine et al., 2006; Syeda et al., 2008). Niyonsaba and colleagues (2007) observed that β-defensins (hBD-2, -3, -4, but not -1) upregulated a variety of genes in keratinocytes including IL-6, IL-10, and MCP-1 via an unidentified Gi-protein and activation of phospholipase C.

An effect of defensins on cell proliferation was reported as long ago as 1991, when Bateman and colleagues noted that high concentrations of neutrophil defensins increased DNA synthesis in HL60 (promyelocytic leukemia) cells. Soon after, both human and rabbit neutrophil defensins were reported to be mitogenic for lens epithelial cells (and also NIH 3T3 fibroblasts) (Murphy et al., 1993). Nishimura and colleagues (2004) observed that low concentrations of HNP-1 but not hBD-1, -2, or -3 stimulated periodontal ligament epithelial cell proliferation. Recently, Niyonsaba and colleagues (2007) reported that β-defensins stimulated keratinocyte proliferation and also migration. They also observed that the β-defensins stimulated phosphorylation of the epidermal growth factor receptor (EGFR) and signal transducer and activator of transcription (STAT)-1 and -3, which are known to be important signaling molecules for keratinocyte proliferation and migration. High expression of hBD-1 has also been shown to promote keratinocyte differentiation (Frye et al., 2001).

That defensins can modulate epithelial cell cytokine/chemokine production, proliferation, and migration has led to the suggestion that they may modulate wound healing responses. While this has yet to be proven *in vivo*, several lines of evidence support this hypothesis. We have shown that hBD-2 is upregulated in regenerating corneal epithelial cells in organ culture (McDermott et al., 2001). The amount of human neutrophil defensins was increased in the tear film of patients after ocular surface surgery (Zhou et al., 2004) and the level of α-defensins in tears correlated with the time course of healing in a rabbit model of a corneal epithelial scrape injury (Zhou et al., 2007). Also Aarbiou and colleagues (2004) reported that HNP1–3 increased epithelial wound closure in a model using NCI-H292 airway epithelial cells. This involved activation of the MAP kinase ERK1/2.

LL-37 has been found to have similar effects on epithelial cells as the defensins and the receptors and signaling pathways involved have been better characterized. In airway epithelial cells, Tjabringa and colleagues (2003) reported that LL-37 activated ERK and stimulated IL-8 secretion. This occurred via transactivation of EGFR. Thus they proposed that LL-37 activates a metalloproteinase (possibly an ADAM), which in turn cleaves membrane-anchored EGFR ligands that activate the EGFR and so promote cytokine gene transcription via the ERK pathway. MAPK pathway (ERK and/or p38MAPK) activation in response to LL-37 was also confirmed for airway epithelial cells by others (Bowdish et al., 2004; Shaykhiev et al., 2005). In a model of airway epithelial injury, Shaykhiev and colleagues (2005) observed that LL-37 stimulated migration and cell proliferation via EGFR and an unidentified G-protein coupled receptor (not FPRL-1). Niyonsaba and colleagues (2005) showed that LL-37 stimulates cytokine (IL-18, IL-8, and IL-20) production by keratinocytes via MAPK, and Tokumaru and colleagues (2005) observed that EGFR mediates LL-37-stimulated keratinocyte migration. Interestingly, Braff and colleagues (2005b) reported that LL-37 stimulated keratinocyte IL-8 secretion, but while this was mediated via EGFR and an unidentified G-protein coupled receptor, activation of these receptors appeared to be the result of membrane perturbation.

In our laboratory we have shown that LL-37 is able to stimulate migration and cytokine secretion (IL-1β, IL-6, IL-8, and TNFα) by human corneal epithelial cells (Huang et al., 2006). Using receptor antagonists and a series of intracellular signaling pathway inhibitors we established that these effects involve activation of MAPK (ERK1/2 and jun-N-terminal kinase, JNK), PI3K, and tyrosine kinases. Activation of FPRL-1 was found to lead to increased cell migration, whereas transactivation of EGFR led to both increased migration and stimulation of cytokine secretion.

As has been proposed by many authors, these effects on various epithelial cells are in keeping with a role for LL-37 in modulating in wound healing. Heilborn and colleagues (2003) showed that LL-37 expression is increased in skin after wounding and in an *ex vivo* wound healing model with human skin observed that antibodies to LL-37 inhibited re-epithelialization. Also, rCRAMP (the rat homolog of LL-37) has been shown to promote healing of gastric ulcers in rats (Yang et al., 2006) and adenoviral transfer of LL-37 into excisional wounds of ob/ob mice significantly improved re-epithelialization and granulation tissue formation (Carretero et al., 2008). Additionally, Koczulla and colleagues (2003) reported that LL-37 stimulates proliferation of vascular endothelial cells and neovascularization via activation of FPRL-1. While this would be beneficial for the majority of tissues, this effect would be detrimental to healing of a corneal wound where maintaining avascularity is essential to good vision.

In summary, defensins and cathelicidin are capable of modulating the activities of a range of human cells. Here the focus was on immune/inflammatory and epithelial cells, but actions on other cells types including fibroblasts and smooth muscle cells have been reported. Both AMP families utilize receptors, often linked to Gi proteins, once thought specific for other molecules. Of the receptors identified to date, defensins utilize CCR6, TLR 1, 2 and 4, and P2Y6, whereas LL-37 uses FPRL-1, EGFR, and P2X7. In turn, activation of these receptors frequently leads to initiation of MAPK pathways. Through modulation of these pathways, these AMPs are believed to play important roles in modulating immune responses and regulating wound healing *in vivo*.

16.3.3 Inhibition/Inactivation of AMP Activity

A number of endogenous molecules have been identified to bind to or otherwise inactivate AMP activity. Panyutich and Ganz (1991) reported that α2-macroglobulin bound HNP-1, possibly via thiol–disulfide exchange, and suggested that it may act as a scavenger for defensins in inflamed tissues. Several serine proteinase inhibitors (serpins) such as α1-proteinase inhibitor and α1-antichymotrypsin have also been shown to bind defensins and inhibit their cytotoxic activity (Panyutich et al., 1995). Notably, in a reciprocal effect, the defensins inactivated the activity of the serine proteinases. Dermatan sulfate also binds α-defensins and neutralizes their antimicrobial activity (Schmidtchen et al., 2001). This glycosaminoglycan is released by the action of bacterial proteases on proteoglycans and so may represent a mechanism used by organisms to circumvent AMP activity. Glycosaminoglycans have also been shown to inhibit the antibacterial activity of LL-37 (Baranska-Rybak et al., 2006). ADP-ribosylation of HNP-1 was found to inhibit the antimicrobial and cytotoxic effects of this AMP but not its ability to chemoattract T cells or induce epithelial cell IL-8 secretion (Paone et al., 2002). Interestingly, the authors also reported that ADP-ribosylated HNP-1 was detected in bronchiolar lavage fluid from smokers but not nonsmokers, and in patients with asthma and idiopathic pulmonary fibrosis (Paone et al., 2002, 2006). Apolipoprotein A-I was observed to inhibit the antimicrobial activity of LL-37 (Wang et al., 1998). Also mucins, large anionic glycoproteins, have also been reported to bind and inhibit the antimicrobial activity of both LL-37 (Felgentreff et al., 2006) and hBD-2 (Huang et al., 2007c). It has also been shown that hBD-2 and -3 are degraded and inactivated by the cysteine proteases cathepsins B, L, and S (Taggart et al., 2003).

These studies suggest a range of possible mechanisms to regulate the biological activities of various AMPs. However, it remains to be determined which of these mechanisms effectively terminate AMP action *in vivo*. Also, while some mechanisms may serve to scavenge or sequester AMPs and so regulate their potential proinflammatory (i.e., cytokine/chemokine secretion and so on) and cytotoxic activities, inappropriate inactivation of AMPs may lead to increased susceptibility to infection.

16.4 PATHOLOGICAL MODE OF ACTION

16.4.1 Cytotoxicity

As eluded to earlier (Section 16.3.1.1), although AMPs show a preference for prokaryotic cells, at high concentrations many also become cytotoxic to eukaryotic cells. As reviewed by Lehrer and colleagues (1993), early studies showed that the cytotoxic effects were concentration- and time-dependent and could be detected as early as three hours after beginning exposure to the peptide. AMPs were observed to first bind to the membrane of the target cell through weak electrostatic interactions. Notably this phase could be prevented by the presence of serum, suggesting that the peptides would only have significant toxic effects at low-serum tissue sites. Subsequently, the AMPs then permeabilized the membrane, at which point the toxic effect became insensitive to serum. Studies also indicated a third event involving DNA injury (Gera and Lichtenstein, 1991). Indeed many studies have shown AMPs can enter mammalian cells (Sandgren et al., 2004; Lau et al., 2005; Bandholtz et al., 2006; Lande et al., 2007) and AMP induction of apoptosis has been confirmed. For example, HNPs have been shown to induce caspase-mediated apoptosis in Jurkat T cells (Aarbiou et al., 2006) and lung epithelial cells (Liu et al., 2007). LL-37 has been shown to stimulate apoptosis in airway epithelial cells although while one group (Lau et al., 2006) suggested this was caspase-3 mediated the findings of another group (Aarbiou et al., 2006) were not in keeping with this.

The reader should note that there is significant variability in cytotoxicity depending upon the particular AMPs and cells being tested. For example, whereas α-defensins at high concentrations were toxic to porcine periodontal ligament epithelial cells and human gingival fibroblasts, β-defensins were not (Nishimura et al., 2004). On the other hand, hBD-1 had strong cytotoxicity to NIH3t3 fibroblasts (Zucht et al., 1998) and hBD-2 was toxic to human bronchial epithelial cells (Sakamoto et al., 2005).

In general, high local concentrations of AMPs, whether from endogenous or exogenous sources, have the potential to cause significant tissue damage. Mechanisms exist to protect cells that produce AMPs from their potentially damaging effects. For example, Valore and colleagues (1996) reported that during synthesis and processing the anionic propiece of HNP-1 inhibits its cytotoxicity. Also, LL-37 is released from its parent, noncytotoxic, molecule hCAP18 extracellularly (Sorensen et al., 2001). Additionally, as discussed here, serum and several specific molecules (Section 16.3.3) have been identified that can inactivate AMPs and thus may serve to limit their cytotoxicity. It follows then, if these mechanisms are compromised or are inadequate due to excessive AMP secretion, then the cytotoxic effects may prevail. To what extent this may contribute to disease processes is largely unknown. It has been reported that HNPs reach cytotoxic concentrations in individuals with α(1)-antitrypsin deficiency and that this may contribute to the lung damage that occurs in these people (Spencer et al., 2004; Wencker and Brantly, 2005). Intriguingly, loss of an apoptotic effect of hBD-1 has been implicated in prostate cancers (Bullard et al., 2008). Also, as discussed later (Section 16.6) the cytotoxic effects of AMPs may prove useful in developing therapeutic versions of these peptides for cancer treatment.

16.4.2 AMPs and Disease

Many studies over recent years have begun to explore a link between various AMPs and human diseases. Altered AMP expression and function has been identified in several disease states although, as is often the case, it is not always readily apparent if this is cause or effect. Given that AMPs induce many cells to secrete proinflammatory cytokines and chemokines that attract immune and inflammatory cells, it is reasonable to propose that they may contribute to exacerbation of inflammatory processes in general and hence tissue damage in a number of situations. Associations between AMPs and a variety of different diseases and conditions are discussed below.

One of the first conditions linked to AMPs was susceptibility to infection in cystic fibrosis (CF) patients. Here it was suggested that increased salt content in respiratory tract fluids led to inactivation of locally produced AMPs, which contributed to microbial colonization and the development of frequent infections. Although this initial salt-dependent hypothesis of inactivation may not be entirely accurate, there is strong evidence (reviewed by Laube et al., 2006) to support lack of AMP activity in models of CF.

Crohn's disease is a chronic inflammatory disease of the intestinal mucosa. Two forms are recognized and altered AMP expression is found in both: ileal Crohn's is associated with diminished expression of Paneth cell defensins whereas Crohn's disease of the colon is associated with impaired expression of inducible β-defensins in enterocytes (Wehkamp et al., 2005). This reduction in antimicrobial protection is suggested to weaken the intestinal barrier function to pathogens, thus leading to a chronic inflammatory response. Reduced hBD-2, -3, and LL-37 expression, a consequence of increased levels of IL-4, IL-10, and IL-13, has also been implicated in the development of skin infections in patients with atopic dermatitis (Howell, 2007).

Morbus Kostmann is a severe congenital neutropenia (Putsep et al., 2002). The neutrophil count in these patients can be normalized by treatment with granulocyte colony stimulating factor. The cells have normal respiratory burst function and lactoferrin levels but are deficient in LL-37 and have reduced amounts of HNP1–3. This reduction has been linked to periodontal disease in these patients (Putsep et al., 2002; Carlsson et al., 2006).

AMPs have also been linked to atherosclerosis. Defensins, HNP1–3, are present at high concentrations in atheromatous plaques in the coronary circulation and have been shown to stimulate the binding of lipoprotein (a) to vascular endothelial cells and smooth muscle cells (Barnathan et al., 1997; Higazi et al., 1997). They also form complexes with lipoprotein (a) and low-density lipoproteins that accumulate in the extracellular space (Bdeir et al., 1999). These interactions are believed to predispose to the development of atherosclerosis, in part by facilitating lipoprotein exposure to harmful modifications such as oxidation (Kougias et al., 2005). Recently, LL-37 has also been shown to be present in atherosclerotic lesions and is capable of causing death of vascular smooth muscle cells (Ciornei et al., 2006).

Patients with acne rosacea, an inflammatory skin disease characterized by erythema, papulopustules, and telangiectasia, have been found to have abnormally high levels of LL-37 and novel proteolytically processed forms due to an increased epidermal expression of stratum corneum tryptic enzyme (Yamasaki et al., 2007). Also a very interesting study describes the ability of LL-37 to transport self-DNA into plasmacytoid dendritic cells where it then activates TLR9, leading to activation of the cells, production of interferon-γ, and consequent activation of auto-T-lymphocytes, which are implicated in the pathogenesis of psoriasis, an autoimmune disease of the skin (Lande et al., 2007).

As noted earlier (Section 16.2) several AMPs show copy number polymorphisms and links between this and specific diseases are just beginning to be investigated. For example, Fellermann and colleagues (2006) observed that while healthy individuals had a median of four hBD-2 gene copies per genome, patients with colonic Crohn's disease had a median of only three. This lower number was associated with reduced mucosal hBD-2 expression and correlates with the findings mentioned above that colonic Crohn's is associated with impaired expression of inducible β-defensins. Hollox and colleagues (2008) recently showed an association between psoriasis and an increased genomic copy number in a β-defensin gene cluster (which included the genes for hBD-2, -3, and -4) from chromosome 8.

DEFB1, the gene for hBD-1, does not show the same copy number variation as the other β-defensins, but several single nucleotide polymorphisms have been identified in this gene (Dork and Stuhrmann, 1998). A number of studies have now linked DEFB1 polymorphisms to a range of pathological conditions including chronic obstructive pulmonary disease (Matsushita et al., 2002; Hu et al., 2004), although a more recent study questions this association (Hersh et al., 2006), susceptibility to oral candidiasis (Jurevic et al., 2003), susceptibility to HIV-1 infection (Braida et al., 2004), asthma (Levy et al., 2005; Leung et al., 2006), atopic dermatitis (Prado-Montes de Oca et al., 2007), susceptibility to and fatal outcome of severe sepsis (Chen et al., 2007), and colonization of the airways with *P. aeruginosa* in CF patients (Tesse et al., 2008).

A final example for AMPs linked with disease is the association with various cancers. This is particularly interesting as in some malignancies AMPs are implicated in promoting tumor cell proliferation and development, whereas in others loss of AMP expression is implicated as a contributing factor to tumor growth.

Muller and colleagues (2002) observed that renal cell carcinomas expressed HNP-1, -2, and -3 and that these α-defensins stimulated proliferation of renal cell carcinoma cell lines suggesting a possible influence on tumor proliferation *in vivo*. Interestingly, they also observed that the peptides became cytotoxic to the cell lines at high concentrations and necrotic areas were observed in carcinoma tissues where there were also high levels of the defensins. Holterman and colleagues (2006) noted that the expression of α-defensins in bladder cancer increased with tumor invasiveness and that exogenously added α-defensin increased proliferation and motility/invasiveness of bladder cancer cell lines. This led them to conclude that α-defensins may play an important role in facilitating bladder cancer invasiveness. Coffelt and colleagues (2008) have reported a similar effect of LL-37 on ovarian tumors. They observed a high level of LL-37 in malignant ovarian tissue and found that exogenously added AMP stimulated proliferation, chemotaxis, invasion, and matrix metalloproteinase expression in ovarian cancer cell lines and so concluded that LL-37 contributes to ovarian tumorigenesis. LL-37 has been found to be highly expressed by breast carcinomas and promoted epithelial cell proliferation (Heilborn et al., 2005). Also, von Haussen and colleagues (2008) reported that adenocarcinoma and squamous cell carcinoma of the lung express LL-37 and found that the peptide stimulated proliferation of lung cancer cell lines. Thus LL-37 may also act as a growth factor promoting the development of lung and breast cancers.

In contrast to the above examples where AMP expression appears to promote tumor growth, hBD-1 has been identified as a candidate tumor suppressor gene. Donald and colleagues (2003) reported the loss of hBD-1 expression in 90% of renal clear cell carcinomas and 82% of malignant prostate tissue samples, whereas high levels were present in normal and benign tissues. Subsequently, Sun and colleagues (2006) showed that hBD-1 inhibited proliferation of a bladder cancer cell line and induced apoptosis when overexpressed. Furthermore they identified several polymorphisms including one -688 bases upstream of the start codon that reduces hBD-1 promoter activity. Bullard and colleagues (2008) also observed that hBD-1 expression reduced the growth and could induce caspase-mediated apoptosis of prostate cancer cell lines. These studies suggest a role for hBD-1 in prostate cancer cell death and support Donald and colleagues' original hypothesis that this defensin may be a tumor suppressor.

16.5 STRUCTURAL ANALOGS

A wide range of AMP analogs are reported in the literature. For the most part, these have been created in order to help understand mechanisms of action, to establish where the functionality of the peptide resides within the sequence, and to create modified versions in an attempt to eliminate some of the less desirable AMP effects, particularly cytotoxicity, and enhance the positive effects, thus paving the way for the development of AMPs with therapeutic potential.

With regard to defensins, analog studies have revealed that the disulfide bonding of hBD-3, while required for the chemotactic effects of this AMP, is not required for antimicrobial activity (Wu et al., 2003; Kluver et al., 2005). Also, a series of linear hBD-3 analogs were found to have reduced cytotoxicity to mammalian cells while retaining strong antimicrobial activity (Liu et al., 2008). Two recent studies have also shown that truncated analogs of HNP-1 lacking disulfide bridges retained broad-spectrum antimicrobial activity (Varkey and Nagaraj, 2005; Lundy et al., 2008). Such variants are easier, and hence cheaper, to synthesize, so are more attractive in terms of therapeutic potential.

Modulations of the cathelicidin structure and sequence include the following. [L(1,8)]-CRAMP-18, an analog of CRAMP-18 (amino-acid residues 16–33 of the mature peptide) with Leu substitution showed a dramatic increase in antibacterial activity but no increase in hemolytic activity (Kang et al., 2002). Similarly, Nagaoka and colleagues (2005) observed that amino-acid substitutions to increase hydrophobicity and cationicity enhanced the antibacterial activity of an 18-mer peptide (residues 15 to 32) of LL-37. However, these substitutions were also shown to be associated with significant hemolysis (Ciornei et al., 2005). On the other hand, removing the N-terminal hydrophobic amino acids from LL-37 did not compromise its antimicrobial activity or ability to neutralize LPS but did reduce its cytotoxicity and sensitivity to serum (Ciornei et al., 2005). It has also been noted that an LL-37 fragment (residues 17–32) containing D- rather than L-amino acids had similar antibacterial activity to the L-diastereomer but lost its cytotoxicity to mammalian cells (Li et al., 2006). An LL-37 fragment (residues 14–34) deduced *in silico* was observed to have antimicrobial activity comparable to the parent compound and be less cytotoxic (Sigurdardottir et al., 2006). Again these studies are informative regarding the relative importance of different parts of the peptide sequence and help identify suitable candidate analogs for further development.

Other examples of interesting AMP analogs include a cecropin A–melittin hybrid, which has the first 13 amino acids of each AMP and has very potent antimicrobial activity but lacks hemolytic activity (Boman et al., 1989). Pexiganan (MSI-78) is a synthetic 22-amino-acid peptide constructed through a series of amino acid substitutions and deletions of the Xenopus peptide magainin 2. It was the first AMP-based product to be rigorously tested clinically and is discussed further along with other AMP analogs in clinical and preclinical development in Section 16.6.

16.6 CLINICAL APPLICATIONS

16.6.1 Advantages and Disadvantages of Therapeutic AMP Development

The clinical potential of antimicrobial peptides is obvious. These molecules show promise not only as novel antimicrobial agents capable of direct microbial killing, but also, given evidence for their effects on mammalian cell function, may find a place as immune-modulators and/or as mediators of wound healing. However, to date, although a number of pharmaceutical companies have undertaken the task, no AMP-based product has been successfully brought to market. Here the advantages of AMP-based pharmaceuticals, the problems associated with their development, and past and present clinical trials are discussed.

The primary features of AMPs that make them attractive as commercial antimicrobial agents are reduced development of resistance, rapid killing activity, and broad spectrum of activity. From early on, AMPs were expected to have low resistance potential due to their mechanism of action, which, as discussed previously, is dependent upon an interaction with biological membranes, an essential part of an organism that is difficult to reconfigure without inducing detrimental effects. Also, some peptides appear to have multiple cellular targets, meaning that resistance to one facet of peptide action may not be sufficient to prevent peptide effects completely. Furthermore, in evolutionary terms AMPs are an ancient line of defense against pathogens, and they have remained effective against bacteria for at least 10^8 years despite pathogens being continually exposed to them. It is not that resistance does not exist, as noted earlier (Section 16.3.1.4) several AMP resistance mechanisms have been identified, rather resistance is less likely to develop than with conventional antibacterials, and because there is such a wide range of AMPs an organism is particularly unlikely to become resistant to them all. Also, resistance when it has been detected experimentally typically represents only a two- to fourfold increase. For example 30 passages of *P. aeruginosa* in subminimal inhibitory concentration (MIC) AMP peptides increased resistance two- to fourfold where for comparison there was a 190-fold increase in resistance to the traditional aminoglycoside gentamycin (Steinberg et al., 1997; Zhang et al., 2005).

AMP action is rapid and can involve multiple cellular targets. Interaction with the microbial cell membrane is the first step but then, as noted previously, some peptides are able to translocate across the membrane and affect a variety of intracellular processes (Brogden, 2005). Furthermore, AMPs are typically bacteriocidal, whereas many conventional antibacterials are only bacteriostatic and a functional immune system is required for actual killing of the invading pathogen. Also, in contrast to many conventional antimicrobials, AMPs typically have an unusually broad spectrum of activity that may extend to fungi and viruses in addition to both Gram-positive and -negative bacteria. Also adding to their attractive qualities, AMPs have frequently been found to have activity against bacterial strains resistant to many conventional antibiotics and some such as LL-37 and its derivatives have the additional ability to bind and neutralize LPS (endotoxin) so may help reduce endotoxemia (Larrick et al., 1991, 1995).

There are of course several problems associated with developing AMPs as therapeutic agents. One significant issue is the previously discussed (Section 16.4.1) cytotoxic effect of many AMPs. To date, this has largely limited AMPs to topical applications, but as more is learnt about the nature of the cytotoxicity and where it resides within specific AMP sequences, derivatives with minimal systemic toxicity can likely be developed. Other issues to overcome, particularly for systemic use, include susceptibility to proteases. Several novel approaches to circumvent this are being adapted such as using D- rather than L-amino acids (McPhee et al., 2005). Perhaps the biggest hurdle is the cost of manufacturing. In this regard elucidating the minimal amino acid sequence for optimal activity and development of cheaper methodologies for synthesis by traditional or recombinant means is essential. Recent reviews by Marr and colleagues (2006) and Hancock and Sahl (2006) provide more detailed discussion of the pros and cons of developing AMPs as therapeutic agents.

16.6.2 AMPs in Clinical Development: Past and Present

A number of AMP-based products have progressed significantly through the various stages of the long process to achieve regulatory authority approval, with some reaching phase III trials; however, none has yet achieved the ultimate goal of being marketed successfully. More often than not, AMPs are tested as topical formulations as this circumvents many of the issues previously raised that may compromise their activity. Some of the antimicrobial peptide-based pharmaceuticals that are currently under various stages of development are summarized in Table 16.3 and those that are the most advanced are discussed in more detail in this section along with a brief review of some of the products that were developed but did not achieve approval. Where possible appropriate references are given; other information is based on the company website, press releases, and other documents in the public domain.

The first peptide to undergo significant commercial development was pexiganan (also known as MSI-78 or by the trade name Locilex). This was developed by Magainin Pharmaceuticals Inc. (Pennsylvania, USA), a company now known as Genaera (www.genaera.com). Magainins were discovered by Zasloff in 1987 in the skin of the African clawed frog *Xenopus laevis* (Zasloff, 1987).

TABLE 16.3
AMP Based Therapeutics Under Development

Company	Drug	Stage	Indication
AM-Pharma (Netherlands)	hLF1-11, derivative of human lactoferrin	Phase IIa	Prophylaxis in stem cell transplants; nosocomial *S. epidermidis*; candidiasis in neutropenic patients
Migenix (Canada)	Omiganan (MX-226, omigard 1% topical gel), 12aa analogue of indolicidin	Phase IIIb/II	Central venous catheter related local infections
	MX-594AN (formerly MBI 594AN)	Phase IIa and b (Seeking development partner)	Acne
Pacgen (Canada)	PAC-113, 12aa derivative of histatin 5	Phase IIb	Oral candidiasis in immunocompromised patients
BioLineRx (Israel)	BL-2060, peptidimimetic of acyl-lysine oligomers	Lead optomization	Nosocomial multi-drug resistant bacterial infections (Radzishevsky et al., 2007)
Ceragenix (USA)	CSA-13, cationic aminosterol steroid (lead compound)	Preclinical	Multi-drug resistant P.aeruginosa; MRSA skin infection
	CSA-54	Preclinical	Anti-viral
			Other uses being tested: coating for indwelling devices, wound dressings, contact lens disinfectant solutions. (Bucki et al., 2007; Chin et al., 2007)
HelixBioMedix (USA)	HB50, a lipohexapeptide mimetic of cecropin (lead compound)	Preclinical	Skin and wound infections, prophylaxis of MRSA skin infection
	HB107, 19aa fragment of cecropin B		Wound healing (Lee et al., 2004)
Inimex Pharm. Inc (Canada)	IMX942, 5 aa peptide	Lead optomization	An innate defence regulator. Infections associated with cancer chemotherapy-induced immunosuppression (Scott et al., 2007)
Lytix Biopharma (Norway)	"Ultra broad spectrum synthetic antimicrobial peptidomimetics" SAMPs		
	LTX 100	Preclinical	Topical antibacterial
	LTX 200	Preclinical	Topical antifungal
	LTX 300, 400	Preclinical	Anti-cancer
Migenix (Canada)	MX-2401, a lipopeptide	Preclinical	Serious Gram +ve infections

continued

TABLE 16.3 (continued)

Company	Drug	Stage	Indication
Novacta Biosystems (UK)	Lantibiotic (Gram +ve bacterial AMP) derivatives		
	NOV203	Preclinical	Nosocomial infections
	NOV206	Preclinical	Skin infections
	NOV207	Preclinical	*C. difficile*
	NOV209	Preclinical	Acne
NovoBiotic (USA)	Identifying novel antimicrobial agents produced by microorganisms using a patented culture system to grow previously "unculturable" organisms (Kaeberlein et al., 2002)		
Novozymes (Denmark)	Plecstatin (fungal defensin) variants	Preclinical	Systemic anti-Gram +ve, especially *Pneumococcal* and *Streptococcal* strains
PepTX (USA)	PTX002 (Prosep) a 33mer peptide	Discovery	Broad spectrum antimicrobial and anti-endotoxin activity. 005 has improved activity over 002. 006 has better Gram positive activity than 005. 007 has no anti-endotoxin activity.
	PTX005 (12mer derivative of 002)	Discovery	
	PTX006 (acylated version of 005)	Discovery	
	PTX007 (non-peptidic analogue of 005)	Discovery	
Polymedix (USA)	Peptidomimetic polymers	Discovery	Antiseptic hand lotion, antimicrobial coating materials for surfaces

Note: Those most advanced in the process are listed first. Information is based on recent review articles (Hancock and Sahl, 2006; Zhang and Falla, 2006), company web sites and other documentation in the public domain.

Pexiganan, a synthetic 22-amino-acid analog of magainin 2, was found to have broad-spectrum *in vitro* activity against 3109 bacterial clinical isolates. Furthermore, resistant mutants could not be generated following repeated passage with subinhibitory concentrations (Ge et al., 1999). Development of this very promising drug candidate progressed all the way to phase III trials in patients with infected diabetic foot ulcers. In these trials topical pexiganan acetate 1% achieved clinical cure or improvement in 90% of patients, comparable to the standard treatment of oral ofloxacin 800 mg/day. Eradication of pathogens at the end of therapy was achieved in 66% of the pexiganan recipients compared to 82% of the ofloxacin recipients (Lamb and Wiseman, 1998). However, FDA approval was denied in 1999 as pexiganan was deemed to be no more effective than other antibiotics used to treat foot ulcers (Moore, 2003). Following this somewhat controversial decision, Genaera discontinued its development.

Iseganan (IB-367) is a synthetic peptide derived from protegrin 1, an antimicrobial peptide expressed by porcine leucocytes (Panyutich et al., 1997). This peptide was developed by Intrabiotics Pharmaceuticals (California, USA) who have since morphed into Ardea Biosciences (www.ardeabio.com). Iseganan was initially developed as a mouthwash to prevent the development of ulcerative mucositis in high-risk patients, such as those receiving chemo- or radiotherapy. Despite promising earlier results, iseganan failed to prevent or reduce stomatitis, ulcerative oral mucositis, or its clinical sequelae relative to a placebo in one phase III trial of patients receiving stomatotoxic chemotherapy (Giles et al., 2004). In a second phase III trial of patients receiving radiotherapy for head and neck cancer, iseganan combined with standard-of-care oral hygiene demonstrated no benefit over a placebo combined with standard-of-care or the latter alone (Trotti et al., 2004). In addition to these disappointing results, a phase III trial of the use of aerosolized iseganan for the prevention of ventilator-related pneumonia was stopped in 2004 by the data monitoring committee when it was determined that there was a higher rate of pneumonia and mortality in the iseganan treatment group compared to the control group. Although phase IIa trials to explore the safety and efficacy of aerosolized iseganan for decreasing bacterial burden in the lungs in CF patients with chronic respiratory tract infections were planned, the failure of the phase III trails ultimately led the company to abandon development of iseganan.

XOMA 629 (previously known as XMP.629) is a small synthetic peptide derived from bactericidal/permeability-increasing protein (BPI), which is a 456-amino-acid cationic protein produced by neutrophils (Weiss et al., 1978). XOMA 629 was developed by Xoma (California, USA, www.xoma.com) and is a 9-amino-acid peptide that was created by reverse synthesis of the native 9mer peptide using D amino acids and substitution of two residues with napthyl–alanine. Preclinical studies of a dermatological formulation showed XOMA 629 to have activity (MIC range 0.5 to 4 μg/mL) against *Propionibacterium acnes*, including strains resistant to some traditional antibiotics. Results of a phase I trial showed a reduction in lesions and no significant skin irritation or systemic absorption. However, results released in 2004 regarding a phase II trial were inconclusive as to the clinical benefit of XOMA 629 versus placebo. XOMA also developed an injectable formulation of recombinant BPI ($rBPI_{21}$ or opebacan, trade name Neuprex). This molecule is a modified 193-amino-acid fragment of the native BPI protein in which the cysteine at position 12 is replaced with an alanine to improve stability (Giroir et al., 1997) and has comparable properties to the native compound, including efficacy against Gram-negative bacteria (particularly *Neisseria meningitidis*), neutralization of endotoxin, and inhibition of angiogenesis. In a phase I/II open-label trial in children with severe meningococcal sepsis, $rBPI_{21}$ was demonstrated to be safe and its use was associated with a higher survival rate compared with matched historical control subjects and predictive indices of mortality (Giroir et al., 1997). In a phase III trial performed in the U.K. and U.S. in children with severe meningococcemia, adjunctive parenteral therapy with Neuprex failed to significantly reduce mortality compared to the placebo control (Levin et al., 2000). However, the trial was determined to be substantially underpowered to detect a statistically significant difference in this the primary end point. Significant benefits in prospectively defined secondary end points of reduction in morbidity and improved functional outcome were observed (Levin et al., 2000; Giroir et al., 2001). Other indications for Neuprex that have been under investigation include treatment of burns and reduction of inflammatory complications associated with pediatric open-heart surgery patients. As of the end of 2007, Neuprex was available for licensing as per the company web site.

AM-Pharma (The Netherlands, www.AM-Pharma.com) is developing a small peptide, referred to as hLF1-11, which consists of the first 11 amino acids of the N-terminus of human lactoferrin. Currently the peptide is being tested for its ability to prevent infections in patients undergoing hematopoetic stem cell transplants. To date, a phase IIa study showed that there were no serious drug-related safety issues when hLF1–11 was given as a single dose in a small number of patients. hLF1–11 is also in phase IIa "proof of concept" trials for the treatment of nosocomial *Staphylococcus epidermidis* (including methicillin-resistant strains) infections and for the treatment of systemic *Candida albicans* infection in neutropenic patients.

Migenix (formerly Microbiologix, British Columbia, Canada, www.migenix.com) has developed a number of novel topical antibacterial compounds including omiganan pentahydrochloride (MX-226, formerly known as MBI-226, trade name Omigard and partnered with Cadence Pharmaceuticals), which is under investigation in the form of a 1% aqueous gel for the prevention of central venous catheter-related infections. Omiganan is a 12-amino-acid synthetic analog (also known as a bactolysin) of indolicidin, a bovine antimicrobial peptide (Selsted et al., 1992) and has rapid bactericidal and fungicidal activity of prolonged duration against microorganisms commonly found on skin, including methicillin-resistant strains of *Staphylococcus aureus*. Meaningful resistance to omiganan could not be induced in the laboratory and development of cross-resistance to other antimicrobials was not observed (Sader et al., 2004). The results of a large U.S.-based phase III trial released in 2004 showed that topical treatment with omiganan did not lead to a statistically significant reduction in central venous catheter-related bloodstream infections compared to povidone–iodine as the control. However, while the primary efficacy end point was not reached, two secondary efficacy end points did achieve statistical significance: reduction in catheter colonization and reduction in catheter-related local infections. A confirmatory multinational phase III trial was launched in 2005 with the primary end point being the reduction of local catheter site infections. Preliminary data analysis indicates a statistically significant reduction of local catheter site infections of approximately 42% in the omiganan treatment group compared to the povidone–iodine treatment control and a reduction in overall local infection rate. An increase in the number of patients to be enrolled in the confirmatory study was announced in July 2007 and, assuming a positive outcome, Cadence Pharmaceuticals expect to submit a new drug application for omiganan (Omigard) in 2009.

Pacgen Biopharmaceuticals Corporation (British Columbia, Canada, www.pacgenbiopharm.com) is developing a compound called PAC-113. This was originally developed by Periodontix (Massachusetts, USA), then acquired by Demgen (Pennsylvania, USA) in 2001 and by Pacgen in 2005. PAC-113 (originally referred to as P113) is a 12-amino-acid peptide based on histatins, specifically histatin 5, which are naturally occurring antimicrobial peptides found in saliva (Oppenheim et al., 1988). It has potent antifungal activity against *Candida albicans*, including isolates from drug-resistant HIV patients. Its MIC against Candida species is in the range 1.4 to 4 μg/mL, an improvement over conventional antifungal agents such as ketoconazole (MIC 4 μg/mL) and fluconazole (MIC 16 to 32 μg/mL). PAC-113 is formulated as a sugar-free nonviscous mouth rinse and is being developed for the treatment of oral candidiasis in immunocompromised patients and patients with salivary dysfunction. Results from a phase I/II trial released in early 2007 showed that PAC-113 is generally safe and well tolerated. Complete or partial responses after 14 days of treatment were observed in 95% of patients receiving PAC-113 and 87% receiving Nystatin, a widely used topical antifungal mouth rinse. Thirty-seven percent of PAC-113 patients and 36% of Nystatin treated patients were clinically cured after 14 days of treatment. A Phase IIb dose-ranging clinical trial in which different doses of PAC-113 were compared to Nystatin in seropositive patients with oral candidiasis was completed in June 2008. The results showed that PAC-113 is effective in treating oral candidiasis and compares favourably to Nystatin.

16.6.3 Other Possible Clinical Applications

The majority of investigations to date have focused on using AMP-based products for their direct antimicrobial effects, although a few have exploited additional effects of AMPs such as endotoxin neutralization by the BPI-based product Neuprex. However, the goal of one company, Inimex (www.inimexpharma.com) is to harness the immunomodulatory effects of AMP-based products to treat infectious disease, cancer and inflammatory disease. Inimex are developing a series of so-called innate defense regulators (IDR), which selectively trigger the body's innate defenses without causing inflammation. A recently published paper (Scott et al., 2007) describes a prototype IDR, "IDR-1", which is a 13-amino-acid synthetic peptide with no direct antibacterial activity, but which protected mice from experimental systemic *Staphylococcus aureus* infection. The latter was mediated by an effect on monocyte/macrophage function. Inimex's lead compound is IMX942, which is in preclinical development for prevention of infections associated with cancer-chemotherapy-induced immunosuppression.

Other possible uses of AMPs include applications as novel disinfectants in contact lens cleaning/storage solutions (Sousa et al., 1996). The bovine peptide bactenecin has been used as a component of organ preservation media that was associated with better graft survival than the standard media (Ambiru et al., 2004). Also, radiolabeled AMPs are under investigation as tools to facilitate the detection, diagnosis, and monitoring of infection (Lupetti et al., 2003; Brouwer et al., 2007).

The early observation that AMPs are cytotoxic to mammalian cells led several researchers in the field to suggest they have potential as anticancer agents. Indeed numerous *in vitro* studies have confirmed that AMPs are cytotoxic to a wide range of tumor cells. However, a major stumbling block in their development as anticancer agents is the fact that AMPs also show toxicity to normal cells. When a cell becomes malignant it sometimes shows membrane alterations that may increase its susceptibility to AMP activity, for example an increase in negative charge. The challenge then is to identify specific AMPs, or derivatives thereof, that show selectivity for malignant rather than normal cells. Promising candidates include magainin 2, which kills several different cancer cell lines at concentrations that are five to ten times lower that those that are cytotoxic to normal cells (Jacob and Zasloff, 1994). Indeed this peptide was recently shown to kill bladder cancer cells without toxicity to normal cells (Lehmann et al., 2006). Several excellent reviews (Papo and Shai, 2005; Mader and Hoskin, 2006; Hoskin and Ramamoorthy, 2008) provide in-depth discussion of the potential of AMPs in the treatment of various cancers.

AMPs have also been suggested as possible markers to facilitate the diagnosis of malignant tumors. Arimura and colleagues (2004) observed that serum levels of hBD-1, and -2 were higher in patients with lung cancer than in normal subjects or those with pneumonia, and suggested that the serum hBD-1 level can be used as an auxiliary tool for lung cancer diagnosis. Albrethsen and colleagues (2006) identified serum/tissue levels of HNPs as a potential marker for prognostic assessment, monitoring progression and treatment effects in colorectal cancer. Zou and colleagues (2007) investigated the possibility of quantitating HNPs in stool samples for the detection of colorectal cancers, but found that these defensins were nonspecifically elevated and thus unsuitable as a diagnostic tool. Also, Nam and colleagues (2005) reported that serum HD-6 may be a possible marker for adenocarcinoma of the colon.

16.7 COMMON METHODOLOGIES AND ASSAY SYSTEMS

In terms of commercial products for investigating the peptides and their biological activities, recombinant versions of several human and murine peptides are available, as are several antibodies. Specialist companies will also undertake solid-phase synthesis of peptides. Such products are much cheaper than the recombinant versions but the user must take care to utilize the services of a reputable company that will guarantee a suitable level of purity and in the case of the defensins, utilize appropriate methodology to ensure correct connectivity of the disulfide bonds. A number of companies also now market enzyme immunoassay kits for the detection of hBD-1, 2, 3, α-defensins, and LL-37 in human cell, tissue, and body fluid samples.

In vitro, the antimicrobial effectiveness of the peptides (isolated from samples, or manufactured by synthetic or recombinant means) has been investigated using standard microbial testing such as colony-forming unit assays and MIC determinations (Weigand et al., 2008). These assays are easy to perform and, excluding the peptides themselves, the reagents are inexpensive. The broth dilution method for MIC determination has also been adapted for multiwell plates. These assays provide easy comparison of effectiveness between different AMPs and AMPs and conventional antibiotics and facilitate the study of synergistic/additive effects among the peptides and other antimicrobial molecules such as lysozyme. The effect of other molecules, such as NaCl or body fluids, for example tears, on peptide efficacy can also be examined in a quantitative manner.

A small number of transgenic or knockout animal models have also been developed and used to investigate the antimicrobial properties of AMPs *in vivo*. Inactivation of the gene for matrilysin, an enzyme required for processing mouse intestinal α-defensins to their active form, increased the susceptibility of the animals to infection with *Salmonella typhimurium* (Wilson et al., 1999). Knocking out the mouse β-defensin-1 gene led to delayed clearance of *Haemophilus influenzae* from the lung (Moser et al., 2002) and an increase in Staphylococcus colonization in the bladder (Morrison et al., 2002). Salzman and colleagues (2003) reported that transgenic mice expressing the gene for human intestinal α-defensin HD-5 were much more resistant to an oral challenge with *S. typhimurium* than their wild-type counterparts. Also, mice expressing hBD-2 were found to be more resistant to infection with *S. aureus* (Zhang et al., 2006). A knockout mouse line deficient in the gene Cnlp that codes for cathelicidin-related antimicrobial peptide (CRAMP), the mouse homolog of LL-37, has been very useful in showing the importance of this peptide in defense against a number of organisms. CRAMP knockout animals were found to be much more susceptible to skin infection caused by group A Streptococcus (Nizet et al., 2001), skin infections due to vaccinia virus (Howell et al., 2004), intestinal colonization by the murine pathogen *Citrobacter rodentium* (Iimura et al., 2005), urinary tract infection caused by *E.coli* (Chromek et al., 2006), meningococcal infection (Bergman et al., 2006) and keratitis induced by *Pseudomonas aeruginosa* (Huang et al., 2007b). Additionally, mice expressing PR-39, a porcine cathelicidin, were found to be more resistant to group A Streptococcus skin infection (Lee et al., 2005).

Expression of AMPs by human cells and tissues has been studied using standard techniques such as RT-PCR for gene expression and immunostaining for protein by western blotting and fluorescence microscopy. How AMP expression is modulated has been studied using chemical inhibitors of a variety of cellular signaling pathways, particularly those that affect MAPK pathways such as SB203590 (p38-MAPK inhibitor), SP600125 (JNK inhibitor), and PD98059 (ERK inhibitor). Some of the same inhibitors have also been employed to elucidate the mechanisms by which AMPs exert their effects on mammalian cell behavior. The nature of the cellular receptors mediating the effects of AMPs has largely been deduced based on assays utilizing known receptor inhibitors, competition with agonists, or observation of positive effects in response to the presence of the peptide in cell types transfected with a particular candidate receptor. Other studies have used classic receptor binding assays with radiolabeled peptide to investigate receptor sites for LL-37 on a variety of cell types (e.g., Niyonsaba et al., 2001). Some of the aforementioned knockout animals have also been used to gain insight into the nonmicrobicidal roles of the peptides *in vivo* and confirm that some of the effects seen *in vitro* have physiological relevance. For example, utilizing CRAMP knockout mice Koczulla and colleagues (2003) showed a role for CRAMP in angiogenesis during wound healing.

16.8 FUTURE DEVELOPMENTS

Interest in AMPs has expanded rapidly over recent years, with almost 2000 papers on defensins alone being published in last 10 years. Indeed, each month reveals new findings linking AMPs to previously unidentified functions, new diseases, and unexpected traits. For example, a mutated β-defensin gene was recently identified to be associated with black coat color in domestic dogs (Candille et al., 2007). This veritable explosion of interest reflects the realization of the importance of the innate immune system in defense against pathogens and the potential to exploit AMPs' as novel therapeutics, in particular to help circumvent the issue of ever growing bacterial resistance to conventional antibiotics.

Yet, despite this intensive research, there is still much to learn about AMPs. We know little of how their expression is regulated, knowledge essential if we are to correct excesses/deficits related to disease processes. Also, conclusive evidence of their roles as modulators of immunity and wound healing *in vivo* is still lacking. Undoubtedly as the number of groups around the world working in the field of AMP research increases, some of these gaps in our knowledge will be filled in upcoming years. Similarly, from a therapeutic and commercial point of view, although great strides have been made toward developing AMP-based pharmaceuticals, actual approval of a product still eludes the various companies involved. Development of AMP-based products will be enhanced by a better understanding of the mechanisms of AMP action on both pro- and eukaryotic cells and development of novel AMP mimetics and delivery systems that help circumvent issues of AMP inactivation, degradation, and toxicity. As discussed there are a number of AMP-based products on the horizon with many more in development, so it is realistic to suggest the appearance of one of more on the market within the next five years.

ACKNOWLEDGMENTS

The author thanks Ms. Kim Thompson of the University of Houston College of Optometry audiovisual services for drawing Figures 16.1 and 16.2. The author is supported by grants from the National Institutes of Health (EY13175) and National Science Foundation (DMR-0706627).

REFERENCES

Aarbiou, J. et al., Neutrophil defensins enhance lung epithelial wound closure and mucin gene expression *in vitro*, *Am. J. Respir. Cell Mol. Biol.*, 30, 193–201, 2004.

Aarbiou, J. et al., Mechansims of cell death induced by the neutrophil antimicrobial peptides alpha-defensins and LL-37, *Inflamm. Res.*, 55, 119–127, 2006.

Agerberth, B. et al., Amino acid sequence of PR-39. Isolation from pig intestine of a new member of the family of proline-arginine-rich antibacterial peptides, *Eur. J. Biochem.*, 202, 849–854, 1991.

Agerberth, B.et al., FALL-39, a putative human peptide antibiotic, is cysteine-free and expressed in bone marrow and testis, *Proc. Natl Acad. Sci. USA*, 92, 195–199, 1995.

Agerbeth, B. et al., The human antimicrobial and chemotactic peptides LL-37 and alpha-defensins are expressed by specific lymphocyte and monocyte populations, *Blood*, 96, 3086–3093, 2000.

Albrethsen, J. et al., Human neutrophil peptides 1, 2 and 3 are biochemical markers for metastatic colorectal cancer, *Eur. J. Cancer*, 42, 3057–3064, 2006.

Aldred, P.M.R., Hollox, E.J., and Armour, J.A.L., Copy number polymorphism and expression level variation of the human α-defensin genes DEFA1 and DEFA3, *Hum. Mol. Genet.*, 14, 2045–2052, 2005.

Ambiru, S. et al., Improved survival of orthotopic liver allograft in swine by addition of trophic factors to University of Wisconsin solution, *Transplantation*, 77, 302–319, 2004.

Arimura, Y. et al., Elevated serum beta-defensins concentrations in patients with lung cancer, *Anticancer Res.*, 24, 4051–4057, 2004.

Arnljots, K., Sorensen, O., Lollike, K., and Borregaard, N., Timing, targeting, and sorting of azurophil granule proteins in human myeloid cells, *Leukemia*, 12, 1789–1795, 1998.

Ayabe, T. et al., Secretion of microbicidal alpha-defensins by intestinal Paneth cells in response to bacteria, *Nat. Immunol.*, 1, 113–118, 2000.

Ayabe, T. et al., Modulation of mouse Paneth cell alpha-defensin secretion by mIKCa1, a Ca2+-activated, intermediate conductance potassium channel, *J. Biol. Chem.*, 277, 3793–3800, 2002.

Baev, D. et al., Human salivary histain 5 causes disordered volume regulation and cell cycle arrest in *Candida albicans*, *Infect Immunol.*, 70, 4777–4784, 2002.

Baev, D. et al., The TRK1 potassium transporter is the critical effector for killing of *Candida albicans* by the cationic protein, histatin 5, *J. Biol. Chem.*, 279, 55060–55072, 2004.

Bainton, D.F., Distinct granule populations in human neutrophils and lysosomal organelles identified by immuno-electron microscopy, *J. Immunol. Methods*, 232, 153–168, 1999.

Ballana, E., Gonzalez, J.R., Bosch, N., and Estivill, X., Inter-population variability of DEFA3 gene absence: Correlation with haplotype structure and population variability, *BMC Genomics*, 8, 14, 2007.

Bals, R., Wang, X., Zasloff, M., and Wilson, J.M., The peptide antibiotic LL-37/hCAP18 is expressed in epithelia of the human lung where it has broad spectrum antimicrobial activity at the airway surface, *Proc. Natl Acad. Sci. USA*, 95, 9541–9546, 1998.

Bals, R. and Wilson, J.M., Cathelicidins—a family of multifunctional antimicrobial peptides, *Cell Mol. Life Sci.*, 60, 711–720, 2003.

Bandholtz, L. et al., Antimicrobial peptide LL-37 internalized by immature dendritic cells alters their phenotype, *Scand. J. Immunol.*, 63, 410–419, 2006.

Baranska-Rybak, W., Sonesson, A., Nowicki, R., and Schmidtchen, A., Glycosaminoglycans inhibit the antibacterial activity of LL-37 in biological fluids, *J. Antimicrob. Chemother.*, 57, 260–265, 2006.

Barnathan, E.S. et al., Immuohistochemical localization of defensins in human coronary vessels, *Am. J. Pathol.*, 150, 1009–1020, 1997.

Bastian, A. and Schafer, H., Human alpha-defensin-1 (HNP-1) inhibits adenoviral infection *in vitro*, *Regul. Pept.*, 101, 157–161, 2001.

Bateman, A., Singh, A., Congote, L.F., and Solomon, S., The effect of HP-1 and related neutrophil granule peptides on DNA synthesis in HL60 cells, *Regul. Pept.*, 35, 135–143, 1991.

Bdeir, K. et al., Defensin promotes the binding of lipoprotein (a) to vascular matrix, *Blood*, 94, 2007–2019, 1999.

Bechinger, B. and Lohner, K., Detergent-like actions of linear amphipathic cationic antimicrobial peptides, *BBA Biomembranes*, 1758, 1529–1539, 2006.

Befus, A.D. et al., Neutrophil defensins induce histamine secretion from mast cells: Mechanisms of action, *J. Immunol.*, 163, 947–953, 1999.

Bensch, K.W.et al., hBD-1: A novel beta-defensin from human plasma, *FEBS Lett.*, 368, 331–335, 1995.

Bergman, P. et al., Induction of the antimicrobial peptide CRAMP in the blood–brain barrier and meninges after meningococcal infection, *Infect. Immunol.*, 74, 6982–6992, 2006.

Bergman, P. et al., The antimicrobial peptide LL-37 inhibits HIV-1 replication, *Curr. HIV Res.*, 5, 410–415, 2007.

Bhat, S., Song, Y.H., Lawyer, C., and Milner, S.M., Modulation of the complement system by human beta-defensin 2, *J. Burns Wound*, 5, e10, 2007.

Biragyn, A. et al., Mediators of innate immunity that target immature, but not mature, dendritic cells induce antitumor immunity when genetically fused with nonimmunogenic tumor antigens, *J. Immunol.*, 167, 6644–6653, 2001.

Biragyn, A. et al., Toll-like receptor-4-dependent activation of dendritic cells by beta-defensin-2, *Science*, 298, 1025–1029, 2002.

Boman, H.G. et al., Antibacterial and antimalarial properties of peptides that are cecropin-melittin hybrids, *FEBS Lett.*, 259, 103–106, 1989.

Boman, H.G., Agerberth, B., and Boman, A., Mechanism of action on *Escherichia coli* of cecropin Pa and PR-39, two antibacterial peptides from pig intestine, *Infect Immunol.*, 62, 2978–2984, 1993.

Boniotto, M. et al., Human beta-defensin 2 induces a vigorous cytokine response in peripheral blood mononuclear cells, *Antimicrob. Agents Chemother.*, 50, 1433–1441, 2006.

Borregaard, N. et al., Human neutrophil granules and secretory vesicles, *Eur. J. Haematol.*, 51, 187–198, 1993.

Bowdish, D.M., Davidson, D.J., Speert, D.P., and Hancock, R.E.W., The human cationic peptide LL-37 induces activation of the extracellular signal-regulated kinase and p38 kinase pathways in primary human monocytes, *J. Immunol.*, 172, 3758–3765, 2004.

Braff, M.H., Di Nardo, A., and Gallo, R.L., Keratinocytes store the antimicrobial peptide cathelicidin in lamellar bodies, *J. Invest Dermatol.*, 124, 394–400, 2005a.

Braff, M.H. et al., Structure–function relationships among human cathelicidin peptides: Dissociation of antimicrobial properties from host immunostimulatory activities, *J. Immunol.*, 174, 4271–4278, 2005b.

Braida, L. et al., A single-nucleotide polymorphism in the human beta-defensin 1 gene is associated with HIV-1 infection in Italian children, *AIDS*, 18, 1598–1600, 2004.

Brogden, K.A., Ackermann, M., McCray, P.B., and Tack, B.E., Antimicrobial peptides in animals and their role in host defences, *Int. J. Antimicrob. Agents*, 22, 465–478, 2003.

Brogden, K.A., Antimicrobial peptides: Pore formers or metabolic inhibitors in bacteria? *Nat. Rev. Microbiol.*, 3, 238–250, 2005.

Brotz, H. et al., The lantibiotic mersacidin inhibits peptidoglycan synthesis by targeting lipid II, *Antimicrob. Agents Chemother.*, 42, 154–160, 1998.

Brouwer, C.P.J.M., Wulferink, M., and Welling, M.M., The pharmacology of radiolabeled cationic antimicrobial peptides, *J. Pharm. Sci.*, e-pub ahead of print, 2007.

Buck, C.B. et al., Human alpha-defensins block papillomavirus infection, *Proc. Natl Acad. Sci. USA*, 103, 1516–1521, 2006.

Bucki, R. et al., Resistance of the antibacterial agent ceragenin CSA-13 to inactivation by DNA or F-actin and its activity in cystic fibrosis sputum, *J. Antimicrob. Chemother.*, 60, 535–545, 2007.

Bulet, P. and Stocklin, R., Insect antimicrobial peptides: Structures, properties and gene regulation, *Protein Pept. Lett.*, 12, 3–11, 2005.

Bullard, R.S. et al., Functional analysis of the host defense peptide human beta defensin-1: New insight into its potential role in cancer, *Mol. Immunol.*, 45, 839–848, 2008.

Candille, S.I. et al., A β-defensin mutation causes black coat color in domestic dogs, *Science*, 318, 1418–1423, 2007.

Carlsson, G. et al., Peridontal disease in patients from the original Kostmann family with severe congenital neutropenia, *J. Peridontol.*, 77, 744–751, 2006.

Carretero, M. et al., *In vitro* and *in vivo* wound healing-promoting activities of human cathelicidin LL-37, *J. Invest. Dermatol.*, 128, 223–236, 2008.

Chaly, Y.V. et al., Neutrophil alpha-defensin human neutrophil peptide modulates cytokine production in monocytes and adhesion molecule expression in endothelial cells, *Eur. Cytokine Netw.*, 11, 257–266, 2000.

Chang, T.L., Vargas, J., DelPortillo, A., and Klotman, M.E., Dual role of alpha-defensins-1 in anti-HIV innate immunity, *J. Clin. Invest.*, 115, 765–773, 2005.

Chavakis, T. et al., Regulation of neovascularisation by human neutrophil peptides (alpha-defensins): A link between inflammation and angiogenesis, *FASEB J.*, 18, 1306–1308, 2004.

Chen, Q.X. et al., Genomic variations within DEFB1 are associated with the susceptibility to and the fatal outcome of severe sepsis in Chinese Han population, *Genes Immunol.*, 8, 439–443, 2007.

Chertov, O. et al., Identification of defensin-1, defensin-2 and CAP37/azurocidin as T-cell chemoattractant proteins released from IL-8 stimulated neutrophils, *J. Biol. Chem.*, 271, 2935–2940, 1996.

Chin, J.N., Rybak, M.J., Cheung, C.M., and Savage, P.B., Antimicrobial activities of ceragenins against clinical isolates of resistant *Staphylococcus aureus*, *Antimicrob. Agents Chemother.*, 51, 1268–1273, 2007.

Chromek, M. et al., The antimicrobial peptide cathelicidin protects the urinary tract against invasive bacterial infection, *Nature. Med.*, 12, 636–641, 2006.

Ciornei, C.D., Sigurdardottir, T., Schmidtchen, A., and Bodelsson, M., Antimicrobial and chemoattractant activity, lipopolysaccharide neutralization, cytotoxicity, and inhibition by serum of analogs of human cathelicidin LL-37, *Antimicrob. Agents Chemother.*, 49, 2845–2850, 2005.

Ciornei, C.D. et al., Human antimicrobial peptide LL-37 is present in atherosclerotic plaques and induces death of vascular smooth muscle cells: A laboratory study, *BMC Cardiovasc. Disord.*, 6, 49, 2006.

Cirioni, O. et al., LL-37 protects rats against lethal sepsis caused by gram-negative bacteria. *Antimicrob. Agents Chemother.*, 50, 1672–1679, 2006.

Clark, D.P., Durell, S., Maloy, W.L., and Zasloff, M., Ranalexin. A novel antimicrobial peptide from bullfrog (*Rana catesbeiana*) skin, structurally related to the bacterial antibiotic, polymixin, *J. Biol. Chem.*, 269, 10849–10855, 1994.

Coffelt, S.B. et al., Ovarian cancers overexpress the antimicrobial protein hCAP-18 and its derivative LL-37 increases ovarian cancer cell proliferation and invasion, *Int. J. Cancer*, 122, 1030–1039, 2008.

Cole, A.M., Weis, P., and Diamond, G.J., Isolation and characterization of pleurocidin, an antimicrobial peptide in the skin secretions of winter flounder, *J. Biol. Chem.*, 272, 12008–12013, 1997.

Cole, A.M. et al., Cutting edge: IFN-inducible ELR-CXC chemokines display defensin-like antimicrobial activity, *J. Immunol.*, 167, 623–627, 2001.

Conejo-Garcia, J.R. et al., Tumor-infiltrating dendritic cell precursors recruited by a beta-defensin contribute to vasculogenesis under the influence of Vegf-A, *Nat. Med.*, 10, 950–958, 2004.

Cowland, J., Johnsen, A., and Borregaard, N., hCAP-18, a cathelin/pro-bactenecin-like protein of human neutrophil specific granules, *FEBS Lett.*, 368, 173–176, 1995.

Cowland, J.B. and Borregaard, N., The individual regulation of granule protein mRNA levels during neutrophil maturation explains the heterogeneity of neutrophil granules, *J. Leukoc. Biol.*, 66, 989–995, 1999.

Daher, K.A., Selsted, M.E., and Lehrer, R.I., Direct inactivation of viruses by human granulocyte defensins, *J. Virol.*, 60, 1068–1074, 1986.

Daher, K.A., Lehrer, R.I., Ganz, T., and Kronenberg, M., Isolation and characterization of human defensin cDNA clones, *Proc. Natl Acad. Sci. USA*, 85, 7327–7331, 1988.

Davidson, D.J. et al., The cationic antimicrobial peptide LL-37 modulates dendritic cell differentiation and dendritic cell-induced T cell polarization, *J. Immunol.*, 172, 1146–1156, 2004.

den Hertog, A.L. et al., Candidacidal effects of two antimicrobial peptides: Histatin 5 causes small membrane defects, but LL-37 causes massive disruption of the cell membrane, *Biochem. J.*, 388, 689–695, 2005.

De Yang, Chen, Q., Schmidt, A.P. et al., LL-37, the neutrophil granule- and epithelial cell-derived cathelicidin, utilizes formyl peptide receptor-like 1 (FPRL-1) as a receptor to chemoattract human peripheral blood neutrophils, monocytes and T cells, *J. Exp. Med.*, 192, 1069–1074, 2000.

Diamond, G. et al., Tracheal antimicrobial peptide, a cysteine-rich peptide from mammalian tracheal mucosa: Peptide isolation and cloning of a cDNA, *Proc. Natl Acad. Sci. USA*, 88, 3952–3956, 1991.

Diamond, G. et al., Transcriptional regulation of beta-defensin gene expression in tracheal epithelial cells, *Infect. Immunol.*, 68, 113–119, 2000.

Di Nardo, A., Vitiello, A., and Gallo, R.L., Mast cell antimicrobial activity is mediated by expression of cathelicidin antimicrobial peptide, *J. Immunol.*, 170, 2274–2278, 2003.

Di Nardo, A. et al., Cathelicidin antimicrobial peptides block dendritic cell TLR4 activation and allergic contact sensitization, *J. Immunol.*, 178, 1829–1834, 2007.

Donald, C. et al., Cancer specific loss of β-defensin 1 in renal and prostatic carcinomas, *Lab. Invest.*, 83, 501–505, 2003.

Dork, T. and Stuhrmann, M., Polymorphisms of the human beta-defensin-1 gene, *Mol. Cell Probes*, 12, 171–173, 1998.

Dorschner, R.A. et al., The mammalian ionic environment dictates microbial susceptibility to antimicrobial defense peptides, *FASEB J.*, 20, 35–42, 2006.

Duits, L.A. et al., Expression of beta-defensin 1 and 2 mRNA by human monocytes, macrophages and dendritic cells, *Immunology*, 106, 517–525, 2002.

Durr, U.H.N., Sudheendra, U.S., and Ramamoorthy, A., LL-37, the only human member of the cathelicidin family of antimicrobial peptides, *Biochim. Biophys. Acta (Biomembranes)*, 1578, 1408–1425, 2006.

Economopoulou, M. et al., Inhibition of pathologic retinal neovascularisation by alpha-defensins, *Blood*, 106, 3831–3838, 2005.

Ehrlich, P., *Farbenanalytische untersuchungen zur histology und klinik des blutes*, Hirschwald, Berlin, 1891.

Ehrlich, P. and Lazarus, A., *Die anamie I. Normale und patologische histology des blutes*, Holder, Wein (revised and re-published 1909), 1898.

Elssner, A., Duncan, M., Gavrilin, M., and Wewers, M.D., A novel P2X7 receptor activator, the human cathelicidin-derived peptide LL37, induces IL-1 beta processing and release, *J. Immunol.*, 172, 4987–4994, 2004.

Ericksen, B., Wu, Z., Lu, W., and Lehrer, R.I., Antibacterial activity and specificity of the six human alpha-defensins, *Antimicrob. Agents Chemother.*, 49, 269–274, 2005.

Faurschou, M. et al., Defensin-rich granules of human neutrophils: Characterization of secretory properties, *Biochim. Biophys. Acta*, 1591, 29–35, 2002.

Felgentreff, K. et al., The antimicrobial peptide cathelicidin interacts with airway mucus, *Peptides*, 27, 3100–3106, 2006.

Fellermann, K. et al., A chromosome 8 gene-cluster polymorphism with low human beta-defensin 2 gene copy number predisposes to Crohn disease of the colon, *Am. J. Hum. Genet.*, 79, 439–448, 2006.

Feng, Z. et al., Human beta-defensins: Differential activity against Candidal species and regulation by *Candida albicans*, *J. Dent. Res.*, 84, 445–450, 2005.

Fernandez de Caleya, R., Gonzalez-Pascual, B., Garcia-Olmedo, F., and Carbonero, P., Susceptibility of phytopathogenic bacteria to wheat purothionins *in vitro*, *Appl. Microbiol.*, 23, 998–1000, 1972.

Fleming, A., On a remarkable bacteriolytic element found in tissues and secretions, *Proc. R. Soc.*, 93, 306–317, 1922.

Frank, R.W. et al., Amino acid sequences of two proline-rich bactenecins. Antimicrobial peptides of bovine neutrophils, *J. Biol. Chem.*, 265, 18871–18874, 1990.

Frohm, M. et al., The expression of the gene coding for the antibacterial peptide LL-37 is induced in human keratinocytes during inflammatory disorders, *J. Biol. Chem.*, 272, 15258–15263, 1997.

Frohm-Nilsson, M. et al., The human cationic antimicrobial protein (hCAP18), a peptide antibiotic, is widely expressed in human squamous epithelia and colocalizes with interleukin-6, *Infect Immunol.*, 67, 2561–2566, 1999.

Frye, M. et al., Expression of human alpha-defensin 5 (HD5) mRNA in nasal and bronchial epithelial cells, *J. Clin. Pathol.*, 53, 770–773, 2000.

Frye, M., Bargon, J., and Gropp, R., Expression of human β-defensin-1 promotes differentiation of keratinocytes, *J. Mol. Med.*, 79, 275–282, 2001.

Funderburg, N. et al., Human-defensin-3 activates professional antigen-presenting cells via Toll-like receptors 1 and 2, *Proc. Natl Acad. Sci. USA*, 104, 18631–18635, 2007.

Furci, L. et al., Alpha-defensins block the early steps of HIV-1 infection: Interference with the binding of gp120 to CD4, *Blood*, 109, 2928–2935, 2007.

Gallo, R.L. et al., Identification of CRAMP, a cathelin-related antimicrobial peptide expressed in the embryonic and adult mouse, *J. Biol. Chem.*, 272, 13088–13093, 1997.

Ganz, T. et al., Defensins. Natural peptide antibiotics of human neutrophils, *J. Clin. Invest.*, 76, 1427–1435, 1985.

Ganz, T., Extracellular release of antimicrobial defensins by human polymorphonuclear leukocytes, *Infect. Immunol.*, 55, 568–571, 1987.

Ganz, T. et al., The structure of the rabbit macrophage defensin genes and their organ-specific expression, *J. Immunol.*, 143, 1358–1365, 1989.

Ganz, T., Defensins: Antimicrobial peptides of innate immunity, *Nat. Rev. Immunol.*, 3, 710–720, 2003.

Ganz, T., Hepcidin—a peptide hormone at the interface of innate immunity and iron metabolism, *Curr. Top. Microbiol. Immunol.*, 306, 183–198, 2006.

Garcia, J.R. et al., Identification of a novel, multifunctional beta-defensin (human beta-defensin 3) with specific antimicrobial activity. Its interaction with plasma membranes of Xenopus oocytes and the induction of macrophage chemoattraction, *Cell Tissue Res.*, 306, 257–264, 2001a.

Garcia, J.R. et al., Human beta-defensin 4: A novel inducible peptide with a salt-sensitive spectrum of antimicrobial activity, *FASEB J.* 15, 1819–1821, 2001b.

Gazit, E., Boman, A., Boman, H., and Shai, Y., Interaction of the mammalian antibacterial peptide cecropin P1 with phospholipids vesicles, *Biochemistry*, 34, 11479–11488, 1995.

Ge, Y. et al., *In vitro* antibacterial properties of pexiganan, an analog of magainin, *Antimicrob. Agents Chemother.*, 43, 782–788, 1999.

Gennaro, R., Skerlavaj, B., and Romeo, D., Purification, composition, and activity of two bactenecins, antibacterial peptides of bovine neutrophils, *Infect. Immunol.*, 57, 3142–3146, 1989.

Gera, J.F. and Lichtenstein, A., Human neutrophil peptide defensins induce single strand DNA breaks in target cells, *Cell Immunol.*, 138, 108–120, 1991.

Ghosh, D. et al., Paneth cell trypsin is the processing enzyme for human defensin-5, *Nat. Immunol.*, 3, 583–590, 2002.

Giles, F.J. et al., A phase III, randomized, double-blind, placebo controlled, study of iseganan for the reduction of stomatitis in patients receiving stomatotoxic chemotherapy, *Leuk. Res.*, 28, 559–565, 2004.

Giroir, B.P. et al., Preliminary evaluation of recombinant amino-terminal fragment of human bactericidal/permeability-increasing protein in children with severe meningococcal sepsis, *Lancet*, 350, 1439–1443, 1997.

Giroir, B.P., Scannon, P.J., and Levin, M., Bactericial/permeability-increasing protein—lessons learned from the phase III, randomized, clinical trial of $rBPI_{21}$ for adjunctive treatment of children with severe meningococcemia, *Crit. Care Med.*, 29, S130–S135, 2001.

Gordon, Y.J. et al., Human cathelicidin (LL-37), a multifunctional peptide, is expressed by ocular surface epithelia and has potent antibacterial and antiviral activity, *Curr. Eye Res.*, 30, 1–10, 2005.

Grigat, J. et al., Chemoattraction of macrophages, T lymphocytes, and mast cells is evolutionarily conserved within the human alpha-defensin family, *J. Immunol.*, 179, 3958–3965, 2007.

Groeneveld, T.W. et al., Human neutrophil peptide-1 inhibits both the classical and the lectin pathway of complement activation, *Mol. Immunol.*, 44, 3608–3614, 2007.

Gropp, R., Frye, M., Wagner, T.O.F., and Bargon, J., Epithelial defensins impair adenoviral infection: Implication for adenovirus-mediated gene therapy, *Human Gene Ther.*, 10, 957–964, 1999.

Grutkoski, P.S. et al., Alpha-defensin 1 (human neutrophil protein 1) as an antichemotactic agent for human polymorphonuclear leukocytes, *Antimicrob. Agents Chemother.*, 47, 2666–2668, 2003.

Gudmundsson, G.H. et al., The human gene FALL39 and processing of the cathelin precursor to the antibacterial peptide LL-37 in granulocytes, *Eur. J. Biochem.*, 238, 325–332, 1996.

Guo, C.-J. et al., Alpha-defensins inhibit HIV infection of macrophages through upregulation of CC-chemokines, *AIDS*, 18, 1217–1228, 2004.

Hancock, R.E.W. and Sahl, H.-G., Antimicrobial and host-defense peptides as new anti-infective therapeutic strategies, *Nature Biotechnol.*, 24, 1551–1557, 2006.

Harder, J., Bartels, J., Christophers, E., and Schroder, J.M., A pepide antibiotic from human skin, *Nature*, 387, 861, 1997.

Harder, J. et al., Mucoid *Pseudomonas aeruginosa*, TNF alpha, and IL-beta, but not IL-6, induce human beta-defensin-2 in repiratory epithelia, *Am. J. Respir. Cell Mol. Biol.*, 22, 714–721, 2000.

Harder, J., Bartels, J., Christophers, E., and Schroder, J.M., Isolation and characterization of human-beta-defensin-3, a novel inducible peptide antibiotic, *J. Biol. Chem.*, 276, 5707–5713, 2001.

Harwig, S.S., Park, A.S., and Lehrer, R.I., Characterization of defensin precursors in mature human neutrophils, *Blood*, 79, 1532–1537, 1992.

Harwig, S.S. et al., Gallinacins L cysteine-rich antimicrobial peptides of chicken leukocytes, *FEBS Lett.*, 342, 281–285, 1994.

Hattenbach, L.O., Gumbel, H., and Kippenberger, S., Identification of beta-defensins in human conjunctiva, *Antimicrob. Agents Chemother.*, 42, 3332, 1998.

Haynes, R.J., Tighe, P.J., and Dua, H.S., Antimicrobial defensin peptides of the human ocular surface, *Br. J. Ophthalmol.*, 83, 737–741, 1999.

Hazrati, E. et al., Human alpha- and beta-defensins block multiple steps in herpes simplex virus infection, *J. Immunol.*, 177, 8658–8666, 2006.

He, K., Ludtke, S.J., Heller, W.T., and Huang, H.W., Mechanism of alamethicin insertion into lipid bilayers, *Biophys. J.*, 71, 2669–2679, 1996.

Heilborn, J.D. et al., The cathelicidin anti-microbial peptide LL-37 is involved in re-epithelialization of human skin wounds and is lacking in chronic ulcer epithelium, *J. Invest. Dermatol.*, 120, 379–389, 2003.

Heilborn, J.D. et al., Antimicrobial protein hCAP18/LL-37 is highly expressed in breast cancer and is a putative growth factor for epithelial cells, *Int. J. Cancer*, 114, 713–719, 2005.

Helmerhorst, E.J., Troxler, R.F., and Oppenheim, F.G., The human salivary peptide histatin 5 exerts its antifungal activity through formation of reactive oxygen species, *Proc. Natl Acad. Sci. USA*, 98, 14637–14642, 2001.

Henzler Wildman, K.A., Lee, D.K., and Ramamoorthy, A., Mechanism of lipid bilayer disruption by the human antimicrobial peptide LL-37, *Biochemistry*, 42, 6545–6558, 2003.

Hersh, C.P. et al., Genetic linkage and association analysis of COPD-related traits on chromosome 8p, *COPD*, 3, 189–194, 2006.

Higazi, A.A.-R. et al., Defensin stimulates the binding of lipoprotein (a) to human vascular endothelial and smooth muscle cells, *Blood*, 89, 4290–4298, 1997.

Hirsch, J.G., Phagocytin: A bactericidal substance from polymorphonuclear leucocytes, *J. Exp. Med.*, 103, 589–611, 1956.

Hollox, E.J., Armour, J.A.L., and Barber, J.C.K., Extensive normal copy number variation of a β-defensin antimicrobial-gene cluster, *Am. J. Hum. Genet.*, 73, 591–600, 2003.

Hollox, E.J. et al., Psoriasis is associated with increased beta-defensin genomic copy number, *Nat. Genet.*, 40, 23–25, 2008.

Holterman, D.A. et al., Overexpression of alpha-defensin is associated with bladder cancer invasiveness, *Urol. Oncol.*, 24, 97–108, 2006.

Hoover, D.M. et al., The structure of human beta-defensin-2 shows evidence of higher order oligomerization, *J. Biol. Chem.*, 275, 32911–32918, 2000.

Hoover, D.M., Wu, Z., Tucker, K., Lu, W., and Lubkowski, J., Antimicrobial characterization of human β-defensin 3 derivatives, *Antimicrob. Agents Chemother.*, 47, 2804–2809, 2003.

Hoskin, D.W. and Ramamoorthy, A., Studies on anticancer activities of antimicrobial peptides, *Biochim. Biophys. Acta*, 1778, 357–375, 2008.

Howell, M.D. et al., Selective killing of vaccinia virus by LL-37: Implications for eczema vaccinatum, *J. Immunol.*, 172, 1763–1767, 2004.

Howell, M.D., The role of human beta defensins and cathelicidins in atopic dermatitis, *Curr. Opin. Allergy Clin. Immunol.*, 7, 413–417, 2007.

Hu, R.C. et al., Correlation of HDEFB1 polymorphism and susceptibility to chronic obstructive pulmonary disease in Chinese Han population, *Chin. Med. J. (England)*, 117, 1637–1641, 2004.

Huang, H.W., Action of antimicrobial peptides: Two-state model, *Biochemistry*, 39, 8347–8352, 2000.

Huang, L.C. et al., Multifunctional roles of human cathelicidin (LL-37) at the ocular surface, *Invest. Ophthalmol. Vis. Sci.*, 47, 2369–2380, 2006.

Huang, L.C. et al., Ocular surface expression and *in vitro* activity of antimicrobial peptides, *Curr. Eye Res.*, 32, 595–609, 2007a.

Huang, L.C., Reins, R.Y., Gallo, R.L., and McDermott, A.M., Cathelicidin-deficient (Cnlp-/-) mice show increased susceptibility to *Pseudomonas aeruginosa* keratitis, *Invest Ophthalmol. Vis. Sci.*, 48, 4498–4508, 2007b.

Huang, L.C. et al., *In vitro* activity of human beta-defensin 2 against *Pseudomonas aeruginosa* in the presence of tear fluid, *Antimicrob. Agents Chemother.*, 51, 3853–3860, 2007c.

Hubert, P. et al., Defensins induce the recruitment of dendritic cells in cervical human papillomavirus-associated (pre)neoplastic lesions formed *in vitro* and transplanted *in vivo*, *FASEB J.*, 21, 2765–2775, 2007.

Ihi, T., Nakazato, M., Mukae, H., and Matsukura, S., Elevated concentrations of human neutrophil peptides in plasma, blood, and body fluids from patients with infections, *Clin. Infect. Dis.*, 25, 1134–1140, 1997.

Iimura, M. et al., Cathelicidin mediates innate intestinal defense against colonization with epithelial adherent bacterial pathogens, *J. Immunol.*, 174, 4901–4907, 2005.

Jacob, L. and Zasloff, M., Potential therapeutic applications of magainins and other antimicrobial agents of animal origin, *Ciba Found Symp.*, 186, 197–216, 1994.

Jia, H.P. et al., Abundant human beta-defensin-1 expression in milk and mammary gland epithelium, *J. Pediatr.*, 138, 109–112, 2001.

Jin, T. et al., *Staphylococcus aureus* resists human defensins by producing staphylokinase, a novel bacterial evasion mechanism, *J. Immunol.*, 172, 1169–1176, 2004.

Johansson, J. et al., Conformation-dependent antibacterial activity of the naturally occurring human peptide LL-37, *J. Biol. Chem.*, 273, 3718–3724, 1998.

Jones, D.E. and Bevins, C.L., Paneth cells of the human small intestine express an antimicrobial peptide gene, *J. Biol. Chem.*, 267, 23216–23225, 1992.

Jones, D.E. and Bevins, C.L., Defensin-6 mRNA in human Paneth cells: Implications for antimicrobial peptides in host defense of the human bowel, *FEBS Lett.*, 315, 187–192, 1993.

Jurevic, R.J. et al., Single-nucleotide polymorphisms (SNPs) in human beta-defensin 1: High-throughput SNP assays and association with Candida carriage in type I diabetics and nondiabetic controls, *J. Clin. Microbiol.*, 41, 90–96, 2003.

Kaeberlein, T., Lewis, K., and Epstein, S.S., Isolating "uncultivable" microorganisms in pure culture in a simulated natural environment, *Science*, 296, 1127–1129, 2002.

Kandler, K. et al., The anti-microbial peptide LL-37 inhibits the activation of dendritic cells by TLR ligands, *Int. Immunol.*, 18, 1729–1736, 2006.

Kang, S.W. et al., CRAMP analog having potent antibiotic activity without hemolytic activity, *Protein Pept. Lett.*, 9, 275–282, 2002.

Kao, C.Y., Chen, Y., Zhao, Y.H., and Wu, R., ORFeome-based search of airway epithelial cell-specific novel human beta-defensin genes, *Am. J. Resp. Cell Mol. Biol.*, 29, 71–80, 2003.

Kaplan, S.S., Heine, R.P., and Simmons, R.L., Defensins impair phagocytic killing by neutrophils in biomaterial-related infection, *Infect. Immunol.*, 67, 1640–1645, 1999.

Kavanagh, K. and Dowd, S., Histatins: Antimicrobial peptides with therapeutic potential, *J. Pharm. Pharmacol.*, 56, 285–289, 2004.

Khine, A.A. et al., Human neutrophil peptides induce interleukin-8 production through the P2Y6 signaling pathway, *Blood*, 107, 2936–2942, 2006.

Klotman, M.E. and Chang, T.L., Defensins in innate antiviral immunity, *Nat. Rev. Immunol.*, 6, 447–456, 2006.

Kluver, E. et al., Structure–activity relation of human beta-defensin 3: Influence of disulfide bonds and cysteine substitution on antimicrobial activity and cytotoxicity, *Biochemistry*, 44, 9804–9816, 2005.

Koczulla, R. et al., An angiogenic role for the human peptide antibiotic LL-37/hCAP-18, *J. Clin. Invest.*, 111, 1665–1672, 2003.

Kokryakov, V.N. et al., Protegrins: Leukocyte antimicrobial peptides that combine features of corticostatic defensins and tachyplesins, *FEBS Lett.*, 327, 231–236, 1993.

Koshlukova, S.E., Lloyd, T.L., Araujo, M.W., and Edgerton, M., Salivary histatin 5 induces nonlytic release of ATP from *Candida albicans* leading to cell death, *J. Biol. Chem.*, 274, 18872–18879, 1999.

Kougias, P. et al., Defensins and cathelicidins: Neutrophil peptides with roles in inflammation, hyperlipidemia and atherosclerosis, *J. Cell Mol. Med.*, 9, 3–10, 2005.

Kragol, G. et al., The antibacterial peptide pyrrhocoricin inhibits the ATPase actions of DnaK and prevents chaperone-assisted protein foldin, *Biochemistry*, 40, 3016–3026, 2001.

Kraus, D. and Peschel, A., Molecular mechanisms of bacterial resistance to antimicrobial peptides, *Curr. Top. Microbiol. Immunol.*, 306, 231–250, 2006.

Krause, A. et al., LEAP-1, a novel highly disulfide-bonded human peptide, exhibits antimicrobial activity, *FEBS Lett.*, 480, 147–150, 2000.

Krause, A. et al., Isolation and biochemical characterization of LEAP-2, a novel blood peptide expressed in the liver, *Protein Sci.*, 12, 143–152, 2003.

Kurosaka, K. et al., Mouse cathelin-related antimicrobial peptide chemoattracts leukocytes using formyl peptide receptor-like 1/mouse formyl peptide receptor-like 2 as the receptor and acts an as immune adjuvant, *J. Immunol.*, 174, 6257–6265, 2005.

Lai, R., Liu, H., Hui Lee, W., and Zhang, Y., An anionic antimicrobial peptide from toad *Bombina maxima*, *Biochem. Biophys. Res. Commun.*, 295, 796–799, 2002.

Lamb, H.M. and Wiseman, L.R., Pexiganan acetate, *Drugs*, 56, 1047–1052; discussion 1053–1044, 1998.

Lande, R. et al., Plasmacytoid dendritic cells sense self-DNA coupled with antimicrobial peptide, *Nature*, 449, 564–569, 2007.

Larrick, J.W. et al., Complementary DNA sequence of rabbit CAP18—a unique lipopolysaccharide binding protein, *Biochem. Biophys. Res. Commun.*, 179, 170–175, 1991.

Larrick, J. et al., Human CAP 18: A novel antimicrobial lipopolysaccharide-binding protein, *Infect. Immunol.*, 63, 1291–1297, 1995.

Lau, Y.E. et al., Interaction and cellular localization of the human host defense peptide LL-37 with lung epithelial cells, *Infect. Immunol.*, 73, 583–591, 2005.

Lau, Y.E. et al., Apoptosis of airway epithelial cells: Human serum sensitive induction by the cathelicidin LL-37, *Am. J. Respir. Cell Mol. Biol.*, 34, 399–409, 2006.

Laube, D.M. et al., Antimicrobial peptides in the airway, *Curr. Top. Microbiol. Immunol.*, 306, 153–182, 2006.

Lay, F.T., Brugliera, F., and Anderson, M.A., Isolation and properties of floral defensins from ornamental tobacco and petunia, *Plant Physiol.*, 131, 1283–1293, 2003.

Lee, J.Y. et al., Antibacterial peptides from pig intestine: Isolation of a mammalian cecropin, *Proc. Natl Acad. Sci. USA*, 86, 9159–9162, 1989.

Lee, P.H. et al., HB-107, a nonbacteriostatic fragment of the antimicrobial peptide cecropin B, accelerates murine wound repair, *Wound Repair Regen.*, 12, 351–358, 2004.

Lee, P.H. et al., Expression of an additional cathelicidin antimicrobial peptide protects against bacterial skin infection, *Proc. Natl Acad. Sci. USA*, 102, 3750–3755, 2005.

Lehmann, O.J., Hussain, I.R., and Watt, P.J., Investigation of beta-defensin gene expression in the ocular anterior segment by semiquantitative RT-PCR, *Br. J. Ophthalmol.*, 84, 523–526, 2000.

Lehmann, J. et al., Antitumour effect of the antimicrobial peptide magainin II against bladder cancer cell lines, *Eur. Urol.*, 50, 141–147, 2006.

Lehrer, R.I. et al., Interaction of human defensins with *Escherichia coli.* Mechanism of bactericidal activity, *J. Clin. Invest.*, 84, 553–561, 1989.

Lehrer, R.I., Lichtenstein, A.K., and Ganz, T., Defensins: Antimicrobial and cytotoxic peptides of mammalian cells, *Annu. Rev. Immunol.*, 11, 105–128, 1993.

Leikina, E. et al., Carbohydrate-binding molecules inhibit viral fusion and entry by crosslinking membrane glycoproteins, *Nat. Immunol.*, 6, 995–1001, 2005.

Leonova, L. et al., Circular minidefensins and posttranslational generation of molecular diversity, *J. Leukoc. Biol.*, 70, 461–464, 2001.

Leung, T.F. et al., Asthma and atopy are associated with DEFB1 polymorphisms in Chinese children, *Genes Immunol.*, 7, 59–64, 2006.

Levin, M. et al., Recombinant bactericial/permeability-increasing protein (rBPI21) as adjunctive treatment for children with severe meningococcal sepsis: A randomized trial, *Lancet*, 356, 961–967, 2000.

Levy, H. et al., Association of defensin beta-1 gene polymorphisms with asthma, *J. Allergy Clin. Immunol.*, 115, 252–258, 2005.

Li, X.S., Reddy, M.S., Baev, D., and Edgerton, M., *Candida albicans* Ssa1/2p is the cell envelope binding protein for salivary histatin 5, *J. Biol. Chem.*, 278, 28553–28561, 2003.

Li, X. et al., Solution structures of human LL-37 fragments and NMR-based identification of a minimal membrane-targeting antimicrobial and anticancer region, *J. Am. Chem. Soc.*, 128, 5776–5785, 2006.

Li, Q., Kumar, A., Gui, J.F., and Yu, F.S., *Staphylococcus aureus* lipoproteins trigger human corneal epithelial innate response through toll-like receptor-2, *Microb. Pathog.*, 44, 426–434, 2008.

Linzmeier, R., Michaelson, D., Liu, L., and Ganz, T., The structure of neutrophil defensin genes, *FEBS Lett.*, 321, 267–273, 1993.

Linzmeier, R., Ho, C.H., Hoang, B.V., and Ganz, T., A 450-kb contig of defensin genes on human chromosome 8p23, *Gene*, 233, 205–211, 1999.

Linzmeier, R.M. and Ganz, T., Human defensin gene copy number polymorphisms: Comprehensive analysis of independent variation in α and β-defensin regions at 8p22–p23, *Genomics*, 86, 423–430, 2005.

Linzmeier, R.M. and Ganz, T., Copy number polymorphisms are not a common feature of innate immune genes, *Genomics*, 88, 122–126, 2006.

Liu, L. and Ganz, T., The pro region of human neutrophil defensin contains a motif that is essential for normal subcellular sorting, *Blood*, 85, 1095–1103, 1995.

Liu, L., Zhao, C., Heng, H.H., and Ganz, T., The human beta-defensins-1 and alpha defensins are encoded by adjacent genes: Two peptide families with differing disulfide topology share a common ancestry, *Genomics*, 43, 316–320, 1997.

Liu, L. et al., Structure and mapping of the human beta-defensin HBD-2 gene and its expression at sites of inflammation, *Gene*, 222, 237–244, 1998.

Liu, C.Y. et al., The concentration-dependent chemokine release and pro-apoptotic effects of neutrophil-derived alpha-defensin-1 on human bronchial and alveolar epithelial cells, *Life Sci.*, 80, 749–758, 2007.

Liu, S. et al., Linear analogs of human beta-defensin 3: Concepts for design of antimicrobial peptides with reduced cytotoxicity to mammalian cells, *Chembiochem.*, 9, 964–972, 2008.

Lopez-Garcia, B., Lee, P.H., Yamasaki, K., and Gallo, R.L., Anti-fungal activity of cathelicidins and their potential role in *Candida albicans* skin infection, *J. Invest. Dermatol.*, 125, 108–115, 2005.

Lundy, F.T. et al., Antimicrobial activity of truncated alpha-defensins (human neutrophil peptide (HNP)-1) analogues without disulfide bridges, *Mol. Immunol.*, 45, 190–193, 2008.

Lupetti, A., Welling, M.M., Pauwels, E.K., and Nibbering, P.H., Radiolabelled antimicrobial peptides for infection detection, *Lancet Infect. Dis.*, 3, 223–229, 2003.

Mader, J.S. and Hoskin, D.W., Cationic antimicrobial peptides as novel cytotoxic agents for cancer treatment, *Expert Opin. Investig. Drugs*, 15, 933–946, 2006.

Mallow, E.B. et al., Human enteric defensins. Gene structure and developmental expression, *J. Biol. Chem.*, 271, 4038–4045, 1996.

Marr, A.K., Gooderham, W.J., and Hancock, R.E.W., Antibacterial peptides for therapeutic use: Obstacles and realistic outlook, *Curr. Opin. Pharmacol.*, 6, 468–472, 2006.

Mars, W.M. et al., Inheritance of unequal numbers of the genes encoding the human neutrophil defensins HP-1 and HP-3, *J. Biol. Chem.*, 270, 30371–30376, 1995.

Matsushita, I. et al., Genetic variants of human beta-defensin-1 and chronic obstructive pulmonary disease, *Biochem. Biophys. Res. Commun.*, 291, 17–22, 2002.

Matsuzaki, K., Sugishita, K., Fujii, N., and Miyajima, K., Molecular basis for membrane selectivity of an antimicrobial peptide, magainin 2, *Biochemistry*, 34, 3423–3429, 1995.

Matsuzaki, K., Murase, O., Fujii, N., and Miyajima, K., An antimicrobial peptide, magainin 2, induced rapid flip-flop of phospholipids coupled with pore formation and peptide translocation, *Biochemistry*, 35, 11361–11368, 1996.

Matsuzaki, K., Magainins as paradigm for the mode of action of pore forming polypeptides, *Biochim. Biophys. Acta*, 1376, 391–400, 1998.

McCray, P.B. and Bentley, L., Human airway epithelia express a beta-defensin, *Am. J. Resp. Cell Mol. Biol.*, 16, 343–349, 1997.

McDermott, A.M., Redfern, R.L., and Zhang, B., Human β-defensin 2 is up-regulated during re-epithelialization of the cornea, *Curr. Eye Res.*, 22, 64–67, 2001.

McDermott, A.M. et al., Defensin expression by the cornea: Multiple signaling pathways mediate IL-1beta stimulation of hBD-2 expression by corneal epithelial cells, *Invest Ophthalmol. Vis. Sci.*, 44, 1859–1865, 2003.

McIntosh, R.S. et al., The spectrum of antimicrobial peptide expression at the ocular surface, *Invest. Ophthalmol. Vis. Sci.*, 46, 1379–1385, 2005.

McNamara, N.A., Van, R., Tuchin, O.S., and Fleiszig, S.M.J., Ocular surface epithelia express mRNA for human beta defensin-2, *Exp. Eye Res.*, 69, 483–490, 1999.

McPhee, J.B., Scott, M.G., and Hancock, R.E.W., Design of host defence peptides for antimicrobial and immunity enhancing activities, *Comb. Chem. High Throughput Screen*, 8, 257–272, 2005.

Miteva, M., Anderson, M., Karshikoff, A., and Otting, G., Molecular electroporation: A unifying concept for the description of membrane pore formation by antibacterial peptides, exemplified with NK-lysin, *FEBS Lett.*, 462, 155–158, 1999.

Miyata, T. et al., Antimicrobial peptides, isolated from horseshoe crab hemocytes, tachyplesin II, and polyphemusins I and II: Chemical structures and biological activity, *J. Biochem. (Tokyo)*, 106, 663–668, 1989.

Mookherjee, N. et al., Modulation of the TLR-mediated inflammatory response by the endogenous human host defense peptide LL-37, *J. Immunol.*, 176, 2455–2464, 2006.

Moore, A., The big and small of drug discovery. Biotech versus pharma: Advantages and drawbacks in drug development, *EMBO Rep.*, 4, 114–117, 2003.

Morikawa, N., Hagiwara, K., and Nakajima T., Brevinin-1 and -2, unique antimicrobial peptides from the skin of the frog, *Rana brevipoda porsa*, *Biochem. Biophys. Res. Commun.*, 189, 184–190, 1992.

Morrison, G., Kilanowski, F., Davidson, D., and Dorin, J., Characterisation of the mouse β-defensin 1, Defb1, mutant mouse model, *Infect. Immunol.*, 70, 3053–3060, 2002.

Moser, C. et al., β-Defensin 1 contributes to pulmonary innate immunity in mice, *Infect. Immunol.*, 70, 3068–3072, 2002.

Muller, C.A. et al., Human alpha-defensins HNPs-1, -2, -3 in renal cell carcinoma: Influences on tumour cell proliferation, *Am. J. Pathol.*, 160, 1311–1324, 2002.

Murakami, M. et al., Cathelicidin anti-microbial peptide expression in sweat, an innate defence system for the skin, *J. Invest. Dermatol.*, 119, 1090–1095, 2002.

Murakami, M. et al., Postsecretory processing generates multiple cathelicidins for enhanced topical antimicrobial defense, *J. Immunol.*, 172, 3070–3077, 2004.

Murphy, C.J. et al., Defensins are mitogenic for epithelial cells and fibroblasts, *J. Cell Physiol.*, 155, 408–443, 1993.

Mygind, P.H. et al., Plectasin is a peptide antibiotic with therapeutic potential from a saprophytic fungus, *Nature*, 437, 975–980, 2005.

Nagaoka, I. et al., Synergistic actions of antibacterial neutrophil defensins and cathelicidins, *Inflamm. Res.*, 49, 73–79, 2000.

Nagaoka, I. et al., Augmentation of the bactericidal activities of human cathelicidin CAP18/LL-37-derived antimicrobial peptides by amino acid substitutions, *Inflamm. Res.*, 5, 66–73, 2005.

Nagaoka, I., Tamura, H., and Hirata, M., An antimicrobial cathelicidin peptide, human CAP18/LL-37 suppresses neutrophil apoptosis via activation of formyl-peptide receptor-like 1 and P2X7, *J. Immunol.*, 176, 3044–3052, 2006.

Nam, M.J., Kee, M.K., Kuick, R., and Hanash, S.M., Identification of defensin alpha 6 as a potential biomarker in colon adenocarcinoma, *J. Biol. Chem.*, 280, 8260–8265, 2005.

Narayanan, S., Miller, W.L., and McDermott, A.M., Expression of human β-defensins in conjunctival epithelium: Relevance to dry eye disease, *Invest Ophthalmol. Vis. Sci.*, 44, 3795–3801, 2003.

Neville, F. et al., Lipid headgroup discrimination by antimicrobial peptide LL-37: Insight into mechanism of action, *Biophys. J.*, 90, 1275–1287, 2006.

Nguyen, T.X., Cole, A.M., and Lehrer, R.I., Evolution of primate theta-defensins: A serpentine path to a sweet tooth, *Peptides*, 24, 1647–1654, 2003.

Nishimura, M. et al., Effect of defensin peptides on eukaryotic cells: Primary epithelial cells, fibroblasts and squamous cell carcinoma cell lines, *J. Dermatol. Sci.*, 36, 87–95, 2004.

Niyonsaba, F. et al., Evaluation of the effects of peptide antibiotics human beta-defensins-1/-2 and LL-37 on histamine release and prostaglandin D(2) production from mast cells, *Eur. J. Immunol.*, 31, 106675, 2001.

Niyonsaba, F. et al., Epithelial cell-derived human beta-defensin-2 acts as a chemotaxin for mast cells through a pertussis toxin-sensitive and phospholipase C-dependent pathway, *Int. Immunol.*, 14, 421–426, 2002a.

Niyonsaba, F. et al., A cathelicidin family of human antibacterial peptide LL-37 induces mast cell chemotaxis, *Immunology*, 106, 20–26, 2002b.

Niyonsaba, F., Ogawa, H., and Nagaoka, I., Human beta-defensin-2 functions as a chemotactic agent for tumour necrosis factor-alpha-treated human neutrophils, *Immunology*, 111, 273–281, 2004.

Niyonsaba, F. et al., The human beta-defensins (-1, -2, -3, -4) and cathelicidin LL-37 induce IL-18 secretion through p38 and ERK MAPK activation in primary human keratinocytes, *J. Immunol.*, 175, 1776–1784, 2005.

Niyonsaba, F. et al., Antimicrobial peptides human beta-defensins stimulate epidermal keratinocyte migration, proliferation and production of proinflammatory cytokines and chemokines, *J. Invest. Dermatol.*, 127, 594–604, 2007.

Nizet, V. et al., Innate antimicrobial peptide protects the skin from invasive bacterial infection, *Nature*, 414, 454–457, 2001.

Obata-Onai, A. et al., Comprehensive gene expression analysis of human NK cella and CD8(+) T lymphocytes, *Int. Immunol.*, 14, 1085–1098, 2002.

Oh, J. et al., Cationic peptide antimicrobials induce selective transcription of micF and osmY in *Escherichia coli*, *Biochim. Biophys. Acta*, 1463, 43–54, 2000.

O'Neil, D.A., Porter E.M., and Elewaut, D., Expression and regulation of the human beta-defensins hBD-1 and hBD-2 in intestinal epithelium, *J. Immunol.*, 163, 6718–6724, 1999.

Ooi, E.H. et al., Fungal allergens induce cathelicidin LL-37 expression in chronic rhinosinusitis patients in a nasal explant model, *Am. J. Rhinol.*, 21, 367–372, 2007.

Oppenheim, F.G. et al., Histatins, a novel family of histidine-rich proteins in human parotid secretion. Isolation, characterization, primary structure, and fungistatic effects on *Candida albicans*, *J. Biol. Chem.*, 263, 7472–7477, 1988.

Oren, Z. et al., Structure and organization of the human antimicrobial peptide LL-37 in phospholipids membranes: Relevance to the molecular basis for its non-cell-selective activity, *Biochem. J.*, 341, 501–513, 1999.

Oren, A., Ganz, T., Liu, L., and Meerloo, T., In human epidermis, β-defensin 2 is packaged in lamellar bodies, *Exp. Mol. Pathol.*, 74, 180–182, 2003.

Otvos, L. Jr O, I., Rogers, M.E. et al., Interaction between heat shock proteins and antimicrobial peptides, *Biochemistry*, 39, 14150–14159, 2000.

Ouellette, A.J. et al., Developmental regulation of cryptdin, a corticostatin/defensin precursor mRNA in mouse small intestinal crypt epithelium, *J. Cell Biol.*, 108, 1687–1695, 1989.

Ouellette, A.J. and Lualdi, J.C., A novel mouse gene family coding for cationic, cysteine-rich peptides. Regulation in small intestine and cells of myeloid origin, *J. Biol. Chem.*, 265, 9831–9837, 1990.

Panyutich, A. and Ganz, T., Activated α2-macroglobulin is a principal defensin-binding protein, *Am. J. Respir. Cell Mol. Biol.*, 5, 101–106, 1991.

Panyutich, A.V., Hiemstra, P.S., van Wetering, S., and Ganz, T., Human neutrophil defensin and serpins form complexes and inactivate each other, *Am. J. Respir. Cell Mol. Biol.*, 12, 351–357, 1995.

Panyutich, A. et al., Porcine polymorphonuclear leukocytes generate extracellular microbicidal activity by elastase-mediated activation of secreted proprotegrins, *Infect. Immunol.*, 65, 978–985, 1997.

Paone, G. et al., ADP ribosylation of human neutrophil peptide-1 regulates its biological activity, *Proc. Natl Acad. Sci. USA*, 99, 8231–8235, 2002.

Paone. G. et al., ADP-ribosyltransferase-specific modification of human neutrophil peptide-1, *J. Biol. Chem.*, 128, 17054–17060, 2006.

Papo, N. and Shai, Y., Host defense peptides as new weapons in cancer treatment, *Cell Mol. Life Sci.*, 62, 784–790, 2005.

Park, C.H., Valore, E.V., Waring, A.J., and Ganz, T., Hepcidin, a urinary antimicrobial peptide synthesized in the liver, *J. Biol. Chem.*, 276, 7806–7810, 2001.

Patil, A.A. et al., Cross-species analysis of the mammalian beta-defensin gene family: Presence of syntenic clusters and preferential expression in the male reproductive tract, *Physiol. Genomics*, 23, 5–17, 2005.

Patrzykat, A. et al., Sublethal concentrations of pleurocidin-derived antimicrobial peptides inhibit macromolecular synthesis in *Escherichia coli*, *Antimicrob. Agents Chemother.*, 46, 605–614, 2002.

Paulsen, F.P. et al., Detection of natural peptide antibiotics in human nasolacrimal ducts, *Invest. Ophthalmol. Vis. Sci.*, 42, 2157–2163, 2001.

Petterson, A., Ueber die bakterisiziden leukocytenstoffe und ihre beziehung zur immuninitat, *Centr. Bakteriol. Parsitenk. Abt. I*, 39, 423–437, 1905.

Pkorny, A. and Almeida, P.F.F., Kinetics of dye efflux and lipid flip-flop induced by delta-lysin in phosphatidylcholine vesicle and the mechanism of graded release by amphipathic, alpha-helical peptides, *Biochemistry*, 43, 8846–8857, 2004.

Poyart, C. et al., Attenuated virulence of *Streptococcus agalactiae* deficient in D-alanyl-liopteichoic acid is due to an increased susceptibility to defensins and phagocytic cells, *Mol. Microbiol.*, 49, 1615–1625, 2003.

Prado-Montes de Oca, E. et al., Association of beta-defensin 1 single nucleotide polymorphisms with atopic dermatitis, *Int. Arch. Allergy. Immunol.*, 142, 211–218, 2007.

Premratanachai, P. et al., Expression and regulation of novel human beta-defensins in gingival keratinocytes, *Oral Microbiol. Immunol.*, 19, 111–117, 2004.

Prohaszka, Z. et al., Defensins purified from human granulocytes bind C1q and activate the classical complement pathway like the transmembrane glycoprotein gp41 of HIV, *Mol. Immunol.*, 34, 809–816, 1997.

Putsep, K., Carlsson, G., Boman, H., and Andersson, M., Deficiency of antibacterial peptides in patients with morbus Kostmann: An observational study, *Lancet*, 360, 1144–1149, 2002.

Quayle, A.J. et al., Gene expression, immunolocalization and secretion of human defensin-5 in female reproductive tract, *Am. J. Pathol.*, 152, 1247–1258, 1998.

Quinones-Mateu, M.E. et al., Human epithelial beta-defensins 2 and 3 inhibit HIV-replication, *AIDS*, 17, F39–F48, 2003.

Radek, K. and Gallo, R.L., Antimicrobial peptides: Natural effectors of the innate immune system, *Semin. Immunopathol.*, 29, 27–43, 2007.

Radzishevsky, I.S. et al., Improved antimicrobial properties based on acyl-lysine oligomers, *Nature Biotechnol.*, 25, 657–659, 2007.

Rapaport, D. and Shai, Y., Interaction of fluorescently labeled pardaxin and its analogues with lipid bilayers, *J. Biol. Chem.*, 266, 23769–23775, 1991.

Ritonja, A., Kopitar, M., Jerala, R., and Turk, V., Primary structure of a new cysteine proteinase inhibitor from pig leukocytes, *FEBS Lett.*, 255, 211–214, 1989.

Rodriguez-Garcia, M. et al., Human immature monocyte-derived dendritic cells produce and secrete alpha defensins 1–3, *J. Leukoc. Biol.*, 82, 1143–1146, 2007.

Rodriguez-Jimenez, F.J. et al., Distribution of new human beta-defensin genes clustered on chromosome 20 in functionally different segments of epididymis, *Genomics*, 81, 175–183, 2003.

Romeo, D., Skerlavaj, B., Bolognesi, M., and Gennaro, R., Structure and bactericidal activity of an antibiotic dodecapeptide purified from bovine neutrophils, *J. Biol. Chem.*, 263, 9573–9575, 1988.

Ryan, L.K. et al., Detection of HBD1 peptide in peripheral blood mononuclear cell subpopulations by intracellular flow cytometry, *Peptides*, 24, 1785–1794, 2003.

Sader, H.S. et al., Omiganan pentahydrochloride (MBI 226), a topical 12 amino-acid-cationic peptide: Spectrum of antimicrobial activity and measurements of bactericidal activity, *Antimicrob. Agents Chemother.*, 48, 3112–3118, 2004.

Sakamoto, N. et al., Differential effects of α- and β-defensin on cytokine production by cultured human bronchial epithelial cells, *Am. J. Physiol. Lung Cell Mol. Physiol.*, 288, L508–L513, 2005.

Salvatore, M. et al., Alpha-defensin inhibits influenza virus replication by cell-mediated mechanism(s), *J. Infect. Dis.*, 196, 835–843, 2007.

Salzman, N.H. et al., Protection against enteric salmonellosis in transgenic mice expressing a human intestinal defensin, *Nature*, 422, 522–526, 2003.

Samakovlis, C. et al., The andropin gene and its products, a male-specific antibacterial peptide in *Drosophila melanogaster*, *EMBO J.*, 10, 163–169, 1991.

Sandgren, S. et al., The human antimicrobial peptide LL-37 transfers extracellular DNA plasmid to the nuclear compartment of mammalian cells via lipid rafts and proteogycan-dependent endocytosis, *J. Biol. Chem.*, 279, 17951–17956, 2004.

Schibli, D.J. et al., The solution structures of the human beta-defensins lead to a better understanding of the potent bactericidal activity of hBD-3 against *Staphylococcus aureus*, *J. Biol. Chem.*, 277, 8279–8289, 2002.

Schittek, B. et al., Dermcidin: A novel antibiotic peptide secreted by sweat glands, *Nature Immunol.*, 2, 1133–1137, 2001.

Schmidtchen, A., Frick, I.-M., and Bjork, L., Dermatan sulphate is released by proteinases of common pathogenic bacteria and inactivates antibacterial α-defensin, *Mol. Microbiol.*, 39, 708–713, 2001.

Schutte, B.C. et al., Discovery of five conserved beta-defensin gene clusters using a computational search strategy, *Proc. Natl Acad. Sci. USA*, 99, 2129–2133, 2002.

Scott, M.G. et al., The human antimicrobial peptide LL-37 is a multifunctional modulator of innate immune responses, *J. Immunol.*, 169, 3883–3891, 2002.

Scott, M.G. et al., An anti-infective peptide that selectively modulates the innate immune response, *Nature Biotechnol.*, 25, 465–472, 2007.

Selsted, M.E., Szklarek, D., and Lehrer, R.I., Purification and antibacterial activity of antimicrobial peptides of rabbit granulocytes, *Infect. Immunol.*, 45, 150–154, 1984.

Selsted, M.E. et al., Primary structures of six antimicrobial peptides of rabbit peritoneal neutrophils, *J. Biol. Chem.*, 260, 4579–4584, 1985a.

Selsted, M.E. et al., Primary structures of three human neutrophil defensins, *J. Clin. Invest.*, 76, 1436–1439, 1985b.

Selsted, M.E. et al., Indolicidin, a novel bactericidal tridecapeptide amide from neutrophils, *J. Biol. Chem.*, 267, 4292–4295, 1992.

Selsted, M.E. et al., Purification, primary structures, and antibacterial activities of beta-defensins, a new family of antimicrobial peptides from bovine neutrophils, *J. Biol. Chem.*, 268, 6641–6648, 1993.

Sevcsik, E., Pabst, G., Jilek, A., and Lohner, K., How lipids influence the mode of action of membrane active peptides, *Biochim. Biophys. Acta*, 1768, 2586–2595, 2007.

Shafer, W.M., Qu, X, Waring, A.J., and Lehrer, R.L., Modulation of *Neisseria gonorrhoeae* susceptibility to vertebrate antibacterial peptides due to a member of the resistance/modulation/division efflux pump family, *Proc. Natl Acad. Sci. USA*, 95, 1829–1833, 1998.

Shai, Y., Mode of action of membrane active antimicrobial peptides, *Biopolymers* (*Pept. Sci.*), 66, 236–248, 2002.

Shaykhiev, R. et al., Human endogenous antibiotic LL-37 stimulates airway epithelial cell proliferation and wound closure, *Am. J. Physiol. Lung Cell Mol. Physiol.*, 289, L842–L848, 2005.

Sieprawska-Lupa, M. et al., Degradation of human antimicrobial peptide LL-37 by *Staphylococcus aureus*-derived proteinases, *Antimicrob. Agents Chemother.*, 48, 4673–4679, 2004.

Sigurdardottir, T. et al., *In silico* identification and biological evaluation of antimicrobial peptides based on human cathelicidin LL-37, *Antimicrob. Agents Chemother.*, 50, 2983–2989, 2006.

Singh, P.K., Tack, B.F., McCray, P.B., and Welsh, M.J., Synergistic and additive killing by antimicrobial factors found in human airway surface liquid, *Am. J. Physiol. Lung Cell Mol. Physiol.*, 279, L799–L805, 2000.

Skerlavaj, B., Romeo, D., and Gennaro, R., Rapid membrane permeabilization and inhibition of vital functions gram-negative bacteria by bactenecins, *Infect. Immunol.*, 58, 3724–3730, 1990.

Sorensen, O.E. et al., Human cathelicidin, hCAP18, is processed to the antimicrobial peptide LL-37 by extracellular cleavage with proteinase 3, *Blood*, 97, 3951–3959, 2001.

Soruri, A. et al., Beta-defensins chemoattract macrophages and mast cells but not lymphocytes and dendritic cells: CCR6 is not involved, *Eur. J. Immunol.*, 37, 2474–2486, 2007.

Sousa, L.B. et al., The use of synthetic Cecropin (D5C) in disinfecting contact lens solutions, *CLAO J.*, 22, 114–117, 1996.

Spencer, L.T. et al., Role of human neutrophil peptides in lung inflammation associated with alpha1-antitrypsin deficiency, *Am. J. Physiol. Lung Cell Mol. Physiol.*, 286, L514–L520, 2004.

Spitznagel, J.K. and Chi, H.Y., Cationic proteins and antibacterial properties of infected tissues and leukocytes, *Am. J. Pathol.*, 43, 679–711, 1963.

Steinberg, D.A. et al., Protegrin-1: A broad spectrum, rapidly microbicidal peptide with *in vivo* activity, *Antimicrob. Agents Chemother.*, 41, 1738–1742, 1997.

Steiner, H. et al., Sequence and specificity of two antibacterial proteins involved in insect immunity, *Nature*, 292, 246–248, 1981.

Subbalakshmi, C. and Sitaram, N., Mechanism of antimicrobial action of indolicidin, *FEMS Microbiol. Lett.*, 160, 91–96, 1998.

Sun, L. et al., Human beta-defensins suppress human immunodeficiency virus infection: Potential role in mucosal protection, *J. Virol.*, 79, 14318–14329, 2005.

Sun, C.Q. et al., Human β-defensin-1, a potential chromosome 8p tumor suppressor: Control of transcription and induction of apoptosis in renal cell carcinoma, *Cancer Res.*, 66, 8542–8549, 2006.

Syeda, F. et al., Differential signaling mechanisms of HNP-induced IL-8 production in human lung epithelial cells and monocytes, *J. Cell Physiol.*, 214, 820–827, 2008.

Taggart, C.C. et al., Inactivation of human beta-defensins 2 and 3 by elastolytic cathepsins, *J. Immunol.*, 171, 931–937, 2003.

Tan, B.H. et al., Macrophages acquire neutrophil granules for antimicrobial activity against intracellular pathogens, *J. Immunol.*, 177, 1864–1871, 2006.

Tang, Y.Q. et al., A cyclic antimicrobial produced in primate leukocytes by the ligation of two truncated alpha-defensins, *Science*, 286, 498–502, 1999.

Tang, Y.Q., Yeaman, M.R., and Selsted, M.E., Antimicrobial peptides from human platelets, *Infect. Immunol.*, 277, 37647–37654, 2002.

Terras, F.R. et al., Small cysteine-rich antifungal proteins from radish: Their role in host defense, *Plant Cell*, 7, 573–588, 1995.

Territo, M.C., Ganz, T., Selsted, M.E., and Lehrer, R.I., Monocyte-chemotactic activity of defensins from human neutrophils, *J. Clin. Invest.*, 84, 2017–2020, 1989.

Tesse, R. et al., Association of beta-defensin-1 gene polymorphisms with *Pseudomonas aeruginosa* airway colonization in cystic fibrosis, *Genes Immunol.*, 9, 57–60, 2008.

Thevissen, K. et al., Fungal membrane responses induced by plant defensins and thionins, *J. Biol. Chem.*, 271, 15018–15025, 1996.

Thouzeau, C. et al., Spheniscins, avian beta-defensins in preserved stomach contents of the king penguin, *Aptenodytes patagonicus*, *J. Biol. Chem.*, 278, 51053–51058, 2003.

Tjabringa, G.S. et al., The antimicrobial peptide LL-37 activates innate immunity at the airway epithelial surface by transactivation of the epidermal growth factor receptor, *J. Immunol.*, 171, 6690–6696, 2003.

Tjabringa, G.S. et al., Human cathelicidin LL-37 is a chemoattractant for eosinophils and neutrophils that acts via formyl-peptide receptors, *Int. Arch. Allergy Immunol.*, 140, 103–112, 2006.

Tokumaru, S. et al., Induction of keratinocyte migration via transactivation of the epidermal growth factor recepor by the antimicrobial peptide LL-37, *J. Immunol.*, 175, 4662–4668, 2005.

Tran, D. et al., Homodimeric theta-defensins from Rhesus macaque leukocytes: Isolation, synthesis, antimicrobial activities, and bacterial binding properties of the cyclic peptides, *J. Biol. Chem.*, 277, 3079–3084, 2002.

Trotti, A. et al., A multinational, randomized phase III trial of iseganan HCL oral solution for reducing the severity of oral mucositis in patients receiving radiotherapy for head-and-neck malignancy, *Int. J. Radiat. Oncol. Biol. Phys.*, 58, 674–681, 2004.

Tytler, E.M. et al., Molecular basis for prokaryotic specificity of magainin-induced lysis, *Biochemistry*, 34, 4393–4401, 1995.

Valore, E.V. and Ganz, T., Posttranslational processing of defensins in immature human myeloid cells, *Blood*, 79, 1538–1544, 1992.

Valore, E.V., Martin, E., Harwig, S.S., and Ganz, T., Intramolecular inhibition of human defensin HNP-1 by its propiece, *J. Clin. Invest.*, 97, 1624–1629, 1996.

Valore, E.V. et al., Human beta-defensin-1: An antimicrobial peptide of urogenital tissues, *J. Clin. Invest.*, 101, 1633–1642, 1998.

van den Berg, R.H. et al., Inhibition of activation of the classical pathway of complement by human neutrophil defensins, *Blood*, 92, 3898–3903, 1998.

van't Hof, W., Veerman, E.C.I., Helmerhorst, E.J., and Amerongen, A.V., Antimicrobial peptides: Properties and applicability, *Biol. Chem.*, 382, 597–619, 2001.

van Wetering, S. et al., Effect of defensins on interleukin-8 synthesis in airway epithelial cells, *Am. J. Physiol.*, 272, L888–L896, 1997.

van Wetering, S., Mannesse-Lazeroms, S.P.G., van Sterkenburg, M.A., and Hiemstra, P.S., Neutrophil defensins stimulate the release of cytokines by airway epithelial cells: Modulation by dexamethasone, *Inflamm. Res.*, 51, 8–15, 2002.

Varkey, J. and Nagaraj, R., Antibacterial activity of human neutrophil defensin HNP-1 analogs without cysteines, *Antimicrob. Agents Chemother.*, 49, 4561–4666, 2005.

Verkleij, A.J. et al., The asymmetric distribution of phospholipids in the human red cell membrane. A combined study using phospholipases and freeze-etch electron microscopy, *Biochim. Biophys. Acta*, 323, 178–193, 1973.

von Haussen, J. et al., The host defence peptide LL-37/hCAP-18 is a growth factor for lung cancer cells, *Lung Cancer*, 59, 12–23, 2008.

Vylkova, S., Li, X.S., Berner, J.C., and Edgerton, M., Distinct antifungal mechanisms: Beta-defensins require *Candida albicans* Ssa1 protein, while Trk1p mediates activity of cysteine-free cationic peptides, *Antimicrob. Agents Chemother.*, 50, 324–331, 2006.

Vylkova, S., Nayyar, N., Li, W., and Edgerton, M., Human beta-defensins kill *Candida albicans* in an energy-dependent and salt-sensitive manner without causing membrane disruption, *Antimicrob. Agents Chemother.*, 51, 154–161, 2007.

Wada, A. et al., *Helicobacter pylori*-mediated transcriptional regulation of the human beta-defensin-2 gene requires NF-κB, *Cell Microbiol.*, 3, 115–123, 2001.

Wang, Y. et al., Apolipoprotein A-I binds and inhibits the human antibacterial/cytotoxic peptide LL-37, *J. Biol. Chem.*, 273, 33115–33118, 1998.

Wang, X. et al., Airway epithelia regulate expression of human beta-defensin-2 through Toll-like receptor 2, *FASEB J.*, 17, 1727–1729, 2003.

Wang, T.-T. et al., Cutting edge: 1,25-dihydroxyvitamin D_3 is a direct iducer of antimicrobial peptide gene expression, *J. Immunol.*, 173, 2909–2912, 2004a.

Wang, W. et al., Activity of alpha- and theta-defensins against primary isolates of HIV-1, *J. Immunol.*, 173, 515–520, 2004b.

Wehkamp, J. et al., Mechanisms of disease: Defensins in gastrointestinal diseases, *Nat. Clin. Pract. Gastroenterol. Hepatol.*, 2, 406–415, 2005.

Wehkamp, J. et al., Paneth cell antimicrobial peptides: Topographical distribution and quantification in human gastrointestinal tissues, *FEBS Lett.*, 580, 5344–5350, 2006.

Weigand, I., Hilpert, K., and Hancock, R.E.W., Agar and broth dilution methods to determine the minimal inhibitory concentration (MIC) of antimicrobial substances, *Nature Protoc.*, 3, 163–175, 2008.

Weiss, J., Elsbach, P., Olsson, I., and Odeberg, H., Purification and characterization of a potent bactericidal and membrane active protein from the granules of human polymorphonuclear leukocytes, *J. Biol. Chem.*, 253, 2664–2672, 1978.

Wencker, M. and Brantly, M.L., Cytotoxic concentrations of alpha-defensins in the lungs of individuals with alpha(1)-antitrypsin deficiency and moderate to severe lung disease, *Cytokine*, 32, 1–6, 2005.

Wilde, C.G. et al., Purification and characterization of human neutrophil peptide-4, a novel member of the defensin family, *J. Biol. Chem.*, 264, 11200–11203, 1989.

Wilson, C.L. et al., Regulation of intestinal α-defensin activation by the metalloproteinase matrilysin in innate host defence, *Science*, 286, 113–117, 1999.

Wimley, W.C., Selsted, M.E., and White, S.H., Interactions between human defensins and lipid bilayers: Evidence for formation of multimeric pores, *Protein Sci.*, 3, 1362–1373, 1994.

Wu, Z. et al., Engineering disulfide bridges to dissect antimicrobial and chemotactic activities of human beta-defensin-3, *Proc. Natl Acad. Sci. USA*, 100, 8880–8885, 2003.

Wu, Z. et al., Human neutrophil alpha-defensin 4 inhibits HIV-1 infection *in vitro*, *FEBS Lett.*, 579, 162–166, 2005.

Wu, Z. et al., Impact of pro segments on the folding and function of human neutrophil alpha-defensins, *J. Mol. Biol.*, 368, 537–549, 2007.

Yamaguchi, S. et al., Orientation and dynamics of an antimicrobial peptide in the lipid bilayer by solid-state NMR spectroscopy, *Biophys. J.*, 81, 2203–2214, 2001.

Yamaguchi, Y. et al., Identification of multiple novel epididymis specific beta-defensin isoforms in humans and mice, *J. Immunol.*, 169, 2516–2523, 2002.

Yamasaki, K. et al., Kallikrein-mediated proteolysis regulates the antimicrobial effects of cathelicidins in skin, *FASEB J.*, 20, 2068–2080, 2006.

Yamasaki, K. et al., Increased serine protease activity and cathelicidin promotes skin inflammation in rosacea, *Nat. Med.*, 13, 904–906, 2007.

Yan, H. and Hancock, R.E., Synergistic interactions between mammalian antimicrobial defense peptides, *Antimicrob. Agents Chemother.*, 45, 1558–1560, 2001.

Yang, D. et al., Beta-defensins: Linking innate and adaptive immunity through dendritic and T cell CCR6, *Science*, 286, 525–528, 1999.

Yang, D., Chen, Q., Chertov, O., and Oppenheim, J.J., Human neutrophil defensins selectively chemoattract naïve T and immature dendritic cells, *J. Leukoc. Biol.*, 68, 9–14, 2000.

Yang, L. et al., Barrel-stave model or toroidal model? A case study on melittin pores, *Biophys. J.*, 81, 1475–1485, 2001.

Yang, D. et al., Many chemokines including CCL20/MIP-3alpha display antimicrobial activity, *J. Leukoc. Biol.*, 74, 448–455, 2003.

Yang, Y.H. et al., The cationic host defense peptide rCRAMP promotes gastric ulcer healing in rats, *J. Pharmacol. Exp. Ther.*, 318, 547–554, 2006.

Yasin, B. et al., Evaluation of the inactivation of infectious herpes simplex virus by host-defense peptides, *Eur. J. Clin. Microbiol. Infect. Dis.*, 19, 187–194, 2000.

Yenugu, S. et al., The androgen-regulated epididymal sperm-binding protein, human beta-defensin 118 (DEFB118) (formerly ESC42) is an antimicrobial beta-defensin, *Endocrinology*, 145, 3165–3173, 2004.

Zaiou, M., Nizet, V., and Gallo, R.L., Antimicrobial and protease inhibitory functions of the human cathelicidin (hCAP18/LL-37) prosequence, *J. Invest. Dermatol.*, 120, 810–816, 2003.

Zanetti, M., Gennaro, R., and Romeo, D., Cathelicidins: A novel protein family with a common proregion and a variable C-terminal antimicrobial domain, *FEBS Lett.*, 374, 1–5, 1995.

Zasloff, M., Magainins, a class of antimicrobial peptides from Xenopus skin: Isolation, characterization of two active forms, and partial cDNA sequence of a precursor, *Proc. Natl Acad. Sci. USA*, 84, 5449–5453, 1987.

Zeya, H.I. and Spitznagel, J.K., Antibacterial and enzymatic basic proteins from leukocyte lysosomes: Separation and identification, *Science*, 142, 1085–1087, 1963.

Zeya, H.I. and Spitznagel, J.K., Cationic proteins of polymorphonuclear leukocyte lysosomes II. Composition, properties and mechanism of antibacterial action, *J. Bacteriol.*, 91, 755–762, 1966.

Zeya, H.I. and Spitznagel, J.K., Arginine-rich proteins of polymorphonuclear leukocyte lysosomes, *J. Exp. Med.*, 127, 927–941, 1968.

Zeya, H.I. and Spitznagel, J.K., Cationic protein-bearing granules of polymorphonuclear leukocytes: Separation from enzyme rich granules, *Science*, 163, 1069–1071, 1969.

Zhang, L. et al., Contribution of human alpha-defensins 1, 2, and 3 to the anti-HIV-1 activity of CD8 antiviral factor, *Science*, 298, 995–1000, 2002.

Zhang, L. et al., Antimicrobial peptide therapeutics for cystic fibrosis, *Antimicrob. Agents Chemother.*, 49, 2921–2927, 2005.

Zhang, S. et al., Effect of human beta defensin 2 on *Staphylococcus aureus* infection: *In vivo* study with transgenic mice, *Zhonghua Yi Xue Za Zhi*, 86, 1834–1836, 2006.

Zhang, L. and Falla, T.J., Antimicrobial peptides: Therapeutic potential, *Exp. Opin. Pharmacother.*, 7, 653–663, 2006.

Zheng, Y. et al., Cathelicidin LL-37 induces the generation of reactive oxygen species and release of human-alpha defensins from neutrophils, *Br. J. Dermatol.*, 157, 1124–1131, 2007.

Zhou, L. et al., Proteomic analysis of human tears: Defensin expression after ocular surface surgery, *J. Proteome Res.*, 3, 410–416, 2004.

Zhou, L. et al., Proteomic analysis of rabbit tear fluid: Defensin levels after an experimental corneal wound are correlated to wound closure, *Proteomics*, 7, 3194–3206, 2007.

Zou, H. et al., Detection of colorectal disease by stool defensin assay: An exploratory study, *Clin. Gastroenterol. Hepatol.*, 5, 865–868, 2007.

Zucht, H.D. et al., Human β-defensin-1: A urinary peptide present in variant molecular forms and its putative functional implication, *Eur. J. Med. Res.*, 3, 315–323, 1998.

Part IV

Venom Components and Toxins

17 Mast Cell Degranulating Peptides

Joseph A. Price

CONTENTS

Make everything as simple as possible, but not simpler.

—Albert Einstein (1879–1955)

17.1 INTRODUCTION

Unlike many of the other chapters in this publication, which focus on a peptide or family of closely related peptides, here our scope of discussion is functionally and source defined. We will only consider those venom protein/(poly)peptides with the particular functional property that they cause the activation of mast cells, leading to "degranulation," the exocytosis of preformed granules. This scope immediately leads to many interesting questions, whose interdisciplinary answers lie in the nature of venom, the physiology of mast cells, and in peptide chemistry. By examining these together we may gain the needed perspective to understand the phenomena of mast cell degranulating peptides. It is an area in which the data have been confusing, contradictory, and confounding.

One might begin by asking "What is venom, and how do histamine releasing peptides play a key role in the function of venom?" "What are mast cells, and what is their physiological and medical significance?" "Which venom peptides are mast cell secretagogues?" "By what mechanism(s) do these peptides achieve this effect?" Of course, what is known so far in response to these questions will open further questions.

17.2 SOURCES OF PEPTIDES—VENOM

Venoms are produced by a wide variety of animals, including bees, wasps, ants, snakes, fish, lizards, scorpions, and spiders, to name but a few (Minton, 1974; Schmidt, 1986; Meier and White, 1995). And with the exception of certain ants, venoms are mainly composed of proteins and peptides (Devi, 1968; Brown, 1973; Tu, 1977, 1982, 1996; Eaker and Wadström, 1980; Russell, 1983; Meier et al., 1995). Historically, bee, wasp, scorpion, and snake venoms have been more frequently studied due to their medical effects. Others have been examined subsequently, not only to better understand their biology and chemistry, but also to investigate the interesting pharmacological activities found in venoms. Snake venoms have been relatively more extensively studied because of the large number of fatalities and cases of severe injury by envenomation, the relative ease in collecting convenient volumes of venoms, and the diversity of active agents found within the various snake venoms.

Many animals produce poisonous substances—toxins—which act as deterrents, and make them distasteful and/or sickening to predators. However, when animals inject a mixture of toxins, usually by fang or stinger, we consider that a venom. Why have venom?

The useful effect of venom to an animal varies. Venom is commonly used on prey, for example, by jellyfish, scorpions, spiders, wasps, and snakes. As an example, snake venom is a derivation of the snake's salivary secretion, and its first useful effect is to quickly incapacitate the victim, so the potential meal will not harm the snake, and to prevent the envenomed meal from fleeing before it dies. Paralytic neurotoxins and hypotensive agents (which lead to shock with loss of consciousness) would be useful for incapacitation, and in fact are amply present in venoms. Recently, a nicely integrated model was proposed suggesting that many of the enzymes and purines present in snake venom itself and those generated from prey tissues by venoms may contribute to the incapacitating effects of hypotension and neurotoxicity through a variety of means (Aird, 2002). Similarly, some predatory insects and spiders store their paralyzed live prey alive, for later consumption. Finally, because many venomous animals do not masticate prey, but rather drink the body fluids or swallow the prey whole, it can be helpful if the venom begins the digestive process within the prey. This makes the degradative enzymes found in many venoms very helpful. Venoms can also be potent defensively. Insects stinging livestock or humans who have disturbed a nest is a familiar example.

Venom composition varies widely between species, from simple to very complex mixtures. Insect venom, for example, bee venom, contains only a few peptides in any significant proportions, and a few more in trace amounts, although minor variations may occur in composition (Habermann, 1972; Franklin et al., 1975; Rader et al., 1987; Hider, 1988). Although many venoms have relatively limited numbers of toxic components, snake venoms are complex (Devi, 1968), often containing any of over two dozen agents, including enzyme activities, kinins, nerve growth factor, purines, salts, and so on (Tu, 1977, 1982; Eaker et al., 1980). Even within a snake species the composition of the venom may vary, as has been shown for Crotalids, where the composition varies with age and other factors (Glenn et al., 1983, 1994; Glenn and Straight, 1985; Adame et al., 1990; Straight et al., 1991, 1992; French et al., 2004). Given that the food source for a snake may be insects when the animal is young, then birds or mammals as it matures, the venom must provide a wide variety of active agents, and/or agents with widely effective activities.

We can roughly divide the snake venoms into two groupings by the pathology of envenomation, which is in turn a reflection of composition. The first is typified by massive tissue destruction, an example of which are the Crotalids. These have venoms rich in proteases, phospholipases, and other enzymes. The second group includes the neurotoxic venoms, exemplified by Elapids, which are a rich source of peptide neurotoxins that while lethal, often by respiratory failure, do not generate massive tissue damage. These generally have much less degradative enzymes.

Venoms are a very successful tool for causing pharmacological changes in prey (or attackers), and as such are Nature's storehouse of biologically active peptides. Venoms have been successfully mined for prototypic enzymes and toxins, yet many peptides found chromatographically are yet uncharacterized. One wonders whether they are all merely degradation products, or whether they have some activities yet to be discovered.

17.3 BIOLOGY OF MAST CELLS

To ask "What then would be the involvement of venom peptides with mast cells?", let us consider the cell that defines the activity we are examining. Mast cells (Erlich, 1877) are located in mucosa and dispersed in other tissues throughout the body. Histology separated the mucosal/serosal and connective tissue mast cells, and revealed basophils, which release similar granules. From early histological studies, interest in mast cells turned to cell biology and functions. Basophils are present in low numbers in blood and have been less well studied, particularly in pure culture. In contrast, rat peritoneal mast cells are more easily obtainable. Analysis of granulocyte granules revealed that basophils (Falcone et al., 2000) are much like mast cells functionally, releasing histamine, the major medically important biogenic amine, from the granules. But basophils migrate through blood and tissues. One may in general consider mast cells and basophils together as potent histamine releasers, while remaining mindful of the constantly growing list of specific functional differences between mast cells and basophils indicating too that their lineages are distinct (Galli and Hammel, 1994; Costa et al., 1997; Galli, 2000; Wedemeyer et al., 2000a). For example, c-kit is a major growth factor for and able to regulate secretion by mast cells. In contrast, basophils can develop without c-kit, and it has a much less regulatory effect upon them. Several mast cell degranulating agents do not have this effect on basophils, as we will soon see.

Activated mast cells are instrumental in driving acute inflammation. Inflammation (acute; Kumar et al., 2005) is an ubiquitous, mechanistically stereotypic response to tissue injury present in pathologies from allergies to arthritis. It is characterized by increased vascular permeability, redness of skin (if the site is cutaneous), swelling, and a prompt cellular infiltrate into the site of damage. Acute inflammation is also a normal part of innate immunity, an important prelude to healing, and focuses innate and adaptive immunity on the site of likely infection. However, the response can be extreme in magnitude, or occur without protecting the host, actually causing damage. Although tissue injury is the common elicitation, inflammation can also be elicited directly by stimulating key reactive cells, particularly mast cells, as any allergy sufferer can attest.

Once activated, mast cells (and basophils) (Abbas, 2005) are key sources of primary mediators found preformed in granules (Lemanske et al., 1983) and newly synthesized mediators (Galli et al., 1993; Galli and Costa, 1995; Galli 1993, 2000) that collectively drive inflammation, among other effects (Table 17.1). Primary mediators include biogenic amines such as histamine, which causes increased smooth muscle contraction and vascular permeability, and secretion by bronchial, gastric, and nasal glands. Neutral proteases (chymase and tryptase) activate Complement, generate kinins, and damage tissues. Proteoglycans present bind many of the components; heparin is an anticoagulant. Secondary mediators include cytokines, which regulate many types of cells. And finally phospholipase A2 is activated and generates arachidonic acid, which is in turn the parent compound for the lipoxygenase and cyclooxygenase pathways that produce the leukotrienes, prostaglandins, and platelet activating factor, all of which have multiple proinflammatory roles. If large numbers of mast cells/basophils are systemically activated to activate and degranulate, this large response can be responsible for severe hypotensive shock, an incapacitating event. The key role of mast cells in inflammation from allergy to autoimmunity (Wasserman, 1984; Marone, 1989; Kaliner, 1989; Maurer et al., 2003; Theoharides and Kalogeromitros, 2006; Blank et al., 2007; Eklund, 2007; Kinet, 2007; Maruotti et al., 2007; Dawicki et al., 2007) has sustained the study of mast cells, their eliciting agents, and the products they produce.

What useful functions do mast cells serve? The inflammatory response, as mentioned above, helps stimulate both innate and adaptive immunity to the infection typically occurring at the site of traumatic tissue injury, and is a physiological prelude to healing. Moreover, helminth infestations are a way of life for most populations of wild animals and many peoples. One may expect natural

TABLE 17.1
Products Released by Activated Mast Cells

Class of Product	Products
Preformed	Histamine, serotonin (in rodents), heparin and/or chondroitin sulfates, tryptase, chymase, major basic protein, cathepsin, carboxypeptidase-A
Lipid-derived	PGD2, PGE2, LTB4, LTC4, PAF
Cytokines and growth factors	GM-CSF[ab], IFN-α[ab], IFN-β[c], IFN-γ[a], IL-1α[bc], IL-1β[ab], IL-1R antagonist[b], IL-2[a], IL-3[ab], IL-4[ab], IL-5[ab], IL-6[ab], IL-8 (CXCL8)[b], IL-9[ab], IL-10[ab], IL-11[b], IL-12[ad], IL-13[ab], IL-14[d], IL-15[d], IL-16[bc], IL-17E (IL-25)[a], IL-17F[c], IL-18[d], IL-22 (IL-TIF)[c], LIF[d], LTβ[d], M-CSF[cd], MIF[d], SCF[b], TGF-β1[ab], TNF[ab], TSLP[c],bFGF[ab], EGF[b], IGF-1[a], NGF[a], PDGF-AA[d], PDGF-BB[b], VEGF[ab]
Chemokines	CCL1 (TCA3/I309)[bc], CCL2 (MCP-1)[ab], CCL3 (MIP-1α)[ad], CCL3L1 (LD78β)[d], CCL4 (MIP-1β)[cd], CCL5 (RANTES)[ab], CCL7 (MCP-3)[cd], CCL8 (MCP-2)[d], CCL11 (eotaxin)[d], CCL13 (MCP-4)[d], CCL16 (LEC/HCC-4)[d], CCL17 (TARC)[ad], CCL20 (LARC)[d], CCL22 (MDC)[ad], CXCL1 (Groα/KC)[ad], CXCL2 (Groβ/MIP-2)[ad], CXCL3 (Groγ)[d], CXCL10 (IP-10)[bc], CXCL11 (I-TAC)[d], XCL1 (lymphotactin)[bc]
Free radicals	Nitric oxide[ef], superoxide[ef]
Others	Corticotropin-releasing factor[b], urocortin[b], substance P[a]

Source: Adapted from Galli, S.J., Nakae, S., Tsai, M., *Nature Immunology* 6 (2), 135–142, 2005.

Note: Certain cytokines and growth factors, including TNF and VEGF, can be released from both preformed and newly synthesized pools. Many others have been localized to mast cell cytoplasmic granules by immunohistochemistry. Some of these cytokines, growth factors, and chemokines have been detected only at the mRNA level, only in studies of *in vitro* derived mast cells, and/or only from mast cells from a single species. For these products, the following apply: [a,b] Protein detected by ELISA or by immunohistochemistry; [c,d] mRNA expression; [a,c,e] Rodent; [b,d,f] Human.

selection should have encouraged a variety of protective systems, just as the parasites have evolved to persist. Mast cells accumulate in response to helminth infections, where they offer some innate protection (Bell, 1996; Maizels et al., 1998; Else et al., 1998; Onah et al., 2000; Ferreira et al., 2002). That may have been their origin, but their utility now is perceived as broader. The current view of mast cells is that they play varied key roles in adaptive and innate immunity (Galli, 1993; Costa et al., 1997; Galli et al., 1999, 2002, 2005; Williams et al., 2000; Galli, 2000; Wedemeyer et al., 2000a, 2000b; Boyce, 2004), even to suppressing chronic inflammation (Metz et al., 2007).

On a wider scale, it has been suggested that perhaps allergy, which is driven by mast cells, is an evolutionary advantage for dealing with toxins (Profet, 1991). This may at first seem counterintuitive. Although many venoms cause histamine release (Clark and Higginbotham, 1971; Cormier, 1981; Flowers and Hessinger, 1981; Chen et al., 1984; Chiu et al., 1989; Price and Sanny, 1996; Chai et al., 2001), mast cells may also offer protection from the effects of venom proteins/peptides (Higginbotham, 1965; Metz et al., 2006; Marx, 2006; Rivera, 2006). Thus the mast cell degranulating peptides in venom stimulate the mast cells to activate and degranulate, which seems to neutralize some effects of the venom. The mechanism(s) of protection is unclear. Their granules include degradative enzymes (Nadel, 1991; 1992), which degrade proteins, and perhaps the toxins including the mast cell degranulating peptides. Granules contain proteoglycans such as heparin, which can complex with and inactivate cationic peptides (Higginbotham and Karnella, 1971). Mast cells also release cytokines, which might be involved in regulating host responses.

When the analysis of mast cell degranulation showed that a major pharmacological agent produced in granules was histamine, the assay of released histamine became the usual gold standard for the study of degranulation reactions of mast cells, both to better understand this response, and for the search for drugs to inhibit histamine release. Many substances were found to be mast cell secretagogues, including many simple chemicals such as compound 48/80 (Paton, 1951) and the calcium ionophore A23187. The subject of mast cell secretagogues is extensive and has often been reviewed (Paton, 1957; Rothschild, 1966; Lagunoff et al., 1983; Metcalfe, 1984; Pearce, 1989; White et al., 1989; Cochrane, 1990; Kaplan et al., 1991; Church et al., 1991; Marone et al., 1993; Nosal, 1994; Mousli et al., 1994a; Church and Levi, 1997; Ferry et al., 2002). These histamine releasing factors (HRFs) include peptides such as the nasal allergens, venom peptides, and endogenous peptides such as Complement peptides (C3a, C5a).

Although some HRFs are endogenous peptides, do they occur at sufficient levels for this effect to be biologically meaningful? Peptides seem to bind quickly to tissues *in vivo*, and are fairly soon eliminated from circulation (Ferreira and Vane, 1967). Some endogenous HRF peptides, such as Complement peptide C3a, may occur at sufficient levels in circulation to have an effect, as discussed by Mousli and colleagues (1994a). It has been argued that others do not. However, the actual concentrations of endogenous HRFs in the microenvironment of tissues are unclear, and might in some situations be sufficient for stimulation. Thus conclusions based on circulating levels of endogenous HRF peptides may be an underestimate of meaningful interactions, and thus the question of which of these endogenous HRF activities are physiologically important is still open ended.

Reports continue to appear showing the activity of various peptides from disparate sources, including defensins, the clotting cascade, and microorganisms. By the 1980s it was clear that unrelated, even randomly chosen peptides and other agents were active secretagogues. During that time reports showed the activity of certain HRFs and the inactivity of the same agent, leading to confusion. Eventually a picture emerged that explained most of the confounding results, incorporating them into a coherent model. Simply put, aside from occasional technical issues, the disparity in the data was because the various experimental model systems for detecting histamine release being used do in fact behave differently. Substantial physiological differences in the responses of responding cells are seen not only (a) between species, but also (b) between the serosal rat peritoneal mast cells, which are similar to mast cells from human skin, and also when compared to (c) mucosal mast cells and (d) basophils, even of the same species (Keller, 1968; Pearce, 1982; Ennis, 1982; Pearce et al., 1982; Befus et al., 1982a, 1982b; Bienenstock et al., 1983; Barrett and Pearce, 1983b; Leung and Pearce, 1984; Shanahan et al., 1984; Wasserman, 1984; Casolaro et al., 1989; Ishizaka et al., 1991; Tainsh et al., 1992; Fureder et al., 1995). Such functional diversity of mast cells may even be seen (Kitamura, 1989) in the microenvironment within an organ (Patella et al., 1995). Thus the relative secretagogue activity of a peptide depends very much on which cell/species it is tested, to the point that in some cases an active peptide could show essentially no HRF activity. A summary of some of this early work is presented in Table 17.2.

Was this functional diversity an issue of different mechanisms of triggering in different cells? How do secretagogues activate mast cells and basophils? Is there a physical basis for their activity, a receptor?

17.4 MOLECULAR MECHANISMS OF ACTION

There are three distinctly different, fairly well-known mechanisms by which peptides may cause degranulation of mast cells, and probably more that remain obscure. These peptides may come from any source in nature, including venom.

Nonspecific direct cytotoxicity leading to release of contents and cell death can be the result of exposure to any of the various toxins that disrupt membranes. Venoms contain significant amounts of degradative enzymes including lipases and proteases. Many also contain cytotoxins (e.g., mellitin, a PLA2 activator), which are destructive to membranes of most types of cells, and will nonspecifically release histamine by cell lysis. However, discussions of mast cell secretagogues generally center on nonlytic histamine release.

A second common biological mechanism is allergy, a type I hypersensitivity (atopic reactions) resulting in inflammation (Kumar et al., 2005). Allergies may be nasal, ocular, gastrointestinal, or cutaneous, with diverse symptoms depending on exposure to any of a wide variety of environmental

TABLE 17.2
Early Work Showing Mast Cell Diversity in Responsiveness to Secretagogues

Species	Sensitive Cells	Nonsensitive Cells
Guinea pig		Basophils[b]
		Heart[c]
	Lung-np[a]	Lung[c]
		Mesentary[c]
		Skin-np[b]
Human	Skin[f]	Basophils[bg]
		BAL-np[g]
		Intestine[f]
		Lung[fg]
Monkey	Lung[e]	Lung[b]
		Intestine[e]
Rat	Cocultured RBL-2H3 (basophilic)[h]	RBL-2H3 (basophilic)[h]
	Intestine[d]	Intestine[g]
	Mesentary[b]	Heart[g]
	Peritoneal cells[b]	
	Lung[ag]	
	Skin[g]	

Source: Adapted from Mousli, M., Hugli, T.E., Landry, Y., and Bronner, C., *Immunopharmacology*, 27, 1–11, Table 1, 1994a.

Note: BAL, bronchoalveolar lavage cells; RBL, rat basophilic leukemia cells; np, nonpurified.

[a] Barrett, K.E., Ennis, M., and Pearce, F.L., Mast cells isolated from guinea-pig lung: Characterization and studies on histamine secretion, *Agents Actions*, 13, 122–126, 1983a.

[b] Pearce, F.L., Mast cell heterogeneity, *Trends Pharmacol.* Sci., 4, 165–167, 1983.

[c] Ali, H. and Pearce, F.L., Isolation and properties of cardiac and other mast cells from the rat and guinea-pig, *Agents Actions*, 16, 138–140, 1985.

[d] Shanahan, F. et al., Mast cell heterogeneity: Effects of neuroenteric peptides on histamine release, *J. Immunol.*, 135, 1331–1337, 1985.

[e] Barrett, K.E., Szucs, E.F., and Metcalfe, D.D., Mast cell heterogeneity in higher animals: A comparison of the properties of autologous lung and intestinal mast cells from nonhuman primates, *J. Immunol.*, 137, 2001–2008, 1986.

[f] Lawrence, I.D. et al., Purification and characterization of human skin mast cells. Evidence for human mast cell heterogeneity, *J. Immunol.*, 139, 3062–3069, 1987.

[g] Pearce, F.L., Mast cell heterogeneity: The problem of nomenclature, *Agents Actions*, 23, 125–128, 1988.

[h] Swieter, M., Hamawy, M.M., Siraganian, R.P., and Mergenhagen, S.E., Mast cells and their microenvironment: The influence of fibronectin and fibroblasts on the functional repertoire of rat basophilic leukemia cells, *J. Periodontol.*, 64, Suppl. 5, 492–496, 1993.

substances. Thus air-borne allergens trigger hay fever, food allergens gastrointestinal distress, and so on. Life-threatening allergic responses, for example, to Hymenoptera venom (bees, wasps, hornets) are not uncommon.

During an initial exposure to the allergen, a primary adaptive immune response occurs, usually with minimal sensitization. For reasons not fully understood and which include the genetics of the responder (Dreskin, 2006; Finkelman and Vercelli, 2007) but no obvious structural feature of the allergen (Aalberse, 2000, 2006; Platts-Mills et al., 2004; Aalberse and Platts-Mills, 2004), some antigens result in an antigenic priming and antibody response with not only the usual isotype switch in the production of antibodies from IgM to IgA and IgG, but also in appreciable amounts of IgE. This critical isotype switching is dependent on the balance of Th1 and Th2 lymphocyte effects, and control of this is poorly understood but of great interest. Some peptides are more likely to elicit IgE than others, and venom peptides can be potent allergens. A functional requirement for the peptide to serve as an allergen includes a minimal molecular weight to be antigenic (at least 5 to 10 kDa).

The mast cells' high-affinity Fc-epsilon RI receptors bind the constant region of immunoglobulin epsilon chains of IgE, thus sensitizing the cells. In a later exposure, allergens bind to mast cells via the IgE. As a result, the immune specificity of the IgE defines the reactivity of the mast cell to allergens (type I hypersensitivities). Receptor aggregation then leads to signal transduction via multiple pathways including Lyn and Syk tyrosine kinases (Stanworth and Ghaderi, 1990; Turner and Kinet, 1999; Parravicini et al., 2002; Rivera et al., 2002; Furumoto et al., 2004; Gilfillan and Tkaczyk, 2006), with resultant mast cell activation and degranulation. This reaction may be mimicked by crosslinking the receptor with lectins. Clinical preparations to test patients for such allergies use whole venom, a mixture of peptides, in part because the allergic response of patients may vary to different antigenic peptides.

The third well-established mechanism of mast cell degranulation, a peptidergic means, is totally irrespective of the Fc-epsilon RI receptor (Bloom et al., 1967a; Goth et al., 1971). Soon after noting the hypotensive action of certain compounds, these polycationic drugs were found to be nonlytic active secretagogues for connective tissue mast cells, human skin and rat peritoneal mast cells, but not mucosal mast cells or blood basophils (Mousli et al., 1990, 1991b, 1994a). The ability of these HRFs to stimulate mast cells was found to be independent of allergic desensitization, or of IgE blocking. It is very rapid (10 to 20 sec) compared to IgE mediated reactions, which take minutes to complete. Peptidergic responses also do not require the large influx of extracellular Ca^{2+} needed by IgE or calcium ionophore mediated reactions. When the simultaneous inhibition by pertussis toxin of compound 48/80 induction of histamine secretion, inositol phospholipid breakdown, and arachidonic acid release was found in mast cells (Nakamura and Ui, 1985), and basophils as well (Warner et al., 1987), work led to the GTP-binding regulatory proteins (G proteins) pathway of cell activation by various peptides, including venom peptides (Mousli et al., 1990, 1991b).

There are two inhibitors of this pathway. Benzalkonium chloride, a quaternary ammonium invert soap, inhibits histamine release (Read and Kiefer, 1979; Read et al., 1982; Seebeck et al., 2000) by polyamines but not the calcium ionophore A23187 (discussed by Mousli in Mousli et al., 1994a) by selectively interacting with heterotrimeric G proteins of the G(i) type via an unclear mechanism. The pertussis toxin inhibits mast cell histamine release by the cationic compound 48/80 (and others) but not by the calcium ionophore A23187 or IgE mediated release from basophils (Nakamura and Ui, 1985), and targets other Gi proteins (Bueb et al., 1990).

Then what is particular about the very active mast cell degranulating peptides discovered that defines a mast cell degranulating peptide? It is that they are in fact very cationic (basic) peptides of various sequences and structures, often of high prevalence. Even some allergens can elicit IgE-independent mast cell degranulation (Machado et al., 1996). Some peptides are more potent than others, which led to studies of most effective structures, as illustrated in Table 17.3. Thus mast cell degranulating peptides became tools to study signal transduction.

Mast cell degranulation secretagogues have recently been reviewed from the perspective of G protein dependence (Ferry et al., 2002). Models for the peptidergic and IgE-mediated pathways (Reischl et al., 1999; Tkaczyk and Gilfillan, 2001; Frossi et al., 2004; Kopec et al., 2006; Galli et al., 2008; Rivera and Olivera, 2008) are shown in Figure 17.1. Basophil responses differ from mast cells, and most attention has focused on mast cells. There are a number of endogenous peptides that have receptors inside appropriate target cells of various types. However, when these peptides enter serosal mast cells they act as receptor mimetics and bind to G proteins, and act as secretagogues. Peptidergic responses occur at higher concentrations than usual for typical receptor-mediated responses. That specific receptors are not involved is also indicated by negative receptor cloning attempts, apparent absence of saturable binding of ligands, lack of identifiable receptor mRNA, and lack of effect of known antagonists of these receptors to influence the peptidergic mast cell degranulation. In turn, this peptidergic stimulation pathway may be a future target for inhibiting these responses (Druey, 2003).

TABLE 17.3
Structure of Compounds that Trigger the Peptidergic Pathway of Mast Cells with Corresponding Biologic Potencies

Compound	Structure	EC_{50} (μM)	Reference
Natural Polyamines			
Spermin	NH_2–$(CH_2)_3$–NH–$(CH_2)_4$–NH–$(CH_2)_3$–NH_2	1000	Bueb et al., 1992
Spermidin	NH_2–$(CH_2)_4$–NH–$(CH_2)_3$–NH_2	4000	Bueb et al., 1992
Neuropeptides and Analogs			
Substance $P_{1\text{-}11}$ (SP)	**R**-P-**K**-P-Q-Q-F-F-G-L-M	2.6	Repke et al., 1987b
$SP_{1\text{-}4}$	**R**-P-**K**-P	300	Repke et al., 1987b
$SP_{1\text{-}4}C_{12}$	**R**-P-**K**-P-N**H**-$(\mathbf{CH_2})_{11}$-$\mathbf{CH_3}$	0.047	Repke et al., 1987b
$[DTrp^{7,9}]SP_{1\text{-}11}$	**R**-P-**K**-P-Q-Q-DW-F-DW-L-M	0.21	Foreman and Jordan, 1983
$[DTrp^{7,9,10}]$ $SP_{1\text{-}11}$	**R**-P-**K**-P-Q-Q-DW-F-DW-DW-M	0.062	Devillier et al., 1988
Neuropeptide $Y_{1\text{-}36}$ (NPY)	Y°P-S-**K**-P-D-N-P-G-E-D-A-P-A-E-D-M->	3.0	Mousli et al., 1994b
$NPY_{18\text{-}36}$	A-**R**-Y-Y-S-A-L-**R**-**H**-Y-I-N-L-I-T-**R**-Q-**R**-Y	0.018	Mousli et al., 1994b
Hormone Peptides and Analogs			
Bradykinin	**R**-P-P-G-F-S-P-F-**R**	19.8	Bueb et al., 1990
$[Thi^{5,8}$, $DPhe^7]$bradykinin	**R**-P-P-G-β-(2-thienyl)-A-F-S-P-DF-β-(2-thienyl)-A-**R**	2.3	Bueb et al., 1990
Kallidin	**K**-**R**-P-P-G-F-S-P-F-**R**	6.1	Devillier et al., 1988
$[Thi^{6,9}$, $DPhe^7]$kallidin	**K**-**R**-P-P-G-β-(2-thienyl)-A-F-S-P-DF-β-(2-thienyl)-A-**R**	0.054	Devillier et al., 1988
Anaphylatoxins			
C3a	<-C-N-Y-I-T-E-L-**R**-**R**-Q-**H**-A-**R**-A-S-**H**-L-G-L-A-**R**	3.3	Mousli et al., 1992
$C3a_{des\ Arg}$	<-C-N-Y-I-T-E-L-**R**-**R**-Q-**H**-A-**R**-A-S-**H**-L-G-L-A	2.2	Mousli et al., 1992
Venom Peptides			
Mastoparan	I-N-L-**K**-A-L-A-A-L-A-**K**-**K**-I-L	0.3	Hirai et al., 1979
MCD	I-**K**-C-N-C-**K**-**R**-**H**-V-I-**K**-P-**H**-I-C-**R**-**K**-I-C-G-**K**-N	4.8	Mousli et al., 1991a
Mellitin	G-I-G-A-V-L-**K**-V-L-T-T-G-L-P-A-L-I-S-W-I-**K**-**R**-**K**-**R**-Q-Q	0.7	Jasani et al., 1979

Source: Reprinted from Mousli, M., Hugli, T.E., Landry, Y., and Bronner, C., *Immunopharmacology*, 27, 1–11, 1994a. With permission from Elsevier.

Note: Characters in bold indicate the basic amino-acid residues.

The mechanism for many peptides, particularly venom peptides, to enter cells with varying effects is still unclear, as Ferry has commented (Ferry et al., 2002); see Figure 17.2. A hydrophobic region is thought helpful for associating with the cell membrane (Repke and Bienert, 1987a, 1988; Repke et al., 1987b). Binding is neuraminidase-sensitive, so an initial interaction with the surface sialyic acid begins the process. For the basic secretagogues there may be an external binding event followed by translocation. When sepharose beads were bound to either polymyxin B (Morrison et al., 1975) covalently, or compound 48/80 with albumen linkage (Hino et al., 1977), these agarose beads, which are larger than the cells and thus could not be internalized, caused statistically significantly more histamine release than with control bead treatment. Morrison had a linear response to the amount of beads used in a 30-min exposure reflected by the release of labeled serotonin, whereas soluble HRF dose responses are generally nonlinear. Hino found the coupled beads as an affinity column retained 67% of the mast cells compared to 22% by a control column. Addition of control beads led to 18% release, while coupled beads caused 53% release. However, one might like

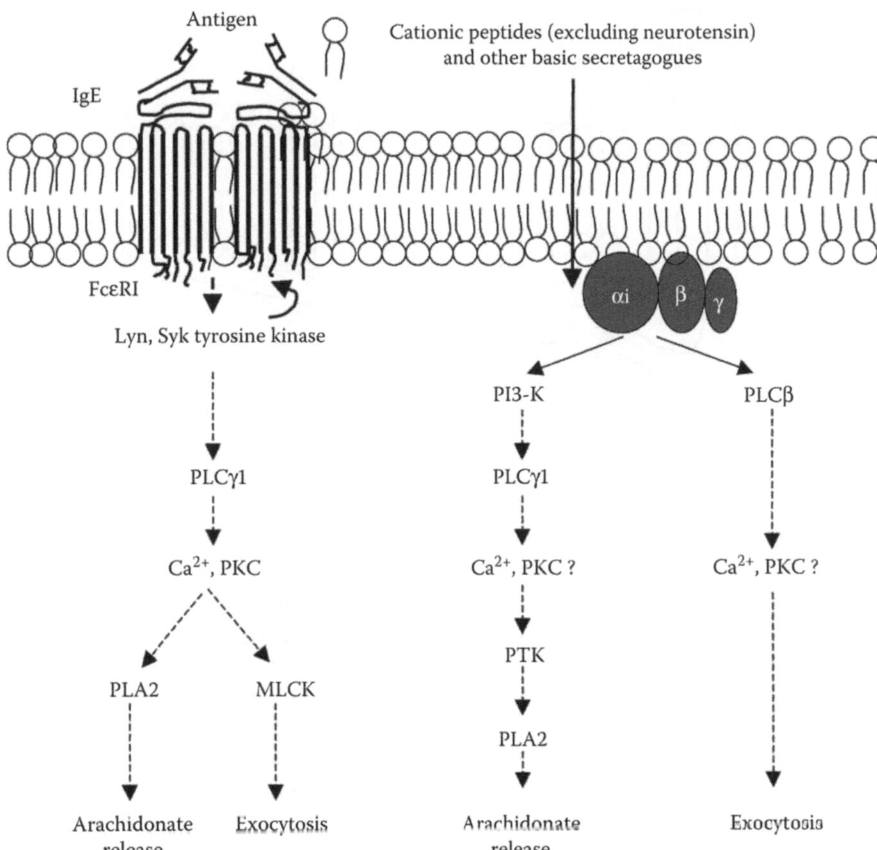

FIGURE 17.1 The two main pathways for serosal (connective tissue) mast cell activation. Antigens crosslink selective IgE bound to FcεRI receptors leading to the activation of Lyn and Syk tyrosine kinases (Kinet, 1999; Nadler et al., 2000). Most basic secretagogues directly activate G_{i2} and G_{i3} proteins leading through βγ subunits to the activation of phosphatidylinositol 3-kinase (PI3K) and phospholipase Cβ (PLCβ) (Ferry et al., 2001). In contrast, neurotensin binds to its selective receptor coupled to G proteins (Barrocas et al., 1999). Alternative hypotheses for the role of the plasma membrane in the effect of basic secretagogues are shown in Figure 17.2. (Reprinted from Ferry, X., Brehin, S., Kamel, R., and Landry, Y., *Peptides*, 23, 1507–1515, 2002. With permission from Elsevier.)

to see lower background in the data with 48/80 (Hino et al., 1977), and one wonders if the degranulation in these models was via the assumed pathway, given the unusual means of stimulation. Further study is required to understand this data better, and to ask when and how nonsecretagogue peptides can potentiate responses, possibly by a more widely distributed binding and translocation mechanism(s) than previously suspected. Binding without stimulation, but with possible potentiation can occur with basophils (Tobin et al., 1986). It is not always clear to what extent binding studies have revealed the transport system, instead of the putative intracellular receptor.

Many snake (Clark and Higginbotham, 1971; Cormier, 1981; Flowers Hessinger, 1981; Chen et al., 1984; Chiu et al., 1989; Price and Sanny, 1996; Chai et al., 2001) and other venoms cause degranulation of mast cells. But are all accounts of nonlytic, nonallergic stimulation of mast cells by venom really a peptidergic reaction? Cytotoxicity controls have not always been done in published studies. But, apart from that, the magnitude of peptidergic responses may be biased. The significant level of purines in many venoms, the recent identification of adenosine A3 receptors in mast cells (Ramkumar et al., 1993), and the release of vasoactive products by adenosine and inosine via A3

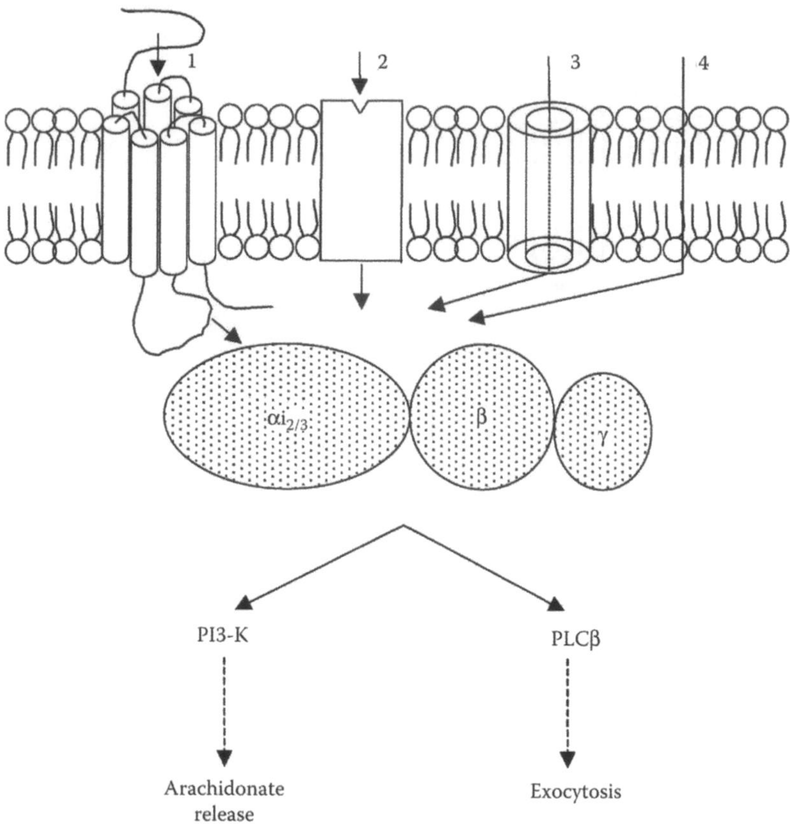

FIGURE 17.2 Proposed role of the plasma membrane for the effect of basic secretagogues on serosal (connective tissue) mast cells. A preliminary step is the interaction of triggers with sialic acid residues at the cell surface (not shown). Then, four possibilities may be proposed: (1) the interaction with a selective receptor, as demonstrated for neurotensin (Miller et al., 1995; Barrocas et al., 1999); (2) interaction with a nonselective transmembrane receptor coupled to G proteins; (3) entry through a membrane transporter; (4) translocation across the plasma membrane, a poorly defined process termed protein transduction for cell-penetrating peptides. (Reprinted from Ferry, X., Brehin, S., Kamel, R., and Landry, Y., *Peptides*, 23, 1507–1515, 2002. With permission from Elsevier.)

receptors (Tilley et al., 2000) indicate a reappraisal (Aird, 2002) of our views. Purines may contribute to titrations of whole venom histamine release activity. And while the reality of the peptidergic pathway is secure, perhaps passive adherence of purines to some peptides isolated from venom may have covertly enhanced the appearance of their secretagogue activity. It may be useful to reevaluate quantitative activities of certain venom peptides with synthetic rather than isolated peptides, for those for which it has not been done.

17.5 STRUCTURAL STUDIES

17.5.1 Specific Venom Peptides

Of the many basic peptides from diverse sources that are mast cell secretagogues, many are only lightly studied, while several venom peptides have been useful models to study other processes. Mastoparan is a potent histamine-releasing peptide, but is considered elsewhere in this publication, and so will not be discussed here. Phospholipase A2 is found in venoms, but its role is complex, is related to a variety of systems, has a complicated history of its relationship to mast cells and basophils, and although worthy of its own review (yet again), is tangential to the focus of this chapter and will not be explored here. The other venom peptides that have been studied relatively intensively as mast cell degranulating peptides are the mast cell degranulating peptide (MCD), and to a lesser extent with mast cells than other systems, nerve growth factor (NGF).

17.5.2 Mast Cell Degranulating Peptide

MCD has been the most studied prototype mast cell degranulating agonist and probe of mast cell receptors for structural studies. It is also a neurotoxin (reviewed by Ziai et al., 1990). It was resolved that although patients could be allergic to bee venom, it was also clear that bee venom induced non-allergic histamine release. Although compositional analysis (reviewed by O'Conner et al., 1967; Hider, 1988) revealed that several peptides in the venom had this activity, one was particularly active (Fredholm, 1966; Fredholm and Haegermark, 1967a, 1967b; Breithaupt and Habermann, 1968; Assem and Atkinson, 1973; Hanson et al., 1974; Gauldie et al., 1976, 1978; Gushchin et al., 1977). This best studied nonlytic highly active peptide was "peptide 401," later usually referred to as mast cell degranulating peptide (MCDP or MCD), and most often now as MCD.

MCD was used with other agonists in early studies revealing differences between species and type of responsive cell within species for degranulation in response to secretagogues including peptides, and later when it was found representative of peptidergic (G protein) degranulation (discussed above) (Pearce et al., 1979; Mousli et al., 1991a, 1991b, 1994a; Emadi-Khiav et al., 1995). Suppression of degranulation by pertussis toxin has shown the role of the G protein pathway in histamine release by MCD (Fujimoto et al., 1991; Mousli et al., 1991a, 1991b).

A long and extensive history of studies to understand its structure–function relationships still remains incomplete, but began with structure. Various initial studies reported its amino acid sequence (Hartter, 1976, 1977, 1980). This led to a model of its structure in solution based on sequence and circular dichroism (CD) spectroscopy (Hider and Ragnarsson, 1981; Lomize and Popov, 1983; Dotimas et al., 1987; Kumar et al., 1988). It is a spherical 22-amino-acid bicyclic peptide with two intramolecular disulfide bonds, and a high number (ten) of positive charges; see Figure 17.3. NMR analysis (Steinmetz et al., 1994) suggests a 2:1 ratio of conformers in equilibrium. Cationic clusters at one side of the surface were concluded to be more important in activating the G protein than the α-helix conformation (Fujimoto et al., 1991). This was explored more directly using reconstituted phospholipid vesicles measuring the steady-state rate of GTP hydrolysis, D and L forms of different toxins, and pertussis inhibition, also finding that the cationic clusters at one side of the α-helical surface appeared more important in the direct activation of G proteins than a specific, α-helical structure (Tomita et al., 1991). A variety of studies showed that positive charges were needed for histamine release activity. Various peptides, their activity, and the interaction with G proteins are discussed by Mousli and colleagues (1994a, 1994b). Genes for degranulating peptides have been cloned (Gmachl and Kreil, 1995) from bee and wasp (Zhang et al., 2003). The bee MCD precursor cDNA shows three exons plus one shared with apamin. Related but not always directly comparable approaches were investigated with MCD in the brain model (Ziai et al., 1990).

A large number of variants of MCD have been synthesized as probes of its interaction with mast cells to elucidate more detail in the structure–function relationships. Jassani compared various histamine-releasing peptides (Jasani et al., 1979) as a step to understanding the functional structure, and discussed relationships. Gushchin began but did not continue histamine release and activity studies with MCD analogs (Gushchin et al., 1981). Buku synthesized and purified the peptide, and confirmed the product was identical to the naturally occurring peptide in terms of physical, chemical, and biological properties (Buku et al., 1989). Using three analogs with markedly lowered histamine-releasing activity, the ability to release superoxide from neutrophils varied, thus showing variation of function with structural changes (Buku et al., 1992). Although the helicity of four analogs determined by CD was not significantly different from the native MCD peptide, the two analogs with C-terminal deletions lost up to tenfold histamine-releasing activity (Buku et al., 1994). Analogs losing

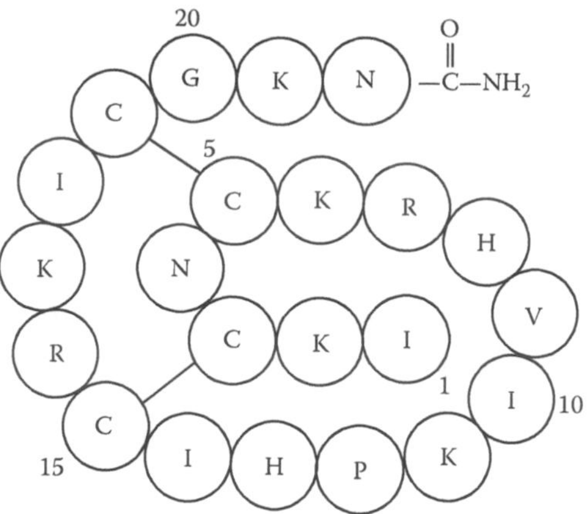

FIGURE 17.3 Primary structure of the MCD peptide with the two disulfide bonds formed at positions 3, 15, and 5, 19, respectively. (Reprinted from Buku, A., *Peptides*, 20, 415–420, 1999. With permission from Elsevier.)

histamine-releasing activity were made and these relationships reviewed, showing that the cyclic structure was needed for activity (Buku et al., 1998, 2004; Buku and Price, 2001a; Buku, 1999). Buku then sought to develop MCD variants that bound to mast cells but did not stimulate degranulation, which led to a model for a second target for MCD binding to cells.

In addition to evidence showing that mast cell degranulation by MCD can be inhibited by pertussis toxin (Fujimoto et al., 1991; Mousli et al., 1991a, 1991b), and is apparently a G protein stimulation, there is evidence that MCD may bind to a surface receptor, the Fc-ε-RI receptor, on the rat basophilic leukemia cells (RBL-2H3). Increasing levels of IgE suppressed histamine release by MCD-peptide-stimulated rat peritoneal mast cells, while the structure of MCD was argued to be consistent with a ligand for a specific receptor (Buku, 1999). Fluorescent and biotinylated analogs of MCD peptide were synthesized labeled at position 1, retaining histamine-releasing activity (Buku et al., 2001b). These bound to the surface of rat basophilic leukemia cells (RBL-2H3) with dose saturation responses as seen in Figure 17.4, and competed with the binding of FITC–IgE as seen in Figure 17.5. Microscopy revealed patchy binding and internalization in vesicles, as expected for endocytosis. The results of those studies and the MCD structure argued against it behaving as a cell-penetrating peptide (Langel, 2007). A series of analogs were prepared with varied efficacy (Buku et al., 2003). The analog [Ala^{12}]MCD was 120-fold less potent then MCD in histamine-releasing activity with rat peritoneal cells, and fivefold more potent in binding affinity to RBL-2H3 mast cell receptors than the parent MCD peptide, with a 50% reduction in IgE binding and hexosaminidase release (Buku et al., 2005). Buku suggested that MCD occupies the disulfide hinge regions of adjacent cellular IgE molecules, causing conformational changes that prevent successful signal transduction (Buku, 1999). It would be interesting to continue the binding studies with rat peritoneal mast cells and human basophils. This approach remains unfinished.

There is always a concern in generalizing results from one experimental model. And with the known cell type and species differences in mast cell degranulation experimental systems, data using RBL-2H3 must be interpreted carefully, although it is very valuable. RBLs are from a chemically derived tumor cell line that is histologically characterized as mucosal, not as connective mast cells (Seldin et al., 1985). They also may show a flexible phenotype, as histamine release is enhanced by adherence to fibronectin, and were reported not to be responsive to 48/80 unless cocultured with fibroblasts (Swieter et al., 1993). It remains to be seen what is the importance of the effect of enhanced release when adhered to fibronectin on interpretations of data from other *in vitro* models. It is also worth considering that basophils vary in their binding and response to the peptides that stimulate serosal mast cells compared to mast cells, and it seems among species too.

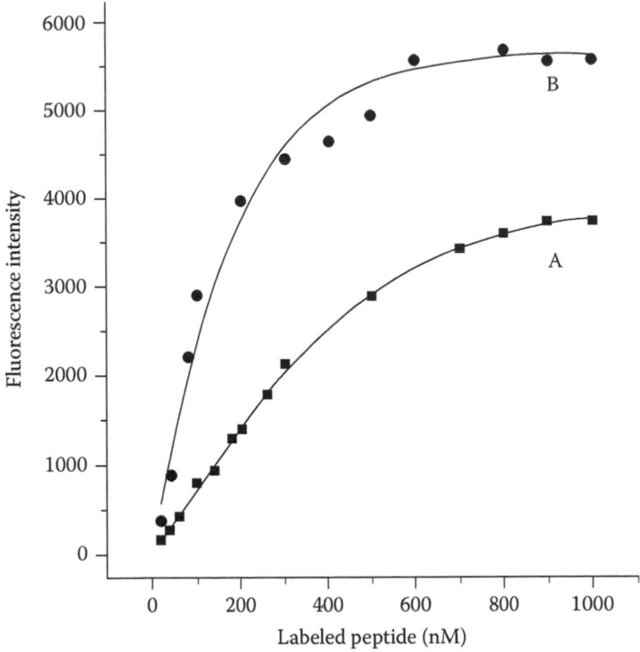

FIGURE 17.4 Saturation of binding of the FITC–MCD ligand (A) and bio-MCD ligand (B) to RBL-2H3 cells. Cells were incubated at 22°C with increasing concentrations of A and B for 30 min. Cells incubated with B were additionally incubated with FITC–avidin for 30 min. The fluorescence intensity of the bound ligands was measured by flow cytometry. Data points represent means of four independent experiments. (Reprinted from Buku, A., Price, J.A., Mendlowitz, M., and Masur, S., *Peptides*, 22, 1993–1998, 2001b. With permission from Elsevier.)

17.5.3 Nerve Growth Factor

Another interesting mast cell degranulating peptide with varied effects is nerve growth factor (NGF). NGF has a variety of roles and is found not only in many animals but in venoms as well (Kostiza and Meier, 1996). Its neurological functions will not be considered here. It stimulates the growth of mast cells and also their degranulation with the release of mediators. The mode of action is incompletely understood. Bruni suggested there may be a specific receptor based on its effect (Bruni et al., 1982). It is at least as potent an HRF as compound 48/80; external calcium is not required for binding to mast cells, but is required for degranulation (Pearce and Thompson, 1986) in a slow, nonlytic release for which the venom NGF requires (lyso)phosphotidylserine. This release is not inhibited by competition with IgE or IgG, but antibodies to NGF in the reaction did block the release. Glucose polymers and specific quaternary ammonium salts that block dextran and polyamine stimulation were ineffective. Unlike IgE-mediated responses, there was no fluctuation in cyclic AMP levels, activation was more persistent, and there was no cross-sensitization. This suggested a receptor-mediated response, but not the Fc-ε R1 receptor active in atopy, and not a peptidergic response seen with polycations. NGF receptors are found on many nerve and other types of cells (Chao et al., 1992). Among neurotrophins, NGF caused mast cell degranulation (Horigome et al., 1993), but brain derived neurotrophic factor (BDNF) or neurotrophin 3 (NT3) did not. Using Northern and Western blot analysis, the receptor for NGF on mast cells is TrkA, and not the low-affinity neurotrophin receptor p75, although this is an NGF receptor on other types of cells. How this relates to other signal transduction systems in mast cells remains to be seen.

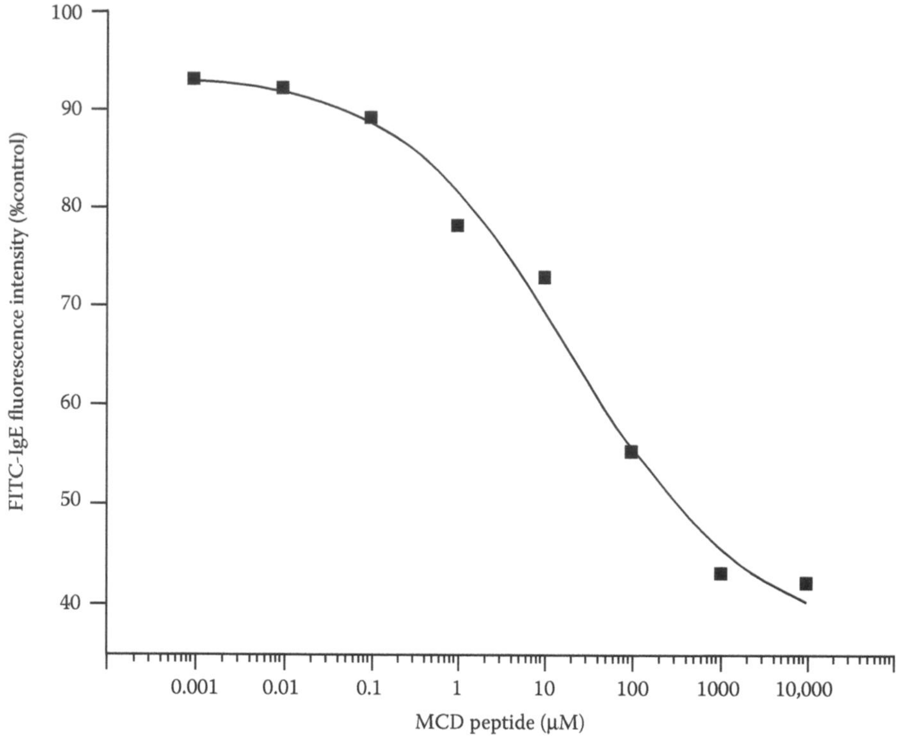

FIGURE 17.5 Inhibition of FITC–IgE binding to RBL-2H3 cells by increasing concentrations of MCD peptide. After preincubation of cells with MCD peptide, fluorescence intensity was measured upon addition of 20nM FITC–IgE. Results are expressed as the percentage of remaining FITC–IgE fluorescence bound to the cells compared to the control. The control value was the fluorescence intensity of 20nM FITC–IgE in the absence of peptide corrected for nonspecific binding detected in the presence of 2μM rat IgE. Data points represent means of four independent experiments. (Reprinted from Buku, A., Price, J.A., Mendlowitz, M., and Masur, S., *Peptides*, 22, 1993–1998, 2001b. With permission from Elsevier.)

17.6 CLINICAL APPLICATIONS

Various peptides from venoms have found clinical use, but not in connection to their properties as mast cell degranulating agents or interactions with mast cells. However, bee venom has a long anecdotal history of use as an anti-inflammatory substance (Beck, 1935). The mechanism is unclear. Higginbotham and colleagues (1971) showed that heparin, released in mast cell granules, reduces the cytolytic and lethal activities of venom, and suggested that it may do so upon mast cell degranulation. Thus mast cell granule contents may bind cationic peptides, which would be ingested by local mononuclear cells and then degraded, which they speculated might be a protective mechanism.

MCD constitutes only about 2% of the venom by weight (Breithaupt and Habermann, 1968; Gauldie et al., 1976), but Billingham and colleagues (1973) found that in bee venom only MCD of the bee venom peptides contributed significant anti-inflammatory action, and activity was much stronger than hydrocortisone in the rat paw edema model. Using a joint protein accumulation assay, MCD was highly protective, as were several drugs and to a lesser extent the cationic compound 48/80. Cutaneously injected low doses of MCD stimulated vascular permeability (assayed by blue dye extravasation) that is abrogated by mepyramine and methysergide. Injection of much higher doses of MCD before challenge with any of a variety of inflammatory agents was protective, but 48/80 had no effect in this assay, and minimal effect in the rat paw system. Mellitin, a cytolytic histamine releaser, was not protective. Thus release of granules *per se* seemed not protective. MCD was also very protective in the adjuvant arthritis model, a result also reported by Chang and colleagues (1979).

MCD derivatives lost degranulation activity concurrently with anti-inflammatory effect (Banks et al., 1980), again consistent with some process concurrent with mast cell activation that is protective. Somerfield's review (1984) discussed possible modes of action, which included that protection by MCD was not inhibited by antihistamines, denervation, or adrenalectomy, and differed from steroid effects but was similar to those of cyclophosphamide. MCD may interfere with PGE2 formation (Banks et al., 1976), and possibly other arachidonic metabolism involvement.

Somewhat in contrast, Banks and colleagues (1990) found that compound 48/80 and allergic challenge both reduced inflammation in the rat carrageenin-induced edema model. Further, MCD and 48/80 were both anti-inflammatory, although less so when skin reactions were suppressed by mepyramine and methysergide. Pretreatment with 48/80 reduced protective effects of MCD or 48/80. It is unclear why these results have somewhat differed from those of Billingham. Perhaps these questions will be revisited with newer methods. As discussed by Ziai and colleagues (1990) the mechanism of high-dose MCD inhibition of mast cells is as yet unclear.

17.7 COMMON METHODOLOGIES FOR THE ASSAY OF HISTAMINE RELEASE

Originally, histamine release was assayed *in vitro* with tissues (e.g., as in Shore et al., 1959; Higginbotham and Karnella, 1971; Ferreira et al., 1973). Various types of cells *ex vivo* and the *in vitro* cell line RBL have been used. Radionuclide-labeled serotonin-treated cells have been used *in vitro*, as well as hexaminidase or tryptase release as evidence of mast cell degranulation. Improvements in the histamine assay came with tube assays using the amount released and the amount residual in cells to calculate percent release. A simple method of assay was developed by Shore using the O-phthaldialdehyde reaction at alkaline pH (Shore, 1971). Many have used the assay without the extraction step. This assay was later modified (Brody et al., 1972; Hakanson et al., 1972) although rarely cited as such. Lebel discussed studies of assay conditions and technologies of the time while presenting another automated assay system (Lebel, 1983). Later, Price made two contributions. First, based on approaches used to isolate lymphocytes from blood (Boyum, 1977, 1983) using varying density and osmolarity, a method was developed to isolate rat peritoneal mast cells in reasonable yield and high purity with a one-step discontinuous gradient (Price, 1997b). The centrifugational separation used media that was endotoxin free and thus less likely to activate the mast cells or others in the preparation and bias the *in vitro* reactions of the mast cells. Secondly, the modified assay was adapted to 96-well microplates, with the same reaction but minor variations in reactant solutions as used by Hakanson and colleagues (1972). This assay uses a single set of experiment-wide controls for unreleased (total) histamine (Price, 1997a). Later we conducted a statistical analysis that provided background for improved experimental designs of histamine release experiments, and provided a basis for an informed choice of parametric versus nonparametric methods of statistical analysis (Coberly and Price, 2005). This histamine assay remains the most cost-effective histamine assay, and is an accurate measure of mast cell degranulation.

Subsequently, an ELISA-based histamine assay kit has been marketed by AMAC (Immunotech International, AMAC, Westbrook, ME, USA). Chemicon (Chemicon, Millipore, MA, USA) offers a mast cell degranulation kit to measure the tryptase released when the cells degranulate. A flow cytometry method in which granules from exocytosis may be stained and measured is available (Demo et al., 1999). Other indirect measurements of cell activation are possible and may prove useful if validated by appropriate controls.

17.8 FUTURE DEVELOPMENTS

Indications of unfinished questions and areas to be explored have been included in context above, but a few ideas may be worth consideration here. One can expect that efforts to find inhibitors of peptidergic stimulation, the G protein mechanism, will continue. The details of the signal transduction pathways, relationships between them, control systems, and receptor blocking will be interesting areas to watch. The possible alternative pathways for MCD and NGF as HRFs could bear examination. The intersection of a better understanding of the cell entry of peptides and these pathways will be useful. Just how high-dose MCD acts therapeutically should be clarified. And for the assayist, perhaps rather than measuring the aftermath of degranulation, in the future some uniquely specific probe will become available to quantitatively measure *in situ* the irreversible commitment of mast cells to degranulation on a cell-by-cell basis as well as for the population.

REFERENCES

Aalberse, R.C., Structural biology of allergens, *J. Allergy Clin. Immunol.*, 106, 228–238, 2000.

Aalberse, R.C., Structural features of allergenic molecules, *Chem. Immunol. Allergy*, 91, 134–146, 2006.

Aalberse, R.C. and Platts-Mills, T.A., How do we avoid developing allergy: Modifications of the TH2 response from a B-cell perspective, *J. Allergy Clin. Immunol.*, 113, 983–986, 2004.

Abbas, A.K., Diseases of Immunity, in *Robins and Cotran Pathologic Basis of Disease*, Kumar, V., Abbas, A.K., and Fausto, N., Eds., Elsevier Saunders, Philadelphia, 2005, 193–267.

Adame, B.L. et al., Regional variation of biochemical characteristics and antigeneity in Great Basin rattlesnake (*Crotalus viridis lutosus*) venom, *Comparative Biochem. Physiol. B: Comparative Biochem.*, 97, 95–101, 1990.

Aird, S.D., Ophidian envenomation strategies and the role of purines, *Toxicon*, 40, 335–393, 2002.

Ali, H. and Pearce, F.L., Isolation and properties of cardiac and other mast cells from the rat and guinea-pig *Agents Actions*, 16, 138–140, 1985.

Assem, E.S. and Atkinson, G., Histamine release by MCDP (401), a peptide from the venom of the honey bee, *Br. J. Pharmacol.*, 48, 337P–338P, 1973.

Banks, B.E. et al., Anti-inflammatory activity of bee venom peptide 401 (mast cell degranulating peptide) and compound 48/80 results from mast cell degranulation *in vivo*, *Br. J. Pharmacol.*, 99, 350–354, 1990.

Banks, B.E., Dempsey, C.E., Vernon, C.A., and Yamey, J., The mast cell degranulating peptide from bee venom, *J. Physiol.*, 308, Suppl., 95p–96p, 1980.

Banks, B.E., Rumjanek, F.D., Sinclair, N.M., and Vernon, C.A., Possible thereapeutic use of a peptide from bee venom, *Bulletin de L'Institute Pasteur*, 74, 137–144, 1976.

Barrett, K.E., Ennis, M., and Pearce, F.L., Mast cells isolated from guinea-pig lung: Characterization and studies on histamine secretion, *Agents Actions*, 13, 122–126, 1983a.

Barrett, K.E. and Pearce, F.L., A comparison of histamine secretion from isolated peritoneal mast cells of the mouse and rat, *Int. Arch. Allergy Appl. Immunol.*, 72, 234–238, 1983b.

Barrett, K.E., Szucs, E.F., and Metcalfe, D.D., Mast cell heterogeneity in higher animals: A comparison of the properties of autologous lung and intestinal mast cells from nonhuman primates, *J. Immunol.*, 137, 2001–2008, 1986.

Barrocas, A.M., Cochrane, D.E., Carraway, R.E., and Feldberg, R.S., Neurotensin stimulation of mast cell secretion is receptor-mediated, pertussis-toxin sensitive and requires activation of phospholipase C, *Immunopharmacology*, 41, 31–137, 1999.

Beck, B., *Bee Venom Therapy*, Appleton-Century, New York, 1935.

Befus, A.D. et al., Mucosal mast cells. I. Isolation and functional characteristics of rat intestinal mast cells, *J. Immunol.*, 128, 2475–2480, 1982a.

Befus, A.D., Pearce, F.L., Goodacre, R., and Bienenstock, J., Unique functional characteristics of mucosal mast cells, *Adv. Exp. Med. Biol.*, 149, 521–527, 1982b.

Bell, R.G., IgE, allergies and helminth parasites: A new perspective on an old conundrum, *Immunol. Cell Biol.*, 74, 337–345, 1996.

Bienenstock, J. et al., Mast cell heterogeneity, *Monogr. Allergy*, 18, 124–128, 1983.

Billingham, M.E. et al., Letter: An anti-inflammatory peptide from bee venom, *Nature*, 245, 163–164, 1973.

Blank, U. et al., Mast cells and inflammatory kidney disease, *Immunol. Rev.*, 217, 79–95, 2007.

Bloom, G.D., Fredholm, B., and Haegermark, O., Studies on the time course of histamine release and morphological changes induced by histamine liberators in rat peritoneal mast cells, *Acta Physiologica Scandinavica*, 71, 270–282, 1967a.

Bloom, G.D. and Haegermark, O., Studies on morphological changes and histamine release induced by bee venom, n-decylamine and hypotonic solutions in rat peritoneal mast cells, *Acta Physiologica Scandinavica*, 71, 257–269, 1967b.

Boyce, J.A., The biology of the mast cell, *Allergy & Asthma Proceedings*, 25, 27–30, 2004.

Boyum, A., Separation of lymphocytes, lymphocyte subgroups and monocytes: A review, *Lymphology*, 10, 71–76, 1977.

Boyum, A., Isolation of human blood monocytes with Nycodenz, a new non-ionic iodinated gradient medium, *Scandinavian J. Immunol.*, 17, 429–436, 1983.

Breithaupt, H. and Habermann, E., MCD-peptide from bee venom: Isolation, biochemical and pharmacolgical properties. [German], *Naunyn-Schmiedebergs Archiv fur Experimentelle Pathologie und Pharmakologie*, 261, 252–270, 1968.

Brody, M.J., Hakanson, R., Owman, C., and Sundler, F., An improved method for the histochemical demonstration of histamine and other compounds producing fluorophores with o-phthaldialdehyde, *J. Histochem. Cytochem.*, 20, 945–948, 1972.

Brown, J.H., *Toxicology and Pharmacology of Venoms from Poisonous Snakes*, Charles C. Thomas, Springfield Ill, 1973.

Bruni, A. et al., Interaction between nerve growth factor and lysophosphatidylserine on rat peritoneal mast cells, *FEBS Lett.*, 138, 190–192, 1982.

Bueb, J.L., Da Silva, A., Mousli, M., and Landry, Y., Natural polyamines stimulate G-proteins, *Biochem. J.*, 282, Part 2, 545–550, 1992.

Bueb, J.L. et al., Activation of Gi-like proteins, a receptor-independent effect of kinins in mast cells, *Mol. Pharmacol.*, 38, 816–822, 1990.

Buku, A., Mast cell degranulating (MCD) peptide: A prototypic peptide in allergy and inflammation, *Peptides*, 20, 415–420, 1999.

Buku, A., Blandina, P., Birr, C., and Gazis, D., Solid phase synthesis and biological activity of mast cell degranulating (MCD) peptide: A component of bee venom, *Int. J. Pept. Protein Res.*, 33, 86–93, 1989.

Buku, A., Condie, B.A., Price, J.A., and Mezei, M., [Ala12]MCD peptide: A lead peptide to inhibitors of immunoglobulin E binding to mast cell receptors, *J. Pept. Res.*, 66, 132–137, 2005.

Buku, A., Maulik, G., and Hook, W.A., Bioactivities and secondary structure of mast cell degranulating (MCD) peptide analogs, *Peptides*, 19, 1–5, 1998.

Buku, A., Mendlowitz, M., Condie, B.A., and Price, J.A., Histamine-releasing activity and binding to the FcepsilonRI alpha human mast cell receptor subunit of mast cell degranulating peptide analogues with alanine substitutions, *J. Med. Chem.*, 46, 3008–3012, 2003.

Buku, A., Mendlowitz, M., Condie, B.A., and Price, J.A., Partial alanine scan of mast cell degranulating peptide (MCD): Importance of the histidine- and arginine residues, *J. Pept. Sci.*, 10, 313–317, 2004.

Buku, A., Mirza, U., and Polewski, K., Circular dichroism (CD) studies on biological activity of mast cell degranulating (MCD) peptide analogs, *Int. J. Pept. Protein Res.*, 44, 410–413, 1994.

Buku, A. and Price, J.A., Further studies on the structural requirements for mast cell degranulating (MCD) peptide-mediated histamine release, *Peptides*, 22, 1987–1991, 2001a.

Buku, A., Price, J.A., Mendlowitz, M., and Masur, S., Mast cell degranulating peptide binds to RBL-2H3 mast cell receptors and inhibits IgE binding, *Peptides*, 22, 1993–1998, 2001b.

Buku, A. et al., Mast cell degranulating (MCD) peptide analogs with reduced ring structure, *J. Protein Chem.*, 11, 275–280, 1992.

Casolaro, V. et al., Human basophil/mast cell releasability. V. Functional comparisons of cells obtained from peripheral blood, lung parenchyma, and bronchoalveolar lavage in asthmatics, *Am. Rev. Respir. Dis.*, 139, 1375–1382, 1989.

Chai, O.H. et al., Histamine release induced by dendroaspis natriuretic peptide from rat mast cells, *Peptides*, 22, 1421–1426, 2001.

Chang, Y.H. and Bliven, M.L., Anti-arthritic effect of bee venom, *Agents Actions*, 9, 205–211, 1979.

Chao, M.V., Battleman, D.S., and Benedetti, M., Receptors for nerve growth factor, *Int. Rev. Cytol.*, 137B, 169–180, 1992.

Chen, I., Chiu, H., Huang, H., and Teng, C., Edema formation and degranulation of mast cells by Trimeresurus mucrosquamatus snake venom, *Toxicon*, 22, 17–28, 1984.

Chiu, H.F., Chen, I.J., and Teng, C.M., Edema formation and degranulation of mast cells by a basic phospholipase A2 purified from Trimeresurus mucrosquamatus snake venom, *Toxicon*, 27, 115–125, 1989.

Church, M.K. and Levi, S.F., The human mast cell, *J. Allergy Clin. Immunol.*, 99, 155–160, 1997.

Church, M.K., Okayama, Y., and el-Lati, S., Mediator secretion from human skin mast cells provoked by immunological and non-immunological stimulation, *Skin Pharmacol.*, 4, Suppl. 1, 15–24, 1991.

Clark, J.M. and Higginbotham, R.D., Cottonmouth moccasin venom: Fractionation of toxic and allergenic components and interaction with tissue mast cells, *Tex. Rep. Biol. Med.*, 29, 181–192, 1971.

Coberly, W. and Price, J.A., Analysis of histamine release assays using the bootstrap, *J. Immunol. Methods*, 296, 103–114, 2005.

Cochrane, D.E., Peptide regulation of mast-cell function, *Progr. Med. Chem.*, 27, 143–188, 1990.

Cormier, S.M., Physalia venom mediates histamine release from Mast cells, *J. Exp. Zool.*, 218, 117–120, 1981.

Costa, J.J., Weller, P.F., and Galli, S.J., The cells of the allergic response: Mast cells, basophils, and eosinophils, *JAMA*, 278, 1815–1822, 1997.

Dawicki, W. and Marshall, J.S., New and emerging roles for mast cells in host defence, *Curr. Opin. Immunol.*, 19, 31–38, 2007.

Demo, S.D. et al., Quantitative measurement of mast cell degranulation using a novel flow cytometric annexin-V binding assay, *Cytometry*, 36, 340–348, 1999.

Devi, A., The protein and nonprotein constituents of snake venoms, in *Venomous Animals and their Venoms Vol. I*, Bucherl, W., Buckley, E.E., and Deulofeu, V., Eds., Academic Press, New York, 1968, pp. 119–165.

Devillier, P., Renoux, M., Drapeau, G., and Regoli, D., Histamine release from rat peritoneal mast cells by kinin antagonists, *Eur. J. Pharmacol.*, 149, 137–140, 1988.

Dotimas, E.M., Hamid, K.R., Hider, R.C., and Ragnarsson, U., Isolation and structure analysis of bee venom mast cell degranulating peptide, *Biochimica et Biophysica Acta*, 911, 285–293, 1987.

Dreskin, S.C., Genetics of food allergy, *Curr. Allergy Asthma Rep.*, 6, 58–64, 2006.

Druey, K.M., Regulators of G protein signalling: Potential targets for treatment of allergic inflammatory diseases such as asthma, *Exp. Opin. Therap. Targets*, 7, 475–484, 2003.

Eaker, D. and WadstrÖm, T., Eds., *Natural Toxins*, Pergamon Press, New York, 1980.

Eklund, K.K., Mast cells in the pathogenesis of rheumatic diseases and as potential targets for anti-rheumatic therapy, *Immunol. Rev.*, 217, 38–52, 2007.

Else, K.J. and Finkelman, F.D., Intestinal nematode parasites, cytokines and effector mechanisms, *Int. J. Parasitol.*, 28, 1145–1158, 1998.

Emadi-Khiav, B., Mousli, M., Bronner, C., and Landry, Y., Human and rat cutaneous mast cells: Involvement of a G protein in the response to peptidergic stimuli, *Eur. J. Pharmacol.*, 272, 97–102, 1995.

Ennis, M., Histamine release from human pulmonary mast cells, *Agents Actions*, 12, 60–63, 1982.

Erlich, P., Beitrage zur kenntniss der quilinfarbunger und ihrer verivendung in der mikroskopischen technik, *Alch. Mikros. Anat.*, 13, 263, 1877.

Falcone, F.H., Haas, H., and Gibbs, B.F., The human basophil: A new appreciation of its role in immune responses, *Blood*, 96, 4028–4038, 2000.

Ferreira, M.B., da Silva, S.L., and Carlos, A.G., Atopy and helminths, *Allergie et Immunologie*, 34, 10–12, 2002.

Ferreira, S.H., Ng, K.K., and Vane, J.R., The continuous bioassay of the release and disappearance of histamine in the circulation, *Br. J. Pharmacol.*, 49, 543–553, 1973.

Ferreira, S.H. and Vane, J.R., Half-lives of peptides and amines in the circulation, *Nature*, 215, 1237–1240, 1967.

Ferry, X., Brehin, S., Kamel, R., and Landry, Y., G protein-dependent activation of mast cell by peptides and basic secretagogues, *Peptides*, 23, 1507–1515, 2002.

Ferry, X., Eichwald, V., Daeffler, L., and Landry, Y., Activation of betagamma subunits of G(i2) and G(i3) proteins by basic secretagogues induces exocytosis through phospholipase Cbeta and arachidonate release through phospholipase Cgamma in mast cells, *J. Immunol.*, 167, 4805–4813, 2001.

Finkelman, F.D. and Vercelli, D., Advances in asthma, allergy mechanisms, and genetics in 2006, *J. Allergy Clin. Immunol.*, 120, 544–550, 2007.

Flowers, A.L. and Hessinger, D.A., Mast cell histamine release induced by Portuguese man-of-war (Physalia) venom, *Biochem. Biophys. Res. Commun.*, 103, 1083–1091, 1981.

Foreman, J.C. and Jordan, C.C., Histamine release and vascular changes induced by neuropeptides, *Agents Actions*, 13, 105–106, 1983.

Franklin, R. and Baer, H., Comparison of honeybee venoms and their components from various sources, *J. Allergy Clin. Immunol.*, 55, 285–298, 1975.

Fredholm, B., Studies on a mast cell degranulating factor in bee venom, *Biochem. Pharmacol.*, 15, 2037–2043, 1966.

Fredholm, B. and Haegermark, O., Histamine release from rat mast cell granules induced by bee venom fractions, *Acta Physiologica Scandinavica*, 71, 357–367, 1967a.

Fredholm, B. and Haegermark, O., Histamine release from rat mast cells induced by a mast cell degranulating fraction in bee venom, *Acta Physiologica Scandinavica*, 69, 304–312, 1967b.

French, W.J. et al., Mojave toxin in venom of Crotalus helleri (Southern Pacific Rattlesnake): Molecular and geographic characterization, *Toxicon*, 44, 781–791, 2004.

Frossi, B., De, C.M., and Pucillo, C., The mast cell: An antenna of the microenvironment that directs the immune response, *J. Leukoc. Biol.*, 75 (4), 579–585, 2004.

Fujimoto, I. et al., Mast cell degranulating (MCD) peptide and its optical isomer activate GTP binding protein in rat mast cells, *FEBS Lett.*, 287, 15–18, 1991.

Fureder, W. et al., Differential response of human basophils and mast cells to recombinant chemokines, *Annals Hematol.*, 70, 251–258, 1995.

Furumoto, Y. et al., Rethinking the role of Src family protein tyrosine kinases in the allergic response: New insights on the functional coupling of the high affinity IgE receptor. [Review] [67 refs], *Immunolog. Res.*, 30, 241–253, 2004.

Galli, S.J., New concepts about the mast cell, *New Engl. J. Med.*, 328, 257–265, 1993.

Galli, S.J., Mast cells and basophils, *Curr. Opin.Hematol.*, 7, 32–39, 2000.

Galli, S.J. and Costa, J.J., Mast-cell-leukocyte cytokine cascades in allergic inflammation, *Allergy*, 50, 851–862, 1995.

Galli, S.J., Gordon, J.R., and Wershil, B.K., Mast cell cytokines in allergy and inflammation, *Agents Actions—Supplements*, 43, 209–220, 1993.

Galli, S.J. and Hammel, I., Mast cell and basophil development, *Curr. Opin.Hematol.*, 1, 33–39, 1994.

Galli, S.J., Maurer, M., and Lantz, C.S., Mast cells as sentinels of innate immunity, *Curr. Opin. Immunol.*, 11, 53–59, 1999.

Galli, S.J., Nakae, S., and Tsai, M., Mast cells in the development of adaptive immune responses, *Nature Immunol.*, 6, 135–142, 2005.

Galli, S.J., Tsai, M., and Piliponsky, A.M., The development of allergic inflammation, *Nature*, 454 (7203), 445–454, 2008.

Galli, S.J., Wedemeyer, J., and Tsai, M., Analyzing the roles of mast cells and basophils in host defense and other biological responses, *Int. J. Hematol.*, 75, 363–369, 2002.

Gauldie, J. et al., The peptide components of bee venom, *Eur. J. Biochem.*, 61, 369–376, 1976.

Gauldie, J., Hanson, J.M., Shipolini, R.A., and Vernon, C.A., The structures of some peptides from bee venom, *Eur. J. Biochem.*, 83, 405–410, 1978.

Gilfillan, A.M. and Tkaczyk, C., Integrated signalling pathways for mast-cell activation, *Nature Rev. Immunol.*, 6, 218–230, 2006.

Glenn, J.L. and Straight, R.C., Venom properties of the rattlesnakes (Crotalus) inhabiting the Baja California region of Mexico, *Toxicon*, 23, 769–775, 1985.

Glenn, J.L., Straight, R.C., Wolfe, M.C., and Hardy, D.L., Geographical variation in *Crotalus scutulatus scutulatus* (Mojave rattlesnake) venom properties, *Toxicon*, 21, 119–130, 1983.

Glenn, J.L., Straight, R.C., and Wolt, T.B., Regional variation in the presence of canebrake toxin in *Crotalus horridus* venom, *Comp. Biochem. Physiol. Pharmacol. Toxicol. & Endocrinol.*, 107 (3), 337–346, 1994.

Gmachl, M. and Kreil, G., The precursors of the bee venom constituents apamin and MCD peptide are encoded by two genes in tandem which share the same 3'-exon, *J. Biol. Chem.*, 270, 12704–12708, 1995.

Goth, A., Adams, H.R., and Knoohuizen, M., Phosphatidylserine: Selective enhancer of histamine release, *Science*, 173, 1034–1035, 1971.

Gushchin, I.S., Miroshnikov, A.I., and Martynov, V.I., Histamine-liberating action of MCD-peptide from bee venom, *Biull. Eksp. Biol. Med.*, 84, 78–80, 1977.

Gushchin, I.S., Miroshnikov, A.I., Martynov, V.I., and Sviridov, V.V., Histamine releasing and anti-inflammatory activities of MCD-peptide and its modified forms, *Agents Actions*, 11, 69–71, 1981.

Habermann, E., Bee and wasp venoms, *Science*, 177, 314–322, 1972.

Hakanson, R., Ronnberg, A.L., and Sjolund, K., Fluorometric determination of histamine with OPT: Optimum reaction conditions and tests of identity, *Anal. Biochem.*, 47, 356–370, 1972.

Hanson, J.M., Morley, J., and Soria-Herrera, C., Anti-inflammatory property of 401 (MCD-peptide), a peptide from the venom of the bee Apis mellifera (L.), *Br. J. Pharmacol.*, 50, 383–392, 1974.

Hartter, P., [Basic peptides in bee venom, II. Synthesis of two pentapeptides from the sequence of the mast-cell-degrading peptide (author's transl)]. [German], *Hoppe-Seylers Zeitschrift fur Physiologische Chemie*, 357, 1683–1693, 1976.

Hartter, P., [Basic peptides in bee venom, III. Synthesis of peptide fragments from the sequence of the mast-cell-degranulating peptide (author's transl)]. [German], *Hoppe-Seylers Zeitschrift fur Physiologische Chemie*, 358, 331–337, 1977.

Hartter, P., Basic peptides from bee venom, IV. Synthesis of the mast cell-degranulating peptide by liquid-phase fragment condensation (author's transl). [German], *Hoppe-Seylers Zeitschrift fur Physiologische Chemie*, 361, 503–513, 1980.

Hider, R.C., Honeybee venom: A rich source of pharmacologically active peptides, *Endeavour*, 12, 60–65, 1988.

Hider, R.C. and Ragnarsson, U., A comparative structural study of apamin and related bee venom peptides, *Biochimica et Biophysica Acta*, 667, 197–208, 1981.

Higginbotham, R.D., Mast cells and local resistance to Russell's viper venom, *J. Immunol.*, 95, 867–875, 1965.

Higginbotham, R.D. and Karnella, S., The significance of the mast cell response to bee venom, *J. Immunol.*, 106, 233–240, 1971.

Hino, R.H., Lau, C.K., and Read, G.W., The site of action of the histamine releaser compound 45/80 in causing mast cell degranulation, *J. Pharmacol. Exp. Therapeut.*, 200, 658–663, 1977.

Hirai, Y. et al., A new mast cell degranulating peptide "mastoparan" in the venom of Vespula lewisii, *Chem. Pharmaceut. Bull.*, 27, 1942–1944, 1979.

Horigome, K., Pryor, J.C., Bullock, E.D., and Johnson, E.M., Jr., Mediator release from mast cells by nerve growth factor. Neurotrophin specificity and receptor mediation, *J. Biol. Chem.*, 268, 14881–14887, 1993.

Ishizaka, T., Furitsu, T., and Inagaki, N., *In vitro* development and functions of human mast cells, *Int. Arch. Allergy Appl. Immunol.*, 94, 116–121, 1991.

Jasani, B., Kreil, G., Mackler, B.F., and Stanworth, D.R., Further studies on the structural requirements for polypeptide- mediated histamine release from rat mast cells, *Biochem. J.*, 181, 623–632, 1979.

Kaliner, M., Asthma and mast cell activation, *J. Allergy Clin. Immunol.*, 83, 510–520, 1989.

Kaplan, A.P., Baeza, M., Reddigari, S., and Kuna, P., Histamine-releasing factors, *Int. Arch. Allergy Appl. Immunol.*, 94, 148–153, 1991.

Keller, R., Interrelations between different types of cells. II. Histamine- release from the mast cells of various species by cationic polypeptides of polymorphonuclear leukocyte lysosomes and other cationic compounds, *Int. Arch. Allergy Appl. Immunol.*, 34, 139–144, 1968.

Kinet, J.P., The high-affinity IgE receptor (Fc epsilon RI): From physiology to pathology, *Annu. Rev. Immunol.*, 17, 931–972, 1999.

Kinet, J.P., The essential role of mast cells in orchestrating inflammation, *Immunolog. Rev.*, 217, 5–7, 2007.

Kitamura, Y., Heterogeneity of mast cells and phenotypic change between subpopulations, *Annu. Rev. Immunol.*, 7, 59–76, 1989.

Kopec, A., Panaszek, B., and Fal, A.M., Intracellular signaling pathways in IgE-dependent mast cell activation, *Arch. Immunol Ther. Exp (Warsz)*, 54 (6), 393–401, 2006.

Kostiza, T. and Meier, J., Nerve growth factors from snake venoms: Chemical properties, mode of action and biological significance, *Toxicon*, 34, 787–806, 1996.

Kumar, N.V., Wemmer, D.E., and Kallenbach, N.R., Structure of P401 (mast cell degranulating peptide) in solution, *Biophys. Chem.*, 31, 113–119, 1988.

Kumar, V., Abbas, A.K., and Fausto, N., Eds., *Robins and Cotran Pathologic Basis of Disease*, Elsevier Saunders, Philadelphia, PA, 2005.

Lagunoff, D., Martin, T.W., and Read, G., Agents that release histamine from mast cells, *Annu. Rev. Pharmacol. Toxicol.*, 23, 331–351, 1983.

Langel, U., Ed., *Handbook of Cell-Penetrating Peptides*, CRC Press, Atlanta, GA, 2007.

Lawrence, I.D. et al., Purification and characterization of human skin mast cells. Evidence for human mast cell heterogeneity, *J. Immunol.*, 139, 3062–3069, 1987.

Lebel, B., A high-sampling-rate automated continuous-flow fluorometric technique for the analysis of nanogram levels of histamine in biological samples, *Anal. Biochem*, 133, 16–29, 1983.

Lemanske, R.F., Jr., Joiner, K., and Kaliner, M., The biologic activity of mast cell granules, *J. Immunol.*, 130, 1881–1884, 1983.

Leung, K.B. and Pearce, F.L., A comparison of histamine secretion from peritoneal mast cells of the rat and hamster, *British J. Pharmacology*, 81, 693–701, 1984.

Lomize, A.L. and Popov, E.M., Theoretical conformation analysis of MCD peptide. [Russian], *Molekuliarnaia Biologiia*, 17, 1212–1219, 1983.

Machado, D.C. et al., Potential allergens stimulate the release of mediators of the allergic response from cells of mast cell lineage in the absence of sensitization with antigen-specific IgE, *Eur. J. Immunology*, 26, 2972–2980, 1996.

Maizels, R.M. and Holland, M.J., Parasite immunology: Pathways for expelling intestinal helminths, *Current Biology*, 8, R711–R714, 1998.

Marone, G., The role of mast cell and basophil activation in human allergic reactions, *Eur. Respiratory J.—Supplement*, 6, 446s–455s, 1989.

Marone, G., Stellato, C., Mastronardi, P., and Mazzarella, B., Mechanisms of activation of human mast cells and basophils by general anesthetic drugs, *Annales Francaises d Anesthesie et de Reanimation*, 12, 116–125, 1993.

Maruotti, N. et al., Mast cells in rheumatoid arthritis, *Clinical Rheumatology*, 26, 1–4, 2007.

Marx, J., Immunology. Mast cells defang snake and bee venom, *Science*, 313, 427, 2006.

Maurer, M. et al., What is the physiological function of mast cells? *Experimental Dermatology*, 12, 886–910, 2003.

Meier, J. and White, J., Eds., *Handbook of Clinical Toxicology of Animal Venoms and Poisons*, CRC Press, Boca Raton, 1995.

Metcalfe, D.D., Mast cell mediators with emphasis on intestinal mast cells, *Annals of Allergy*, 53, 563–575, 1984.

Metz, M. et al., Mast cells in the promotion and limitation of chronic inflammation, *Immunological Reviews*, 217, 304–328, 2007.

Metz, M. et al., Mast cells can enhance resistance to snake and honeybee venoms, *Science*, 313, 526–530, 2006.

Miller, L.A., Cochrane, D.E., Carraway, R.E., and Feldberg, R.S., Blockade of mast cell histamine secretion in response to neurotensin by SR 48692, a nonpeptide antagonist of the neurotensin brain receptor, *British J. Pharmacology*, 114, 1466–1470, 1995.

Minton, S.A., *Venom Diseases*, Charles C. Thomas, Springfield Ill, 1974.

Morrison, D.C., Roser, J.F., Cochrane, C.G., and Henson, P.M., Two distinct mechanisms for the initiation of mast cell degranulation, *Int. Archives Allergy Applied Immunology*, 49, 172–178, 1975.

Mousli, M., Bronner, C., Bueb, J.L., and Landry, Y., Evidence for the interaction of mast cell-degranulating peptide with pertussis toxin-sensitive G proteins in mast cells, *Eur. J. Pharmacology*, 207, 249–255, 1991a.

Mousli, M. et al., G protein activation: A receptor-independent mode of action for cationic amphiphilic neuropeptides and venom peptides, *Trends Pharmacol. Sci.*, 11, 358–362, 1990.

Mousli, M. et al., G-proteins as targets for non-immunological histamine releasers, *Agents Actions*, 33, 81–83, 1991b.

Mousli, M., Hugli, T.E., Landry, Y., and Bronner, C., A mechanism of action for anaphylatoxin C3a stimulation of mast cells, *J. Immunol.*, 148, 2456–2461, 1992.

Mousli, M., Hugli, T.E., Landry, Y., and Bronner, C., Peptidergic pathway in human skin and rat peritoneal mast cell activation, *Immunopharmacology*, 27, 1–11, 1994a.

Mousli, M. and Landry, Y., Role of positive charges of neuropeptide Y fragments in mast cell activation, *Agents Actions*, 41, C41–C42, 1994b.

Nadel, J.A., Biology of mast cell tryptase and chymase, *Annals New York Academy Sciences*, 629, 319–331, 1991.

Nadel, J.A., Biologic effects of mast cell enzymes, *Am. Review Respiratory Disease*, 145, S37–S41, 1992.

Nadler, M.J., Matthews, S.A., Turner, H., and Kinet, J.P., Signal transduction by the high-affinity immunoglobulin E receptor Fc epsilon RI: Coupling form to function, *Advances Immunology*, 76, 325–355, 2000.

Nakamura, T. and Ui, M., Simultaneous inhibitions of inositol phospholipid breakdown, arachidonic acid release, and histamine secretion in mast cells by islet-activating protein, pertussis toxin. A possible involvement of the toxin-specific substrate in the Ca2+-mobilizing receptor-mediated biosignaling system, *J. Biological Chemistry*, 260, 3584–3593, 1985.

Nosal, R., Pharmacology of histamine liberation. Cationic amphiphilic drugs and mast cells, *J. Physiol. Pharmacol.*, 45, 377–386, 1994.

O'Conner, R. et al., *The venom of the Honeybee (Apis mellifera) I. General character*, in Animal Toxins, Russell, F.E. and Saunders, P.R., Eds. Pergamon Press, Oxford, 1967, 17–22.

Onah, D.N. and Nawa, Y., Mucosal immunity against parasitic gastrointestinal nematodes, *Korean J. Parasitology*, 38, 209–236, 2000.

Parravicini, V. et al., Fyn kinase initiates complementary signals required for IgE-dependent mast cell degranulation, *Nature Immunology*, 3, 741–748, 2002.

Patella, V. et al., Human heart mast cells: A definitive case of mast cell heterogeneity, *Int. Arch. Allergy Immunol.*, 106, 386–393, 1995.

Paton, W.D., Compound 48/80: A potent histamine liberator, *British J. Pharmacology Chemotherapy*, 6, 499–508, 1951.

Paton, W.D.M., Histamine release by compounds of simple chemical structure, *Pharmacol. Reviews*, 9, 269–321, 1957.

Pearce, F.L., Functional heterogeneity of mast cells from different species and tissues, *Klinische Wochenschrift*, 60, 954–957, 1982.

Pearce, F.L., Mast cell heterogeneity, *Trends Pharmacological Sciences*, 4, 165–167, 1983.

Pearce, F.L., Mast cell heterogeneity: The problem of nomenclature, *Agents Actions*, 23, 125–128, 1988.

Pearce, F.L., Non-IgE-mediated mast cell stimulation, *Ciba Foundation Symposium*, 147, 74–87, 1989.

Pearce, F.L., Atkinson, G., and Ennis, M., Studies on histamine release induced by compound 48/80 and peptide 401, *Agents Actions*, 9, 63–64, 1979.

Pearce, F.L., Befus, A.D., Gauldie, J., and Bienenstock, J., Mucosal mast cells, *J. Immunology*, 128, 2481–2486, 1982.

Pearce, F.L. and Thompson, H.L., Some characteristics of histamine secretion from rat peritoneal mast cells stimulated with nerve growth factor, *J. Physiology*, 372, 379–393, 1986.

Platts-Mills, T.A., Woodfolk, J.A., Erwin, E.A., and Aalberse, R., Mechanisms of tolerance to inhalant allergens: The relevance of a modified Th2 response to allergens from domestic animals. [Review] [39 refs], *Springer Seminars Immunopathology*, 25, 271–279, 2004.

Price, J.A., Microplate assay for measurement of histamine release from mast cells, *BioTechniques*, 22, 958–962, 1997a.

Price, J.A., Two-layer gradient isolation of rat peritoneal mast cells, *BioTechniques*, 22, 616–618, 1997b.

Price, J.A. and Sanny, C.G., Histamine Releasing Factors in Venom, *Res. Commun. Pharmacology Toxicology*, 1, 127–136, 1996.

Profet, M., The function of allergy: Immunological defense against toxins, *Quarterly Review Biology*, 66, 23–62, 1991.

Rader, K. et al., Characterization of bee venom and its main components by high-performance liquid chromatography, *J. Chromatogr. A*, 408, 341–348, 1987.

Ramkumar, V., Stiles, G.L., Beaven, M.A., and Ali, H., The A3 adenosine receptor is the unique adenosine receptor which facilitates release of allergic mediators in mast cells, *J. Biological Chemistry*, 268, 16887–16890, 1993.

Read, G.W., Hong, S.M., and Kiefer, E.F., Competitive inhibition of 48/80-induced histamine release by benzalkonium chloride and its analogs and the polyamine receptor in mast cells, *J. Pharmacology Experimental Therapeutics*, 222, 652–657, 1982.

Read, G.W. and Kiefer, E.F., Benzalkonium chloride: Selective inhibitor of histamine release induced by compound 48/80 and other polyamines, *J. Pharmacology Experimental Therapeutics*, 211, 711–715, 1979.

Reischl, I.G., Coward, W.R., and Church, M.K., Molecular consequences of human mast cell activation following immunoglobulin E-high-affinity immunoglobulin E receptor (IgE-FcepsilonRI) interaction, *Biochem Pharmacol,* 58 (12), 1841–1850, 1999.

Repke, H. and Bienert, M., Mast cell activation—a receptor-independent mode of substance P action? *FEBS Lett.*, 221, 236–240, 1987a.

Repke, H. and Bienert, M., Structural requirements for mast cell triggering by substance P- like peptides, *Agents Actions*, 23, 207–210, 1988.

Repke, H., Piotrowski, W., Bienert, M., and Foreman, J.C., Histamine release induced by Arg-Pro-Lys-Pro(CH2)11CH3 from rat peritoneal mast cells, *J. Pharmacol. Exp. Ther.*, 243, 317–321, 1987b.

Rivera, J., Snake bites and bee stings: The mast cell strikes back, *Nature Medicine*, 12, 999–1000, 2006.

Rivera, J. and Olivera, A., A current understanding of Fc epsilon RI-dependent mast cell activation, *Curr. Allergy Asthma Rep.,* 8 (1), 14–20, 2008.

Rivera, J. et al., Macromolecular protein signaling complexes and mast cell responses: A view of the organization of IgE-dependent mast cell signaling. [Review] [22 refs], *Molecular Immunology*, 38, 1253–1258, 2002.

Rothschild, A., Histamine release by basic compounds, *Handbook Exp. Pharmacol.*, 28, 386–430, 1966.

Russell, F., *Snake Venom Poisoning*, Scholium International, Great Neck, NY, 1983.

Schmidt J.O., *Chemistry, Pharmacology, and Chemical Ecology of Ant Venoms*, in Venoms of the Hymenoptera, Piek, T., Ed., Academic Press Inc., London, 1986, 425–508.

Seebeck, J., Krebs, D., and Ziegler, A., Influence of salmeterol and benzalkonium chloride on G-protein-mediated exocytotic responses of rat peritoneal mast cells, *Eur. J. Pharmacology*, 397, 19–24, 2000.

Seldin, D.C. et al., Homology of the rat basophilic leukemia cell and the rat mucosal mast cell. *Proceedings of the National Academy of Sciences of the United States of America,* 82 (11), 3871–3875, 1985.

Shanahan, F., Denburg, J.A., Bienenstock, J., and Befus, A.D., Mast cell heterogeneity, *Canadian J. Physiology Pharmacology*, 62, 734–737, 1984.

Shanahan, F. et al., Mast cell heterogeneity: Effects of neuroenteric peptides on histamine release, *J. Immunol.*, 135, 1331–1337, 1985.

Shore, P.A., Fluorometric assay of histamine, *Methods in Enzymology*, 17, 842–845, 1971.

Shore, P.A., Burkhalter, A., and Cohn, V.H., Jr., A method for the fluorometric assay of histamine in tissues, *J. Pharmacology Experimental Therapeutics*, 127, 182–186, 1959.

Somerfield, S.D., The concept of anti-inflammatory peptides, *New Zealand Medical J.*, 97, 298–300, 1984.
Stanworth, D.R. and Ghaderi, A.A., The role of high and low affinity IgE receptors in cell signalling processes, *Molecular Immunology*, 27, 1291–1296, 1990.
Steinmetz, W.E., Bianco, T.I., Zollinger, M., and Pesiri, D., Characterization of the multiple forms of mast cell degranulating peptide by NMR spectroscopy, *Peptide Research*, 7, 77–82, 1994.
Straight, R.C., Glenn, J.L., Wolt, T.B., and Wolfe, M.C., Regional differences in content of small basic peptide toxins in the venoms of *Crotalus adamanteus* and *Crotalus horridus*, *Comparative Biochemistry Physiology—B: Comparative Biochemistry*, 100, 51–58, 1991.
Straight, R.C., Glenn, J.L., Wolt, T.B., and Wolfe, M.C., North-south regional variation in phospholipase A activity in the venom of Crotalus ruber, *Comparative Biochemistry Physiology—B: Comparative Biochemistry*, 103, 635–639, 1992.
Swieter, M., Hamawy, M.M., Siraganian, R.P., and Mergenhagen, S.E., Mast cells and their microenvironment: The influence of fibronectin and fibroblasts on the functional repertoire of rat basophilic leukemia cells, *J. Periodontology*, 64, 5 Suppl., 492–496, 1993.
Tainsh, K.R. et al., Mast cell heterogeneity in man: Unique functional properties of skin mast cells in response to a range of polycationic stimuli, *Immunopharmacology*, 24, 171–180, 1992.
Theoharides, T.C. and Kalogeromitros, D., The critical role of mast cells in allergy and inflammation, *Annals New York Academy Sciences*, 1088, 78–99, 2006.
Tilley, S.L. et al., Adenosine and inosine increase cutaneous vasopermeability by activating A(3) receptors on mast cells, *J. Clin. Investigation*, 105, 361–367, 2000.
Tkaczyk, C. and Gilfillan, A.M., Fc(epsilon)Ri-dependent signaling pathways in human mast cells, *Clin. Immunol,* 99 (2), 198–210, 2001.
Tobin, M.C., Karns, B.K., Anselmino, L.M., and Thomas, L.L., Potentiation of human basophil histamine release by protamine: A new role for a polycation recognition site, *Molecular Immunology*, 23, 245–253, 1986.
Tomita, U. et al., Direct activation of GTP-binding proteins by venom peptides that contain cationic clusters within their alpha-helical structures, *Biochem. Biophys. Res. Commun.*, 178, 400–406, 1991.
Tu, A.T., Ed., *Venoms: Chemistry and Molecular Biology*, John Wiley & Sons, New York, 1977.
Tu, A.T., Ed., *Rattlesnake Venoms*, Marcel Dekker Inc., New York, 1982.
Tu, A.T., Overview of snake venom chemistry, *Adv. Exp. Med. Biol.*, 391, 37–62, 1996.
Turner, H. and Kinet, J.P., Signalling through the high-affinity IgE receptor Fc epsilonRI, *Nature*, 402, 6760 Suppl, B24–B30, 1999.
Warner, J.A., Yancey, K.B., and MacGlashan, D.W., Jr., The effect of pertussis toxin on mediator release from human basophils, *J. Immunol.*, 139, 161–165, 1987.
Wasserman, S.I., The human lung mast cell, *Environmental Health Perspectives*, 55, 259–269, 1984.
Wedemeyer, J. and Galli, S.J., Mast cells and basophils in acquired immunity, *British Medical Bulletin*, 56, 936–955, 2000a.
Wedemeyer, J., Tsai, M., and Galli, S.J., Roles of mast cells and basophils in innate and acquired immunity, *Current Opinion Immunology*, 12, 624–631, 2000b.
White, M.V., Kowalski, M.L., and Kaliner, M.A., Mast cell secretagogues, *Progress Clin. Biological Res.*, 297, 83–100, 1989.
Williams, C.M. and Galli, S.J., The diverse potential effector and immunoregulatory roles of mast cells in allergic disease, *J. Allergy Clin. Immunology*, 105, 847–859, 2000.
Zhang, S.F., Shi, W.J., Cheng, J.A., and Zhang, C.X., Cloning and characterization analysis of the genes encoding precursor of mast cell degranulating peptide from 2 honeybee and 3 wasp species, *I Chuan Hsueh Pao—Acta Genetica Sinica*, 30, 861–866, 2003.
Ziai, M.R. et al., Mast cell degranulating peptide: A multi-functional neurotoxin, *J. Pharmacy Pharmacology*, 42, 457–461, 1990.

18 Mastoparans

Sarah Jones and John Howl

CONTENTS

The tetradecapeptide mastoparan (MP) translocates cell membranes as an amphipathic α-helix to regulate the activity of heterotrimeric G proteins. Besides its most widely acknowledged role as an allosteric modulator of these proteins, MP and its structurally related analogs bind additional intracellular targets and modulate a plethora of biological phenomena including mast cell degranulation, insulin secretion, intracellular calcium mobilization, and cell death. Initially isolated from the venom of *Vespula lewisii*, many endogenous variants continue to be discovered and this prototypic peptide has generated numerous structural and chimeric analogs. Some of the most significant of these facilitate mitochondrial apoptosis and serve as cell-penetrating peptides.

18.1 INTRODUCTION

Mastoparans comprise a group of mast cell degranulating amphipathic helical peptides found in wasp venom. Often, if not relentlessly, mastoparans are referred to as "remarkable tetradecapeptides." Although this is somewhat of an unconventional scientific introduction, it is none other than the truth. Before introducing and evaluating the biological effects of these peptides, a process that only peptide scientists may find captivating, one must emphasize the unique and ever forthcoming biological opportunities that these peptides have to offer. To date, MP analogs have served as novel inducers of apoptosis, cell-penetrant delivery vectors, modulators of secretory events, and regulators of intracellular calcium homeostasis, to name but a few.

18.2 SOURCE OF PEPTIDES

Initially isolated from the venom of *Vespula lewisii* (Hirai et al., 1979) the C-terminally amidated sequence, *H*-INLKALAALAKKIL-NH_2, has served as a prototype for numerous synthetic analogs, peptide chimeras in addition to being the forerunner to the discovery of a plethora of endogenous mastoparans found in the venoms of numerous other species (Table 18.1). Amongst the most recently isolated endogenous mastoparans are the myotoxic peptide polybia-MP1 isolated from the venom of *Polybia paulista* (Rocha et al., 2007), three novel biologically active peptides from the venom of

TABLE 18.1
Endogenous Mastoparans

Mastoparans	Sequence	Species Origin	Reference
Mastoparan	*H*-INLKALAALAKKIL-NH_2	*Vespula lewisii*	Hirai et al., 1979
Mastoparan X	*H*-INWKGIAAMAKKLL-NH_2	*Vespa xanthoptera*	Higashijima et al., 1983
Mastoparan M	*H*-INLKAIAALAKKLL-NH_2	*Vespa mandarinia*	Wu et al., 1999
Mastoparan A	*H*-IKWKAILDAVKKVL-NH_2	*Vespa analis*	Murata et al., 2006
Mastoparan T	*H*-INLKAIAAFAKKLL-NH_2	*Vespa tropica*	Murata et al., 2006
Mastoparan II	*H*-INLKALAALVKKVL-NH_2	*Vespa orientalis*	Murata et al., 2006
Mastoparan C	*H*-INLKALLAVAKKIL-NH_2	*Vespa crabro*	Argiolas and Pisano, 1984
Mastoparan B	*H*-LKLKSIVSWAKKVL-NH_2	*Vespa basalis*	Murata et al., 2006
Polistes Mastoparan	*H*-VDWKKIGQHILSVL-NH_2	*Polistes jadwagae*	Nakajima et al., 1985
Protopolybia MPI	*H*-INWLKLGKKVSAIL-NH_2	*Protobolybia exigua*	Mendes et al., 2005
Protopolybia MPII	*H*-INWKAIIEAAKQAL-NH_2	*Protobolybia exigua*	Mendes et al., 2005
Protopolybia MPIII	*H*-INWLKLGKAVIDAL-NH_2	*Protobolybia exigua*	Mendes et al., 2005
Polistes-mastoparan-R1	*H*-INWLKLGKKILGAI-NH_2	*Polistes rothneyi iwatai*	Murata et al., 2006
Polistes-mastoparan-R2	*H*-LNFKALAALAKKIL-NH_2	*Polistes rothneyi iwatai*	Murata et al., 2006
Polistes-mastoparan-R3	*H*-INWLKLGKQILGAL-NH_2	*Polistes rothneyi iwatai*	Murata et al., 2006
Pronectarina mastoparan	*H*-INWKALLDAAKKVL-NH_2	*Pronectarina sylveirae*	Murata et al., 2006
Polybia-MPI	*H*-IDWKKLLDAAWQIL-NH_2	*Polybia paulista*	de Souza et al., 2004
Agelaia-MP	*H*-INWLKLGKAIIDAL-NH_2	*Agelaia pallipes pallipes*	Murata et al., 2006
Unnamed	*H*-INWAKLGKLALEVI-NH_2	*Parapolybia indica*	Murata et al., 2006
Unnamed	*H*-INWSKLLSMAKEVI-NH_2	*Ropalidia sp.* (Papua New Guinea)	Murata et al., 2006
Eumenine mastoparan-AF	*H*-INLLKIAGKIIKSL-NH_2	*Anterhynchium flavomarginatum micado*	dos Santos Cabrera et al., 2004

Polistes rothneyi iwatai, polistes-mastoparan-R1, 2, and 3, with potent mast cell degranulating properties (Murata et al., 2006), and the antimicrobial peptide eumenine mastoparan-AF isolated from the venom of *Anterhynchium flavomarginatum micado* (dos Santos Cabrera et al., 2004). The reader is advised to consult Table 18.1 for an extensive reference list of endogenous MPs isolated to date.

Strangely, the biosynthetic pathway and the biological precursors of the endogenous MPs have not yet been identified owing to an inability to solve the putative precursor gene sequence. It was not known whether MP was a product of the genome or nonribosomally produced. Recently, however, one seminal study by Xu and coworkers (2006) screened the venom gland cDNA library of the wasp *Vespula magnifica* and identified the first biological precursor of MP, the "mastoparanogen." The cDNA of mastoparanogen encodes a polypeptide of 40 amino acids. The N-terminal is composed of multiple acidic residues with multiple tandem repeats of AXP (X is aspartic acid or glycine). The C-terminal is composed of a MP peptide (INLKAIAALAKKLL) and extended with a glycine, as the donator of NH_4 for the amidation of the C-terminal leucine.

18.3 MOLECULAR MECHANISMS OF ACTION

Consensus has long dictated that the formation of an amphipathic α-helix is a major molecular and structural determinant for the biological activity of MP and its analogs. In aqueous solution the structure of MP is largely disordered, but in the presence of lipid and bilayer mimetic environments (Higashijima et al., 1983; Hori et al., 2001) and when interacting with heterotrimeric G proteins, particularly pertussis toxin-sensitive (PTX) G_i and G_o (Sukumar and Higashijima, 1992), MP adopts an amphipathic α-helical structure. Moreover, the ability of MP and related analogs to translocate cell membranes as amphipathic α-helices allows these peptides to present a cationic hydrophilic face on the inner side of the plasma membrane. It is for this very reason that MP has become most widely acknowledged as a receptor-independent allosteric regulator of heterotrimeric G proteins because cationic charge, provided by Lysyl residues 4, 11, and 12 together with the amino terminus, mimics the G-protein binding domain of G protein-coupled receptors (GPCRs) (Figure 18.1). Once bound to G protein α subunits, MP catalyses guanosine diphosphate (GDP)/guanosine triphosphate (GTP) exchange and stimulates GTPase activity, in a manner similar to the activated receptor (Higashijima et al., 1988, 1990; Weingarten et al., 1990; Oppi et al., 1992).

Recently, two orientational states of MP α-helices in a lipid bilayer have been observed distributed 90% in plane and 10% transmembrane (Hori et al., 2001). Interestingly, the first crystal structure of an endogenous MP isolated from *Polistes jadwagae* (Table 18.1) adopts an amphipathic α helical structure even in the absence of detergents and thus its amphipathicity is not restricted to biological membranes (Lui et al., 2007). This feature presents numerous and as of yet unidentified biological targets for this MP analog, because, interaction with other molecules may contribute to the formation of its "biologically active" amphipathic helical structure.

To date, a plethora of studies have used MP merely as a means to establish the role of heterotrimeric G proteins in a variety of biological systems. However, there is increasing evidence to suggest that MP demonstrates a distinct biological activity besides the modulation of heterotrimeric G proteins. Additional targets comprise those that modulate intracellular Ca^{2+} concentration $[Ca^{2+}]_i$ including, Ca^{2+}-ATPases (Longland et al., 1998), calmodulin (Malencik and Anderson, 1983), and glycogen phosphorylase (Hirata et al., 2003), in addition to the small monomeric GTP binding proteins rho and rac (Koch et al., 1991) and mitochondrial membranes (Nicolay et al., 1994).

18.4 BIOLOGICAL AND PATHOLOGICAL MODE OF ACTION

18.4.1 SECRETION

MP and related analogs stimulate exocytosis in various cell types to include the secretion of histamine, serotonin (5-HT), insulin, β-hexoseaminidase, and catecholamines. Although the regulation of

(a)

(b)

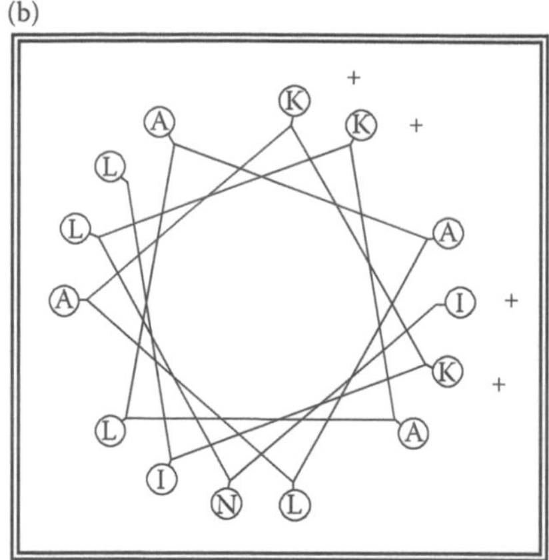

FIGURE 18.1 Predicted secondary structure of MP. (a) The predicted conformation of the MP in lipid bilayers. The cylinder diagram was compiled by the RCSB (Research Collaboratory for Structural Bioinformatics) Protein Data Bank (http://www.rcsb.org/pdb/) and is based on data obtained from ^{1}H-NMR spectroscopy in detergent micelles, solid-state ^{2}H-NMR and ^{15}N-NMR spectroscopy in lipid bilayers. (Adapted from Hori, Y. et al., *Eur. J. Biochem.*, 268, 302–309, 2001.) (b) The predicted helical wheel projection for MP (*H*-INLKALAALAKKIL-NH_2) with localization of cationic charge provided by lysyl residues and the amino terminus. The cationic hydrophilic face is considered to make contact with G proteins and contributes to the amphipathicity of this helical peptide. The helical wheel projection was produced using the Interactive Java Helical Wheel Applet at Berkley (http://marqusee9.berkeley.edu/kael/helical.htm). (Adapted from Jones, S. and Howl, J., *Regul. Pept.*, 121, 121–128, 2004.)

exocytosis involves a complexity of signal transduction proteins, common intracellular targets for MP are also important components of stimulus–secretion coupling, such as PTX-sensitive G proteins and targets that regulate $[Ca^{2+}]_i$.

18.4.1.1 MP and Secretion from Mast Cells

It has long been established that cationic amphipathic peptides including venom peptides stimulate secretion of pharmacological mediators from mast cells via a receptor-independent activation of PTX-sensitive G-proteins (Mousli et al., 1990). A more detailed analysis of the secretory mechanisms reveals that cationic peptides such as MP interact with G_{i2} and G_{i3} heterotrimeric G proteins to facilitate dissociation of the βγ subunits that activate the β isoform of phospholipase C (PLC) (Ferry et al., 2001). The subsequent hydrolysis of phophatidylinositol 4,5-bisphosphate (PIP_2) by PLCβ to inositol 1,4,5-trisphophate (IP_3) and diacylglycerol (DAG) stimulates a rise in $[Ca^{2+}]$ through IP_3 binding to its cognate receptor located on the endoplasmic reticulum (ER). The resultant efflux of Ca^{2+} into the cytoplasm triggers exocytotic events including activation of phospholipase D (PLD), an enzyme implicated in membrane ruffling and mast cell secretion.

RBL-2H3 (rat basophilic leukemia) cells have served as a prototypic model for the study of mast cell degranulation and exocytotic phenomena and have been widely used to elucidate the mechanisms of MP-induced secretion of pharmacological mediators such as β-hexoseaminidase and 5-HT. Moreover, in this cell line, secretion of β-hexoseaminidase by MP is most likely to also involve the activation of the small monomeric G proteins of the Rho family, Rac and Cdc42, with a concomitant stimulation of PLC and PLD (Chahdi et al., 2004). Thus, a dual activation of both heterotrimeric and monomeric G proteins appears to be an integral component of exocytotic events stimulated by MP.

MP is a strong stimulator of PLD in intact cells (Farquhar et al., 2002), and to introduce further complexities to the effects of MP on secretory events, MP has also been shown to selectively activate the plasma-membrane-bound PLD2 isoenzyme in RBL-2H3 cells (Chahdi et al., 2003). Unlike PLD1, which is located in cytoplasmic secretory granules, activation of plasma membrane PLD2 is only dependent upon intracellular production of PIP_2 and independent of both monomeric and heterotrimeric G proteins.

It is of particular importance to note that novel chimeric MP secretagogues have also served as valuable tools to establish differential mechanisms of secretion for different pharmacological mediators present within mast cells (Farquhar et al., 2002). A more comprehensive and detailed account of which, shall be presented in Section 18.5.2.

18.4.1.2 MP as an Insulin Secretagogue

Yokokawa et al., (1989) first identified MP as a potent facilitator of insulin secretion from pancreatic β cells. This action was found to be both PTX- and phospholipase (PLA_2)-dependent. In contrast, Eddlestone et al. (1995) demonstrated that the exocytotic release of insulin by MP in the insulin-secreting cell lines RINm5F and HIT is independent of PTX-sensitive G proteins but dependent upon an elevation in $[Ca^{2+}]_i$. Such observations involve a MP-induced suppression of ATP-sensitive K^+ channel activity, leading to cellular depolarization and activation of voltage-dependent Ca^{2+} channels with a subsequent entry of Ca^{2+}. However, this finding was contested by Komatsu and coworkers (1993) because the increase in $[Ca^{2+}]_i$ induced by MP was not found to be involved in the ability of MP to stimulate insulin secretion, an observation that was later confirmed in human pancreatic islets by Straub and colleagues (1998). Putative targets for MP include the exocytotic stimulatory G protein G_e (Komatsu et al., 1993; Straub et al., 1998). Further investigations by this group demonstrated that MP stimulates insulin release by the activation of signaling pathways that can be augmented by nutrients such as glucose, a phenomenon that is heavily dependent upon the intracellular concentration of GTP (Straub et al., 1998), yet independent of PTX treatment.

Debate continued until an alternative intracellular locus for MP was recently identified that would partly account for the above discrepancies. The most recent molecular mechanism of action, involved in MP-induced insulin release, shows direct activation by MP of the Rho family of small monomeric GTP binding proteins, particularly Rac (Amin et al., 2003). MP facilitated the translocation of cytosolic Rac to the membrane whereupon a strong colocalization of rac with insulin was observed in β cells treated with MP. Interestingly, MP has also been shown to stimulate a novel protein kinase in β cells that selectively uses GTP as a phosphoryl donor. These novel protein kinases specifically phosphorylate proteins at histidine residues, such as the β subunit of heterotrimeric G proteins (Kowluru et al., 2002).

It is of particular interest that MP potentiates the action of insulin by facilitating the uptake of glucose in rat adipocytes by enhancing the fusion of glucose transporter (GLUT4)-containing vesicles to the plasma membrane (Omata et al., 2005).

18.4.2 MP as a Modulator of $[Ca^{2+}]_i$

MP and various MP analogs interact with proteins that regulate intracellular Ca^{2+} homeostasis. More specifically, MP and chimeric analogs mobilize a concentration-dependent (1 to 20 μM) release of Ca^{2+} from both porcine cerebellar microsomes and rabbit skeletal muscle sarcoplasmic reticulum (SR) (Longland et al., 1998). Further analysis demonstrated that MP inhibits Ca^{2+}-ATPases and activates the ryanodine receptor (RyR) to promote Ca^{2+}-efflux from subcellular vesicles (Longland et al., 1998). Moreover, the molecular interaction of MP with the SR Ca^{2+}-ATPase abolishes cooperative Ca^{2+} binding and is a consequence of the ability of MP to adopt an amphiphilic helical conformation in membranes (Longland et al., 1999). Interestingly, a 97-kDa mastoparan-binding protein, first identified in rabbit muscle SR (Hirata et al., 2000) has more recently been identified as glycogen phosphorylase (GP) (Hirata et al., 2003). Moreover, there is evidence that GP negatively regulates Ca^{2+} release via the RyR. Hence, the binding of MP to GP may dissociate GP from the SR to relieve inhibition of RyR activity by GP.

Activation of PLC, with a subsequent phosphoinositide breakdown, to produce the ER-binding second messenger IP_3 is yet another mechanism by which MP promotes release of Ca^{2+} from intracellular stores (Lin et al., 1997; Schnabel et al., 1997). Furthermore, MP induces suppression of ATP-sensitive K^+ channel activity, which in turn promotes cellular depolarization and activation of voltage-dependent Ca^{2+} channels with a subsequent influx of extracellular Ca^{2+} (Eddlestone et al., 1995).

18.4.3 MP as a Cytotoxin

The mechanism(s) by which MP and structural analogs mediate cell death remains unresolved. The ability of MP and its analogs to interact and insert into biological membranes has prompted a more nonspecific necrotic model of MP-induced cell death (Suh et al., 1996). Accordingly, and particularly at higher concentrations, MP facilitates random pore formation in the plasma membrane, leading to the increased permeability of extracellular cations and subsequent cell lysis (Epand et al., 1995). Conversely, cell free models using isolated mitochondria have shown that MP induces apoptosis through interaction with mitochondrial membranes (Nicolay et al., 1994) and opening of mitochondrial permeability transition pores (Pfeiffer et al., 1995; He and Lemasters, 2003). This in turn promotes the release of apoptogenic proteins such as cytochrome *c*, which consequentially leads to activation of the caspase cascade (Ellerby et al., 1997). Most recently, novel target-selective chimeric MP analogs have demonstrated the ability to induce mitochondrial apoptosis in living cellular systems and shall be discussed in Section 18.6.2. Furthermore, the cytotoxic actions of MP have also been attributed to MP sharing homology with the amphipathic helix forming death domains of the tumor necrosis factor receptor-1 (TNFR-1) and Fas (Chapman, 1995).

Given the central role of Ca^{2+} to the orchestration of apoptotic events, the release of calcium from intracellular stores cannot be discounted as a mechanism for mediating the cytotoxic effects of MP. Moreover, MP-induced apoptosis of cultured cerebellar granule neurons is initiated by activation of PLC and a subsequent increase in $[Ca^{2+}]_i$ (Lin et al., 1997). However, Lin and coworkers postulate that disruption to plasma membrane integrity and nonspecific entry of extracellular Ca^{2+}, either as a secondary or parallel process, may ultimately be responsible for MP-induced cell death of these cells, particularly at higher concentrations.

18.4.4 Additional Biological Activities of MP

Many reports indicate that MP stimulates PLA_2 (Argiolas and Pisano, 1983; Gusovsky et al., 1991) with a consequential liberation of arachidonic acid. One such study established that the MP-induced release of arachidonic acid stimulates mitogenesis of Swiss 373 cells (Gil et al., 1991). This observation is in stark contrast to the numerous reports concerning the cytotoxic nature of MP. Interestingly, MP and a range of structural analogs promote ERK1/2 (alternatively designated p42/p44 mitogen-activated protein kinase, MAPK) phosphorylation in U373MG astrocytoma cells (Jones and Howl, 2004), although the consequential activation of ERK1/2 has no proliferative effect and the biological outcome of it remains unresolved. The highly potent structural analog Mas-7 (Table 18.2) has also been reported to activate ERK1/2, although on this occasion a consequential activation of mitogenesis was observed (McKillop et al., 1999). In contrast, Mas-7 inhibited proliferation in nontransformed hepatocytes. Single reports also claim that MP modulates the stress-activated MAPKs and include inhibition of p38 MAPK in human monocytes and peripheral blood mononuclear cells (Solomon et al., 1998) and activation of c-Jun N-terminal kinase in human embryonic

TABLE 18.2
Synthetic Structural Analogs

Analog	Sequence	Properties	Reference
Mastoparan 7	*H*-INLKALAALAKALL-NH_2	Highly potent synthetic MP analog	Higashijima et al., 1990
Mastoparan 17	*H*- INLKALAALAKKLL-*OH*	Inactive MP analog	Higashijima et al., 1990
Mitoparan	*H*-INLKKLAKL(Aib)KKIL-NH_2; Aib = α-aminoisobutyric acid	Potent mitochondriotoxic CPP	Jones and Howl, 2004
Mastoparan S	*H*-INWKGIASMAibRQVL-NH_2	Stimulates G_s	Sukumar et al., 1997
Mastoparan 11	*H-INLKALAALKKKLL*-NH_2	Non-lytic MP analog	Mukai et al., 2007

kidney cells (Yamauchi et al., 2000). For a further appraisal of the biological effects of MP, the reader is advised to consult the work of Soomets and coworkers (1997).

18.5 STRUCTURAL ANALOGS

18.5.1 Structure-Activity Relationship (SAR) Studies

Relatively few studies have specifically focused upon the SARs of MP. Investigations that provide SAR data have used different bioassays to assess the activities of MP analogs. Of particular interest are studies of the stimulation of heterotrimeric G proteins (Higashijima et al., 1990) and inhibition of NADPH oxidase activation (Tisch-Indelson et al., 2001) using purified and reconstituted proteins in cell free systems. However, studies utilizing intact cellular systems are perhaps a more rigorous assessment of bioactivity and require plasma membrane translocation of MP analogs, subsequent to protein binding and activation. One such study by Jones and Howl (2004) used secretory and cytotoxicity efficacies to evaluate the impact of charge delocalization within the hydrophilic face of MP. As previously mentioned, the amphipathic nature of MP presents lysyl residues at positions 4, 11, and 12, which, together with the amino terminus, contribute to a cationic hydrophilic face that contacts G proteins (Figure 18.1). Using living cellular systems, synthetic analogs were designed to address the functional significance of these residues. Using arginine substitution to delocalize the cationic charge provided by lysyl residues, three mono-arginine substituted analogs were synthesized: $[Arg]^4$MP, $[Arg]^{11}$MP, and $[Arg]^{12}$MP.

Charge delocalized MP analogs activated intracellular signaling pathways, leading to β-hexoseaminidase secretion, and induced a concentration-dependent decrease in cellular viability. All of the active analogs studied were better secretagogues than MP. Given that cationic peptide-induced secretion from mast cells is predominantly mediated by heterotrimeric G proteins, these findings indicated that charge delocalization positively enhances the ability of MP analogs to reach, bind, or stimulate heterotrimeric G proteins. However, a very different rank order was observed when comparing the cytotoxicity indices of MP analogs and further indicates that additional intracellular protein targets, besides heterotrimeric G proteins, modulate MP-induced cytotoxicity.

Of particular interest was the synthesis and evaluation of an MP analog in which positive charge provided by lysyl residues was more dramatically delocalized by the substitution of additional lysyl residues together with a known helix promoter and stabilizer α-aminoisobutyric acid (Aib) to produce [Lys^5, Lys^8, Aib^{10}]MP (mitoparan), Table 18.2. This novel MP analog exhibited enhanced secretory and cytotoxicity efficacies. The design of [Lys^5, Lys^8, Aib^{10}]MP incorporated sequence permutations that were predicted to be compatible with the formation of an amphipathic helix. Moreover, Lys^8, together with lysyl residues at positions 4, 5, 11, and 12, is located within an extremely hydrophilic face, contributing to the increased amphipathicity of this peptide and offering a simple explanation for the enhanced bioactivities of this peptide. Although this rather simplistic observation may account for an enhanced plasma membrane translocation, intriguingly, further investigations demonstrated rather more specific molecular interactions of this MP analog. Moreover, [Lys^5, Lys^8, Aib^{10}]MP crosses plasma membranes, subsequently redistributes to specifically interact with mitochondria and ultimately induces mitochondrial-dependent apoptosis, features that cannot simply be explained by the enhanced amphipathicity of this molecule (Jones et al., 2008). Interestingly, the design of the potent MP analog Mas 7 (Higashijima et al., 1990) incorporated elimination of the cationic charge at position 12 of MP with a mono-substitution of Ala (Table 18.2), thus further indicating the complex biology of this peptide besides its ability to form an amphipathic structure.

To advance these findings, [Lys^5, Lys^8, Aib^{10}]MP was extended by simple N-terminal acylation with peptidyl *address* motifs to produce novel apoptotic chimeric analogs, with further enhanced cytotoxic potencies compared to [Lys^5, Lys^8, Aib^{10}]MP alone (see Section 18.6.2).

As previously mentioned, at higher concentrations MP and the majority of its structural analogs promote random pore formation of the plasma membrane. This nonspecific activity is of obvious concern when investigating the biological actions of this peptide. A most recent development has been the synthesis of a nonlytic MP analog, Mas-11 ([Lys^{10}, Leu^{13}]MP) (Mukai et al., 2007) (Table 18.2). Whilst retaining and even exceeding the secretory potency of MP, Mas-11 is nonlytic up to and including concentrations of 100 μM.

It is noteworthy that the MP analog Mas-17 ([Lys^6, Leu^{13}]MP) (Table 18.2) is essentially inactive (Higashijima et al., 1990) and incompatible with the formation of an amphipathic α-helix. As such, it is extensively used as a negative control in numerous MP-related investigations.

18.5.2 Chimeric MP Analogs

Chimeric combinations of MP with GPCR ligands have produced some of the most valuable research tools for the study of the biological and molecular mechanisms of MP. Moreover, this chimeric strategy has produced novel compounds that exhibit quite different biological activities from that of their components.

The novel peptide galparan, consisting of the amino terminal of the neuropeptide galanin (Gal^{1-13}, GWTLNSAGYLLGP) and MP (Langel et al., 1996) (Table 18.3) is the best-characterized chimeric MP analog that combines a GPCR ligand with MP. Galparan is a ligand for the rat hypothalamic galanin receptor ($K_d = 5.8$ nM) although, intriguingly, a similar chimera combining Gal^{1-13} and the inactive MP analog Mas-17 (*H*-GWTLNSAGYLLGPINLKAKAALAKKIL-NH_2) displays even higher affinity ($K_d = 0.6$ nM) for the same receptor (Pooga et al., 1998a). In addition to binding GPCRs, galparan also exhibits a variety of biological activities that are independent of GPCR activation and quite distinct from its composite parts. For instance, galparan promotes the release of acetylcholine from rodent brain frontal cortex and activates the Na^+, K^+-ATPase (Langel et al., 1996). Galparan is also a potent insulin secretagogue that acts at a distal site in the secretory pathway of B cells that is distinct from that of MP (Östenson et al., 1997). This secretory action is mostly independent of $[Ca^{2+}]_o$ and $[Ca^{2+}]_i$ and of similar magnitude in normal rat islets and those from diabetic GK rats.

Chimeric MP constructs have also been synthesized using ligands for the vasopressin (AVP) receptor. M375 and M391 (Table 18.3) combine the sequence of a linear V_{1a}-selective antagonist [$Phaa_D Tyr(Me)^2 Arg^6 Tyr^9$]AVP; (Howl et al., 1996) with MP and display selective, high-affinity binding to the V_{1a} receptor, M375 ($K_d = 3.8$ nM), and M391 ($K_d = 9.0$ nM) (Hällbrink et al., 1999). Both M375 and M391 are also potent insulin secretagogues and, like galparan, apparently act at a site in the secretory pathway that is distinct from that modulated by MP. Like MP however, M375 and M391 also interact with proteins involved in intracellular Ca^{2+} homeostasis and include inhibition of ER and SR Ca^{2+}-ATPases and promotion of Ca^{2+} release by activating the RyR *in vitro* (Longland et al., 1998).

This chimeric strategy has also produced constructs that modulate mast cell secretion. Studies by Farquhar and coworkers (2002) have identified differential peptidyl secretagogues that selectively

TABLE 18.3
Synthetic Chimeric Analogs

Analog	Sequence	Properties	Reference
Galparan Gal$^{1-13}$–MP	*H*–GWTLNSAGYLLGPINLK ALAALAKKIL–NH_2	Ach release Activates Na^+K^+-ATPases Inhibitor of GTPase activity Insulin secretagogue	Langel et al., 1996
Gal $^{1-13}$–MP17	*H*–GWTLNSAGYLLGP INLKALAALAKKLL–*OH*	Galanin receptor high-affinity ligand	Pooga et al., 1998a
Transportan Gal$^{1-12}$–Lys–MP	*H*–GWTLNSAGYLLGKINLK ALAALAKKIL–NH_2	CPP	Pooga et al., 1998b
Transportan10 Gal$^{7-12}$–Lys–MP	*H*–AGYLLGKINLKALA ALAKKIL–NH_2	Biologically inert CPP	Soomets et al., 2000
M375 [Phaa$_D$Tyr(Me)^{2}Arg^{6}Tyr^9] AVP(1–9)–MP	Phaa–$_D$Y(Me)FQNRPRYINLK ALAALAKKIL–NH_2	Insulin secretagogue	Howl et al., 1996
M391 [Phaa$_D$Tyr(Me)^{2}Arg^{6}Tyr^9] AVP(1–9)–εAhx–MP	Phaa–$_D$Y(Me)FQNRPRY εAhx INLKALAALAKKIL–NH_2	Insulin secretagogue	Howl et al., 1996
M436 MP–CGRP	*H*–INLKALAALAKKILVTHR LAGLLSRVPTNVGSKAF–NH_2	Potent 5-HT secretagogue	Farquhar et al., 2002
Z–Gly–RGD*f*–mitP	Z–Gly–RGD*f* INLKKLAKL(Aib) KKIL–NH_2	Highly potent mitochondriotoxic CPP	Jones et al., 2008

Note: CPP, cell-penetrating peptide.

stimulate the secretion of 5-HT and β-hexoseaminidase from RBL-2H3 cells. Chimeric constructs included N-terminal extensions of MP with small peptide hormones, neuropeptides, or peptides known to modulate secretion and many of these chimeric constructs were more potent secretagogues than MP itself. Of 70 screened peptides, MP-S (Table 18.2) was the most potent and selective activator of β-hexoseaminidase secretion, whereas M436 (MP-CGRP$^{8-18,28-37}$; Table 18.3) a chimeric construct combining MP at the amino-terminal with sequences of human calcitonin gene-related peptide (CGRP), was the most selective and potent 5-HT secretagogue.

18.5.3 Differential Modulation of Heterotrimeric G Proteins by Structural MP Analogs

The central dogma dictates that MP binds to G protein α subunits, predominantly G_i and G_o, to catalyse GDP/GTP exchange and stimulate GTPase activity. Conversely, the synthetic analog MP-S (Table 18.2) preferentially stimulates GTP hydrolysis by G_s although it weakly stimulates G_i and G_o (Sukumar et al., 1997). Intriguingly, some sequence permutations of MP, including the peptide ALAIKLINNLKAKA, inhibit the GTPase activity of recombinant $G_o\alpha$ (Oppi et al., 1992). Studies with chimeric MP analogs including galparan have further indicated a differential regulation of GTPase activity in rat brain cortical membranes (Zorko et al., 1998). Thus, galparan noncompetitively inhibits GTPase activity without affecting GTPγS binding, an effect reversed by MP. Moreover, this novel action of galparan is probably due to the selective allosteric modulation of G_s rather than G_i/G_o (Zorko et al., 1998). Other chimeric, amino-terminal extended MP analogs selectively activate the GTPase activity of G_i/G_o but probably suppress that of G_s (Hällbrink et al., 1999).

18.6 CLINICAL APPLICATIONS

18.6.1 Cell-Penetrating Peptides

The membrane translocating properties of MP are no doubt a contributory biochemical feature accounting for the cell-penetrating properties of some MP-containing peptides and giving rise to some of the most exciting and potential clinical applications for this peptide.

The discovery of cell-penetrating peptides (CPP) has pioneered a feasible alternative to conventional viral methodologies for the intracellular delivery of nonpermeable compounds. Indeed, the pharmaceutical potential of CPP delivery is demonstrated by the fact that most peptide and nucleic acid-based drugs are poorly taken up into cells and are therefore of limited application as therapeutic agents.

Pioneered by Ulo Langel's group at the University of Stockholm, transportan (galanin(1-12)-Lys-MP) is a chimeric amino-terminal extended analog of MP with efficient cell-penetrating properties (Pooga et al., 1998b) (Table 18.3). The design of it incorporates a reactive ε-amino group at Lys^{13} to provide a convenient point of attachment for cargo molecules (Pooga et al., 1998b; Soomets et al., 2000). Consequently, transportan has been employed as a relatively inert vector for the intracellular delivery of cargoes including peptides, proteins, oligonucleotides, short interfering RNAs (siRNAs), and peptide nucleic acids (PNAs).

The pharmaceutical potential of transportan-mediated delivery has been firmly verified by the following: (a) intracellular delivery of an antisense PNA construct, corresponding to a region of a phosphotyrosine phosphatase overexpressed in pancreatic islet cells that enhances glucose-induced insulin secretion (Östenson et al., 2000); (b) intracellular delivery of an antisense PNA construct of the human type I galanin receptor that modifies pain transmission *in vivo* (Pooga et al., 1998c); and (c) an antiviral PNA–transportan conjugate that decreases the production of HIV virions *in vitro* (Kaushik et al., 2002).

More recent studies have utilized the cell-penetrating peptide vector transportan-10 (TP10, *H*-AGYLLGKINLKALAALAKKIL-NH_2) for the intracellular delivery of a range of bioactive peptide cargoes that modulate signal transduction (Howl et al., 2003; Jones et al., 2005). TP10 (galanin(7-12)-Lys-MP) is a deletion analog of transportan (Table 18.3). Although TP10 retains efficient translocating properties, it does not exert an inhibitory effect on GTPase activity, a trait of most transportan analogs (Soomets et al., 2000). One distinct benefit of this TP10 delivery system is that, following translocation, bioactive cargoes are liberated by the intracellular reduction of cystine (Figure 18.3). Therefore, by utilizing this inert, nontoxic system for the intracellular translocation of peptide cargoes that correspond to partial sequences of the regulatory domains of intracellular signaling proteins, it is now possible to modify biological responses, as has been shown with the modulation of the secretion of pharmacological mediators from mast cells and the dual phosphorylation of p42/44 mitogen-activated protein kinases in astrocytic tumor cells (Howl et al., 2003; Jones et al., 2005). Moreover, the ability of MP-containing cell-penetrant peptides to modulate signal transduction heralds numerous pharmacological applications such as identification of intracellular therapeutic loci and intervention of signal transduction pathways aberrant in various pathological states.

18.6.2 Tumor Cell Death

The ability of MP to facilitate mitochondrial permeability in cell free systems has led to proposals that MP could be of utility in tumor therapeutics provided that it conferred features of cellular penetration and mitochondrial localization (Yamada et al., 2005). Moreover, the highly potent MP analog $[Lys^{5,8}Aib^{10}]$MP (Jones et al., 2008) specifically promotes apoptosis of human cancer cells, which has been confirmed by common methodologies for the detection of apoptotic events such as *in situ* TUNEL staining, activation of caspase-3, and translocation of phosphatidylserine to the outer cell membrane without the necrotic features of plasma membrane breakdown. $[Lys^{5,8}Aib^{10}]$MP penetrates

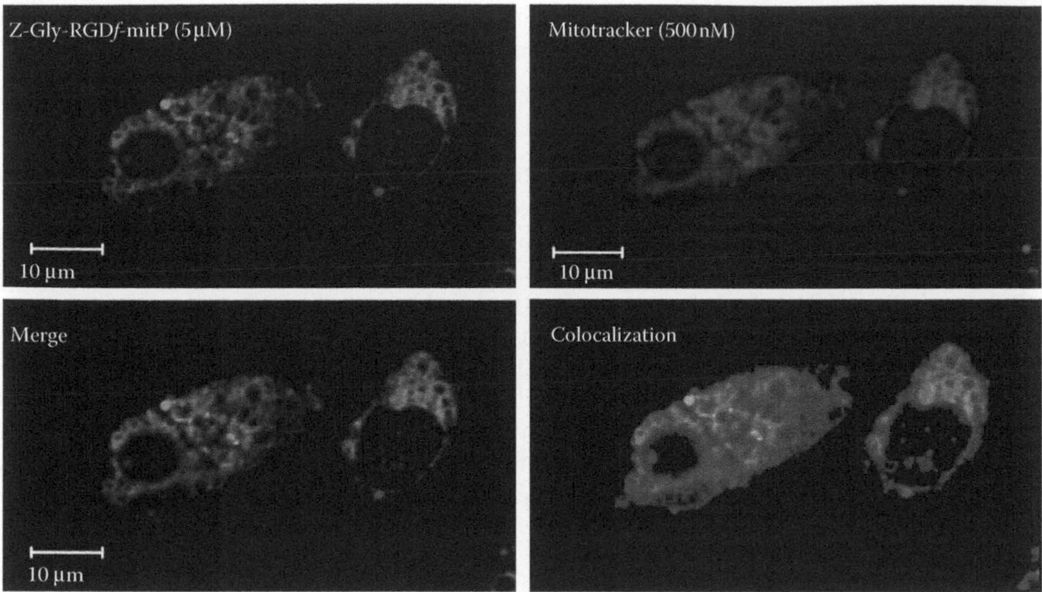

FIGURE 18.2 **(See color insert following page 176.)** The chimeric mitP analog Z-Gly-RGD*f*-mitP demonstrates strong mitochondrial localizing properties. Live confocal cell imaging was used to visualize the colocalization of fluorescein-labeled Z-Gly-RGD*f*-mitP with mitochondria in ECV304 cells, costained with Mitotracker™ (Molecular Probes). Using Carl Zeiss imaging software, colocalized fluorophores are shown.

plasma membranes and redistributes to colocalize with mitochondria, so we have consequently renamed this intriguing MP analog mitoparan (mitP). Complementary studies, using isolated mitochondria, further demonstrate that mitP, through cooperation with a protein of the permeability transition pore complex, voltage-dependent anion channel (VDAC), induces swelling and permeabilization of mitochondria, leading to the release of the apoptogenic factor cytochrome *c*.

An expanding field of peptide and cell-penetrating peptide (CPP) research has focused on the selective targeting of tumors by engineering constructs that incorporate cell-specific or tumor-specific peptidyl *address* motifs. Peptidyl *address* motifs could enhance the selectivity of drug delivery whilst the improved cellular uptake offered by CPP undoubtedly enhances bioavailability. Thus, as a potential therapeutic strategy, we designed target-specific mitP analogs (Jones et al., 2008). One particular analog incorporated the integrin-specific *address* motif RGD by simple N-terminal acylation of mitP, to produce the novel tandem-linked chimeric peptide, Z-Gly-RGD*f*-mitP. Previous reports (Yamada et al., 2005) that advocate MP as a potential antitumor therapeutic have delivered MP in a target-selective cell-penetrant liposomal complex. Such constructs are effective at concentrations of 25 μM. In contrast, Z-Gly-RGD*f*-mitP demonstrates an enhanced cytotoxic potency of 1.4 μM, a concentration readily achievable *in vivo*. Figure 18.2 shows a strong colocalization of Z-Gly-RGD*f*-mitP with mitochondria in ECV304 bladder cancer carcinoma cells. Thus, mitP is a highly potent mitochondriotoxic CPP and a structure that easily incorporates into a tissue-specific construct without impacting on the bioactivity of the parent cytotoxin.

18.7 COMMON METHODOLOGIES AND ASSAY SYSTEMS

There are a multitude of technologies for evaluating the biological effects of MP and its structural analogs. Therefore, routine assay systems used in our laboratory and methods adopted for the synthesis of these peptides shall be described. Biological phenomena upon which MP impacts include mast cell degranulation and cell death. Thus, routine assays for the assessment of mast cell secretion, cell viability, and the induction of apoptosis or necrosis shall be presented.

18.7.1 Measurement of β-Hexoseaminidase from RBL-2H3 Cells

Secretion of β-hexoseaminidase from the RBL-2H3 mast cell line is a relatively simple assay system used to evaluate the impact of MPs on mast cell secretion.

RBL-2H3 cells are maintained in DMEM in a humidified atmosphere of 5% CO_2 at 37°C. The medium contains L-glutamine (0.1 mg/mL) and is supplemented with 10% (w/v) fetal bovine serum (FBS), penicillin (100 U/mL) and streptomycin (100 μg/mL).

Cells are cultured in 24-well plates and transferred to labeling medium consisting of Hams medium supplemented with 0.1% (w/v) BSA and 10 mM HEPES, pH 7.4. Cells are subsequently incubated with MPs for the designated time periods at 37°C. Secreted β-hexoseaminidase is assayed by incubating samples of medium with 1 mM p-nitrophenyl N-acetyl-β-D-glucosamide in 0.1 M sodium citrate buffer for 1 h at 37°C. Addition of 0.1 M Na_2CO_3/$NaHCO_3$ buffer (pH 10.5) allows determination of β-hexoseaminidase activity by colorimetric analysis at 405 nm using a microtiter plate reader.

18.7.2 Measurement of Cell Viability

Cell viability is routinely measured by 3-(4,5-dimethylthiazol-2-yl)-2,5-diphenyl tetrazolium bromide (MTT) conversion. MTT is reduced, by metabolically active cells only, to a colored water-insoluble formazan salt. Following solubilization and colorimetric measurement, only viable cells are therefore detected. Many similar methods using the reduction of tetrazolium salts are also available. Cells are cultured as above in 96-well plates and treated with medium containing MPs or vehicle (medium alone) for the designated time periods at 37°C. Stimulation medium is removed and the cells are incubated in medium containing MTT (0.5 mg/mL) for 3 h at 37°C. Medium is aspirated and the insoluble formazan product is solubilized with DMSO. MTT conversion is determined by colorimetric analysis at 540 nm. Cell viability is expressed as a percentage of cells treated with vehicle (medium) alone.

18.7.3 Measurement of Apoptosis and Necrosis

As previously mentioned, MP can be engineered to produce chimeric constructs that specifically induce apoptosis (Jones et al., 2008). Given that MP, particularly at higher concentrations, induces cell death by random pore formation, it is of the utmost importance to distinguish between necrotic and apoptotic mechanisms.

18.7.3.1 Determination of Annexin V and Propidium Iodide Staining Using Flow Cytometry

Phosphatidylserine translocation to the outer plasma membrane is a feature of early apoptosis and can be measured by annexin-V staining and flow cytometry. Because necrotic cells also expose phosphatidylserine due to loss of membrane integrity, propidium iodide exclusion needs to be carried out to discriminate between apoptosis and necrosis. Annexin V staining can be measured using many available kits such as the Annexin-V-FLUOS staining kit (Roche, UK). With these, 10^6 cells are treated with peptide, washed in Hanks balanced salt solution, trypsinized and centrifuged at 200*g* for 5 min. The cell pellet is resuspended in 100 μL Annexin-V-FLUOS labeling solution containing annexin-V-fluorescein and propidium iodide and incubated for 15 min at 15 to 25°C. Following the addition of 0.5 mL HEPES incubation buffer, fluorescence of annexin-V-fluorescein and propidium iodide is detected in FL-1 and FL-3 channels, respectively. Unstained and untreated cells are used to establish instrument settings and FL-1 and FL-3 cell populations are isolated to the first log decade. Stained and untreated cells are used to establish quadrant settings for dot plot analysis.

18.7.3.2 Detection of DNA Fragmentation *in Situ* by TUNEL Assay

Nuclear DNA fragmentation, a characteristic of apoptosis, can be detected by Tdt (terminal deoxynucleotidyl transferase)-mediated dUTP nick end labeling (TUNEL). Many convenient kits are available including the "In Situ Cell Death Detection kit," TMR red, Roche, UK. Cells are grown to subconfluence on cover slips in 6-well plates, washed and treated with peptides or vehicle alone (medium) for 2 h in medium. Cells are washed with phosphate-buffered saline (PBS), pH 7.4 and fixed with 4% (wt/vol) formaldehyde in PBS for 1 h at room temperature. Fixed cells are permeabilized with 0.1% (vol/vol) Triton X-100 in 0.1% (wt/vol) sodium citrate at 4°C for 2 min, then incubated for 1 h at 37°C in a humidified atmosphere in the dark with TUNEL reaction mixture containing Tdt and TMR red-dUTP to label free 3′OH ends in the DNA. Cover slips are washed in PBS, air-dried and mounted on slides with Vectashield™ (Vector Laboratories Inc, Peterborough, U.K.) containing 4′6′-diamidino-8-phenylindole dihydrochloride (DAPI) to counterstain double-stranded DNA in the nuclei. Samples are analyzed using fluorescence or confocal microscopy. Positive controls are achieved by incubating fixed and permeabilized cells with DNase 1, grade 1 (3000 U/ml in 50 mM Tris-HCl, pH 7.5, 1 mg/mL BSA) for 10 min at room temperature to induced DNA strand breaks, prior to labeling procedure. Negative controls are achieved by incubating fixed and permeabilized cells in TUNEL reaction mixture without Tdt.

18.7.4 Synthesis of MP Peptides

MPs are relatively small peptides and, as such, are routinely synthesized manually in our laboratory. Nevertheless, semiautomated or automated synthetic approaches could be used. MP peptides are synthesized (0.1 mmol scale) on Rink amide methylbenzhydrylamine (MBHA) resin employing an N-α-Fmoc protection strategy with 2-(1-H-benzotriazole-1-yl)-1,1,3,3-tetramethyluronium hexafluorophosphate/N-hydroxybenzotriazole) (HBTU/HOBt) activation. Crude peptides are purified to apparent homogeneity by semipreparative scale HPLC (Howl and Wheatley, 1993). To engineer fluorescent peptides for the determination of cellular penetration and intracellular colocalization with various organelles, peptides are synthesized by N-terminal acylation with 6-carboxy-tetramethylrhodamine or 5-carboxy-fluorescein (Novabiochem, Beeston, U.K.). The predicted masses of all peptides used (average $M + H^+$) are always confirmed to an accuracy of ±1 by matrix-assisted laser desorption ionization (MALDI) time of flight mass spectrometry (Kratos Kompact Probe operated in positive ion mode).

18.7.4.1 Synthesis and Conjugation of Peptide Cargoes to the Cell-Penetrant Peptide TP10

TP10 and peptide cargoes can also be manually synthesized (0.1 mmol scale) on Rink amide methylbenzhydrylamine resin as above. Figure 18.3 shows the synthetic route to the generation of TP10 conjugated to a bioactive peptide cargo. The method was pioneered by Ursel Soomets and is detailed by Howl et al. (2003) and Jones and Howl (2006a). Step [i] indicates selective removal of the 4-methyltrityl group of Lys^7 with trifluoracetic acid (TFA) (3% (vol/vol) in dichloromethane (DCM; 2 × 10 min), acylation with Boc-Cys(Npys) (2 equiv.) and cleavage with TFA/H_2O/triisopropylsilane (95:2.5:2.5%) to yield the fully deprotected [$Lys^7N^{\varepsilon Cys(Npys)}$]TP10. Disulfide bond formation, to generate a TP10–peptide chimera, is achieved by dissolving [$Lys^7N^{\varepsilon Cys(Npys)}$]TP10 and individual cargoes (at a twofold molar ratio) in a minimum volume of DMF/DMSO/$C_2H_3O_2Na$ (0.1 M) pH 5 (3:1:1) and mixing overnight [ii]. Cargoes are liberated from the TP10 vector by reduction of the disulfide bond following cellular penetration.

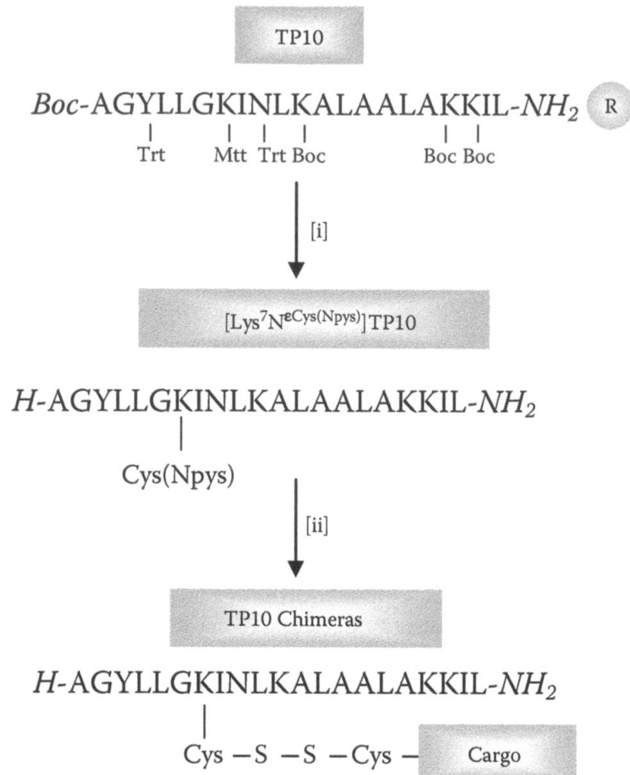

FIGURE 18.3 The synthetic route of TP10 conjugation to a bioactive peptide cargo. Full details of this synthetic route can be found in Howl, J., Jones, S., and Farquhar, M., *Chembiochem.*, 4, 1312–1316, 2003 and Jones, S. and Howl, J. *Handbook of Cell Penetrating Peptides*, 2nd ed., Langel, U., Ed. CRC Press, Washington, 2006, pp. 273–291. Also, see text for an explanatory précis.

18.8 FUTURE DEVELOPMENTS

In the authors' opinion, one of the most encouraging developments to date is the emergence of cell-penetrant MP analogs. Moreover, CPP technologies are now breaking new ground and entering clinical trials. In the laboratory, TP10 is an established and inert vector for the delivery of many bioactive moieties. Typically, CPP design has demanded that these peptidyl vectors be biologically inactive. However, MP analogs such as mitP are both cell penetrant and "intrinsically" biologically active. Thus, mitP and its target-selective analogs possess the dual functions of cellular penetration and induction of mitochondrial apoptosis, a feature that eliminates the requirement for conjugation to an inert CPP.

One consideration, however, for the therapeutic utility of MP analogs would be their ability to induce mast cell degranulation. Thus, it would be an achievement to generate analogs with a low secretory efficacy, whilst still retaining the desired biological effects. Further SAR studies are warranted to identify essential pharmacophors and structural determinants for this process.

One apparent drawback to the utility of peptides as therapeutics is their susceptibility to proteolysis. However, this apparent disadvantage would appear to be advantageous for the treatment of aggressive malignancies where *in situ* drug delivery is commonly practiced, offering features of low toxicity to nearby tissue and avoidance of multiple drug resistance. In addition, tumor-specific targeting through the use of peptidyl *address* motifs could enhance the selectivity of drug delivery leading to a reduction in nonselective side-effects.

The susceptibility of peptides to proteolysis is, however, currently not a restriction to their use as therapeutics. Successful strategies for the design of protease-resistant peptides are routinely used and include retro-inverso transformation, the incorporation of D amino acids, and an array of effective blocking groups.

Quite simply, as my old PhD supervisor would say, in a pronounced yet dignified West Midlands accent, "There is no end of possibilities to this remarkable tetradecapeptide."

REFERENCES

Amin, R.H. et al., Mastoparan-induced insulin secretion from insulin-secreting betaTC3 and INS-1 cells: Evidence for its regulation by Rho subfamily of G proteins, *Endocrinology*, 144, 4508–4518, 2003.

Argiolas, A. and Pisano, J.J., Facilitation of phospholipase A_2 activity by mastoparans, a new class of mast cell degranulating peptides from wasp venom, *J. Biol. Chem.*, 258, 13697–13702, 1983.

Argiolas, A. and Pisano, J.J., Isolation and characterization of two new peptides, mastoparan C and crabrolin, from the venom of the European hornet, *Vespa crabro*, *J. Biol. Chem.*, 259, 10106–10111, 1984.

Chahdi, A. et al., Mastoparan selectively activates phospholipase D2 in cell membranes, *J. Biol. Chem.*, 278, 12039–12045, 2003.

Chahdi, A. et al., The Rac/Cdc42 guanine nucleotide exchange factor β_1Pix enhances mastoparan-activated G_i-dependent pathway in mast cells, *Biochem. Biophys. Res. Commun.*, 317, 384–389, 2004.

Chapman, B.S., A region of the 75 kDa neurotrophin receptor homologous to the death domains of TNFR-I and Fas, *FEBS Lett.*, 374, 216–220, 1995.

Eddlestone, G.T. et al., Mastoparan increases the intracellular free calcium concentration in two insulin-secreting cell lines by inhibition of ATP-sensitive potassium channels, *Mol. Pharmacol.*, 47, 787–797, 1995.

Ellerby, H.M. et al., Establishment of a cell-free system of neuronal apoptosis: Comparison of premitochondrial, mitochondrial, and postmitochondrial phases, *J. Neurosci.*, 17, 6165–6178, 1997.

Epand, R.M. et al., Mechanisms for the modulation of membrane bilayer properties by amphipathic helical peptides, *Biopolymers*, 37, 319–338, 1995.

Farquhar, M. et al., Novel mastoparan analogs induce differential secretion from mast cells, *Chem. Biol.*, 9, 63–70, 2002.

Ferry, X. et al., Activation of betagamma subunits of G_{i2} and G_{i3} proteins by basic secretagogues induces exocytosis through phospholipase Cβ and arachidonate release through phospholipase Cγ in mast cells, *J. Immunol.*, 167, 4805–4813, 2001.

Gil, J., Higgins, T., and Rozengurt, E., Mastoparan, a novel mitogen for Swiss 3T3 cells, stimulates pertussis toxin-sensitive arachidonic acid release without inositol phosphate accumulation, *J. Cell Biol.*, 113, 943–950, 1991.

Gusovsky, F., Soergel, D.G., and Daly, J.W., Effects of mastoparan and related peptides on phosphoinositide breakdown in HL-60 cells and cell-free preparations, *Eur. J .Pharmacol.*, 206, 309–314, 1991.

Hällbrink, M., et al., Effects of vasopressin–mastoparan chimeric peptides on insulin release and G-protein activity, *Regul. Pept.*, 82, 45–51, 1999.

He, L. and Lemasters, J.J., Heat shock suppresses the permeability transition in rat liver mitochondria, *J. Biol. Chem.*, 278, 16755–16760, 2003.

Higashijima, T. et al., Conformational change of mastoparan from wasp venom on binding with phospholipid membrane, *FEBS Lett.*, 152, 227–230, 1983.

Higashijima, T. et al., 1988. Mastoparan, a peptide toxin from wasp venom, mimics receptors by activating GTP-binding regulatory proteins (G proteins), *J. Biol. Chem.*, 263, 6491–6494, 1988.

Higashijima, T., Burnier, J., and Ross, E., Regulation of G_i and G_o by mastoparan, related amphiphilic peptides, and hydrophobic amines. Mechanism and structural determinants of activity, *J. Biol. Chem.*, 265, 14176–14186, 1990.

Hillaire-Buys, D. et al., Insulin releasing effects of mastoparan and amphiphilic substance P receptor antagonists on RINm5F insulinoma cells, *Mol. Cell. Biochem.*, 109, 133–138, 1992.

Hirai, Y. et al., A new mast cell degranulating peptide "mastoparan" in the venom of *Vespula lewisii*, *Chem. Pharm. Bull. (Tokyo)*, 27, 1942–1944, 1979.

Hirata, Y., Nakahata, N., and Ohizumi, Y., Identification of a 97-kDa mastoparan-binding protein involving in Ca^{2+} release from skeletal muscle sarcoplasmic reticulum, *Mol. Pharmacol.*, 57, 1235–1242, 2000.

Hirata, Y. et al., Mastoparan binds to glycogen phosphorylase to regulate sarcoplasmic reticular Ca^{2+} release in skeletal muscle, *Biochem. J.*, 371, 81–88, 2003.

Hori, Y. et al., Interaction of mastoparan with membranes studied by ^{1}H-NMR spectroscopy in detergent micelles and by solid-state ^{2}H-NMR and ^{15}N-NMR spectroscopy in oriented lipid bilayers, *Eur. J. Biochem.*, 268, 302–309, 2001.

Howl, J., Jones, S., and Farquhar, M., Intracellular delivery of bioactive peptides to RBL-2H3 cells induces beta-hexosaminidase secretion and phospholipase D activation, *ChemBioChem*, 4, 1312–1316, 2003.

Howl, J. and Wheatley, M., V1a vasopressin receptors: Selective biotinylated probes, *Methods Neurosci.*, 13, 281–296, 1993.

Howl, J. et al., Probing the V1a vasopressin receptor binding site with pyroglutamate-substituted linear antagonists, *Neuropeptides*, 30, 73–79, 1996.

Jones, S. et al., Intracellular translocation of the decapeptide carboxyl terminal of $G_i3\alpha$ induces the dual phosphorylation of p42/p44 MAP kinases, *Biochim. Biophys. Acta*, 1745, 207–214, 2005.

Jones, S. and Howl, J. Charge delocalisation and the design of novel mastoparan analogs: Enhanced cytotoxicity and secretory efficacy of [Lys^5, Lys^8, Aib^{10}]MP, *Regul. Pept.*, 121, 121–128, 2004.

Jones, S. and Howl, J., Applications of cell-penetrating peptides as signal transduction modulators, in *Handbook of Cell Penetrating Peptides*, 2nd ed., Langel, U., Ed., CRC Press, Washington, 2006a, p. 273–291.

Jones, S. and Howl, J., Biological applications of the receptor mimetic peptide mastoparan, *Curr. Protein Pept. Sci.*, 7, 501–508, 2006b.

Jones, S. et al., Mitoparan and target-selective chimeric analogues: Membrane translocation and intracellular redistribution induces mitochondrial apoptosis, *Biochim. Biophys. Acta Mol. Cell Res.*, 1783, 849–863, 2008.

Kaushik, N. et al., Anti-TAR polyamide nucleotide analog conjugated with a membrane-permeating peptide inhibits human immunodeficiency virus type 1 production, *J. Virol.*, 76, 3881–3891, 2002.

Koch, G. et al., Interaction of mastoparan with the low molecular mass GTP-binding proteins rho/rac, *FEBS Lett.*, 291, 336–340, 1991.

Komatsu, M., McDermott, A.M., and Gillison, S.L., Mastoparan stimulates exocytosis at a Ca^{2+}-independent late site in stimulus-secretion coupling. Studies with the RINm5F beta-cell line, *J. Biol. Chem.*, 268, 23297–23306, 1993.

Kowluru, A., Identification and characterization of a novel protein histidine kinase in the islet beta cell: Evidence for its regulation by mastoparan, an activator of G-proteins and insulin secretion, *Biochem. Pharmacol.*, 63, 2091–20100, 2002.

Langel, Ü. et al., A galanin–mastoparan chimeric peptide activates the Na^+,K^+-ATPase and reverses its inhibition by ouabain, *Regul. Pept.*, 62, 47–52, 1996.

Lin, S.Z., Yan, G.M., and Koch, K.E., Mastoparan-induced apoptosis of cultured cerebellar granule neurons is initiated by calcium release from intracellular stores, *Brain. Res.*, 771, 184–195, 1997.

Liu, S., Wang, F., and Tang, L., Crystal structure of mastoparan from *Polistes jadwagae* at 1.2 Å resolution, *J. Struct. Biol.*, 160, 28–34, 2007.

Longland, C.L. et al., Biochemical mechanisms of calcium mobilisation induced by mastoparan and chimeric hormone–mastoparan constructs, *Cell Calcium*, 24, 27–34, 1998.

Longland, C.L., Mezna, M., and Michelangeli, F., The mechanism of inhibition of the Ca^{2+}-ATPase by mastoparan. Mastoparan abolishes cooperative Ca^{2+} binding, *J. Biol. Chem.*, 274, 14799–14805, 1999.

Malencik, D.A. and Anderson, S.R., High affinity binding of the mastoparans by calmodulin, *Biochem. Biophys. Res. Commun.*, 114, 50–56, 1983.

McKillop, I.H. et al., Inhibitory guanine nucleotide regulatory protein activation of mitogen-activated protein kinase in experimental hepatocellular carcinoma *in vitro*, *Eur. J. Gastroenterol. Hepatol.*, 11, 761–768, 1999.

Mendes, M.A., de Souza, B.M., and Palma, M.S., Structural and biological characterization of three novel mastoparan peptides from the venom of the neotropical social wasp *Protopolybia exigua* (Saussure), *Toxicon*, 45, 101–106, 2005.

Mousli, M. et al., G protein activation: A receptor-independent mode of action of cationic amphiphilic neuropeptides and venom peptides, *Trends Pharmacol. Sci.*, 11, 358–362, 1990.

Mukai, H. et al., A mastoparan analog without lytic effects and its stimulatory mechanisms in mast cells, *Biochem. Biophys. Res. Commun.*, 362, 51–55, 2007.

Murata, K. et al., Novel biologically active peptides from the venom of *Polistes rothneyi iwatai*, *Biol. Pharm. Bull.*, 29, 2493–2497, 2006.

Nakajima. T. et al., Wasp venom peptides; wasp kinins, new cytotrophic peptide families and their physicochemical properties, *Peptides*, 3, 425–430, 1985.

Nicolay, K., Laterveer, F.D., and van Heerde, W.L., Effects of amphipathic peptides, including presequences, on the functional integrity of rat liver mitochondrial membranes, *J. Bioenerg. Biomembr.*, 26, 327–334, 1994.

Omata, W. et al., Duality in the mastoparan action on glucose transport in rat adipocytes, *Endocrin. J.*, 52, 395–405, 2005.

Oppi, C. et al., Attenuation of GTPase activity of recombinant G(o) alpha by peptides representing sequence permutations of mastoparan, *Proc. Natl Acad. Sci. USA*, 89, 8268–8272, 1992.

Östenson, C.-G. et al., Galparan: A powerful insulin-releasing chimeric peptide acting at a novel site, *Endocrinology*, 138, 3308–3313, 1997.

Östenson, C.G. et al., Overexpression of protein–tyrosine phosphatase PTP sigma is linked to impaired glucose-induced insulin secretion in hereditary diabetic Goto-Kakizaki rats, *Biochem. Biophys. Res. Commun.*, 291, 945–950, 2002.

Pfeiffer, D.R. et al., The peptide mastoparan is a potent facilitator of the mitochondrial permeability transition, *J. Biol. Chem.*, 270, 4923–4932, 1995.

Pooga, M. et al., Novel galanin receptor ligands, *J. Peptide Res.*, 51, 65–74, 1998a.

Pooga, M. et al., Cell penetration by transportan, *FASEB J.*, 12, 67–77, 1998b.

Pooga, M. et al., Cell penetrating PNA constructs regulate galanin receptor levels and modify pain transmission *in vivo*, *Nature Biotechnol.*, 16, 857–861, 1998c.

Raynor, R.L. Zheng, B., and Kuo, J.F., Membrane interactions of amphiphilic polypeptides mastoparan, melittin, polymyxin B and cardiotoxin. Differential inhibition of protein kinase C, Ca^{2+}/calmodulin-dependent protein kinase II and synaptosomal membrane Na, K-ATPase and Na^+ pump and differentiation of HL60 cells, *J. Biol. Chem.*, 266, 2753–2758, 1991.

Rocha, T. et al., Myotoxic effects of mastoparan from *Polybia paulista* (Hymenoptera, Epiponini) wasp venom in mice skeletal muscle, *Toxicon*, 50, 589–599, 2007.

dos Santos Cabrera, M.P. et al., Conformation and lytic activity of eumenine mastoparan: A new antimicrobial peptide from wasp venom, *J. Pept. Res.*, 64, 95–103, 2004.

Schnabel, P. et al., G protein-independent stimulation of human myocardial phospholipase C by mastoparan, *Br. J. Pharmacol.*, 122, 31–36, 1997.

Solomon, K.R. et al., Heterotrimeric G proteins physically associated with the lipopolysaccharide receptor CD14 modulate both *in vivo* and *in vitro* responses to lipopolysaccharide, *J. Clin. Invest.*, 102, 2019–2027, 1998.

Soomets, U. et al., From galanin and mastoparan to galparan and transportan, *Curr. Top. Pept. Protein Res.*, 2, 229–274, 1997.

Soomets, U. et al., Deletion analogues of transportan, *Biochim. Biophys. Acta*, 1467, 165–176, 2000.

de Souza, B.M. et al., Mass spectrometric characterization of two novel inflammatory peptides from the venom of the social wasp *Polybia paulista*, *Rapid Commun. Mass Spectrom.*, 18, 1095–1102, 2004.

Straub, S.G. et al., Glucose augmentation of mastoparan-stimulated insulin secretion in rat and human pancreatic islets, *Diabetes*, 47, 1053–1057, 1998.

Suh, B.C. et al., Induction of cytosolic Ca^{2+} elevation mediated by Mas-7 occurs through membrane pore formation, *J. Biol. Chem.*, 271, 32753–32759, 1996.

Sukumar, M. and Higashijima, T., G protein-bound conformation of mastoparan-X, a receptor-mimetic peptide, *J. Biol. Chem.*, 267, 21421–21424, 1992.

Sukumar, M., Ross, E.M., and Higashijima, T., A G_s-selective analog of the receptor-mimetic peptide mastopoan binds to Gsα in a kinked helical conformation, *Biochemistry*, 36, 3632–3639, 1997.

Tisch-Indelson, D. et al., Structure–function relationship in the interaction of mastoparan analogs with neutrophil NADPH oxidase, *Biochem. Pharmacol.*, 61, 1063–1071, 2001.

Weingarten, R. et al., Mastoparan interacts with the carboxyl terminus of the alpha subunit of G_i, *J. Biol. Chem.*, 265, 11044–11049, 1990.

Wu, T.M. and Li, M.L., The cytolytic action of all-D mastoparan M on tumor cell lines, *Int. J. Tissue React.*, 21, 35–42, 1999.

Xu, X. et al., The mastoparanogen from wasp, *Peptides*, 27, 3053–3057, 2006.

Yamada, Y. et al., Mitochondrial delivery of mastoparan with transferrin liposomes equipped with a pH-sensitive fusogenic peptide for selective cancer therapy, *Int. J. Pharm.*, 303, 1–7, 2005.

Yamauchi, J. et al., G_i-dependent activation of c-Jun N-terminal kinase in human embryonal kidney 293 cells, *J. Biol. Chem.*, 275, 7633–7640, 2000.

Yokokawa, N. et al., Mastoparan, a wasp venom, stimulates insulin release by pancreatic islets through pertussis toxin sensitive GTP-binding protein, *Biochem. Biophys. Res. Commun.*, 158, 712–716, 1989.

Zorko, M. et al., Differential regulation of GTPase activity by mastoparan and galparan, *Arch. Biochem. Biophys.*, 349, 321–328, 1998.

19 Spider Venom and Hemolymph-Derived Cytolytic and Antimicrobial Peptides

Lucia Kuhn-Nentwig, Christian Trachsel, and Wolfgang Nentwig

CONTENTS

It is only within the last decade that 30 cytolytic and antimicrobial peptides have been detected in the venom glands of five spider species. Due to their high species number, spiders are predestined to exhibit a very high molecular diversity of such peptides, which still remain to be identified. In spite of different amino acid sequences, all of these peptides form an amphipathic α-helix and show lytic activity against cell membranes. This process is still not well understood but the fact that several highly similar peptides with slightly differing lytic properties are usually found within one spider species offers a unique chance to analyse it.

19.1 INTRODUCTION

Cytolytic and antimicrobial peptides play a decisive role in a spider's life. These animals (and arthropods in general) have evolved an innate immune system consisting of mechanisms involving humoral as well as cellular responses to different bacteria, fungi, and protozoa. In this system, antimicrobial peptides (AMPs) play an important role in destroying such aggressors (Iwanaga and Lee, 2005; Fukuzawa et al., 2008).

Simultaneously, many spiders have developed in their venom glands, besides other neurotoxic components, a great diversity of cytolytic peptides designed to paralyze or kill prey rapidly, and also to attack aggressors. These cytolytic peptides in the venom have a number of functions: they can depolarize excitable cells, lyse cell membranes, support the activity of neurotoxins synergistically, and perform other functions that are not yet fully understood. In consequence of their constitutive presence in the venom, cytolytic peptides can, as a positive side effect, protect the venom glands from microbial infections. Very recently, a spider species with a venom strategy based mainly on cytolytic peptides was identified for the first time (Vassilevski et al., 2008).

19.2 SPIDERS AS A SOURCE OF CYTOLYTIC AND ANTIMICROBIAL PEPTIDES

Spiders colonize, with high species diversity, all ecological niches in all possible terrestrial habitats. From 108 families, a total of 40,024 spider species have been described to date (Platnick, 2008). With the exception of a few specialists, spiders are polyphagous predators and have optimized their venom to paralyse a wide variety of different prey types. The main prey groups are insects, but other arthropods, including spiders, are frequently taken. The high molecular diversity of prey items could help to explain the high diversity of different neurotoxic and/or cytolytic peptides in the venom of even one single spider species. Bactericidal activity has been verified for the venom of about ten species from six different spider families: Amaurobiidae, Ctenidae, Lycosidae, Oxyopidae, Thomisidae, and Zodariidae. These families represent 17% of the spider species described so far. Additionally, cytotoxic activity due to venom components not yet further analysed has been demonstrated for nine species from eight spider families: Ctenizidae, Theraphosidae, Filistatidae, Loxoscelidae, Plectreuridae, Clubionidae, Theridiidae, and Salticidae which comprise another ~18% of all described spider species (Cohen and Quistad, 1998; Kuhn-Nentwig, 2003; Villegas and Corzo, 2005). Meanwhile, nearly 30 different peptides with antimicrobial and/or cytolytic activities have been purified from spider venom. If these single species records can be generalized, many thousands of spider species can be predicted to possess such peptides. It can also be hypothesized that each species produces several slightly different peptides.

The innate immune system of spiders is a further source of antimicrobial peptides. To date, we only have information concerning the mygalomorph spider *Acanthoscurria gomesiana*. Furthermore, two different antimicrobial peptides, acanthoscurrin and gomesin, have been purified from the hemocytes isolated from the spider's hemolymph (Silva et al., 2000; Lorenzini et al., 2003a).

The relevance and function of such peptides in the venom are still a matter of debate, ranging from immune defence to a tool for prey capture. In this review, peptides derived from the spider innate immune system are named antimicrobial peptides. Peptides derived from spider venom are entitled cytolytic peptides when their cytolytic activities include more than antimicrobial activity.

19.3 ARANEOMORPH SPIDER VENOM

The function of venom is to paralyse or kill prey rapidly or to deter aggressors with components that have been optimized during evolution in response to the neuronal diversity of the potential prey spectrum. A highly active venom composition can be achieved by applying different molecular strategies, ranging from a few highly specific components to a composition of highly diverse components with different targets and modes of action, including synergisms between them. In addition, venom compositions range from the absence of cytotoxic/cytolytic peptides to mixed compositions with neurotoxins and/or polyamines and, as recently reported for the zodariid spider *Lachesana tarabaevi*, even to venoms only containing cytolytic peptides as paralytic components (Vassilevski et al., 2008).

Cytolytic peptides have manifold functions in the venom: they act simultaneously as neurotoxin synergists with additional neurotoxic and cytolytic effects (Kuhn-Nentwig et al., 2002a; Wullschleger et al., 2005), they dissipate ion and voltage gradients by pore formation in membranes of excitable and nonexcitable cells (Yan and Adams, 1998; Corzo et al., 2002; Vassilevski et al., 2008) and, in a general way, they act cytolytically as a spreading factor, giving the neurotoxins better access to their targets (Corzo et al., 2002; Kuhn-Nentwig et al., 2002a; Kuhn-Nentwig, 2003). Owing to the absence of cytolytic peptides in several spider venoms, it is supposed that in general the role of cytolytic peptides may be similar to neurotoxins and other substances focusing on membrane-depolarizing activity (Villegas and Corzo, 2005; Estrada et al., 2007). The protective antimicrobial effect of these peptides in the venom gland is not in doubt (Yan and Adams, 1998; Corzo et al., 2002; Kuhn-Nentwig et al., 2002a), but should be assessed in the light of very recent reports of latarcins and cyto-insectotoxins more as a positive side effect (Kozlov et al., 2006; Vassilevski et al., 2008).

To date, all linear α-helical cytolytic peptides from spider venom have been identified mainly from the venom of araneomorph spiders. Common to these amphipathic and cationic peptides is the absence of cysteine and the ability to adopt an α-helical structure in the presence of membranes or membrane-mimicking solutions. Determination of the α-helicity by circular dichroism spectroscopy or by secondary structure consensus prediction programs reveals a high α-helical character of 62 to 69% for the peptides. Interestingly, the peptides are composed of between 38 and 49% hydrophobic amino acids and display net charges of between +3 and +14, mainly due to lysine (Kuhn-Nentwig, 2003; Vassilevski et al., 2008).

In addition, three peptides with antibacterial or antimalarial activities and exhibiting the inhibitory cystine knot (ICK) fold have been reported from mygalomorph spiders (Choi et al., 2004; Jung et al., 2006).

19.3.1 Lycotoxins and Lycocitins: Structure and Biological Activity

The first report on antimicrobially acting peptides in venom referred to the wolf spider *Lycosa singoriensis* (Xu et al., 1989). Meanwhile, further linear cytolytic peptides have been characterized from two other lycosids: two lycotoxins from *Lycosa carolinensis* (valid taxonomic name *Hogna carolinensis*) (Yan and Adams, 1998), and three lycocitins and two further short possibly cytolytic peptides from *Lycosa singoriensis* (Table 19.1) (Budnik et al., 2004). The peptide masses range between 1959.1 and 3206.9 Da, corresponding to 18 to 27 amino-acid residues. The linear peptides exhibit isoelectric points above pH 10. With 18 amino-acid residues, lycocitin 1 and 2 are among the smallest cytolytic peptides detected in spider venom gland homogenate. Lycocitin 3 differs from the earlier purified lycotoxin II only in the C-terminal part by the absence of one glycine. Characteristic for lycotoxin I and II and lycocitin 3 is the four- to fivefold repeat of four to five amino acid residues, always starting with lysine.

Lycotoxin I and II (~5 to 150 μM) and the lycocitins 1 and 2 (~1.6 to 10 μM) are active against Gram-positive as well as Gram-negative bacteria. In concentrations above 100 μM, both peptides show significant hemolytic activity toward rabbit erythrocytes. In addition, a fungicidal activity in the micromolar range has been demonstrated for lycotoxin I and II and lycocitin 2. A leishmanicidal activity against *Leishmania pifanoi* (amastigotes) and *L. donovani* (promastigotes) in the micromolar range was verified for lycotoxin I and II and their shortened synthesized forms LycoI 1-19,

TABLE 19.1
Amino Acid Sequences and Peptide Parameters of Cytolytic Peptides and Analogs (in bold) from the Venom of *Hogna carolinensis, Lycosa singoriensis, Cupiennius salei,* and *Oxyopes takobius*

Peptide	Amino Acid Sequence	Mass (Da)[b]	aa[c]	pI[d]
Hogna carolinensis				
LycoI 1-15	--IWLTALKFLGKHAAK*	1696.1	15	10.3
LycoI 1-19	--IWLTALKFLGKHAAKHLAK*	2145.6	19	10.5
LycoII 1-17	KIKWFKTMKSIAKFIAK*	2067.6	17	10.7
LycoII 1-21	KIKWFKTMKSIAKFIAKEQMK*	2456.0	21	10.3
Lycotoxin I	--IWLTALKFLGKHAAKHLAKQQLSKL*	2843.5	25	10.6
Lycotoxin II	KIKWFKTMKSIAKFIAKEQMKKHLGGE	3206.9	27	10.2
Lycosa singoriensis				
Lycocitin 1	GKLQAFLAKMKEIAAQTL*	1959.1	18	10.3
Lycocitin 2	GRLQAFLAKMKEIAAQTL*	1987.2	18	10.6
Lycocitin 3	KIKWFKTMKSLAKFLAKEQMKKHLG-E	3147.6	26	10.8
2034[a,e]	AGIGKIGDFIKKAIAKYKN	2034.2	19	10.5
2340[a,e]	MoxIASHLAFEKLSKLGSKHTMoxL*	2372.3	21	10.3
Cupiennius salei				
Cupiennin 1a	GFGALFKFLAKKVAKTVAKQAAKQGAKYVVNKQME*	3798.6	35	10.3
Cupiennin 1b	GFGSLFKFLAKKVAKTVAKQAAKQGAKYIANKQME*	3800.6	35	10.3
Cupiennin 1c	GFGSLFKFLAKKVAKTVAKQAAKQGAKYIANKQTE*	3770.5	35	10.3
Cupiennin 1d	GFGSLFKFLAKKVAKTVAKQAAKQGAKYVANKHME*	3795.5	35	10.3
Oxyopes takobius				
Oxyopinin 1	FRGLAKLLKIGLKSFARVLKKVLPKAAKAGKALAKSMADENAIRQQNQ	5221.3	48	11.3
Oxyopinin 2a	GKFSVFGKILRSIAKVFKGVGKVRKQFKTASDLDKNQ	4126.9	37	10.8
Oxyopinin 2b	GKFSGFAKILKSIAKFFKGVGKVRKQFKEASDLDKNQ	4146.9	37	10.3
Oxyopinin 2c	GKLSGISKVLRAIAKFFKGVGKARKQFKEASDLDKNQ	4064.8	37	10.5
Oxyopinin 2d	GKFSVFSKILRSIAKVFKGVGKVRKQFKTASDLDKNQ	4156.9	37	10.8

Note: Cationic amino-acid residues are boxed in black.

[a] Leu/Ile not assigned.

[b] Calculated masses (www.expasy.org, Peptide Mass).

[c] aa, number of amino-acid residues.

[d] Theoretical isoelectric point calculated (www.expasy.org, ProtParam).

[e] Mox, methinonine oxidized.

[*] Amidated C-terminus.

LycoII 1-17 and LycoII 1-21 (Table 19.1) (Luque-Ortega et al., 2006). Interestingly, further shortening of lycotoxin I (LycoI 1-15) resulted in a peptide less active towards *L. pifanoi*, but showed the least hemolytic activity towards sheep erythrocytes. This peptide adopts an α-helical structure in the presence of membranes and interacts more with negatively charged lipid bilayers than with neutral ones (Adão et al., 2008).

The capability of lycotoxin I to dissipate ion and voltage gradients of excitable cells (body wall muscles of prepupal *Musca domestica*) suggests that cytolytic peptides in the venom might be important for prey paralysis, in addition to the neurotoxins (Yan and Adams, 1998).

19.3.2 Oxyopinins: Structure and Biological Activity

Oxyopinins are large amphipathic peptides isolated from the venom of the lynx spider *Oxyopes kitabensis* (Table 19.1) (valid taxonomic name *Oxyopes takobius*) (Corzo et al., 2002). With 48 amino-acid residues, of which approximately 50% are hydrophobic amino acids, oxyopinin 1 is the largest cytolytic peptide from this venom. Comparable to lycotoxins and cupiennins, oxyopinin 1 and oxyopinins 2a–d are characterized by a six- to ninefold repeat of three to four amino-acid residues, starting with lysine or arginine. Oxyopinins 2a–d are nearly identical, feature high isoelectric points above pH 10.3 and a total charge of +7 to +10, and are composed of approximately 38% hydrophobic amino acids. Oxyopinin 1 and oxyopinin 2b are comparably bactericidal against Gram-positive and Gram-negative bacteria and exhibit different hemolytic activities in dependence on the erythrocyte source, for example, differences in their phosphatidylcholine to sphingomyelin ratio (Belokoneva et al., 2003). Patch clamp experiments with oxyopinins performed on insect Sf9 pupal ovary cells (*Spodoptera frugiperda*) show that the peptides reduce membrane resistance by opening nonselective ion channels and in higher concentrations lyse the cells. Although their insecticidal activity on *Spodoptera litura* larvae, with LD_{50} concentrations of 166 nmol/g (Oxki1) and 500 nmol/g (Oxki 2b), seems to be 33 to 98 times less potent than the neurotoxin OxyTx1 (isolated from the same venom), the cytolytic peptides synergistically facilitate paralytic activity when co-injected with this neurotoxin (Corzo et al., 2002).

19.3.3 Cupiennins

19.3.3.1 Structure and Biological Activity

Mass spectrometric investigations of the venom of the ctenid spider *Cupiennius salei* reveal that peptidic venom components with molecular masses between 1.5 and 4.2 kDa are 2.2 times more abundant in the venom than peptides with masses between 4.8 and 8.3 kDa (Kuhn-Nentwig et al., 2004). A family of cytolytic peptides was first reported as cupiennins 1a–d (Table 19.1) (Kuhn-Nentwig et al., 2002a).

Cupiennins are composed of 35 amino-acid residues, and exhibit a very high percentage (43 to 49%) of hydrophobic residues and a net charge of +8. After a hydrophobic N-terminal stretch, a sixfold repeat of four residues, always starting with lysine, is followed by a more hydrophilic amidated C-terminus. The cytolytic activity is determined by the N-terminal chain segment, whereas the polar C-terminus modulates the accumulation at negatively charged cell surfaces via electrostatic interactions (Kuhn-Nentwig et al., 2002b). C-terminal amidation of cupiennins 1a and 1d does not seem to be important for its cytolytic and insecticidal activity. The peptides are active against bacteria, without Gram specificity, in the submicromolar range (0.08 to 5 μM), and hemolytic against human erythrocytes in micromolar concentrations. In a bioassay on *Drosophila melanogaster*, their insecticidal activity, with LD_{50} concentrations between 5 and 8 nmol/g fly, indicates their importance for prey capture. A 65% enhancement of neurotoxin efficacy by synergistic interactions of cupiennin 1a has been demonstrated for CSTX-1 and CSTX-9 (Wullschleger et al., 2005).

Remarkably, cupiennin 1a also forms complexes with Ca^{2+}-calmodulin, thus inhibiting the formation of NO by neuronal nitric oxide synthase (Pukala et al., 2007a), as previously reported for the amphibian antimicrobial peptides caerins 1 and splendipherin (Pukala et al., 2006).

19.3.3.2 Spatial Structure of Cupiennin 1a

The solution structure of cupiennin 1a was determined by nuclear magnetic resonance (NMR) spectroscopy, exhibiting a helix–hinge–helix structure (Pukala et al., 2007b). This structural motif has been frequently identified in cationic antimicrobial peptides composed of more than 20 amino-acid residues. The flexible hinge is often enriched with the helix-breaking residues Gly or Pro (Oh et al., 2000).

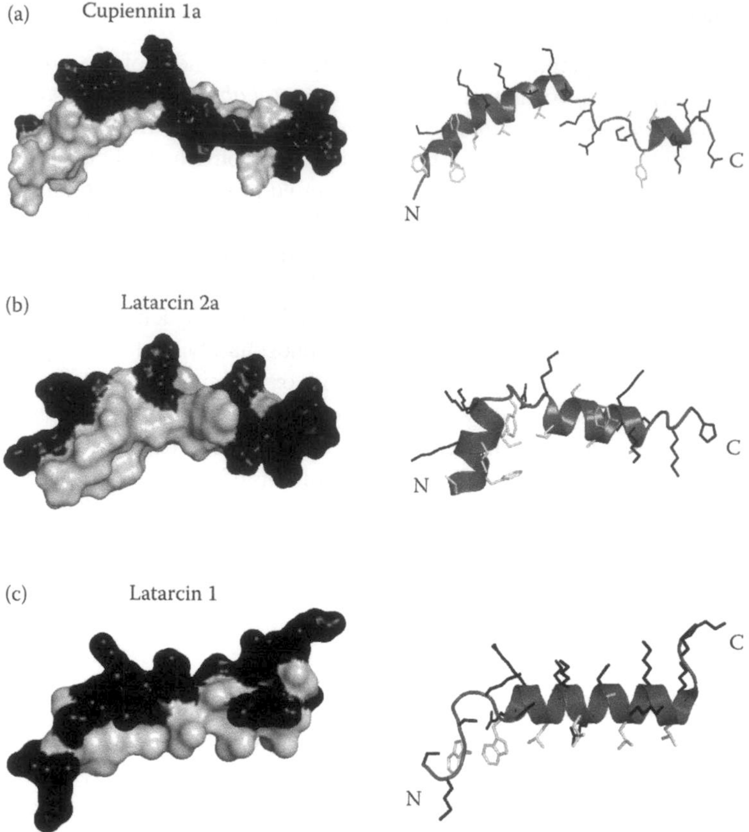

FIGURE 19.1 Three-dimensional structures of cationic α-helical peptides isolated from the venom of the spiders (a) *Cupiennius salei* and (b, c) *Lachesana tarabaevi*. The spatial structure of the peptides shows the distribution of the hydrophobic and charged/polar amino-acid residues on the surfaces of (a) cupiennin 1a, (b) latarcin 2a, and (c) latarcin 1. Hydrophobic amino acids are shown in gray and charged/polar residues in black. The secondary structures of the peptides are illustrated on the right side and the N- and C-termini of the peptides are indicated by N and C, respectively. The Protein Data Bank codes of the peptides are 2PCO (latarcin 1), 2G9P (latarcin 2a), and 2K38 (cupiennin 1a). The figures were produced with PyMOL (DeLano, 2002).

Cupiennin 1a adopts a predominantly α-helical structure in a membrane-mimicking trifluoroethanol solution or in the presence of phospholipid vesicles. The residues Gly^{3}-Ala^{21} and Tyr^{28}-Lys^{32} of the peptide are well-defined helices and it is supposed that Gly^{24} could initiate the hinge region separating both helical domains (Figure 19.1a). The N-terminal-located lysine residues are arranged in an arc of 140°, which enables this region to adopt a strong amphipathic conformation with phenylalanine residues on the other side of the helix. Furthermore, the hinge region, which is responsible for the observed flexibility, seems to allow both helices to orientate independently, corresponding to their differing influence on the peptide's cytolytic activity (Kuhn-Nentwig et al., 2002b). The N-terminal helix of cupiennin 1a, with a length of ~30 Å is able to span the membranes of erythrocytes and bacteria in a transmembrane orientation, thus presumably inducing pores of the toroidal type comparable to melittin (Yang et al., 2001). It is supposed that only subtle differences are responsible for the cell membrane destruction of neutral or negatively charged membranes. In the case of negatively charged membranes, the cell surface disruption could be due to interactions of the more hydrophilic C-terminal helix with the anionic lipid head groups of the outer leaflet. However, the driving force for pore formation and disruption of neutral membranes is assumed to occur by insertion of the N-terminal helix within the core of the membrane bilayer (Pukala et al., 2007b).

19.3.4 Latarcins and Cyto-Insectotoxins

For the first time, a venom composition based mainly on cytolytic peptides has been reported for the Central Asian zodariid spider *Lachesana tarabaevi* (Kozlov et al., 2006; Vassilevski et al., 2008). These peptides have been named latarcins and cyto-insectotoxins. In contrast to the other spider species mentioned here, this family is known to comprise several monophagous spider species feeding mostly on ants. However, limited field observations indicate that *L. tarabaevi* feeds on ants (*Messor* spp.) and also on woodlice (Zonstein and Ovtchinnikov, 1999).

19.3.4.1 Precursor mRNA Structure

Latarcin and cyto-insectotoxin precursor proteins can be divided into simple precursors, binary precursors, and complex precursors (Kozlov et al., 2006; Vassilevski et al., 2008). The structural organization of simple precursors includes signal peptides of 20 (cyto-insectotoxins) and 22 (latarcins) amino-acid residues, acidic fragments comprising 36 to 42 (latarcins) or 40 (cyto-insectotoxins) amino-acid residues, ending C-terminally with the processing quadruplet motif (PQM) and the mature peptides. Posttranslational processing occurs at the PQM yielding the mature latarcins 1, 2a, 2b, 3a, 3b, 5, 7, and the cyto-insectotoxins 1 to 16.

The two identified binary precursor proteins encoding latarcins 6b and 6a or 6c and 6a are composed of a signal peptide (22 amino-acid residues) followed by an acidic propeptide (22 amino-acid residues) ending with a first PQM. Arranged behind this, in one case latarcin 6b and, separated by a second PQM, latarcin 6a are encoded. In the second precursor protein latarcin 6c followed by a second PQM and subsequently latarcin 6a are encoded.

The complex precursors are characterized by a signal peptide (22 residues) followed by an acid fragment (22 residues) with a C-terminal PQM. Subsequently, four or five highly repetitive polypeptide elements are separated by a further four or five posttranslational processing sites (PQMs). The precursor protein is terminated by the mature latarcins 4a or 4b. The repetitive polypeptide elements do not share any homology with the various mature latarcins (Kozlov et al., 2006; Vassilevski et al., 2008).

19.3.4.2 Latarcins: Structure and Biological Activity

Latarcins 1, 2a, 3a, 3b, 4a, 4b, and 5 have been isolated from the venom and are rather small cytolytic peptides with molecular masses between 2424.3 and 3427.9 Da, corresponding to 20 to 28 amino-acid residues. Analysis of the venom gland expressed sequence tag database of *L. tarabaevi* led to the further identification of the putative cytolytic latarcins 2b, 6a, 6b, 6c, and 7. Except for latarcin 1 and 2a, which possess a free C-terminal carboxyl group, all other purified latarcins exhibit a C-terminal amidation. Remarkably, latarcin 1 seems to be further posttranslationally processed by the cleavage of the C-terminal lysine present in the peptide precursor.

Isoelectric points from pH 10 to 11.8 are characteristic for latarcins, as are high net charges of +5 to +10, taking the amidation into account (Table 19.2). Synthesized latarcins 1, 2a, 3a, 3b, 4a, 4b, and 5 are active against Gram-positive and Gram-negative bacteria in the low micromolar range, but differ in their effectiveness (0.5 to more than 45 μM). Except for latarcin 5, which is not fungicidal up to a concentration of 37 μM, all other peptides are active in concentrations of 6.7 to more than 35 μM against two different yeast species. Interestingly, latarcins 1, 2a, and 5 exhibit hemolytic activity in the range 6 to 80 μM. All other latarcins are not hemolytic up to a concentration of 120 μM.

In contrast to cupiennins, which seem to be more toxic, latarcin 2a produces reversible paralytic effects on flesh fly larvae and adult flies (*Sarcophaga carnaria*) when applied in concentrations of ~35 nmol/g fly (Kozlov et al., 2006; Vassilevski et al., 2008).

Most synthetic latarcins are able, in micromolar concentrations, to disrupt planar bilayer membranes, mimicking eukaryotic or prokaryotic membranes only at negative potentials (corresponding to the negative potentials of living cells). In contrast, latarcins 1 and 7 cause conductance changes in membranes but may only form pores without membrane disruption (Kozlov et al., 2006).

TABLE 19.2
Amino-Acid Sequences and Peptide Parameters of Latarcins and Cyto-Insectotoxins (CIT) from the Venom of *Lachesana tarabaevi*

Peptide	Amino Acid Sequence	Mass (Da)[a]	aa[b]	pI[c]
Latarcin 1	SMWSGMWRRKLKKLRNALKKKLKGE	3071.8	25	11.8
Latarcin 2a	GLFGKLIKKFGRKAISYAVKKARGKH	2900.8	26	11.3
Latarcin 2b	GLFGKLIKKFGRKAISYAVKKARGKN	2877.8	26	11.3
Latarcin 3a	SWKSMAKKLKEYMEKLKQRA*	2481.4	20	10.1
Latarcin 3b	SWASMAKKLKEYMEKLKQRA*	2424.3	20	10.0
Latarcin 4a	GLKDKFKSMGEKLKQYIQTWKAKF*	2900.6	24	10.0
Latarcin 4b	SLKDKVKSMGEKLKQYIQTWKAKF*	2882.6	24	10.0
Latarcin 5	GFFGKMKEYFKKFGASFKRRFANLKKRL*	3427.9	28	11.2
Latarcin 6a	QAFQTFKPDWNKIRYDAMKMQTSLGQMKKKFNL	4049.1	33	10.3
Latarcin 6b	QAFKTFTPDWNKIRNDAKRMQDNLEQMKKRFNLNL	4338.2	35	10.2
Latarcin 6c	QAFKTFTPDWNKIRNDAKRMQDNLEQMKKKFNLNL	4310.2	35	10.0
Latarcin 7	GETFDKLKEKLKTFYQKLVEKAEDLKGDLKAKLS	3940.2	34	9.1
CIT 1a	GFFGNTWKKIKGKADKIMLKKAVKIMVKKEGISKEEAQAKVDAMSKKQIRLYLLKYYGKKALQKASEKL	7905.6	69	10.2
CIT 1b	GFFGNTWKKIKGKADKIMLKKAVKLMVKKEGISKEEAQAKVDAMSKKQIRLYLLKYYGKKALQKASEKL	7905.6	69	10.2
CIT 1c	GFFGNTWKKIKGKADKIMLKKAVKIMVKKEGISKEEAQAKVDAMSKKQIRLYVLKYYGKKALQKASEKL	7891.6	69	10.2
CIT 1d	GFFGNTWKKIKGKADKIMLKKAVKIMVKKEGITKEEAQAKVDAMSKKQIRLYLLKYYGKKALQKASEKL	7919.7	69	10.2
CIT 1e	GFFGNTWKKIKGKSDKIMLKKAVKIMVKKEGISKEEAQAKVDAMSKKQIRLYLLKYYGKKALQKASEKL	7921.6	69	10.2
CIT 1f	GFFGNTWKKIKGKADKIMLKKAVKIMVKKEGISKEEAQAKVDAMSKKQIRLYLLKHYGKKALQKASEKL	7879.6	69	10.2
CIT 1 g	GFFGNTWKKIKGKADKIMLKKAVKIMVKKEGISKEEAQAKVDAMSKKQIRLYVLKHYGKKALQKASEKL	7865.6	69	10.2
CIT 1 h	GFFGNAWKKIKGKAEKFFRKKAAKIIAKKEGITKEEAEAKVDTMSKKQIKVYLLKHYGKKALQKASEKL	7885.5	69	10.2
CIT 1-6	GFFGNTWKKIKGKADKIMLKKAVKIMVKKEGIFKEEAQAKVDAMSKKQIRLYLLKYYGKKALQKASEKL	7965.7	69	10.2
CIT 1-9	GFFGNTWKKIKGKTDKIMLKKAVKIMVKKEGISKEEAQAKVDAMSKKQIRLYLLKHYGKKALQKASEKL	7909.6	69	10.2
CIT 1-10	GFFGNTWKKIKGKADKIMLKKAVKIMVKKEGISKEEAQAKVDAMSKKQIRLYVLKHYGKKALQKVSEKL	7893.6	69	10.2
CIT 1-12	GFFGNAWKKIKGKAEKFFRKKAAKIIAKKEGITKEEAEAKVDPMSKKQIKVYLLKHYGKKALQKASEKL	7881.5	69	10.2
CIT 1-13	GFFGNTWKKIKGKADKIMLKKAVKIMVKKEGISKEEAQAKVDAMSKKQIRLYLLKHYGKKLFKKRPKNCDQ	8268.1	71	10.2
CIT 1-14	GFFGNTWKKIKGKADKIMLKKAVKIMVKKEGISKEEAQAKVDAMSKKQIRLYLLKYYGKKLFKKRPKNCDQ	8294.1	71	10.2
CIT 1-15	GFFGNTWKKIKGKADKIMLKKAVKLMVKKEGISKEEAQAKVDAMSKKQIRLYLLKYYGKKSSSKSVRKIVISKSF	8571.4	75	10.4
CIT 1-16	GFFGNTWKKIKGKADKIMLKKAVKIMVKKEGISKEEAQAKVDAMSKKQIRLYVLKHYGKKSSSKSFRKIVISKSF	8579.4	75	10.4

Note: Putative peptides are in italics and cationic amino-acid residues are boxed in black.

[a] Calculated masses (www.expasy.org, Peptide Mass): latarcins = monoisotopic masses and CIT = average masses.

[b] aa, number of amino-acid residues.

[c] Theoretical isoelectric point calculated (www.expasy.org, ProtParam).

[*] Amidated C-terminus.

19.3.4.3 Spatial Structure of Latarcins 1 and 2a

The three-dimensional structure of latarcins 1 and 2a have been determined by means of different NMR spectroscopic techniques (Dubovskii et al., 2006; 2008). Comparable to cupiennin 1a, latarcin 2a exhibits a helical conformation in membrane-like environments. The two identified α-helical domains are located from Phe^{3}-Lys^{9} and Lys^{13}-Lys^{21}. The hinge region separating both helices could be initiated by Gly^{11}. The helix–hinge–helix structural motif of latarcin 2a is characterized by a distinct (N-terminal) and feebly marked (C-terminal) amphipathicity of both helices. Nevertheless, the N-terminal helix is amphipathic with the hydrophobic sidechains of the residues Phe^{3}, Leu^{6}, and Ile^{7} at one side of the helix. In contrast, the C-terminal helix of the peptide is characterized by the absence of hydrophobic patches on the surface due to the uniformly distributed apolar and polar amino-acid residues between the helix sides, leading to a hydrophobicity gradient. Results from ^{31}P NMR and fluorescence spectroscopy suggested that latarcin 2a may disrupt the membrane by a carpet-like mechanism (Figure 19.1b) (Dubovskii et al., 2006). The interaction of synthetic latarcin 3b-G, latarcin 2a and latarcin 1-K with artificial membranes highlights that at low peptide concentrations the formation of unstable short-lived pores may also be possible (Vassilevski et al., 2007).

In contrast to cupiennin 1a and latarcin 2a, latarcin 1 exhibits a consistent α-helical structure from Trp^{3} to Lys^{23} flanked by a simultaneously sparsely structured N- and C-terminus. The hydrophobic N-terminal region passes into a highly amphipathic rigid α-helix with a thin hydrophobicity stretch on the surface and ends in a short hydrophilic C-terminal area. It is supposed that latarcin 1 acts in a membrane-potential-dependent fashion towards Gram-negative bacteria. After attraction and formation of an α-helix, latarcin 1 may intercalate into the lipid matrix and be placed parallel to the membrane plane. The peptide may adopt a transmembrane orientation, thus forming voltage-dependent conductivity lesions in the membrane (Figure 19.1c) (Dubovskii et al., 2008).

19.3.4.4 Cyto-Insectotoxins

Cyto-insectotoxins are a novel class of equally potent cytolytic and insecticidal, unexpectedly long peptides isolated from the venom of *L. tarabaevi*. They are considered to comprise approximately 20% of the total venom protein. From a total of 16 described peptides, eight have been purified from the venom, and sequence information on a further eight putative peptides has been obtained by analysis of the venom gland expressed sequence tag database of the spider (Table 19.2) (Vassilevski et al., 2008).

Cyto-insectotoxin (CIT) 1a to 1 h are linear, highly cationic (~30% lysine) peptides composed of 69 amino-acid residues with molecular masses between 7865.6 and 7921.6 Da and nearly identical amino-acid sequences. Furthermore, the peptides are characterized by a nonamidated C-terminus, a high net charge of +14 and pI > 10. Circular dichroism (CD) spectroscopy of CIT 1a revealed that the peptide has a random coiled structure in water and adopts an α-helical structure in the presence of 50% trifluoroethanol and sodium dodecyl sulfate (SDS) micelles.

Antimicrobial activity has been determined with synthetic CIT 1a on diverse Gram-positive and Gram-negative bacteria. Comparable to short-chain cytolytic peptides, CIT 1a is active in low micromolar concentrations without Gram specificity (0.5 to greater than 30 µM). Similarly, CIT 1a, in micromolar concentrations, lyses human erythrocytes (EC_{50}, 6 µM), human leucocytes (EC_{50}, 3 µM), and ovarian insect Sf9 cells (EC_{50}, 1 µM). In contrast to latarcin 2a, which is approximately 2.4 times shorter than CIT 1a, CIT 1a is more insecticidal towards flesh fly larvae (*Sarcophaga carnaria*) (~2.5 nmol/g) and adult flesh flies (~0.63 nmol/g). Artificial membranes mimicking eukaryotic and prokaryotic membranes are disrupted by CIT 1a in submicromolar concentrations in a voltage- and concentration-dependent manner.

Of extreme importance is the sequence organization of the cyto-insectotoxins. The authors suggest that the peptides exhibit a modular character, where two short linear cytolytic peptides are combined into one peptide in a "head-to-tail" orientation, connected by a short linker sequence EEAQ or EEAE. It is assumed that the quadruplet could be a cryptic or mutated PQM, missing an arginine residue (Vassilevski et al., 2008).

19.4 MYGALOMORPH SPIDER VENOM

To date, no cytolytic linear cationic peptides from mygalomorph spider venom have been described. However, one peptide from *Grammostola spatulata*—GsMTx-4—and two very similar peptides from the venom of *Psalmopoeus cambridgei* (both Theraphosidae)—psalmopeotoxin I and II—exhibit antimicrobial activities.

19.4.1 GsMTx-4: Precursor, Structure, and Biological Activity

GsMTx-4 (4093.9 Da) was first described as an inhibitor of cationic mechanosensitive channels identified in nonspecialized eukaryotic tissues. The gene encoding the cDNA sequence of the peptide is translated into a putative signal peptide of 21 residues and subsequently into the 25-residue-containing propeptide with a C-terminal arginine as processing site for the release of the mature peptide (34 amino-acid residues). The last two amino acids of the GsMTx-4 precursor, Gly-Lys, are removed during C-terminal amidation. The peptide contains three disulfide bridges and occurs in the venom at a concentration of approximately 2 mM (Figure 19.2a). (Suchyna et al., 2000; Ostrow et al., 2003). Solution structure determination of GsMTx-4 by NMR spectroscopy exhibits the inhibitor cystine knot (ICK) motif, which has been identified in many invertebrate neurotoxins (Oswald et al., 2002; Corzo and Escoubas, 2003; Escoubas, 2006). Importantly, the peptide is amphipathic due to hydrophilic and hydrophobic surfaces. The hydrophobic surface is surrounded by mostly positive charges. It is hypothesized that this hydrophobic patch may be involved in binding to membranes, resulting in penetration into the lipid bilayer and thereby altering the channel gating (Figure 19.2b) (Oswald et al., 2002; Posokhov et al., 2007a, 2007b). A bactericidal activity of GsMTx-4 against Gram-positive bacteria (0.5 to 8 μM) and to a lesser extent against Gram-negative

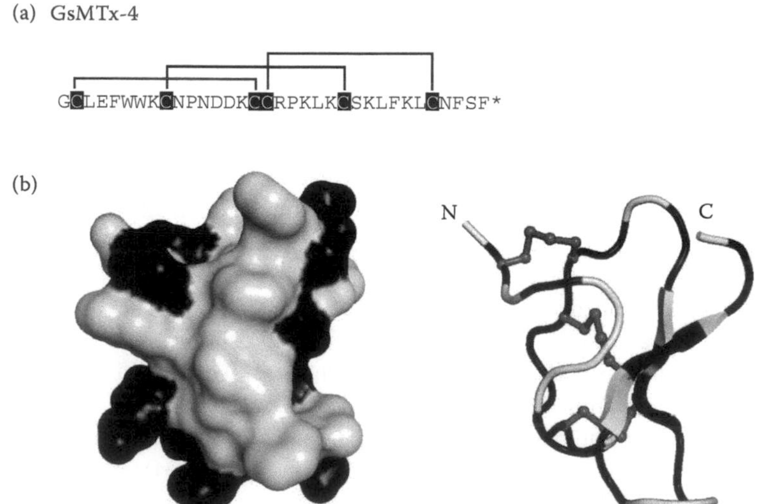

FIGURE 19.2 GsMTx-4, an inhibitor of mechanosensitive channels with antimicrobial activity isolated from the venom of the spider *Grammostola spatulata*. (a) Amino acid sequence of GsMTx-4. Disulfide bridges are represented by lines, the corresponding ICK motif cysteine residues are in black and C-terminal amidation is denoted with an asterisk. (b) The three-dimensional structure of GsMTx-4 shows the distribution of the hydrophobic and charged/polar amino acid residues on the peptide surface. Hydrophobic amino acids are shown in gray, and charged/polar residues shaded in black. The secondary structure of the peptide is illustrated on the right side, and the N- and C-termini of the peptide are indicated by N and C, respectively. The three disulfide bonds are shown as balls and sticks (dark gray). The Protein Data Bank code of GsMTx-4 is 1LU8. The figures were produced with PyMOL (DeLano, 2002).

(a) Psalmopeotoxin I and II

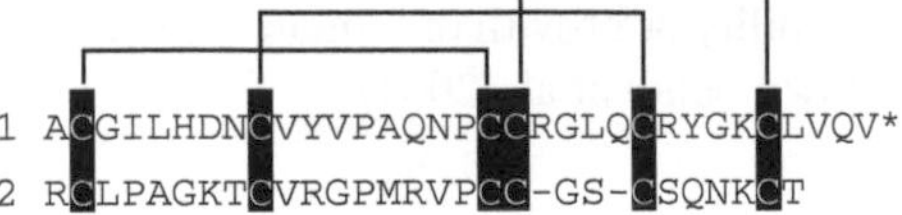

(b)

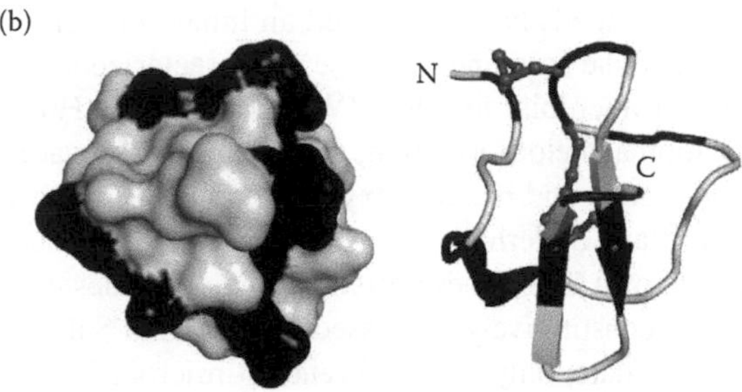

FIGURE 19.3 Antimalarial peptides isolated from the venom of *Psalmopoeus cambridgei*. (a) Amino acid sequence alignment of (1) psalmopeotoxin I (PcFK1) and (2) psalmopeotoxin II (PcFK2). Disulfide bridges are represented by lines, the corresponding ICK motif cysteine residues are in black, and C-terminal amidation is denoted with an asterisk. (b) The three-dimensional structure of PcFK1 shows the distribution of the hydrophobic and charged/polar amino acid residues on the peptide surface. Hydrophobic amino acids are shown in gray and charged/polar residues shaded in black. The secondary structure of the peptide is illustrated on the right side, and the N- and C-termini of the peptide are indicated by N and C, respectively. The three disulfide bonds are shown as balls and sticks (dark gray). The Protein Data Bank code of PcFK1 is 1X5V. The figures were produced with PyMOL (DeLano, 2002).

bacteria (8 to 64 μM) has been reported. The observed antimicrobial activity might be due to modulating the gating of the bacterial mechanosensitive channels or to altering the membrane phospholipid packing, leading to cell membrane disruption (Jung et al., 2006).

19.4.2 Psalmopeotoxin I and II: Precursor, Structure, and Biological Activity

Psalmopeotoxin I (PcFK1, 3615.6 Da) is composed of 33 amino-acid residues and Psalmopeotoxin II (PcFK2, 2948.3 Da) of 28 residues. Both peptides are characterized by three disulfide bridges, exhibiting the ICK motif (Figure 19.3a). Analysis of the cDNAs encoding the peptides reveals a peptide precursor structure composed of signal peptides (21 and 19 residues) followed by the acidic propeptides (30 and 19 residues) with the basic doublet processing site (Lys/Arg or Arg/Arg) and the mature peptides. In contrast to PcFK2, mature PcFK1 exhibits two extra C-terminal residues, Gly-Arg, which are removed after posttranslational C-terminal amidation (Choi et al., 2004).

Determination of the solution structure of recombinant PcFK1 by ^{1}H two-dimensional NMR spectroscopy reveals that the peptide can be classified as a member of the ICK superfamily (Pimentel et al., 2006). The peptide shares structural determinants common to other ion channel effectors, such as HWTX-I (Qu et al., 1997) and HNTX-I (Li et al., 2003). These three peptides share similarities of their molecular surfaces, characterized by the association of basic with hydrophobic amino-acid residues. Adsorption to erythrocytes could occur through hydrophobic or electrostatic interactions caused by this basic/hydrophobic patch (Figure 19.3b) (Pimentel et al., 2006).

Both PcFK1 and PcFK2 inhibit the growth of *Plasmodium falciparum* in low micromolar concentrations with IC_{50} values of 1.6 μM (PcFK1) and 1.2 μM (PcFK2). PcFK1 adsorbed to infected and noninfected human erythrocytes. In contrast, PcFK2 binds selectively to infected erythrocytes, suggesting that the two peptides may have different targets. In concentrations up to 10 μM, both peptides neither cause hemolysis nor affect neuromuscular transmission of isolated mouse and frog preparations. Furthermore, in concentrations up to 20 μM neither peptide exhibits a bactericidal or fungicidal activity nor affects the viability or growth of human epithelial cells. The exact molecular target of both peptides is still unknown (Choi et al., 2004).

19.5 SPIDER HEMOLYMPH

As a first line of defence, arthropods have developed an innate immune system involving cellular and humoral components. Cellular immune response is characterized by phagocytosis, nodule formation, and encapsulation of microbial invaders (Williams, 2007). Humoral components involve wound healing by coagulation and clot formation, melanization of invading microbes mediated by the phenoloxidase cascade system and reactive oxygen and nitrogen intermediates (Li et al., 2002; Kumar et al., 2003; Cerenius and Söderhäll, 2004). Besides these defence molecules, an important humoral component consists of AMPs. These often cationic peptides can be induced by microbes and their compounds, or are constitutively expressed (Vizioli and Salzet, 2002; Iwanaga and Lee, 2005; Kurata et al., 2006). To date, only two different antimicrobial peptides, acanthoscurrin and gomesin, have been identified from spider hemolymph, namely, from the mygalomorph spider *A. gomesiana*. Both peptides are constitutively expressed in hemocytes and can be stored in the same or in different granules. Hemocyte migration to the place of infection is affected by yeast and microbial components such as lipopolysaccharides. Destruction of invaders is mediated mainly by hemocyte degranulation, for example, due to the release of both peptides. In addition, the secretion of components from the coagulation system into the hemolymph supports this process (Fukuzawa et al., 2008).

19.5.1 Gomesin

19.5.1.1 Spatial Structure and Precursor

Gomesin is a short (2270.4 Da) cationic peptide of 18 amino-acid residues with two disulfide bridges (Cys^2-Cys^{15} and Cys^6-Cys^{11}), forming a well-defined β-hairpin structure, in which a two-stranded antiparallel β sheet ($pGlu^1$-Tyr^7 facing Arg^{10}-Arg^{16}) is connected by a noncanonical β turn (Tyr^7-Arg^{10}) as compiled by NMR determination. The peptide is characterized by two posttranslational modifications: an N-terminal pyroglutamic acid (pGlu) and an amidated C-terminal arginine. Gomesin is highly amphipathic with two distinct hydrophilic and positively charged domains (N-terminal Arg^3, Arg^4; C-terminal Arg^{16}, Arg^{18} and in the turn Lys^8, Gln^9, and Arg^{10}) and one hydrophobic region on the concave surface of the peptide (Leu^5, Tyr^7, Val^{12}, and Tyr^{14}). The opposite surface is shaped by the charged sidechain of Arg^4 and the polar sidechain of Thr^{13}, which is delimited by the two apolar disulfide bridges (Figure 19.4a,b) (Silva et al., 2000; Mandard et al., 2002). Interestingly, gomesin shares several common features in the distribution of hydrophilic and hydrophobic domains with other AMPs, such as androctonin (scorpion hemolymph) (Mandard et al., 1999), protegrin-1 (porcine leucocytes) (Kokryakov et al., 1993), tachyplesin I and polyphemusin II (horse shoe crab hemolymph) (Miyata et al., 1989; Kawano et al., 1990), which are essential for their antimicrobial activity (Silva et al., 2000; Mandard et al., 2002).

Analysis of preprogomesin cDNA revealed a putative signal peptide of 23 amino-acid residues, followed by the mature peptide and the anionic C-terminal region of 43 amino-acid residues comprising the propeptide. The posttranslational modifications of the N- and C-terminus of mature gomesin are confirmed by the presence of Gln^{24} leading to pyroglutamic acid and Gly^{42} in the peptide precursor, which act as the NH_4 donator for the C-terminal amidation of arginine (Lorenzini et al., 2003b). It is supposed that gomesin is stored as progomesin in granules of the hemocytes and that processing could take place intracellularly within the granules or in the plasma after peptide secretion (Lorenzini et al., 2003b; Fukuzawa et al., 2008). Analysis of expressed sequence tags from

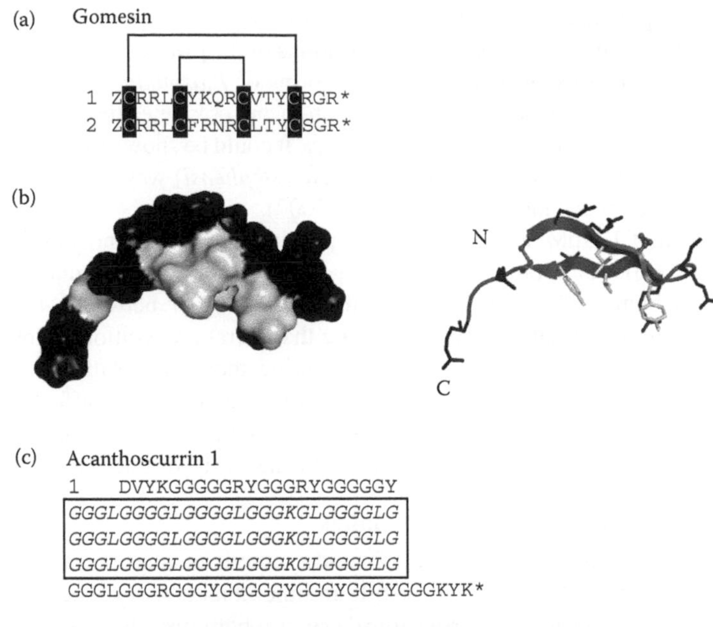

FIGURE 19.4 Antimicrobial peptides isolated from hemocytes of the spider *Acanthoscurria gomesiana*. (a) Amino acid sequence alignment of (1) gomesin and (2) a gomesin isoform. Disulfide bridges are represented by lines, the corresponding cysteine residues are shaded in black, and C-terminal amidation is denoted with an asterisk. (b) The three-dimensional structure of gomesin shows the distribution of the hydrophobic and charged/polar amino acid residues on the peptides surface. Hydrophobic amino acids are shown in gray and charged/polar residues in black. The secondary structure of the peptide is illustrated on the right side, and the N- and C-termini of the peptide are indicated by N and C, respectively. The two disulfide bonds are shown as balls and sticks (dark gray). The Protein Data Bank code of gomesin is 1KFP. The figures were produced with PyMOL (DeLano, 2002). (c) Amino acid sequence alignment of (1) acanthoscurrin 1 and (2) the C-terminal part of acanthoscurrin 2. The three sequence repeats within the peptides are written in italics and are boxed. C-terminal amidation is denoted with an asterisk.

hemocytes has allowed the identification of a gomesin precursor isoform that differs from the gomesin precursor by four substitutions in the precursor regions and five substitutions within the mature peptide region (Figure 19.4a) (Lorenzini et al., 2006).

19.5.1.2 Biological Activity

Gomesin exhibits broad antimicrobial activity in concentrations of between 0.2 and 12.5 μM without Gram specificity to various bacteria. In addition, a fungicidal activity of between 0.2 and 25 μM is shown for diverse filamentous fungi and yeasts (Silva et al., 2000). The peptide is able to cross the capsular barrier of the yeast-like pathogen *Cryptococcus neoformans*; it binds at the cell surface and induces cell permeabilization in micromolar concentrations (2 μM), leading to cell death. Furthermore, gomesin affects the capsule expression of *C. neoformans*. Remarkably, in concentrations of 0.1 to 1 μM, the peptide and the antifungal drug fluconazole (1 μM) act synergistically in killing *C. neoformans* cells. Administered together, both also enhance the antimicrobial activity of human brain phagocytes. Nevertheless, at a concentration of 10 μM both compounds are toxic to phagocytes (Barbosa et al., 2007). In a comparison of synthetic amidated gomesin with the nonamidated synthesized form, no difference in their antimicrobial activity could be determined, suggesting that amidation does not influence the peptide's activity (Silva et al., 2000).

Gomesin is also highly active against parasitic protozoans. The peptide reduces the viability of *Leishmania amazonensis* promastigotes in concentrations of 2.5 μM (Silva et al., 2000) and inhibits the *in vitro* growth of the asexual intraerythrocytic forms of *Plasmodium falciparum* (IC_{50}, 76 to 87 μM). Gomesin also inhibits male gamete exflagellation and ookinetes formation of mature gametocytes as demonstrated in *in vitro* cultures of *P. berghei*. It could be shown that for both *Plasmodium* species, the number of oocysts in mosquitos (*Anopheles stephensi*) was reduced dose dependently after a gomesin-enriched blood meal (Moreira et al., 2007).

At a concentration of 100 μM, gomesin causes approximately 40% hemolysis of human erythrocytes (Fázio et al., 2006). The peptide exerts a direct cytotoxic activity on human and murine tumor cells as well as on human endothelial cells, with IC_{50} values of between1.4 and 8.1 μM. The enantiomer D-gomesin was equally active, indicating that chiral recognition is not required for the toxic activity. Topical treatment of subcutaneous murine melanoma with gomesin (4 μg/20 mg cream) resulted in significantly delayed melanoma development, without affecting the peripheral healthy skin of the treated mice. These results indicate that gomesin could become a useful topical drug in the treatment of intraepithelial and intradermal cancers (Rodrigues et al., 2008).

19.5.1.3 Conformational and Functional Studies

Structure–activity relationship studies of gomesin and synthetic analogs revealed that the peptide's secondary structure is crucially important for full antimicrobial activity. The hemolytic and antimicrobial activities are correlated properties. Furthermore, both disulfide bridges are required for optimal peptide conformation, full antimicrobial activity, and high serum stability. Variants with only one of the disulfide bridges exhibit distinct antimicrobial activity but reduced serum stability. With bicyclo(2-15, 6-11)[Glu^{2}, $Cys^{6,11}$, Lys^{15}]-Gm, a stable and potent analog has been developed (Fázio et al., 2006).

The design of several linear gomesin analogs, characterized by the removal of both disulfide bridges and the incorporation of a D- or L-Pro, led to two encouraging candidates with improved therapeutic indices, that is, less hemolytic but slightly reduced activity against bacteria (Gram-positive and Gram-negative bacteria) when compared with gomesin. The analog [D-$Thr^{2,6,11,15}$, Pro^{9}]-D-Gm was additionally analysed by NMR spectroscopy and revealed a conformation very similar to native gomesin (Fázio et al., 2007).

Recently, functional and conformational studies of further gomesin analogs in the presence of different solvent environments mimicking biological membranes have been reported. For fluorescence spectroscopy, tryptophan-labeled gomesin analogs and their corresponding linear analogs have been synthesized. The introduction of the paramagnetic amino acid 2,2,6,6-tetramethylpiperidine-1-oxyl-4-amino-4-carboxylic acid (TOAC) as the N-terminal residue of gomesin and its linear form permits electron paramagnetic resonance (EPR) spectroscopic investigations. Moreover, CD spectroscopy has been performed for all analogs. As a result, it is suggested that the first step in membrane binding of gomesin is the result of electrostatic interactions with the lipid bilayer of bacteria but that hydrophobic interactions are also important. Nevertheless, the mechanism by which gomesin and its analogs disrupt cell membranes is still unknown (Moraes et al., 2007).

19.5.2 Acanthoscurrin: Structure, Precursor, and Biological Activity

Identification and amino acid sequence determination from two isoforms of acanthoscurrin (1 and 2) were mediated by the combination of mass spectroscopy, Edman degradation, and cDNA cloning strategies. Acanthoscurrin 1 and 2 are glycine-rich peptides (72 to 73%) with high molecular masses of 10,111 and 10,225 Da, and differ only by the presence of two additional glycines (Gly^{111} and Gly^{116}) in acanthoscurrin 1. Both isoforms are C-terminally amidated and are composed of, besides glycine, only a few amino acids, such as Arg (2%), Asp (1%), Leu (12%), Lys (4 to 5%), Tyr (7%), and Val (1%). A remarkable feature is a threefold repeat of 26 amino acids within both acanthoscurrins (Figure 19.4c).

The putative signal peptides contain 23 amino-acid residues and are followed by proacanthoscurrin 1 comprising 132 amino-acid residues, for example, proacanthoscurrin 2 with 130 amino-acid residues. The C-terminal Gly^{133} (Gly^{131}) of the precursors is used as an NH_4 donator for C-terminal amidation and is followed by the stop codon (Lorenzini et al., 2003a). In contrast to gomesin, neither of the acanthoscurrins exhibits an additional C-terminal propeptide in its precursors (Lorenzini et al., 2003b).

Acanthoscurrin 1 in the micromolar range is active against Gram-negative bacteria (*E. coli*, 2.3 to 5.6 μM) and yeast (*Candida albicans*, 1.15 to 2.3 μM), but up to a concentration of 5.6 μM is inactive against Gram-positive bacteria (*Micrococcus luteus*) (Lorenzini et al., 2003a).

19.6 CONCLUSIONS AND FUTURE DEVELOPMENTS

The diversity of identified cytolytic peptides in spider venom is striking. Species from several araneomorph families have evolved an array of different linear cationic peptides that affect prey in the absence of, or in combination with, further neurotoxically acting components. These cationic peptides adopt an α-helical conformation in the presence of diverse eukaryotic and prokaryotic cell membranes. They are present in the venom in millimolar concentrations, differ in shape (rod form versus the helix–hinge–helix conformation), and exhibit an approximately 3.7-fold variation in peptide length and in the arrangement of their amphipathicity due to the pattern of hydrophobic and charged/polar amino acids within the peptide chain. It is peculiar that all these peptides belong to the same class of linear cationic α-helical peptides and have obviously been streamlined during the course of their evolution to disturb nearly all types of membranes (Kuhn-Nentwig, 2003; Kozlov et al., 2006; Vassilevski et al., 2008).

In explanations for the existence of cytolytic and pore-forming peptides in spider venom, three hypotheses have been put forward: (1) they could be a part of an antimicrobial defence system in venom glands, including a disinfecting and possibly preserving role in prey conservation (e.g., Kozlov et al., 2006); (2) they might synergistically support the activity of neurotoxins (e.g., Wullschleger et al., 2005); (3) they have evolved to act directly as insecticidal substances (e.g., Vassilesvki et al., 2008). An antimicrobial defence function is possible, but so far has never been shown for spider venom glands under natural conditions. However, an inducible antibacterial response of the venom gland from the scorpion *Buthus martensii* has been demonstrated: a short 18-residue α-helical antimicrobial peptide (BmKb1) was upregulated at the transcriptional level after venom gland challenge (Gao et al., 2007).

The synergistic support of neurotoxins by cationic peptides has been demonstrated for two spider species and several neurotoxins (Corzo et al., 2002; Kuhn-Nentwig et al., 2004; Wullschleger et al., 2005). Corzo and coworkers observed that not all spider venoms exhibit bactericidal and/or cytotoxic activity and that of two *Oxyopes* species only one possesses cytolytic activity (Villegas and Corzo, 2005). They reasoned that cytolytic peptides in the venom have a membrane-depolarizing activity toward different cell types of a given prey item, thus functioning similarly to neurotoxins or other neurotoxic components in spider venom. Cytolytic peptides, therefore, would represent an alternative to the classical neurotoxins. This line of argument is supported by the very recent detection involving the venom of *Lachesana tarabaevi*, which is exclusively based on cytolytic peptides (Kozlov et al., 2006; Vassilevski et al., 2008).

The increasing resistance of bacteria to many traditional antibiotics is indeed a driving force in developing new antibiotics (e.g., Verma et al., 2007). Gomesin, an AMP derived from the innate immune system of a mygalomorph spider seems to be a potential lead in the development of new antibiotic drugs (Fázio et al., 2006, 2007; Rodrigues et al., 2008). Cytolytic peptides from spider venom, however, might not be the first choice for drug development because of their broad cytolytic activity. Furthermore, the awareness that cupiennin 1a exhibits not only insecticidal and cytolytic activities but also functions as a very potent inhibitor of the formation of nitric oxide by neuronal nitric oxide synthase suggests that such peptides in general may have additional, but up to now unidentified cellular targets (Pukala et al., 2007a). These targets could account for a further indisposition of imperfectly paralysed prey items, because venom from one spider species differs considerably in its insecticidal potency towards different prey species (Kuhn-Nentwig et al., 1998).

To date, our knowledge of the structural features of cytolytic and antimicrobial peptides that determine specific or broad membrane lytic activity is still very limited. In contrast to the high diversity of cytolytic peptides across spider species, the majority of such peptides found within a species are rather similar. These subtle differences in structure can provide clues to identifying the prerequisites for cytolytic and pore-forming activity specific to a particular membrane type (Dubovskii et al., 2008).

ACKNOWLEDGMENT

We thank Dr. H. Murray for critical comments on the manuscript.

REFERENCES

Adão, R. et al., Membrane structure and interactions of a short Lycotoxin I analogue, *J. Pept. Sci.*, 14, 528–534, 2008.

Barbosa, F.M. et al., Gomesin, a peptide produced by the spider *Acanthoscurria gomesiana*, is a potent anticryptococcal agent that acts in synergism with fluconazole, *FEMS Microbiol. Lett.*, 274, 279–286, 2007.

Belokoneva, O.S. et al., The hemolytic activity of six arachnid cationic peptides is affected by the phosphatidylcholine-to-sphingomyelin ratio in lipid bilayers, *Biochim. Biophys. Acta*, 1617, 22–30, 2003.

Budnik, B.A. et al., *De novo* sequencing of antimicrobial peptides isolated from the venom glands of the wolf spider *Lycosa singoriensis*, *J. Mass Spectrom.*, 39, 193–201, 2004.

Cerenius, L. and Söderhäll, K., The prophenoloxidase-activating system in invertebrates, *Immunol. Rev.*, 198, 116–126, 2004.

Choi, S-J. et al., Isolation and characterization of Psalmopeotoxin I and II: Two novel antimalarial peptides from the venom of the tarantula *Psalmopoeus cambridgei*, *FEBS Lett.*, 572, 109–117, 2004.

Cohen, E. and Quistad, G.B., Cytotoxic effects of arthropod venoms on various cultured cells, *Toxicon*, 36, 353–358, 1998.

Corzo, G. and Escoubas, P., Pharmacologically active spider peptide toxins, *Cell. Mol. Life Sci.*, 60, 2409–2426, 2003.

Corzo, G. et al., Oxyopinins, large amphipathic peptides isolated from the venom of the wolf spider *Oxyopes kitabensis* with cytolytic properties and positive insecticidal cooperativity with spider neurotoxins, *J. Biol. Chem.*, 277, 23627–23637, 2002.

DeLano, W.L., The PyMOL molecular graphics system, available at http://www.pymol.org, 2002.

Dubovskii, P.V. et al., Spatial structure and activity mechanism of a novel spider antimicrobial peptide, *Biochemistry*, 45, 10759–10767, 2006.

Dubovskii, P.V. et al., Three-dimensional structure/hydrophobicity of latarcins specifies their mode of membrane activity, *Biochemistry*, 47, 3525–3533, 2008.

Escoubas, P., Molecular diversification in spider venoms: A web of combinatorial peptide libraries, *Mol. Divers.*, 10, 545–554, 2006.

Estrada, G., Villegas, E., and Corzo, G., Spider venoms: A rich source of acylpolyamines and peptides as new leads for CNS drugs, *Nat. Prod. Rep.*, 24, 145–161, 2007.

Fázio, M.A. et al., Structure–activity relationship studies of gomesin: Importance of the disulfide bridges for conformation, bioactivities, and serum stability, *Biopolymers*, 84, 205–218, 2006.

Fázio, M.A. et al., Biological and structural characterization of new linear gomesin analogues with improved therapeutic indices, *Biopolymers*, 88, 386–400, 2007.

Fukuzawa, A.H. et al., The role of hemocytes in the immunity of the spider *Acanthoscurria gomesiana*, *Dev. Comp. Immunol.*, 32, 716–725, 2008.

Gao, B., Tian, C., and Zhu, S., Inducible antibacterial response of scorpion venom gland, *Peptides*, 28, 2299–2305, 2007.

Iwanaga, S. and Lee, B.L., Recent advances in the innate immunity of invertebrate animals, *J. Biochem. Mol. Biol.*, 38, 128–150, 2005.

Jung, H.J. et al., Lipid membrane interaction and antimicrobial activity of GsMTx-4, an inhibitor of mechanosensitive channel, *Biochem. Biophys. Res. Commun.*, 340, 633–638, 2006.

Kawano, K. et al., Antimicrobial peptide, tachyplesin I, isolated from hemocytes of the horseshoe crab (*Tachypleus tridentatus*). NMR determination of the beta-sheet structure, *J. Biol. Chem.*, 265, 15365–15367, 1990.

Kokryakov, V.N. et al., Protegrins: Leukocyte antimicrobial peptides that combine features of corticostatic defensins and tachyplesins, *FEBS Lett.*, 327, 231–236, 1993.

Kozlov, S.A. et al., Latarcins, antimicrobial and cytolytic peptides from the venom of the spider *Lachesana tarabaevi* (Zodariidae) that exemplify biomolecular diversity, *J. Biol. Chem.*, 281, 20983–20992, 2006.

Kuhn-Nentwig, L., Antimicrobial and cytolytic peptides of venomous arthropods, *Cell. Mol. Life Sci.*, 60, 2651–2668, 2003.

Kuhn-Nentwig, L., Bücheler, A., Studer, A., and Nentwig, W., Taurine and histamine: Low molecular compounds in prey hemolymph increase the killing power of spider venom, *Naturwissenschaften*, 85, 136–138, 1998.

Kuhn-Nentwig, L. et al., Cupiennin 1, a new family of highly basic antimicrobial peptides in the venom of the spider *Cupiennius salei* (Ctenidae), *J. Biol. Chem.*, 277, 11208–11216, 2002a.

Kuhn-Nentwig, L. et al., Cupiennin 1d*: The cytolytic activity depends on the hydrophobic N-terminus and is modulated by the polar C-terminus, *FEBS Lett.*, 527, 193–198, 2002b.

Kuhn-Nentwig, L., Schaller, J., and Nentwig W., Biochemistry, toxicology and ecology of the venom of the spider *Cupiennius salei* (Ctenidae), *Toxicon*, 43, 543–553, 2004.

Kumar, S. et al., The role of reactive oxygen species on Plasmodium melanotic encapsulation in *Anopheles gambiae*, *Proc. Natl Acad. Sci. USA*, 100, 14139–14144, 2003.

Kurata, S., Ariki, S., and Kawabata, S., Recognition of pathogens and activation of immune responses in *Drosophila* and horseshoe crab innate immunity, *Immunobiology*, 211, 237–249, 2006.

Li, D. et al., Insect hemolymph clotting: Evidence for interaction between the coagulation system and the prophenoloxidase activating cascade, *Insect Biochem. Mol. Biol.*, 32, 919–928, 2002.

Li, D. et al., Function and solution structure of hainantoxin-I, a novel insect sodium channel inhibitor from the Chinese bird spider *Selenocosmia hainana*, *FEBS Lett.*, 555, 616–622, 2003.

Lorenzini, D.M. et al., Acanthoscurrin: A novel glycine-rich antimicrobial peptide constitutively expressed in the hemocytes of the spider *Acanthoscurria gomesiana*, *Dev. Comp. Immunol.*, 27, 781–791, 2003a.

Lorenzini, D.M. et al., Molecular cloning, expression analysis and cellular localization of gomesin, an antimicrobial peptide from hemocytes of the spider *Acanthoscurria gomesiana*, *Insect Biochem. Mol. Biol.*, 33, 1011–1016, 2003b.

Lorenzini, D.M. et al., Discovery of immune-related genes expressed in hemocytes of the tarantula spider *Acanthoscurria gomesiana*, *Dev. Comp. Immunol.*, 30, 545–556, 2006.

Luque-Ortega, J.R. et al., Leishmanicidal activity of antimicrobial peptides from the wolf spider *Lycosa carolinensis*, *J. Pept. Sci.*, 12, Suppl. S1, 527, 2006.

Mandard, N. et al., Androctonin, a novel antimicrobial peptide from scorpion *Androctonus australis*: Solution structure and molecular dynamics simulations in the presence of a lipid monolayer, *J. Biomol. Struct. Dyn.*, 17, 367–380, 1999.

Mandard, N. et al., The solution structure of gomesin, an antimicrobial cysteine-rich peptide from the spider, *Eur. J. Biochem.*, 269, 1190–1198, 2002.

Miyata, T. et al., Antimicrobial peptides, isolated from horseshoe crab hemocytes, tachyplesin II, and polyphemusins I and II: Chemical structures and biological activity, *J. Biochem. (Tokyo)*, 106, 663–668, 1989.

Moraes, L.G. et al., Conformational and functional studies of gomesin analogues by CD, EPR and fluorescence spectroscopies, *Biochim. Biophys. Acta*, 1768, 52–58, 2007.

Moreira, C.K. et al., Effect of the antimicrobial peptide gomesin against different life stages of *Plasmodium* spp., *Exp. Parasitol.*, 116, 346–353, 2007.

Oh, D. et al., Role of the hinge region and the tryptophan residue in the synthetic antimicrobial peptides, cecropin A(1-8)-magainin 2(1-12) and its analogues, on their antibiotic activities and structures, *Biochemistry*, 39, 11855–11864, 2000.

Ostrow, K.L. et al., cDNA sequence and in vitro folding of GsMTx4, a specific peptide inhibitor of mechanosensitive channels, *Toxicon*, 42, 263–274, 2003.

Oswald, R.E. et al., Solution structure of peptide toxins that block mechanosensitive ion channels, *J. Biol. Chem.*, 277, 34443–34450, 2002.

Pimentel, C. et al., Solution structure of PcFK1, a spider peptide active against *Plasmodium falciparum*, *Protein Sci.*, 15, 628–634, 2006.

Platnick, N.I., The World Spider Catalog, version 8.5, American Museum of Natural History, available at http://research.amnh.org/entomology/spiders/catalog/index.html, 2008.

Posokhov, Y.O. et al., Is lipid bilayer binding a common property of inhibitor cysteine knot ion-channel blockers?, *Biophys. J.*, 93, L20–L22, 2007a.

Posokhov, Y.O., Gottlieb, P.A., and Ladokhin, A.S., Quenching-enhanced fluorescence titration protocol for accurate determination of free energy of membrane binding, *Anal. Biochem.*, 362, 290–292, 2007b.

Pukala, T.L. et al., Host-defence peptides from the glandular secretions of amphibians: Structure and activity, *Nat. Prod. Rep.*, 23, 368–393, 2006.

Pukala, T.L. et al., Cupiennin 1a, an antimicrobial peptide from the venom of the neotropical wandering spider *Cupiennius salei*, also inhibits the formation of nitric oxide by neuronal nitric oxide synthase, *FEBS J.*, 274, 1778–1784, 2007a.

Pukala, T.L. et al., Solution structure and interaction of cupiennin 1a, a spider venom peptide, with phospholipid bilayers, *Biochemistry*, 46, 3576–3585, 2007b.

Qu, Y. et al., Proton nuclear magnetic resonance studies on huwentoxin-I from the venom of the spider *Selenocosmia huwena*: 2. Three-dimensional structure in solution, *J. Protein Chem.*, 16, 565–574, 1997.

Rodrigues, E.G. et al., Effective topical treatment of subcutaneous murine B16F10-Nex2 melanoma by the antimicrobial peptide gomesin, *Neoplasia*, 10, 61–68, 2008.

Silva, P.I. Jr, Daffre, S., and Bulet, P., Isolation and characterization of gomesin, an 18-residue cysteine-rich defense peptide from the spider *Acanthoscurria gomesiana* hemocytes with sequence similarities to horseshoe crab antimicrobial peptides of the tachyplesin family, *J. Biol. Chem.*, 275, 33464–33470, 2000.

Suchyna, T.M. et al., Identification of a peptide toxin from *Grammostola spatulata* spider venom that blocks cation-selective stretch-activated channels, *J. Gen. Physiol.*, 115, 583–598, 2000.

Vassilevski, A.A. et al., Cyto-insectotoxins, a novel class of cytolytic and insecticidal peptides, *Biochem. J.*, 411, 687–696, 2008.

Vassilevski, A.A. et al., Synthetic analogues of antimicrobial peptides from the venom of the Central Asian spider *Lachesana tarabaevi*, *Bioorg Khim*, 33, 405–412, 2007.

Verma, C. et al., Defensins: Antimicrobial peptides for therapeutic development, *Biotechnol. J.*, 2, 1353–1359, 2007.

Villegas, E. and Corzo, G., Pore-forming peptides from spiders, *Tox. Rev.*, 24, 1–13, 2005.

Vizioli, J. and Salzet, M., Antimicrobial peptides from animals: Focus on invertebrates, *Trends Pharmacol. Sci.*, 23, 494–496, 2002.

Williams, M.J., *Drosophila* hemopoiesis and cellular immunity, *J. Immunol.*, 178, 4711–4716, 2007.

Wullschleger, B., Nentwig, W., and Kuhn-Nentwig, L., Spider venom: Enhancement of venom efficacy mediated by different synergistic strategies in *Cupiennius salei*, *J. Exp. Biol.*, 208, 2115–2121, 2005.

Xu, K., Ji, Y., and Qu, X., Purification and characterization of an antibacterial peptide from venom of *Lycosa singoriensis*, *Acta Zool. Sinica*, 35, 300–305, 1989.

Yan, L. and Adams, M.E., Lycotoxins, antimicrobial peptides from venom of the wolf spider *Lycosa carolinensis*, *J. Biol. Chem.*, 273, 2059–2066, 1998.

Yang, L. et al., Barrel-stave model or toroidal model? A case study on melittin pores, *Biophys. J.*, 81, 1475–1485, 2001.

Zonstein, S.L. and Ovtchinnikov, S.V., A new Central Asian species of the spider genus *Lachesana* Strand, 1932 (Araneae, Zodariidae: Lachesaninae), *Tethys Entomol. Res.*, 1, 59–62, 1999.

Index

B

C

D

O

P

T

U

V

W

X

Y

Z

Milton Keynes UK
Ingram Content Group UK Ltd.
UKHW050456071024
449327UK00015B/409